Introduction to electromagnetic and microwave e
23001
TK7876 .K35 1998
Karmel, Paul R.

DATE	ISSUED TO

Introduction to Electromagnetic and Microwave Engineering

WILEY SERIES IN MICROWAVE AND OPTICAL ENGINEERING

KAI CHANG, Editor
Texas A&M University

A complete list of the titles in this series appears at the end of this volume.

Introduction to Electromagnetic and Microwave Engineering

PAUL R. KARMEL
GABRIEL D. COLEF
RAYMOND L. CAMISA

A WILEY-INTERSCIENCE PUBLICATION
JOHN WILEY & SONS, INC.
NEW YORK / CHICHESTER / WEINHEIM / BRISBANE / SINGAPORE / TORONTO

This book is printed on acid-free paper. ♾

Published simultaneously in Canada.

Library of Congress Cataloging-in-Publication Data:

Karmel, Paul R.
Introduction to electromagnetic and microwave engineering/Paul R. Karmel, Gabriel D. Colef, and Raymond L. Camisa.
p. cm. — (Wiley series in microwave and optical engineering)
"A Wiley-Interscience publication."
Includes index.
ISBN 0-471-17781-4 (cloth: alk. paper)
1. Microwave devices. 2. Microwaves. 3. Microwave transmission lines. I. Colef, Gabriel D. II. Camisa, Raymond L. III. Title. IV. Series.
TK7876.K35 1997
621.3—dc21 97-16542

Printed in the United States of America.

10 9 8 7 6 5 4 3 2 1

Contents

Preface

This text is an outgrowth of undergraduate and graduate courses given at the City College of New York and is intended to serve as a two-semester text in electromagnetics.

For the undergraduate student and first-year graduate student this text provides a comprehensive foundation in the field of electromagnetic theory and microwave engineering. Practicing engineers will also find this text useful to update their knowledge in the field.

Considerable thought and effort were devoted to the pedagogy and clarity of presentation. Throughout the text, illustrative examples are presented, design problems are discussed, and the use of simulation software is emphasized. In addition, great care has been exercised in the development of approximately 300 homework problems. These exercises serve to elucidate the text and in some cases extend the discussion by including analysis, synthesis, design, and computer-aided problem solving.

The first chapter explains the importance of electromagnetics and its role in the microwave engineering and telecommunication fields. Chapter 2 provides the basic vector mathematics needed for the remainder of the text. It is assumed that the reader has some previous exposure to linear, passive circuit theory. Hence this chapter can be omitted in a graduate course.

Chapters 3, 4, and 5 define the electromagnetic fields and introduce Maxwell's equations. This approach permits the reader to understand the interrelation between electromagnetic fields and their sources. Chapters 6, 7, and 8 show that statics are a special case of electromagnetics.

Chapters 9 and 10 discuss transverse electromagnetic (TEM) waves guided by transmission lines and Chapters 11 and 12 discuss the propagation of waves in unbounded media and transverse electric (TE) and transverse magnetic (TM) waves transmitted through waveguides. Chapters 13 and 14 introduce the notion of microwave circuits and networks as well as radiation of waves using antennas. These six chapters can form a second semester (undergraduate or graduate) course and are of particular value to the practicing engineer. These topics are not often found in senior level texts; however, they have been included here because of their relevance in communications, radar, and electronic warfare systems.

This text was written by Professor Paul R. Karmel, who, unfortunately, passed away before he could complete the final editing. We were Professor Karmel's students and also his colleagues. We want to thank Professor Karmel's family for giving us the opportunity to complete his excellently written text. We also want to thank Professor Donald Schilling for his encouragement during this project.

We also want to thank Midshipmen Thomas Ritter, John Long, and Gregory Sachs, from the U.S. Merchant Marine Academy, for their help with some of the figures.

Special thanks go to our families for supporting us throughout the project.

Gabriel D. Colef

Associate Professor, Engineering Department
U.S. Merchant Marine Academy
Kings Point, New York

Raymond L. Camisa

Director of Microwave and Optoelectronics Laboratory
David Sarnoff Research Center
Princeton, New Jersey

CHAPTER ONE

Introduction

1.1 ELECTROMAGNETICS

Electromagnetics is the study of the interaction of electric and magnetic fields. This interaction is described mathematically in a set of formulas known as Maxwell's equations. A primary purpose of this text is to foster an understanding of the physical fields and of the mathematics that represents them.

An understanding of these fields is necessary to describe physical electromagnetic components from resistors to rotating machines. It is needed to understand the motion of charged particles in a cathode ray tube, a cyclotron, or a space plasma. It is required to analyze the transmission of electromagnetic signals through space, along wires or other guiding structures, and in material media such as fiber optical cables. It is required to explain radiation and propagation of electromagnetic energy through space and to analyze and design antenna systems.

1.2 MICROWAVE ENGINEERING

Basic circuit theory analyzes voltages and currents in circuits consisting of lumped elements, such as resistors, capacitors, and mutual inductors, using Kirchhoff's laws. The circuits and physical components are small in comparison with the wavelength of the signals, and the lengths of the wires that connect components are also small. Wavelength is the speed of light, $c = 3 \times 10^8$ m/s ($c = 11.8 \times 10^9$ in./s) in free space, divided by frequency, f, so that a frequency of 1 GHz (1×10^9 Hz) corresponds to a wavelength of 0.3 m. As a general guide we can take 1 GHz as the upper end of the frequency range where basic circuit theory applies.

Optics involves physical components that are much larger than the wavelength of the signals. In the visible light band around 300 THz (300×10^{12} Hz) the wavelength is 1 μm (1×10^{-6} m) and most components, such as lenses, are much larger. As a general guide we can take 100 GHz with a wavelength of 0.003 m as the lower end of the optical band.

In between, from 1 to 100 GHz, we have the range that is generally called the microwave frequency band. In this band components and circuits are on the same length scale as the wavelength. Microwave circuits and systems find wide use in communications systems, radar, commercial and consumer heating and drying ovens, and medical applications.

A second purpose of this text is to introduce the theory of transmission lines and waveguides and the techniques for analyzing them. We will also introduce concepts of voltage, current, and impedance as applied to microwaves, techniques for matching a load to a source, applications to microwave filters, and the elements of microwave circuit descriptions using scattering concepts. Many of these ideas will rely on basic circuit theory and many are also applicable to optical systems.

1.3 DIMENSIONS AND UNITS

All physical quantities have dimensions. The choice of which and how many dimensions are considered fundamental is somewhat arbitrary but has been standardized by international convention (in 1960) in the Système International d'Unités (SI) or International System of Units. The SI chooses seven fundamental dimensions: length, mass, time, electric current, temperature, luminous intensity, and the amount of a substance present. The same convention defines fundamental units to express each of these quantities. These units are, respectively, the meter (m), kilogram (kg), second (s), ampere (A), kelvin (K), candela (ca), and the mole (mol).

In our study of electromagnetics only the first four quantities are of interest. The dimensions of all the physical quantities we will study can be expressed in terms of length (L), mass (M), time (T), and current (I). As examples, velocity is distance per unit of time, hence [velocity] $=$ L/T; acceleration is change in velocity per unit of time, hence [acceleration] $= \mathrm{L/T^2}$; force is mass multiplied by acceleration, hence [force] $= \mathrm{ML/T^2}$; energy is force multiplied by distance, hence [energy] $= \mathrm{ML^2/T^2}$; energy is power multiplied by time, hence [power] $= \mathrm{ML^2/T^3}$; power is also voltage multiplied by current, hence [voltage] $= \mathrm{ML^2/IT^3}$; voltage is resistance multiplied by current, hence [resistance] $= \mathrm{ML^2/I^2T^3}$; voltage is also inductance multiplied by the change in current per unit time, hence [inductance] $=$ [voltage $\times$ time/current] or [inductance] $= \mathrm{ML^2/I^2T^2}$; magnetic flux is inductance multiplied by current, hence [magnetic flux] $= \mathrm{ML^2/IT^2}$; and so on. Table 1.1 lists some electromechanical quantities and their dimensions.

All equations must be balanced dimensionally; that is, the dimensions of the quantities on the left side of the equation must always equal the dimensions of the quantities on the right side. Correctness of dimensionality should always be checked after any complicated manipulation of equations. If the dimensions are not balanced, the result is certainly wrong; if the dimensions are balanced, the result may be correct or incorrect.

Units are different from dimensions. For example, a given length can be expressed in units of inches, miles, meters, or kilometers. A length of 1 mile (U.S. statute) $=$ 63,360 inches $=$ 160.9347219 meters $=$ 1.609347219 kilometers. These all have different units, but the dimension of each is length. The SI system of units chooses the meter, kilogram, second, and ampere as fundamental units and defines all other units in terms of these. Thus, for example, the unit of voltage, the volt, is equivalent to a kilogram meter squared per ampere second cubed, $[\mathrm{V}] = \mathrm{kg\text{-}m^2/A\text{-}s^3}$. Table 1.1 also lists the standard units and their symbols.

TABLE 1.1 Quantities, Symbols, Units, and Dimensions

Quantity	Symbol	SI Unit	Abbreviations	Dimensions
Admittance; conductance; susceptance	Y; G; B	siemens (mho)	S	I^2T^3/ML^2
Attenuation constant	α	neper/meter	Np/m, m^{-1}	1/L
Capacitance	C	farad	F	I^2T^4/ML^2
Charge; point charge	Q; q	coulomb	C	IT
Charge density, volume	ρ	coulomb/meter³	C/m^3	IT/L^3
Charge density, surface	ρ_s	coulomb/meter²	C/m^2	IT/L^2
Conductivity	σ	siemens/meter	S/m	I^2T^3/ML^3
Current; line current	I; i	ampere	A	I
Current density, surface	$\vec{J}_s$	ampere/meter	A/m	I/L
Current density, volume	$\vec{J}$	ampere/meter²	A/m^2	I/L^2
Electric dipole moment	$\vec{p} = q\vec{l}$	coulomb-meter	C-m	ILT
Electric field intensity	$\vec{E}$	volt/meter	V/m	ML/IT^3
Electromotive force (emf); voltage; electric potential	V	volt	V	ML^2/IT^3
Electric flux density	$\vec{D}$	coulomb/meter²	C/m^2	IT/L^2
Electric flux	Ψ_e	coulomb	C	IT
Energy; work	W	joule	J	ML^2/T^2
Energy density, electric; magnetic	w_e; w_m	joule/meter³	J/m^3	M/LT^2
Frequency	f	hertz	Hz	1/T
Frequency, angular	ω	radian/second	rad/s, 1/s	1/T
Impedance; resistance; reactance	Z; R; X	ohm	Ω	ML^2/I^2T^3
Inductance	L	henry	H	ML^2/I^2T^2
Length; wavelength	$L, l, a, \ldots \lambda$ λ	meter	m	L
Line charge density	ρ_l	coulomb/meter	C/m	IT/L
Magnetic dipole moment	$m = IS$	ampere-meter²	$A\text{-}m^2$	IL^2
Magnetic field intensity	$\vec{H}$	ampere/meter	A/m	I/L
Magnetic flux density	$\vec{B}$	tesla, weber/meter²	T, Wb/m^2	M/IT^2
Magnetic flux	Φ	weber	Wb	ML^2/IT^2
Magnetomotive force (mmf)	V_m	ampere	A	I
Mass	M, m	kilogram	kg	M
Permeability, relative	μ_r			
Permeability; permeability of free space	μ; μ_0	henry/meter	H/m	ML/I^2T^2
Permittivity, relative	ε_r			
Permittivity; permittivity of free space	ε; ε_0	farad/meter	F/m	I^2T^4/ML^3
Phase constant	β	radian/meter	rad/m, m^{-1}	1/L
Poynting vector, instantaneous; time average	$\vec{S}$; $\vec{S}_c$	watt/meter²	W/m^2	M/T^3
Propagation constant	γ	1/meter	m^{-1}	1/L
Reluctance	$\mathscr{R}$	1/henry	H^{-1}	I^2T^2/ML^2
Resistance, surface	R_s	ohm	Ω	ML^2/I^2T^3
Skin depth	δ_s	meter	m	L
Time; time constant	t, T; τ	second	s	T

In order to compare the size of two quantities they must have the same dimensions and be expressed in the same units. You cannot say the distance to the moon is greater than a pound of cheese. These have different dimensions and cannot have the same units. You can say the distance to the sun, 1 AU (astronomical unit), is greater than the length of a marathon race, 26 miles 385 yards. These both have the dimensions of length and can be converted to the same units.

In this text we use the rationalized SI system of units. This means that the dimensionless factor 4π does not appear in Maxwell's equations. It does, however, turn up in other relations.

In rationalized SI units the permeability of free space is defined as $\mu_0 = 4\pi \times 10^{-7}$ H/m, exactly. As we will see, the speed of light in vacuum equals the reciprocal of the square root of the permeability of free space multiplied by the permittivity of free space, $c = 1/\sqrt{\mu_0 \varepsilon_0}$. The speed of light in vacuum has accurately been measured as $c = 2.997925 \times 10^8 \approx 3.00 \times 10^8$ m/s. As a consequence the permittivity of free space is $\varepsilon_0 = 8.854185 \times 10^{-12} \approx (1/36\pi) \times 10^{-9}$ F/m.

The same convention defined a system of preferred prefixes, which represent multipliers by factors of 10^3 or 10^{-3}. These prefixes are listed, along with their proper pronunciation, in Table 1.2.

1.4 THE FREQUENCY SPECTRUM

Maxwell's equations describe the behavior of electric and magnetic fields at all frequencies, from zero on up (at least as far as any experiment has been performed). Different portions of the frequency spectrum, however, have different engineering uses and have been assigned different names. Historically, the engineering use of the

TABLE 1.2 Multipliers[a]

Prefix	Abbreviation	Multiplier	Pronunciation[b]
exa	E	10^{18}	eks'â
peta	P	10^{15}	pet'â
tera	T	10^{12}	ter'â
giga	G	10^{9}	ji'gâ
mega	M	10^{6}	meg'â
kilo	k	10^{3}	kil'â
milli	m	10^{-3}	mil'é
micro	μ	10^{-6}	mi'kró
nano	n	10^{-9}	ná'nâ
pico	p	10^{-12}	pí'kó
femto	f	10^{-15}	fém'tó
atto	a	10^{-18}	at'tó

[a] The prefix cent (c) for 10^{-2} is common.

[b] â as in alone; á as in able; é as in equal; í as in ice; ó as in over.

frequency spectrum has been pushed ever higher as we have learned to generate, modulate, guide, transmit, receive, and demodulate ever higher frequencies. The reason for this is that the quantity of information that can be transmitted in a given time depends on the bandwidth (frequency width) of the signal. As we wish to transmit more and more information, we must move to ever higher frequencies. For example, the real-time bandwidth of a telephone conversation containing voice information only is about 10 kHz, while for a television channel containing picture and sound

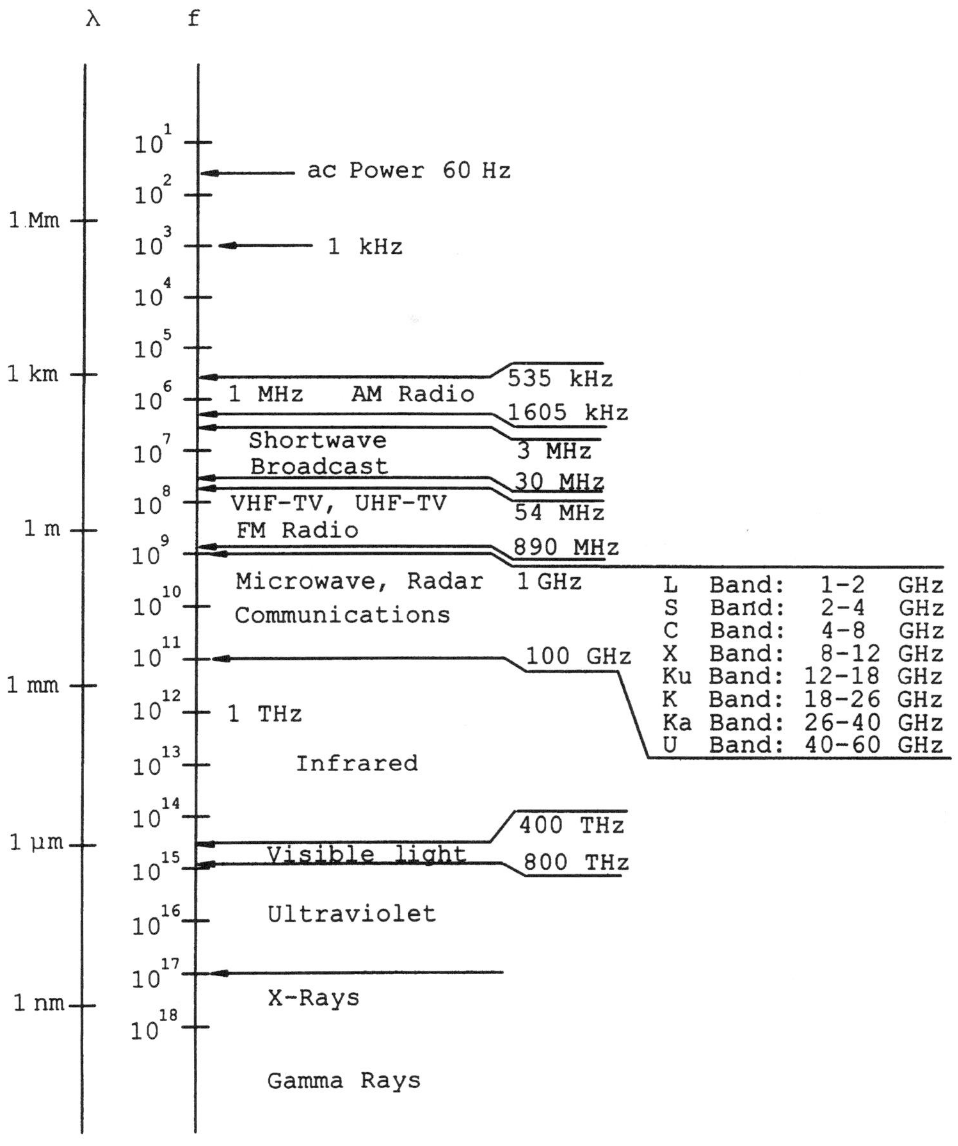

FIGURE 1.1 Frequency and wavelength electromagnetic spectrum, band designations, and associated applications.

information the bandwidth is 6 MHz, 600 times as much. In order to keep the fractional bandwidth the same, the television signal must modulate a carrier at 600 times as high a frequency as the telephone signal's carrier.

We have already indicated that the microwave frequency band extends from about 1 to 100 GHz. Figure 1.1 presents the electromagnetic spectrum and the names and uses of various frequency ranges.

1.5 BIOGRAPHICAL DATA

The study of electric and magnetic phenomena may be traced to ancient Greece, but the modern study of electricity begins in the 18th century. Table 1.3 lists those engineers and scientists who have had SI units named in recognition of their contributions to electrical science. Any general or biographical encyclopedia will include the significant names. Particular attention is drawn to the following, which includes

TABLE 1.3 Brief Biography of SI Unit Honorees

Name, Dates, Quantity	Electromagnetic Achievement
COULOMB, Charles Augustin de, 1736–1806 Charge	French military engineer. Invented a delicate torsion balance and used it to measure the force of attraction and repulsion between harges. His measurements showed that force is proportional to the product of the charges and inversely proportional to the distance between them.
VOLTA, Alessandro, 1745–1827 Electric potential, voltage (volt)	Italian physicist. Invented the electrophorus—a device for building up charge on a plate—and the voltaic cell or battery.
AMPÈRE, André Marie, 775–1836 Current	French mathematician and physicist. Described the direction of the magnetic field around a current-carrying wire. Introduced the concept of lines of force. Showed that the force acting on parallel wires is attractive/repulsive when the currents are in the same/opposite direction.
FARADAY, Michael, 1791–1867 Capacitance (farad)	English physicist and chemist. A meticulous experimenter, he developed the concept of magnetic lines of force and discovered electrical induction, whereby a current is induced in a closed circuit subjected to a changing magnetic field. Invented the first electrical generator.
OHM, Georg Simon, 1787–1854 Resistance, impedance	German physicist. By experimenting with different wires he found that for a constant electric potential difference the current was inversely proportional to the length of the wire and proportional to the area. Discovered Ohm's law.

TABLE 1.3 (*Continued*)

Name, Dates, Quantity	Electromagnetic Achievement
HENRY, Joseph, 1797–1878 Inductance	American physicist. Designed and built an improved electromagnet. Invented the telegraph with a system of repeater relays to permit long-distance transmission. Discovered electrical induction independently of Faraday.
JOULE, James Prescott, 1818–1889 Energy	English physicist. Carried out extensive measurements of conversion of mechanical and electrical energy to heat, leading eventually to the law of conservation of energy. Showed that the heat produced by a resistor is proportional to the square of the current passing through it.
WEBER, Wilhelm Eduard, 1804–1891 Magnetic flux	German physicist. Worked on terrestrial magnetism and a practical telegraph. Introduced a logical system of units for magnetism, now replaced by SI units.
SIEMENS, Ernst Werner von, 1861–1892 Conductance, admittance	German electrical engineer. Invented an electric dynamo. Constructed the first telegraph line and built the first electric railway in Germany.
HERTZ, Heinrich Rudolf, 1857–1894 Frequency	German physicist. Generated hertzian or radio waves by producing an oscillating spark between two metal balls. Detected the waves at a distance, showed they involved both electric and magnetic fields, and measured the wavelength, thus experimentally verifying Maxwell's equations.
TESLA, Nikola, 1856–1943 Magnetic flux density	American electrical engineer (Serbian, born in Croatia). Pioneered alternating current systems, inventing the high-voltage transformer making long-distance power transmission practical. Invented ac motors and generators and made contributions to wireless communications. Designed the Niagra Falls generating station.

delightfully readable descriptions of scientific achievements and interesting life stories: Isaac Asimov, *Asimov's Biographical Encyclopedia of Science and Technology*, Doubleday & Co., New York, 1964, 1972.

Unfortunately, a listing with this restriction leaves out the most significant contributor, James Clerk Maxwell (1831–1879), the great Scottish mathematician and physicist. The maxwell is a unit of magnetic flux in the Gaussian system of units, which is frequently used in atomic physics. The equivalence is 1 weber = 10^8 maxwell. Maxwell made significant contributions to mathematics, astronomy, and thermodynamics. His crowning achievement was to place Faraday's experimental observations on magnetic lines of force into mathematical form. In this way he obtained a few simple equations that described the electrical and magnetic observations of previous scientists. His equations showed that electric and magnetic phenomena were inseparably linked and they predicted radiation due to oscillating charges, which would spread out with the speed of light. He therefore induced that light is part of the same electromagnetic phenomena and that there should be a full spectrum of electromagnetic waves from the lowest to the highest frequency. Maxwell's equations remain the foundation of electromagnetic science and engineering and their validity is still completely intact.

CHAPTER TWO

Vector Mathematics I

2.1 INTRODUCTION

The variables in circuit or system analysis courses are generally functions of time. For example, in circuit problems we deal with voltage $v(t)$ and current $i(t)$. The mathematics needed for the description of circuits usually involves algebra, ordinary differential equations, linear algebra, and, for discrete systems, difference equations. For the study of switching systems, you are required to learn Boolean algebra.

In the study of electromagnetic fields and waves we are required to use a description of physical quantities that accounts for their dependence on position in space as well as time. Also, some quantities have direction. For these reasons we must use partial differential equations and vector calculus. This chapter introduces vector mathematics for vector field quantities. It covers coordinate systems, vector arithmetic, and some integral operations. A later chapter will deal with differential operations on scalar and vector fields.

2.2 SCALAR FIELDS

A field is any physical quantity that varies with space and time. There are two kinds of fields, depending on the nature of the quantity described.

A scalar field describes a physical quantity that has amplitude (size) only. For example, the temperature in the room in which you are sitting is a function of position throughout the room and of time. It is completely specified by giving its size or value, a single number in units of degrees Celsius, at each point in space and at each instant of time. If we use the symbol T to represent the temperature, then we write

$$T = T(x, y, z, t) \tag{2.1}$$

to indicate that the temperature is a scalar function of position, given by the coordinates x, y, and z and the time t.

Other examples of scalar fields are mass density $\rho_m(x, y, z, t)$ and electric voltage $V(x, y, z, t)$.

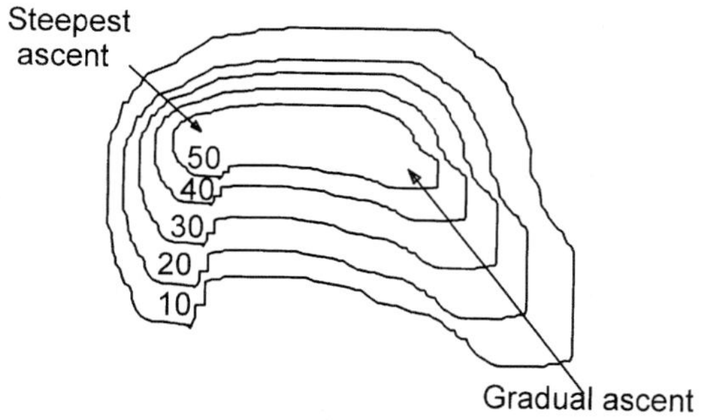

FIGURE 2.1 Topographical map. A graphical representation of a two-dimensional, time-independent scalar field.

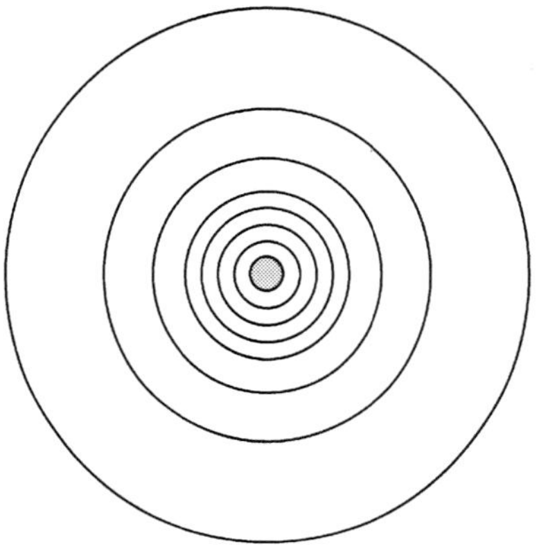

FIGURE 2.2 Lines of equal gravitational potential for a uniform mass density of radius a.

An example of a two-dimensional scalar field that is time independent, at least over short time spans, is the elevation of ground level above sea level $h(x, y)$. If we make a diagram of lines connecting points at regular elevation intervals, for example, every 10 feet, then we have a topographic map of the terrain (Fig. 2.1). The places where the lines are closest together are where the height is changing most quickly and the normal to these lines is the direction in which the ascent or descent would be steepest.

A similar figure can be drawn for a three-dimensional field, but it is difficult to display the lines on a two-dimensional page. We usually draw the lines for a slice taken through some representative plane. For example, the lines of equal gravitational potential for a uniform mass density of radius a are shown in Fig. 2.2 for any plane passing through the center of the mass.

2.3 VECTOR FIELDS

A vector field describes a physical quantity that has amplitude and direction. For example, the velocity of the air particles in your room is a function of position and

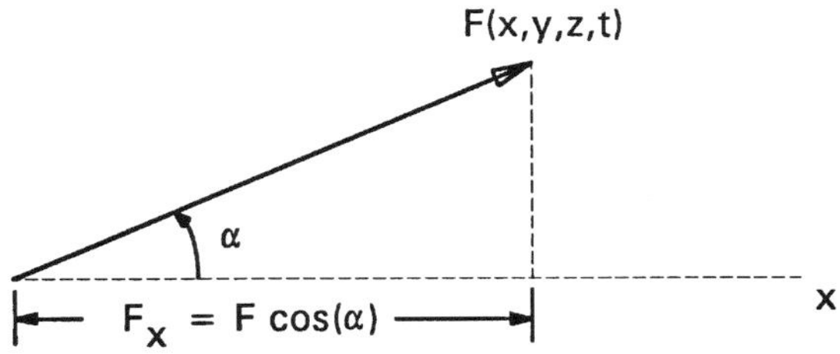

FIGURE 2.3 Vector field.

time. In order to completely describe the velocity of a small volume of air, it is necessary to give the speed of the volume, perhaps in meters per second, and to give the direction in which the volume is traveling. If we use the symbol $\vec{u}$ to represent the velocity, then we write

$$\begin{aligned} \vec{u} &= \vec{u}(x, y, z, t) \\ &= \vec{a}_x u_x(x, y, z, t) + \vec{a}_y u_y(x, y, z, t) + \vec{a}_z u_z(x, y, z, t). \end{aligned} \tag{2.2}$$

The arrow symbol designates the quantity as a vector possessing direction as well as size.† The $\vec{a}$ designates a unit vector, meaning a vector with unity size or value. The subscript x, y, or z indicates the direction of the unit vector. Hence $\vec{a}_x$ is a unit vector in the direction of the x axis. The scalar fields u_x, u_y, and u_z represent the components of velocity in the x, y, and z directions, respectively. These give the speed (a scalar quantity) with which the air element is moving in the specified direction as a function of position and time.

Other examples of vector fields are force $\vec{F}(x, y, z, t)$ and electric field intensity $\vec{E}(x, y, z, t)$.

A vector field at a point in space can be represented by a line with an arrowhead (Fig. 2.3). The length of the line represents the value of the field and the direction of the arrow gives the field direction. The projection of the arrow in some particular direction gives the component of the field in that direction. For example, from the same figure, the component of $\vec{F}$ in the x direction is $F_x = F \cos \alpha$.

A vector field may also be displayed graphically with lines tangent to the direction of the field. Figure 2.4 shows the lines of gravitational force surrounding a point mass. The lines are radial inward with the greatest density of lines close to the point mass where the field is strongest. Figure 2.5 shows the gravitational force field around two point masses m_1 and m_2, where $m_1 = 2m_2$. Note that there are twice as many lines inward to the larger mass m_1 and that close to the point masses the lines are radial as for isolated point masses. Far from the masses they would appear as if they were a single point mass and the lines would appear radial like those around a mass $3m_2$.

† Many texts use boldface type to differentiate vector quantities from scalar quantities. This book uses the arrow symbol over the quantity to encourage you, the student, to use the same symbol whenever you write a vector quantity. While experienced engineers may omit this, the beginner will find it improves understanding to keep the distinction between vectors and scalars clearly represented in this way.

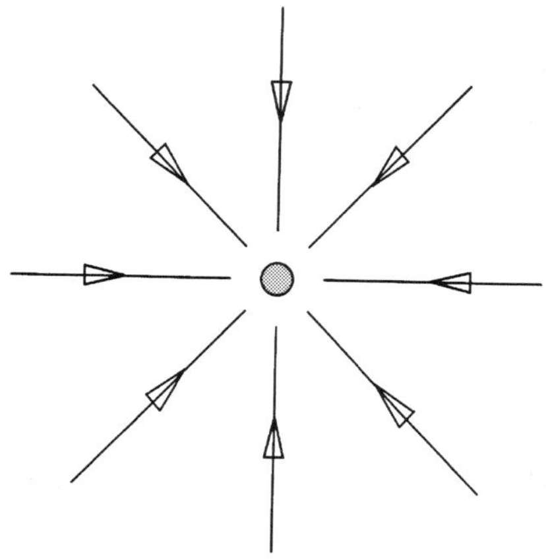

FIGURE 2.4 Lines of gravitational force surrounding a point mass.

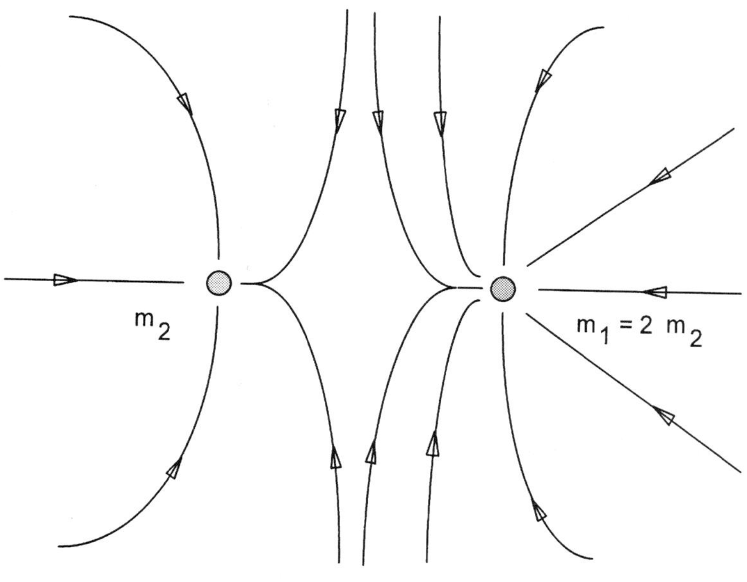

FIGURE 2.5 Gravitational force field around two point masses m_1 and m_2 ($m_1 = 2m_2$).

2.4 COORDINATE SYSTEMS

Fields, whether scalar or vector, are functions of position and time. In order to clearly specify the position of a point in space we must agree on a coordinate system. There are many established coordinate systems and everyone is free to make up their own, but the most useful and the simplest are the Cartesian, cylindrical, and spherical coordinate systems.

2.4.1 Cartesian Coordinates

In the Cartesian or rectangular coordinate system we establish three mutually orthogonal axes, which are fixed in space and specify the position of a point in space by giving its distance from the origin along each axis. Figure 2.6 shows the axes labeled x, y, and z. The order of labeling the axes is such that if the fingers of the right hand are used to turn the x axis toward the y axis, the thumb will point in the direction of the z axis. The x, y, z ordered system is therefore said to be a right-hand system.

When you specify the x coordinate of a point, you indicate that it lies somewhere on a plane perpendicular to the x axis the specified number of units away (positive or negative) from the plane of the y and z axes (Fig. 2.7). Similarly, the y or z coordinate indicates the location of a plane perpendicular to the y or z axis, respectively. The intersection of any two of these planes is a straight line. For example, the x and y coordinates determine planes that intersect in a line parallel to the z axis. The intersection of this line with the plane specified by z is at a point $P(x, y, z)$.

To locate any point in all of three-dimensional space, the range of x, y, and z is from $-\infty$ to $+\infty$.

The position vector is a vector from the origin to a point in space. Designating $\vec{a}_x$, $\vec{a}_y$, and $\vec{a}_z$ as unit vectors along the x, y, and z axes, the position vector $\vec{R}$ from the origin to the point at (x, y, z) is given by

$$\vec{R} = \vec{a}_x x + \vec{a}_y y + \vec{a}_z z. \tag{2.3}$$

The length of the position vector $\vec{R}$ represents the distance from the point in space to the origin of the coordinate system and is calculated as follows:

$$|\vec{R}| = \sqrt{x^2 + y^2 + z^2}. \tag{2.4}$$

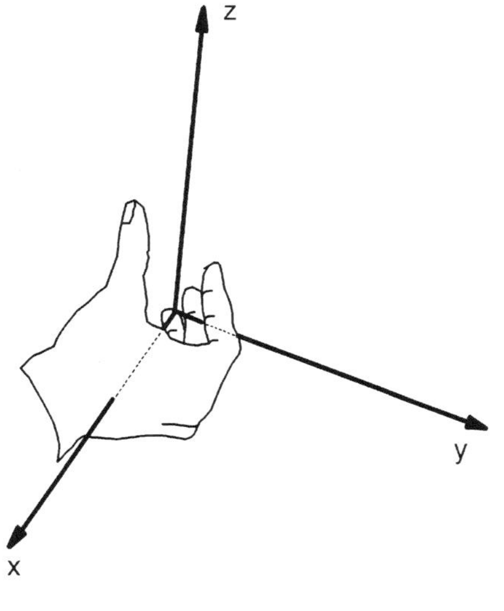

FIGURE 2.6 The x, y, and z axes of the Cartesian coordinate system.

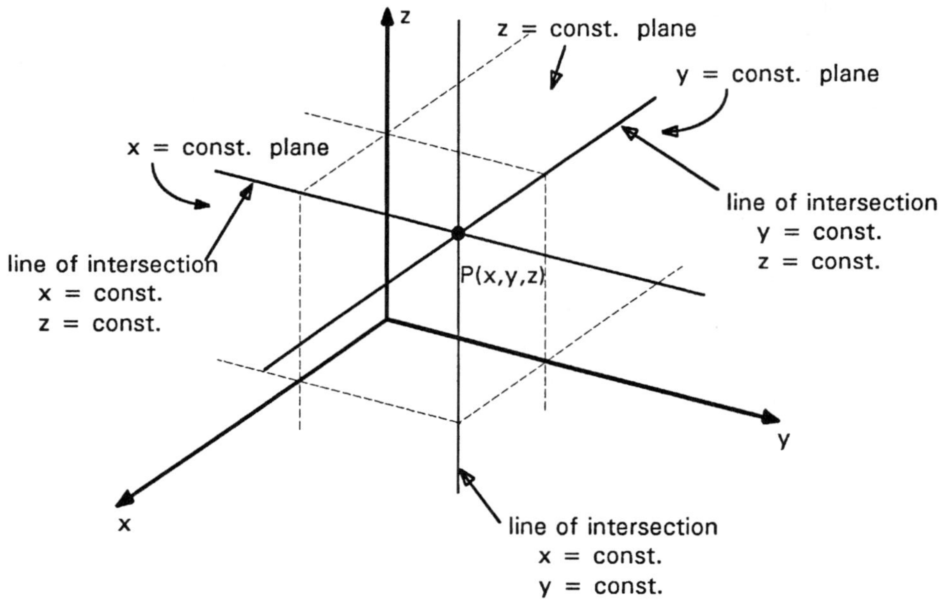

FIGURE 2.7 A point $P(x, y, z)$ in the Cartesian coordinate system.

A unit vector in the direction of the position vector can also be defined by

$$\vec{a}_R = \frac{\vec{R}}{|\vec{R}|} = \vec{a}_x \frac{x}{\sqrt{x^2 + y^2 + z^2}} + \vec{a}_y \frac{y}{\sqrt{x^2 + y^2 + z^2}} + \vec{a}_z \frac{z}{\sqrt{x^2 + y^2 + z^2}}. \tag{2.5}$$

Example 2.1 Determine the position vector for the point $(x, y, z) = (3, 4, 12)$ as well as the unit vector in the direction of that position vector.

Solution The position vector is simply $\vec{R} = \vec{a}_x 3 + \vec{a}_y 4 + \vec{a}_z 12$. The length of the position vector is then calculated to be

$$|\vec{R}| = \sqrt{3^2 + 4^2 + 12^2} = 13.$$

The unit vector in the direction of the position vector is evaluated to be

$$\vec{a}_R = \vec{a}_x \frac{3}{13} + \vec{a}_y \frac{4}{13} + \vec{a}_z \frac{12}{13}.$$

Referring to Fig. 2.8 we see that if the x, y, and z coordinates change by small amounts dx, dy, and dz, then the position vector $d\vec{l}$ from the original point (x, y, z) to the new point $(x + dx, y + dy, z + dz)$ is given by

$$d\vec{l} = \vec{a}_x\, dx + \vec{a}_y\, dy + \vec{a}_z\, dz. \tag{2.6}$$

This is the differential length vector, in Cartesian coordinates.

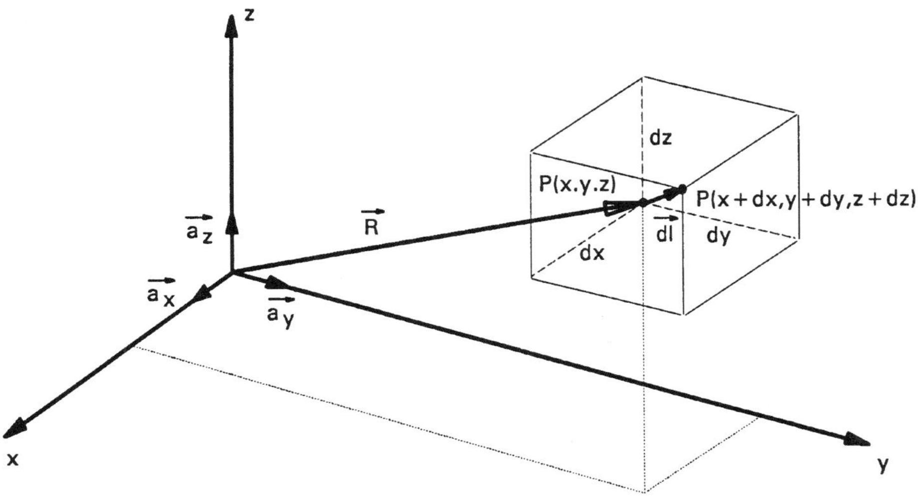

FIGURE 2.8 Position vector and differential length vector in Cartesian coordinates.

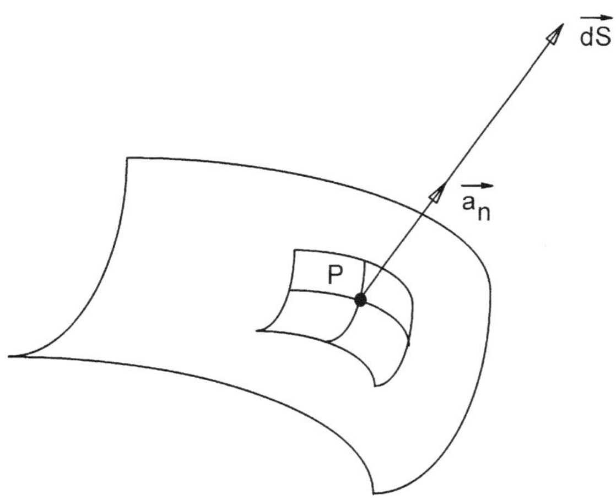

FIGURE 2.9 Smooth curved surface with normal vector to the surface at point *P*.

Figure 2.9 shows a smooth curved surface. We can draw several lines that lie in the surface and pass through a common point P. In a very small region surrounding P the surface appears flat and we can draw a vector $\vec{a}_n$ that is normal to all the lines passing through P. Then $\vec{a}_n$ is defined as the normal to the surface at P. If dS is a small element of surface area, then the vector differential surface is $d\vec{S} = \vec{a}_n\, dS$.

The differential surface formed by changing x by dx and y by dy is $\vec{a}_z\, dx\, dy$ (Fig. 2.10). If all three coordinates experience small changes, then the differential vector area is

$$d\vec{S} = \vec{a}_x\, dy\, dz + \vec{a}_y\, dx\, dz + \vec{a}_z\, dx\, dy. \tag{2.7}$$

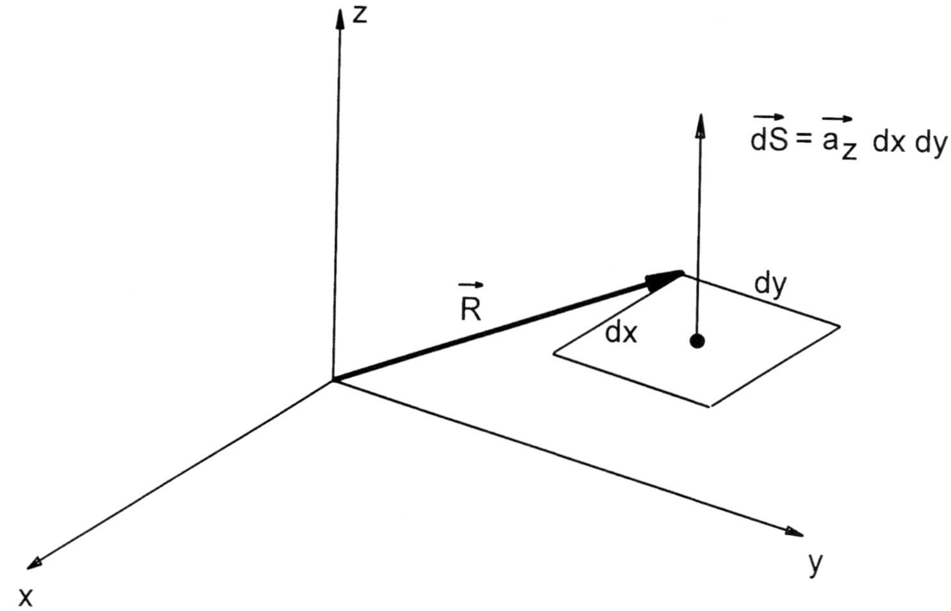

FIGURE 2.10 Differential surface in the z direction.

The differential volume formed by the changes dx, dy and dz is

$$dV = dx\, dy\, dz. \tag{2.8}$$

Volume is always a scalar quantity.

Note that the dimension of each term in the position vector $\vec{R}$ and differential length $d\vec{l}$ is length. The dimensions of each term in the differential area $d\vec{S}$ and of the volume dV are length squared and length cubed, respectively.

2.4.2 Cylindrical Coordinates

In cylindrical coordinates we specify the position of a point in space by giving the radius r from the z axis, the angle ϕ around the z axis measured from the x axis, and the coordinate z above the xy plane. Thus the three coordinates required to specify the position of a point are r, ϕ, and z.

In detail, referring to Fig. 2.11, when $r = r_0$ the point lies somewhere on the circular cylindrical surface of radius r_0, which is coaxial with the z axis. When $\phi = \phi_0$ the point lies somewhere on the half-plane, which is hinged like a door on the z axis and is rotated from the x axis by the angle ϕ_0. The rotation is in the direction of the fingers of the right hand when the thumb points along the z axis. The intersection of the cylinder $r = r_0$ and the half-plane $\phi = \phi_0$ is a straight line parallel to the z axis. The z coordinate is identical to that of the Cartesian coordinate system. When $z = z_0$ we specify a plane perpendicular to the z axis at a height z_0 above the xy plane. The intersection of this plane with the cylinder $r = r_0$ is a circle; its intersection with the half-plane

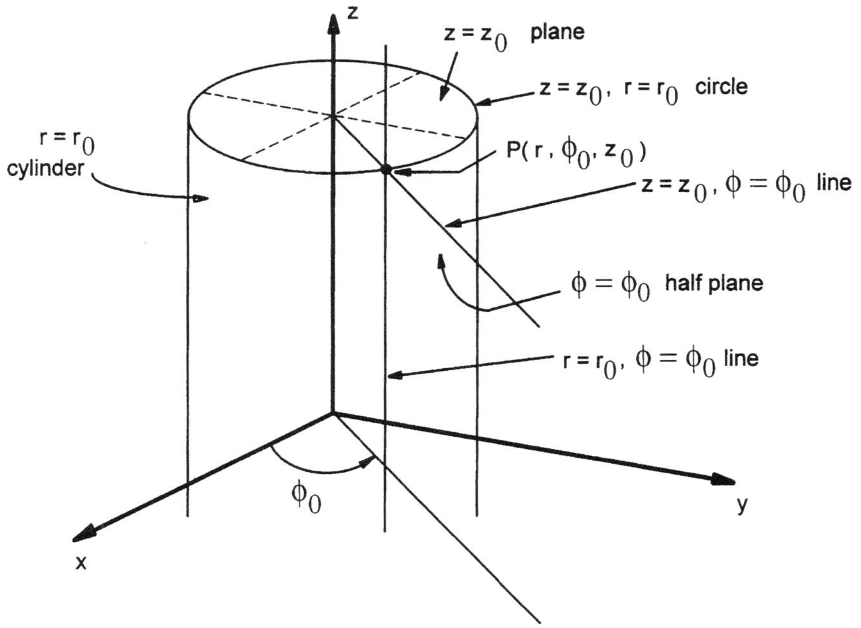

FIGURE 2.11 A point $P(r_0, \phi_0, z_0)$ in the cylindrical coordinate system.

$\phi = \phi_0$ is a straight line radially outward from the z axis. When all three coordinates are given the three surfaces intersect in a point $P(r_0, \phi_0, z_0)$.

To specify any point in all of the space and to not repeat any points, the range of each variable is given by

$$0 \le r < \infty, \quad 0 \le \phi < 2\pi, \quad -\infty < z + \infty. \tag{2.9}$$

When r changes by a small amount dr the point moves as shown in Fig. 2.12a, from P to P_1. The unit vector $\vec{a}_r$ is defined in this direction. When ϕ changes by a small amount $d\phi$ the point moves from P to P_2 (Fig. 2.12b). The length of the arc from P to P_2 is $r\,d\phi$ and the unit vector $\vec{a}_\phi$ is defined as the tangent to the arc in the direction of increasing angle. When z changes by dz the point moves from P to P_3 and the unit vector $\vec{a}_z$ is defined in this direction (Fig. 2.12c). The three unit vectors $\vec{a}_r$, $\vec{a}_\phi$, and $\vec{a}_z$ form a mutually orthogonal right-hand set when taken in this order.

In cylindrical coordinates the position vector is

$$\vec{R} = \vec{a}_r r + \vec{a}_z z. \tag{2.10}$$

If the three coordinates each change by small amounts, so that the point moves from (r, ϕ, z) to $(r + dr, \phi + d\phi, z + dz)$, then the position vector from the original point to the new point is

$$d\vec{l} = \vec{a}_r\, dr + \vec{a}_\phi r\, d\phi + \vec{a}_z\, dz. \tag{2.11}$$

This is the differential length in cylindrical coordinates.

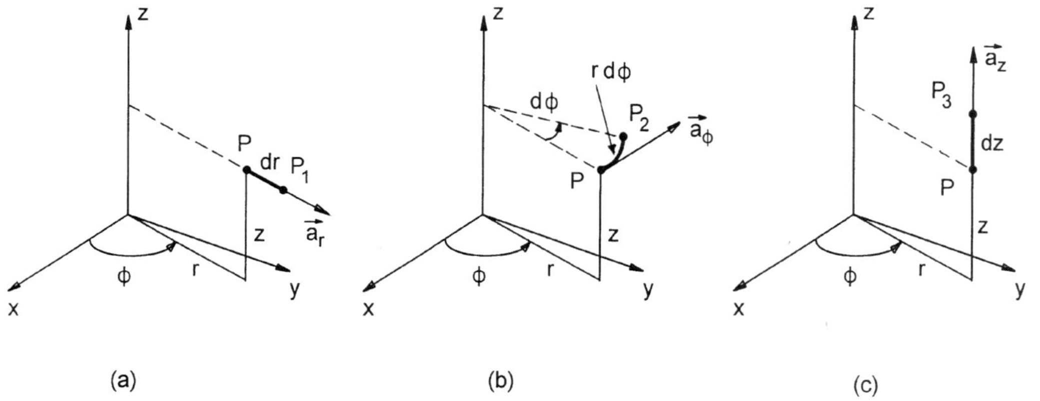

FIGURE 2.12 Unit vectors and differential distances in cylindrical coordinates.

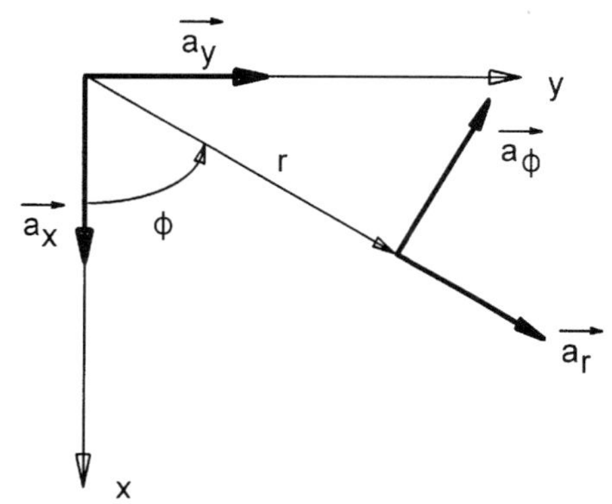

FIGURE 2.13 Unit vectors in Cartesian and cylindrical coordinate systems.

In Eq. (2.8) the position vector $\vec{R}$ appears to be a function of the variables r and z only. This is misleading. It is important to recognize that both unit vectors $\vec{a}_r$ and $\vec{a}_\phi$ are functions of the ϕ coordinate. Figure 2.13 is a view with the z axis pointing directly out of the page. The cylindrical unit vectors can be written in terms of the fixed Cartesian unit vectors:

$$\vec{a}_r = \vec{a}_x \cos\phi + \vec{a}_y \sin\phi, \tag{2.12}$$

$$\vec{a}_\phi = -\vec{a}_x \sin\phi + \vec{a}_y \cos\phi. \tag{2.13}$$

The position vector is now seen to be an explicit function of r and z and an implicit function of ϕ.

Figure 2.14 shows the differential areas in cylindrical coordinates. The differential surface area formed when r changes by dr and ϕ changes by $d\phi$ is $\vec{a}_z\,dr(r\,d\phi) = \vec{a}_z r\,dr\,d\phi$. When r and z change incrementally, we get $\vec{a}_\phi\,dr\,dz$ for the differential

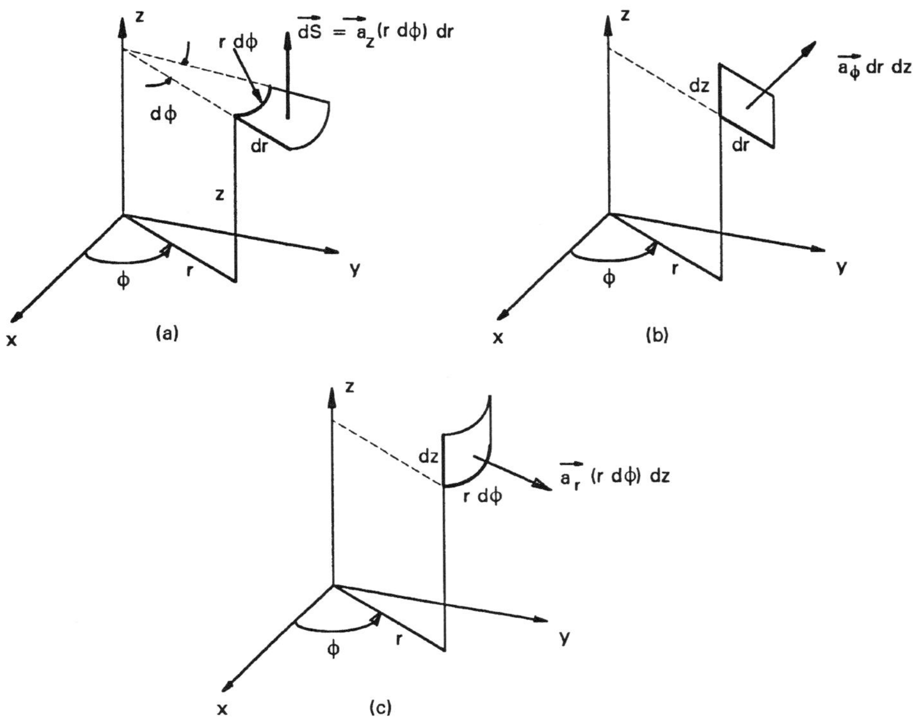

FIGURE 2.14 Surface elements in cylindrical coordinates.

surface area. When ϕ and z change, we get $\vec{a}_r r\ d\phi\ dz$. The differential vector surface area is therefore

$$d\vec{S} = \vec{a}_r r\ d\phi\ dz + \vec{a}_\phi\ dr\ dz + \vec{a}_z r\ d\phi\ dr. \tag{2.14}$$

The differential volume in cylindrical coordinates is

$$dV = dr\,(r\,d\phi)\,dz = r\ dr\ d\phi\ dz. \tag{2.15}$$

Note that the dimension of each term in the position vector $\vec{R}$ and differential length $d\vec{l}$ is length. The dimensions of each term in the differential area $d\vec{S}$ and volume dV are length squared and length cubed, respectively.

2.4.3 Spherical Coordinates

In spherical coordinates we specify the position of a point in space by giving the radius R from the origin, the polar angle θ from the z axis, and the angle ϕ from the x axis. The three coordinates required to specify a point are therefore R, θ, and ϕ.

Referring to Fig. 2.15, when $R = R_0$ the point lies on a sphere of radius R_0 concentric with the origin. When $\theta = \theta_0$ the point lies on a cone of angle θ_0 from

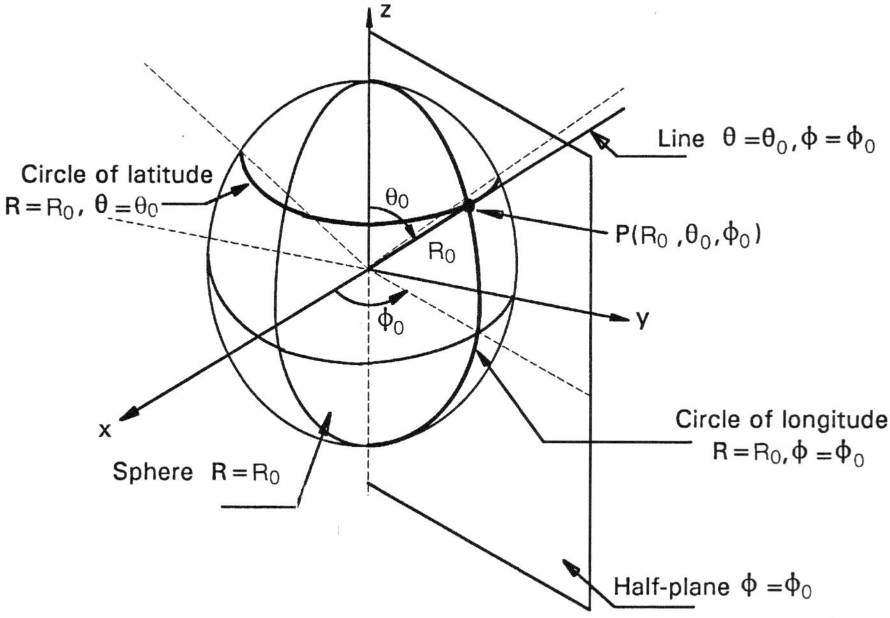

FIGURE 2.15 Sphere of radius R_0 in spherical coordinates.

the positive z axis. The angle ϕ is the same as the cylindrical coordinate and $\phi = \phi_0$ gives the same half-plane hinged on the z axis and rotated around from the x axis by ϕ_0. The intersection of the sphere $R = R_0$ and the cone $\theta = \theta_0$ is a circle, like a line of latitude on a globe. The intersection of the sphere $R = R_0$ and the hinged plane $\phi = \phi_0$ is a semicircle, like a line of longitude on a globe. The cone $\theta = \theta_0$ and the hinged plane $\phi = \phi_0$ intersect in a straight line extending outward from the origin. When all three coordinates are given the three surfaces intersect in a point $P(R_0, \theta_0, \phi_0)$.

To specify any point in all of space and to not repeat any points, the range of each variable is given by

$$0 \le R < \infty, \quad 0 \le \theta \le \pi, \quad 0 \le \phi < 2\pi. \tag{2.16}$$

When R changes by a small amount dR, the point moves from P to P_1, as shown in Fig. 2.16a. The unit vector $\vec{a}_R$ is defined in the direction from P to P_1. When θ changes by $d\theta$, the point changes from P to P_2 and sweeps out a small arc on the surface of the sphere along the line of longitude (Fig. 2.16b). The length of arc from P to P_2 is $R\,d\theta$. The unit vector tangent to the arc from P to P_2 is $\vec{a}_\theta$. When ϕ changes by $d\phi$ the point moves from P to P_3 sweeping out an arc on a circle of latitude (Fig. 2.16c). The radius of this circle is $R \sin\theta$ and hence the length of arc is $R \sin\theta\, d\phi$. The unit vector tangent to the arc from P to P_3 is $\vec{a}_\phi$, just as in cylindrical coordinates. The three unit vectors $\vec{a}_R$, $\vec{a}_\theta$, and $\vec{a}_\phi$ form a mutually orthogonal right-hand set in the order given.

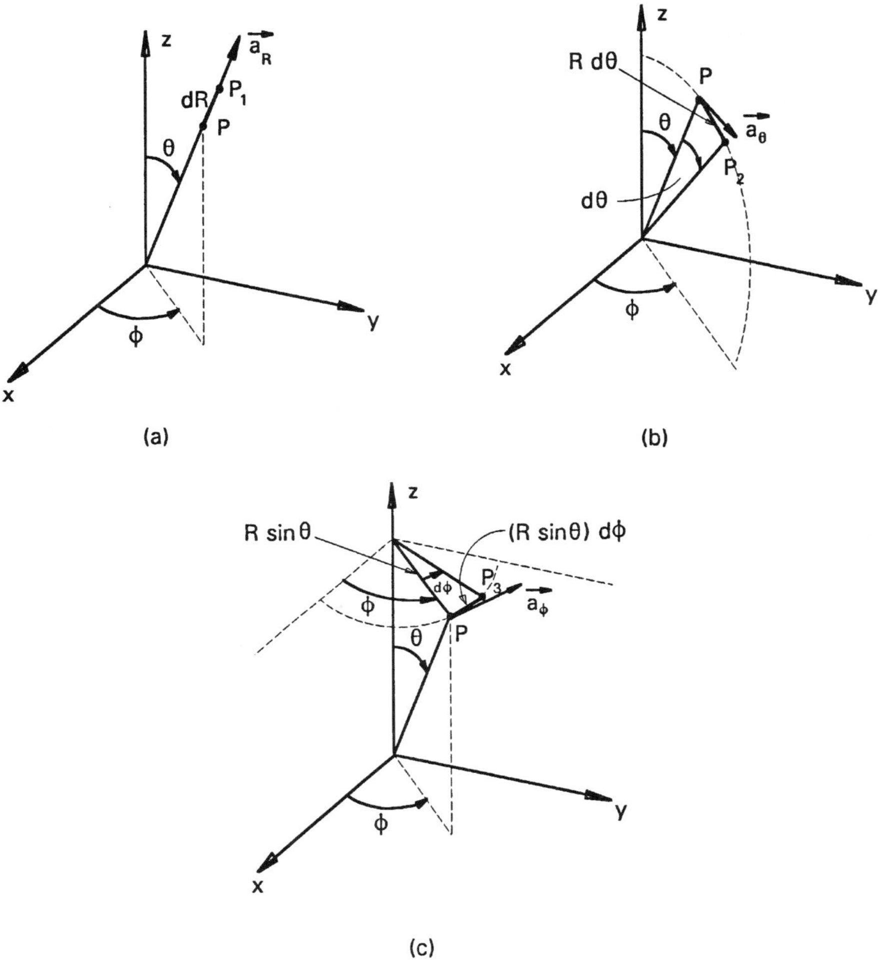

FIGURE 2.16 Unit vectors and differential distances in spherical coordinates.

The position vector in spherical coordinates is given by the deceptively simple formula

$$\vec{R} = \vec{a}_R R. \tag{2.17}$$

If the three coordinates each change by small amounts, so that the point moves from (R, θ, ϕ) to $(R + dR, \theta + d\theta, \phi + d\phi)$, then the position vector from the original point to the new point is

$$d\vec{l} = \vec{a}_R \, dR + \vec{a}_\theta \, R \, d\theta + \vec{a}_\phi R \sin\theta \, d\phi. \tag{2.18}$$

Equation (2.17) is, in fact, not as simple as it appears. This is because the unit vector $\vec{a}_R$ is a function of the θ and ϕ coordinates of the point. In fact, so is $\vec{a}_\theta$ and $\vec{a}_\phi$. To show the dependence we must write the unit vectors in terms of the constant Cartesian unit vectors. We can do this in stages by first going to the cylindrical unit vectors and then to the Cartesian:

$$\vec{a}_R = \vec{a}_r \sin\theta + \vec{a}_z \cos\theta \tag{2.19}$$

$$= \vec{a}_x \sin\theta \cos\phi + \vec{a}_y \sin\theta \sin\phi + \vec{a}_z \cos\theta, \tag{2.20}$$

$$\vec{a}_\theta = \vec{a}_r \cos\theta - \vec{a}_z \sin\theta \tag{2.21}$$

$$= \vec{a}_x \cos\theta \cos\phi + \vec{a}_y \cos\theta \sin\phi - \vec{a}_z \sin\theta, \tag{2.22}$$

$$\vec{a}_\phi = -\vec{a}_x \sin\phi + \vec{a}_y \cos\phi. \tag{2.23}$$

Equation (2.13) has been substituted in Eqs. (2.19) and (2.21) to obtain Eqs. (2.20) and (2.22), respectively. Equation (2.23) is the same as Eq. (2.13) and is repeated here for completeness.

Example 2.2 The point A is specified in rectangular coordinates as $(x, y, z) = (3, 4, 12)$. Give the position vector in Cartesian, cylindrical, and spherical coordinates.

Solution The position vector in Cartesian coordinates is immediately expressed as:

$$\vec{R} = \vec{a}_x 3 + \vec{a}_y 4 + \vec{a}_z 12.$$

In order to find the expression of the position vector in cylindrical coordinates, the r coordinate of point A has to be evaluated: $r = \sqrt{x^2 + y^2} = \sqrt{3^2 + 4^2} = 5$. Hence the position vector of point A in cylindrical coordinates is $\vec{R} = \vec{a}_r 5 + \vec{a}_z 12$.

The position vector in spherical coordinates is simply $\vec{R} = \vec{a}_R R$. The R coordinate of point A is evaluated to be $R = |\vec{R}| = \sqrt{x^2 + y^2 + z^2} = \sqrt{3^2 + 4^2 + 12^2} = 13$. The position vector of point A in spherical coordinates is $\vec{R} = \vec{a}_R 13$.

The spherical differential areas shown in Fig. 2.17. When θ and ϕ change by small amounts the small surface on the sphere has area $(R\,d\theta) \times (R \sin\theta\, d\phi)$ and the normal to this surface is in the radial direction, $\vec{a}_R$. When R and ϕ change, the small surface on the cone has area $(dR) \times (R \sin\theta\, d\phi)$ and the normal to this surface is in the θ direction, $\vec{a}_\theta$. When R and θ change, the surface on the hinged plane has area $(dR) \times (R\, d\theta)$ and the normal to the surface is in the ϕ direction, $\vec{a}_\phi$. Combining these results, the vector differential area in spherical coordinates is

$$d\vec{S} = \vec{a}_R R^2 \sin\theta\, d\theta\, d\phi + \vec{a}_\theta R \sin\theta\, dR\, d\phi + \vec{a}_\phi R\, dR\, d\theta. \tag{2.24}$$

The differential volume formed when all three coordinates change is

$$dV = R^2 \sin\theta\, dR\, d\theta\, d\phi.$$

Note that the dimension of each term in the position vector $\vec{R}$ and differential length $d\vec{l}$ is length. The dimensions of each term in the differential area $d\vec{S}$ and volume dV are length squared and length cubed, respectively.

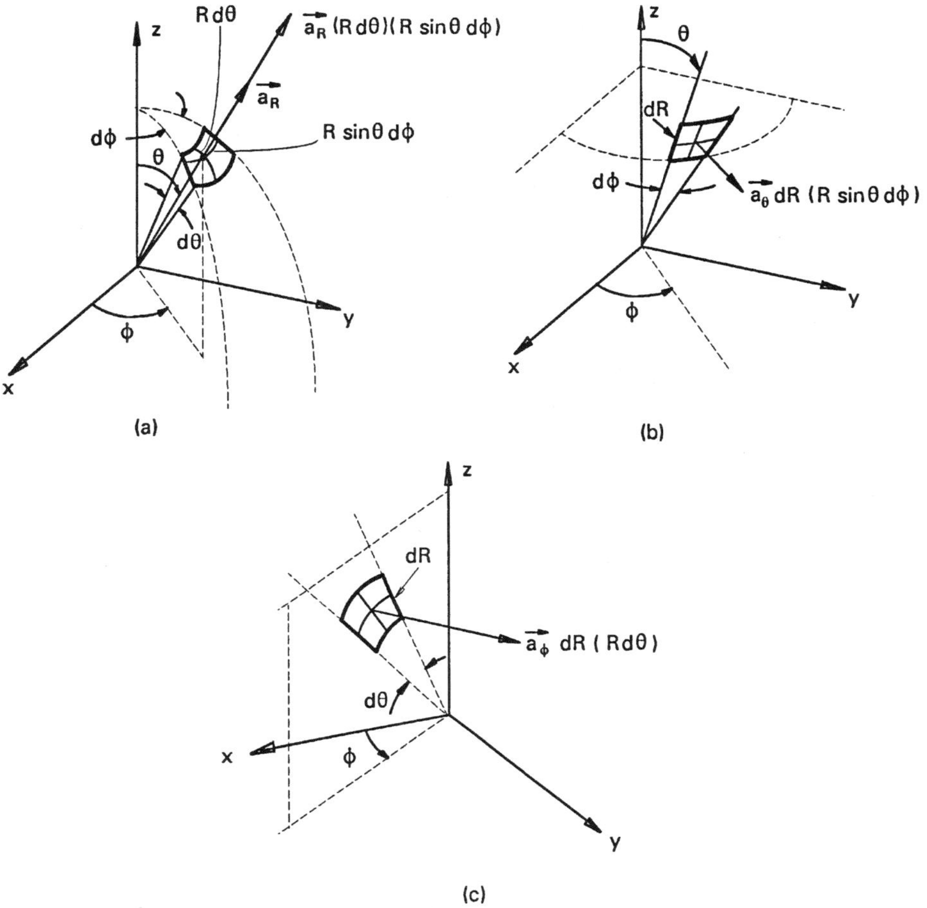

FIGURE 2.17 Surface elements in spherical coordinates.

2.5 VECTOR ARITHMETIC

Just as there are rules for performing arithmetic with scalar quantities, there are rules for arithmetic with vector quantities. Of course, these rules are more complicated than the rules for scalars. Probably you are already familiar with the rules, but they are presented here for review.

In order to reduce unnecessary duplication of equations, when the formulas that follow are applicable to all three of the coordinate systems discussed, we will use the notation u, v, w to represent x, y, z or r, ϕ, z or R, θ, ϕ.

2.5.1 Magnitude

The magnitude of a vector is the strength of the field quantity without regard to direction. In geometric terms it is the length of the line used to represent the vector.

The magnitude of the vector $\vec{A}$ is written as $|\vec{A}|$ or, simply, A. If

$$\vec{A} = \vec{a}_u A_u + \vec{a}_v A_v + \vec{a}_w A_w, \tag{2.25}$$

then

$$|\vec{A}| = A = +\sqrt{A_u^2 + A_v^2 + A_w^2}. \tag{2.26}$$

Since $\vec{a}$ represents a unit vector, $|\vec{a}| = 1$.

2.5.2 Addition

Vector addition is accomplished by the scalar addition of identical components of the vectors to be added. For example, the vector addition of the vectors $\vec{A}$ and $\vec{B}$ to give the sum $\vec{C}$ is given as follows: If

$$\vec{A} = \vec{a}_u A_u + \vec{a}_v A_v + \vec{a}_w A_w \tag{2.27}$$

and

$$\vec{B} = \vec{a}_u B_u + \vec{a}_v B_v + \vec{a}_w B_w, \tag{2.28}$$

then

$$\vec{C} = \vec{A} + \vec{B} = \vec{a}_u (A_u + B_u) + \vec{a}_v (A_v + B_v) + \vec{a}_w (A_w + B_w). \tag{2.29}$$

Figure 2.18 shows how vector addition is accomplished graphically.

The fact that the unit vectors are not constant vectors in the cylindrical and spherical coordinate systems means that we must be very careful in adding vectors expressed in those coordinate systems. For example, if the velocity of an object is given in spherical coordinates at two different points in space and we wish to add these velocities together (perhaps as a step in finding the average velocity), then we cannot simply add the R, θ, and ϕ components because these components are in different directions at the two different points. The addition can only be done by first changing the spherical coordinates to Cartesian coordinates.

Example 2.3 Inside a cathode ray tube (used for an oscilloscope), an electron is subjected to a force $\vec{F}_1 = \vec{a}_z 24$, while traveling in the z direction. As it passes through the deflection zone, two additional forces $\vec{F}_2 = \vec{a}_x 8$ and $\vec{F}_3 = \vec{a}_y 6$ act on the electron.

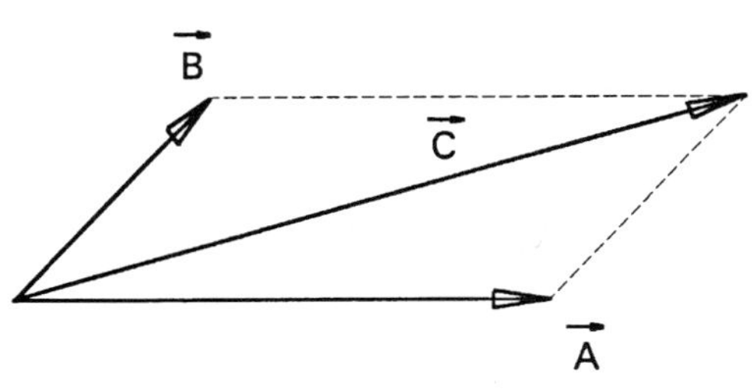

FIGURE 2.18 Addition of two vectors.

Determine the total force that is acting on the electron, as the sum of the three individual forces.

Solution The total force that is acting on the electron will be the sum of the forces:

$$\vec{F}_T = \vec{F}_1 + \vec{F}_2 + \vec{F}_3 = \vec{a}_z 24 + \vec{a}_y 6 + \vec{a}_x 8.$$

Rearranging terms, we get

$$\vec{F}_T = \vec{a}_x 8 + \vec{a}_y 6 + \vec{a}_z 24.$$

Example 2.4 Three balls located at $(R, \theta, \phi) = (10, 0, 0), (10, \pi/2, 0)$, and $(10, \pi/2, \pi/2)$ each have velocity $\vec{u} = \vec{a}_R 5$. Express their velocities in rectangular coordinates and give the sum of their velocities.

Solution The first ball is on the z axis, moving in the radial or z direction. Hence $\vec{u}_1 = \vec{a}_z 5$. The second ball is on the x axis, moving in the radial or x direction. Hence $\vec{u}_2 = \vec{a}_x 5$. The third ball is on the y axis, moving in the radial or y direction. Hence $\vec{u}_3 = \vec{a}_y 5$. The sum of the three velocities is $\vec{u} = 5(\vec{a}_x + \vec{a}_y + \vec{a}_z)$.

2.5.3 Multiplication by a Scalar

To multiply a vector by a scalar we simply multiply each component of the vector by the scalar. If

$$\vec{A} = \vec{a}_u A_u + \vec{a}_v A_v + \vec{a}_w A_w \tag{2.30}$$

then

$$\alpha\vec{A} = \vec{a}_u \alpha A_u + \vec{a}_v \alpha A_v + \vec{a}_w \alpha A_w. \tag{2.31}$$

2.5.4 Scalar or Dot Product

The scalar product is an operation that is performed often times in the study of electromagnetics. It is encountered in the operation of line and surface integrals, in the integral and differential formulations of Maxwell's equations, and in the expression of two of the boundary conditions that are applied to electromagnetic fields, to name only a few of the more frequent usages.

The scalar product of two vectors $\vec{A}$ and $\vec{B}$ is given by

$$\vec{A}\cdot\vec{B} = |\vec{A}||\vec{B}| \cos\theta_{AB} = AB \cos\theta_{AB}, \tag{2.32}$$

where θ_{AB} is the angle from the first vector to the second (Fig. 2.19).

An application of the scalar product is the evaluation of the work done in moving a body for a distance L, when the force applied to the body is not necessarily in the direction of the motion (Fig. 2.20). The work done to move the body from point A to point B is

$$W = \vec{F}\cdot\vec{L} = |\vec{F}||\vec{L}| \cos\theta_{FL} = FL \cos\theta_{FL}$$

where $\vec{L}$ is a vector in the direction of motion and has an amplitude equal to L.

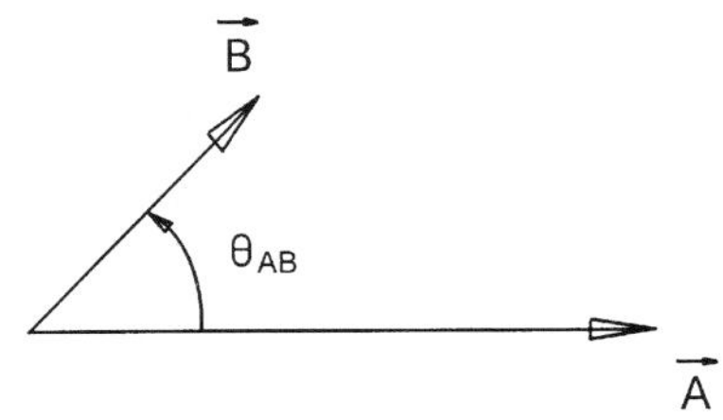

FIGURE 2.19 Two vectors, whose scalar product is to be taken.

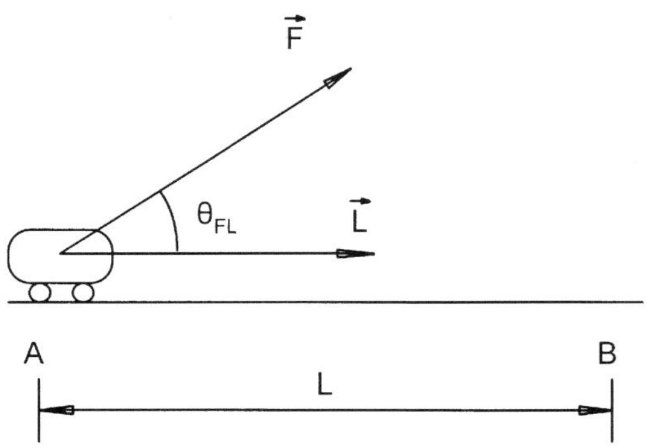

FIGURE 2.20 Force acting on a body to be moved from point A to point B.

Several properties of the scalar product are worth stating. If the two vectors are in the same direction, $\theta_{AB} = 0^\circ$, then $\vec{A}\cdot\vec{B} = AB$; if they are in opposite directions, $\theta_{AB} = 180^\circ$, then $\vec{A}\cdot\vec{B} = -AB$. If the two vectors are orthogonal, $\theta_{AB} = \pm 90^\circ$, then $\vec{A}\cdot\vec{B} = 0$. The scalar product is commutative, $\vec{A}\cdot\vec{B} = \vec{B}\cdot\vec{A}$. Finally, if one of the vectors is a unit vector $\vec{a}_s$, then the dot product gives the component of the other vector in the direction of the unit vector, $\vec{a}_s\cdot\vec{A} = A\cos\theta = A_s$ (Fig. 2.21).

The name "scalar product" is given to this operation because the result is a scalar. The popular terminology "dot product" comes from the dot $(\cdot)$ symbol used to represent this type of product.

The scalar product of two vectors is obtained by taking the product of corresponding components of the vectors and then adding the products together. Thus if

$$\vec{A} = \vec{a}_u A_u + \vec{a}_v A_v + \vec{a}_w A_w \tag{2.33}$$

and

$$\vec{B} = \vec{a}_u B_u + \vec{a}_v B_v + \vec{a}_w B_w, \tag{2.34}$$

then

$$\vec{A}\cdot\vec{B} = A_u B_u + A_v B_v + A_w B_w. \tag{2.35}$$

Example 2.5 A cart is being pulled with a force $\vec{F} = \frac{1}{2}\vec{a}_x + (\sqrt{3}/2)\vec{a}_y$ along a straight road in the x direction for a distance of 100 meters. Determine the work done. Determine the angle that the force $\vec{F}$ is making with the road.

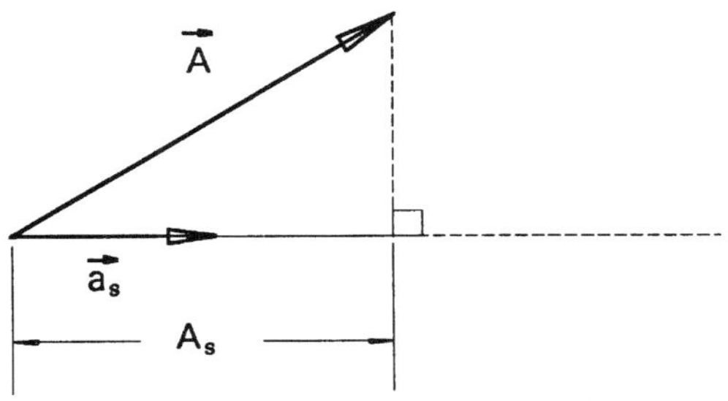

FIGURE 2.21 Projection of a vector on a line obtained as the scalar product of the vector and a unit vector along that line.

Solution Since the road is in the (x) direction, $\vec{L} = 100\,\vec{a}_x$. The work done is then calculated as

$$W = \vec{F}\cdot\vec{L} = \left(\frac{1}{2}\vec{a}_x + \frac{\sqrt{3}}{2}\vec{a}_y\right)\cdot(100\,\vec{a}_x) = \frac{1}{2}\cdot 100 + 0.0 + \frac{\sqrt{3}}{2}\cdot 0 = 50 \text{ Joules}.$$

In order to determine the angle θ between the force $\vec{F}$ and $\vec{L}$ we can use

$$W = \vec{F}\cdot\vec{L} = F\cdot L\cdot\cos\theta.$$

But

$$F = |\vec{F}| = \sqrt{\left(\frac{1}{2}\right)^2 + \left(\frac{\sqrt{3}}{2}\right)^2} = 1 \quad \text{and} \quad L = |\vec{L}| = 100.$$

Thus

$$\cos\theta = \frac{W}{F\cdot L} = \frac{50}{1\cdot 100} = \frac{1}{2}$$

and

$$\theta = \cos^{-1}\left(\frac{1}{2}\right) = \pm\, 60^\circ.$$

2.5.5 Vector or Cross Product

The vector product is an operation that is often encountered in the study of electromagnetics. It is found in the differential formulation of Maxwell's equations and in the expression of two of the boundary conditions that are applied to electromagnetic fields, to name only a couple of the more frequent usages.

A practical example of the use of the vector product is found in calculating the torque produced by a force acting on a rotating body (Fig. 2.22). The torque $\vec{T}$ developed by a force $\vec{F}$ acting at point A characterized by a position vector $\vec{r}$ from the rotation axis is calculated as $\vec{T} = \vec{r}\times\vec{F}$.

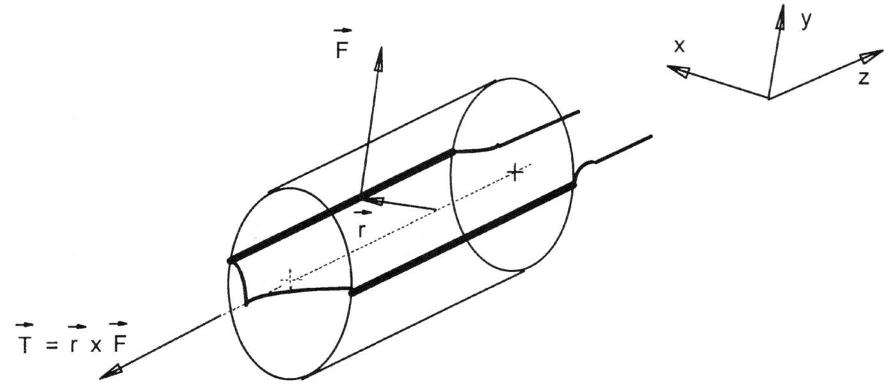

FIGURE 2.22 Torque due to the force acting on the conductors of a rotating machine.

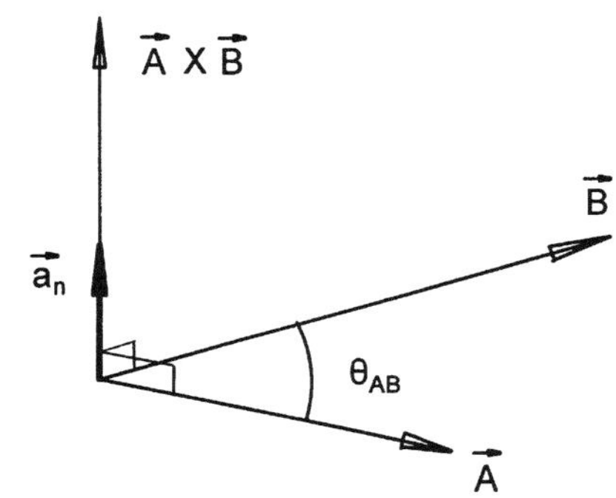

FIGURE 2.23 Cross product of two vectors.

The vector product of two vectors $\vec{A}$ and $\vec{B}$ is given by

$$\vec{A} \times \vec{B} = \vec{a}_n |\vec{A}||\vec{B}| \sin \theta_{AB} = \vec{a}_n AB \sin \theta_{AB}, \tag{2.36}$$

where θ_{AB} is the angle from the first vector to the second (Fig. 2.23), and $\vec{a}_n$ is the unit vector that is perpendicular to both vectors in the direction of the thumb when the fingers of the right hand turn $\vec{A}$ toward $\vec{B}$. The resultant vector is perpendicular to the two original vectors.

There are several important properties of the vector product. If the two vectors are collinear, $\theta_{AB} = 0^\circ$ or 180°, then $\vec{A} \times \vec{B} = 0$. If the two vectors are orthogonal, $\theta_{AB} = 90^\circ$, then $\vec{A} \times \vec{B} = \vec{a}_n AB$. The vector product is not commutative. In fact, $\vec{A} \times \vec{B} = -\vec{B} \times \vec{A}$. If the two vectors are orthogonal unit vectors, then the vector product is a mutually orthogonal unit vector in the direction defined by the right-hand rule. In particular, for the three coordinate systems we are interested in, we have

$$\vec{a}_z = \vec{a}_x \times \vec{a}_y, \tag{2.37}$$

$$\vec{a}_z = \vec{a}_r \times \vec{a}_\phi, \tag{2.38}$$

$$\vec{a}_\phi = \vec{a}_R \times \vec{a}_\theta. \tag{2.39}$$

Rather than memorize these relations, you should sketch the coordinate system and use your right hand to obtain them. There are six additional relations (e.g., $\vec{a}_\theta = \vec{a}_\phi \times \vec{a}_R$) that can be obtained in the same way.

The name "vector product" is given to this operation because the result is a vector. The popular terminology "cross product" comes from the ($\times$) symbol used to represent this type of product.

The formula for taking the vector product of two vectors represented in the same coordinate system is best remembered as a determinant. If

$$\vec{A} = \vec{a}_u A_u + \vec{a}_v A_v + \vec{a}_w A_w \tag{2.40}$$

and

$$\vec{B} = \vec{a}_u B_u + \vec{a}_v B_v + \vec{a}_w B_w, \tag{2.41}$$

then

$$\vec{A} \times \vec{B} = \begin{vmatrix} \vec{a}_u & \vec{a}_v & \vec{a}_w \\ A_u & A_v & A_w \\ B_u & B_v & B_w \end{vmatrix} \tag{2.42}$$

$$= \vec{a}_u(A_v B_w - A_w B_v) + \vec{a}_v(A_w B_u - A_u B_w) + \vec{a}_w(A_u B_v - A_v B_u). \tag{2.43}$$

Example 2.6 Figure 2.24 shows the force acting on a conductor inside a rotating machine. The distance from the conductor to the axis of rotation is $d = 12$ cm. Assume the force $\vec{F}$ is acting at point A and is given by $\vec{F} = 50\vec{a}_\phi$ N. Determine the torque developed by the given force.

Solution The position vector of point A is $\vec{d} = 0.12\vec{a}_r$ m in cylindrical coordinates. The torque is then calculated as

$$\vec{T} = \vec{d} \times \vec{F} = (0.12\vec{a}_r) \times (50\vec{a}_\phi) = 6\vec{a}_z \text{ N-m}.$$

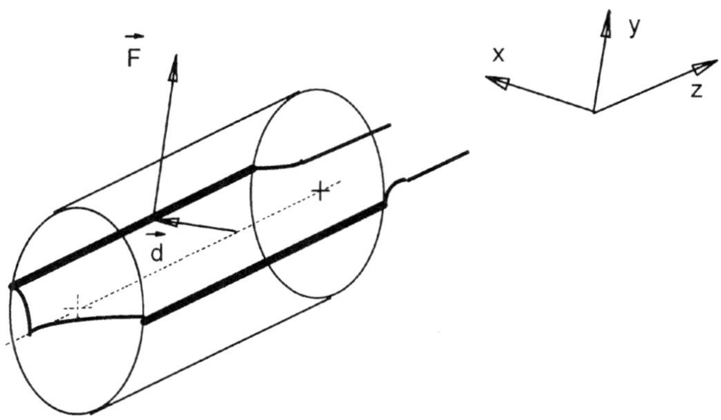

FIGURE 2.24 The force acting on the conductors of the rotating machine in Example 2.6.

In evaluating the torque, Eq. (2.36) was modified as

$$\vec{a}_r \times \vec{a}_\phi = \vec{a}_z.$$

2.5.6 Vector Identities

There are two vector identities that we will use. The first is

$$\vec{A} \cdot (\vec{B} \times \vec{C}) = \vec{C} \cdot (\vec{A} \times \vec{B}) = \vec{B} \cdot (\vec{C} \times \vec{A}). \tag{2.44}$$

This can be proved by substitution of the three vectors in component form. More interestingly, we can show that if $\vec{B}$ and $\vec{C}$ form the sides of a parallelogram, then $\vec{D} = \vec{B} \times \vec{C}$ is the vector area of the parallelogram (Fig. 2.25). Furthermore, $\vec{A} \cdot \vec{D} = \vec{A} \cdot (\vec{B} \times \vec{C})$ gives the volume of the parallelepiped. The other two forms also represent this volume and hence the identity holds.

The second identity is

$$\vec{A} \times (\vec{B} \times \vec{C}) = \vec{B}(\vec{A} \cdot \vec{C}) - \vec{C}(\vec{A} \cdot \vec{B}). \tag{2.45}$$

This can also be proved by substitution of the three vectors in component form. A more elegant proof is tedious and adds no insights of importance. An interesting proof is given in Ref. [1].

2.5.7 Vector Arithmetic Using MathCad

MathCad is a mathematics program that has a number of interesting features, which include regular real and complex arithmetic, vectors and matrices, symbolic manipulation of expressions, two- and three-dimensional graphics, and other higher level

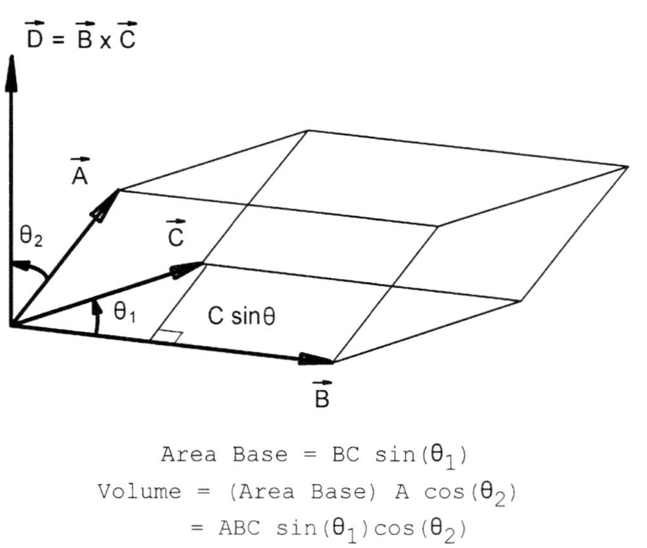

FIGURE 2.25 Physical significance of the cross and dot products.

math functions. As far as this section is concerned, the vector arithmetic capabilities will be explored. It is assumed that the student has some familiarity with MathCad, at least with entering expressions. Appendix B offers a brief introduction to MathCad for the novice user.

2.5.7.1 Vector Definition Vectors are defined in terms of their coordinates. The three coordinates are listed in a column, from top to bottom, surrounded by round or square brackets:

$$\vec{F} = F_x\vec{a}_x + F_y\vec{a}_y + F_z\vec{a}_z = \begin{pmatrix} F_x \\ F_y \\ F_z \end{pmatrix}.$$

MathCad recognizes this format and vectors are defined using the matrix function. The vectors that we will be using are defined as matrices with three rows and one column. The names of vectors do not have arrows on top. They do look like regular variables. A vector is defined in MathCad as follows:

$$D: = \begin{pmatrix} 1 \\ 2 \\ 3 \end{pmatrix}.$$

The name of the variable is entered and then the assign operator is entered by typing ":". The matrix can be created two ways. Select Math from the top menu. Then select Matrices from the submenu. Enter 3 for the number of rows and 1 for the number of columns; then press Enter or select the Create button. The other way is to select the second palette of buttons of math functions on the left side of the screen. Then, select the fourth button from the top from that palette (it looks like a matrix with little black squares as its elements). That brings you to the matrix window, where you have to enter the number of rows and columns.

Example 2.7 Define, in MathCad, the vector $\vec{F} = 3\vec{a}_x + 4\vec{a}_y + 12\vec{a}_z$.

Solution The vector $\vec{F}$ is defined in MathCad as

$$F: = \begin{pmatrix} 3 \\ 4 \\ 12 \end{pmatrix}.$$

2.5.7.2 Magnitude of a Vector Using MathCad The magnitude of a vector can be obtained very easily in MathCad using the function "$|x|$", which can be located as the 8th button in the first palette on the left side of the screen. The magnitude of the vector can be assigned to another variable or can simply be evaluated by typing the " = " sign after the absolute value of the vector.

Example 2.8 Determine the magnitude of the vector $\vec{F}$ defined in the previous example using MathCad.

Solution Once the vector is defined, place the cursor under the vector (MathCad evaluates expressions from the top of the page to the bottom) and type $|F|$ followed by the equal sign. MathCad will display the answer right after the equal sign:

$$|F| = 13.$$

2.5.7.3 Multiplication by a Scalar Using MathCad The multiplication by a scalar is accomplished using the regular multiplication sign.

Example 2.9 Determine the unit vector in the direction of the vector $\vec{F}$ from the previous example using MathCad.

Solution The unit vector in the direction of $\vec{F}$ is obtained by dividing the coordinates of the vector by the magnitude of the vector; in other words,

$$\vec{u}_F = \frac{\vec{F}}{|\vec{F}|}.$$

In MathCad, the unit vector is found as follows:

$$U := \frac{F}{|F|}.$$

The value is displayed by typing U followed by the equal sign:

$$U = \begin{pmatrix} 0.2308 \\ 0.3077 \\ 0.9231 \end{pmatrix}.$$

2.5.7.4 Dot Product Using MathCad The scalar or dot product of two vectors is performed in MathCad using the regular multiplication sign.

Example 2.10 Solve Example 2.5 using MathCad.

Solution From Example 2.5, the force that is acting on the cart is given by $\vec{F} = \frac{1}{2}\vec{a}_x + (\sqrt{3}/2)\vec{a}_y$. The vector $\vec{L}$ is defined as $\vec{L} = 100\vec{a}_x$. The two vectors will be defined in MathCad as

$$F := \begin{pmatrix} \frac{1}{2} \\ \frac{\sqrt{3}}{2} \end{pmatrix} \qquad L := \begin{pmatrix} 100 \\ 0 \end{pmatrix}.$$

The two vectors can be defined as having three coordinates, with the third coordinate being zero. The dot product, which gives the work done in pulling the cart along that distance, is calculated as $W := F \cdot L$, and the numerical result is displayed by typing W followed by the equal sign: $W = 50$.

The angle between the two vectors is obtained using the equation

$$\theta = \cos^{-1}\left(\frac{\vec{F}\cdot\vec{L}}{|\vec{F}||\vec{L}|}\right).$$

The MathCad operation is described below:

$$\theta := a\cos\left(\frac{F\cdot L}{|F|\cdot|L|}\right).$$

The Greek letter θ is obtained by locating it on palette 4 (palettes 4 and 5 contain Greek symbols) on the left side of the screen, and clicking on it with the mouse. The numerical value of θ is obtained by retyping the variable followed by the equal sign. The result will be displayed in radians. To convert the result to degrees, click with the mouse on the placeholder (a small black rectangle visible right after the numerical result is obtained) at the right of the numerical result, and type "deg". The result is converted to degrees when the Enter key is pressed. The screen will look like this: $\theta = 60$ deg.

2.5.7.5 Cross Product Using MathCad The cross product of two vectors can be performed in MathCad by using the $\vec{x}\times\vec{y}$ button from the second palette on the left side of the screen. As soon as the cross product button is selected with the mouse, two placeholders separated by an "X" appear on the working area in front of you. The two vectors have to be entered at the two placeholders. The result can be assigned to a variable or can be displayed directly by typing the equal sign. For example,

$$A := \begin{pmatrix} 1 \\ 0 \\ 0 \end{pmatrix}, \quad B := \begin{pmatrix} 0 \\ 1 \\ 0 \end{pmatrix},$$

$$D := A \times B,$$

$$D = \begin{pmatrix} 0 \\ 0 \\ 1 \end{pmatrix}.$$

MathCad allows the symbolic evaluation of algebraic expressions. The vector algebra operations can also be performed symbolically in MathCad.

Example 2.11 Evaluate the cross product of the following vectors symbolically using MathCad:

$$\vec{V}_1 = 2xy\cdot\vec{a}_x + x^2z\cdot\vec{a}_y + 6yz^2\cdot\vec{a}_z,$$

$$\vec{V}_2 = 3z^2\cdot\vec{a}_x + 2y^2\cdot\vec{a}_y + 5x^2\cdot\vec{a}_z.$$

Solution The two vectors are explicitly entered in the two placeholders of the cross product. The expression is then selected by clicking the cross product and using the up-arrow key until the whole expression is surrounded by a blue rectangle. Select Symbolic from the top menu, then select Evaluate and Evaluate Symbolically from the submenus:

$$\begin{pmatrix} 2xy \\ x^2z \\ 6yz^2 \end{pmatrix} \times \begin{pmatrix} 3z^2 \\ 2y^2 \\ 5x^2 \end{pmatrix}.$$

After the symbolic evaluation, the result is

$$\begin{pmatrix} 5x^4z - 12y^3z^2 \\ 18yz^4 - 10x^3y \\ 4xy^3 - 3x^2z^3 \end{pmatrix}.$$

2.6 LINE INTEGRALS

The line integration is an operation that is widely used in the study of electromagnetic phenomena, from the evaluation of the electric potential and the current due to a surface current density, to the definition and applications of Ampère's and Faraday's laws (which will be introduced and studied thoroughly in later chapters).

Line integration consists of calculating the component of a vector field that is along a small element of the path, multiplying the component by the element length and then adding up these products over the entire path length.

For example, we might want to find the work done by the force field $\vec{F}$ in moving an object along a particular path L going from point P_1 to point P_2 (Fig. 2.26). At the point P a small element of path length is $d\vec{l}$, taken in the direction of the tangent to the path. Then the work dW to move along the element of length is found by taking

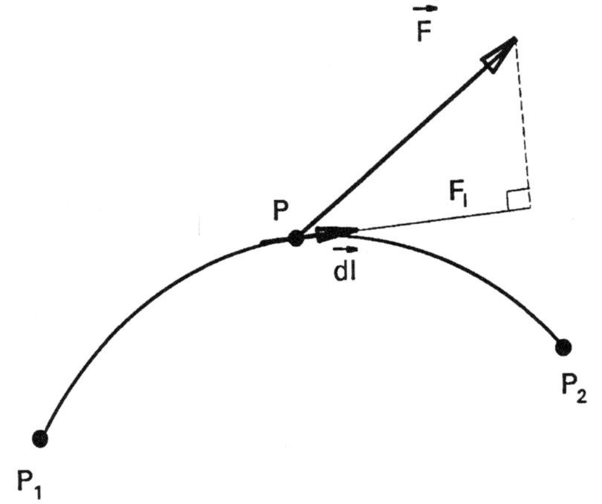

FIGURE 2.26 Vector field to be integrated over a line.

the component of force in that direction and multiplying it by the path length:

$$dW = F_l\, dl = \vec{F}\cdot d\vec{l}. \tag{2.46}$$

To find the total work over the path we add up or integrate the work for each element of the path:

$$W = \int_L \vec{F}\cdot d\vec{l}. \tag{2.47}$$

This is the line integral of the vector field $\vec{F}$ along the path L. It is important to recognize that the result depends on both the field and the particular path.

Example 2.12 Calculate the line integral of

$$\vec{F} = \vec{a}_x 2xy + \vec{a}_y x^2 yz + \vec{a}_z x^3 yz^2 \quad \text{(N)} \tag{2.48}$$

from P_1 to P_2 along the paths (a) L_1 and L_2 and (b) L_3, all lying in the $z = 4$ plane (Fig. 2.27).

Solution We will use Cartesian coordinates because the field is given that way and because it is easy to describe the paths shown in that system. The differential length is $d\vec{l} = \vec{a}_x\, dx + \vec{a}_y\, dy + \vec{a}_z\, dz$ so that

$$\begin{aligned} W &= \int_L \vec{F}\cdot d\vec{l} \\ &= \int_L 2xy\, dx + \int_L x^2 yz\, dy + \int_L x^3 yz^2\, dz. \end{aligned} \tag{2.49}$$

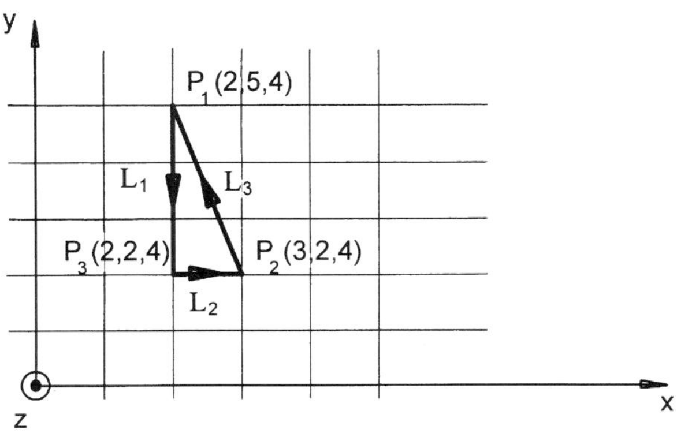

FIGURE 2.27 Paths for line integrals in Example 2.12.

(a) Along the path L_1, dx and dz are zero, so the first and third integrals are zero. The second integral is then

$$W_1 = \int_{y=5}^{2} (2^2)y(4)\,dy = -168 \quad \text{(J)}. \tag{2.50}$$

In this integrand we have substituted $x = 2$ and $z = 4$, which is the equation of the line L_1.

Along the path L_2, dy and dz are zero, so the second and third integrals in Eq. (2.49) are zero. The first integral is

$$W_2 = \int_{x=2}^{3} 2x(2)\,dx = 10 \quad \text{(J)}. \tag{2.51}$$

In this integrand we have substituted $y = 2$ and $z = 4$, which is the equation of the line L_2. The total work along the lines L_1 and L_2 is the sum of the work over each line:

$$W_{1,2} = W_1 + W_2 = -158 \quad \text{(J)}.$$

(b) To evaluate the line integral along the path L_3 we must first obtain the equation of the sloped straight line. Letting $y = mx + b$, substituting the end points (2,5) and (3,2), and solving for m and b, we obtain $y = -3x + 11$. The third integral in Eq. (2.49) is again zero ($dz = 0$). Thus

$$W_3 = \int_{x=2}^{3} 2xy\,dx + \int_{y=5}^{2} x^2yz\,dy.$$

In the first integral we substitute $-3x + 11$ for y. In the second integral we could solve $y = -3x + 11$ for x and substitute (along with $z = 4$). Alternately, we can change the variable of integration in the second integrand from y to x. Since $y = -3x + 11$, $dy = -3\,dx$ and we can write for the two integrals

$$W_3 = \int_{x=2}^{3} 2x(-3x + 11)\,dx + \int_{x=2}^{3} x^2(-3x + 11)(4)(-3\,dx)$$

$$= \int_{2}^{3} (36x^3 - 138x^2 + 22x)\,dx = -234 \quad \text{(J)}.$$

Note that the result along the path L_1–L_2 is not the same as the result along the path L_3. The line integral between two points is generally dependent on the path taken between the points. Later on, we will learn what must be true about the vector field $\vec{F}$ for the line integral to be independent of the path.

2.6.1 Line Integral Using MathCad

MathCad allows the calculation and symbolic evaluation of integrals. It is therefore appropriate to investigate the application of MathCad to line integration. The previous example will be solved using MathCad. In setting up the line integral, the definition of the unit vectors in the x, y, and z directions is required. These vectors are defined as

$$\vec{a}_x \rightarrow a_x := \begin{pmatrix} 1 \\ 0 \\ 0 \end{pmatrix},$$

$$\vec{a}_y \rightarrow a_y := \begin{pmatrix} 0 \\ 1 \\ 0 \end{pmatrix},$$

$$\vec{a}_z \rightarrow a_z := \begin{pmatrix} 0 \\ 0 \\ 1 \end{pmatrix}.$$

The vector $\vec{F}$ is defined in MathCad as

$$\vec{F} = \vec{a}_x 2xy + \vec{a}_y x^2 yz + \vec{a}_z x^3 yz^2 \rightarrow F(x, y, z) := \begin{pmatrix} 2xy \\ x^2 yz \\ x^3 yz^2 \end{pmatrix}.$$

The line integral along line L_1 is then calculated as

$$W1 := \int_5^2 F(2, y, 4) \cdot a_y \, dy$$

$$W1 = -168.$$

The integral symbol is found in one of the palettes on the left side of the screen (just like the cross product and the matrix buttons).

In a similar way, the integral along line L_2 is found to be

$$W2 := \int_2^3 F(x, 2, 4) \cdot a_x \, dx,$$

$$W2 = 10.$$

The line integral along the second path P_2 is calculated by substituting for the proper variable (x or y) in the function $F(x, y, z)$.

$$W3 := \int_2^3 F(x, -3x + 11, 4) \cdot a_x \, dx + \int_5^2 F\left(\frac{11 - y}{3}, y, 4\right) \cdot a_y \, dy$$

$$W3 = -234.$$

2.7 SURFACE INTEGRALS

Figure 2.28 shows a vector field $\vec{u}$ and an imaginary surface S. If we want to know the quantity of the field that passes through the surface, then we must evaluate the surface integral of the field.

Let us assume that the field $\vec{u}$ represents the velocity of a fluid flowing in a channel as a function of space and time, and we want to know the volume of the fluid that flows through the surface per unit of time. Consider a small element of the area $d\vec{S} = \vec{a}_n\, dS$. The volume flow rate through the element of area will equal the product of the velocity component in the direction of the normal to the area with the size of the differential area. To evaluate the component of velocity normal to the surface element we use the scalar product. The volume flow rate will therefore be

$$dQ = u_n\, dS = \vec{u} \cdot \vec{a}_n\, dS = \vec{u} \cdot d\vec{S}. \tag{2.52}$$

To find the total volume flow rate through the complete surface we add up the contributions of each element of area or integrate over the surface:

$$Q = \int_S dQ = \int_S \vec{u} \cdot d\vec{S}. \tag{2.53}$$

It is always a good idea to check units. Velocity, area, and volume flow rate have units of m/s, m^2, and m^3/s, respectively, so the product of the first two gives the third.

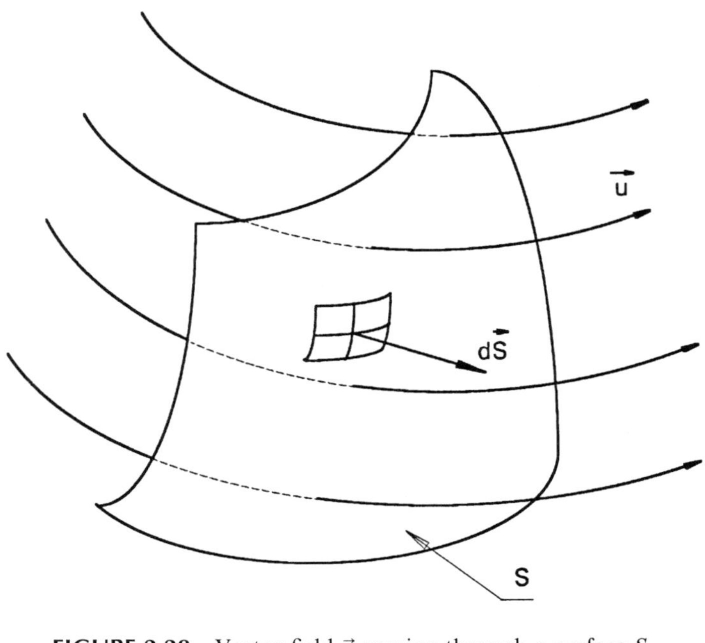

FIGURE 2.28 Vector field $\vec{u}$ passing through a surface S.

The surface integral will be encountered in the expression of Gauss's law for electric and magnetic fields, as well as the Poynting theorem, which gives the total power through a surface, to name only a few of its applications.

In general, for a vector field $\vec{A}$ and a surface S, the quantity of the field or flux Ψ of $\vec{A}$ through the surface S is

$$\Psi = \int_S \vec{A} \cdot d\vec{S}. \tag{2.54}$$

A closed surface is a surface that completely encloses a volume of space. When S is a closed surface we add a little circle to the integral symbol. The flux of $\vec{A}$ through the closed surface S is then

$$\Psi = \oint_S \vec{A} \cdot d\vec{S}. \tag{2.55}$$

For closed surfaces we take the normal to the surface to be directed outward from the enclosed volume.

Example 2.13 Find the flux of the field

$$\vec{A} = \vec{a}_R \frac{C}{R^2}, \quad C = \text{constant}$$

through a surface consisting of a polar spherical cap of radius R_0 and polar angle θ_0 (Fig. 2.29).

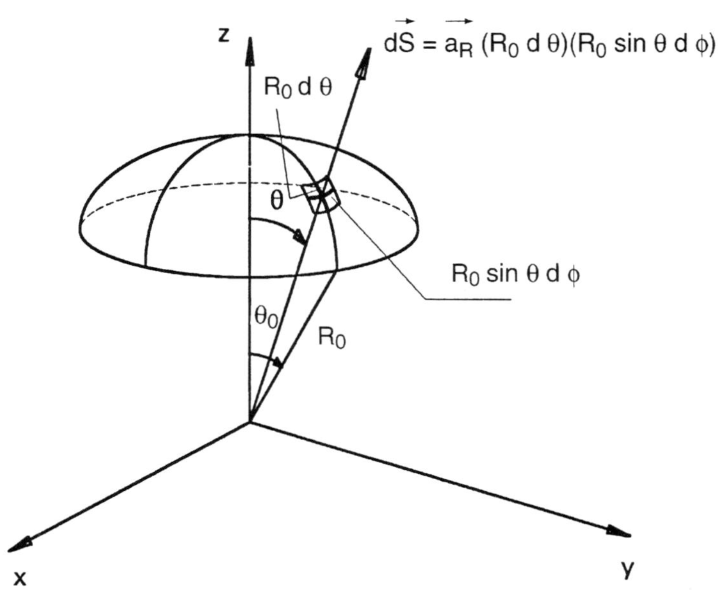

FIGURE 2.29 Spherical cap surface used to determine the flux of the vector field in Example 2.13.

Solution The solution will proceed in spherical coordinates because both the surface and the field are described that way. On the surface at radius R_0 the field is $\vec{a}_R C/R_0^2$ and the element of surface is $\vec{a}_R R_0^2 \sin\theta \, d\theta \, d\phi$. Note that the specific radius of the sphere R_0 has been substituted for the general radius R because that describes the particular surface over which we are integrating. The solution is then

$$\begin{aligned}\Psi &= \int_S \vec{A}\cdot d\vec{S} \\ &= \int_{\phi=0}^{2\pi}\int_{\theta=0}^{\theta_0}\left(\vec{a}_R \frac{C}{R_0^2}\right)\cdot(\vec{a}_R R_0^2 \sin\theta \, d\theta \, d\phi) = 2\pi C(-\cos\theta)|_0^{\theta_0} \\ &= 2\pi C(1-\cos\theta_0). \end{aligned} \tag{2.56}$$

Example 2.14 Find the flux of the same field $\vec{A}$ through the flat circular disk that forms the bottom surface of the polar cap in the previous example. See Fig. 2.30.

Solution We now have a mixed coordinate system problem. The field is in spherical coordinates and the disk is described most easily in cylindrical coordinates. It is easiest to convert the field to cylindrical coordinates and then proceed. From the geometry of the figure, the disk is at height $z_0 = R_0 \cos\theta_0$ above the xy plane. The spherical radius R is given in terms of the cylindrical radius r as $R = \sqrt{z_0^2 + r^2}$. Using Eq. (2.17) we have

$$\begin{aligned}\vec{A} &= \vec{a}_R \frac{C}{R^2} \\ &= (\vec{a}_r \sin\theta + \vec{a}_z \cos\theta)\frac{C}{R^2},\end{aligned}$$

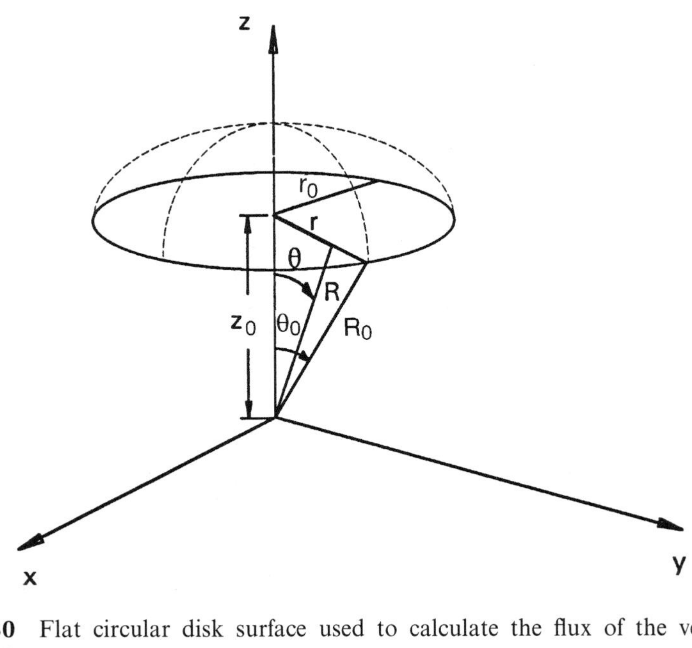

FIGURE 2.30 Flat circular disk surface used to calculate the flux of the vector field in Example 2.14.

where

$$\sin\theta = \frac{r}{R} = \frac{r}{\sqrt{z_0^2 + r^2}}$$

and

$$\cos\theta = \frac{z_0}{R} = \frac{z_0}{\sqrt{z_0^2 + r^2}}.$$

An element of area on the disk is $d\vec{S} = \vec{a}_z r\, d\phi\, dr$. The disk has an outer radius $r_0 = R_0 \cos\theta_0$. The flux through the surface is

$$\begin{aligned}\Psi &= \int_S \vec{A} \cdot d\vec{S} \\ &= \int_{r=0}^{r_0} \int_{\phi=0}^{2\pi} \left[\left(\vec{a}_r \frac{r}{\sqrt{z_0^2 + r^2}} + \vec{a}_z \frac{z_0}{\sqrt{z_0^2 + r^2}} \right) \frac{C}{z_0^2 + r^2} \right] \cdot \vec{a}_z r\, d\phi\, dr \\ &= 2\pi C z_0 \int_0^{r_0} \frac{r\, dr}{(z_0^2 + r^2)^{3/2}} = 2\pi C z_0 \left(\frac{1}{z_0} - \frac{1}{\sqrt{z_0^2 + r_0^2}} \right) \\ &= 2\pi C (1 - \cos\theta_0).\end{aligned} \tag{2.57}$$

The result is the same as Eq. (2.56) for the previous example. Later on we will learn how to test whether the result depends on the particular surface, or just on the outer boundary of the surface. Note that the integral of the field $\vec{A}$ over the closed surface consisting of the polar cap and the circular disk is

$$\begin{aligned}\oint_S \vec{A} \cdot d\vec{S} &= \int_{\text{cap}} \vec{A} \cdot d\vec{S} - \int_{\text{disk}} \vec{A} \cdot d\vec{S} \\ &= 0.\end{aligned}$$

The second integral is subtracted because for the disk we took the normal to the area to be directed upward, $\vec{a}_n = \vec{a}_z$. For the closed surface the outward normal over the disk is downward, $\vec{a}_n = -\vec{a}_z$, and the direction of positive flow reverses.

2.7.1 Surface Integral Using MathCad

MathCad allows the calculation and symbolic evaluation of double integrals. It is therefore appropriate to investigate the application of MathCad to surface integration. If the surface integral is to be numerically evaluated, then the integrant field and $d\vec{S}$ can be entered implicitly under the integral sign. If the symbolic evaluation of the surface integral is required, then both the field and the $d\vec{S}$ vector have to be entered explicitly in the expression of the integral. Example 2.13 will be solved using MathCad.

In order to evaluate the surface integral numerically, the vector $\vec{F}$ and the $d\vec{S}$ vector will be defined as follows:

$$F(R) := \begin{pmatrix} \dfrac{C}{R^2} \\ 0 \\ 0 \end{pmatrix} \qquad dS(R,\theta) := \begin{pmatrix} R^2 \sin(\theta) \\ 0 \\ 0 \end{pmatrix}.$$

The surface integral can then be evaluated as the following function:

$$\Psi(R, \theta, \phi) := \int_0^{\phi} \int_0^{\theta} F(R) \cdot dS(R, \theta)\, d\theta\, d\phi.$$

The function can then be evaluated for different values of R, θ, and ϕ. The value of C has to be defined prior to the definition of the function $\Psi(R, \theta, \phi)$. The numerical value of the function can be obtained for different values of the function arguments. For example, the value of the flux of $\vec{F}$ through a surface characterized by $R = 1$, $\theta = \pi/2$, and $\phi = 2\pi$, which is the upper-half of a sphere of radius 1, is obtained:

$$\Psi(1, \pi/2, 2\pi) = 6.283.$$

The same surface integral can be evaluated symbolically. All expressions involved in the integral have to be declared explicitly:

$$\int_0^{2\pi} \int_0^{\theta_0} \frac{C}{R_0^2} \cdot R_0^2 \cdot \sin(\theta)\, d\theta\, d\phi.$$

The next operation is to select the entire surface integral, and then select from the Symbolic menu the Evaluate function. Select the Evaluate Symbolically operation from the submenu of the Evaluate function. The result is then displayed by MathCad right under the surface integral and looks like this:

$$-2 \cdot C \cdot (\cos(\theta_0) - 1) \cdot \pi.$$

The result is identical to the one obtained in Example 2.13. The numerical value obtained above can very easily be checked by substituting $R_0 = 1$, $C = 1$, and $\theta_0 = \pi/2$.

PROBLEMS

2.1 A vector field is given by $\vec{A} = \vec{a}_x x^2 \sin \pi y + \vec{a}_z x^2 \cos \pi y$. Evaluate the components at the point $(x, y, z) = (2, \frac{1}{2}, 3)$.
Ans. $A_x = 4, \quad A_y = 0, \quad A_z = 0.$

2.2 **a.** Give the position vector $\vec{R}$ of the point $(x, y, z) = (2, 3, 4)$.

b. Give the position vector $\vec{R}_{AB}$ of point A at (5, 7, 9) relative to point B at (3, 4, 5).

Ans. a. $\vec{R} = \vec{a}_x 2 + \vec{a}_y 3 + \vec{a}_z 4$; b. $\vec{R}_{AB} = \vec{a}_x 2 + \vec{a}_y 3 + \vec{a}_z 4$.

2.3 Point A is specified in cylindrical coordinates as $(3, \pi/3, 4)$.

a. Give the position vector in cylindrical coordinates.

b. Give the position vector in Cartesian coordinates.

Ans. a. $\vec{R} = \vec{a}_r 3 + \vec{a}_z 4$; b. $\vec{R} = \vec{a}_x 3/2 + \vec{a}_y 3\sqrt{3}/2 + \vec{a}_z 4$.

2.4 Point A is specified in spherical coordinates as $[5, \tan^{-1}(3/4), \pi/3]$.

a. Give the position vector in spherical coordinates.

b. Give the position vector in cylindrical coordinates.

c. Give the position vector in Cartesian coordinates.

Ans. a. $\vec{R} = \vec{a}_R 5$; b. $\vec{R} = \vec{a}_r 3 + \vec{a}_z 4$; c. $\vec{R} = \vec{a}_x 3/2 + \vec{a}_y 3\sqrt{3}/2 + \vec{a}_z 4$.

2.5 Let $\vec{R}$ be the position vector of the point $P(R, \theta, \phi)$ and $\vec{R}'$ be the position vector of the point $P'(R', \theta', \phi')$.

a. Find the position vector $\vec{R}_d$ of point P relative to point P'.

b. Show that $|\vec{R}_d| = |\vec{R} - \vec{R}'| = \sqrt{R^2 + (R')^2 - 2RR' \cos\alpha}$, where $\cos\alpha = \sin\theta \sin\theta' \cos(\phi - \phi') + \cos\theta \cos\theta'$.

c. Assume that $R > R'$ and expand $|\vec{R} - \vec{R}'|^{\nu}$ in a Maclaurin series, to terms of order $(R'/R)^2$.

2.6 Find the magnitude of $\vec{A} = \vec{a}_x 5 + \vec{a}_y 12 + \vec{a}_z 13$, $\vec{B} = \vec{a}_r 5 + \vec{a}_\phi 12 + \vec{a}_z 13$, and $\vec{C} = \vec{a}_R 5 + \vec{a}_\theta 12 + \vec{a}_\phi 13$.

Ans. $|\vec{A}| = |\vec{B}| = |\vec{C}| = 13\sqrt{2}$.

2.7 Find the magnitude of the vectors in Problem 2.6 using MathCad.

2.8 Find the magnitude of $\vec{A} = \vec{a}_x x^2 \sin \pi y + \vec{a}_z x^2 \cos \pi y$ and evaluate the magnitude at $(x, y, z) = (2, \frac{1}{2}, 3)$.

Ans. $|\vec{A}| = 4$.

2.9 Determine symbolically the magnitude of the vector in Problem 2.8 using MathCad. Evaluate the magnitude numerically as required in Problem 2.8, using MathCad.

2.10 **a.** Find the sum of $\vec{A} = \vec{a}_x 2 + \vec{a}_y 3 + \vec{a}_z 4$ and $\vec{B} = \vec{a}_x 4 - \vec{a}_y 3 + \vec{a}_z 2$.

b. Find $|\vec{A}|$, $|\vec{B}|$, and $|\vec{A} + \vec{B}|$.

c. Under what conditions does $|\vec{A}| + |\vec{B}| = |\vec{A} + \vec{B}|$?

d. Prove that $|\vec{A}| + |\vec{B}| \geq |\vec{A} + \vec{B}|$.

Ans. b. 5.39, 5.39, 8.49.

2.11 Solve Problem 2.10 using MathCad.

2.12 Given that $\vec{A} = \vec{a}_R R^2 \sin^2\theta$ and $\vec{B} = \vec{a}_r 2r^2 \cos^2\phi$, evaluate $\vec{A} + \vec{B}$ at the point $(x, y, z) = (5, 0, 0)$.

Ans. $\vec{a}_x 75$.

2.13 If $\vec{A} = \vec{a}_x 2 + \vec{a}_y 3 + \vec{a}_z 4$ and $\vec{B} = -\vec{a}_x + \vec{a}_y 2 - \vec{a}_z$, show that $\vec{A}$ and $\vec{B}$ are orthogonal.

2.14 Solve Problem 2.13 using MathCad.

2.15 Given the fields $\vec{A} = \vec{a}_x 2x^2y + \vec{a}_y 3y^2z + \vec{a}_z 4z^2x$ and $\vec{B} = \vec{a}_x 3y^2z + \vec{a}_y 4xz^2 + \vec{a}_z 5x^2y$, evaluate the component of $\vec{A}$ in the direction of $\vec{B}$ at the point (1, 2, 3).

Ans. 34.7.

2.16 Solve Problem 2.15 using MathCad.

2.17 Evaluate $\vec{A} \times \vec{B}$ at the point (1, 2, 3) for the vector fields given in Problem 2.15.

Ans. $-\vec{a}_x 936 + \vec{a}_y 1256 - \vec{a}_z 1152$.

2.18 Solve Problem 2.17 using MathCad.

2.19 The forces on the two sides of the winding in Fig. 2.31 are 50 N each and are acting as shown. The distance from the winding to the axis of rotation is $d = 12$ cm. Determine the total torque developed by this elementary motor.

Ans. $\vec{T} = 12\vec{a}_z$ N-m.

2.20 Three vectors $\vec{A}$, $\vec{B}$, and $\vec{C}$ form a triangle with opposite angles a, b, and c.

a. Show that the area of the triangle is given by

$$\tfrac{1}{2}|\vec{A} \times \vec{B}| = \tfrac{1}{2}|\vec{B} \times \vec{C}| = \tfrac{1}{2}|\vec{C} \times \vec{A}|.$$

b. Derive the law of cosines: $C^2 = A^2 + B^2 - 2AB \cos c$.

2.21 Find the line integral of $\vec{B} = \vec{a}_\phi C(1 + \cos\theta)/R \sin\theta$ around the path $R = R_0$, $\theta = \theta_0$.

Ans. $2\pi C(1 + \cos\theta_0)$.

2.22 Solve Problem 2.21 using MathCad.

2.23 Find the line integral of $\vec{F} = \vec{a}_x kxy^2 + \vec{a}_y(x^2 + 2)y$ over the paths L_1, L_2, and L_3 and around the closed path that they form (Fig. 2.32). For what value of k is the closed line integral equal to zero?

Ans. Over L_1, -16; L_2, 0; L_3, $-64k + 80$; $k = 1$.

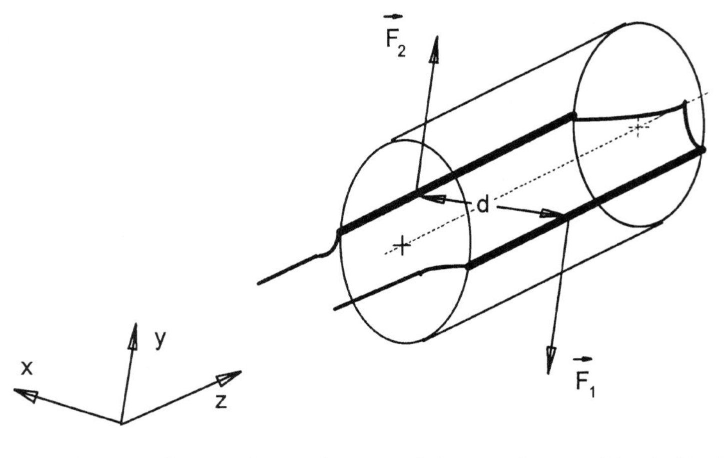

FIGURE 2.31 Forces acting on the conductors of the rotating machine in Problem 2.19.

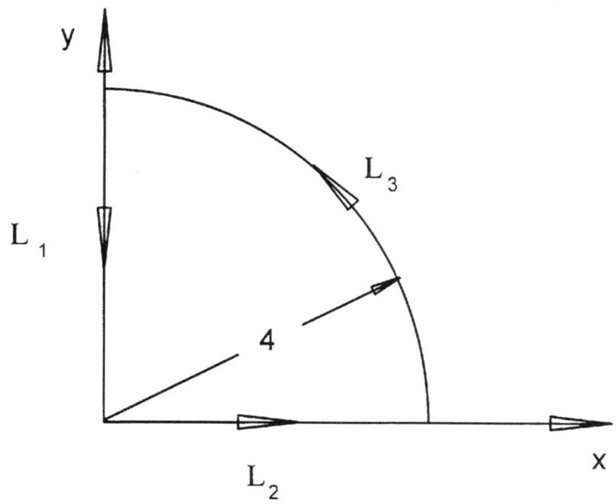

FIGURE 2.32 Closed path for the line integral in Problem 2.23.

2.24 Solve Problem 2.23 using MathCad.

2.25 Find the flux of the field $\vec{A} = \vec{a}_R(C/R^2)$ through the south polar cap of radius R and angle $\theta_0 < \theta < \pi$.
Ans. $2\pi C(1 + \cos\theta_0)$.

2.26 Solve Problem 2.25 using MathCad.

2.27 Given the field $\vec{A} = C(\vec{a}_R 2\cos\theta + \vec{a}_\theta \sin\theta)/R^3$:
a. Find the flux through a polar cap of radius R_0 and angle $0 < \theta < \theta_0$.
b. Find the flux through a cone of angle θ_0 and radius $R_1 < R < R_2$.
Ans. a. $(2\pi C/R_0)\sin^2\theta_0$; b. $2\pi C[(1/R_1) - (1/R_2)]\sin^2\theta_0$.

2.28 Solve Problem 2.27 using MathCad.

2.29 Find the flux of the field $\vec{A} = \vec{a}_\phi(C/2\pi r)$ through the surface $\phi = \phi_0$ extending over $r_1 < r < r_2$ and $z_0 < z < z_0 + L$.
Ans. $(CL/2\pi)\ln(r_2/r_1)$.

2.30 Solve Problem 2.29 using MathCad.

REFERENCES

[1] M. Javid and P. Brown, *Field Analysis and Electromagnetics*, McGraw-Hill, New York, 1963, p. 6.

CHAPTER THREE

Electromagnetic Field Definitions

3.1 INTRODUCTION

In this chapter we will introduce the major field quantities needed in electromagnetics and microwave engineering. It is not our purpose at this time to give complete methods of calculating the fields; this will be done in subsequent chapters. Here we will define the field quantities and briefly describe their properties.

3.2 ELECTRIC CHARGE

Electric charge consists of electrons, the smallest unit of negative charge, and protons, the smallest unit of positive charge. The proton and the electron have equal but opposite polarity charge. In the International System of Units (SI system) the unit of charge is the coulomb, C. The charge of one electron is approximately -1.602×10^{-19} C and hence it takes about $\frac{5}{8} \times 10^{19}$ electrons (or protons) to make 1 coulomb of charge.

Experiments have shown that the magnitude of the force between two charged bodies is given by

$$F_e = k_e \frac{q_1 q_2}{d_2}, \tag{3.1}$$

where q_1 and q_2 are the charges, d is the distance between them, and k_e is a constant that depends on the system of units. The force is along the line connecting the charges. When both charges have the same polarity (both positive or both negative), then the force is directed so as to push them apart. When the charges have opposite polarity (one positive and one negative), then the force is directed so as to pull them together.

In the SI system of units the constant of proportionality is about $k_e = 8.988 \times 10^9$ N-m^2/C^2. The force between two 1 C charges separated by 1 meter is therefore about 9×10^9 N or 2×10^9 pounds. This is a tremendous force and so we conclude that 1 coulomb is an extremely large quantity of charge.

The force of gravitational attraction between two masses m_1 and m_2 separated by a distance d is

$$F_g = k_m \frac{m_1 m_2}{d^2}, \tag{3.2}$$

which is of the same form as Eq. (3.1). The constant is $k_m = 6.670 \times 10^{-11}$ N-m^2/kg^2. The mass of an electron is 9.107×10^{-31} kg. Using Eqs. (3.1) and (3.2) we find that the electric force of repulsion between two electrons is about 4×10^{42} times as large as the gravitational force of attraction. The gross difference in the magnitude of these types of forces means that it is possible to separate and identify them. A proton has the same magnitude charge as an electron, but its mass is about 1834 times larger than an electron's. The force of electric repulsion between two protons is still larger than that of gravitational attraction by a factor of 1.2×10^{36}.

3.2.1 Volume Charge Density

Usually we are not concerned with individual electrons or protons, but with charges distributed throughout some region of space. We describe the location and quantity of this charge by giving the volume charge density ρ (in units of C/m^3). Volume charge density is analogous to the mass density of a body given by ρ_m (in units of kg/m^3). Charge density will, in general, be a function of position and time:

$$\rho = \rho(u, v, w, t). \tag{3.3}$$

Figure 3.1 represents a region with volume charge density, which may be considered as a fluid spread throughout the volume. When we look at a differential volume dV we assume that even though this volume is very small it is still large enough to contain

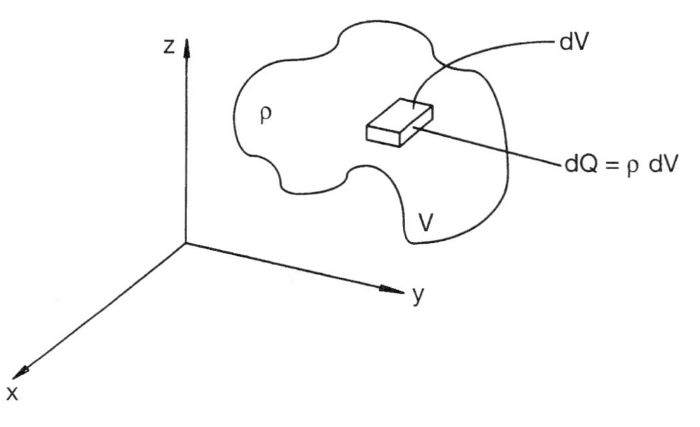

FIGURE 3.1 Region with volume charge density.

many charged particles. We do not have to consider each particle separately but can express the differential charge in this volume as

$$dQ = \rho \, dV. \tag{3.4}$$

The total charge present in the volume is then

$$Q = \int_V dQ = \int_V \rho \, dV. \tag{3.5}$$

This is a volume integral and hence the "integral over V" symbol represents a triple integral.

Example 3.1 The charge density in the rectangular parallelepiped shown in Fig. 3.2 is

$$\rho = \begin{cases} Axy^2z^3, & 0 \le x < 4, \quad 0 \le y < 3, \quad 0 \le z < 2 \\ 0, & \text{elsewhere} \end{cases} \quad (\mathrm{C/m^3}).$$

Find the total charge in the region.

Solution Cartesian coordinates are the most suitable for solving the problem since both the charge density and the region geometry are expressed in this system. Therefore $dV = dx\, dy\, dz$ and for a volume in the charged region

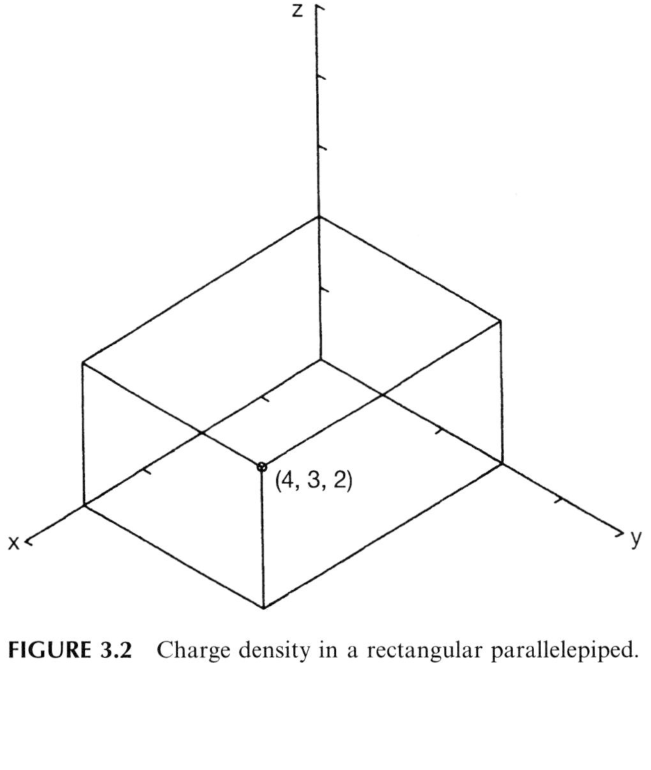

FIGURE 3.2 Charge density in a rectangular parallelepiped.

$dQ = (Axy^2z^3)(dx\,dy\,dz)$. Integrating over the region containing the charge, we have

$$Q = \int_{z=0}^{2}\int_{y=0}^{3}\int_{x=0}^{4} Axy^2z^3\,dx\,dy\,dz$$
$$= A\int_0^4 x\,dx\int_0^3 y^2\,dy\int_0^2 z^3\,dz = A\left(\frac{x^2}{2}\right)\Bigg|_0^4\left(\frac{y^3}{3}\right)\Bigg|_0^3\left(\frac{z^4}{4}\right)\Bigg|_0^2$$
$$= 288A \quad \text{(C)}.$$

3.2.2 Surface Charge Density

Sometimes we wish to describe a charge distribution that consists of charges lying in such a thin layer that we say the charge lies on or forms a surface. Let us suppose we take the volume charge distribution shown in Fig. 3.1 and squeeze it together until the same total charge lies in an extremely thin layer. As the layer gets thinner, the volume occupied by the charge gets smaller and the charge density ρ gets larger. When the charge lies in a true surface the volume occupied by the charge is zero and the volume charge density is infinite or, in mathematical jargon, singular. The result of squeezing all the charge onto a surface is surface charge described by its surface charge density ρ_s (C/m^2).

The charge on a small element of surface dS (Fig. 3.3) is

$$dQ = \rho_s\,dS. \tag{3.6}$$

The total charge on a surface is

$$Q = \int_S dQ = \int_S \rho_s\,dS. \tag{3.7}$$

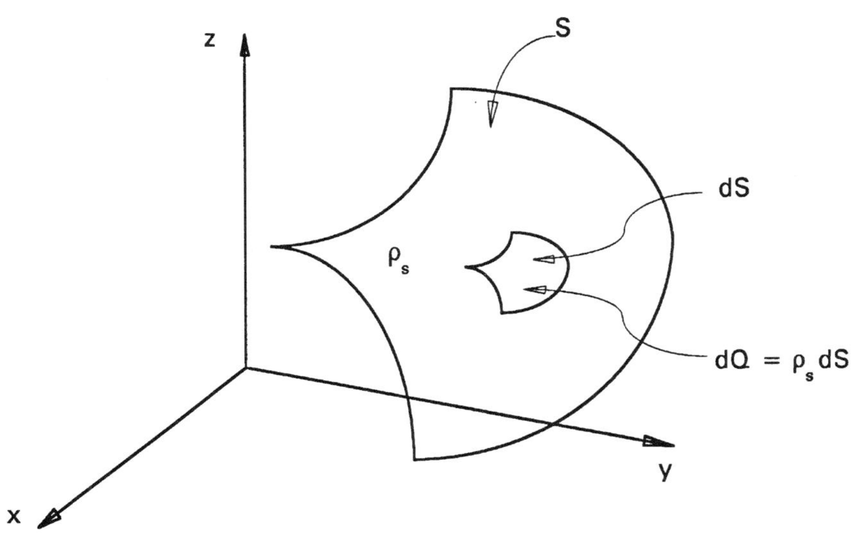

FIGURE 3.3 Charge on a small element of surface dS.

This is an integral over a surface and hence the "integral over S" symbol represents a double integration.

Example 3.2 The charge on the surface of a sphere of radius a is given by $\rho_s = \rho_{s0} \cos^2 \phi$ C/m^2. Find the total charge on the surface of the sphere.

Solution The differential area on the surface of the sphere is $dS = a^2 \sin\theta \, d\theta \, d\phi$ (Section 2.4.3). Hence

$$\begin{aligned} Q &= \int_{\phi=0}^{2\pi} \int_{\theta=0}^{\pi} (\rho_{s0} \cos^2 \phi) a^2 \sin\theta \, d\theta \, d\phi \\ &= \rho_{s0} a^2 \int_{\phi=0}^{2\pi} \cos^2 \phi \, d\phi \int_{\theta=0}^{\pi} \sin\theta \, d\theta = \rho_{s0} a^2 (\pi)(2) \\ &= 2\pi a^2 \rho_{s0} \quad \text{(C)}. \end{aligned}$$

3.2.3 Line Charge Density

If the total charge on a surface is squeezed together so that it now lies along a line, then the surface area occupied by the charge is zero (as is the volume). The surface charge density ρ_s is infinite (singular) and, of course, so is the volume charge density ρ. The charge is described as a line charge with line charge density ρ_l (C/m).

The charge on a small length of line dl (Fig. 3.4) is

$$dQ = \rho_l \, dl \tag{3.8}$$

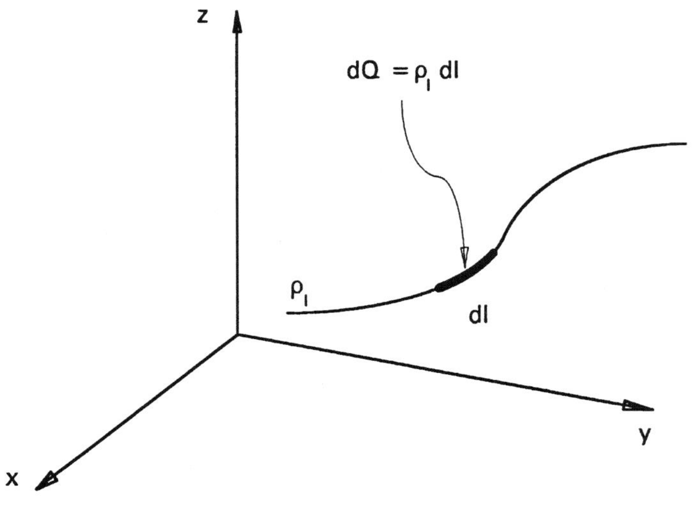

FIGURE 3.4 Charge on a small length of line dl.

and the total charge on the line is

$$Q = \int_L \rho_l \, dl. \tag{3.9}$$

Since this is an integral along a line, it involves only a single integration.

Example 3.3 A line charge $\rho_l = \rho_{l0}(|z|/L)$ C/m lies along the z axis from $z = -L$ to $+L$ m. Find the total charge on the line.

Solution Along the z axis the differential length is dz. Hence

$$Q = \int_{z=-L}^{L} \rho_{l0} \frac{|z|}{L} \, dz = \frac{\rho_{l0}}{L} \left[\int_{-L}^{0} (-z) \, dz + \int_{0}^{L} z \, dz \right]$$

$$= \rho_{l0} L \quad \text{(C)}.$$

3.2.4 Point Charge

Squeezing the charge along the line together until the entire charge is compressed to a single point produces a point charge q (C). The point has no volume, so volume charge density ρ is singular; no area, so surface charge density ρ_s is singular; and no length, so line charge density ρ_l is singular.

3.2.5 Charge Equivalences

As a final note, it is worth repeating the equivalence of charge located in a small volume, area, length, or at a point.

$$\begin{aligned} dQ &= \rho \, dV \quad && \textit{for volume charge density,} \\ dQ &= \rho_s \, dS \quad && \textit{for surface charge density,} \\ dQ &= \rho_l \, dl \quad && \textit{for line charge density,} \\ dQ &= q \quad && \textit{for a point charge.} \end{aligned} \tag{3.10}$$

The SI unit for each term is the coulomb, C.

3.3 ELECTRIC CURRENT

The flow of electric charge constitutes an electric current I (A). In an electronic vacuum tube, such as the picture tube in your television set, there are free electrons moving in space and these form a current. In conducting materials there are also free electrons, which move under the influence of an electric field, thereby constituting a current. As the individual electrons move within the conductor they change velocity and direction, but the electrons as a whole, or on the average, have a flow direction and speed that yields the current. When we considered charge density we did not concern ourselves with individual particles of charge but thought of the aggregates of charge as a continuous fluid. When charge moves we take the same view and consider the average motion of the "charge fluid" as a whole.

Current (often referred to as conventional current) is defined in terms of the flow of positive charge. Hence if electrons, which are negative charge, flow in one direction the current is in the opposite direction.

The current crossing a surface is defined as the charge per unit time that crosses the surface. The relations between charge and current are therefore

$$I = \frac{dQ}{dt} \quad \text{and} \quad Q = \int^{t} I\, dt. \tag{3.11}$$

The coulomb is therefore equal to an ampere-second.

3.3.1 Volume Current Density

Consider a tube of charge (Fig. 3.5) with cross-sectional area ΔS flowing in the x direction with speed u_x. In time Δt the charge will move a distance $\Delta x = u_x\, \Delta t$ and the charge that will cross the surface is

$$\Delta Q = \rho\, \Delta V = \rho\, \Delta x\, \Delta S = \rho u_x\, \Delta t\, \Delta S.$$

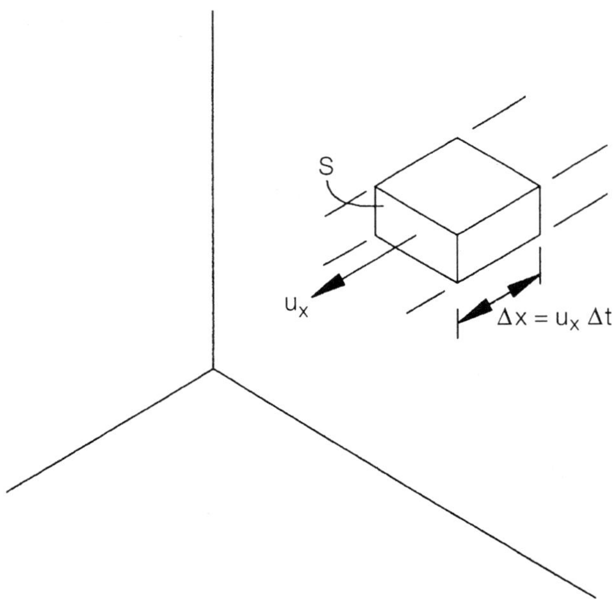

FIGURE 3.5 Tube of charge with cross-sectional area ΔS.

The charge per unit time crossing the surface is the x-directed current. The current per unit area of surface is the x component of the current density

$$J_x = \frac{\Delta I}{\Delta S} = \frac{\Delta Q}{\Delta t\, \Delta S}$$

$$= \rho u_x. \tag{3.12}$$

If the charge density is moving in the y and z directions also, then there would be y and z components of current density too. The total volume current density $\vec{J}$ (A/m^2) is

$$\vec{J} = \vec{a}_x \rho u_x + \vec{a}_y \rho u_y + \vec{a}_z \rho u_z$$

$$= \rho \vec{u}. \tag{3.13}$$

The last equation is independent of the coordinate system. The current has direction as well as magnitude and hence current density is a vector quantity. Note that this quantity is called a volume current density, because it is distributed throughout a volume, but its dimensions are current per unit area.

To find the total current crossing a surface S (Fig. 3.6), we first note that the current crossing the small element of surface $d\vec{S}$ is

$$dI = \vec{J} \cdot d\vec{S}.$$

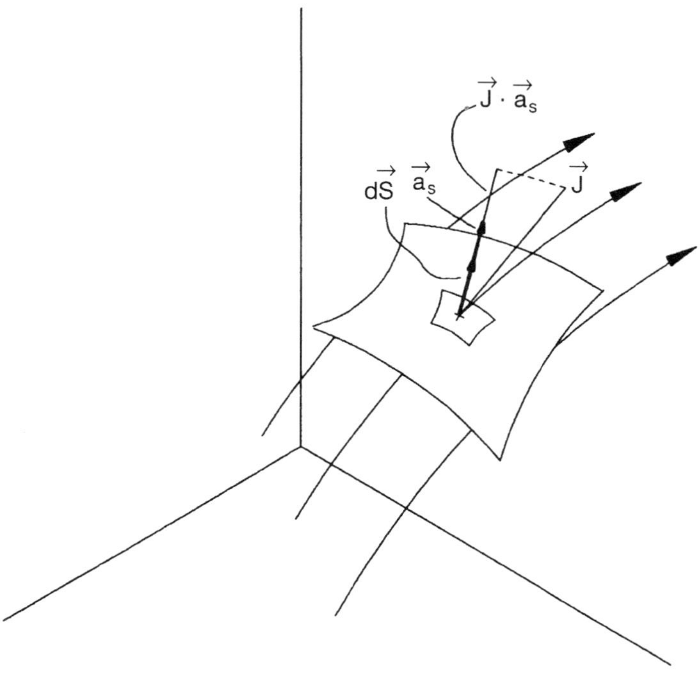

FIGURE 3.6 Current crossing a small element of surface dS.

Then the total current is obtained by adding up or integrating these contributions over the whole surface:

$$I = \int_S dI = \int_S \vec{J} \cdot d\vec{S}. \tag{3.14}$$

The scalar product in this equation eliminates the component of current that flows along or tangent to the surface and hence does not cross it. In terms of the general surface integral, Eq. (2.52), I is the flux of $\vec{J}$ through the surface S.

Example 3.4 The current density is given by $\vec{J} = \vec{a}_z J_0 \exp(-k^2 r^2)$ A/m^2. Find the current through a circle of radius a centered on the z axis and lying parallel to the xy plane (Fig. 3.7).

Solution The current density is given in cylindrical coordinates and the circle is easily described in these coordinates, so the problem should be solved in the cylindrical system. The element of area in the plane of the circle is $d\vec{S} = \vec{a}_z r\, dr\, d\phi$. The current through this element is then

$$dI = \vec{J} \cdot d\vec{S} = J_0 e^{-k^2 r^2}\, r\, dr\, d\phi,$$

and the total current is

$$\begin{aligned} I &= \int_{\phi=0}^{2\pi} \int_{r=0}^{a} J_0 e^{-k^2 r^2}\, r\, dr\, d\phi \\ &= J_0 \int_0^{2\pi} d\phi \int_0^a e^{-k^2 r^2}\, r\, dr = J_0 (2\pi) \left(-\frac{1}{2k^2} e^{-k^2 u} \Big|_0^{a^2} \right) \\ &= \frac{\pi}{k^2} J_0 (1 - e^{-k^2 a^2}) \quad \text{(A)}. \end{aligned}$$

In performing the integration over r we have substituted $u = r^2$, $du = 2r\, dr$. The units of J_0 are A/m^2 and those of k^2 are 1/m^2, so the unit of the answer is $(A/\text{m}^2)/(1/m^2) = \text{A}$, as it should be. If $a \to \infty$, the result is the total current flowing in all space, $(\pi J_0/k^2)$ A.

3.3.2 Surface Current Density

If the current flowing in a region is squeezed together until it flows in an infinitesimally thin sheet, then the result is a surface current $\vec{J}_s$ (A/m). Since this is a current flowing through zero cross-sectional area, the current density $\vec{J}$ is infinite or singular.

Although we have not yet studied resistance you will remember from elementary physics that, as the cross-sectional area of a resistor gets smaller, the resistance gets larger. The resistance to a current flowing on a surface would therefore be infinite. From this we conclude that surface currents can only exist on the surface of a perfect conductor. A physical example is a superconductor. Even when we are not dealing

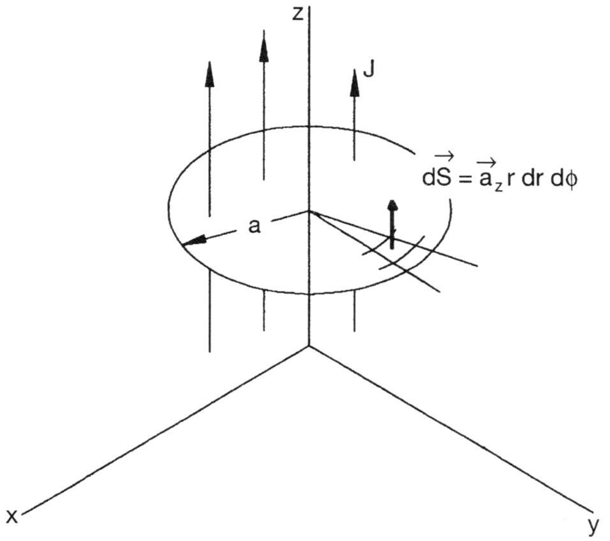

FIGURE 3.7 Current through a circle of radius a due to the current density in Example 3.4.

with superconductors we will sometimes model a very good conductor as a perfect conductor in order to simplify the problem and make it easier or possible to find a solution.

Figure 3.8 shows a surface current $\vec{J}_s$ and a line L in the surface on which $\vec{J}_s$ is flowing. To find the current crossing the line we must integrate the component of $\vec{J}_s$ that is perpendicular to the line over the length of the line. Hence

$$I = \int_L J_{sn}\, dl, \tag{3.15}$$

where J_{sn} is the component of the surface current density that is perpendicular to the line.

Example 3.5 A surface current $\vec{J}_s = -\vec{a}_x 3 + \vec{a}_y 7$ A/m flows on the $z = 0$ plane. Find the current crossing the line of $y = 5, z = 0$ extending from $x = 2$ to $x = 8$ m (Fig. 3.9).

Solution The line is of length 6 m parallel to the x axis. Therefore $J_{sn} = J_{sy} = 7$ A/m and $I = (7)(6) = 42$ A.

3.3.3 Line Current

If a sheet or surface of current is squeezed together so that the total current flows along a line, then the surface current density is infinite or singular (as is the current density), and the resultant current flow is called a line current i (or I) A. Line current has the direction of the line along which the current is flowing. A line current models a current flowing in a thin wire.

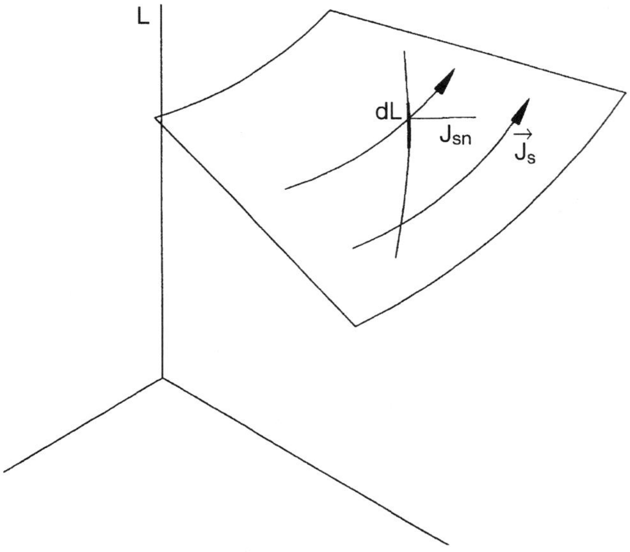

FIGURE 3.8 Surface current density J_s and a line L in the surface on which it is flowing.

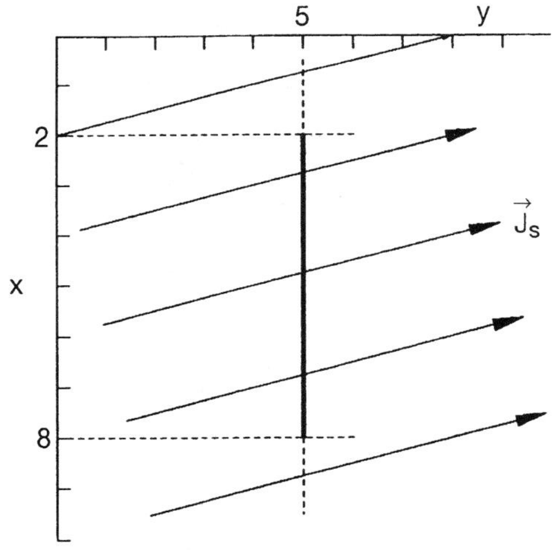

FIGURE 3.9 Surface current density in Example 3.5.

3.3.4 Current Equivalences

The following equivalences apply for current densities:

$$\begin{aligned} dI &= \vec{J} \cdot d\vec{S} && \textit{for volume current density,} \\ dI &= J_{sn}\, dl && \textit{for surface current density,} \\ dI &= i && \textit{for line current,} \end{aligned} \tag{3.16}$$

all with dimensions of current, A; and

$$\vec{J}\,dv \Leftrightarrow \vec{J}_s\,dS \Leftrightarrow i\,d\vec{l}, \tag{3.17}$$

with dimensions of current times length, A-m, and where the symbol $\Leftrightarrow$ means equivalent. The second set of equivalences will be useful in formulas to be developed later.

3.4 ELECTRIC FIELD INTENSITY

Electric field intensity $\vec{E}$(V/M) is defined in terms of the force on a stationary test charge q_t. Let us picture the following "thought experiment."[†]

A small test charge is attached to a scale or "force meter" and held at a fixed point in space. The scale indicates the direction and magnitude of the force $\vec{F}_t$ on the test charge. The measured force is divided by the test charge to compute a ratio, which is the force per unit of test charge. The test charge is reduced in size until this ratio is found to stabilize at a particular value. The vector value of this ratio is defined as the electric field intensity.

Mathematically, we can represent this measurement and define the electric field intensity as

$$\vec{E} = \lim_{q_t \to 0} \frac{\vec{F}_t}{q_t}. \tag{3.18}$$

It is necessary to reduce the stationary test charge (mathematically, to take the limit shown) to zero to ensure that the test charge does not affect the original electric field intensity we are measuring by causing its sources to move or vary from what they would be in the absence of the test charge.

From this definition it is clear that the force on a stationary charge q, located at a point where the electric field intensity is $\vec{E}$, is given by

$$\vec{F} = q\vec{E}. \tag{3.19}$$

This is also the force on a moving charge in the absence of a magnetic field, which is defined later in this chapter.

Practising engineers will often speak of the "electric field" when they mean the electric field intensity. In this book we will use the full name, in order to avoid confusion with other electric field quantities.

Consider a test charge q_t and a point charge q in otherwise empty space. Then, from Eq. (3.1), the force on the test charge will be

$$\vec{F}_t = \vec{a}_d k_e \frac{q_t q}{d^2},$$

[†] A thought experiment is an experiment performed in the mind. It should represent a physically possible experiment in order to be useful, but we do not need to concern ourselves with the exact details of the experiment or the accuracy and resolution of the equipment.

where $\vec{a}_d$ is along the line joining the two charges, directed from q toward q_t. The electric field intensity due to the charge q is therefore

$$\vec{E} = \frac{\vec{F}_t}{q_t} = \vec{a}_d k_e \frac{q}{d^2}. \tag{3.20}$$

This vector field is shown in Fig. 3.10. The closeness of the lines near the point charge represents the great strength of the field near the charge. Even if other charges are present in the space, when we get close enough to a point charge the field of that charge will dominate over the field of the remaining charge distribution. The fact that the lines spread out radially is due to the spherical symmetry of the single point charge. The point charge itself has no orientation or difference in shape or size in any particular direction. Hence, if we picture the point charge at the origin of our coordinate system, there is no reason to choose any particular orientation of the axes. One direction is just like any other.

Lines of electric field start from positive charge. Figure 3.10 has been drawn for a positive charge. For negative charge the arrows, representing the direction of the field, would be reversed. Figure 3.11 represents the field surrounding two equal but opposite polarity charges. Note that the lines start from positive charge and end on negative charge and that close to each point charge the lines are radial like those in Fig. 3.10 for the isolated point charge.

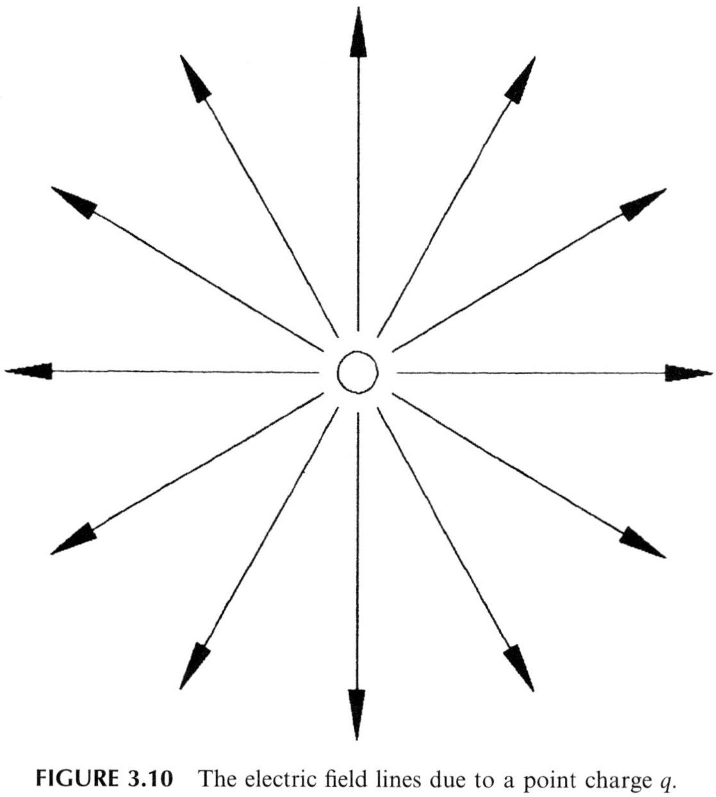

FIGURE 3.10 The electric field lines due to a point charge q.

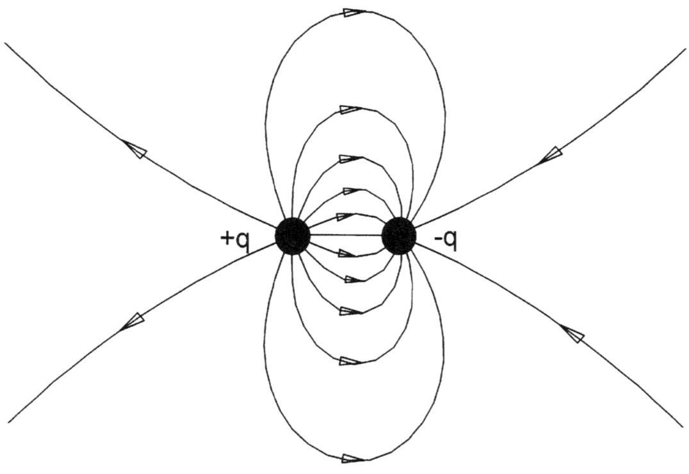

FIGURE 3.11 The field lines surrounding two equal but opposite polarity charges.

In Chapter 4 we will obtain the electric field due to various highly symmetric charge distributions. In Chapter 6 we will learn general methods for calculating the electric field for a known charge distribution. Other methods are introduced as needed.

3.5 ELECTROMOTIVE FORCE (EMF) OR VOLTAGE

The electromotive force (emf) or voltage V(V) along a line (Fig. 3.12) is the line integral of the electric field intensity:

$$\text{emf} = V_{\text{AB}} = \int_{L_{\text{AB}}} \vec{E} \cdot d\vec{l} \tag{3.21}$$

$$= \int_{L_{\text{AB}}} E_l \, dl. \tag{3.22}$$

In general, the voltage depends on the particular path or line between the end points A and B. The name electromotive force comes from the function of potential difference in driving the electrons through electric circuits. Voltage is considered as the "driving force" behind the electrons in circuits, hence the name electromotive force. In later chapters we will discover the different sources of electromotive force other than batteries, which provide potential difference between their terminals as a result of an internal chemical reaction. At this point, the definition of emf is left as presented by Eq. (3.21), and it is general enough to account for voltage rises and voltage drops in conventional electric circuits.

3.6 ELECTRIC FLUX DENSITY

Electric flux density $\vec{D}$(C/m^2) can also be defined in terms of a thought experiment. Let us consider a pair of thin circular conducting metal plates, which are held together

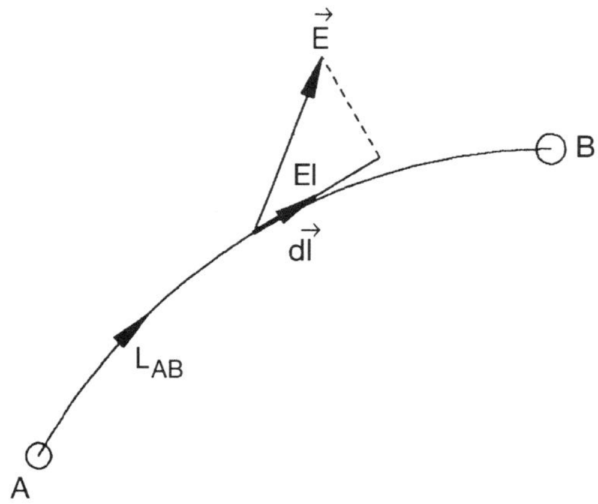

FIGURE 3.12 A path between points A and B in an electric field used to calculate emf or voltage.

in an electric field (Fig. 3.13a). The area of the plates ($\Delta S = \pi a^2$) is very small and the thickness of the plates δt is much smaller than the diameter of the plates, $\delta t \ll 2a$. If an electric field is present, electric charge will be displaced and form a surface charge on the plates. The displaced charge will arrange itself so that the original field plus the field due to the surface charge will add up to give a null or zero total field inside the plates. The plates are now separated (Fig. 3.13b), trapping the displaced charge on each plate. The plates are next removed from the field and a charge meter (e.g., a ballistic galvanometer) is connected to the plates (Fig. 3.13c). The previously trapped charge flows through the charge meter, giving a reading of the total charge on each plate. The experiment is performed many times, each time with a different orientation of the plates in the original field. For some particular orientation, the displaced charge, and hence the reading of the charge meter, will be a maximum. The electric flux density is equal to the ratio of the maximum charge displaced to the surface area of the plates and is in the direction of the normal to the surface of the positively charged plate when the plates are oriented so as to maximize the displaced charge.

Mathematically, if $\vec{D}$ represents the electric flux density and $\vec{a}_n$ is the normal directed from the positive plate to the negative plate when the maximum charge ΔQ is displaced, then

$$\vec{D} = \vec{a}_n \left[\lim_{\Delta S \to 0} \frac{\Delta Q}{\Delta S} \right]_{\max}. \tag{3.23}$$

$\vec{D}$ is called a flux density because it gives the field lines or flux per unit area. The symbol "*D*" recalls that this field quantity used to be called the electric *D*isplacement field. The derivation of this name is related to the displaced charge on the conducting plates of our thought experiment.

Electric flux density may be pictured by drawing field lines, as we did for electric field intensity. In fact, as we shall see, in simple media, the pictures of electric flux density around charge would be identical to those of field intensity.

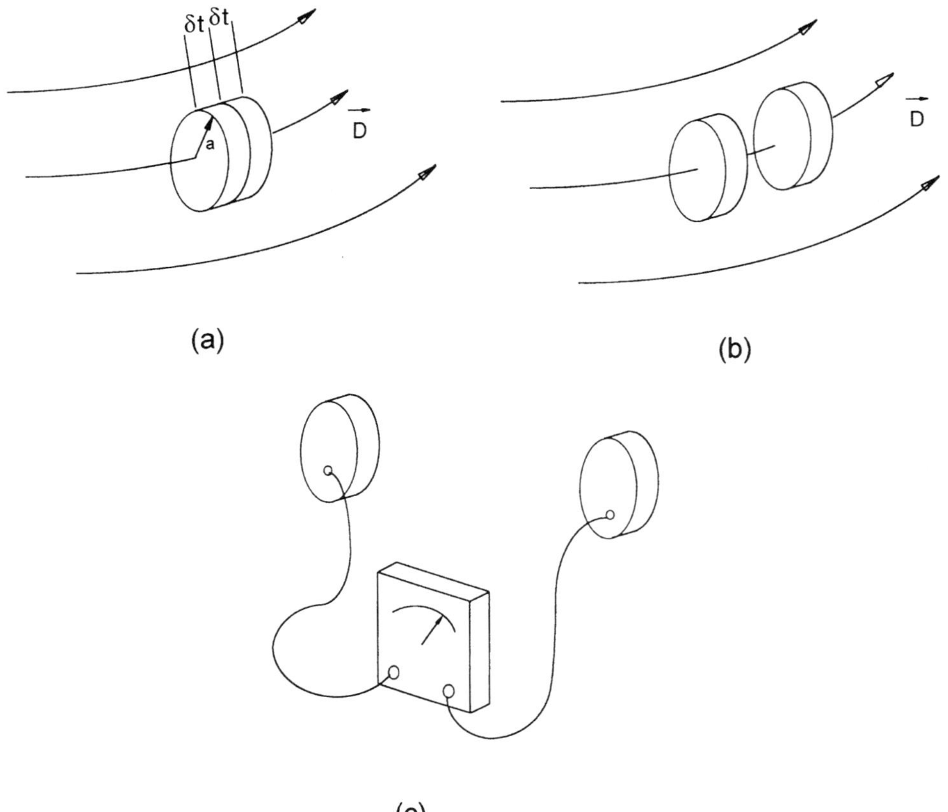

FIGURE 3.13 A pair of thin circular conducting metal plates in an electric field.

3.7 MAGNETIC FLUX DENSITY

Magnetic flux density $\vec{B}$ (T or Wb/m^2) can be defined in either of two ways, which represent the two aspects of this field quantity. Two different thought experiments are described to demonstrate these definitions.

The first definition is in terms of the force on a moving test charge. Let us assume we move a test charge q_t with constant velocity $\vec{u}$ through a region in space and, at each point, measure the total force $\vec{F}_t$ on the charge (Fig. 3.14). At the same time, we use a stationary test charge at the point to measure the electric force $\vec{F}_e$ due to whatever electric field may be present. At each point we could then find a force $\vec{F}_m = \vec{F}_t - \vec{F}_e$, which we attribute to the magnetic flux density $\vec{B}$. A little experimentation with different velocities (direction and magnitude) would show that

$$\vec{F}_m = q_t \vec{u} \times \vec{B}. \tag{3.24}$$

This equation can be considered a definition of magnetic flux density $\vec{B}$.

The second definition is in terms of the voltage developed across a loop inserted into the field. Let us assume we move a small loop of wire with area ΔS from a distant

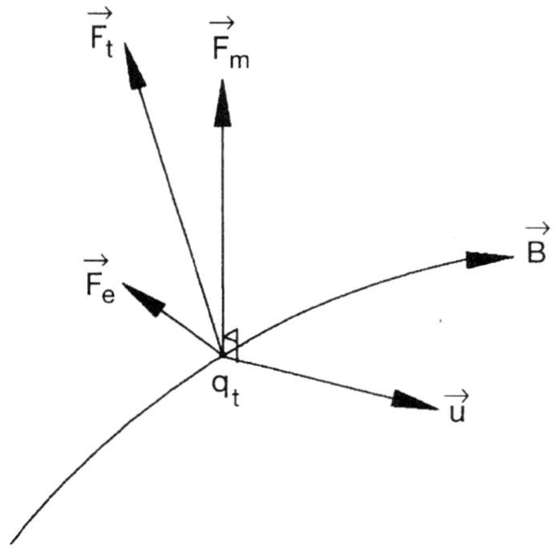

FIGURE 3.14 Forces acting on a charge moving with constant velocity inside an electromagnetic field.

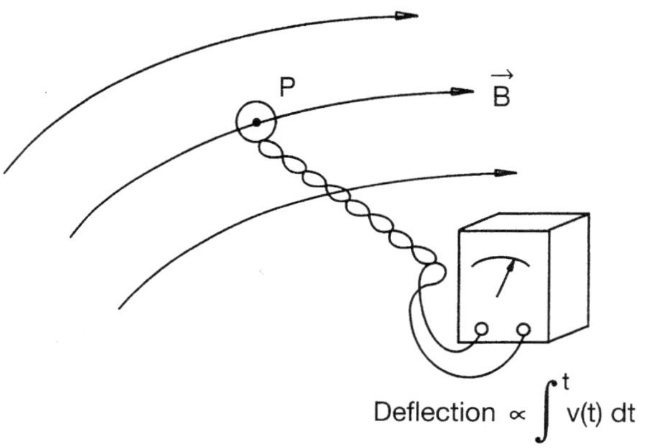

FIGURE 3.15 Small loop of wire connected to an integrating voltmeter, which is used to define magnetic flux density.

point, where there is no field, to the point P where we are measuring the field (Fig. 3.15). The loop is connected to an integrating voltmeter, which gives a deflection or scale reading proportional to the time integral of the voltage at the terminals of the loop. The orientation of the loop is adjusted at P until the integrating voltmeter gives a maximum value. The unit vector $\vec{a}_n$ is in the direction of the normal to the plane of the loop and hence to the area ΔS when this maximum is obtained. Then the magnetic flux density can be defined as

$$\vec{B} = \vec{a}_n \lim_{\Delta S \to 0} \left[\frac{\int^t v(t)\, dt}{\Delta S} \right]_{\max}. \tag{3.25}$$

The first definition, in terms of the force on a moving test charge, is similar to the definition for electric field intensity, which was given in terms of the force on a stationary test charge. This might make you feel the magnetic flux density is misnamed and ought to be called a field intensity also. The second definition, however, is in terms of a per unit area measurement and justifies the name of flux density. This is similar to the definition for electric flux density, which was also defined in terms of a flux per unit area, Eq. (3.23). The force equation,

$$\vec{F} = q(\vec{E} + \vec{u} \times \vec{B}), \tag{3.26}$$

is important in calculating the motion of charges in electromagnetic fields. Nevertheless, it is the aspect of $\vec{B}$ displayed in the second definition, Eq. (3.25), which is most important in our studies.

3.8 MAGNETIC FIELD INTENSITY

Another thought experiment is used to define the magnetic field intensity $\vec{H}$ (A/m). This experiment uses a solenoid, which is a coil of closely wound wire in the shape of a cylinder (Fig. 3.16a). The particular solenoid we need has a short length ΔL, which is still very long compared to the diameter of the cross section, $\Delta L \gg 2a$. There are n turns of wire per unit length. We connect a current source to the wires to produce a current I in the wire. The current per unit length circulating around the solenoid is therefore nI. An idealized equivalent of this solenoid is a short, very thin cylinder with surface current $J_{s0} = nI/\Delta L$ flowing round the cylinder, as shown in Fig. 3.16b.

A small bar magnet is now placed within the solenoid and the torque on the magnet is measured. It is found that the intensity of the torque is uniform within the solenoid (as long as we stay away from the ends) and that it is proportional to the current I. The torque is in such a direction as to align the bar magnet with the axis of the solenoid.

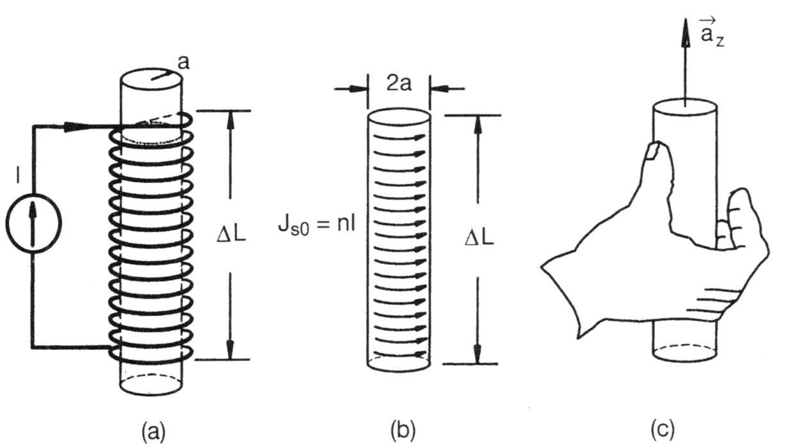

FIGURE 3.16 Solenoid with closely wound wire, which is used to define the magnetic field intensity.

We define the magnitude of the magnetic field intensity interior to the solenoid as being equal to the surface current and we define its direction as being parallel to the axis of the solenoid in the direction of the thumb when the fingers of the right hand turn around the solenoid in the direction of the current flow (Fig. 3.16c). Thus

$$\vec{H} = \vec{a}_s J_{s0}. \tag{3.27}$$

where $\vec{a}_s$ is given by the right-hand rule.

To measure the magnetic field intensity at any point P in space, we place our little solenoid at the point and adjust the current and the orientation of the solenoid until the torque on the bar magnet is zero. Then the total magnetic field intensity inside the solenoid must be zero. Since this total field is the sum of the original field, which we are trying to measure, and the field due to the current in the solenoid, the original field must be equal to but in the opposite direction from the solenoid field. This measurement can work within material media if we imagine (in our thought experiment) a thin tunnel with the solenoid fitted inside.

3.9 MAGNETOMOTIVE FORCE (MMF)

The magnetomotive force, mmf (in ampere units, A), along a line (Fig. 3.17) is the line integral of the magnetic field intensity:

$$\text{mmf} = U_{\text{AB}} = \int_{L_{\text{AB}}} \vec{H} \cdot d\vec{l} \tag{3.28}$$

$$= \int_{L_{\text{AB}}} H_l \, dl. \tag{3.29}$$

In general, the mmf depends on the particular path or line between the end points A and B. The magnetomotive force represents, just like the emf, the "driving force"

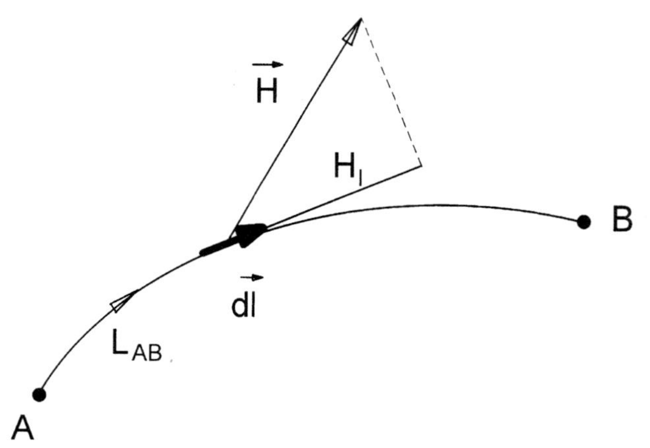

FIGURE 3.17 A path between points A and B in magnetic field used to calculate mmf.

behind the magnetic flux in magnetic circuits. The concept of magnetomotive force becomes important and very useful in the study of electromagnetic relays, transformers, and rotating machines, which operate on the existence of a time-varying magnetic field inside a magnetic circuit. In conventional magnetic circuits, the mmf is produced by a solenoid excited by an electric current. This kind of mmf is considered as a magnetic potential rise. As a result of the presence of the magnetic flux in the magnetic circuit, one could define a magnetic potential drop along different sections of the magnetic circuit (very much like the definition of voltage drop in electric circuits). The above definition is general enough to be applicable to both magnetic potential rise (mmf) as well as magnetic potential drops.

3.10 CONSTITUTIVE EQUATIONS

In this chapter, several field quantities have been defined independently. We are interested in the relationship between these fields in different materials, such as air, glass, copper, or iron. The relations that depend on the properties of the material are called the constitutive equations. These relations link together the two electric field quantities, the two magnetic field quantities, and, in conductors, the electric field intensity and the current density.

Our purpose in this textbook is to study the interrelationships of the electromagnetic field quantities, not the internal structure of materials. Hence a discussion of the atomic or molecular causes of the material properties will not be given here.

3.10.1 Dielectric Materials and Permittivity

Electric field intensity $\vec{E}$ and electric flux density $\vec{D}$ have been defined independently. These two electric field quantities are related by a multiplicative factor ε called the permittivity, which depends on the electric properties of the material.

In free space,

$$\vec{D} = \varepsilon_0 \vec{E}, \tag{3.30}$$

where ε_0 (F/m) is the permittivity of free space. The value of this constant is

$$\varepsilon_0 = 8.854 \cdots \times 10^{-12} \text{ F/m} \tag{3.31}$$

$$\approx \frac{1}{36\pi} \times 10^{-9} \text{ F/m}. \tag{3.32}$$

The form given in Eq. (3.32) is easy to remember and convenient to use in many of the formulas we will derive.

Many materials have a different permittivity from that of free space, but the permittivity is still a constant value. These materials are linear and isotropic. The permittivity of linear and isotropic materials could be a function of frequency, especially at higher frequencies. In these materials the relation between electric flux density and electric field intensity is

$$\vec{D} = \varepsilon\vec{E} \tag{3.33}$$

$$= \varepsilon_0\varepsilon_r\vec{E}. \tag{3.34}$$

The constant ε (F/m) is the permittivity of the material. The constant ε_r is the relative permittivity or the dielectric constant. It is a dimensionless quantity. A short table of the relative permittivity of different materials is given in Appendix A. The range of values is from air, $\varepsilon_r = 1.0$, to Teflon, $\varepsilon_r = 2.1$, to quartz, $\varepsilon_r = 3.8$, to distilled water, $\varepsilon_r = 80$.

In nonlinear materials the permittivity is dependent on the magnitude of the applied electric field intensity. The relation is then given by

$$\vec{D} = \varepsilon(E)\vec{E}. \tag{3.35}$$

In anisotropic materials the relation depends on the direction of the electric field intensity. In this case a matrix is required to show the relationship:

$$\begin{bmatrix} D_x \\ D_y \\ D_z \end{bmatrix} = \varepsilon_0 \begin{bmatrix} \varepsilon_{xx} & \varepsilon_{xy} & \varepsilon_{xz} \\ \varepsilon_{yx} & \varepsilon_{yy} & \varepsilon_{yz} \\ \varepsilon_{zx} & \varepsilon_{zy} & \varepsilon_{zz} \end{bmatrix} \begin{bmatrix} E_x \\ E_y \\ E_z \end{bmatrix}. \tag{3.36}$$

In these materials the electric field intensity and the electric flux density are in different directions.

In this text we will deal only with linear isotropic dielectric materials.

3.10.2 Magnetic Materials and Permeability

Magnetic flux density $\vec{B}$ and magnetic field intensity $\vec{H}$ have been defined independently. These two magnetic field quantities are related by a multiplicative factor μ called the permeability, which depends on the magnetic properties of the material.

In free space

$$\vec{B} = \mu_0\vec{H}, \tag{3.37}$$

where μ_0(H/m) is the permeability of free space. The value of this constant is

$$\mu_0 = 4\pi \times 10^{-7}\ \text{H/m}. \tag{3.38}$$

This value is exact and is determined by the independent absolute standards used to define the meter, kilogram, second, and ampere.

Many materials have a different permeability from that of free space, but the permeability is still a constant. In these linear isotropic materials the relation between the magnetic fields becomes

$$\vec{B} = \mu\vec{H} \tag{3.39}$$

$$= \mu_0\mu_r\vec{H}. \tag{3.40}$$

The constant μ is the permeability of the material with units of H/m. The constant μ_r is the relative permeability, which is a dimensionless quantity. Most linear materials have relative permeabilities that are very close to unity. For example, the relative permeability of copper is 0.99999 and for aluminum it is 1.00002. For all practical purposes, the value of relative permeability for these materials can be taken as 1.

In nonlinear magnetic materials, such as iron, the permeability depends on the applied magnetic field intensity. The relation is represented mathematically by

$$\vec{B} = \mu(H)\vec{H} \tag{3.41}$$

and graphically as in Fig. 3.18. The solid curve represents saturation: as H becomes very large, the rate of increase in B becomes smaller, approaching that of free space. The dashed curve is called a hysteresis curve. Starting at the origin, as H increases to H_m, decreases to $-H_m$, and then increases to H_m again, the path OABCDEFA is followed. Materials such as iron, which have this property, have memory, meaning that the value of B for any given H depends on the past history of the field. Note that at points B and E the magnetic field intensity is zero but the magnetic flux density is not. This represents a residual magnetic field in the material. For unsaturated pure iron the relative permeability is about 4000; for Mumetal it is as high as 100,000.

Magnetic materials such as ferrites exhibit anisotropic properties resulting in a relation like that of Eq. (3.36).

In this text we will mainly be concerned with linear isotropic magnetic materials. A short table of relative permeabilities for various materials is given in Appendix A.

3.10.3 Conductors and Conductivity

Current density and electric field intensity have been defined independently. In some materials these field quantities are related by a multiplicative constant σ(S/m) called the conductivity.

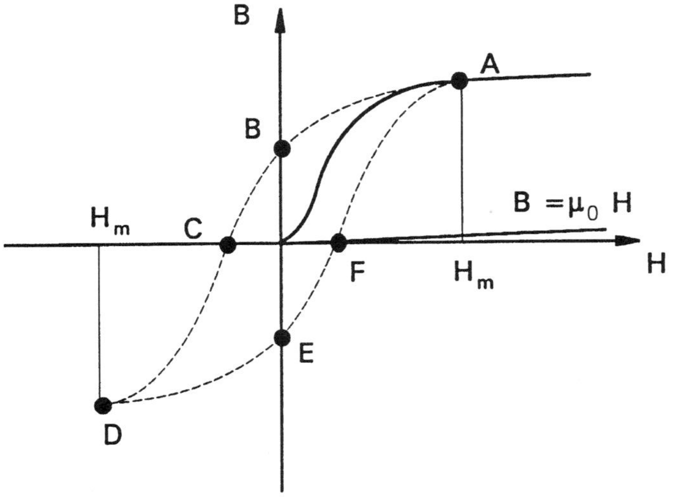

FIGURE 3.18 A plot of B versus H for a nonlinear magnetic material that exhibits hysteresis.

In linear isotropic conducting media the relation is given by

$$\vec{J} = \sigma \vec{E}. \tag{3.42}$$

The reciprocal of conductivity is resistivity, $\rho = 1/\sigma$, which has units of resistance times length, ohm-meter, Ω-m. The siemens is often called the "mho" by practicing engineers and the symbol is often written as an inverted Ω.

The range of conductivity for practical materials is fantastically large. Free space or air has zero conductivity. An insulator or good dielectric, like fused quartz, has a conductivity of less than 2×10^{-17} S/m. A poor conductor like distilled water has 2×10^{-4} S/m. Semiconductors are created by adding impurities to poor conductors like silicon, germanium, or gallium arsenide. Pure or intrinsic silicon has a conductivity of about 0.002 S/m. The conductivity of doped silicon depends on the doping density but is on the order of 1 S/m. Seawater is a better conductor with conductivity of 4 S/m. Metals are the best conductors and copper, for example, has a conductivity of 5.8×10^{7} S/m. From quartz to copper the increase in conductivity is at least 24 orders of magnitude, a factor of 10^{24}. There is no other material property that varies so widely.

A short table of conductivities is given in Appendix A.

3.10.4 Linearity and Superposition

With the exception of saturable magnetic materials, all the media that we will deal with in this text are linear media. Maxwell's equations, which govern all the phenomena we will discuss, are a set of linear integral–differential equations when the media are linear. As a result, the principle of superposition will apply. This has the same meaning for the problems we will encounter as it does in your linear circuits and systems courses. A system is considered linear if it satisfies the additivity and homogeneity conditions.

In detail, what we mean by linearity and superposition is the following. Let $S_1(u, v, w, t)$ represent any source field and $R_1(u, v, w, t)$ represent the response field. Either or both of these might be scalars or vectors. Similarly, let $S_2(u, v, w, t)$ represent any other source field and $R_2(u, v, w, t)$ represent the response field. Then, in linear media, the response to the two sources applied simultaneously is equal to the sum of the responses to each source individually (additivity). In equation form, if

$$R_1(u, v, w, t) = \text{Response } \{S_1(u, v, w, t)\} \tag{3.43}$$

and

$$R_2(u, v, w, t) = \text{Response } \{S_2(u, v, w, t)\} \tag{3.44}$$

then

$$R_1 + R_2 = \text{Response } \{S_1 + S_2\}. \tag{3.45}$$

In linear media the response to a source multiplied by a constant is equal to the product of the constant and the response to that source (homogeneity). In equation form, if

$$R_1(u, v, w, t) = \text{Response}\ \{S_1(u, v, w, t)\} \tag{3.46}$$

then

$$a \cdot R_1 = \text{Response}\ \{a \cdot S_1\}. \tag{3.47}$$

By using the principle of superposition we will be able to find the fields of complicated source distributions by first finding the fields of simpler source distributions.

PROBLEMS

3.1 A hydrogen atom consists of a single proton orbited by a single electron. A simple model of this atom has the proton at the center and the electron in a circular orbit at a distance $R = 5.28 \times 10^{-11}$ m.

a. Evaluate the electric force of attraction between the proton and the electron.

b. Evaluate the gravitational force of attraction.

Ans. a. 8.27×10^{-8}; b. 3.64×10^{-47}.

3.2 Find the charge Q inside a sphere of radius a for each of the following charge distributions:

a. $\rho = \rho_0, \quad R < a.$

b. $\rho = \rho_0, \quad R < b < a.$

c. $\rho = \rho_0(R/a), \quad R < a.$

d. $\rho = \rho_0 \cos\theta, \quad R < a.$

e. $\rho = \rho_0 \cos^2\theta, \quad R < a.$

Ans. a. $Q = \frac{4}{3}\pi a^3 \rho_0$; b. $\frac{4}{3}\pi b^3 \rho_0$; c. $\pi a^3 \rho_0$; d. 0; e. $(4\pi/9)a^3\rho_0$.

3.3 Solve Problem 3.2 using MathCad.

3.4 Find the charge per unit length inside a cylinder of radius a coaxial with the z axis for each of the following charge distributions:

a. $\rho = \rho_0, \quad r < a.$

b. $\rho = \rho_0, \quad b < r < a.$

c. $\rho = \rho_0(r/a), \quad r < a.$

d. $\rho = \rho_0(a/r), \quad r < a.$

Ans. a. $\rho_l = \pi a^2 \rho_0$; b. $\pi b^2 \rho_0$; c. $(2\pi/3)a^2\rho_0$; d. $2\pi a^2 \rho_0$.

3.5 Solve Problem 3.4 using MathCad.

3.6 Surface charge $\rho_s = \rho_{s0} \exp(-kr)$ covers the $z = 0$ plane.

a. Find the charge on a ring of inner radius a and outer radius b.

b. Find the charge on the portion of the plane $a < r$.

c. Find the total charge on the plane.

Ans. a. $Q = (2\pi\rho_{s0}/k^2)\,[(ka + 1)\exp(-ka) - (kb + 1)\exp(-kb)]$;
b. $(2\pi\rho_{s0}/k^2)(ka + 1)\exp(-ka)$; c. $(2\pi\rho_{s0}/k^2)$.

3.7 Use the symbolic evaluation feature of MathCad to solve Problem 3.6.

3.8 **a.** For the current density given in Example 3.4, draw a graph of the z component of current density versus normalized radius kr for $0 \le kr < 2$. Find the slope at $z = 0$ in order to draw the figure properly.

b. Find the average current density inside a circle of radius $a = (1/k)$. Add this value to your sketch with a horizontal line from $kr = 0$ to 1.

Ans. b. $(J_z)_{avg} = J_0[1 - \exp(-1)] = 0.632J_0$.

3.9 A toroid has inner radius a, outer radius b, and height h. The walls are idealized as perfect conductors with total current I_0 flowing up the inner cylinder, radially outward on the upper surface, down the outer cylinder, and radially inward on the bottom surface (see Fig. 3.19, for a cut open view of half the toroid). Find the surface current density on each surface.

Ans. On $r = a$, $J_{sz} = I_0/2\pi a$; on $z = h$, $J_{sr} = I_0/2\pi r$; on

$r = b$, $J_{sz} = -I_0/2\pi b$; on $z = 0$, $J_{sr} = -I_0/2\pi r$.

3.10 The surface current on a sphere of radius a is

$\vec{J}_s = J_{s0}(\vec{a}_\theta \cos\theta \cos^2\phi + \vec{a}_\phi \sin\theta \sin^2\phi)$.

a. Find the current crossing the circle $R = a$, $\theta = \theta_0$.

b. Find the current crossing the semicircle $R = a$, $\phi = \phi_0$.

Ans. a. $I_a = \pi a J_{s0} \cos\theta_0 \sin\theta_0$; b. $I_b = 2aJ_{s0} \sin^2\phi_0$.

3.11 Use the symbolic evaluation feature of MathCad to solve Problem 3.10.

3.12 **a.** Evaluate the electric field intensity at a distance of 1 m from a 2 μC point charge in otherwise empty space.

b. Evaluate the force at this point on a 3 μC point charge.

c. Repeat if the distance is 10 m.

Ans. b. 54×10^{-3} N; c. 0.54×10^{-3} N.

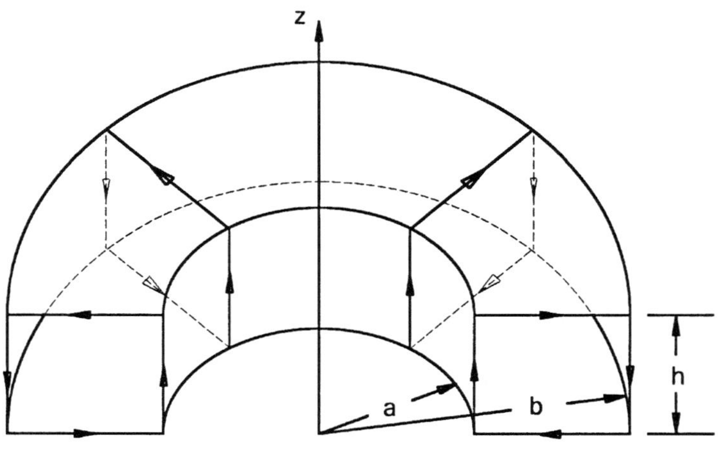

FIGURE 3.19 Problem 3.9.

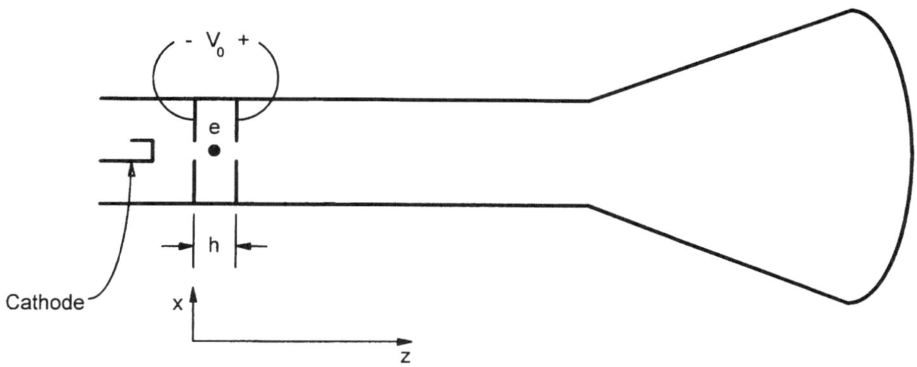

FIGURE 3.20 Problem 3.13.

3.13 Figure 3.20 shows (symbolically) the cathode of a cathode ray tube. Electrons are emitted in a stream with zero initial velocity. They enter a pair of parallel plates through a small hole in the first plate. The field between the plates is $\vec{E} = -\vec{a}_z(v_0/h)$. Find the velocity u_{z0} with which the electrons leave through the hole in the second plate. Find the kinetic energy K.E. $= mu_{z0}^2/2$ of the exiting electron.

Ans. $u_{z0} = \sqrt{2eV_0/m}$, K.E. $= eV_0$.

3.14 The electrons of Problem 3.13 now have velocity $\vec{u} = \vec{a}_z u_{z0}$. They next enter a second pair of parallel plates between which the field is $\vec{E} = -\vec{a}_x E_{x0}(t/T)$. The second pair of plates is d wide. The distance from the end of the second pair of plates to the screen is D (see Fig. 3.21).

a. Consider an electron that enters these plates at time t'. This electron leaves the plates at time $t' + d/u_{z0}$. Find the velocity u_{x0} with which this electron leaves the plates.

b. Assume that $t' \gg d/u_{z0}$. Show that the exit velocity is proportional to t'.

c. Under the assumption of part (b), find the x deflection at the screen.

Ans. a. $u_{x0} = (eE_0/2mT)[(t' + d/u_{z0})^2 - (t')^2]$; b. $u_{x0} \approx (eE_0 d/mTu_{z0})t' = kt'$;
c. $x = u_{x0}(D/u_{z0}) = (eE_0 dD/mTu_{z0}^2)t'$.

Note that a repetitive sweep field (Fig. 3.22a) will yield a periodic deflection (Fig. 3.22b). Furthermore, if a third set of plates is added with a periodic field in the y direction (Fig. 3.22d), a graph of that field versus x deflection will appear on the screen (Fig. 3.22c). This is the basic principle of the oscilloscope.

3.15 The field of a point charge q located at the origin is $\vec{E} = \vec{a}_R k_e(q/R^2)$.

a. Find the voltage along the radial line $\theta = \theta_0$, $\phi = \phi_0$ extending from point A at R_A to point B at $R_B > R_A$.

b. Find the voltage at R_A relative to the voltage at infinity, by letting $R_B \to \infty$.

Ans. a. $V_{AB} = V(R_A) - V(R_B) = k_e q[(1/R_A) - (1/R_B)]$.

3.16 Use the symbolic evaluation feature of MathCad to solve Problem 3.15.

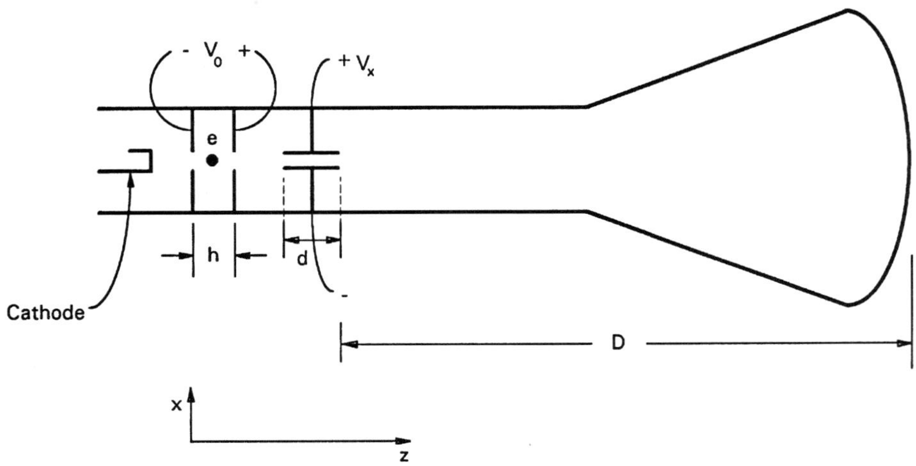

FIGURE 3.21 Problem 3.14.

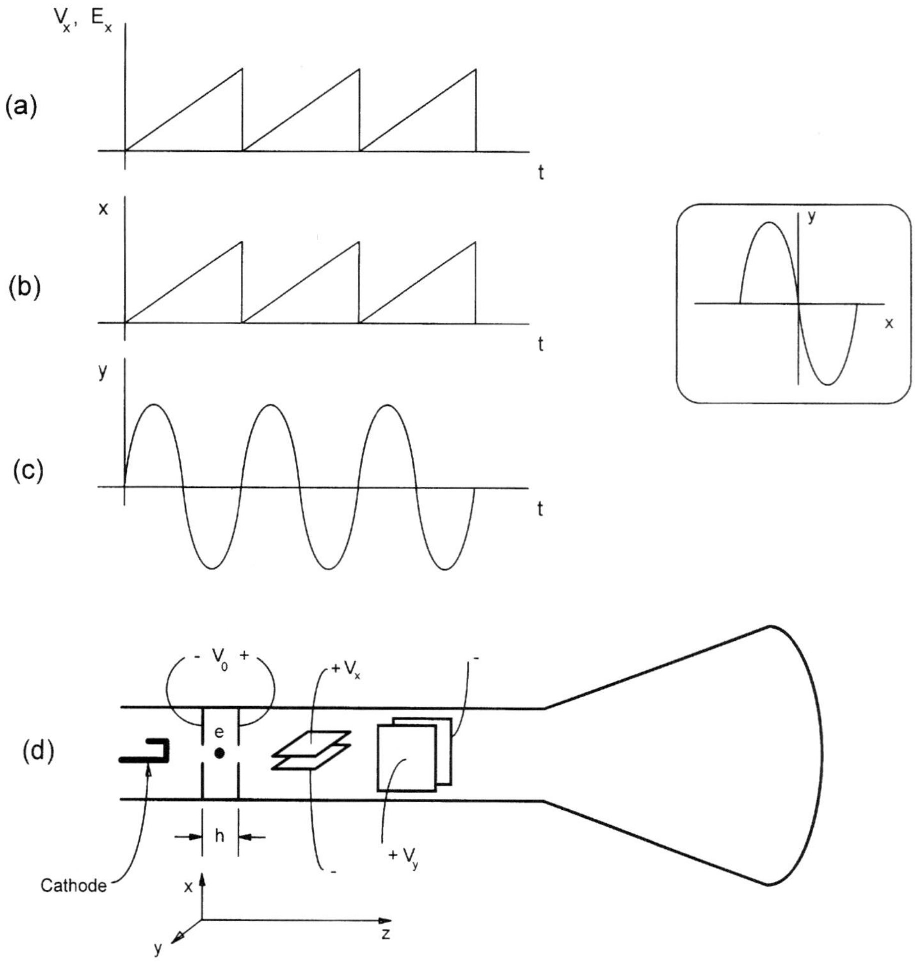

FIGURE 3.22 Basic concept of the operation of an oscilloscope.

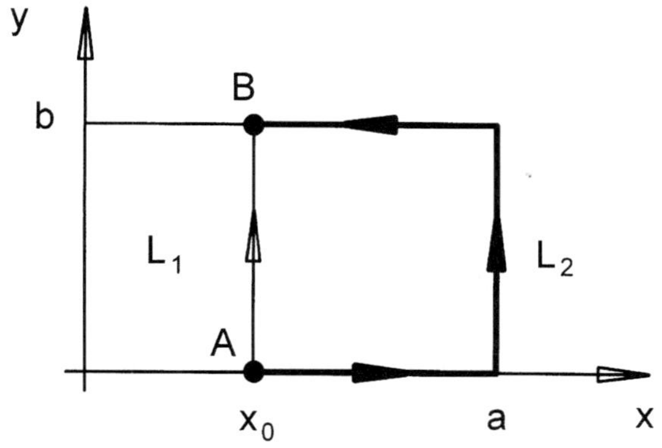

FIGURE 3.23 Problem 3.17.

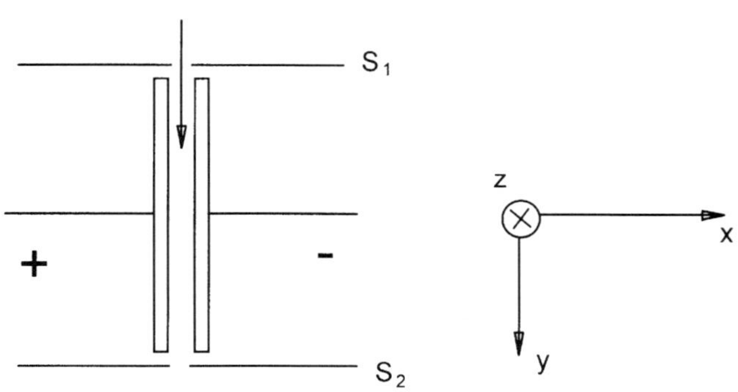

FIGURE 3.24 Problem 3.22.

3.17 The electric field at a particular instant of time is $\vec{E} = \vec{a}_y E_0 \sin(\pi x/a)$.

a. Find the voltage V_{AB} along the path L_1 (Fig. 3.23).

b. Find the voltage V_{AB} along the path L_2 (Fig. 3.23).

Ans. a. $V_{AB} = E_0 b \sin(\pi x_0/a)$; b. $V_{AB} = 0$.

3.18 **a.** An electron is projected into a magnetic field $\vec{B} = \vec{a}_z 10$ T with a velocity $\vec{u} = \vec{a}_x 3 \times 10^7$ m/s. Compute the magnetic force on the electron.

b. Repeat if $\vec{u} = \vec{a}_y 3 \times 10^7$ m/s.

c. Repeat if $\vec{u} = \vec{a}_z 3 \times 10^7$ m/s.

d. Compare this force with the gravitational attraction (weight) of an electron.

Ans. a. $\vec{F} = \vec{a}_y 48$ pN; b. $\vec{F} = -\vec{a}_x 48$ pN; c. $F = 0$; d. $\vec{F}_g = 89 \times 10^{-31}$ N.

3.19 Solve Problem 3.18 using MathCad.

3.20 A charge q is located at the origin and has a velocity $\vec{u} = \vec{a}_x u_{x0}$ at time $t = 0$. The magnetic flux density is $\vec{B} = \vec{a}_z B$.

a. Show that the x and y components of the velocity are related by $du_y/dt = -\omega u_x$ and $du_x/dt = \omega u_y$, where $\omega = qB/m$.

b. Solve for u_x and u_y.

c. Integrate the velocities to obtain the position of the charge, $x = (u_{x0}/\omega)\sin\omega t$ and $y = (u_{x0}/\omega)(\cos\omega t - 1)$. Show that these satisfy

$$x^2 + \left(y + \frac{u_{x0}}{\omega}\right)^2 = \left(\frac{u_{x0}}{\omega}\right)^2$$

and draw the circle that this represents. The charge rotates around the circle with angular frequency $\omega = qB/m\ \mathrm{s}^{-1}$ and period $T = 2\pi m/qB$ s, both independent of u_{x0}. The circle's radius is $R = u_{x0}/qB$.

3.21 **a.** Obtain the path of the charge of Problem 3.20 if $\vec{u} = \vec{a}_x u_{x0} + \vec{a}_z u_{z0}$.

b. Show that at $t = nT$, $n = 1, 2, \ldots$, where $T = 2\pi m/qB$, all charges injected at time $t = 0$ will converge on a single "focused" spot, independent of the value of u_{x0}. What is the coordinate of that spot? This is the principle of magnetic focusing of the beam in a cathode ray tube.

3.22 Figure 3.24 shows a velocity filter. Positive ions of charge q pass through slit S_1 with differing velocities $\vec{u} = \vec{a}_y u$. Between the plates there is an electric field $\vec{E} = \vec{a}_x E_0$ and a magnetic field $\vec{B} = -\vec{a}_z B_0$. Only those ions that are not

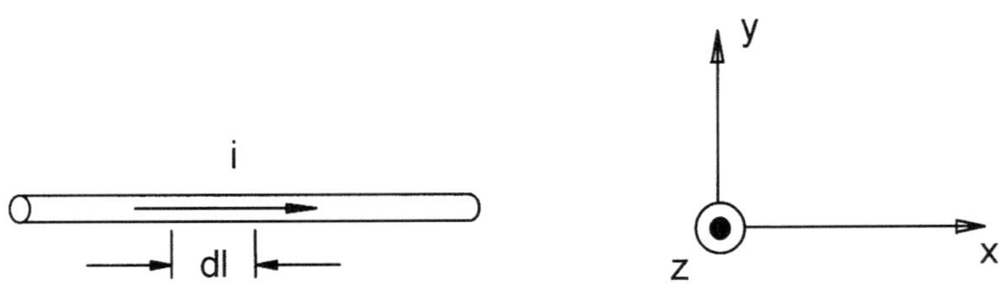

FIGURE 3.25 Problem 3.23.

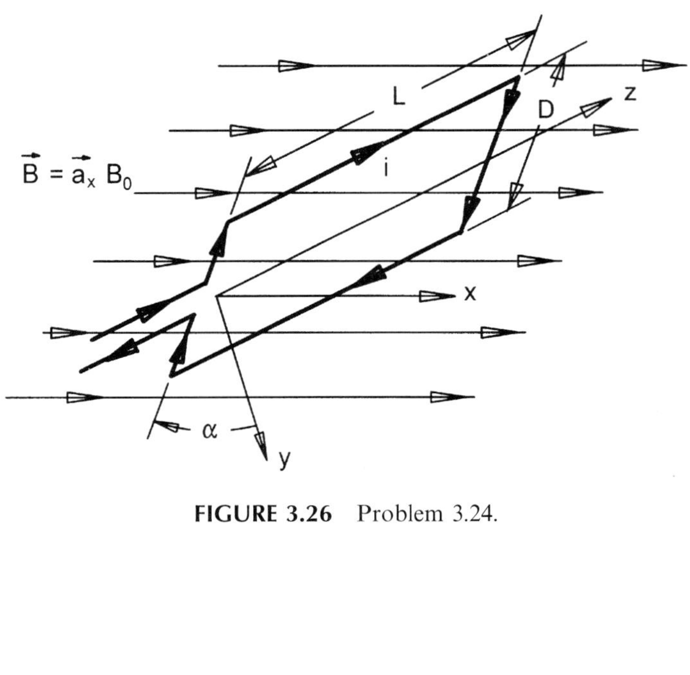

FIGURE 3.26 Problem 3.24.

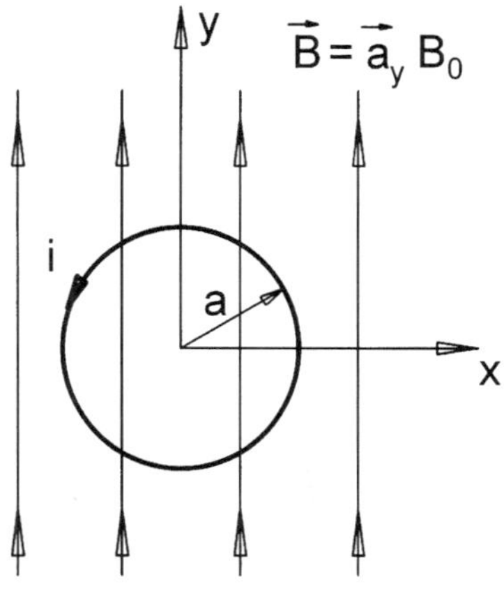

FIGURE 3.27 Problem 3.25.

deflected by the fields will emerge through slit S_2. Show that these ions have the velocity $\vec{u} = \vec{a}_y(E_0/B_0)$.

3.23 **a.** A current $i = \rho A u_x$ flows in the positive x direction in a wire of cross-sectional area A (ρ and u_x are negative). The wire is in a magnetic field $\vec{B} = \vec{a}_y B_0$. See Fig. 3.25. Show that the force $d\vec{F}$ on a section of wire $d\vec{l}$ is $d\vec{F} = \vec{a}_z i\ dl\ B_0$. *Hint*: Find the force on an individual electron and then multiply by the number of electrons.

b. Show that this is a special case of the general result that $d\vec{F} = i\, d\vec{l} \times \vec{B}$. This equation represents the basic operating principle of any motor (dc or ac).

3.24 Use the result of Problem 3.23 to find the torque about the z axis for the rectangular loop shown in Fig. 3.26.

Ans. $\vec{\tau} = -\vec{a}_z i A B_0 \sin\alpha$, $A = DL$.

3.25 **a.** Use the result of Problem 3.23 to find the torque about the x axis for the circular loop shown in Fig. 3.27.

b. Repeat, if the plane of the circle is at angle α with the z axis.

Ans. b. $\vec{\tau} = -\vec{a}_x i A B_0 \sin\alpha$, $A = \pi a^2$.

CHAPTER FOUR

Maxwell's Integral Equations

4.1 INTRODUCTION

In the previous chapter various electromagnetic field quantities were defined. The constituent equations give the interrelationship between some of these quantities. In this chapter Maxwell's equations are introduced in integral form. These equations give the complete interrelation between the four electromagnetic fields and their sources. Several examples are given in which Maxwell's integral equations are used to find the fields so as to illustrate the meaning of the equations.

4.2 GAUSS'S LAW FOR ELECTRIC FLUX DENSITY

Gauss's law for electric flux density states that the total electric flux leaving a closed surface, Ψ_e, equals the total charge enclosed by the surface, Q_{enclosed}:

$$\Psi_e = Q_{\text{enclosed}}. \tag{4.1}$$

The total electric flux leaving a closed surface is the integral of the outward normal component of the electric flux density $\vec{D}$ over the closed surface (Fig. 4.1). The total charge enclosed by the surface is the integral of the charge density ρ throughout the volume enclosed by the surface. Hence the electric flux density is related to the volume charge density by

$$\oint_S \vec{D} \cdot d\vec{S} = \int_V \rho \, dv. \tag{4.2}$$

This is the integral form of Gauss's law. If surface, line, or point charges are present within the volume V, then terms to integrate or sum over these charges must be added to the right side of the equation.

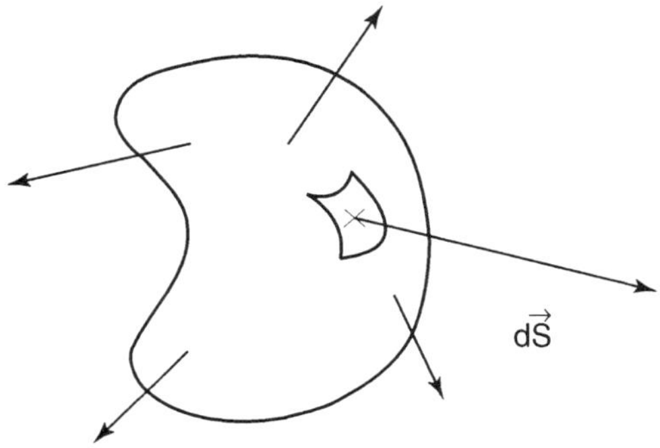

FIGURE 4.1 Electric flux density leaving a closed surface.

Note that the closed surface over which the integral is performed need not be a surface corresponding to some particular physical feature of the problem. The law states that the equality holds for any closed surface that can be imagined. A closed surface over which Gauss's law is applied is frequently called a Gaussian surface.

This integral form of the law relates total electric flux to total charge enclosed. It does not give any information about the relation between the electric flux density and the charge density at a particular point in space. This form of Gauss's law will therefore not generally be useful in finding the flux density due to a charge distribution. There are, however, some problems that can be solved using the integral law and these problems are important in leading to an understanding of many more problems.

The problems that can be solved using Eq. (4.2) are ones in which the charge distribution is very symmetrical. Symmetry in charge distribution leads to symmetry in the flux density. This may mean that the field is constant over some suitably chosen closed surface and hence that the total flux through the surface is just equal to the flux density multiplied by the surface area. In these cases Eq. (4.2) can be used to find the electric flux density at each point.

Another way of stating the same idea is as follows. Equation (4.2) can be interpreted to mean that the average value of electric flux density through a surface is equal to the total charge enclosed by the surface divided by the area of the surface. Thus

$$(D_{\text{normal}})_{\text{average}} = \frac{Q_{\text{enclosed}}}{\text{Area}}. \tag{4.3}$$

If the normal component of the electric flux density is constant over the surface, then the average value given by Eq. (4.3) is equal to that constant value.

The examples that follow will illustrate the calculation of the electric flux density using Gauss's law in integral form when the charge distribution is highly symmetric.

4.2.1 Electric Field of a Point Charge

Consider a sphere of charge of radius a in which the charge is uniformly distributed throughout the sphere (Fig. 4.2). This uniform sphere of charge has complete spherical symmetry with respect to the center point of the sphere. That is, if we choose the center

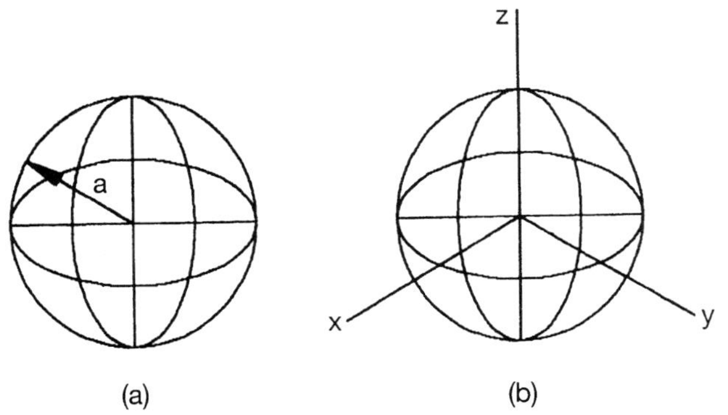

FIGURE 4.2 Sphere of uniformly distributed charge of radius a.

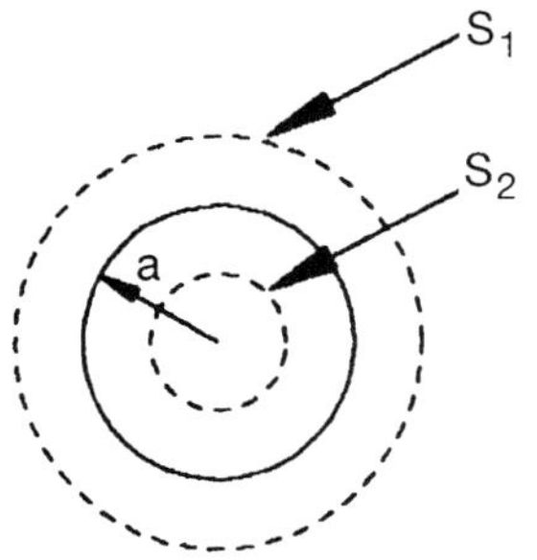

FIGURE 4.3 Gaussian surfaces for the charged sphere of radius a.

point as the origin, then the orientation of the x, y, and z axes is completely arbitrary. The figure would be unchanged no matter how the axes were rotated.

The charge distribution is given by

$$\rho(R, \theta, \phi) = \begin{cases} \rho_0, & R < a, \\ 0 & R > a. \end{cases} \tag{4.4}$$

Since the charge distribution has spherical symmetry and since lines of electric flux emanate from the charge, the electric flux density has only a radial component. This component is independent of θ and ϕ, also due to the spherical symmetry of the charge. Thus $\vec{D}(R, \theta, \phi) = \vec{a}_R D_R(R)$. We now apply Eq. (4.2) over a spherical surface of radius R centered at the center of the spherical charge distribution, that is, at the origin. Figure 4.3 shows two such Gaussian surfaces as dotted lines, S_1 and S_2. For $R < a$,

$$\oint_{S_1} \vec{D} \cdot d\vec{S} = \int_V \rho \, dv$$

$$\int_{\phi=0}^{2\pi} \int_{\theta=0}^{\pi} [\vec{a}_R D_R(R)] \cdot [\vec{a}_R R^2 \sin\theta \, d\theta \, d\phi] = \int_{\phi=0}^{2\pi} \int_{\theta=0}^{\pi} \int_{R=0}^{R} \rho_0 R^2 \sin\theta \, dR \, d\theta \, d\phi$$

$$4\pi R^2 D_R(R) = \tfrac{4}{3}\pi R^3 \rho_0$$

$$D_R(R) = \frac{\tfrac{4}{3}\pi R^3 \rho_0}{4\pi R^2} \tag{4.5}$$

$$D_R(R) = \frac{R}{3}\rho_0, \quad R < a. \tag{4.6}$$

Equation (4.5) is written in the form of Eq. (4.3) and could have been obtained directly without the steps that precede it. For $R > a$,

$$\oint_{S_2} \vec{D} \cdot d\vec{S} = \int_V \rho \, dv$$

$$\int_{\phi=0}^{2\pi} \int_{\theta=0}^{\pi} [\vec{a}_R D_R(R)] \cdot [\vec{a}_R R^2 \sin\theta \, d\theta \, d\phi] = \int_{\phi=0}^{2\pi} \int_{\theta=0}^{\pi} \int_{R=0}^{a} \rho_0 R^2 \sin\theta \, dR \, d\theta \, d\phi$$

$$4\pi R^2 D_R(R) = \tfrac{4}{3}\pi a^3 \rho_0 \tag{4.7}$$

$$D_R(R) = \frac{\tfrac{4}{3}\pi a^3 \rho_0}{4\pi R^2} \tag{4.8}$$

$$D_R(R) = \frac{a^3}{3R^2}\rho_0, \quad R > a. \tag{4.9}$$

On the left side, the integration is identical to the previous case. Only the right side, however, the integration over R extends only to the upper limit $R = a$ because the charge density is zero in the spherical shell from a to $R > a$. Again, Eq. (4.8) is in the form of Eq. (4.3) and could have been written directly without the steps that precede it.

Assuming the medium is free space, the electric field intensity is

$$\vec{E} = \begin{cases} \vec{a}_R \dfrac{R}{3\varepsilon_0}\rho_0, & R < a, \qquad (4.10) \\[2ex] \vec{a}_R \dfrac{a^3}{3\varepsilon_0 R^2}\rho_0, & R > a. \qquad (4.11) \end{cases}$$

The total charge Q is given by the right-hand side of Eq. (4.7), $Q = \tfrac{4}{3}\pi a^3 \rho_0$. Equations (4.9) and (4.11) become

$$D_R(R) = \frac{Q}{4\pi R^2}, \quad E_R(R) = \frac{Q}{4\pi\varepsilon_0 R^2}, \quad R > a. \tag{4.12}$$

If we let $a \to 0$ but keep the total charge constant, then the charge source becomes a point charge, $q = Q$, and the region $R > a$ becomes all of space. The electric flux

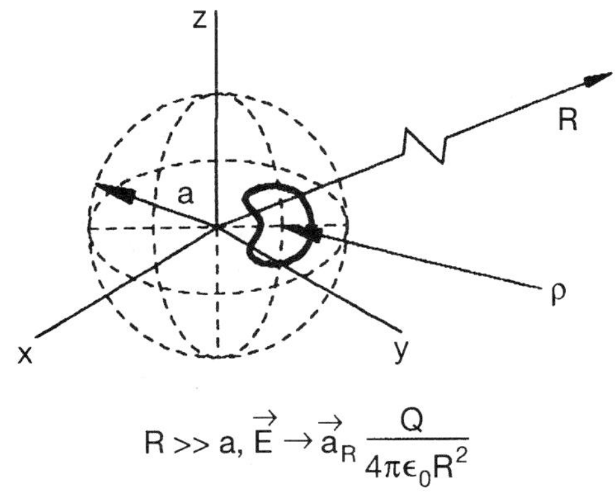

FIGURE 4.4 Field due to a charge density ρ for $R \gg a$.

density and the electric field intensity are then

$$\vec{D} = \vec{a}_R \frac{q}{4\pi R^2}, \tag{4.13}$$

$$\vec{E} = \vec{a}_R \frac{q}{4\pi\varepsilon_0 R^2}, \tag{4.14}$$

for all values of R. These are the free-space fields of a point charge q located at the origin.

The field of a point charge is, as expected, inversely proportional to the square of the distance from the charge, R^2. If we are very far away from any charge distribution, that distribution will appear as if it were a point. The field very far from any distribution (Fig. 4.4) will approach the field of a point charge $q = Q$, where Q is the total charge in the distribution.

Conversely, if $\vec{E}$ is inversely proportional to R^2 as $R \to 0$, then a point charge must be located at the point where $R = 0$.

The electric field intensity results given in this section assume the medium is free pace. For any medium that maintains the spherical symmetry (i.e., where the permittivity is at most a function of the radial coordinate R), the electric flux density is unchanged. The electric field should be found by dividing the electric flux density by the permittivity of the medium.

4.2.2 Electric Field of a Uniform Line Charge

Figure 4.5 shows a line charge with the z axis lying along the line. The line charge has cylindrical symmetry around the z axis; no matter how the x and y axes are rotated around the z axis, the picture is unchanged. The fields due to the line charge must therefore be independent of the ϕ coordinate and must not have a ϕ component. Thus $\vec{D}(r, \phi, z) = \vec{a}_r D_r(r, z) + \vec{a}_z D_z(r, z)$.

Let the line charge density be uniform along the entire z axis. This means that $\rho_l = \rho_{l0}$, a constant, for all values of z from $-\infty$ to $+\infty$. Then the position of the xy

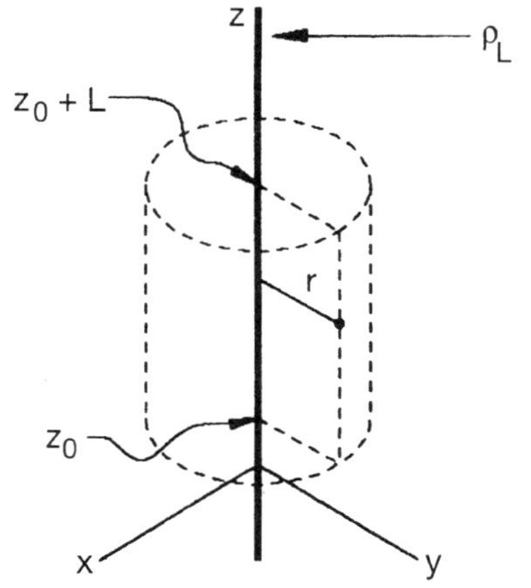

FIGURE 4.5 Line charge along the z axis.

plane (of $z = 0$) is arbitrary: no matter where it is located the picture is unchanged. The fields must also be independent of the z coordinate and must not have a z component. Thus $\vec{D}(r, \phi, z) = \vec{a}_r D_r(r)$.

We now apply Gauss's law by integrating over the Gaussian surface shown in Fig. 4.5.

$$\oint_S \vec{D}\cdot d\vec{S} = \int_V \rho\, dv$$

$$\int_{z=z_0}^{z_0+L}\int_{\phi=0}^{2\pi} [\vec{a}_r D_r(r)]\cdot[\vec{a}_r r\, d\phi\, dz] + \int_{\text{top \& bottom}} \vec{D}\cdot d\vec{S} = \int_{z=z_0}^{z_0+L} \rho_{l_0}\, dz \tag{4.15}$$

$$2\pi r L D_r(r) = \rho_{l_0} L$$

$$\vec{D} = \vec{a}_r \frac{\rho_{l_0}}{2\pi r}. \tag{4.16}$$

The integral over the top and bottom surfaces of the Gaussian cylinder are zero because the normals to these surfaces are in the $\vec{a}_z$ and $-\vec{a}_z$ directions, respectively, and $D_z = 0$.

In the vicinity of any line charge the field varies as $1/r$ for small r. Conversely, if the field varies as $1/r$ for small r, then a line charge must be present at $r = 0$. A practical example of the above situation would be the long lines found in integrated circuits, where the determination of the electric field around the lines is very important in evaluating the capacitance and cross interference between adjacent lines. This problem will be addressed in a later chapter.

The electric flux density is given by this result for any medium that maintains the cylindrical symmetry (i.e., where the permittivity is at most a function of the radial coordinate r). The electric field intensity is found by dividing the electric flux density by the permittivity.

4.2.3 Electric Field of a Uniform Surface Charge

Figure 4.6 shows an infinite plane surface, which may be taken to be the $z = 0$ plane. The surface has a uniform surface charge density ρ_{s0} on it. One point on the surface is exactly like every other point and hence the choice of position for the z axis (the $x = y = 0$ point) is arbitrary. Also, the region above the plane is exactly like the region below the plane. As a result of these symmetries, the electric flux density must be in the z direction, it must not be a function of the x or y coordinates, and $D_z(z) = -D_z(-z)$.

To find the value of the electric flux density we draw a cylindrical Gaussian surface with sides parallel to the z axis and the top and bottom surfaces of area A located at $z = z'$ and $z = -z'$, respectively. Applying Gauss's law for electric flux density we get

$$\int_{\text{top}} \vec{D} \cdot d\vec{S} + \int_{\text{bottom}} \vec{D} \cdot d\vec{S} + \int_{\text{side}} \vec{D} \cdot d\vec{S} = Q_{\text{enclosed}}$$

$$D_z(z')A - D_z(-z')A = \rho_{s0} A$$

$$D_z(z') = \frac{\rho_{s0}}{2}.$$

The integral over the side is zero because the surface of the side is orthogonal to the z-directed flux density.

Since z' can have any value, the field above and below the plane of the surface charge is uniform or independent of z. The result, in vector form, is

$$\vec{D} = \begin{cases} +\vec{a}_z \dfrac{\rho_{s0}}{2}, & z > 0, \\ -\vec{a}_z \dfrac{\rho_{s0}}{2}, & z < 0. \end{cases} \tag{4.17}$$

This electric flux density is unchanged for any medium that maintains the symmetry. That means the permittivity may be an even function of the z coordinate only, with $\varepsilon(z) = \varepsilon(-z)$. The electric field intensity is then found by dividing the electric flux

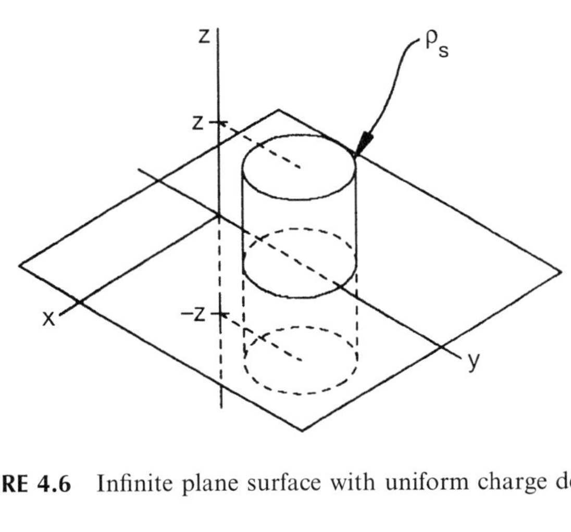

FIGURE 4.6 Infinite plane surface with uniform charge density.

density by the permittivity of the medium. A similar situation occurs when two parallel conductive plates with opposite surface charges are put together. The determination of the electric field between the plates is treated in a later chapter when the concept of capacitance is introduced.

4.3 GAUSS'S LAW FOR MAGNETIC FLUX DENSITY

Gauss's law for magnetic flux density states that the total magnetic flux leaving any closed surface, Ψ_m, equals zero:

$$\Psi_m = 0. \tag{4.18}$$

The total magnetic flux through the closed surface is the integral of the normal component of surface density over the surface (Fig. 4.7). In equation form this gives

$$\oint_S \vec{B} \cdot d\vec{S} = 0. \tag{4.19}$$

These two equations are identical in form with Eqs. (4.1) and (4.2). They differ in the right side, which is zero for magnetic fields. The physical reason for this is that no magnetic charge, the magnetic dual of electric charge, has ever been discovered.

A consequence of Gauss's law for electric flux density is that lines of electric flux density are drawn from positive charge, where they originate, to negative charge, where they terminate. A consequence of Gauss's law for magnetic flux density is that lines of magnetic flux density are continuous, forming closed loops in space.

4.4 CONSERVATION OF CHARGE

Charge cannot be created or destroyed. Current is the flow of charge and, therefore, if there is a net current leaving a region, then the charge remaining in the region must be

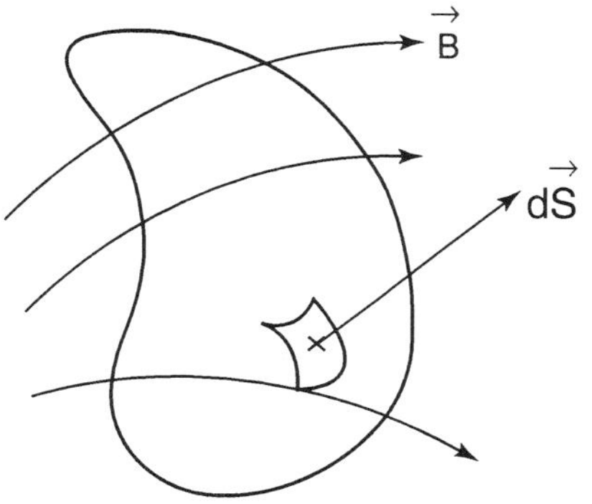

FIGURE 4.7 Magnetic flux density through a closed surface.

decreasing at an appropriate rate. The law of conservation of charge formalizes this statement as follows: the total current I outward through a closed surface is equal to the time rate of decrease of the charge enclosed:

$$I_{\text{outward}} = -\frac{dQ_{\text{enclosed}}}{dt}. \tag{4.20}$$

The total outward current through a closed surface is the outward normal current density integrated over the surface. The charge enclosed by the surface is the integral of the charge density throughout the volume enclosed by the surface. The time rate of decrease of this charge is the negative of the derivative with respect to time. Hence the current density is related to the charge by

$$\oint_S \vec{J} \cdot d\vec{S} = -\frac{d}{dt}\int_V \rho \, dv. \tag{4.21}$$

In evaluating the left side of this equation we must include surface currents and line currents flowing through the surface. On the right side we must include surface charge, line charge, and point charges throughout the enclosed volume V.

4.4.1 Relaxation Time Constant in a Conductor

Let us consider a spherical conductor of radius a (Fig. 4.8). At time $t = 0$ charge is distributed throughout the sphere in a spherically symmetric way, so that the charge density is only a function of the radial coordinate R from the center of the sphere. Let $Q(R, t)$ be the charge inside a spherical surface of radius $R < a$ at time t. Then Gauss's law for electric flux density applied to the closed surface at radius $R < a$ will show, as in Eq. (4.12), that

$$D_R(R, t) = \frac{Q(R, t)}{4\pi R^2}, \quad R < a.$$

Dividing by the permittivity of the conductor we obtain the electric field intensity in the conductor. Multiplying this quotient by the conductivity of the conductor, we

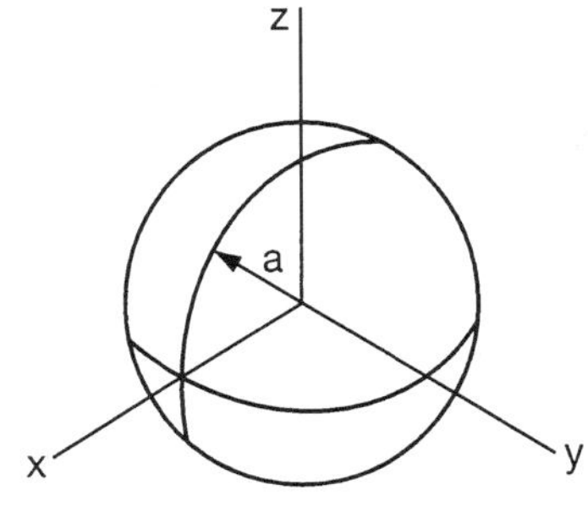

FIGURE 4.8 Spherical conductor of radius a.

obtain the current density

$$J_R(R, t) = \frac{\sigma Q(R, t)}{4\pi\varepsilon R^2}, \quad R < a. \tag{4.22}$$

We now apply the law of conservation of charge, Eq. (4.21), to the same spherical surface, substituting Eq. (4.22) for J_R:

$$\oint_S \vec{J} \cdot d\vec{S} = -\frac{d}{dt}\int_V \rho \, dv$$

$$\int_{\phi=0}^{2\pi} \int_{\theta=0}^{\pi} J_R(R, t) R^2 \sin\theta \, d\theta \, d\phi = -\frac{dQ(R, t)}{dt}$$

$$\frac{\sigma}{\varepsilon} Q(R, t) = -\frac{dQ(R, t)}{dt}, \quad R < a. \tag{4.23}$$

This last equation is a first-order differential equation for the charge. Its solution is

$$Q(R, t) = Q(R, 0)e^{-t/\tau}, \quad \text{where } \tau = \varepsilon/\sigma. \tag{4.24}$$

This equation shows that the charge within a sphere of radius $R < a$ is decreasing at an exponential rate, with a time constant $\tau = \varepsilon/\sigma$ seconds, called the relaxation time constant. This time constant is exactly like the *RC* (resistance multiplied by capacitance) time constant in an *RC* circuit (Fig. 4.9).

The charge originally found distributed throughout the sphere cannot disappear. As the charge within the conductor diminishes, charge is building up as an increasing surface charge on the outer surface of the sphere. At $t = 0$ all the charge $Q_0 = Q(a, 0)$ ies within the sphere. At any time thereafter the charge within the sphere is $Q(a, t) = Q_0 \exp(-t/\tau)$. The difference is the charge on the surface, so the surface

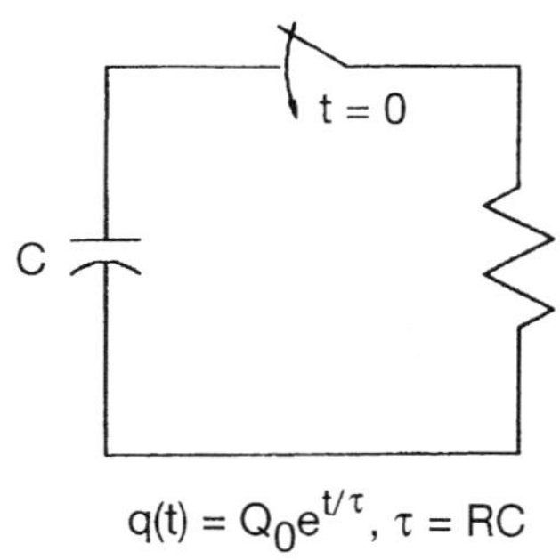

FIGURE 4.9 Equivalent *RC* circuit that models the relaxation time constant of a conductor.

charge density is

$$\rho_s(t) = \frac{Q(a, 0) - Q(a, t)}{4\pi a^2}$$

$$= \frac{Q_0}{4\pi a^2}(1 - e^{-t/\tau}). \tag{4.25}$$

Equations (4.24) and (4.25) are plotted in Fig. 4.10.

The time constant for a good conductor such as copper is on the order of 10^{-19} s, which is very small indeed. For a poor conductor such as distilled water it is on the order of 10^{-6} s. For an excellent insulator such as fused quartz it is on the order of 10^6 s or 12 days. In materials commonly used as conductors (e.g., copper, silver, aluminum) the time constant is so small that we say that the metal cannot contain any net charge in its interior; all the charge goes quickly to the surface as surface charge.

This derivation of the time constant and the expulsion of charge to the surface used a spherical body and a spherically symmetric charge distribution so as to make the mathematics easily manageable. The result holds for any shape of charged conductor and for any initial charge distribution. After a very short transient time, all the charge has flowed to the surface as a surface charge density, ρ_s.

Example 4.1 Figure 4.11 shows a spherical conductor of radius a and a concentric conducting spherical shell of inner radius b and outer radius c. The region between the sphere and the shell is a dielectric with permittivity ε and the region outside the shell is

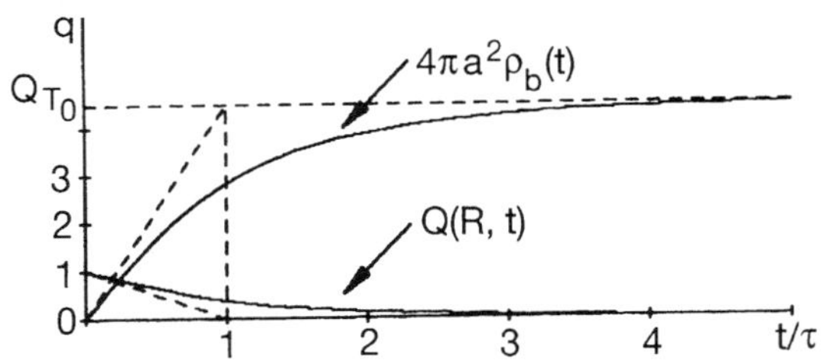

FIGURE 4.10 Plot of surface charge density and charge within the sphere versus time.

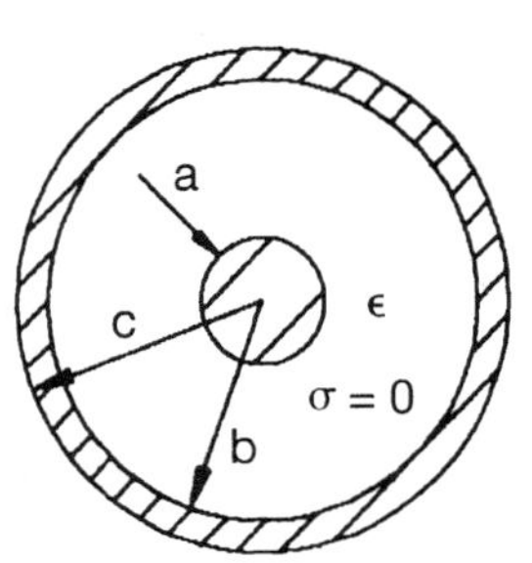

FIGURE 4.11 Spherical conductor with a concentric conducting spherical shell.

free space. A charge Q is placed on the conducting sphere. Calculate the electric flux density and the electric field intensity everywhere.

Solution After a brief transient, the charge on the conductors will reside entirely on the surface. The spherical symmetry of the problem means the fields will be radial and functions of the radial coordinate only: $\vec{D} = \vec{a}_R D_R(R)$ and $\vec{E} = \vec{a}_R E_R(R)$. Using Gauss's law over a spherical surface of radius R,

$$4\pi R^2 D_R = Q_{\text{enclosed}}. \tag{4.26}$$

$$\text{For } R < a, \quad 4\pi R^2 D_R = 0$$

$$D_R = 0, \quad E_R = 0. \tag{4.27}$$

$$\text{For } a < R < b, \quad 4\pi R^2 D_R = Q$$

$$D_R = \frac{Q}{4\pi R^2}, \quad E_R = \frac{Q}{4\pi\varepsilon R^2}. \tag{4.28}$$

The fields within the conducting shell itself must be zero, because if $\vec{E}$ was not zero there would be current and we are interested in the solution after the transient current has disappeared. If $D_R = 0$, $b < R < c$, then, from Eq. (4.2), $Q_{\text{enclosed}} = 0$ and the charge on the $R = b$ surface must be $-Q$, equal but opposite to the charge on the central sphere. Since the spherical shell was originally uncharged, there is no net charge on it and the charge on the $R = c$ surface must be $+Q$. To find the field outside the shell we again use Eq. (4.2).

$$\text{For } c < R, \quad 4\pi R^2 D_R = Q$$

$$D_R = \frac{Q}{4\pi R^2}, \quad E_R = \frac{Q}{4\pi\varepsilon_0 R^2}. \tag{4.29}$$

The surface charge density on each of the three spherical surfaces is

$$\rho_{sa} = \frac{Q}{4\pi a^2}, \quad \text{at } R = a, \tag{4.30}$$

$$\rho_{sb} = \frac{-Q}{4\pi b^2}, \quad \text{at } R = b, \tag{4.31}$$

$$\rho_{sc} = \frac{Q}{4\pi c^2}, \quad \text{at } R = c. \tag{4.32}$$

4.4.2 Electric Field from a Point Current Source

We will now find the field surrounding a long insulated conducting filament lying along the upper half of the z axis, from $z = 0$ to $z = \infty$, as shown in Fig. 4.12. The filament carries current I_0 in the negative z direction. The medium surrounding the

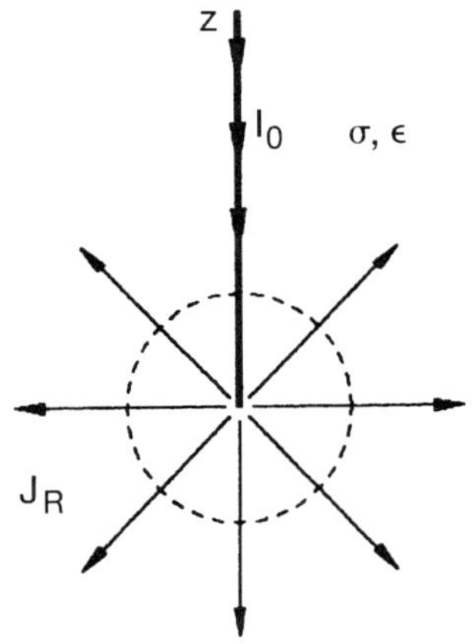

FIGURE 4.12 Field due to a long insulated conducting filament.

filamentary current and extending over all of space has conductivity σ permittivity ε. We assume that the current has been flowing for a long time (compared to $\tau = \varepsilon/\sigma$) so that any transients have disappeared and there is no time variation of any field quantity. The filament is insulated along its entire length so current cannot leave the filament, except at its tip at the origin $z = 0$.†

The tip of the filament is a spherically symmetric point source of current. Since the conducting medium extends outward in all directions, there is no more reason for current to flow in one direction than in another. As a consequence, the current density will be radial and independent of direction, $\vec{J}(R, \theta, \phi) = \vec{a}_R J_R(R)$. Applying the law of conservation of charge to a Gaussian sphere of radius R centered on the origin as shown in the figure, we obtain

$$\oint_S \vec{J} \cdot d\vec{S} = -\frac{d}{dt}\int_V \rho\, dv$$

$$\int_{\phi=0}^{2\pi}\int_{\theta=0}^{\pi} J_R R^2 \sin\theta\, d\theta\, d\phi - I_0 = 0 \tag{4.33}$$

$$4\pi R^2 J_R - I_0 = 0 \tag{4.34}$$

$$J_R = \frac{I_0}{4\pi R^2}. \tag{4.35}$$

The right side of Eq. (4.33) is zero because the charge enclosed by the sphere is constant, $dQ/dt = 0$. The current I_0 is subtracted on the left side of the equation because the net current flowing outward through the surface is the difference between

† A physical situation to approximate the filamentary current would occur if a thin insulated wire carrying current I_0 were lowered far below the surface of the ocean. The fields in that situation would closely approximate the fields calculated here in a region $R < a$, where a is much less than the distance to the ocean surface or the ocean floor.

the integrated current density flowing outward in the medium and the filamentary current flowing inward on the filament. Since the current density in the conducting medium is a constant with respect to the θ and ϕ coordinates, Eq. (4.34) could be written directly without the formal integration shown in Eq. (4.33). The electric field intensity and the electric flux density are given by

$$\vec{E} = \frac{1}{\sigma}\vec{J} = \vec{a}_R \frac{I_0}{4\pi\sigma R^2} \tag{4.36}$$

$$\vec{D} = \varepsilon\vec{E} = \vec{a}_R \frac{\varepsilon I_0}{4\pi\sigma R^2} \tag{4.37}$$

We now apply Gauss's law for electric fields to the same Gaussian sphere of radius R centered at the origin.

$$\oint_S \vec{D}\cdot d\vec{S} = \int_V \rho\, dv$$

$$\int_{\phi=0}^{2\pi}\int_{\theta=0}^{\pi} D_R R^2 \sin\theta\, d\theta\, d\phi = Q_{\text{enclosed}} \tag{4.38}$$

$$\frac{\varepsilon I_0}{4\pi\sigma R^2} 4\pi R^2 = Q_{\text{enclosed}} \tag{4.39}$$

$$Q_{\text{enclosed}} = q = \frac{\varepsilon}{\sigma} I_0. \tag{4.40}$$

Since the charge enclosed by the sphere does not depend on the radius of the sphere, it must all be at the origin as a point charge q on the tip of the filament. This point charge is the source of the electric fields. The charge is exactly large enough to produce the electric field, Eq. (4.36), which causes a current I_0 to flow outward from the origin, so no additional charge accumulates at the origin.

4.5 AMPÈRE–MAXWELL LAW

The Ampère–Maxwell law relates the magnetic field intensity to the currents that are its sources. The law states that the line integral of magnetic field intensity around any closed line equals the total current linked with the closed line (Fig. 4.13). The line integral of magnetic field intensity around a closed line is the magnetomotive force, Eq. (3.28), around that closed path, U_{closed}. The total current linked with the closed line is the total current through any surface that caps the closed path, $(I_t)_{\text{surface}}$. Hence the Ampère–Maxwell law is given by

$$U_{\text{closed}} = (I_t)_{\text{surface}}. \tag{4.41}$$

There are two kinds of current that make up the total current. The first is the current that results from the flow of free charge, either in a conducting medium or as moving electrons, protons, or ions in free space. This is the current discussed in

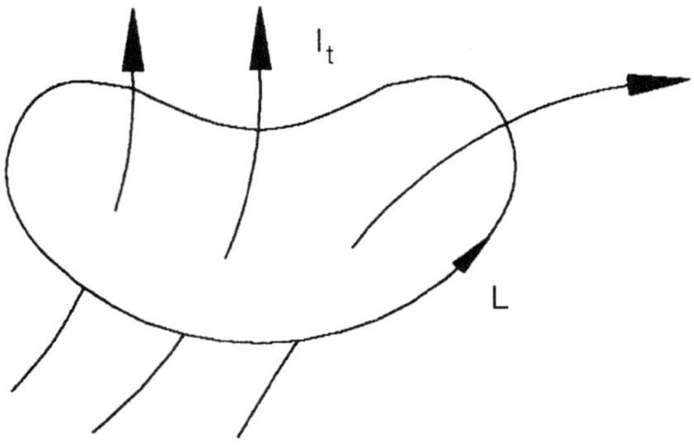

FIGURE 4.13 Current linked by a closed path, which is used to determine the magnetic field intensity.

Section 3.3. The general symbol for this current is I with units of amperes. The current I crossing a surface is the integral of the current density $\vec{J}$ plus the integral of the surface current density $\vec{J}_s$ plus the sum of any line currents i through the surface.

The second component of the total current is unrelated to real or physical charge. This second kind of current is called displacement current, I_d. The displacement current density, $\vec{J}_d$, is equal to the time rate of change of the electric flux density:†

$$\vec{J}_d = \frac{\partial \vec{D}}{\partial t}. \tag{4.42}$$

The displacement current crossing a surface is equal to the integral of the displacement current density normal to the surface.

The dimensions of electric flux density are charge per unit area, $\mathrm{IT/L^2}$, and hence the dimensions of displacement current density are current per unit area, $\mathrm{I/L^2}$. The SI units are $\mathrm{A/m^2}$. These dimensions and units are identical to those of the free current density $\vec{J}$, so the two current densities can be added to give a total current density, $\vec{J}_t$:

$$\vec{J}_t = \vec{J} + \vec{J}_d \tag{4.43}$$

$$= \vec{J} + \frac{\partial \vec{D}}{\partial t}. \tag{4.44}$$

The total current through a surface is the sum of the free current and the displacement current through the surface:

$$(I_t)_{\text{surface}} = I + I_d. \tag{4.45}$$

These currents are the flux of their respective current densities through the surface.

† Recall that the electric flux density used to be called the displacement (Section 3.6).

The Ampère–Maxwell law in integral form is therefore given by

$$\oint_L \vec{H} \cdot d\vec{l} = \int_S \left[\vec{J} + \frac{\partial \vec{D}}{\partial t} \right] \cdot d\vec{S}, \tag{4.46}$$

where S is any surface capping the closed curve L and the positive direction for the normal to the surface is given by the right-hand rule (Fig. 4.14). If surface currents or line currents are present, their integrals through the surface S must be added to the right side.

It is important and useful to understand that we can choose any surface that caps the closed line for the right side of the Ampère–Maxwell law. If we imagine the line to be a steel band and the surface to be a rubber sheet connected to the steel band, then we can stretch the rubber sheet into any shape we want and the result for the surface integral will not change. We will therefore try to pick surfaces for which the integration is easiest. Generally these surfaces are easily described in some coordinate system and/or are normal to the field.

This is an integral form of the law and does not give information about the field at each point. Just as for Gauss's law, we can write the equation in terms of average values of the tangential magnetic field intensity along the curve and the normal component of current density over the surface:

$$(H_{\text{tangential}})_{\text{average}} = [(J_{\text{normal}})_{\text{average}} + (J_{d\ \text{normal}})_{\text{average}}] \times \frac{\text{Area of surface}}{\text{Length of curve}}. \tag{4.47}$$

When the field components are constant over the closed curve and over the surface, then the average values equal those constant values and Eq. (4.47) gives the actual fields.

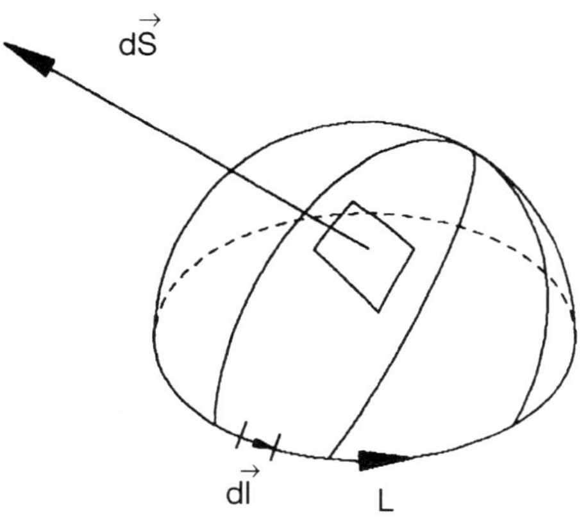

FIGURE 4.14 Surface capping a closed curve L.

To be completely general and allow for the possibility that the closed curve and the capping surface are changing with time, we must modify the Ampère–Maxwell law to be

$$\oint_L \vec{H} \cdot d\vec{l} = \int_S \vec{J} \cdot d\vec{S} + \frac{d}{dt} \int_S \vec{D} \cdot d\vec{S}. \tag{4.48}$$

4.5.1 Magnetic Field of a Current-Carrying Wire

Figure 4.15 shows a solid wire carrying a total constant current I_0 uniformly distributed throughout the cross section of the wire. The wire has radius a and is coaxial with the z axis.

We will apply the Ampère–Maxwell law to two circles of radius r (shown dotted in the figure), one inside the wire and the other outside. The current density has cylindrical symmetry around the z axis (i.e., it is independent of the ϕ coordinate) and hence the magnetic field intensity will also have this symmetry. The source current is independent of the z coordinate and the field is also. This means that $\vec{H}(r, \phi, z) = \vec{a}_\phi H_\phi(r)$. The current density inside the wire is the current divided by the area of the wire, $\vec{J} = \vec{a}_z(I_0/\pi a^2)$. Since only constant (with respect to time) currents are present, the displacement current is zero. For each dotted circle shown in the figure, we choose the surface S to be the flat circular area centered on the z axis lying in the plane of the circle. For this surface, $d\vec{S} = \vec{a}_z r\, dr\, d\phi$.

For the circle with radius $r < a$,

$$\oint_L \vec{H} \cdot d\vec{l} = \int_S \left[\vec{J} + \frac{\partial \vec{D}}{\partial t} \right] \cdot d\vec{S}$$

$$\int_{\phi=0}^{2\pi} [\vec{a}_\phi H_\phi(r)] \cdot [\vec{a}_\phi r\, d\phi] = \int_{\phi=0}^{2\pi} \int_{r=0}^{r} \left[\vec{a}_z \frac{I_0}{\pi a^2} \right] \cdot [\vec{a}_z r\, dr\, d\phi]$$

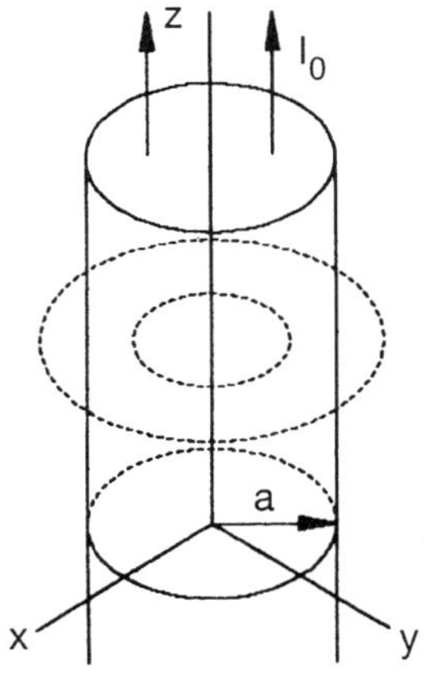

FIGURE 4.15 Solid wire of radius a carrying a uniformly distributed current.

$$2\pi r H_\phi(r) = \frac{I_0}{\pi a^2}\pi r^2$$

$$H_\phi(r) = \left(\frac{I_0}{\pi a^2}\right)\left(\frac{\pi r^2}{2\pi r}\right) \tag{4.49}$$

$$H_\phi(r) = \frac{r}{2\pi a^2} I_0, \quad r < a. \tag{4.50}$$

Equation (4.49) is written in the form of Eq. (4.47) and could be obtained directly without the steps that precede it.

For a circle with radius $r > a$,

$$\oint_L \vec{H}\cdot d\vec{l} = \int_S \left[\vec{J} + \frac{\partial \vec{D}}{\partial t}\right]\cdot d\vec{S}$$

$$\int_{\phi=0}^{2\pi} [\vec{a}_\phi H_\phi(r)]\cdot[\vec{a}_\phi r\, d\phi] = \int_{\phi=0}^{2\pi}\int_{r=0}^{a}\left[\vec{a}_z \frac{I_0}{\pi a^2}\right]\cdot[\vec{a}_z r\, dr\, d\phi] \tag{4.51}$$

$$2\pi r H_\phi(r) = I_0 \tag{4.52}$$

$$H_\phi(r) = \frac{I_0}{2\pi r}, \quad r > a. \tag{4.53}$$

In Eq. (4.51) we integrate over r on the right side up to a limit of a even though the surface of integration extends to some radius $r > a$. This is because the current density is zero outside the wire. Equation (4.52) could have been obtained directly without the integration shown in Eq. (4.51) because H_ϕ is constant around the circle and the current density is constant over the surface.

Assuming both the wire and the surrounding medium have the permeability of free space, the magnetic flux density is

$$\vec{B} = \begin{cases} \vec{a}_\phi \dfrac{r\mu_0 I_0}{2\pi a^2}, & r < a, \\ \vec{a}_\phi \dfrac{\mu_0 I_0}{2\pi r}, & r > a. \end{cases}$$

If we let $a \to 0$ but keep the current constant, the current source becomes a filamentary current $i = I_0$ along the entire z axis. The magnetic field intensity and the magnetic flux density are then

$$\vec{H} = \vec{a}_\phi \frac{I_0}{2\pi r}, \tag{4.54}$$

$$\vec{B} = \vec{a}_\phi \frac{\mu_0 I_0}{2\pi r}, \tag{4.55}$$

for all values of $r > 0$. These are the fields of a line or filamentary current I_0 along the z axis.

In the vicinity of any current filament the field will vary as $1/r$ for small r. Conversely, if the field varies as $1/r$ for small r, then a current filament must be present at $r = 0$.

Note that the magnetic fields make closed curves around the current in accordance with a right-hand rule: point the right thumb in the direction of the current and the right-hand fingers give the direction of the magnetic field intensity.

4.5.2 Magnetic Field of a Uniform Surface Current

Figure 4.16 shows a surface current $\vec{J}_s = \vec{a}_x J_{sx0}$ flowing "out of the page" in the xy plane. Since the surface current is uniform and covers the entire $z = 0$ plane, the location of the origin in that plane is arbitrary and the fields cannot be a function of the x and y coordinates.

Apply the Ampère–Maxwell law around the loop ABCDA, with the line AB located at z' and CD located at $-z'$. Take the capping surface to be the flat surface parallel to the yz plane.

$$U_{\mathrm{AB}} + U_{\mathrm{BC}} + U_{\mathrm{CD}} + U_{\mathrm{DA}} = (I_t)_{\mathrm{surface}}.$$

Symmetry and the requirement that magnetic field lines form closed paths require that $U_{\mathrm{BC}} = U_{\mathrm{DA}} = 0$. The displacement current is zero and therefore the total current crossing the surface is $J_{sx0}L$. Thus

$$U_{\mathrm{AB}} + U_{\mathrm{CD}} = J_{sx0}L$$

$$-H_y(z')L + H_y(-z')L = J_{sx0}L. \tag{4.56}$$

The L's cancel. By symmetry, either $H_y(z') = H_y(-z')$ or $H_y(z') = -H_y(-z')$. The first alternative cannot be true, since it would make the left side of Eq. (4.56) equal to zero while the right side is not zero. Taking the second alternative gives

$$H_y(z') = -\frac{J_{sx0}}{2}.$$

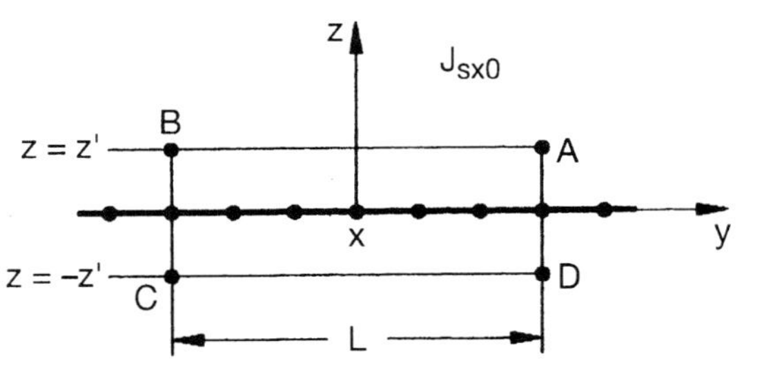

FIGURE 4.16 Uniform surface current in the xy plane.

Since z' can have any value, the field above and below the plane of the surface charge is uniform or independent of z. The result, in vector form, is

$$\vec{H} = \begin{cases} -\vec{a}_y \dfrac{J_{sx0}}{2}, & z > 0, \\ \vec{a}_y \dfrac{J_{sx0}}{2}, & z < 0. \end{cases} \tag{4.57}$$

This magnetic field intensity is unchanged for any medium that maintains the symmetry of the problem. That means the permeability may be an even function of the z coordinate only, with $\mu(z) = \mu(-z)$. The magnetic flux density is found by multiplying the magnetic field intensity by the permeability of the medium.

4.5.3 Magnetic Field of Point Current Source

In Section 4.4.2 we calculated the current density and the electric fields surrounding a point source of current in a conducting medium. Using the Ampère–Maxwell law we can now calculate the magnetic field caused by the current.

Figure 4.17 again shows the constant filamentary current along the positive z axis. We will apply the Ampère–Maxwell law to a circle (shown dotted) located at spherical radius R and cone angle θ. Due to the symmetry of the sources around the z axis, the magnetic field intensity will not be a function of the ϕ coordinate. When θ is small the dotted circle is close to the filament of current and the field must be the same as that given by Eq. (4.54), $H_\phi = -I_0/2\pi r$, where $r = R \sin\theta$. It is therefore reasonable to assume that, for all points in space, $\vec{H} = \vec{a}_\phi H_\phi(R, \theta)$.

In applying Eq. (4.46) we chose the circle shown as the closed curve L over which we integrate the tangential component (ϕ component) of $\vec{H}$ and the spherical cap of radius R as the surface S over which we integrate the normal component (R component) of $\vec{J}$. Since nothing varies with time, there is no displacement current.

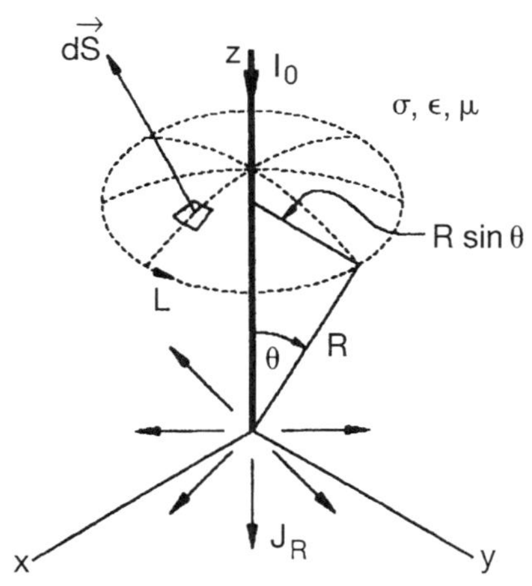

FIGURE 4.17 Constant filamentary current along the positive z axis.

$$\oint_L \vec{H}\cdot d\vec{l} = \int_S \left[\vec{J} + \frac{\partial \vec{D}}{\partial t}\right]\cdot d\vec{S} \tag{4.58}$$

$$\int_{\phi=0}^{2\pi} [\vec{a}_\phi H_\phi]\cdot[\vec{a}_\phi R \sin\theta\, d\phi] = \int_{\phi=0}^{2\pi}\int_{\theta=0}^{\theta}\left[\vec{a}_R \frac{I_0}{4\pi R^2}\right]\cdot[\vec{a}_R R^2 \sin\theta'\, d\theta'\, d\phi] - I_0 \tag{4.59}$$

$$2\pi R \sin\theta\, H_\phi = \frac{I_0}{2}(1-\cos\theta) - I_0$$

$$H_\phi = \frac{I_0(1+\cos\theta)}{4\pi R \sin\theta}. \tag{4.60}$$

The $-I_0$ term comes from the line current, which pierces the spherical cap in an inward direction on the z axis.

Let us check the result for the two limiting cases of $\theta \to 0$ and $\theta \to \pi$. In either case, in the denominator we have $R\sin\theta = r$, the cylindrical radius. In the first case, $\cos\theta \to 1$. In the second case, $\cos\theta \to -1$. Thus

$$H_\phi \to -\frac{I_0}{2\pi r} \quad \text{as } \theta \to 0, \tag{4.61}$$

$$H_\phi \to 0 \quad \text{as } \theta \to \pi. \tag{4.62}$$

The first result correctly matches with the field close to a current filament $-I_0$. The second result is correct since a small circle around the negative z axis does not enclose any current.

The magnetic flux density is found by multiplying by the permeability of the conducting medium.

4.5.4 Magnetic Field of a Point Displacement Current Source

Let us again consider a constant filamentary current $-I_0$ along the positive z axis, but this time we assume that the medium is a perfect dielectric ($\sigma = 0$) (Fig. 4.18). Current cannot flow into the perfect dielectric, so charge $Q = \int I_0\, dt$ accumulates on the tip of the wire.[†]

Consider a Gaussian sphere of radius R around the tip and apply Gauss's law for electric fields to this surface to obtain

$$\vec{D} = \vec{a}_R \frac{Q}{4\pi R^2}$$

$$= \vec{a}_R \frac{\int I_0\, dt}{4\pi R^2}. \tag{4.63}$$

[†] This is a thought experiment. In reality, it would be very difficult to force current into the open circuit represented by the wire tip surrounded by a perfect dielectric. It would require ever-increasing electromotive force (voltage sources) along the wire length.

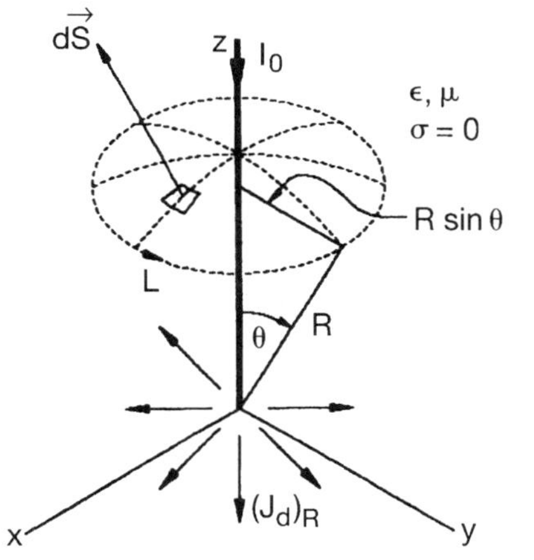

FIGURE 4.18 Constant filamentary current along the positive z axis with a perfect dielectric medium surrounding it.

We again apply the Ampère–Maxwell law to the dotted circle shown in the figure. This time, the current density in the dielectric medium is zero, but the displacement current density is given by

$$\vec{J}_d = \frac{\partial \vec{D}}{\partial t} = \frac{\partial}{\partial t}\left(\vec{a}_R \frac{\int I_0\, dt}{4\pi R^2}\right) = \vec{a}_R \frac{1}{4\pi R^2}\frac{\partial}{\partial t}\int I_0\, dt$$
$$= \vec{a}_R \frac{I_0}{4\pi R^2}. \tag{4.64}$$

This displacement current in the dielectric medium is exactly equal to the conduction current, which was present in the conducting medium case, Eq. (4.35). Applying the Ampère–Maxwell law,

$$\oint_L \vec{H}\cdot d\vec{l} = \int_S \left[\vec{J} + \frac{\partial \vec{D}}{\partial t}\right]\cdot d\vec{S} \tag{4.65}$$

$$\int_{\phi=0}^{2\pi} [\vec{a}_\phi H_\phi]\cdot[\vec{a}_\phi R \sin\theta\, d\phi] = \int_{\phi=0}^{2\pi}\int_{\theta'=0}^{\theta}\left[\vec{a}_R \frac{I_0}{4\pi R^2}\right]\cdot[\vec{a}_R R^2 \sin\theta'\, d\theta'\, d\phi] - I_0 \tag{4.66}$$

$$2\pi R \sin\theta\, H_\phi = \frac{I_0}{2}(1 - \cos\theta) - I_0$$

$$H_\phi = -\frac{I_0(1 + \cos\theta)}{4\pi R \sin\theta}. \tag{4.67}$$

This is the same result as Eq. (4.60) for the conducting medium. The displacement current in the dielectric medium and the conduction current in the conducting medium are identical and they therefore give rise to identical magnetic fields. The total current—equal to the sum of the conduction and displacement currents—in the two examples is the same and therefore the magnetic field is the same.

4.6 FARADAY–MAXWELL LAW

The Faraday–Maxwell law relates the electric field intensity to the time-varying flux, which is its source. The law states that the line integral of electric field intensity around any closed line equals the negative time rate of change of the magnetic flux linked with the closed line (Fig. 4.19). The line integral of electric field intensity around a closed path is the emf or voltage around that closed path, V_{closed}. The negative time rate of change of the magnetic flux linked with the closed line is the negative of the derivative with respect to time of the flux through any surface that caps the closed path, $-d\Phi_{\text{surface}}/dt$. Hence the Faraday–Maxwell law is given by

$$V_{\text{closed}} = -\frac{d\Phi_{\text{surface}}}{dt}. \tag{4.68}$$

The Faraday–Maxwell law in integral form is

$$\oint_L \vec{E} \cdot d\vec{l} = -\frac{d}{dt}\int_S \vec{B} \cdot d\vec{S}, \tag{4.69}$$

where S is any surface capping the closed curve L and the positive direction for the normal to the surface is given by the right-hand rule (Fig. 4.20). It is again important to emphasize that we can choose any surface that caps the closed curve and that we should choose a surface to make the integration on the right side of the equation as easy as possible.

The Faraday–Maxwell law and the Ampère–Maxwell law [Eq. (4.48)] are easily compared. The forms of the left sides are the same. On the right side of the Faraday–Maxwell law there is no equivalent of the current density $\vec{J}$ because there is no magnetic charge or current. The terms for the time rate of change of the flux through S and the time rate of change of the displacement current through S are identical in form but differ in sign.

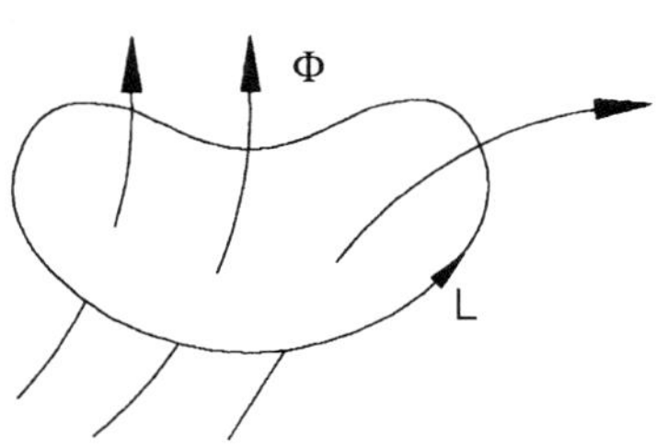

FIGURE 4.19 Magnetic flux linked by a closed path, which is used to determine the electric field intensity.

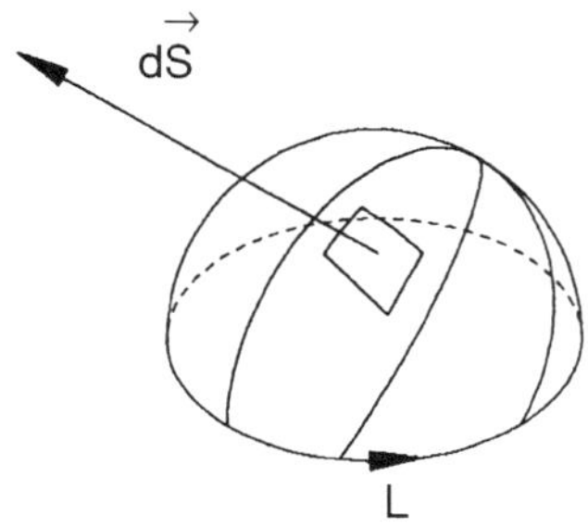

FIGURE 4.20 Surface capping a closed curve L used in Faraday–Maxwell law.

When the closed curve and the capping surface are fixed in space (not changing with time), Eq. (4.69) becomes

$$\oint_L \vec{E} \cdot d\vec{l} = \int_S \left(-\frac{\partial \vec{B}}{\partial t} \right) \cdot d\vec{S}. \tag{4.70}$$

This form is comparable to Eq. (4.46) for the Ampère–Maxwell law for a stationary surface. Here, the time rate of change of the magnetic flux density $\partial \vec{B}/\partial t$ and the time rate of change of the electric flux density $\partial \vec{D}/\partial t$ are comparable terms.

The integral form does not give information about the field at each point. Just as for the Ampère–Maxwell law, we can write Eq. (4.70) in terms of average values of the tangential magnetic field intensity along the curve and the normal component of current density over the surface:

$$(E_{\text{tangential}})_{\text{average}} = \left[\left(-\frac{\partial \vec{B}}{\partial t} \right)_{\text{normal}} \right]_{\text{average}} \times \frac{\text{Area of surface}}{\text{Length of curve}}. \tag{4.71}$$

When the field components are constant over the curve and surface, then the average values equal those constant values and Eq. (4.71) gives the actual fields.

4.6.1 Moving Wire—Constant Magnetic Field

Figure 4.21 shows a wire sliding over conducting rails connected to a resistor. The figure lies in the xy plane and the z component of flux density is $B_z = B_z(x)$, which is constant with time and uniform with respect to the y coordinate. The wire moves in the x direction with a speed $u_x(t)$ and its x position is $x(t)$ with $x(0) = 0$. Of course, $u_x = dx/dt$.

We apply the Faraday–Maxwell law, Eq. (4.69), by integrating counterclockwise around the loop ABCDA, where points A and B move with the wire. We take the surface capping the loop to be the rectangular surface in the xy plane of the figure.

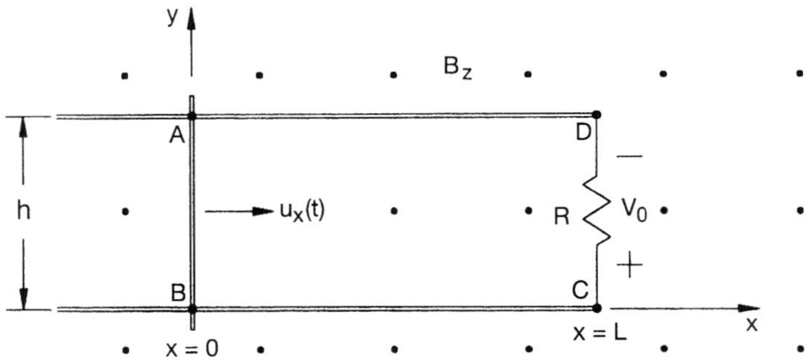

FIGURE 4.21 Wire sliding over conducting rails to illustrate the generation of emf.

Then the magnetic flux through the loop and its time derivative are

$$\Phi = \int_{x'=x(t)}^{L} \int_{y'=0}^{h} B_z(x')\,dy'\,dx' \tag{4.72}$$

$$= h\int_{x'=x(t)}^{L} B_z(x')\,dx' \tag{4.73}$$

and

$$\frac{d\Phi}{dt} = -\frac{dx(t)}{dt} hB_z[x(t)] \tag{4.74}$$

$$= -u_x(t)hB_z[x(t)]. \tag{4.75}$$

The emf or voltage around the loop is therefore

$$V_{\text{closed}} = V_{\text{ABCDA}} = -\frac{d\Phi}{dt}$$

$$= u_x(t)\,hB_z[x(t)]. \tag{4.76}$$

The sign (+ or −) of the result depends on the signs of u_x and B_z at any particular time.

In order to simplify the result and make it easier to understand, assume that $u_x(t) = u_{x0}$ and $B_z(x) = B_{z0}$, both positive constants. The problem then could be stated as the wire moving in the x direction with constant velocity through a uniform constant z-directed magnetic field. The emf is now given by

$$V_{\text{ABCDA}} = V_0 = u_{x0}B_{z0}h. \tag{4.77}$$

Since the emf is positive around the loop in the counterclockwise direction, the current will flow counterclockwise around the loop. A voltage $V_R = V_0$ will be present across the resistor. The device represented in Fig. 4.21 is seen to be a linear direct current generator.

Let us suppose now that the resistor is missing from the circuit (or $R = \infty$), creating an open circuit where the resistor was. The emf around the loop is unchanged and the voltage V_0 will appear across the open circuit.

Let us further assume that the wire is not sliding on the pair of rails but is simply moving through the field B_z with constant speed u_{x0} (Fig. 4.22). The same closed path can be drawn (dotted in the figure) and the Faraday–Maxwell law applied to it, even though no physical wires are present. The results are the same and the voltage

$$V_0 = u_{x0} B_{z0} h \tag{4.78}$$

will appear across the wire. The polarity will be such as to drive the current in the counterclockwise direction if current could flow around the loop.

This same result [Eq. (4.78)] can be obtained from a different viewpoint. Equation (3.26) shows that the force on an electron in the wire is $\vec{F} = -e(\vec{E} + \vec{u} \times \vec{B})$. With the wire moving at a constant speed u_{x0} through a uniform and constant magnetic field B_{z0} a constant voltage and current must be produced. With the wire open circuited, no current can flow in the wire and the total force on charge in the interior of the wire must be zero. This means that the charge will arrange itself on the surface of the wire so as to cause an electric field $\vec{E}$ that is exactly equal and opposite to the field $\vec{u} \times \vec{B}$ inside the wire. For our example, $\vec{E} = -\vec{u} \times \vec{B} = \vec{a}_y u_{x0} B_0$. Then

$$\begin{aligned} V_0 = V_{\text{BA}} = \int_L \vec{E} \cdot d\vec{l} = \int_{y=0}^{h} (\vec{a}_y u_{x0} B_{z0}) \cdot (a_y\, dy) \\ = u_{x0} B_{z0} h, \end{aligned} \tag{4.79}$$

which is the same as Eq. (4.78).

Lenz's law can be derived from these results. Lenz's law states that the polarity of the emf will be such that if current were to flow in the loop, the current would produce magnetic flux so as to oppose the change in magnetic flux. In the previous example, the wire motion causes the total magnetic flux through the loop to decrease in amplitude while pointing in the positive z direction, which means out of the paper as

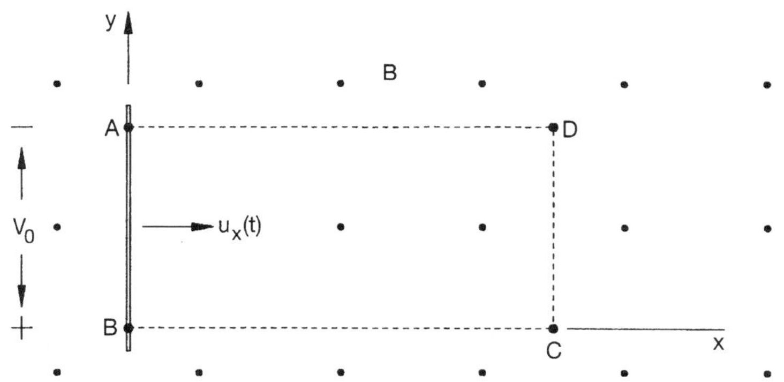

FIGURE 4.22 Voltage induced across a wire that is moving inside a magnetic field.

you look at Fig. 4.21. According to Lenz's law, the induced magnetic flux, produced by the current that would flow through the loop due to the induced emf, will oppose the action that created it (a decrease in the original magnetic flux). In order to oppose a decrease in the original flux, which is pointing in the positive z direction, the induced flux will have to point in the same positive z direction. The direction of the induced current that would flow through the loop is therefore given by the right-hand rule. Pointing the thumb of the right hand in the direction of the induced magnetic flux, which is the positive z direction, the fingers of the right hand will curl in the direction of the induced current, which in our case is counterclockwise.

Lenz's law is verified by conversely applying the results obtained in the previous example. In our example with the wire sliding over the stationary wires with constant speed and uniform flux density, we found that current flows counterclockwise. Placing the right thumb in the direction of current in the wire, the right fingers point out of the page in the interior of the loop. The wire moving to the right results in reduced flux (out of the page) through the loop. Thus the current causes flux that opposes the change in flux due to the motion of the wire.

The Faraday–Maxwell law and Lenz's law, as explained in this section, are very important in understanding the operation of all ac and dc generators and motors. The variation of the magnetic flux through coils in generators is accomplished by rotating the constant magnetic field in front of the coils (ac generator) or by rotating the coils inside a constant magnetic field (dc generators). The following experiment will help you understand how the mere rotation of the coil inside the constant magnetic flux can produce a variation of the net flux through the coil. The magnetic flux can be thought of as the flux of light from a flashlight. The loop and loop surface are the perimeter and the surface of a blank sheet of paper (any size you have at hand). The measure of the flux of light through the loop is the size of the shadow cast by the sheet of paper. If you hold the flashlight steady and start rotating the sheet of paper in front of the flashlight around an axis perpendicular to the light flux (holding the paper from the middle of one side of the sheet), you will see the shadow cast by the rotating paper vary from a maximum, when the paper is perpendicular to the flux of light, to a minimum, when the paper is tangential to the light flux. A depiction of the experiment is shown in Fig. 4.23.

A similar experiment can be performed to illustrate the time-varying flux through a coil produced by rotating a permanent magnet (or a dc excited electromagnet) in front of the coil. The coil is now replaced by a sheet of paper or cardboard with the center cut out, making it look like an empty frame of a picture. Holding the sheet steady in front of a screen or a wall, and rotating the flashlight in front of it, you should be able to notice the varying intensity of the light through the window of the sheet (the open area through which the light passes). The maximum light flux is observed when the flashlight is pointing at the sheet (perpendicular to the sheet), and the minimum flux is obtained when the flashlight is pointing in a tangential direction to the sheet. A depiction of this experiment is given in Fig. 4.24. The effect is more dramatic if the experiment is performed in a dim lighted room.

4.6.2 Stationary Wire—Changing Magnetic Field

We now examine the same loop, Fig. 4.21, but assume the wire is stationary, $u_x = 0$, and the flux density changes with time, $B_z = B_z(x, t)$.

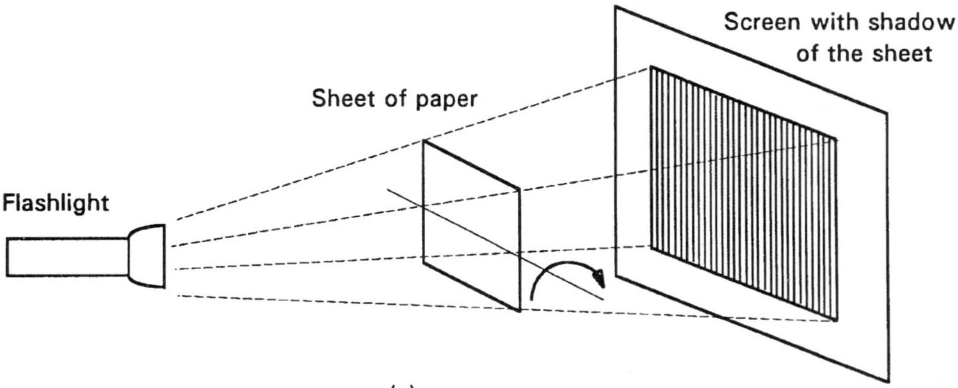

(a)

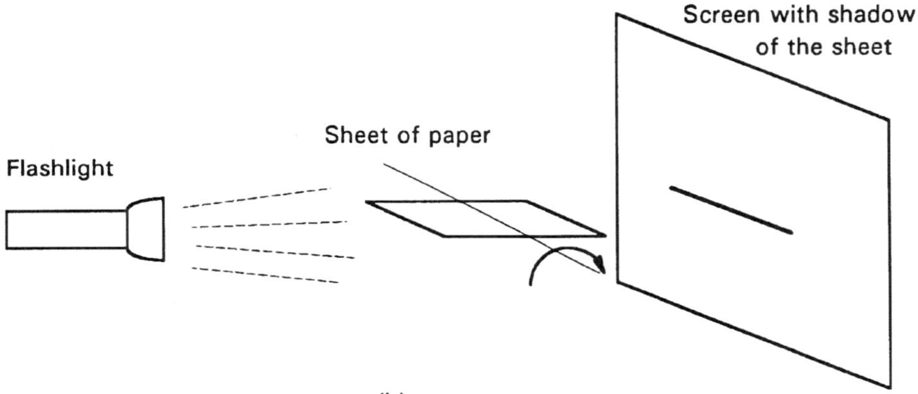

(b)

FIGURE 4.23 Rotating sheet of paper in front of a flashlight to illustrate varying magnetic field through a coil rotated inside a constant magnetic flux for (a) maximum flux and (b) minimum flux.

We again apply the Faraday–Maxwell law, Eq. (4.69), by integrating counterclockwise around the loop ABCDA. Taking the surface capping the loop to be the xy plane of the figure, the magnetic flux through the loop and its time derivative are

$$\Phi = \int_{x'=0}^{L} \int_{y'=0}^{h} B_z(x', t)\, dy'\, dx' \tag{4.80}$$

$$= h \int_{x'=0}^{L} B_z(x', t)\, dx' \tag{4.81}$$

and

$$\frac{d\Phi}{dt} = h \int_{x'=0}^{L} \frac{\partial B_z(x', t)}{\partial t}\, dx'. \tag{4.82}$$

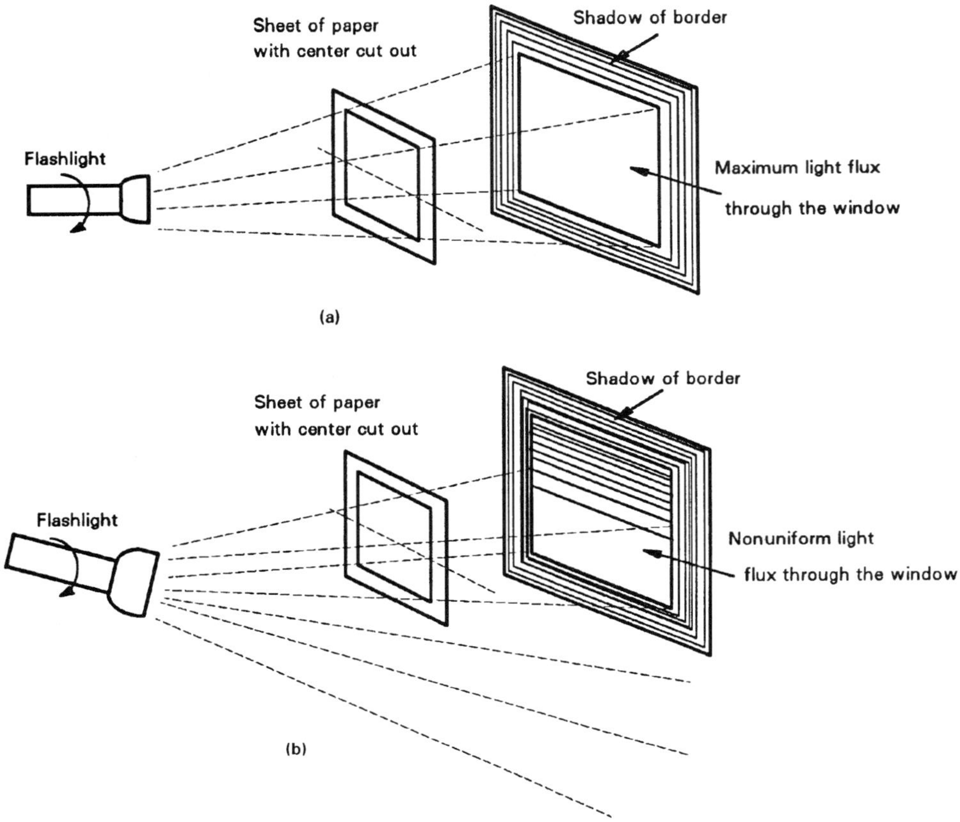

FIGURE 4.24 Rotating flashlight in front of stationary sheet of paper with center cut out to illustrate varying magnetic field through a stationary coil due to a rotating constant magnetic flux in front of it for (a) maximum flux and (b) an intermediate position, other than maximum or minimum.

The emf or voltage around the loop is therefore

$$V_{\text{closed}} = V_{\text{ABCDA}} = -h\int_{x'=0}^{L} \frac{\partial B_z(x', t)}{\partial t}\, dx'. \tag{4.83}$$

We simplify the result by choosing the flux density to be uniform, $B_z = B_z(t)$. Then

$$V_{\text{ABCDA}} = -hL\frac{dB_z(t)}{dt}. \tag{4.84}$$

Assuming that $B_z(t) = (t/T)B_{z0}$, a linearly increasing flux density, then

$$V_{\text{ABCDA}} = -\frac{hL}{T}B_{z0}. \tag{4.85}$$

The result is negative, which means the current will flow in the clockwise direction around the loop. This current would cause flux into the page in the interior of the loop, opposing the change in flux due to the increasing flux density, confirming Lenz's law.

Example 4.2 The magnetic flux through a coil is increased sinusoidally in the upward direction, as shown in Fig. 4.25. Determine the polarity of the induced voltage.

Solution For ease in determining the polarity of the induced voltage, the coil is loaded with a resistor. The induction of a voltage at the terminals of the coil produces an induced current through the coil, which in turn produces an induced magnetic flux. The induced voltage, current, and flux oppose the action that created them. Since the magnetic flux variation, which induces the voltage, is an increase in the upward direction, the induced magnetic flux will have to point in the downward direction, thus opposing the action that created it, as indicated by the dashed arrow in Fig. 4.26. The induced current, which produces the induced flux, will have a direction given by the right-hand rule. Pointing the thumb in the direction of the induced magnetic flux, the fingers will curl in the direction of the induced current. Given the direction of the induced current i_i through the resistor R, the polarity of the voltage across the resistor and hence the polarity of e is as shown in Fig. 4.26. The introduction of the resistor across the terminals of the coil was done only to facilitate the determination of the polarity of the induced voltage e. The induced voltage e will appear and have the same polarity even in the absence of the resistor.

Example 4.3 Repeat the previous example for a magnetic flux through the coil that decreases in strength in the upward direction, as shown in Fig. 4.27.

Solution The coil is loaded with a resistor. Since the magnetic flux variation, which induces the voltage, is a decrease in the upward direction, the induced magnetic flux will have to point in the upward direction, thus opposing the weakening of the original magnetic flux in the upward direction, as indicated by the dashed arrow in

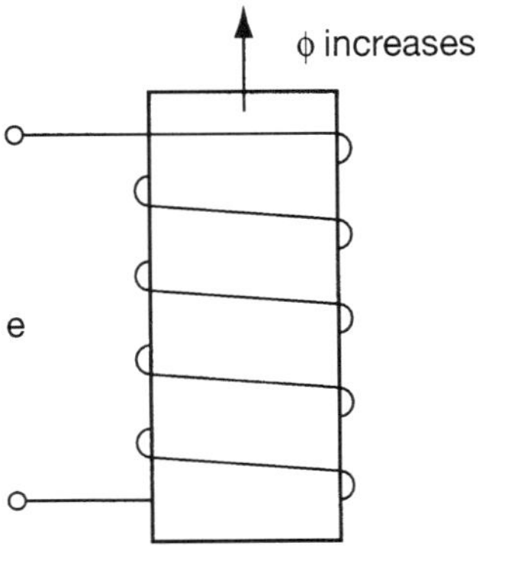

FIGURE 4.25 Coil with a time-varying magnetic flux (increasing upward).

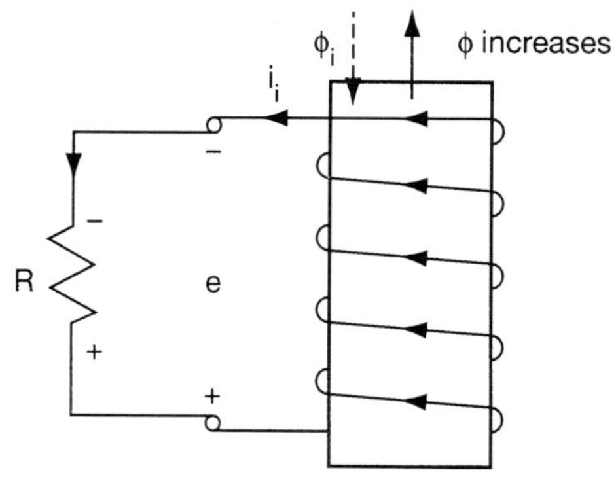

FIGURE 4.26 Coil with resistive loading and a time-varying magnetic flux (increasing upward).

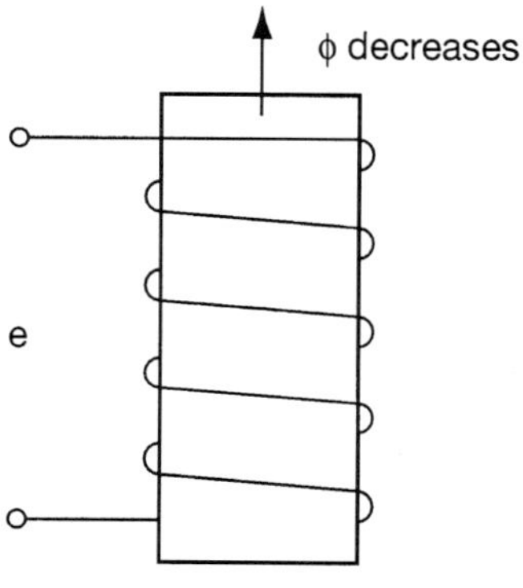

FIGURE 4.27 Coil with a time-varying magnetic flux (decreasing upward).

Fig. 4.28. The induced current, which produces the induced flux, will have a direction given by the right-hand rule, as shown in Fig. 4.28. Given the direction of the induced current i_i through the resistor R, the polarity of the voltage across the resistor and hence the polarity of e is as shown in Fig. 4.28. Note that the polarity of the induced voltage has changed from the previous example.

4.7 BOUNDARY CONDITIONS

Boundary conditions play a very important role in the solution of electromagnetic problems. They give the conditions that must be satisfied at the boundary or interface of two media with different properties or constituent relations.

Let us use temperature as an example of a physical quantity with which we are familiar. It is frequently said, when we want to go swimming, that the water is cold, but the air is warm. We know that, in a literal sense, this cannot be an accurate statement. What is meant is that the water at some distance below the surface is cold and the air

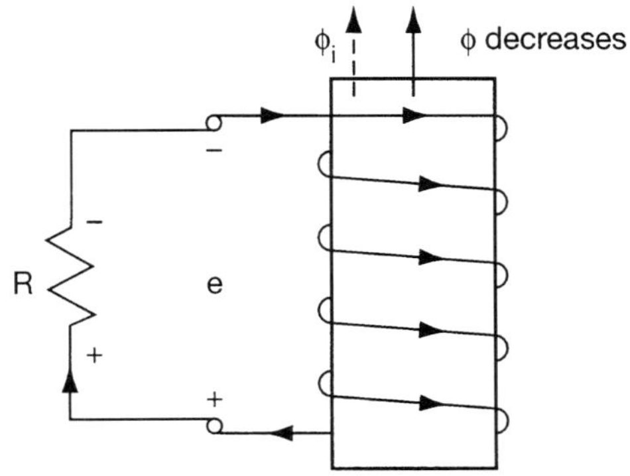

FIGURE 4.28 Coil with resistive loading and a time-varying magnetic flux (decreasing upward).

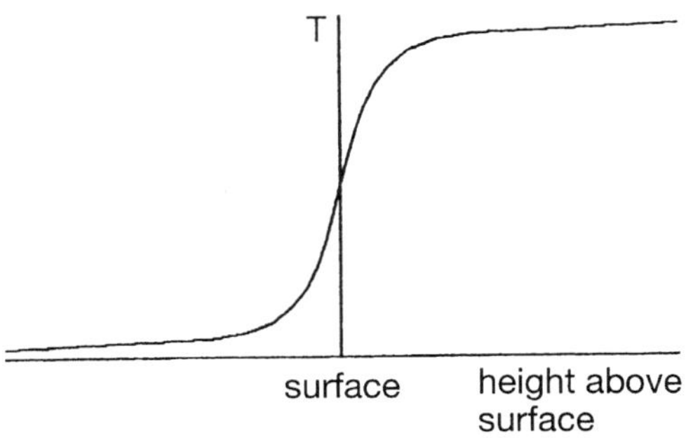

FIGURE 4.29 Temperature as a function of distance from a surface.

at some distance above the surface is warm. In the vicinity of the surface there is a strong gradient (rate of change) of temperature from cold (below) to warm (above). The temperature as a function of distance from the surface must be continuous (Fig. 4.29).

Boundary conditions put strict conditions on field quantities in space at the interface of two media. They have a similarity to continuity conditions in time on voltages and currents in circuit theory. For example, capacitor voltage and inductor current cannot be discontinuous. Similar conditions hold for the field quantities we have introduced. In some instances certain components of the field must be continuous or must be discontinuous in some particular way. This section derives these conditions from the integral laws introduced in this chapter.

4.7.1 Tangential Magnetic and Electric Field Intensities

We will start by using the Ampère–Maxwell integral law to derive the boundary condition on the tangential component of the magnetic field intensity at the boundary between two media with different permeabilities (Fig. 4.30). Medium 1 above the boundary surface in the figure has permeability μ_1. Medium 2, below the boundary

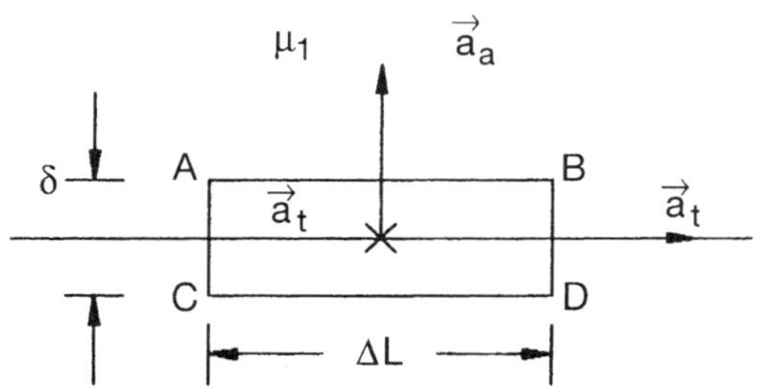

FIGURE 4.30 Boundary between two media with different permeabilities.

surface, has permeability μ_2. The normal to the boundary surface is $\vec{a}_n$ directed from medium 2 to medium 1. Let $\vec{a}_\tau$ be a unit vector tangent to the boundary but otherwise in an arbitrary direction. These two unit vectors are orthogonal and their cross product defines a third mutually orthogonal unit vector,

$$\vec{a}_t = \vec{a}_\tau \times \vec{a}_n, \tag{4.86}$$

which is also tangent to the boundary surface. In the figure, $\vec{a}_t$ is shown in the plane of the page and $\vec{a}_\tau$ is into the page.

We now draw the closed loop ABCDA so that AB is just on one side of the boundary and CD is just on the other. The plane of the loop ABCDA is the plane of $\vec{a}_n$ and $\vec{a}_t$, while $\vec{a}_\tau$ is normal to the plane of the loop. The boundary itself is idealized to be a surface so the length of AD and BC, δ, is vanishingly small. The lengths AB and D, ΔL, are large compared to δ, but still small enough so the field is approximately constant along their lengths.

We now apply the Ampère–Maxwell law around the closed loop ABCDA.

$$U_{\text{ABCDA}} = (I_t)_{\text{surface}}, \tag{4.87}$$

$$\vec{H}_1 \cdot \vec{a}_t \, \Delta L - \vec{H}_{\text{BC}} \cdot \vec{a}_n \delta - \vec{H}_2 \cdot \vec{a}_t \, \Delta L + \vec{H}_{\text{DA}} \cdot \vec{a}_n \delta = \left(\vec{J} + \frac{\partial \vec{D}}{\partial t} \right) \cdot \vec{a}_\tau \, \Delta L \, \delta + \vec{J}_s \cdot \vec{a}_\tau \, \Delta L. \tag{4.88}$$

$\vec{H}_1$ and $\vec{H}_2$ are the fields at the center of ΔL, just above and below the surface, respectively. The terms $\vec{J}$ and $\partial \vec{D}/\partial t$ are per unit area current densities. Since δ is vanishingly small, these terms and the two terms that include δ on the left side of the equation can be neglected. The remaining term on the right side is the surface current term, which continues to flow through the loop even though $\delta \to 0$. Equation (4.88) herefore simplifies to

$$(\vec{H}_1 - \vec{H}_2) \cdot \vec{a}_t \, \Delta L = \vec{J}_s \cdot \vec{a}_\tau \, \Delta L.$$

Canceling ΔL and substituting $\vec{a}_\tau \times \vec{a}_n$ for $\vec{a}_t$, we get

$$(\vec{H}_1 - \vec{H}_2) \cdot (\vec{a}_\tau \times \vec{a}_n) = \vec{J}_s \cdot \vec{a}_\tau \tag{4.89}$$

$$\vec{a}_\tau \cdot [\vec{a}_n \times (\vec{H}_1 - \vec{H}_2)] = [\vec{J}_s] \cdot \vec{a}_\tau. \tag{4.90}$$

The vector identity Eq. (2.42) has been used to get from Eq. (4.89) to Eq. (4.90). The direction of the vector $\vec{a}_\tau$ is tangent to the surface but otherwise is arbitrary. Furthermore, both bracketed expressions are tangent to the surface. Therefore the only way the scalar product of the $\vec{a}_\tau$ and the bracketed expressions can be equal for all $\vec{a}_\tau$ is if the bracketed expressions are equal to each other. This gives us the final result:

$$\vec{a}_n \times (\vec{H}_1 - \vec{H}_2) = \vec{J}_s. \tag{4.91}$$

This equation gives the boundary condition for the tangential components of magnetic field intensity. It involves only the tangential components because of the vector product with the unit vector normal to the surface.

Surface current can only flow on the surface of a perfect conductor. The surface current must therefore be zero unless one of the media is a perfect conductor. Thus

$$\vec{a}_n \times (\vec{H}_1 - \vec{H}_2) = 0, \quad \text{when } \vec{J}_s = 0 \tag{4.92}$$

or

$$H_{1\ \text{tangential}} = H_{2\ \text{tangential}}, \tag{4.93}$$

meaning that the tangential components of magnetic field intensity are continuous or the same on either side of the boundary.

If medium 2 is a perfect conductor, the field in that medium is zero. In that case

$$\vec{a}_n \times \vec{H}_1 = \vec{J}_s, \quad \text{when } \sigma_2 = \infty. \tag{4.94}$$

This means that the component of $\vec{H}_1$ that is tangent to the surface and parallel to the surface current is zero, but the component that is tangent to the surface and normal to the surface current equals the surface current (Fig. 4.31).

We can now deal easily with the boundary condition for the tangential component of electric field intensity at the boundary between media with different permittivities. This is derived from the Faraday–Maxwell integral law. Since the form of that

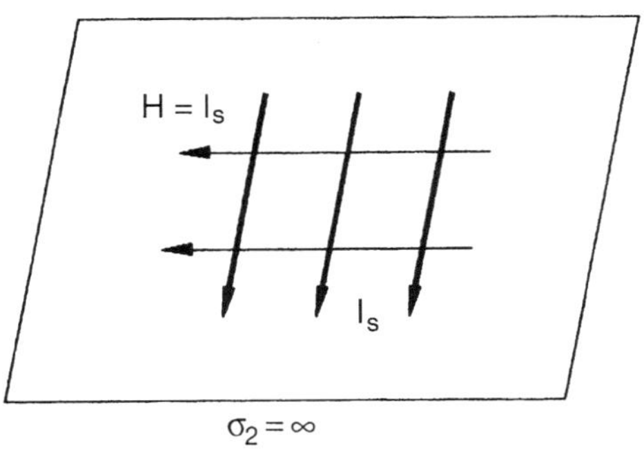

FIGURE 4.31 Surface current density at the interface of two media, one of which is a perfect conductor.

equation and the Ampère–Maxwell equation are identical, the derivation is step-by-step the same. Hence, jumping from the first step to the final result, we have

$$V_{\text{ABCDA}} = -\frac{d\Phi_{\text{surface}}}{dt} \tag{4.95}$$

$$\vec{E}_1 \cdot \vec{a}_t\,\Delta L - \vec{E}_{\text{BC}} \cdot \vec{a}_n \delta - \vec{E}_2 \cdot \vec{a}_t\,\Delta L + \vec{E}_{\text{DA}} \cdot \vec{a}_n\,\delta = -\frac{\partial \vec{B}}{\partial t} \cdot \vec{a}_\tau\,\Delta L\,\delta \tag{4.96}$$

$$\vec{a}_n \times (\vec{E}_1 - \vec{E}_2) = 0. \tag{4.97}$$

This means that the tangential components of electric field intensity are always continuous:

$$E_{1\ \text{tangential}} = E_{2\ \text{tangential}}. \tag{4.98}$$

In an electrostatic problem, where, by definition, the current is zero, the electric field must be zero in a conductor. Thus if medium 2 is a conductor and the problem is electrostatic, then

$$\vec{a}_n \times \vec{E}_1 = 0 \quad \text{or} \quad E_{1\ \text{tangential}} = 0, \quad \sigma_2 \neq 0, \text{ electrostatics}. \tag{4.99}$$

If medium 2 is a perfect conductor, the field in that medium is zero regardless of whether or not the problem is electrostatic. In that case

$$\vec{a}_n \times \vec{E}_1 = 0 \quad \text{or} \quad E_{1\ \text{tangential}} = 0, \quad \text{when } \sigma_2 = \infty. \tag{4.100}$$

4.7.2 Normal Electric and Magnetic Flux Densities and Normal Current Density

To derive the boundary conditions on the normal components of the electric flux density we will use Gauss's law, Eq. (4.2). The closed surface chosen is a small pillbox: a flat box with top and bottom surfaces ΔS and height δh. We may think of the surface as circular with small radius a and then $\delta h \ll a$ (Fig. 4.32).

Medium 1 above the boundary surface has permittivity ε_1 and conductivity σ_1. Medium 2 below the boundary surface has permittivity ε_2 and conductivity σ_2. The pillbox is placed so that the infinitesimal height δh just spans the boundary surface (Fig. 4.33). Let $\vec{a}_n$ be the normal to the boundary surface directed from medium 2 to medium 1. $\vec{D}_1$ is the electric flux density just above the surface and $\vec{D}_2$ is its value just below. Applying Gauss's law for electric fields, Eq. (4.2), gives

$$\int_{\text{top surface}} \vec{D}_1 \cdot d\vec{S} + \int_{\text{bottom surface}} \vec{D}_2 \cdot d\vec{S} + \int_{\text{sides}} \vec{D} \cdot d\vec{S} = Q_{\text{enclosed}}$$

$$\vec{D}_1 \cdot \vec{a}_n\,\Delta S - \vec{D}_2 \cdot \vec{a}_n\,\Delta S + (\vec{D} \cdot \vec{a}_\tau)\,2\pi a \delta h = \rho\,\Delta S\,\delta h + \rho_s\,\Delta S,$$

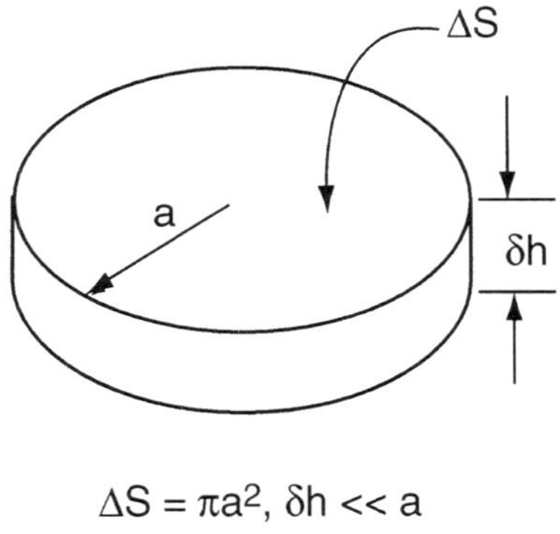

FIGURE 4.32 Pillbox surface used to derive boundary conditions for electric and magnetic flux densities.

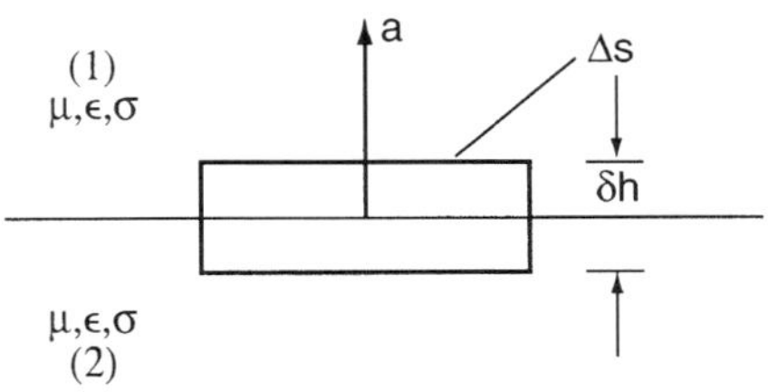

FIGURE 4.33 Side view of the pillbox of Fig. 4.32 showing the interface between the two media.

where ρ is the volume charge density and ρ_s is the surface charge density within the pillbox. Letting $\delta h \to 0$ the term on the left due to outward flux through the side of the pillbox and the term on the right due to volume charge density within the pillbox both go to zero. Hence

$$\vec{D}_1 \cdot \vec{a}_n \, \Delta S - \vec{D}_2 \cdot \vec{a}_n \, \Delta S = \rho_s \, \Delta S$$

$$(\vec{D}_1 - \vec{D}_2) \cdot \vec{a}_n = \rho_s \tag{4.101}$$

or

$$D_{1\ \text{normal}} - D_{2\ \text{normal}} = \rho_s. \tag{4.102}$$

Equations (4.101) and (4.102) both mean that the normal components of electric flux density are discontinuous at the boundary surface between two media by the surface charge density on the surface.

Equation (4.17) gives the electric flux density on either side of a uniform surface charge. Note that these fields satisfy the condition that the discontinuity in the normal component of electric flux density is given by the surface charge density.

Surface charge density can only exist on a conducting surface. Hence if both media are perfect dielectrics, meaning their conductivities are zero, then

$$(\vec{D}_1 - \vec{D}_2) \cdot a_n = 0$$

or

$$D_{1\ \text{normal}} = D_{2\ \text{normal}}, \quad \text{when } \sigma_1 = \sigma_2 = 0. \tag{4.103}$$

Equation (4.103) means that the normal component of electric flux density is continuous at the surface boundary between two perfect dielectrics.

We can now easily derive the boundary condition for the normal component of magnetic flux density at the interface of two media with different permeabilities, μ_1 and μ_2. Gauss's law for magnetic flux density, Eq. (4.19), is identical to the form of auss's law for electric flux density, except that the right side of the magnetic flux density law is always zero. Using Gauss's law for magnetic flux density and following the same steps, we obtain

$$(\vec{B}_1 - \vec{B}_2) \cdot \vec{a}_n = 0 \tag{4.104}$$

or

$$B_{1\ \text{normal}} = B_{2\ \text{normal}}, \tag{4.105}$$

where $\vec{B}_1$ and $\vec{B}_2$ are the magnetic flux density just above and below the surface, respectively. Equations (4.104) and (4.105) mean that the normal components of magnetic flux density are always continuous.

Lastly, we use the law of conservation of charge to find the boundary conditions on the normal components of current density at an interface between two media with different conductivities, σ_1 and σ_2. The form of Eq. (4.21) is again the same as the form of Gauss's law for electric flux density. In this case, the right side of the equation is the time rate of change of the total charge ($\rho\, \Delta S\, \delta h + \rho_s\, \Delta S$) inside the volume of the illbox. As $\delta h \to 0$ the only charge remaining inside the volume will be the surface charge. Following the same steps we obtain

$$(\vec{J}_1 - \vec{J}_2) \cdot \vec{a}_n = -\frac{\partial \rho_s}{\partial t} \tag{4.106}$$

or

$$J_{1\ \text{normal}} - J_{2\ \text{normal}} = -\frac{\partial \rho_s}{\partial t}, \tag{4.107}$$

where $\vec{J}_1$ and $\vec{J}_2$ are the free current density just above and below the surface, respectively. Equations (4.106) and (4.107) mean that the normal component of current density is discontinuous by the time rate of decrease of charge density on the boundary surface. If more current is leaving the surface than is entering it, the charge on the surface must be decreasing to supply the charge for this excess current.

When the fields and charges are constant with time $\partial\rho/\partial t = 0$ and then

$$(\vec{J}_1 - \vec{J}_2)\cdot\vec{a}_n = 0 \tag{4.108}$$

or

$$J_{1\ \text{normal}} - J_{2\ \text{normal}}. \tag{4.109}$$

This means that for steady fields the normal component of current density is continuous.

If medium 2 is a nonconducting region then $\vec{J}_2 = 0$ and $J_{1\ \text{normal}} = 0$.

PROBLEMS

4.1 The charge density and relative permittivity in a spherical region are given by

$$\rho(R) = \begin{cases} \rho_0 \\ 0 \end{cases} \quad \text{and} \quad \varepsilon_r(R) = \begin{cases} R/a, & R < a, \\ 1, & R > a. \end{cases}$$

Find the electric flux density and the electric field intensity for all R.

Ans. $E_R = \rho_0 a/3\varepsilon_0$, $R < a$; $E_R = \rho_0 a^3/3\varepsilon_0 R^2$, $R > a$.

4.2 The charge density in a cylindrical region is

$$\rho(r) = \begin{cases} \rho_0, & r < a, \\ 0, & r > a. \end{cases}$$

a. Assume free space and find the electric flux density and the electric field intensity for all r.

b. Repeat (a) if

$$\varepsilon_r(r) = \begin{cases} r/a, & r < a, \\ 1, & r > a. \end{cases}$$

Ans. b. $E_r = \rho_0 a/2\varepsilon_0$, $r < a$; $E_r = \rho_0 a^2/2\varepsilon_0 r$, $r > a$.

4.3 The electric flux density is $\vec{D} = A(\vec{a}_R \sin\theta - \vec{a}_\theta \cos\theta)/R^2$.

a. Find the charge $Q(R)$ inside a spherical region of radius R.

b. Find the point charge q at the origin. What is the electric flux density due to that point charge alone? Is there an additional charge distribution $\rho(R, \theta)$? What is the electric flux density due to $\rho(R, \theta)$?

Ans. a. $Q(R) = \pi^2 A$.

4.4 The charge density is given by $\rho = (A/R^2)\exp(-kR)$, $k > 0$.

a. Find the electric flux density.

b. Show that as $R \to 0$, $RD_R \to A$ and $D_R \to (A/R)$. Why is this consistent with the fact that the given charge distribution does not include a point charge at the origin?

Ans. a. $D_R = A[1 - \exp(-kR)]/kR^2$.

4.5 For the magnetic flux density $\vec{B} = A(\vec{a}_R 2\cos\theta + \vec{a}_\theta \sin\theta)/R^3$, show by surface integration that $\oint \vec{B}\cdot d\vec{S} = 0$ for the following closed surfaces.

a. A sphere of radius a centered at the origin.

b. A cone of angle θ_0 closed on the top by a spherical cap of radius R_0 and on the bottom by a spherical cap of radius $R_i < R_0$.

4.6 Assume $\varepsilon = \varepsilon_r \varepsilon_0$ everywhere and that a uniform surface charge ρ_{s0} is on the $x = 3$ plane. Find $\vec{D}$ and $\vec{E}$ everywhere.

4.7 Assume the permittivity of all space is $\varepsilon = \varepsilon_r \varepsilon_0$.

a. Find $\vec{D}$ and $\vec{E}$ for a surface charge density ρ_{s0} on the $z = 0$ plane.

b. Find $\vec{D}$ and $\vec{E}$ for a surface charge density $-\rho_{s0}$ on the $z = d > 0$ plane.

c. Assume both surface charges are present simultaneously. Use the principle of superposition to find the total $\vec{D}$ and $\vec{E}$ everywhere.

d. Does the answer to (c) change if the medium outside the two surface charges, $z < 0$ and $z > d$, is changed to free space, $\varepsilon = \varepsilon_0$?

Ans. c. $D_z = \rho_{s0}$, $E_z = \rho_{s0}/\varepsilon$, $0 < z < d$; and zero elsewhere.

4.8 A conductor consists of a solid cone of angle θ_0 coaxial with the positive z axis, a very thin sphere of radius R_0, and a wire along the negative z axis (Fig. 4.34). Current density in the cone is $\vec{J} = \vec{a}_R I_0 / 2\pi R^2 (1 - \cos\theta_0)$. Surface current on

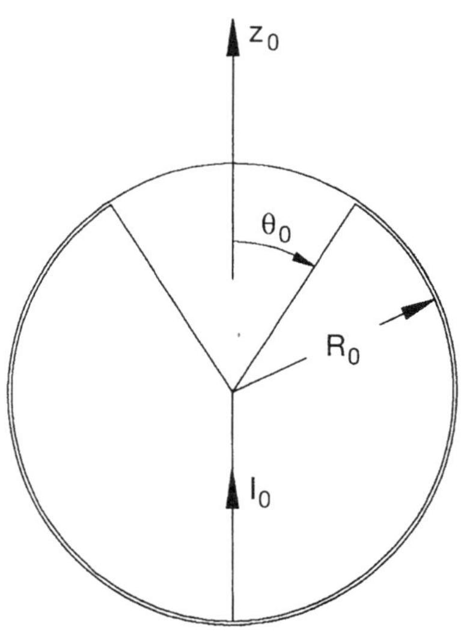

FIGURE 4.34 Conductor consisting of a solid cone of angle θ_0 coaxial with the positive z axis, a thin sphere of radius R_0, and a wire along the negative z axis.

the sphere is $\vec{J}_s = \vec{a}_\theta I_0/2\pi R_0 \sin\theta$. Current I_0 flows in the $+z$ direction in the wire. Use the Ampère–Maxwell law to find $H_\phi(R, \theta)$ if:

a. $R < R_0$, $\theta < \theta_0$ (i.e., within the solid cone).

b. $R < R_0$, $\theta > \theta_0$.

c. $R > R_0$.

Hint: In each case choose the loop to be a circle of radius $R \sin\theta$ centered on the z axis and lying in a z = constant plane and the surface to be the north polar cap over the circle.

Ans. a. $H_\phi = \dfrac{I_0(1-\cos\theta)}{2\pi R(\sin\theta)(1-\cos\theta_0)}$; b. $H_\phi = \dfrac{I_0}{2\pi R \sin\theta}$.

4.9 Assume $\mu = \mu_r\mu_0$ everywhere and that a uniform surface current $\vec{J}_s = \vec{a}_z J_{s0}$ is on the $x = 3$ plane. Find $\vec{H}$ and $\vec{B}$ everywhere.

4.10 Assume the permeability of all space is μ.

a. Find $\vec{H}$ and $\vec{B}$ for a surface current density $\vec{a}_x J_{s0}$ on the $z = 0$ plane.

b. Find $\vec{H}$ and $\vec{B}$ for a surface current density $-\vec{a}_x J_{s0}$ on the $z = d$ plane.

c. Assume both surface currents are present simultaneously. Use the principle of superposition to find the total $\vec{H}$ and $\vec{B}$ everywhere.

d. Does the answer to (c) change if the permeability outside the two surface currents, $z < 0$ and $z > d$, is changed to free space, $\mu = \mu_0$?

Ans. c. $\vec{H} = -\vec{a}_y J_{s0}$, $\vec{B} = -\vec{a}_y \mu J_{s0}$, $0 < z < d$ and zero elsewhere.

4.11 For the conditions leading to Eq. (4.77) find:

a. The power dissipated in the resistor R.

b. The magnetic force $\vec{F}_m$ on the wire.

c. The mechanical force that must be applied to the wire.

d. The mechanical power input to move the wire.

4.12 For a wire sliding over conducting rails (Fig. 4.21), assume that the flux density is constant and uniform, $\vec{B} = \vec{a}_z B_{z0}$, and the velocity of the wire is $\vec{u} = \vec{a}_x u_0 \sin\omega t$. Find the voltage across the resistor.

Ans. $V_0 = u_0 h B_{z0} \sin\omega t$.

4.13 For a wire sliding over conducting rails (Fig. 4.21), assume that the flux density is constant but dependent on position, $\vec{B} = \vec{a}_z B_0 \sin(2\pi x/X_0)$, and the velocity of the wire is constant, $\vec{u} = \vec{a}_x u_0$. Assume that at $t = 0$ the wire is located at $x = 0$. Find the voltage across the resistor.

Ans. $V_0 = u_0 h B_0 \sin(2\pi u_0 t/X_0)$.

4.14 For a wire sliding over conducting rails (Fig. 4.21), assume that the flux density is constant but dependent on position, $\vec{B} = \vec{a}_z B_0 \sin(2\pi x/X_0)$ and the velocity of the wire is $\vec{u} = \vec{a}_x u_0 \sin\omega t$. Assume that at $t = 0$ the wire is located at $x = 0$. Find the voltage across the resistor.

Ans. $V_0 = u_0 h B_0 \sin\omega t \sin[2\pi u_0(1-\cos\omega t)/\omega X_0]$.

4.15 For a wire sliding over conducting rails (Fig. 4.21), assume that the flux density is $\vec{B} = \vec{a}_z B_0 \sin \omega t$ and that the wire is stationary. Find the voltage across the resistor.

Ans. $V_0 = -hLB_0\omega \cos \omega t$.

4.16 For a wire sliding over conducting rails (Fig. 4.21), assume that the flux density is a function of position and time, $\vec{B} = \vec{a}_z B_z(x, t)$, and that the wire position is also a function of time, $x(t)$. Show that the voltage across the resistor is given by

$$V_0 = -h\left\{\int_{x'=x(t)}^{L} \frac{\partial B_z(x', t)}{\partial t}\,dx' - B_z[x(t), t]\,\frac{dx(t)}{dt}\right\}.$$

4.17 A wire is sliding over conducting rails at an angle α (Fig. 4.35). For a constant and uniform magnetic field, $\vec{B} = \vec{a}_z B_{z0}$, and a constant velocity wire, $\vec{u} = \vec{a}_x u_0$, find the voltage across the resistor.

Ans. $V_0 = u_0 h B_0$.

4.18 The magnetic flux through a coil increases in strength in the downward direction, as shown in Fig. 4.36. Determine the polarity of the induced voltage.

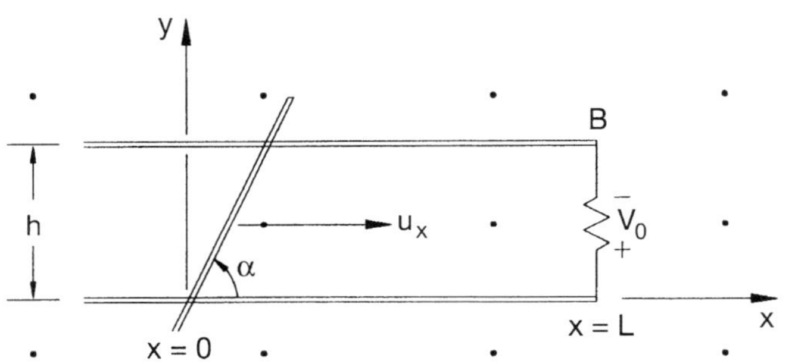

FIGURE 4.35 Wire sliding over conducting rails at an angle α.

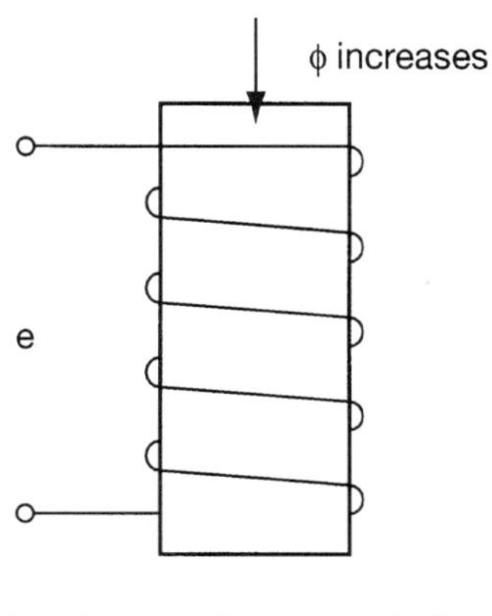

FIGURE 4.36 Coil with a time-varying magnetic flux (increasing downward).

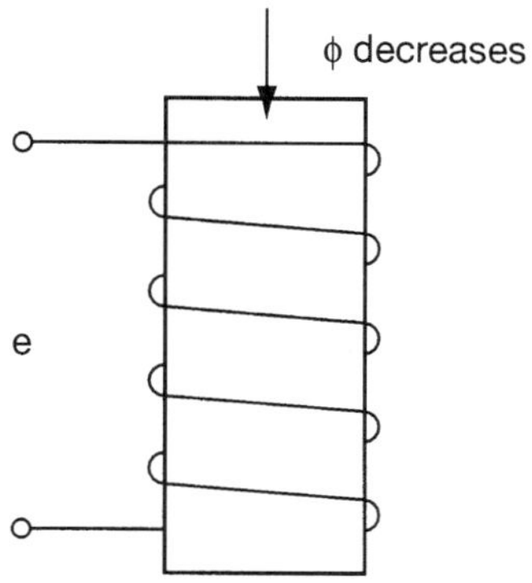

FIGURE 4.37 Coil with a time-varying magnetic flux (decreasing downward).

4.19 The magnetic flux through a coil decreases in strength in the downward direction, as shown in Fig. 4.37. Determine the polarity of the induced voltage. Compare this result with the results of the previous problem and Examples 4.2 and 4.3. Assume now that the flux through the coil is sinusoidal. Comment on the shape and polarity of the induced voltage.

4.20 Assume the region $z < 0$ is free space and the region $z > 0$ is a perfect conductor. The electric and magnetic field in the free space region is given by

$$\vec{E} = \vec{a}_x E_0 \cos\alpha \sin(kz\cos\alpha)\sin(\omega t - kx\sin\alpha)$$
$$- \vec{a}_z E_0 \sin\alpha \cos(kz\cos\alpha)\cos(\omega t - kx\sin\alpha),$$
$$\vec{H} = \vec{a}_y \sqrt{\varepsilon_0/\mu_0} E_0 \cos(kz\cos\alpha)\cos(\omega t - kx\sin\alpha).$$

a. Show that the boundary condition on the tangential electric field intensity is satisfied at $z = 0$.

b. Find the surface charge density $\rho_s(x, t)$ on the $z = 0$ surface.

c. Find the surface current density $\vec{J}_s(x, t)$ on the $z = 0$ surface.

d. Find the value of k such that $\partial J_{sx}/\partial x = -\partial\rho_s/\partial t$.

Ans. d. $k = \omega\sqrt{\mu_0\varepsilon_0}$.

4.21 Repeat Problem 4.20 if the fields in the free-space region $z < 0$ are given by

$$\vec{E} = \vec{a}_y E_0 \sin(kz\cos\alpha)\sin(\omega t - kx\sin\alpha),$$
$$\vec{H} = \vec{a}_x \sqrt{\varepsilon_0/\mu_0}\, E_0 \cos\alpha\cos(kz\cos\alpha)\cos(\omega t - kx\sin\alpha)$$
$$+ \vec{a}_z \sqrt{\varepsilon_0/\mu_0}\, E_0 \sin\alpha\sin(kz\cos\alpha)\sin(\omega t - kx\sin\alpha).$$

For part (d), show that $\partial J_{sy}/\partial y = -\partial\rho_s/\partial t$.

CHAPTER FIVE

Vector Mathematics II and Maxwell's Equations

5.1 INTRODUCTION

In Chapter 2 we studied vector arithmetic and vector integrals. In Chapter 4 we studied problems involving symmetric fields, which could be solved using Maxwell's equations in integral form. To go further and find relations between the field quantities at a point in space, it is necessary to introduce vector differential operations and to convert Maxwell's integral equations to differential form.

5.2 GRADIENT

In introducing scalar fields (Section 2.2) an example of a topographic map (see Fig. 2.1) was given. It was noted that if we start at some particular point on the hill, then the normal to the line of constant height passing through that point represents the greatest grade or slope up or down the hill. It is this concept of the greatest rate of change of a scalar field and the direction in which that maximum occurs that we formalize in the gradient.

5.2.1 Definition

The gradient of a scalar field is a vector field. The magnitude of the gradient at each point is the maximum rate of increase of the scalar field with respect to position at the point. The direction of the gradient is the direction in which that maximum rate of increase occurs.

Let us consider a scalar function of the coordinates and time $f(u, v, w, t)$. At any time the function has a constant value f over some surface and a slightly different value $f + \Delta f$ over an adjacent surface (Fig. 5.1). Let P be a point on the first surface and Q be a point on the second surface and let $\overline{PQ}$ be the distance between these points. Then the ratio $\Delta f/\overline{PQ}$ is the rate of change of the function f at point P in the direction

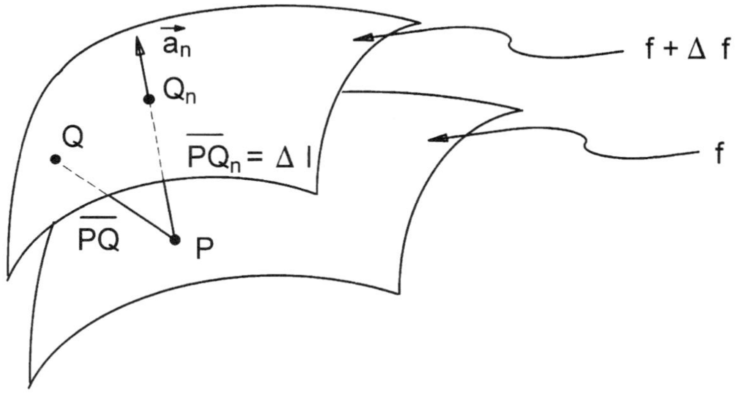

FIGURE 5.1 Surfaces of constant scalar field f used in defining its gradient.

toward Q. To maximize the ratio the distance $\overline{PQ}$ must be made as short as possible. This will occur when point Q coincides with point Q_n along the normal to the first surface. Along the normal let $\overline{PQ_n} = \Delta l$. Then the maximum value of the ratio is $(\Delta f/\overline{PQ})_{maximum} = \Delta f/\Delta l$. The limit of this ratio as $\Delta l \to 0$ is the magnitude of the gradient of f at the point P. The direction of the gradient is the direction in which this maximum ratio occurs. This is the direction of the normal $\vec{a}_n$ to the surface at P. Thus

$$\vec{F} = \text{grad} f = \vec{a}_n \left(\lim_{\Delta l \to 0} \frac{\Delta f}{\Delta l} \right). \tag{5.1}$$

This equation puts the definition in mathematical form. We must next find formulas for the gradient in different coordinate systems.

Note that the gradient involves division of the scalar field by a distance. The units of the gradient are therefore the units of the scalar field divided by length.

5.2.2 Cartesian Coordinates and the Del Operator

The component of the gradient in any direction is the rate of change of the scalar field with respect to position in that direction. In Cartesian coordinates we therefore have

$$F_x = \vec{a}_x \cdot \text{grad} f = \frac{\partial f}{\partial x}, \quad F_y = \vec{a}_y \cdot \text{grad} f = \frac{\partial f}{\partial y}, \quad F_z = \vec{a}_z \cdot \text{grad} f = \frac{\partial f}{\partial z}.$$

We add these components to get the formula for the gradient in Cartesian coordinates:

$$\vec{F}(x, y, z) = \text{grad} f(x, y, z) = \vec{a}_x \frac{\partial f}{\partial x} + \vec{a}_y \frac{\partial f}{\partial y} + \vec{a}_z \frac{\partial f}{\partial z}. \tag{5.2}$$

It is convenient to define a del operator, the symbol for which is ∇. This vector operator is defined as

$$\nabla = \vec{a}_x \frac{\partial}{\partial x} + \vec{a}_y \frac{\partial}{\partial y} + \vec{a}_z \frac{\partial}{\partial z}. \tag{5.3}$$

Since the operator is a vector quantity, it might be logical to write it as $\vec{\nabla}$, but by custom the arrow is omitted.

Multiplying the operator by the scalar field f means that we should perform the indicated operations on f

$$\begin{aligned}\nabla f &= \left(\vec{a}_x \frac{\partial}{\partial x} + \vec{a}_y \frac{\partial}{\partial y} + \vec{a}_z \frac{\partial}{\partial z}\right) f \\ &= \vec{a}_x \frac{\partial f}{\partial x} + \vec{a}_y \frac{\partial f}{\partial y} + \vec{a}_z \frac{\partial f}{\partial z} \\ &= \text{grad}\, f.\end{aligned}$$

Hence we can use the notation ∇f to mean the gradient of f. The notation ∇f is convenient and has the advantage of reminding us that the result is a vector, since the del operator is a vector. As a result, modern notation has replaced the term "grad" with the operator. When reading an equation containing ∇f it is best to say "gradient of f," although it is not uncommon or incorrect to say "del f."

5.2.3 Cylindrical and Spherical Coordinates

The component of the gradient in any direction is the rate of change of the field in that direction. This is the $\lim_{\Delta l \to 0} (\Delta f/\Delta l)$, where Δl is the differential length in the desired direction.

In cylindrical coordinates the differential lengths in the r, ϕ, and z directions are dr, $r\,d\phi$, and dz, respectively. Hence, if $f = f(r, \phi, z, t)$, then

$$F_r = \vec{a}_r \cdot \nabla f = \frac{\partial f}{\partial r}, \quad F_\phi = \vec{a}_\phi \cdot \nabla f = \frac{1}{r}\frac{\partial f}{\partial \phi}, \quad F_z = \vec{a}_z \cdot \nabla f = \frac{\partial f}{\partial z}.$$

The formula for the gradient in cylindrical coordinates is therefore

$$\vec{F}(r, \phi, z, t) = \nabla f(r, \phi, z, t) = \vec{a}_r \frac{\partial f}{\partial r} + \vec{a}_\phi \frac{1}{r}\frac{\partial f}{\partial \phi} + \vec{a}_z \frac{\partial f}{\partial z}. \tag{5.4}$$

In spherical coordinates the differential lengths in the R, θ, and ϕ directions are dR, $R\,d\theta$, and $R \sin\theta\, d\phi$, respectively. Hence, if $f = f(R, \theta, \phi, t)$, then

$$F_R = \vec{a}_R \cdot \nabla f = \frac{\partial f}{\partial R}, \quad F_\theta = \vec{a}_\theta \cdot \nabla f = \frac{1}{R}\frac{\partial f}{\partial \theta}, \quad F_\phi = \vec{a}_\phi \cdot \nabla f = \frac{1}{R \sin\theta}\frac{\partial f}{\partial \phi}.$$

The formula for the gradient in spherical coordinates is therefore

$$\vec{F}(R,\theta,\phi,t) = \nabla f(R,\theta,\phi,t) = \vec{a}_R\frac{\partial f}{\partial R} + \vec{a}_\theta\frac{1}{R}\frac{\partial f}{\partial \theta} + \vec{a}_\phi\frac{1}{R\sin\theta}\frac{\partial f}{\partial \phi}. \tag{5.5}$$

Example 5.1 Find the gradient of the scalar field $f = CR^{-2}\cos\theta$. Evaluate the maximum rate of increase at the point (2, $\pi/3$, $\pi/6$) and the direction in which that maximum occurs.

Solution We use Eq. (5.5) to find the gradient

$$\begin{aligned}\nabla f &= \vec{a}_R(C\cos\theta)\frac{d(R^{-2})}{dR} + \vec{a}_\theta(CR^{-2})\frac{1}{R}\frac{d(\cos\theta)}{d\theta}\\ &= -\frac{C}{R^3}(\vec{a}_R 2\cos\theta + \vec{a}_\theta\sin\theta)\end{aligned}$$

and evaluate this at the point to get $\nabla f = -C(\vec{a}_R 0.125 + \vec{a}_\theta 0.108)$. The maximum rate of change is the magnitude $0.165C$ and the direction is the gradient divided by the magnitude $-(\vec{a}_R 0.756 + \vec{a}_\theta 0.655)$.

5.2.4 MathCad Evaluation of the Gradient

MathCad can be used to evaluate the gradient of a scalar field both numerically and symbolically. The gradient can be evaluated in all types of coordinate systems.

In rectangular coordinates, the gradient of a scalar field $f(x, y, z)$ is defined as follows:

$$\text{grad}(f, x, y, z) := a_x\frac{d}{dx}f(x,y,z) + a_y\frac{d}{dy}f(x,y,z) + a_z\frac{d}{dz}f(x,y,z).$$

The function $f(x, y, z)$ as well as the unit vectors a_x, a_y, and a_z have to be previously defined. The numerical value of the gradient is obtained by typing the grad(f, 1, 1, 1) followed by the equal sign. If the function is

$$f(x,y,z) := \frac{1}{\sqrt{x^2+y^2+z^2}},$$

then the value of the gradient at point (1, 1, 1) is calculated to be

$$\text{grad}(f,1,1,1) = \begin{pmatrix} -0.192 \\ -0.192 \\ -0.192 \end{pmatrix}.$$

The gradient can be evaluated symbolically, if the function is explicitly typed in the expression. As an example, for the same function as above, the gradient is defined as

$$\text{grad}(f, x, y, z) := a_x \frac{d}{dx}\left(\frac{1}{\sqrt{x^2+y^2+z^2}}\right) + a_y \frac{d}{dy}\left(\frac{1}{\sqrt{x^2+y^2+z^2}}\right) + a_z \frac{d}{dz}\left(\frac{1}{\sqrt{x^2+y^2+z^2}}\right).$$

The explicit expression of the gradient (the left side of the assign operator) has to be selected. Using the function "Evaluate Symbolically" from the "Symbolic" menu, will give the symbolic expression of the gradient.

Example 5.2 Use MathCad to evaluate symbolically the gradient in rectangular coordinates of the function $f(x, y, z) = x^2 + \sin(y) + \sqrt{z}$. Evaluate the gradient for $x = 1$, $y = \pi/6$, $z = 36$.

Solution After the definition of the unit vectors, the function has to be defined:

$$f(x, y, z) := x^2 + \sin(y) + \sqrt{z}.$$

In order to evaluate the gradient symbolically, the function has to be explicitly defined in the expression of the gradient:

$$a_x \cdot \frac{d}{dx}(x^2 + \sin(y) + \sqrt{z}) + a_y \cdot \frac{d}{dy}(x^2 + \sin(y) + \sqrt{z}) + a_z \cdot \frac{d}{dz}(x^2 + \sin(y) + \sqrt{z}).$$

Select the expression and choose the Evaluate Symbolically from the Symbolic menu. MathCad outputs the symbolic result as

$$2 \cdot a_x \cdot x + a_y \cdot \cos(y) + \frac{1}{2} \cdot \frac{a_z}{\sqrt{z}}.$$

This result can be assigned to a function of x, y, and z and then evaluated for different values of the arguments.

The gradient of a scalar function can also be evaluated in cylindrical and spherical coordinates. The procedure is similar to the one used for rectangular coordinates.

Example 5.3 Evaluate, using MathCad, the gradient in cylindrical coordinates at the point $(r, \phi, z) = (1, 0, 1)$ of the following function:

$$f(r, \phi, z) = \frac{\cos(\phi)}{r^2}.$$

Solution The unit vectors in cylindrical coordinates have to be defined first:

$$a_r := \begin{pmatrix} 1 \\ 0 \\ 0 \end{pmatrix} \quad a_\phi := \begin{pmatrix} 0 \\ 1 \\ 0 \end{pmatrix} \quad a_z := \begin{pmatrix} 0 \\ 0 \\ 1 \end{pmatrix}.$$

The function definition follows:

$$f(r,\phi,z) := \frac{\cos(\phi)}{r^2}.$$

The gradient can be defined as a function of the function f itself and the variables r, ϕ, and z as follows:

$$\mathrm{grad}_{cyl}(f,r,\phi,z) := a_r \cdot \frac{d}{dr} f(r,\phi,z) + a_\phi \cdot \frac{1}{r} \cdot \frac{d}{d\ \phi} f(r,\phi,z) + a_z \cdot \frac{d}{dz} f(r,\phi,z).$$

The function can now be evaluated at the desired point by typing $\mathrm{grad}_{cyl}(f,1,0,1)$ followed by the equal sign:

$$\mathrm{grad}_{cyl}(f,1,0,1) = \begin{pmatrix} -2 \\ 0 \\ 0 \end{pmatrix}.$$

The gradient can also be evaluated symbolically by following the procedure outlined for the gradient in cylindrical coordinates. The result is

$$\mathrm{grad}_{cyl}(f,r,\phi,z) := -2 \cdot a_r \cdot \frac{\cos(\phi)}{r^3} - \frac{a_\phi}{r^3} \cdot \sin(\phi).$$

Example 5.4 Redo Example 5.1 and evaluate, using MathCad, the gradient in spherical coordinates at the point $(R,\theta,\phi) = (2, \pi/3, \pi/6)$ of the following function:

$$f(R,\theta,\phi) = \frac{\cos(\phi)}{R^2}.$$

Solution The unit vectors in cylindrical coordinates have to be defined first:

$$a_R := \begin{pmatrix} 1 \\ 0 \\ 0 \end{pmatrix} \quad a_\theta := \begin{pmatrix} 0 \\ 1 \\ 0 \end{pmatrix} \quad a_\phi := \begin{pmatrix} 0 \\ 0 \\ 1 \end{pmatrix}.$$

The function definition follows:

$$f(R,\theta,\phi) := \frac{\cos(\phi)}{R^2}.$$

The gradient can be defined as a function of the function f itself and the variables R, θ, and ϕ as follows:

$$\mathrm{grad}_s(f,R,\theta,\phi) := a_R \cdot \frac{d}{dR} f(R,\theta,\phi) + a_\theta \cdot \frac{1}{R} \cdot \frac{d}{d\theta} f(R,\theta,\phi)$$

$$+ a_\phi \cdot \frac{1}{R \cdot \sin(\theta)} \cdot \frac{d}{d\phi} f(R,\theta,\phi).$$

The function can now be evaluated at the desired point as follows by typing $\text{grad}_s(f, 2, \pi/3, \pi/6)$ followed by the equal sign:

$$\text{grad}_s\left(f, 2, \frac{\pi}{3}, \frac{\pi}{6}\right) = \begin{pmatrix} -0.125 \\ -0.108 \\ 0 \end{pmatrix}.$$

The gradient can also be evaluated symbolically by following the procedure outlined for the gradient in cylindrical coordinates. The result is

$$\text{grad}_s(f, R, \theta, \phi) := -2 \cdot a_R \cdot \frac{\cos(\theta)}{R^3} - \frac{a_\theta}{R^3} \cdot \sin(\theta).$$

5.3 DIVERGENCE

The divergence of a vector field is a measure of how much the field spreads out from or diverges from a point in space.

To appreciate what this means we examine the electric flux density surrounding a uniform sphere of charge, which we found in Section 4.2.1. The charge density and the resultant field are repeated here:

$$\rho = \begin{cases} \rho_0 \\ \\ 0 \end{cases} \quad \text{and} \quad D_R = \begin{cases} \dfrac{R}{3}\rho_0, & R < a, \\ \dfrac{a^3}{3R^2}\rho_0, & R > a, \end{cases} \tag{5.6}$$

and Fig. 5.2 shows the field lines. Note that new field lines originate in the region $R < a$ since the region contains charge. If we draw a closed surface in this region, S_1 in

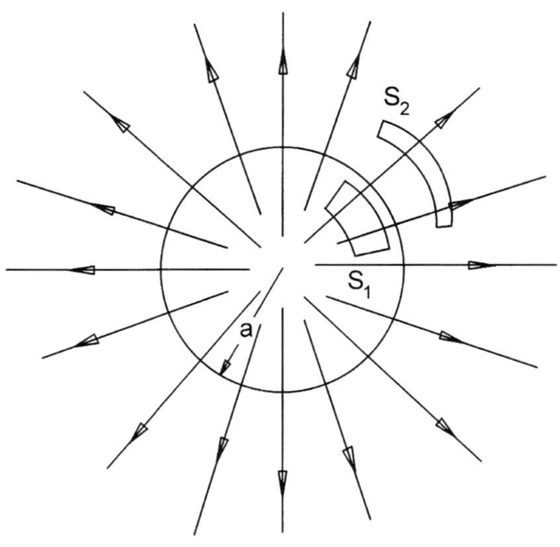

FIGURE 5.2 Field lines due to a change density inside a sphere of radius a.

the figure (the surfaces S_1 and S_2 are formed by revolving the outline shown in the figure around the z axis), more lines of flux will leave the surface than enter it, indicating that the field is diverging from the volume enclosed by the surface. On the other hand, if we draw a closed surface, S_2 in the figure, in the region $R > a$, then all the lines that enter the surface also leave it and no new lines start within the surface. This indicates that the field is not diverging from the volume enclosed by the surface.

Gauss's law, Eq. (4.2), shows that the excess of flux lines leaving the closed surface equals the charge within the volume. The divergence of the electric flux density at a point ought therefore to equal the charge density at the point. To arrive at a charge density, it is necessary to divide the flux of $\vec{D}$ through the closed surface by the volume enclosed. This leads directly to the definition of the divergence.

The divergence of a vector field at a point in space is the ratio of the integral of the field over a closed surface enclosing the point to the volume enclosed by the surface, as the surface and the enclosed volume shrink to the point. In mathematical form, if $\vec{F}(u, v, w, t)$ is the vector field and $f(u, v, w, t)$ is its divergence, then

$$f = \operatorname{div} \vec{F} = \lim_{\Delta v \to 0} \left(\frac{\oint_S \vec{F} \cdot d\vec{S}}{\Delta v} \right). \tag{5.7}$$

We must now use this definition to find formulas for the divergence in the various coordinate systems.

5.3.1 Cartesian Coordinates and the Del Operator

Figure 5.3 shows a rectangular parallelepiped surrounding the point (x, y, z). The corner closest to the origin is located at $(x - \Delta x/2, y - \Delta y/2, z - \Delta z/2)$ and the corner

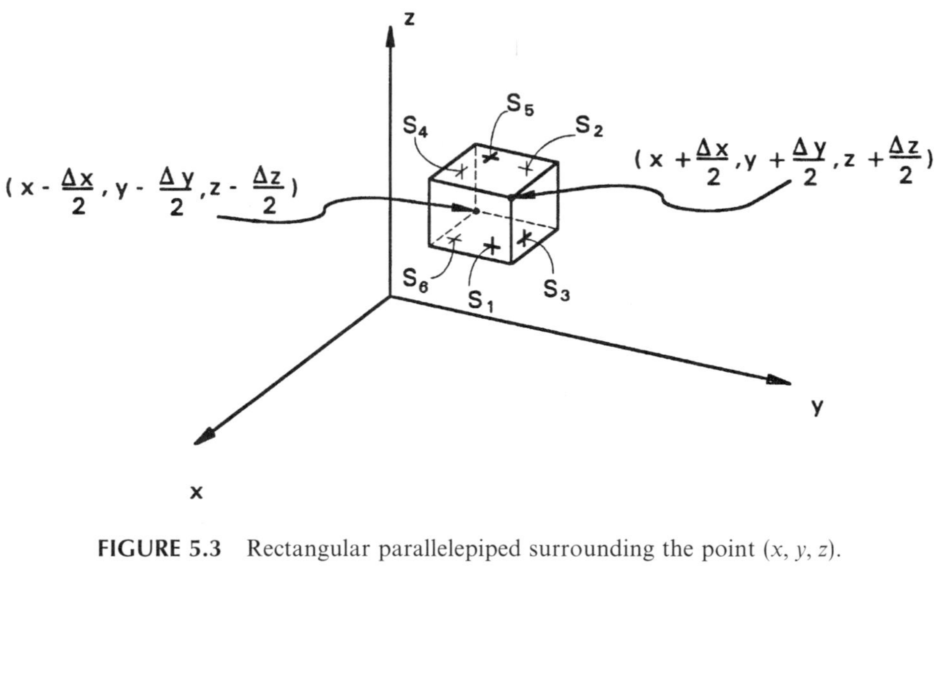

FIGURE 5.3 Rectangular parallelepiped surrounding the point (x, y, z).

furthest from the origin is at $(x + \Delta x/2, y + \Delta y/2, z + \Delta z/2)$. First examine carefully the surface integral over the front surface, S_1, in the plane $(x + \Delta x/2)$.

$$\int_{S_1} \vec{F} \cdot d\vec{S} \approx F_x(x + \Delta x/2, y, z, t)\Delta y\, \Delta z,$$

where $\vec{F}$ has been evaluated at the center of the small surface S_1 to approximate its value over the entire surface. Proceeding in a similar way for the surface S_2 in the plane $(x - \Delta x/2)$, the surface S_3 in the plane $(y + \Delta y/2)$, the surface S_4 in the plane $(y - \Delta y/2)$, the surface S_5 in the plane $(z + \Delta z/2)$, and the surface S_6 in the plane $(z - \Delta z/2)$, we obtain

$$f = \operatorname{div} \vec{F} = \lim_{\Delta v \to 0} \left(\frac{\int_{S_1} + \int_{S_2} + \int_{S_3} + \int_{S_4} + \int_{S_5} + \int_{S_6}}{\Delta x\, \Delta y\, \Delta z} \right)$$

$$= \lim_{\substack{\Delta x \to 0 \\ \Delta y \to 0 \\ \Delta z \to 0}} \left(\frac{\begin{array}{c}[F_x(x + \Delta x/2, y, z, t) - F_x(x - \Delta x/2, y, z, t)]\Delta y\, \Delta z \\ + [F_y(x, y + \Delta y/2, z, t) - F_y(x, y - \Delta y/2, z, t)]\Delta x\, \Delta z \\ + [F_z(x, y, z + \Delta z/2, t) - F_z(x, y, z - \Delta z/2, t)]\Delta x\, \Delta y\end{array}}{\Delta x\, \Delta y\, \Delta z} \right)$$

$$= \lim_{\Delta x \to 0} \left(\frac{F_x(x + \Delta x/2, y, z, t) - F_x(x - \Delta x/2, y, z, t)}{\Delta x} \right)$$
$$+ \lim_{\Delta y \to 0} \left(\frac{F_y(x, y + \Delta y/2, z, t) - F_y(x, y - \Delta y/2, z, t)}{\Delta y} \right)$$
$$+ \lim_{\Delta z \to 0} \left(\frac{F_z(x, y, z + \Delta z/2, t) - F_z(x, y, z - \Delta z/2, t)}{\Delta z} \right).$$

Each of the limits in the last equation is the definition of the partial derivative of the component in the numerator with respect to the variable in the denominator. Hence

$$f(x, y, z, t) = \operatorname{div} \vec{F}(x, y, z, t) = \frac{\partial F_x}{\partial x} + \frac{\partial F_y}{\partial y} + \frac{\partial F_z}{\partial z}. \tag{5.8}$$

This is the formula for the divergence in Cartesian coordinates.

This formula can be written in terms of the del operator, Eq. (5.3). We take the scalar product of the del operator with the vector field $\vec{F}$ to get

$$\begin{aligned} \nabla \cdot \vec{F} &= \left(\vec{a}_x \frac{\partial}{\partial x} + \vec{a}_y \frac{\partial}{\partial y} + \vec{a}_z \frac{\partial}{\partial z} \right) \cdot (\vec{a}_x F_x + \vec{a}_y F_y + \vec{a}_z F_z) \\ &= \frac{\partial F_x}{\partial x} + \frac{\partial F_y}{\partial y} + \frac{\partial F_z}{\partial z} \\ &= \operatorname{div} \vec{F}. \end{aligned} \tag{5.9}$$

Hence we can use the notation $\nabla \cdot \vec{F}$ to mean the divergence of $\vec{F}$. The notation is convenient: the dot product reminds us that the divergence is an operation on a vector field, which results in a scalar field, and the scalar product formally works to give us the correct formula in rectangular coordinates. The notation is also used in other coordinate systems, although the correspondence between the formal dot product and the formula for the divergence is not so simple. When reading an equation containing $\nabla \cdot \vec{F}$ it is best to say "the divergence of F," although it is not uncommon or incorrect to say "del dot F."

5.3.2 Cylindrical and Spherical Coordinates

To find the formula for the divergence in cylindrical coordinates we apply the definition of divergence, Eq. (5.7), to a cylindrical parallelepiped centered on the point (r, ϕ, z) (Fig. 5.4). The surface integral over each face of the integral is approximated by the product of the normal component of $\vec{F}$ at the center of the surface and the surface area of the face. Note that the surface area of the face at $(r + \Delta r/2)$ is different from the surface area of the face at $(r - \Delta r/2)$ and that the F_r component at each face must be multiplied by the proper area. See the first and second terms in the numerator below. Thus

$$\nabla \cdot \vec{F} = \lim_{\substack{\Delta r \to 0 \\ \Delta\phi \to 0 \\ \Delta z \to 0}} \left(\frac{\begin{array}{l}[F_r(r + \Delta r/2, \phi, z, t)(r + \Delta r/2)\,\Delta\phi\,\Delta z \\ \quad - F_r(r - \Delta r/2, \phi, z, t)(r - \Delta r/2)\,\Delta\phi\,\Delta z] \\ + [F_\phi(r, \phi + \Delta\phi/2, z, t) - F_\phi(r, \phi - \Delta\phi/2, z, t)]\,\Delta r\,\Delta z \\ + [F_z(r, \phi, z + \Delta z/2, t) - F_z(r, \phi, z - \Delta z/2, t)]\,\Delta r(r\,\Delta\phi)\end{array}}{\Delta r (r\,\Delta\phi)\Delta z} \right)$$

$$= \lim_{\Delta r \to 0} \left(\frac{F_r(r + \Delta r/2, \phi, z, t)(r + \Delta r/2) - F_r(r - \Delta r/2, \phi, z, t)(r - \Delta r/2)}{r\,\Delta r} \right)$$

$$+ \lim_{\Delta\phi \to 0} \left(\frac{F_\phi(r, \phi + \Delta\phi/2, z, t) - F_\phi(r, \phi - \Delta\phi/2, z, t)}{r\,\Delta\phi} \right)$$

$$+ \lim_{\Delta z \to 0} \left(\frac{F_z(r, \phi, z + \Delta z/2, t) - F_z(r, \phi, z - \Delta z/2, t)}{\Delta z} \right)$$

$$\nabla \cdot \vec{F} = \frac{1}{r}\frac{\partial (rF_r)}{\partial r} + \frac{1}{r}\frac{\partial F_\phi}{\partial \phi} + \frac{\partial F_z}{\partial z}. \tag{5.10}$$

This is the formula for the divergence in cylindrical coordinates. The first term on the right requires these operations: multiply F_r by r; take the derivative of the product with respect to r; divide the result by r.

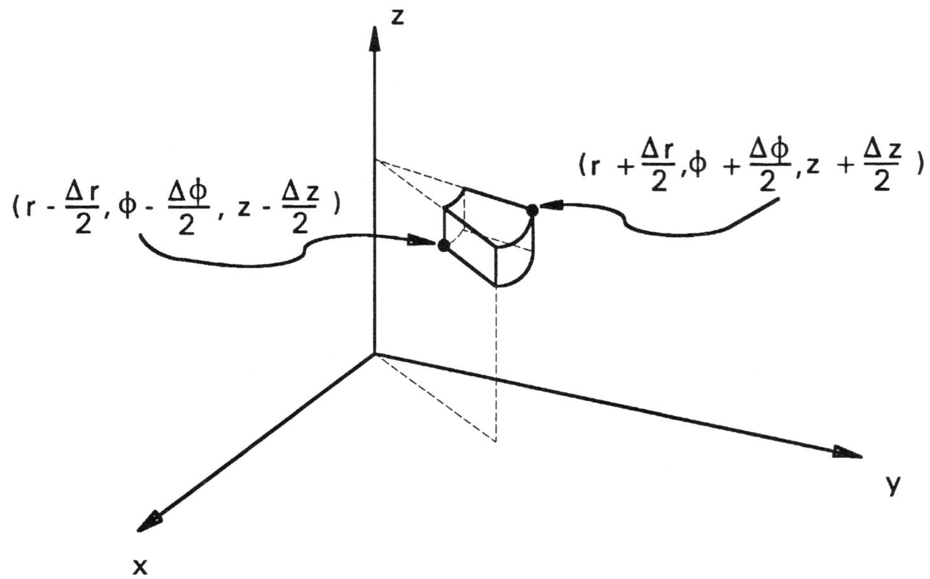

FIGURE 5.4 Cylindrical parallelepiped centered on the point (r, ϕ, z).

Example 5.5 Evaluate the divergence of $\vec{F} = \vec{a}_r 3\phi + \vec{a}_\phi 2\phi^2$ at the point $(2, \pi/6, 14)$.

Solution The divergence is

$$\nabla \cdot \vec{F} = \frac{1}{r}\frac{\partial[r(3\phi)]}{\partial r} + \frac{1}{r}\frac{\partial(2\phi^2)}{\partial \phi}$$

$$= \frac{3\phi}{r} + \frac{4\phi}{r} = \frac{7\phi}{r}.$$

Evaluating at the point we get $(7\pi/12)$.

The formula for the divergence in spherical coordinates is derived in the same way. Figure 5.5 shows a spherical parallelepiped centered around the point (R, θ, ϕ). This time the spherically radial components on the outer and inner spheres are at different radii, $(R + \Delta R/2)$ and $(R - \Delta R/2)$, respectively, so these components must be multiplied by the corresponding areas (the first and second terms in the numerator below). Similarly, the F_θ components must be multiplied by the areas for their faces of the parallelepiped (see the sine factors in the third and fourth terms below). The results are

$$\nabla \cdot \vec{F} = \lim_{\substack{\Delta R \to 0 \\ \Delta\theta \to 0 \\ \Delta\phi \to 0}} \left(\frac{\begin{array}{l} F_R(R + \Delta R/2, \theta, \phi, t)[(R + \Delta R/2)\,\Delta\theta][(R + \Delta R/2)\sin\theta\,\Delta\phi] \\ \quad - F_R(R - \Delta R/2, \theta, \phi, t)[(R - \Delta R/2)\,\Delta\theta][(R - \Delta R/2)\sin\theta\,\Delta\phi] \\ \quad + \{F_\theta(R, \theta + \Delta\theta/2, \phi, t)\,\Delta R\,[R\sin(\theta + \Delta\theta/2)\,\Delta\phi] \\ \quad - F_\phi(R, \theta - \Delta\theta/2, \phi, t)\,\Delta R\,[R\sin(\theta - \Delta\theta/2)\,\Delta\phi]\} \\ \quad + \{[F_\theta(R, \theta, \phi + \Delta\phi/2, t) - F_\phi(R, \theta, \phi - \Delta\phi/2, t)]\,\Delta R\,(R\,\Delta\theta)\} \end{array}}{\Delta R(R\,\Delta\theta)(R\sin\theta\,\Delta\phi)} \right)$$

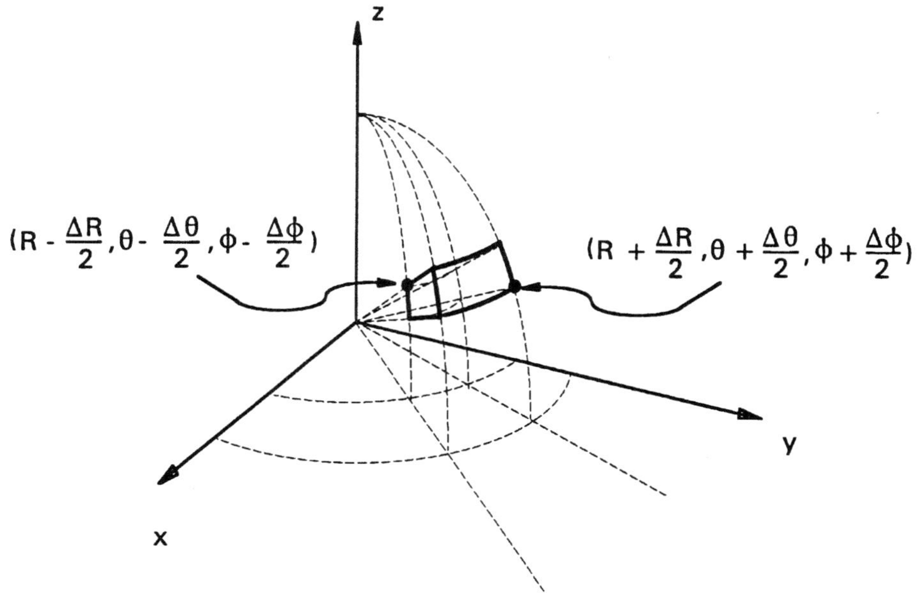

FIGURE 5.5 Spherical parallelepiped centered on the point (R, θ, ϕ).

$$= \lim_{\Delta R \to 0} \left(\frac{F_R(R + \Delta R/2, \theta, \phi, t)(R + \Delta R/2)^2 - F_R(R - \Delta R/2, \theta, \phi, t)(R - \Delta R/2)^2}{R^2 \, \Delta R} \right)$$

$$+ \lim_{\Delta \theta \to 0} \left(\frac{F_\theta(R, \theta + \Delta\theta/2, \phi, t) \sin(\theta + \Delta\theta/2) - F_\theta(R, \theta - \Delta\theta/2, \phi, t) \sin(\theta - \Delta\theta/2)}{R \sin\theta \, \Delta\theta} \right)$$

$$+ \lim_{\Delta \phi \to 0} \left(\frac{F_\phi(R, \theta, \phi + \Delta\phi/2, t) - F_\phi(R, \theta, \phi - \Delta\phi/2, t)}{R \sin\theta \, \Delta\phi} \right)$$

$$\nabla \cdot \vec{F} = \frac{1}{R^2} \frac{\partial(R^2 F_R)}{\partial R} + \frac{1}{R \sin\theta} \frac{\partial(\sin\theta \, F_\theta)}{\partial\theta} + \frac{1}{R \sin\theta} \frac{\partial F_\phi}{\partial\phi}. \tag{5.11}$$

This is the formula for the divergence in spherical coordinates. Again note that the multiplication inside the parentheses must be performed before the derivative of the product is taken.

Example 5.6 Find the divergence of $\vec{F} = \vec{a}_R 3 + \vec{a}_\theta 2 + \vec{a}_\phi \cos\phi$.

Solution

$$\nabla\cdot\vec{F} = \frac{1}{R^2}\frac{\partial(3R^2)}{\partial R} + \frac{1}{R\sin\theta}\frac{\partial(2\sin\theta)}{\partial\theta} + \frac{1}{R\sin\theta}\frac{\partial(\cos\phi)}{\partial\phi}$$
$$= \frac{6}{R} + \frac{2\cos\theta}{R\sin\theta} - \frac{\sin\phi}{R\sin\theta}.$$

5.3.3 The Divergence Theorem

The divergence theorem, also known as Gauss's theorem, follows directly from the definition of the divergence.

Consider a closed surface S surrounding a volume V (Fig. 5.6). The volume is broken up into many small volumes, Δv_k, each of which has an enclosing surface S_k. For each of these small volumes, from Eq. (5.7),

$$\oint_{S_k} \vec{F}\cdot d\vec{S} \approx (\nabla\cdot\vec{F})_{P_k}\,\Delta v_k,$$

where $(\nabla\cdot\vec{F})_{P_k}$ is the value of the divergence of $\vec{F}$ at some point P_k within the volume Δv_k. We add all the small volumes and surfaces to get

$$\sum_k \oint_{S_k} \vec{F}\cdot d\vec{S} \approx \sum_k (\nabla\cdot\vec{F})_{P_k}\,\Delta v_k.$$

The surfaces of the small volumes are of two kinds. The first is an interior surface, such as $\Delta\vec{S}_{\text{interior}}$ in Fig. 5.7, which is a common wall between two small volumes. This surface will occur twice in the summation, once for volume Δv_i and once for the

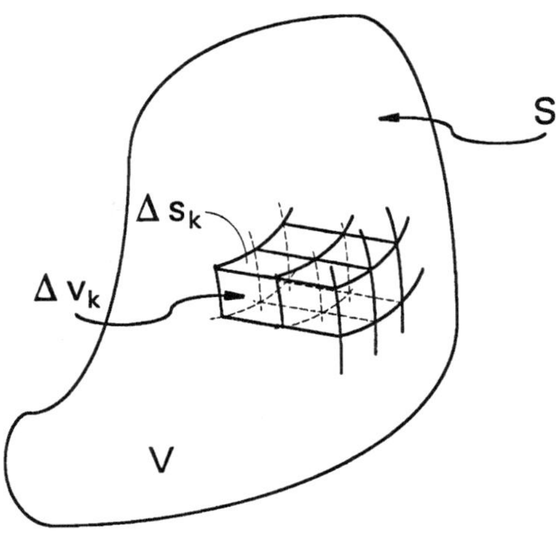

FIGURE 5.6 Closed surface surrounding a volume V, which consists of many small volume elements Δv_k, used in deriving the divergence theorem.

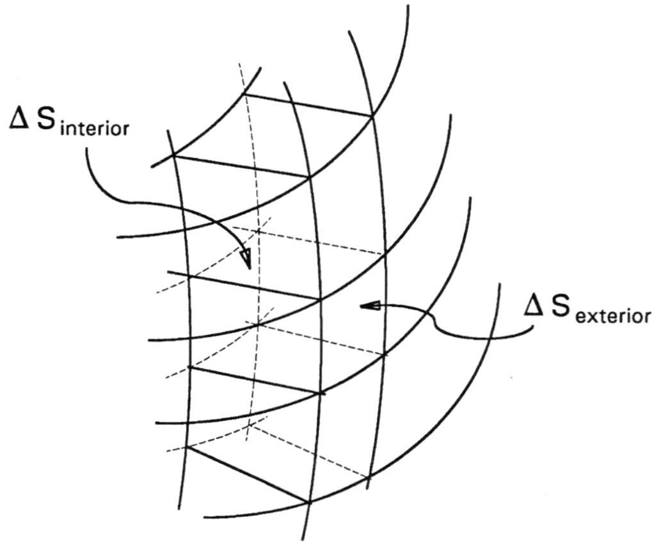

FIGURE 5.7 Details of Fig. 5.6 showing interior and exterior surfaces of the small volumes.

contiguous volume Δv_j. The two occurrences are with opposite sign (if one is positive the other is negative), because the direction of the normal to the surface, always outward from the small volume, reverses. The integrals over all these interior surfaces therefore cancel. The second kind of surface, such as $\Delta\vec{S}_{\text{exterior}}$, occurs only once. Thus the integrals over these surfaces add until the left side of the equation becomes the integral of the normal component of $\vec{F}$ over the entire exterior surface.

Taking the limit as the number of small volumes increases and the size of each small volume decreases, the right side becomes the integral of the divergence of $\vec{F}$ throughout the volume. Thus

$$\oint_S \vec{F}\cdot d\vec{S} = \int_V \nabla\cdot\vec{F}\,dv. \tag{5.12}$$

This is the divergence theorem, which states that the integral of the normal component of a vector field over any closed surface equals the integral of the divergence of that field throughout the volume enclosed by the surface.

Example 5.7 For the field given in Example 5.6, evaluate both sides of the divergence theorem and confirm the equality. Take the surface to be a sphere of radius a centered at the origin.

Solution The integral over the spherical surface $R = a$ is

$$\begin{aligned}\oint_S \vec{F}\cdot d\vec{S} &= \int_{\phi=0}^{2\pi}\int_{\theta=0}^{\pi}(\vec{a}_R 3 + \vec{a}_\theta 2 + \vec{a}_\phi \cos\phi)\cdot\vec{a}_R\, a^2 \sin\theta\, d\theta\, d\phi \\ &= 3a^2\int_0^{\pi}\sin\theta\, d\theta\int_0^{2\pi} d\phi = 3a^2(2)(2\pi) \\ &= 12\pi a^2.\end{aligned}$$

The integral of the divergence throughout the volume is

$$\int_V \nabla\cdot\vec{F}\,dv = \int_{\phi=0}^{2\pi}\int_{\theta=0}^{\pi}\int_{R=0}^{a}\left(\frac{6}{R}+\frac{2\cos\theta}{R\sin\theta}-\frac{\sin\phi}{R\sin\theta}\right)R^2\sin\theta\,dR\,d\theta\,d\phi.$$

The integral of $\cos\theta$ from 0 to π in the second term is zero and the integral of $\sin\phi$ from 0 to 2π in the third term is zero. Continuing with the remaining first term,

$$\int_V \nabla\cdot\vec{F}\,dv = 6\int_0^{2\pi} d\phi\int_0^{\pi}\sin\theta\,d\theta\int_0^a R\,dR = 6(2\pi)(2)(a^2/2)$$
$$= 12\pi a^2.$$

5.3.4 Differential Form of Gauss's Laws

By applying the divergence theorem to Gauss's law for electric flux density we can find the differential or point form of the law.

We repeat Gauss's law for electric flux density, Eq. (4.2), and then apply the divergence theorem to the left side.

$$\oint_S \vec{D}\cdot d\vec{S} = \int_V \rho\,dv,$$
$$\int_V \nabla\cdot\vec{D}\,dv = \int_V \rho\,dv.$$

The first equation is true for any surface and for its enclosed volume. Therefore the second equality holds for any volume and it follows that the integrands must be equal to each other. Thus

$$\nabla\cdot\vec{D} = \rho, \tag{5.13}$$

the divergence of the electric flux density equals the electric charge density at each point in space. This is the differential form of Gauss's law for electric flux density.

Example 5.8 Show that the divergence of the electric flux density found in Section 4.2.1, and repeated as Eq. (5.6), equals the charge density.

Solution For the region $R < a$,

$$\nabla\cdot\vec{D} = \frac{1}{R^2}\frac{\partial[R^2(R\rho_0/3)]}{\partial R} = \frac{\rho_0}{3R^2}\frac{d(R^3)}{dR} = \rho_0,$$

which checks with the given charge distribution for the region.

For the region $R > a$,

$$\nabla\cdot\vec{D} = \frac{1}{R^2}\frac{\partial[R^2(a^3\rho_0/3R^2)]}{\partial R} = \frac{1}{R^2}\frac{d(a^3\rho_0/3)}{dR} = 0,$$

which also checks with the given charge distribution for the region.

The mathematical form of Gauss's law for magnetic flux density, Eq. (4.19), is the same as the form of Gauss's law for electric flux density, except the right side of the equation is zero. Repeating the same steps, we get

$$\nabla \cdot \vec{B} = 0, \tag{5.14}$$

the divergence of the magnetic flux density equals zero at each point in space. This is the differential form of Gauss's law for magnetic flux density.

5.3.5 The Equation of Continuity

The divergence theorem can also be applied to the law of conservation of charge to obtain a differential or point relation between the current density and the charge density.

The law of conservation of charge, Eq. (4.21), can be rewritten for the case of *a* stationary or fixed surface, and hence a constant enclosed volume, as

$$\oint_S \vec{J} \cdot d\vec{S} = \int_V \left(-\frac{\partial \rho}{\partial t} \right) dv. \tag{5.15}$$

By applying the divergence theorem to the left side of this equation and using the same argument to show the equality of the integrands, we obtain

$$\nabla \cdot J = -\frac{\partial \rho}{\partial t}. \tag{5.16}$$

This differential form is known as the equation of continuity. It states that the divergence of the current density equals the time rate of decrease of the charge density at each point in space.

Example 5.9 The charge density at $t = 0$ at a point in a conducting medium is ρ_0. Find the charge density as a function of time. The medium has conductivity σ and permittivity ε.

Solution In the conducting medium $\vec{J} = \sigma \vec{E} = (\sigma/\varepsilon)\vec{D}$. Therefore, using Eq. (5.13), $\nabla \cdot \vec{J} = (\sigma/\varepsilon)\nabla \cdot \vec{D} = (\sigma/\varepsilon)\rho$. Substituting in the equation of continuity we get

$$\frac{\partial \rho}{\partial t} + \frac{\sigma}{\varepsilon}\rho = 0$$

$$\rho = \rho_0 e^{-t/\tau}, \quad \text{where } \tau = \varepsilon/\sigma. \tag{5.17}$$

Note that the time constant τ is the same as the time constant with which charge moves to the surface of the conducting medium (see Section 4.4.1). Thus the charge density at each point in the conducting medium as well as the total charge in the interior of the medium decreases exponentially with this time constant. The charge on the surface increases exponentially at the same rate.

5.3.6 MathCad Evaluation of the Divergence

MathCad can be used to evaluate the divergence of a vector field both numerically as well as symbolically. The divergence can be evaluated in any coordinate system. The process is similar to the evaluation of the gradient. A number of examples in the three coordinate systems are presented.

Example 5.10 Evaluate the divergence of the following vector field numerically at $(x, y, z) = (1, 1, 1)$ and symbolically using MathCad:

$$\vec{F}(x, y, z) = \vec{a}_x 4x^2 y + \vec{a}_y 5xz^4 + \vec{a}_z 2xyz.$$

Solution After defining the unit vectors in rectangular coordinates, the vector field has to be defined as

$$F(x, y, z) := a_x \cdot 4 \cdot x^2 \cdot y + a_y \cdot 5 \cdot x \cdot z^4 + a_z \cdot 2 \cdot x \cdot y \cdot z.$$

The divergence of the vector field is then defined as

$$div_{rec}(F, x, y, z) := \frac{d}{dx}(F(x, y, z) \cdot a_x) + \frac{d}{dy}(F(x, y, z) \cdot a_y) + \frac{d}{dz}(F(x, y, z) \cdot a_z).$$

The divergence can be defined only as a function of the coordinates, but if defined as above, it can be evaluated for any vector field. This only requires the definition of the new vector field. The value of the divergence at the required point is done by typing $div_{rec}(F, 1, 1, 1)$ followed by the equal sign:

$$div_{rec}(F, 1, 1, 1) = 10.$$

In order to evaluate the divergence symbolically, the vector field components have to be explicitly entered in the expression of the divergence:

$$div_{rec}(F, x, y, z) := \frac{d}{dx}(4 \cdot x^2 \cdot y) + \frac{d}{dy}(5 \cdot x \cdot z^4) + \frac{d}{dz}(2 \cdot x \cdot y \cdot z).$$

Following the procedure used in any symbolic evaluation, the result is

$$div_{rec}(F, x, y, z) := 10 \cdot x \cdot y.$$

Example 5.11 Determine the divergence of the following vector field using MathCad:

$$F(r, \phi, z) = \vec{a}_r r^2 z \cos(\phi) + \vec{a}_\phi z^3 \sin(\phi) + \vec{a}_z r^2 \cos(\phi).$$

Evaluate the divergence at the point $(r, \phi, z) = (1, 0, 1)$.

Solution After defining the unit vectors in cylindrical coordinates, the vector field has to be defined as

$$F(r, \phi, z) := a_r \cdot r^2 \cdot z \cdot \cos(\phi) + a_\phi \cdot z^3 \cdot \sin(\phi) + a_z \cdot r^2 \cdot \cos(\phi).$$

The divergence of the vector field is then defined as

$$div_{cyl}(F, r, \phi, z) := \frac{1}{r} \cdot \frac{d}{dr}(r \cdot F(r, \phi, z) \cdot a_r) + \frac{1}{r} \cdot \frac{d}{d\phi}(F(r, \phi, z) \cdot a_\phi) + \frac{d}{dz}(F(r, \phi, z) \cdot a_z).$$

The value of the divergence at the required point is obtained by typing $div_{cyl}(F, 1, 0, 1)$ followed by the equal sign:

$$div_{cyl}(F, 1, 0, 1) = 4.$$

Following the procedure used before, the symbolic expression for the divergence is

$$div_{cyl}(F, r, \phi, z) := 3 \cdot r \cdot z \cdot \cos(\phi) + \frac{1}{r} \cdot z^3 \cdot \cos(\phi).$$

Example 5.12 Determine the divergence of the following vector field using MathCad:

$$F(R, \theta, \phi) = \vec{a}_R \frac{a^4}{R^3} \rho_0.$$

Evaluate the divergence at the point $(R, \theta, \phi) = (1, \pi/4, \pi/4)$.

Solution After defining the unit vectors in cylindrical coordinates, the vector field has to be defined as

$$F(R, \theta, \phi) := a_R \cdot \frac{a^4}{R^3} \cdot \rho_0.$$

The divergence of the vector field is then defined as

$$div_{sph}(F, R, \theta, \phi) := \frac{1}{R^2} \cdot \frac{d}{dR}(R^2 \cdot F(R, \theta, \phi) \cdot a_R)$$

$$+ \frac{1}{R \sin(\theta)} \cdot \frac{d}{d\theta}(\sin(\theta) \cdot F(R, \theta, \phi) \cdot a_\theta) + \frac{1}{R \sin(\theta)} \cdot \frac{d}{d\phi}(F(R, \theta, \phi) \cdot a_\phi).$$

The value of the divergence at the required point is obtained by typing $div_{sph}(F, 1, \pi/4, \pi/4)$ followed by the equal sign:

$$div_{sph}(F, 1, \pi/4, \pi/4) = -1.$$

Following the procedure used before, the symbolic expression for the divergence is

$$div_{sph}(F, R, \theta, \phi) := -\frac{1}{R^4} \cdot a^4 \cdot \rho_0.$$

5.4 LAPLACIAN

The Laplacian of a scalar field is the divergence of the gradient of the scalar field:

$$\text{Laplacian } f(u, v, w, t) = \text{div}[\text{grad } f(u, v, w, t)] \tag{5.18}$$

$$= \nabla \cdot (\nabla f)$$

$$= \nabla^2 f. \tag{5.19}$$

The use of the del operator naturally leads to the notation of ∇^2 to represent the Laplacian. When reading an equation containing $\nabla^2 f$ it is best to say "the Laplacian of *f*," but it is not uncommon or incorrect to say "del squared *f*."

In Cartesian coordinates,

$$\nabla^2 f(x, y, z, t) = \left(\vec{a}_x \frac{\partial}{\partial x} + \vec{a}_y \frac{\partial}{\partial y} + \vec{a}_z \frac{\partial}{\partial z}\right) \cdot \left(\vec{a}_x \frac{\partial f}{\partial x} + \vec{a}_y \frac{\partial f}{\partial y} + \vec{a}_z \frac{\partial f}{\partial z}\right)$$

$$= \frac{\partial^2 f}{\partial x^2} + \frac{\partial^2 f}{\partial y^2} + \frac{\partial^2 f}{\partial z^2}. \tag{5.20}$$

In cylindrical and spherical coordinates, taking the divergence of the gradient establishes the following formulas:

$$\nabla^2 f(r, \phi, z, t) = \frac{1}{r} \frac{\partial}{\partial r}\left(r \frac{\partial f}{\partial r}\right) + \frac{1}{r^2} \frac{\partial^2 f}{\partial \phi^2} + \frac{\partial^2 f}{\partial z^2}, \tag{5.21}$$

$$\nabla^2 f(R, \theta, \phi, t) = \frac{1}{R^2} \frac{\partial}{\partial R}\left(R^2 \frac{\partial f}{\partial R}\right) + \frac{1}{R^2 \sin\theta} \frac{\partial}{\partial \theta}\left(\sin\theta \frac{\partial f}{\partial \theta}\right) + \frac{1}{R^2 \sin^2\theta} \frac{\partial^2 f}{\partial \phi^2}. \tag{5.22}$$

Example 5.13 Evaluate the Laplacian of the following scalar field:

$$f(x, y, z) = \frac{1}{\sqrt{x^2 + y^2 + z^2}}.$$

Solution The Laplacian is evaluated applying the equation for rectangular coordinates:

$$\nabla^2 f(x, y, z) = \frac{d^2}{dx^2} f(x, y, z) + \frac{d^2}{dy^2} f(x, y, z) + \frac{d^2}{dz^2} f(x, y, z).$$

The second derivative of the function with respect to x is evaluated as

$$\frac{d^2}{dx^2}\left(\frac{1}{\sqrt{x^2+y^2+z^2}}\right)=\frac{d}{dx}\left(-\frac{x}{(x^2+y^2+z^2)^{3/2}}\right)$$
$$=-\frac{1}{(x^2+y^2+z^2)^{3/2}}+\frac{3x^2}{(x^2+y^2+z^2)^{5/2}}.$$

Therefore the Laplacian becomes

$$\nabla^2 f(x,y,z)=\frac{2x^2-y^2-z^2}{(x^2+y^2+z^2)^{5/2}}+\frac{2y^2-x^2-z^2}{(x^2+y^2+z^2)^{5/2}}+\frac{2z^2-y^2-x^2}{(x^2+y^2+z^2)^{5/2}}=0.$$

5.5 CURL

The curl of a vector field is a measure of the rotation or circulation of a vector field at a point around an axis that passes through the point. The direction of the axis of rotation gives an orientation to the curl and hence the curl of a vector field is also a vector field.

To appreciate what this means let us look at the field of a current-carrying wire, which was studied in Section 4.5.1. The current density and the magnetic field intensity are repeated below and a top view of the wire is shown in Fig. 5.8.

$$J_z=\begin{cases}\dfrac{I_0}{\pi a^2} \\ 0\end{cases}\qquad H_\phi=\begin{cases}\dfrac{r}{2\pi a^2}I_0, & r<a,\\ \dfrac{I_0}{2\pi r}, & r>a.\end{cases}\tag{5.23}$$

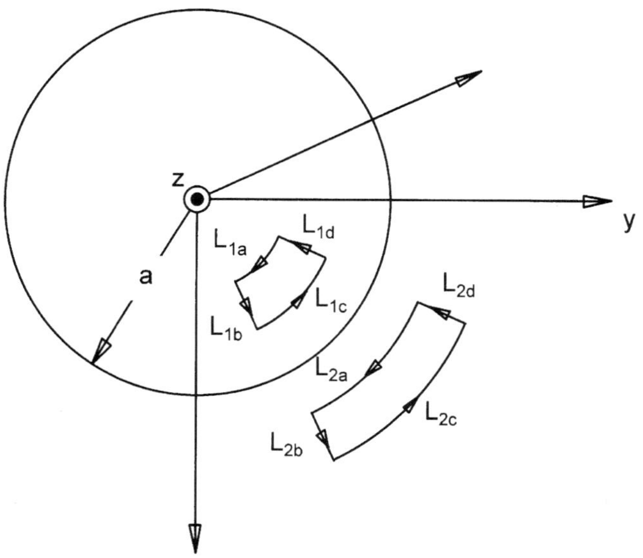

FIGURE 5.8 Sketch of a cross-sectioned current-carrying wire of radius a.

The magnetic field intensity increases in the region of the current density up to the radius of the wire and then decreases where the current density is zero outside the wire. First, examine the line integral $\oint \vec{H} \cdot \vec{dl}$ of the magnetic field intensity around closed loop L_1. Since $H_r = 0$ the line integral over the two ends L_{1b} and L_{1d} are each zero. The line integrals over the sides L_{1a} and L_{1c} are of opposite sign and tend to cancel, but the integral over L_{1c} is larger because on this line segment H_ϕ is slightly greater (proportional to r) and because the line length is also slightly longer (again, proportional to r). As a result, there is a net rotation or circulation of the field around a z-directed axis at any point in the region $r < a$. Second, examine the line integral of the magnetic field intensity around closed loop L_2. The integrals along L_{2b} and L_{2d} are again each equal to zero. The line integrals over L_{2a} and L_{2c} are again of opposite sign. In this region, however, they do cancel because the field over L_{2c} is slightly smaller (proportional to $1/r$) and the line length is slightly longer (proportional to r). As a result, there is no net rotation or circulation of the magnetic field intensity at any point in the region $r > a$.

The Ampère–Maxwell law, Eq. (4.46), shows that net circulation of the magnetic field intensity around a closed curve should equal the total current through a surface capping that curve. The curl at a point ought therefore to equal the total current density at the point on the surface as the surface area shrinks to zero. To arrive at a current density, it is necessary to divide the closed line integral by the capping surface area. This leads directly to the definition of the curl.

The magnitude of the curl of a vector field at a point in space is the maximum ratio of the line integral of the field around a closed curve surrounding the point to the area of the surface capping the line, as the area of the surface and the length of the line shrink to the point. The direction of the curl at a point is the direction normal to the surface when the plane of the line and the surface are oriented so as to produce the maximum magnitude. As always, the direction of the positive normal to the surface is given by the right-hand rule. In mathematical form, if $\vec{F}(u, v, w, t)$ is the vector field and $\vec{G}(u, v, w, t)$ is its curl, then

$$\vec{G} = \text{curl } \vec{F} = \left[\vec{a}_n \lim_{\Delta S \to 0} \frac{\oint_L \vec{F} \cdot \vec{dl}}{\Delta S} \right]_{\max}, \tag{5.24}$$

where $\vec{a}_n$ is in the direction of the normal to the surface when the maximum is achieved.

We must now use this definition to find formulas for the curl in the various coordinate systems.

5.5.1 Cartesian Coordinates and the Del Operator

The component of the curl in any direction is the closed line integral around a curve lying in a plane normal to that direction divided by the capping surface area in that plane. Thus the component in the x direction (see Fig. 5.9a) is

$$G_x = \vec{a}_x \cdot \text{curl } \vec{F} = \lim_{\Delta S_1 \to 0} \left(\frac{\int_{L_a} + \int_{L_b} + \int_{L_c} + \int_{L_d}}{\Delta y \, \Delta z} \right)$$

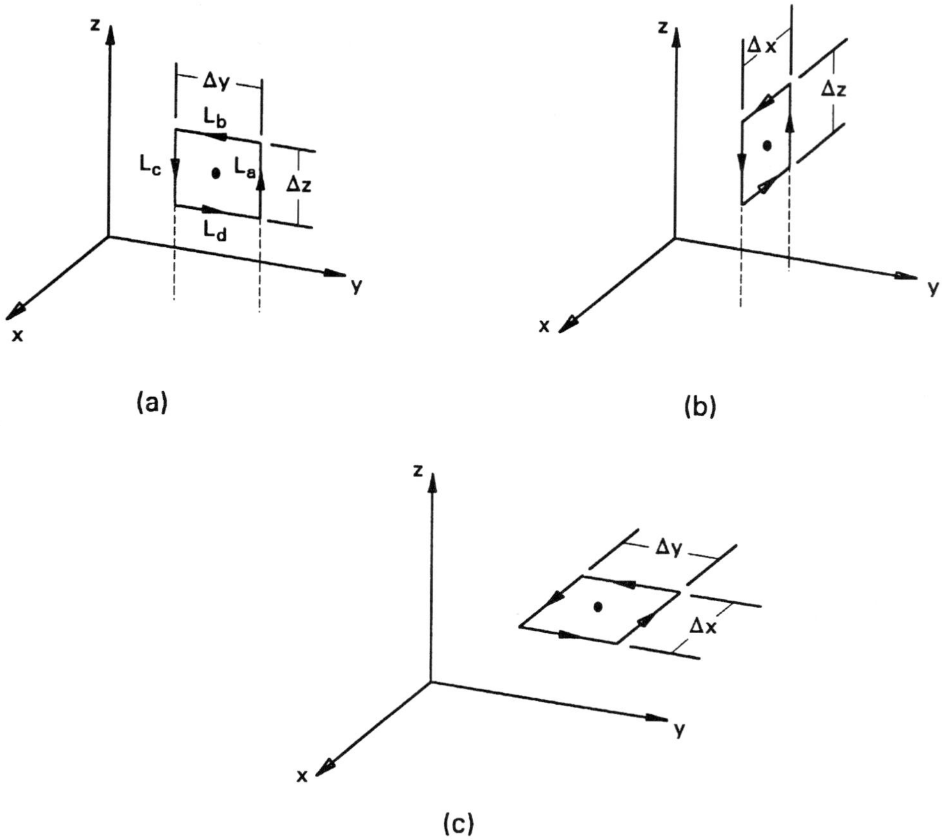

FIGURE 5.9 Closed paths used to calculate the curl operator in Cartesian coordinates.

$$= \lim_{\substack{\Delta y \to 0 \\ \Delta z \to 0}} \left(\frac{\begin{array}{l} F_z(x, y + \Delta y/2, z, t)\, \Delta z \\ \quad - F_y(x, y, z + \Delta z/2, t)\, \Delta y \\ \quad\quad - F_z(x, y - \Delta y/2, z, t)\, \Delta z \\ \quad\quad\quad + F_y(x, y, z - \Delta z/2, t)\, \Delta y \end{array}}{\Delta y\, \Delta z} \right)$$

$$= \lim_{\Delta y \to 0} \left(\frac{F_z(x, y + \Delta y/2, z, t) - F_z(x, y - \Delta y/2, z, t)}{\Delta y} \right)$$

$$- \lim_{\Delta z \to 0} \left(\frac{F_y(x, y, z + \Delta z/2, t) - F_y(x, y, z - \Delta z/2, t)}{\Delta z} \right)$$

$$= \frac{\partial F_z}{\partial y} - \frac{\partial F_y}{\partial z}.$$

This gives the x component of the curl. The derivations for the y and z components are carried out with the aid of Figs. 5.9b and 5.9c. The derivations are identical and the results can be obtained by cyclical replacement of x, y, and z:

$$G_y = \frac{\partial F_x}{\partial z} - \frac{\partial F_z}{\partial x},$$

$$G_z = \frac{\partial F_y}{\partial x} - \frac{\partial F_x}{\partial y}.$$

Putting these results together into a single formula, we obtain

$$\vec{G} = \text{curl } \vec{F} = \vec{a}_x\left(\frac{\partial F_z}{\partial y} - \frac{\partial F_y}{\partial z}\right) + \vec{a}_y\left(\frac{\partial F_x}{\partial z} - \frac{\partial F_z}{\partial x}\right) + \vec{a}_z\left(\frac{\partial F_y}{\partial x} - \frac{\partial F_x}{\partial y}\right), \qquad (5.25)$$

as the formula for the curl in Cartesian coordinates.

This formula can be written in terms of the del operator. Take the vector product of the del operator with the vector field $\vec{F}$,

$$\begin{aligned} \nabla \times \vec{F} &= \begin{vmatrix} \vec{a}_x & \vec{a}_y & \vec{a}_z \\ \dfrac{\partial}{\partial x} & \dfrac{\partial}{\partial y} & \dfrac{\partial}{\partial z} \\ F_x & F_y & F_z \end{vmatrix} \\ &= \vec{a}_x\left(\frac{\partial F_z}{\partial y} - \frac{\partial F_y}{\partial z}\right) + \vec{a}_y\left(\frac{\partial F_x}{\partial z} - \frac{\partial F_z}{\partial x}\right) + \vec{a}_z\left(\frac{\partial F_y}{\partial x} - \frac{\partial F_x}{\partial y}\right) \\ &= \text{curl } \vec{F}. \end{aligned} \qquad (5.26)$$

Hence we can use the notation $\nabla \times \vec{F}$ to mean the curl of $\vec{F}$. The notation is convenient: the cross product reminds us that the curl is an operation on a vector field, which results in a vector field, and the cross product formally works to give us the correct formula in rectangular coordinates. The notation is also used in other coordinate systems, although the correspondence between the formal cross product and the formula for the curl is not so simple. When reading an equation containing $\nabla \times \vec{F}$ it is best to say "the curl of F," although it is not uncommon or incorrect to say "del cross F."

5.5.2 Cylindrical and Spherical Coordinates

To find the formula for the curl in cylindrical coordinates we apply the definition of the curl, Eq. (5.24), to the three curves shown in Fig. 5.10, each centered on the point

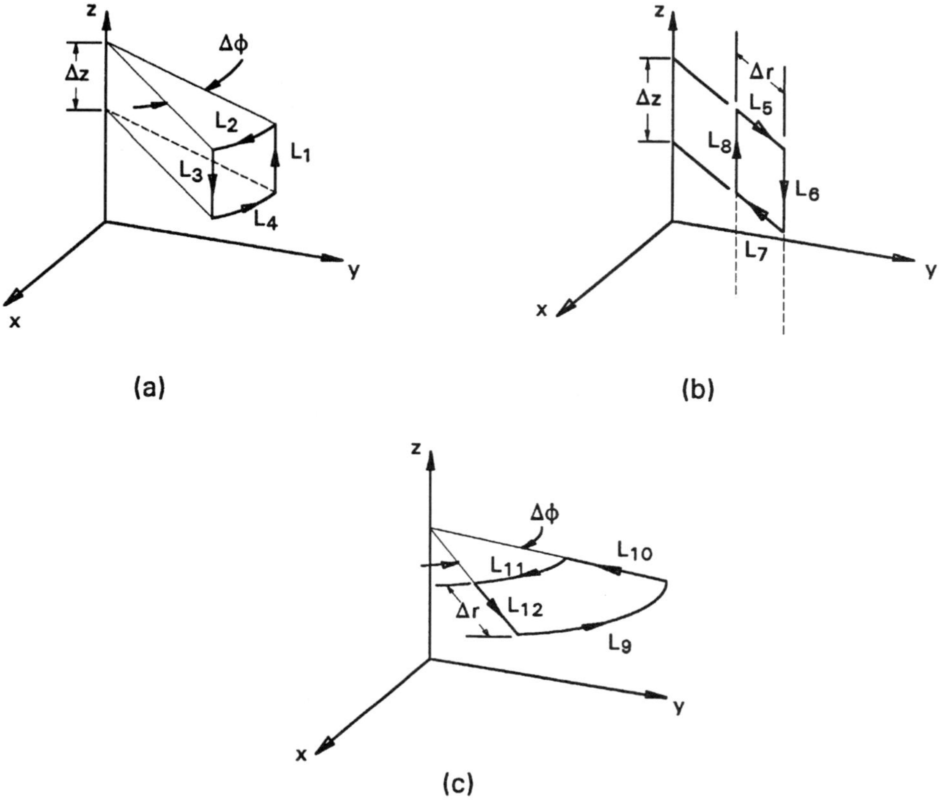

FIGURE 5.10 Closed paths used to calculate the curl operator in cylindrical coordinates.

(r, ϕ, z). Letting $\vec{G} = \nabla \times \vec{F}$, we have from Fig. 5.10a

$$G_r = \lim_{\substack{\Delta\phi \to 0 \\ \Delta z \to 0}} \frac{\int_{L_1} + \int_{L_3} + \int_{L_2} + \int_{L_4}}{r\,\Delta\phi\,\Delta z}$$

$$= \lim_{\substack{\Delta\phi \to 0 \\ \Delta z \to 0}} \left(\frac{\begin{matrix} [F_z(r, \phi + \Delta\phi/2, z, t) - F_z(r, \phi - \Delta\phi/2, z, t)]\,\Delta z \\ -\,[F_\phi(r, \phi, z + \Delta z/2, t) - F_\phi(r, \phi, z - \Delta z/2, t)]r\,\Delta\phi \end{matrix}}{r\,\Delta\phi\,\Delta z} \right)$$

$$= \frac{1}{r}\frac{\partial F_z}{\partial \phi} - \frac{\partial F_\phi}{\partial z}.$$

From Fig. 5.10b, we have

$$G_\phi = \lim_{\substack{\Delta r \to 0 \\ \Delta z \to 0}} \frac{\int_{L_5} + \int_{L_7} + \int_{L_6} + \int_{L_8}}{\Delta r \, \Delta z}$$

$$= \lim_{\substack{\Delta r \to 0 \\ \Delta z \to 0}} \left(\frac{\begin{array}{l}[F_r(r, \phi, z + \Delta z/2, t) - F_r(r, \phi, z - \Delta z/2, t)] \, \Delta r \\ - [F_z(r + \Delta r/2, \phi, z, t) - F_z(r - \Delta r/2, \phi, z, t)] \, \Delta z\end{array}}{\Delta r \, \Delta z} \right)$$

$$= \frac{\partial F_r}{\partial z} - \frac{\partial F_z}{\partial r}.$$

From Fig. 5.10c, we have

$$G_z = \lim_{\substack{\Delta r \to 0 \\ \Delta \phi \to 0}} \frac{\int_{L_9} + \int_{L_{11}} + \int_{L_{10}} + \int_{L_{12}}}{r \, \Delta \phi \, \Delta r}$$

$$= \lim_{\substack{\Delta r \to 0 \\ \Delta \phi \to 0}} \left(\frac{\begin{array}{c}[F_\phi(r + \Delta r/2, \phi, z, t)(r + \Delta r/2) \, \Delta \phi \\ - F_\phi(r - \Delta r/2, \phi, z, t)(r - \Delta r/2) \, \Delta \phi] \\ - [F_r(r, \phi + \Delta \phi/2, z, t) - F_r(r, \phi - \Delta \phi/2, z, t)] \, \Delta r\end{array}}{r \, \Delta \phi \, \Delta r} \right)$$

$$= \frac{1}{r} \left[\frac{\partial (r F_\phi)}{\partial r} - \frac{\partial F_r}{\partial \phi} \right].$$

Putting the three components together we obtain the curl formula in cylindrical coordinates:

$$\nabla \times \vec{F} = \vec{a}_r \left(\frac{1}{r} \frac{\partial F_z}{\partial \phi} - \frac{\partial F_\phi}{\partial z} \right) + \vec{a}_\phi \left(\frac{\partial F_r}{\partial z} - \frac{\partial F_z}{\partial r} \right) + \vec{a}_z \frac{1}{r} \left[\frac{\partial (r F_\phi)}{\partial r} - \frac{\partial F_r}{\partial \phi} \right]. \tag{5.27}$$

Example 5.14 Given that $\vec{F} = \vec{a}_r (z^2/a^2) \sin \omega t + \vec{a}_\phi \cos \omega t$, find $\nabla \times \vec{F}$.

Solution

$$\nabla \times \vec{F} = \vec{a}_\phi \frac{\partial (z^2/a^2) \sin \omega t}{\partial z} + \vec{a}_z \frac{1}{r} \frac{\partial (r \cos \omega t)}{\partial r}$$

$$= \vec{a}_\phi \frac{2z}{a^2} \sin \omega t + \vec{a}_z \frac{1}{r} \cos \omega t.$$

To find the formula for the curl in spherical coordinates we apply the definition of the curl, Eq. (5.23), to the three curves shown in Fig. 5.11, each centered on the point

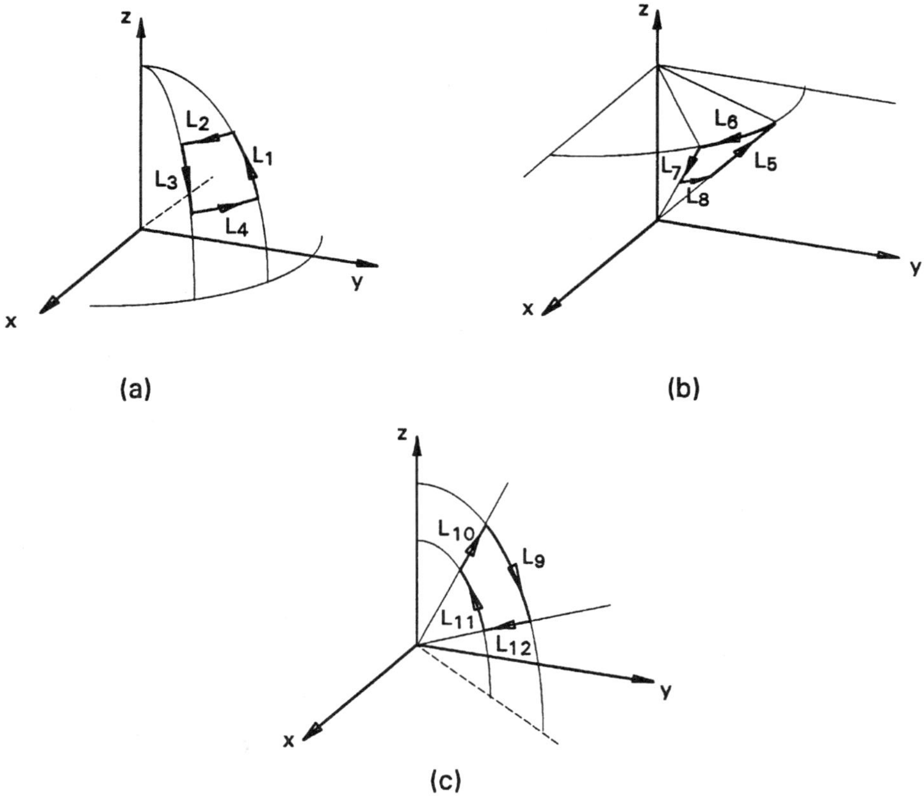

FIGURE 5.11 Closed paths used to calculate the curl operator in spherical coordinates.

(R, θ, ϕ). Letting $\vec{G} = \nabla \times \vec{F}$, we have from Fig. 5.11a

$$G_R = \lim_{\substack{\Delta\theta \to 0 \\ \Delta\phi \to 0}} \frac{\int_{L_1} + \int_{L_3} + \int_{L_2} + \int_{L_4}}{R^2 \sin\theta \, \Delta\theta \, \Delta\phi}$$

$$= \lim_{\substack{\Delta\theta \to 0 \\ \Delta\phi \to 0}} \left(\frac{\begin{array}{c}[F_\phi(R, \theta + \Delta\theta/2, \phi) R \sin(\theta + \Delta\theta/2)\, \Delta\phi \\ - F_\phi(R, \theta - \Delta\theta/2, \phi) R \sin(\theta - \Delta\theta/2)\, \Delta\phi] \\ - [F_\theta(R, \theta, \phi + \Delta\phi/2) - F_\theta(R, \theta, \phi - \Delta\phi/2)]\, R\, \Delta\theta\end{array}}{R^2 \sin\theta \, \Delta\theta \, \Delta\phi} \right)$$

$$= \frac{1}{R \sin\theta} \left[\frac{\partial(F_\phi \sin\theta)}{\partial\theta} - \frac{\partial F_\theta}{\partial\phi} \right].$$

From Fig. 5.11b, we have

$$G_\theta = \lim_{\substack{\Delta R \to 0 \\ \Delta\phi \to 0}} \frac{\int_{L_5} + \int_{L_7} + \int_{L_6} + \int_{L_8}}{R \sin\theta \, \Delta\phi \, \Delta R}$$

$$= \lim_{\substack{\Delta R \to 0 \\ \Delta\phi \to 0}} \left(\frac{\begin{array}{c}[F_R(R, \theta, \phi + \Delta\phi/2) - F_R(R, \theta, \phi - \Delta\phi/2)] \, \Delta R \\ - [F_\phi(R + \Delta R/2, \theta, \phi)(R + \Delta R/2) \sin\theta \\ - F_\phi(R - \Delta R/2, \theta, \phi)(R - \Delta R/2) \sin\theta]\end{array}}{R \sin\theta \, \Delta\phi \, \Delta R} \right)$$

$$= \frac{1}{R}\left[\frac{1}{\sin\theta}\frac{\partial F_R}{\partial\phi} - \frac{\partial(RF_\phi)}{\partial R}\right].$$

From Fig. 5.11c, we have

$$G_\phi = \lim_{\substack{\Delta R \to 0 \\ \Delta\theta \to 0}} \frac{\int_{L_9} + \int_{L_{11}} + \int_{L_{10}} + \int_{L_{12}}}{R \, \Delta R \, \Delta\theta}$$

$$= \lim_{\substack{\Delta R \to 0 \\ \Delta\theta \to 0}} \left(\frac{\begin{array}{c}[F_\theta(R + \Delta R/2, \theta, \phi)(R + \Delta R/2) \, \Delta\theta \\ - F_\theta(R - \Delta R/2, \theta, \phi)(R - \Delta R/2) \, \Delta\theta] \\ - [F_R(R, \theta + \Delta\theta/2, \phi) - F_R(R, \theta - \Delta\theta/2, \phi)] \, \Delta R\end{array}}{R \, \Delta R \, \Delta\theta} \right)$$

$$= \frac{1}{R}\left[\frac{\partial(RF_\theta)}{\partial R} - \frac{\partial F_R}{\partial\theta}\right].$$

Putting the three components together we obtain the curl formula in spherical coordinates:

$$\nabla \times \vec{F} = \vec{a}_R \frac{1}{R\sin\theta}\left[\frac{\partial(\sin\theta \, F_\phi)}{\partial\theta} - \frac{\partial F_\theta}{\partial\phi}\right] + \vec{a}_\theta \frac{1}{R}\left[\frac{1}{\sin\theta}\frac{\partial F_R}{\partial\phi} - \frac{\partial(RF_\phi)}{\partial R}\right] + \vec{a}_\phi \frac{1}{R}\left[\frac{\partial(RF_\theta)}{\partial R} - \frac{\partial F_R}{\partial\theta}\right]. \tag{5.28}$$

Example 5.15 Given that $\vec{F} = -\vec{a}_R 2 + \vec{a}_\theta 3 + \vec{a}_\phi 4$, find $\nabla \times \vec{F}$.

Solution

$$\nabla \times \vec{F} = \vec{a}_R \frac{1}{R\sin\theta}\frac{\partial(4\sin\theta)}{\partial\theta} - \vec{a}_\theta \frac{1}{R}\frac{\partial(4R)}{\partial R} + \vec{a}_\phi \frac{1}{R}\frac{\partial(3R)}{\partial R}$$

$$= \vec{a}_R \frac{4\cos\theta}{R\sin\theta} - \vec{a}_\theta \frac{4}{R} + \vec{a}_\phi \frac{3}{R}.$$

5.5.3 Stokes's Theorem

Stokes's theorem follows directly from the definition of the curl.

Consider a closed curve L capped by any surface S (Fig. 5.12). The surface is broken up into many small surfaces, $\Delta\vec{S}_k$, each of which has a closed boundary L_k. For each of these small surfaces [Eq. (5.24)] we have

$$\oint_{L_k} \vec{F}\cdot dl \approx (\nabla \times \vec{F})_{P_k}\cdot \Delta\vec{S}_k,$$

where $(\nabla\times\vec{F})_{P_k}$ is the curl of $\vec{F}$ at any point P_k on the surface $\Delta\vec{S}_k$. We add all the small surfaces to get

$$\sum_k \oint_{L_k} \vec{F}\cdot d\vec{l} \approx \sum_k (\nabla\times\vec{F})_{P_k}\cdot \Delta\vec{S}_k.$$

The integral over the line segments on the left side of the equation occur twice, except for the portions of the lines that make up the original outside boundary L. The two occurrences are in opposite directions and hence the integral of $\vec{F}\cdot d\vec{l}$ cancels on all the interior portions of L_k. The left side of the equation is therefore equal to the integral over the line L. Taking the limit as the number of small surfaces increases and each small surface area decreases, the right side becomes the integral of the curl of $\vec{F}$ over the entire capping surface. Thus

$$\oint_L \vec{F}\cdot d\vec{l} = \int_S (\nabla\times\vec{F})\cdot d\vec{S}. \tag{5.29}$$

This is Stokes's theorem, which states that the line integral of a vector field around a closed curve is equal to the surface integral of the curl of the field over any surface capping the curve.

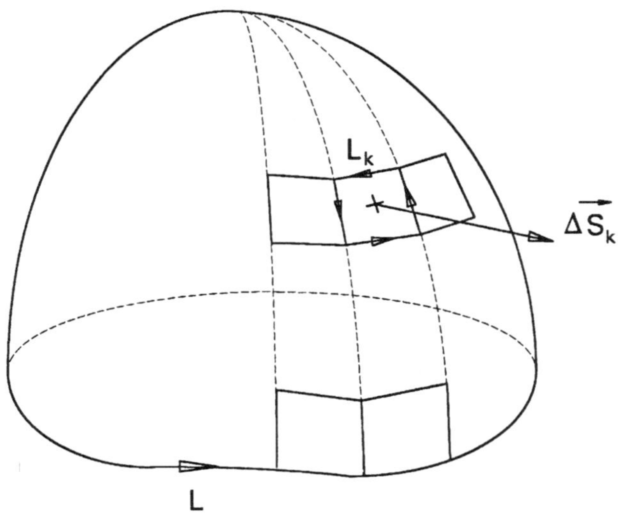

FIGURE 5.12 Closed curve L capped by a surface S, used in deriving Stokes's theorem.

Example 5.16 For the field given in Example 5.15 evaluate both sides of Stokes's theorem and verify the equality. Take the closed curve to be the circle of latitude $R = R_0, \theta = \theta_0$ and the capping surface to be the spherical cap above the circle. Also evaluate for a capping surface consisting of the cone from the origin to the circle of latitude. See Fig. 5.13.

Solution The integral around the closed line is

$$\oint_L \vec{F} \cdot d\vec{l} = \int_{\phi=0}^{2\pi} \vec{F}(R_0, \theta_0, \phi) \cdot (\vec{a}_\phi R_0 \sin\theta_0 \, d\phi)$$
$$= \int_{\phi=0}^{2\pi} 4R_0 \sin\theta_0 \, d\phi$$
$$= 8\pi R_0 \sin\theta_0.$$

The integral of the curl over the spherical cap is

$$\int_{S_1} (\nabla \times \vec{F}) \cdot d\vec{l}_1 = \int_{\theta=0}^{\theta_0} \int_{\phi=0}^{2\pi} (\nabla \times \vec{F})_{R=R_0} \cdot (\vec{a}_R R_0^2 \sin\theta \, d\theta \, d\phi)$$
$$= \int_{\theta=0}^{\theta_0} \int_{\phi=0}^{2\pi} \left(\frac{4\cos\theta}{R_0 \sin\theta}\right) R_0^2 \sin\theta \, d\theta \, d\phi$$
$$= 8\pi R_0 \sin\theta_0.$$

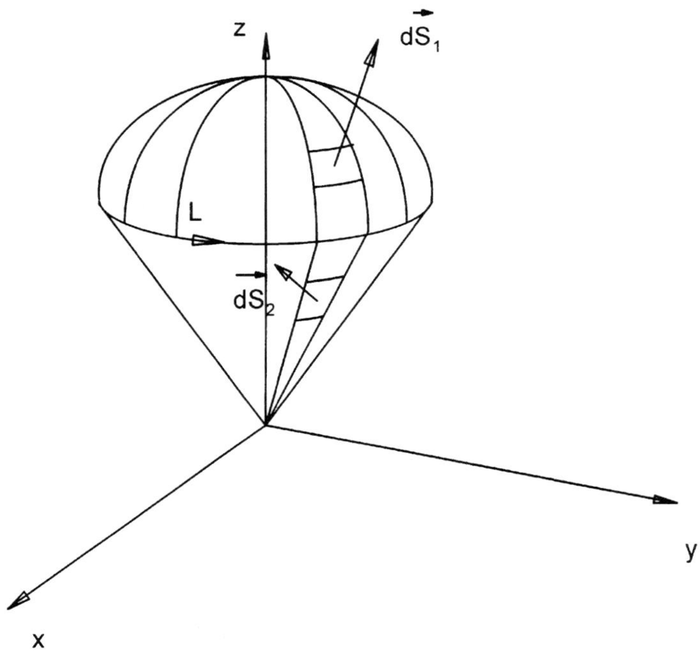

FIGURE 5.13 Spherical cap and capping surface consisting of the cone from the origin to a circle of latitude.

The integral of the curl over the cone is

$$\int_{S_2} (\nabla \times \vec{F}) \cdot d\vec{S}_2 = \int_{\theta=0}^{R_0} \int_{\phi=0}^{2\pi} (\nabla \times \vec{F})_{\theta=\theta_0} \cdot (\vec{a}_\theta R \sin\theta_0 \, d\phi \, dR)$$

$$= \int_{R=0}^{R_0} \int_{\phi=0}^{2\pi} \left(-\frac{4}{R}\right)(-R\sin\theta_0)\, d\phi\, dR$$

$$= 8\pi R_0 \sin\theta_0.$$

Note that the positive normal to the cone, as given by the right-hand rule, is in the $-\vec{a}_\theta$ direction. (If you have difficulty convincing yourself of this, try putting your right-hand fingers around the closed curve L in the wrong direction. Your right thumb then clearly points out through the conical surface in the $\vec{a}_\theta$ direction. With the fingers in the correct direction the normal must be reversed.)

5.5.4 Differential Forms of the Ampère–Maxwell and the Faraday–Maxwell Laws

By applying Stokes's theorem to the integral forms of the Ampère–Maxwell and the Faraday–Maxwell laws we can obtain their differential or point forms.

The Ampère–Maxwell law, Eq. (4.46), is repeated below and then Stokes's theorem is applied to the left side:

$$\oint_L \vec{H} \cdot d\vec{l} = \int_S \left(\vec{J} + \frac{\partial \vec{D}}{\partial t}\right) \cdot d\vec{S},$$

$$\int_S (\nabla \times \vec{H}) \cdot d\vec{S} = \int_S \left(\vec{J} + \frac{\partial \vec{D}}{\partial t}\right) \cdot d\vec{S}.$$

The first equation is true for any line and any capping surface. Therefore the second equality holds for any surface and it then follows that the integrands of the second equation must be equal to each other. Thus

$$\nabla \times \vec{H} = \vec{J} + \frac{\partial \vec{D}}{\partial t}, \tag{5.30}$$

the curl of the magnetic field intensity equals the sum of current density and the displacement current density at each point in space. This is the differential form of the Ampère–Maxwell law.

Example 5.17 Show that the curl of the magnetic fields given in Eqs. (5.23) equals the corresponding current densities.

Solution For $r < a$ and for $r > a$ the magnetic field intensity has only a ϕ component, which is a function of the r coordinate. Hence, using Eq. (5.27),

$\nabla \times \vec{H} = \vec{a}_z(1/r)\partial(rH_\phi)/\partial r$. For $r < a$,

$$\nabla \times \vec{H} = \vec{a}_z \frac{1}{r}\left[\frac{\partial(r^2 I_0/2\pi a^2)}{\partial r}\right] = \vec{a}_z \frac{I_0}{\pi a^2},$$

which checks. For $r > a$,

$$\nabla \times \vec{H} = \vec{a}_z \frac{1}{r}\left[\frac{\partial(I_0/2\pi)}{\partial r}\right] = 0,$$

which also checks.

We can repeat the same derivation for the Faraday–Maxwell law, Eq. (4.70). Thus

$$\oint_L \vec{E}\cdot d\vec{l} = \int_S \left(-\frac{\partial \vec{B}}{\partial t}\right)\cdot d\vec{S}$$

$$\int_S (\nabla \times \vec{E})\cdot d\vec{S} = \int_S \left(-\frac{\partial \vec{B}}{\partial t}\right)\cdot d\vec{S}$$

$$\nabla \times \vec{E} = -\frac{\partial \vec{B}}{\partial t}, \tag{5.31}$$

the curl of the electric field intensity equals the negative time rate of change of the magnetic flux density at each point in space. This is the differential form of the Faraday–Maxwell law.

5.6 VECTOR IDENTITIES

Several vector identities are useful in the vector manipulations encountered in this text. These can be proved by substitution in Cartesian form and obtaining identical results on both sides of the equal sign. Let f and g be any two scalar fields and let $\vec{F}$ and $\vec{G}$ be any two vector fields. Then

$$\nabla(fg) = f\nabla g + g\nabla f, \tag{5.32}$$

$$\nabla\cdot(f\vec{G}) = f\nabla\cdot\vec{G} + \vec{G}\cdot\nabla f, \tag{5.33}$$

$$\nabla\times(f\vec{G}) = f\nabla\times\vec{G} + \nabla f\times\vec{G}, \tag{5.34}$$

$$\nabla\cdot(\vec{F}\times\vec{G}) = \vec{G}\cdot(\nabla\times\vec{F}) - \vec{F}\cdot(\nabla\times\vec{B}), \tag{5.35}$$

$$\nabla\times\nabla f = 0, \tag{5.36}$$

$$\nabla\cdot(\nabla\times\vec{F}) = 0 \tag{5.37}$$

The last two equations are of special interest. Equation (5.36) shows that the curl of the gradient of any scalar field is identically equal to zero. Thus if $\vec{G} = \nabla f$, then from

Stokes's theorem

$$\oint_L \vec{G}\cdot d\vec{l} = \oint_L \nabla f\cdot d\vec{l} = \int_S \nabla\times\nabla f\cdot d\vec{S} = 0.$$

A vector that can be written as the gradient of a scalar has zero curl and its line integral around any closed curve is also zero. The vector field is said to be irrotational. Conversely, if a vector field is irrotational (i.e., its curl is everywhere equal to zero), then that vector field can be written as the gradient of a scalar field. In equation form,

$$\text{If } \nabla\times\vec{G} = 0, \quad \text{then } \vec{G} = \nabla f. \tag{5.38}$$

For a given field $\vec{G}$ the field f is unique except for an additive constant.

Equation (5.37) shows that the divergence of the curl of any vector field is identically equal to zero. Conversely, if the divergence of a vector field is everywhere equal to zero, then that field can be written as the curl of some other vector field. In equation form,

$$\text{If } \nabla\cdot\vec{G} = 0, \quad \text{then } \vec{G} = \nabla\times\vec{F}. \tag{5.39}$$

For a given field $\vec{G}$ the field $\vec{F}$ is not unique: if $\vec{F}' = \vec{F} + \nabla f$, then $\vec{G} = \nabla\times\vec{F} = \nabla\times\vec{F}'$ for any scalar field f.

5.7 VECTOR LAPLACIAN

A vector differential expression that frequently occurs in field problems is

$$\nabla(\nabla\cdot\vec{F}) - \nabla\times(\nabla\times\vec{F}), \tag{5.40}$$

where $\vec{F}$ is any vector field. When the field is written in Cartesian coordinates and these operations are performed, it is found that

$$\nabla(\nabla\cdot\vec{F}) - \nabla\times(\nabla\times\vec{F}) = \vec{a}_x\,\nabla^2 F_x + \vec{a}_y\,\nabla^2 F_y + \vec{a}_z\,\nabla^2 F_z.$$

Note that there are really nine terms on the right side of the equation since

$$\nabla^2 F_x = \frac{\partial^2 F_x}{\partial x^2} + \frac{\partial^2 F_x}{\partial y^2} + \frac{\partial^2 F_x}{\partial z^2}$$

and similarly for $\nabla^2 F_y$ and $\nabla^2 F_z$. Because of the similarity to the scalar Laplacian and for convenience, the operations given in Eq. (5.40) are defined as the vector Laplacian of $\vec{F}$. Thus we define the vector Laplacian as

$$\nabla^2\vec{F} = \nabla(\nabla\cdot\vec{F}) - \nabla\times(\nabla\times\vec{F}). \tag{5.41}$$

In Cartesian coordinates

$$\nabla^2\vec{F} = \vec{a}_x \nabla^2 F_x + \vec{a}_y \nabla^2 F_y + \vec{a}_z \nabla^2 F_z, \tag{5.42}$$

but in other coordinate systems there is no equivalent simple formula. To find the vector Laplacian in cylindrical or spherical coordinates we must calculate $\nabla(\nabla\cdot\vec{F}) - \nabla\times(\nabla\times\vec{F})$, which contains well-defined differential operations in each coordinate system.

5.8 SUMMARY OF MAXWELL'S EQUATIONS

In Chapter 4 we obtained Maxwell's equations in integral form. In this chapter we have derived differential forms of the equations. These are summarized in Table 5.1 for stationary media.

These equations are supplemented by the constitutive equations repeated in Table 5.2.

The equation of continuity can be derived from Maxwell's equations and hence is not an independent relation. Taking the divergence of the Ampère–Maxwell relation and substituting from Gauss' law for electric flux, we obtain

$$\nabla\cdot(\nabla\times\vec{H}) = \nabla\cdot\vec{J} + \frac{\partial\nabla\cdot\vec{D}}{\partial t}$$

$$0 = \nabla\cdot\vec{J} + \frac{\partial\rho}{\partial t},$$

which is the equation of continuity. Note that if the displacement current density term $\partial\vec{D}/\partial t$ were not present in the Ampère–Maxwell law, then the equation of continuity would be contradicted and we would know the Ampère–Maxwell law was wrong.

TABLE 5.1 Maxwell's Equations in Integral and Differential Form

Name	Integral Form	Differential Form	Equation
Faraday–Maxwell law	$\oint_L \vec{E}\cdot d\vec{l} = \int_S \left(-\frac{\partial\vec{B}}{\partial t}\right)\cdot d\vec{S}$	$\nabla\times\vec{E} = -\frac{\partial\vec{B}}{\partial t}$	(4.70) and (5.31)
Ampère–Maxwell law	$\oint_L \vec{H}\cdot d\vec{l} = \int_S \left(\vec{J} + \frac{\partial\vec{D}}{\partial t}\right)\cdot d\vec{S}$	$\nabla\times\vec{H} = \vec{J} + \frac{\partial\vec{D}}{\partial t}$	(4.46) and (5.30)
Gauss's law Electric flux	$\oint_S \vec{D}\cdot d\vec{S} = \int_V \rho\, dv$	$\nabla\cdot\vec{D} = \rho$	(4.2) and (5.13)
Gauss's law Magnetic flux	$\oint_S \vec{B}\cdot d\vec{S} = 0$	$\nabla\cdot\vec{B} = 0$	(4.19) and (5.14)

TABLE 5.2 Constitutive Relations in Linear Isotropic Media

Constitutive Property	General Relation	Free-Space Relation	SI Units
Permittivity	$\vec{D} = \varepsilon\vec{E} = \varepsilon_0\varepsilon_r\vec{E}$	$\vec{D} = \varepsilon_0\vec{E},\ \varepsilon_0 = 8.854\cdots\times 10^{-12} \approx \frac{1}{36\pi}\times 10^{-9}$	F/m
Permeability	$\vec{B} = \mu\vec{H} = \mu_0\mu_r\vec{H}$	$\vec{B} = \mu_0\vec{H},\ \mu_0 = 4\pi\times 10^{-7}$	H/m
Conductivity	$\vec{J} = \sigma\vec{E}$	$\sigma = 0$	S/m

5.9 WAVE EQUATION AND HOMOGENEOUS SOLUTION

We now wish to use Maxwell's equations in differential form to obtain a partial differential equation, called the wave equation. We will then look at a simple special case of this equation and show that the solutions do indeed constitute waves traveling in space. Finally, we will be able to calculate the wave velocity and the wave impedance.

A little manipulation is required in order to obtain the wave equation. In the derivation that follows we assume that μ and ε are constants independent of position and time and that $\sigma = 0$. We start with Eq. (5.31) and take the curl of both sides:

$$\nabla\times(\nabla\times\vec{E}) = -\frac{\partial(\nabla\times\vec{B})}{\partial t}$$

$$\nabla(\nabla\cdot\vec{E}) - \nabla^2\vec{E} = -\frac{\partial}{\partial t}\left[\mu\left(\vec{J} + \frac{\partial\vec{D}}{\partial t}\right)\right],$$

where $\mu\vec{H}$ has been substituted for $\vec{B}$ and the right side of Eq. (5.30) has then been substituted for $\nabla\times\vec{H}$. On the left side we substitute from Eq. (5.12) and on the right we side substitute $\varepsilon\vec{E}$ for $\vec{D}$, to obtain

$$\nabla\left(\frac{1}{\varepsilon}\rho\right) - \nabla^2\vec{E} = -\mu\frac{\partial\vec{J}}{\partial t} - \mu\varepsilon\frac{\partial^2\vec{E}}{\partial t^2}$$

$$\nabla^2\vec{E} - \mu\varepsilon\frac{\partial^2\vec{E}}{\partial t^2} = \mu\frac{\partial\vec{J}}{\partial t} + \frac{1}{\varepsilon}\nabla\rho. \tag{5.43}$$

This is the wave equation for the electric field intensity. The left side involves second-order partial derivatives of $\vec{E}$ with respect to position and time. The right side gives the relation between the electric field intensity and its sources, the current and charge densities.

A similar manipulation starting with $\nabla\times(\nabla\times\vec{H})$ gives the wave equation for the magnetic field intensity:

$$\nabla^2\vec{H} - \mu\varepsilon\frac{\partial^2\vec{H}}{\partial t^2} = -\nabla\times\vec{J}. \tag{5.44}$$

The left side of this second-order partial differential equation is identical in form to the left side of Eq. (5.43). The right side gives the sources of the magnetic field intensity wave.

We will solve a special case of these equations. Before doing that we should first appreciate how compactly the vector notation allows these equations to be written. Each vector equation represents three component equations. In Cartesian coordinates, these are

$$\nabla^2 E_x - \mu\varepsilon \frac{\partial^2 E_x}{\partial^4 t^2} = \mu \frac{\partial J_x}{\partial t} + \frac{1}{\varepsilon}\frac{\partial \rho}{\partial x}, \tag{5.45}$$

$$\nabla^2 E_y - \mu\varepsilon \frac{\partial^2 E_y}{\partial t^2} = \mu \frac{\partial J_y}{\partial t} + \frac{1}{\varepsilon}\frac{\partial \rho}{\partial y}, \tag{5.46}$$

$$\nabla^2 E_z - \mu\varepsilon \frac{\partial^2 E_z}{\partial t^2} = \mu \frac{\partial J_z}{\partial t} + \frac{1}{\varepsilon}\frac{\partial \rho}{\partial z}. \tag{5.47}$$

Even this does not fully expand the equations. The first of these, for example, is

$$\frac{\partial^2 E_x}{\partial x^2} + \frac{\partial^2 E_x}{\partial y^2} + \frac{\partial^2 E_x}{\partial z^2} - \mu\varepsilon \frac{\partial^2 E_x}{\partial t^2} = \mu \frac{\partial J_x}{\partial t} + \frac{1}{\varepsilon}\frac{\partial \rho}{\partial x}, \tag{5.48}$$

where E_x, J_x, and ρ are each functions of x, y, z, and t.

To obtain results that will illustrate the wave nature of the solution without unnecessary complication, we will now make some simplifying assumptions. First, assume that we are seeking a solution in a source-free region of space, $\vec{J} = 0$ and $\rho = 0$. Then the right sides of Eqs. (5.43) through (5.48) are equal to zero and we are left with source-free or homogeneous equations. This is a very important special case. Whenever a wave is launched by a transmitting antenna, it travels in source-free space until it reaches a receiving antenna. Second, we assume that only a single Cartesian component of the field exists and that it is a function of one space coordinate only. Thus, let $\vec{E}(x, y, z, t) = \vec{a}_x E_x(z, t)$. In this case, Eqs. (5.46) and (5.47) are satisfied identically, and Eq. (5.45) or (5.48) becomes

$$\frac{\partial^2 E_x(z,t)}{\partial z^2} - \mu\varepsilon \frac{\partial^2 E_x(z,t)}{\partial t^2} = 0. \tag{5.49}$$

This is a homogeneous second-order partial differential equation in two independent variables z and t.

You are familiar with the homogeneous second-order ordinary differential equation

$$\frac{d^2 G(w)}{dw^2} - k^2 G(w) = 0$$

and know that there are two independent solutions, $G(w) = A \exp(-kw)$ and $G(w) = B \exp(+kw)$. These can be shown to be solutions of the ordinary differential equation by substituting each of them into the differential equation.

For the partial differential equation there are two functional solutions, $E_x(z, t) = f(z - ut)$ and $E_x(z, t) = g(z + ut)$. These will be shown to be solutions of the partial differential equation by substituting each of them into the equation.

Before we do that, however, let us illustrate what is meant by a functional solution by giving some examples.

$$
\begin{aligned}
f_1 &= A \cos[\beta(z - ut)], \\
f_2 &= a_0 + a_1(z - ut) + a_2(z - ut)^2 + a_3(z - ut)^3, \\
f_3 &= B \exp[p(z - ut)^\pi], \\
f_4 &= Cf_1 + Df_2 f_3 .
\end{aligned}
$$

Any function f will do, as long as the independent variables appear together as $(z - ut)$. Similarly, any function g will do, as long as the independent variables appear together as $(z + ut)$.

To show that the function $f(z - ut)$ is a solution of Eq. (5.49), we take the partial derivatives required by that equation. The partial derivative with respect to z is

$$
\frac{\partial E_x}{\partial z} = \frac{df(z - ut)}{d(z - ut)} \frac{\partial(z - ut)}{\partial z} = \frac{df(z - ut)}{d(z - ut)} = \dot{f},
$$

since $\partial(z - ut)/\partial z = 1$. The term $df(z - ut)/d(z - ut)$ is the derivative with respect to the complete argument $(z - ut)$, which we denote by $\dot{f}$. To take the second derivative we continue,

$$
\begin{aligned}
\frac{\partial^2 E_x}{\partial z^2} = \frac{\partial \dot{f}}{\partial z} &= \frac{d\dot{f}}{d(z - ut)} \frac{\partial(z - ut)}{\partial z} = \frac{d\dot{f}}{d(z - ut)} \\
&= \ddot{f},
\end{aligned}
\tag{5.50}
$$

where $\ddot{f}$ denotes the second ordinary derivative with respect to the total argument $(z - ut)$.

Taking the partial derivative with respect to t is similar, except that $\partial(z - ut)/\partial t = -u$. Thus

$$
\begin{aligned}
\frac{\partial E_x}{\partial t} &= \frac{df(z - ut)}{d(z - ut)} \frac{\partial(z - ut)}{\partial t} = -u\dot{f}, \\
\frac{\partial^2 E_x}{\partial t^2} &= -u \frac{\partial \dot{f}}{\partial t} = -u \frac{d\dot{f}}{d(z - ut)} \frac{\partial(z - ut)}{\partial t} \\
&= u^2 \ddot{f}.
\end{aligned}
\tag{5.51}
$$

Substituting Eqs. (5.50) and (5.51) into Eq. (5.49), we find

$$
\begin{aligned}
\ddot{f} - \mu\varepsilon u^2 \ddot{f} &= 0 \\
(1 - \mu\varepsilon u^2)\ddot{f} &= 0.
\end{aligned}
\tag{5.52}
$$

This last equation is satisfied for any $f(z - ut)$ if the coefficient $(1 - \mu\varepsilon u^2) = 0$. With the coefficient set to zero, we obtain

$$u = \frac{1}{\sqrt{\mu\varepsilon}}. \tag{5.53}$$

We have therefore shown that the functional $f(z - ut)$ is a solution to the wave equation, Eq. (5.49), and we have also found the value of u in terms of the constants of the medium in which the wave exists. The dimensions of each term in $(z - ut)$ must be length and the dimensions of u are therefore length per unit time or velocity.

To illustrate that $f(z - ut)$ represents a wave, we plot f as a function of position z with time t as a parameter. Figure 5.14a represents f at some particular time t_1. Figure 5.14b shows f at a time $t_2 = t_1 + \Delta t$, where $\Delta t > 0$. The curve representing f has shifted in the positive z direction by a distance Δz. Corresponding points on the curve at t_1 and t_2, such as A_1 and A_2, represent points where the argument of f has the same value. Hence

$$\begin{aligned} z_1 - ut_1 &= z_2 - ut_2 \\ &= (z_1 + \Delta z) - u(t_1 + \Delta t) \\ 0 &= \Delta z - u\Delta t, \\ u &= \frac{\Delta z}{\Delta t}. \end{aligned} \tag{5.54}$$

As t increases the curve of Fig. 5.14a slides with velocity u in the positive z direction. Thus $f(z - ut)$ represents a wave traveling in the positive z direction with velocity $u = 1/\sqrt{\mu\varepsilon}$.

Using the values of permeability and permittivity for free space (Table 5.2), we calculate the value

$$\begin{aligned} u_0 = c &= \frac{1}{\sqrt{\mu_0\,\varepsilon_0}} \\ c &= 3 \times 10^8 \text{ m/s}, \end{aligned} \tag{5.55}$$

the velocity of light in free space.

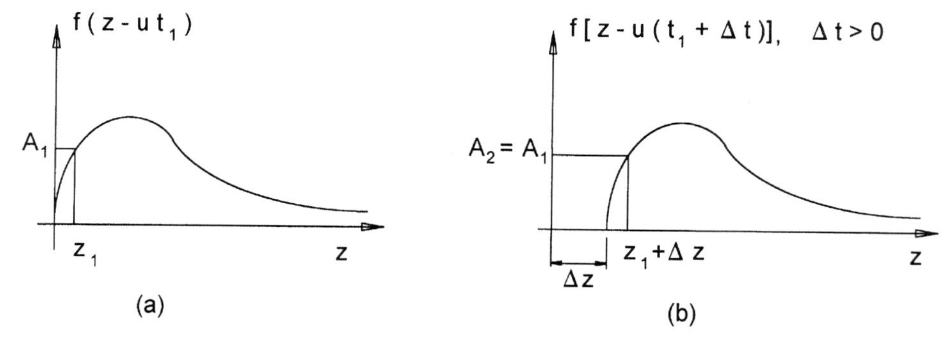

FIGURE 5.14 Plot of a wave $f(z - ut)$ traveling in the positive z direction for two values of time.

There is a corresponding solution for the magnetic field intensity $\vec{H}$. Since $\vec{E}$ has only an x component, which is a function of the z coordinate and time, then Eq. (5.31) becomes

$$\vec{a}_y \frac{\partial E_x}{\partial z} = -\mu \frac{\partial \vec{H}}{\partial t}, \tag{5.56}$$

which shows that $\vec{H}$ has only a y component, which is also a function of the z coordinate and time. Since $E_x = f(z - ut)$, we assume

$$H_y = f(z - ut)/\eta, \tag{5.57}$$

where η is a constant, and substitute in Eq. (5.56) to get

$$\frac{\partial f}{\partial z} = -\mu \frac{\partial (f/\eta)}{\partial t}$$

$$\dot{f} = -\frac{\mu}{\eta}(-u)\dot{f}.$$

This gives us

$$\eta = \mu u = \sqrt{\frac{\mu}{\varepsilon}}. \tag{5.58}$$

The assumption we made for H_y is seen to be correct and the value of the constant η has been found. This constant is the ratio of the positive z traveling electric field (E_x) to the positive z traveling magnetic field (H_y). Its SI unit is the ohm and it is called the intrinsic impedance of the medium. The intrinsic impedance of free space is $\eta_0 = 120\pi \approx 377$ ohms. The term "intrinsic" is used to indicate that this impedance depends only on the properties of the medium.

The second solution $E_x(z, t) = g(z + ut)$ can be shown to satisfy the wave equation, Eq. (5.49), in the same way. The value of u required to satisfy the equation is again $u = 1/\sqrt{\mu\varepsilon}$. As shown in Fig. 5.15, this solution represents a wave traveling in the

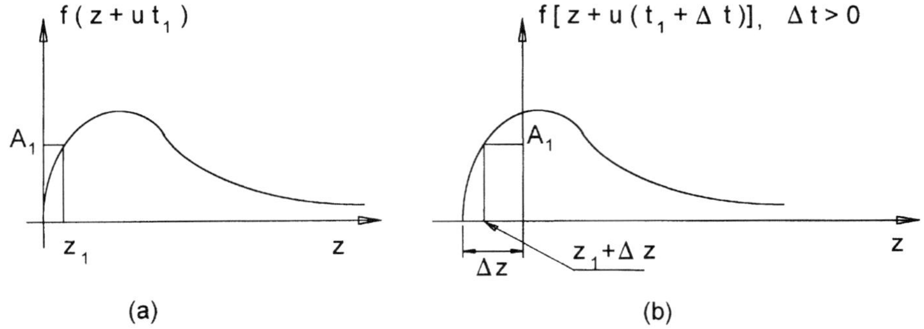

FIGURE 5.15 Plot of a wave $f(z + ut)$ traveling in the negative z direction for two values of time.

negative z direction with velocity u. The corresponding magnetic field intensity is $H_y(z,t) = -g(z+ut)/\eta$.

PROBLEMS

5.1 Find the gradient of:

a. $f_a = 3xy^2z^3$.

b. $f_b = 3rz^2 \cos^2\phi$.

c. $f_c = 4R^3 \cos\theta \sin\phi$.

Ans. a. $\nabla f_a = \vec{a}_x 3y^2z^3 + \vec{a}_y 6xyz^3 + \vec{a}_z 9xy^2z^2$;

b. $\nabla f_b = \vec{a}_r z^2 \cos^2\phi - \vec{a}_\phi 6z^2 \cos\phi \sin\phi + \vec{a}_z 6rz \cos^2\phi$;

c. $\nabla f_c = \vec{a}_R 12R^2 \cos\theta \sin\phi - \vec{a}_\theta 4R^2 \sin\theta \sin\phi + \vec{a}_\phi 4R^2 \cot\theta \cos\phi$.

5.2 Evaluate the gradient of the scalar fields given in the previous problem using MathCad.

5.3 Figure 5.16 shows the value of a two-dimensional scalar electric potential field V in 5 volt steps in any plane parallel to the xy plane. The electric field intensity is given by $\vec{E} = -\nabla V$. Evaluate the electric field intensity at points a, b, and c.

Ans. At a, $\vec{E} \approx -\vec{a}_y 2.5$ V/m; at b, $\vec{E} \approx -(\vec{a}_x 1.4 + \vec{a}_y 1.0)$ V/m; at c, $\vec{E} \approx -(\vec{a}_x 1.3 + \vec{a}_y 3.2)$ V/m.

5.4 Electric field intensity $\vec{E}$ and electric potential V are related by $\vec{E} = -\nabla V$. Find $\vec{E}$ for the following V.

a. $V = q/4\pi\varepsilon_0 R$.

b. $V = qd \cos\theta/4\pi\varepsilon_0 R^2$.

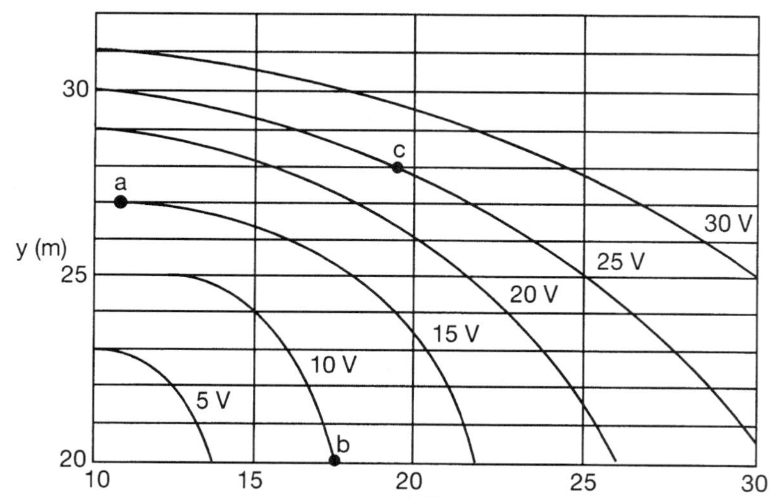

FIGURE 5.16 Plot of a two-dimensional scalar electric potential field.

c. $V = qd^2(3\cos^2\theta - 1)/4\pi\varepsilon_0 R^3$.

d. $V = 3qd^2 \sin^2\theta \cos 2\phi/4\pi\varepsilon_0 R^3$.

Note: These are the fields of (a) a point charge, (b) a dipole, (c) a linear quadrupole (along the z axis), and (d) a square quadrupole (in the xy plane), each located at the origin.

Ans. c. $\vec{E} = (qd^2/4\pi\varepsilon_0 R^4)[\vec{a}_R 3(3\cos^2\theta - 1) + \vec{a}_\theta 6\cos\theta\sin\theta]$;
d. $\vec{E} = (3qd^2/4\pi\varepsilon_0 R^4)(\vec{a}_R 3\sin^2\theta\cos 2\phi - \vec{a}_\theta 2\sin\theta\cos\theta\cos 2\phi + \vec{a}_\phi 2\sin\theta\sin 2\phi)$.

5.5 Electric field intensity $\vec{E}$ and electric potential V are related by $\vec{E} = -\nabla V$. Find $\vec{E}$ for the following V:

a. $V = (\rho_l/2\pi\varepsilon_0)\ln(r_0/r)$.

b. $V = \rho_l d\cos\phi/2\pi\varepsilon_0 r$.

Note: These are the fields of (a) a line charge and (b) a linear dipole, each along the z axis.

Ans. b. $\vec{E} = (\rho_l d/2\pi\varepsilon_0 r^2)(\vec{a}_r\cos\phi + \vec{a}_\phi\sin\phi)$.

5.6 Find the divergence of the following fields:

a. $\vec{F} = \vec{a}_x 3x^2y + \vec{a}_y 4x^2z + \vec{a}_z 4xyz$.

b. $\vec{F} = \vec{a}_r 2r^2z\cos\phi + \vec{a}_\phi z^3\sin\phi + \vec{a}_z 3r^3\cos\phi$.

c. $\vec{F} = (\vec{a}_R\cos\theta + \vec{a}_\theta\sin\theta + \vec{a}_\phi\sin 2\phi)/R^2$.

Ans. a. $\nabla\cdot\vec{F} = 10xy$; b. $\nabla\cdot\vec{F} = z(6r + z^2/r)\cos\phi$; c. $\nabla\cdot\vec{F} = 2(\cos\theta + \cos 2\phi/\sin\theta)/R^3$.

5.7 Evaluate symbolically, using MathCad, the divergence of the fields in the previous problem.

5.8 Electric flux density $\vec{D}$ and electric charge density ρ are related by $\nabla\cdot\vec{D} = \rho$. Find the charge density that is the source of the electric flux density

$$\vec{D} = \begin{cases} \vec{a}_r\rho_0 r/2, & r < a, \\ \vec{a}_r\rho_0 a^2/2r, & r > a. \end{cases}$$

Ans. $\rho = \rho_0$, $r < a$; $\rho = 0$, $r > a$.

5.9 The electric flux density is $\vec{D} = A(\vec{a}_R\sin\theta - \vec{a}_\theta\cos\theta)/R^2$.

a. Find the charge density $\rho = \nabla\cdot\vec{D}$.

b. Using this charge density evaluate Q', the charge within a spherical shell of inner radius R_i and outer radius $R_0 > R_i$ centered at the origin.

c. Use Gauss's law to evaluate $Q(R)$, the charge inside a sphere of radius R centered at the origin.

d. The difference between these results is a point charge q located at the origin. Evaluate q in terms of A.

e. What is the electric flux density $\vec{D}_1$ due to the point charge q? What is the electric flux density $\vec{D}_2$ due to the charge distribution ρ found in part (a)?

Note: The formula $\rho = \nabla \cdot \vec{D}$ cannot easily be used to evaluate charge singularities since the required derivatives of $\vec{D}$ do not exist (are infinite) at the charge singularities.

Ans. a. $\rho = -A \cos(2\theta)/R^3 \sin\theta$; b. $Q' = 0$; c. $Q(R) = A\pi^2$; d. $q = A\pi^2$; e. $\vec{D}_1 = \vec{a}_R A\pi/4R^2$, $\vec{D}_2 = A\{\vec{a}_R[(\sin\theta) - (\pi/4)] - \vec{a}_\theta \cos\theta\}/R^2$.

5.10 The electric flux density is

$$\vec{D} = \vec{a}_R \frac{D_0}{k^2R^2}[1 - (1 + kR)\exp(-kR)].$$

a. Find the charge density $\rho = \nabla \cdot \vec{D}$.

b. D_R appears to be inversely proportional to R^2 as $R \to 0$, implying that a point charge is located at the origin. Show that this is untrue by finding the limit as $R \to 0$ of D_R. *Hint*: Expand the exponential in its Taylor series, $\exp u = 1 + u + u^2/2! + u^3/3! + \cdots$. Be sure to retain a sufficient number of terms.

Ans. a. $\rho = (D_0/R)\exp(-kR)$; b. As $R \to 0$, $D_r \to D_0/2$.

5.11 The magnetic flux density for a current I_0 along the z axis is given by $\vec{B} = \vec{a}_\phi \mu_0 I_0/2\pi r$. Show that $\nabla \cdot \vec{B} = 0$, $r \neq 0$.

5.12 We are given the field $\vec{F} = \vec{a}_r r^2 \cos^2\phi + \vec{a}_\phi rz \cos\phi + \vec{a}_z z^2$. For the closed surface consisting of a cylinder of radius a coaxial with the z axis, the flat top disk at $z = L$ and the bottom disk at $z = 0$:

a. Evaluate $\oint_S \vec{F} \cdot d\vec{S}$.

b. Evaluate $\int_V \nabla \cdot \vec{F}\, dv$ for the enclosed volume.

c. Confirm their equality.

Ans. a and b. $a^3L\pi + a^2L^2\pi$.

5.13 Repeat Problem 5.12 for the wedge-shaped volume $r < a$, $0 < \phi < \phi_0$, $0 < z < L$ and the surface that encloses that volume. Also, show that when $\phi_0 = 2\pi$ the result is the same as that of Problem 5.12.

Ans. a and b. $a^3L[\phi_0(\sin 2\phi_0)/2]/2 - a^2L^2(1 - \cos\phi_0 + 2\phi_0)/4$.

5.14 Find $\nabla \times \vec{F}$ for the following fields:

a. $\vec{F} = \vec{a}_x xy^2z^3 + \vec{a}_y x^3yz^2 + \vec{a}_z x^2y^3z$.

b. $\vec{F} = \vec{a}_r r^2 \cos^2\phi + \vec{a}_\phi rz \cos\phi + \vec{a}_z z^2$.

c. $\vec{F} = \vec{a}_R R^2 \sin^2\theta \cos\phi + \vec{a}_\theta R^2 \sin\phi + \vec{a}_\phi R^2 \cos\theta$.

Ans. a. $\vec{a}_x(3x^2y^2z - 2x^3yz) + \vec{a}_y(3xy^2z^2 - 2xy^3z) + \vec{a}_z(3x^2yz^3 - 2xyz^3)$;
b. $\vec{a}_r(-r\cos\phi) + \vec{a}_z(2rz\cos\phi + 2r^2\cos\phi\sin\phi)/r$;
c. $\vec{a}_R[R^2(\cos^2\theta - \sin^2\theta) - R^2\cos\phi]/R\sin\theta + \vec{a}_\theta(-R^2\sin\theta\sin\phi - 3R^2\cos\theta)/R + \vec{a}_\phi(3R^2\sin\phi - 2R^2\sin\theta\cos\theta\cos\phi)/R$.

5.15 Use the symbolic evaluation feature of MathCad to solve the previous problem.

5.16 Equations (4.60) and (4.67) show that $H_\phi = -I_0(1 + \cos\theta)/4\pi R\sin\theta$. Find $\nabla \times \vec{H}$ and show that it matches with the current density sources, Eqs. (4.35) and

(4.64). Note that H_ϕ is singular along the positive z axis (i.e., $H_\phi \to \infty$ as $\theta \to 0$) and its derivatives do not exist. Thus taking $\nabla \times \vec{H}$ does not easily show the line current $-I_0$ along the positive z axis.
Ans. $\vec{J}$ or $\partial \vec{D}/\partial t = \vec{a}_R I_0/4\pi R^2$.

5.17 Derive Eq. (5.42).

5.18 A wave is excited at angular frequency ω. Show that $E_x(z,t) = E_0 \cos(\omega t - \beta z)$ is a solution to the homogeneous wave equation, Eq. (5.47). Find β in terms of ω, μ, and ε. Writing E_x in the form $E_0 \cos \beta(z - ut)$, show that $\omega/\beta = u = 1/\sqrt{\mu\varepsilon}$.

5.19 Let $E_x = E_0 \cos(\omega t - \beta z)$, where $\omega = 2\pi f$ and $\beta = \omega\sqrt{\mu\varepsilon}$ (see Problem 5.15).

a. For a frequency $f = 100$ MHz, calculate ω.

b. For free space, calculate β.

c. Sketch $E_x(z, t)$ at $t = 0$ for z between -3 and 6 meters.

d. Repeat (c) at $t = 0.25 \times 10^{-8}$.

e. How far (Δz) has the wave moved in time $\Delta t = 0.25 \times 10^{-8}$ s? Calculate $u = \Delta z/\Delta t$.

Ans. a. $\omega = 2\pi \times 10^8\ \text{s}^{-1}$; b. $\beta \equiv 2\pi/3\ \text{m}^{-1}$; e. $u = u_0 = 3 \times 10^8$ m/s.

5.20 **a.** Confirm that $E_x(z, t) = g(z + ut)$ is a solution to the homogeneous wave equation by substituting in Eq. (5.47).

b. Use Eq. (5.54) to show that the corresponding magnetic field intensity is $\vec{H} = -\vec{a}_y g(z + ut)/\eta$, $\eta = \sqrt{\mu/\varepsilon}$.

5.21 **a.** In a conductor $\vec{J} = \sigma\vec{E}$ and $\rho = 0$. Derive the wave equation for $\vec{E}$ and $\vec{H}$ in a conductor.

b. If $\vec{E} = a_x E_x(z, t)$, show that E_x must satisfy the wave equation:

$$\frac{\partial^2 E_x}{\partial z^2} - \mu\sigma \frac{\partial E_x}{\partial t} - \mu\varepsilon \frac{\partial E_x^2}{\partial t^2} = 0.$$

c. Show that $E_x(z, t) = E_0 \exp(-\alpha z) \cos(\omega t - \beta z)$ is a solution and find α and β in terms of ω, μ, ε, and σ.

Ans. a. $\nabla^2 \vec{E} - \mu\sigma \dfrac{\partial \vec{E}}{\partial t} - \mu\varepsilon \dfrac{\partial^2 \vec{E}}{\partial t^2} = 0, \quad \nabla^2 \vec{H} - \mu\sigma \dfrac{\partial \vec{H}}{\partial t} - \mu\varepsilon \dfrac{\partial^2 \vec{H}}{\partial t^2} = 0;$

b. $\alpha = \dfrac{\sigma\sqrt{\mu/\varepsilon}}{\xi}, \quad \beta = (\omega\sqrt{\mu\varepsilon})\xi, \quad \xi = \sqrt{\dfrac{1}{2}\left[1 + \sqrt{1 + \left(\dfrac{\sigma}{\omega\varepsilon}\right)^2}\right]}.$

CHAPTER SIX

Electrostatics

6.1 INTRODUCTION

In this chapter we examine the electric fields when the charge, which is the source of that field, is constant with time and stationary in space and the fields are therefore independent of time. With the fields independent of time, all derivatives with respect to time equal zero ($\partial/\partial t = 0$). Maxwell's equations (see Table 5.1) separate into two independent sets of equations, one for electric field quantities related to stationary charge, and the other for magnetic field quantities related to constant current. It is the first set that we examine in this chapter and that are written below for the electrostatic situation. The integral and differential forms of the equations are

$$\oint_L \vec{E}\cdot d\vec{l} = 0 \qquad \nabla\times\vec{E} = 0, \tag{6.1a,b}$$

$$\oint_S \vec{D}\cdot d\vec{S} = \int_V \rho\, dv \qquad \nabla\cdot\vec{D} = \rho. \tag{6.2a,b}$$

The boundary conditions are repeated as

$$\vec{a}_n \times (\vec{E}_1 - \vec{E}_2) = 0 \quad \text{or} \quad E_{1\ \text{tangential}} = E_{2\ \text{tangential}}, \tag{6.3}$$

$$(\vec{D}_1 - \vec{D}_2)\cdot\vec{a}_n = \rho_s \quad \text{or} \quad D_{1\ \text{normal}} - D_{2\ \text{normal}} = \rho_s. \tag{6.4}$$

The constituent relation is

$$\vec{D} = \varepsilon\vec{E} = \varepsilon_0\varepsilon_r\vec{E}. \tag{6.5}$$

This chapter discusses the solution of electrostatic field problems. When we study the propagation of fields along transmission lines we will find that the electrostatic solution is important for these time-dependent fields as well.

6.2 ELECTRIC FIELD INTENSITY FOR A CHARGE DISTRIBUTION

The electric field of a point charge q located at the origin in free space was found in Section 4.2.1 and is given as Eq. (4.14), repeated here:

$$\vec{E} = \vec{a}_R \frac{q}{4\pi\varepsilon_0 R^2}. \tag{6.6}$$

Since $\vec{a}_R = \vec{R}/R$ this can also be written as

$$\vec{E} = \frac{q\vec{R}}{4\pi\varepsilon_0 R^3}. \tag{6.7}$$

This equation gives the response, the electric field, to an impulsive input, the point charge, located at the origin.

Let us now move the origin of the coordinate system so that the charge is located at some point P′, called the source point. The position vector $\vec{R}'$ gives the location of P′ with respect to the origin (Fig. 6.1). We want to find the field at some point P, called the field point. The position vector $\vec{R}$ gives the location of P with respect to the origin. The difference vector $\vec{R}_d = \vec{R} - \vec{R}'$ gives the location of the field point with respect to the source point. Since the field is not dependent on our choice of the position of the origin, the equation for the electric field will be unchanged, except that $\vec{R}$ should be

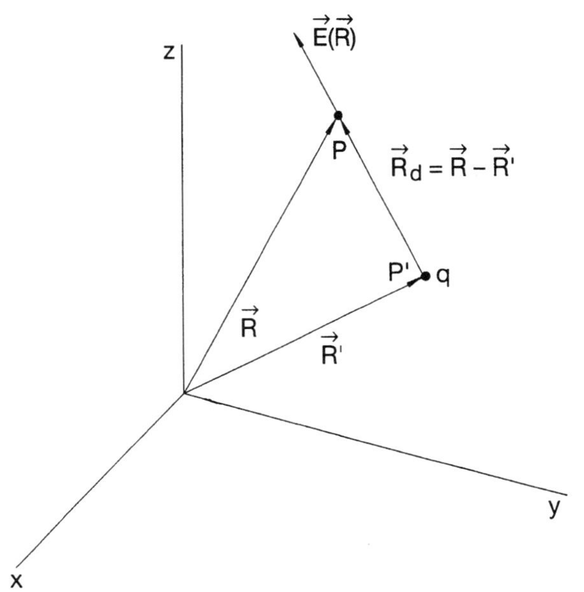

FIGURE 6.1 Electric field ($\vec{E}$) due to a point charge q.

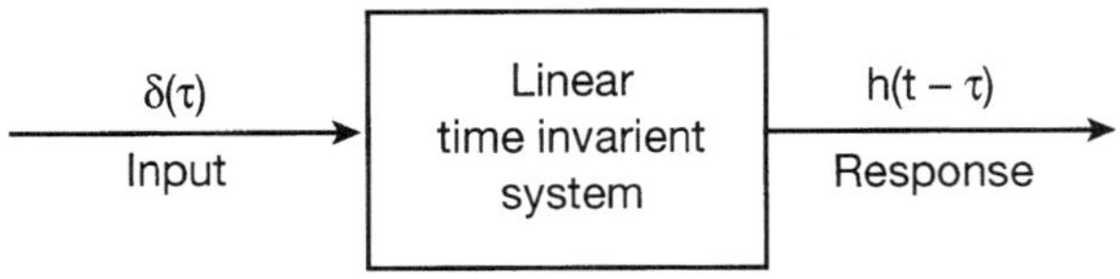

FIGURE 6.2 Linear time invariant system with response to an impulse.

replaced by $\vec{R}_d$. Thus

$$\vec{E}(\vec{R}) = \frac{q\vec{R}_d}{4\pi\varepsilon_0 R_d^3}$$

$$= \frac{q(\vec{R} - \vec{R}')}{4\pi\varepsilon_0 |\vec{R} - \vec{R}'|^3}. \tag{6.8}$$

This last formula gives the response (electric field intensity) at the field point $\vec{R}$ due to an impulsive input (point charge) at the source point $\vec{R}'$.

This function is often called a Green's function. It is similar in concept to the impulse response $h(t - \tau)$ encountered in circuit or system theory courses. The impulse response gives the response (voltage, current, or whatever) at time t due to an impulsive input (of voltage current, or whatever else) applied at time τ (Fig. 6.2). The Green's function gives a response at some position relative to the source position, while the impulse response gives the response at some time relative to the input time. One function is related to response over distance while the other is related to response over time, but the idea of the two functions is the same.

Equation (6.8) is the basic relation for calculating electric field intensity. As we shall see, the relations for any charge distribution are simple variations on this equation.

6.2.1 Multiple Point Charges

We have already noted (Section 3.10.4) that for linear media, which of course includes free space, the principle of superposition will apply. Thus if we have N point sources (Fig. 6.3), $q_1, q_2, \ldots, q_N$, located at the source points $\vec{R}'_1, \vec{R}'_2, \ldots, \vec{R}'_N$, respectively, then the electric field intensity at the field point $\vec{R}$ is given by the sum of the fields due to each charge acting alone. Thus

$$\vec{E}(\vec{R}) = \frac{q_1(\vec{R} - \vec{R}'_1)}{4\pi\varepsilon_0 |\vec{R} - \vec{R}'_1|^3} + \frac{q_2(\vec{R} - \vec{R}'_2)}{4\pi\varepsilon_0 |\vec{R} - \vec{R}'_2|^3} + \cdots + \frac{q_N(\vec{R} - \vec{R}'_N)}{4\pi\varepsilon_0 |\vec{R} - \vec{R}'_N|^3}$$

$$= \sum_{i=1}^{N} \frac{q_i(\vec{R} - \vec{R}'_i)}{4\pi\varepsilon_0 |\vec{R} - \vec{R}'_i|^3}. \tag{6.9}$$

It is important to recognize that the addition indicated by the summation is vector addition.

Example 6.1 A point charge $q_1 = 7\ \mu\text{C}$ is located at (1, 2, 3) m and a second point charge $q_2 = -8\ \mu\text{C}$ is located at (− 4, 5, 6) m. Find the electric field intensity at any point (x, y, z) and evaluate it at the specific point (9, − 10, 11) m.

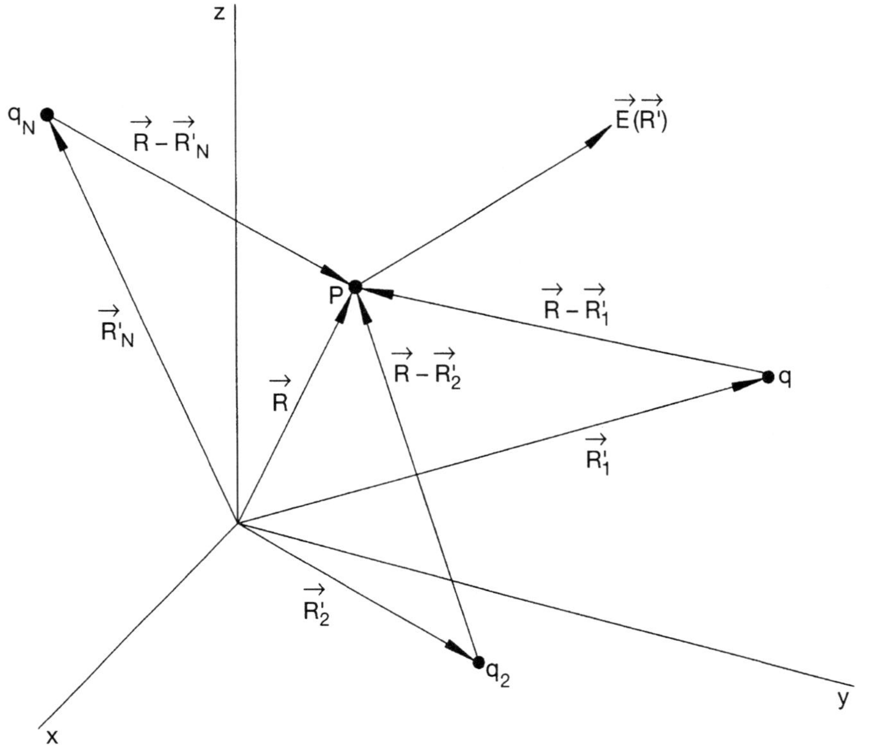

FIGURE 6.3 Electric field due to multiple point charges obtained by superposition.

Solution The two source points are located at $\vec{R}'_1 = \vec{a}_x + \vec{a}_y 2 + \vec{a}_z 3$ and $\vec{R}'_2 = -\vec{a}_x 4 + \vec{a}_y 5 + \vec{a}_z 6$ and the field point is at $\vec{R} = \vec{a}_x x + \vec{a}_y y + \vec{a}_z z$. Then Eq. (6.9) gives

$$\vec{E}(x, y, z) = \frac{10^{-6}}{4\pi\varepsilon_0}\left\{7\frac{\vec{a}_x(x-1) + \vec{a}_y(y-2) + \vec{a}_z(z-3)}{[(x-1)^2 + (y-2)^2 + (z-3)^2]^{3/2}}\right.$$
$$\left. - 8\frac{\vec{a}_x(x+4) + \vec{a}_y(y-5) + \vec{a}_z(z-6)}{[(x+4)^2 + (y-5)^2 + (z-6)^2]^{3/2}}\right\}.$$

To evaluate the field at the given point, substitute 9 for x, -10 for y, and 11 for z. Evaluating the numerator and the denominator of each term, we calculate

$$\vec{E}(9, -10, 11) = \frac{10^{-6}}{4\pi\varepsilon_0}\left[7\frac{(\vec{a}_x 8 - \vec{a}_y 12 + \vec{a}_z 8)}{4486} - 8\frac{(\vec{a}_x 13 - \vec{a}_y 15 + \vec{a}_z 5)}{8577}\right]$$
$$= \frac{10^{-9}}{4\pi\varepsilon_0}(\vec{a}_x 0.358 - \vec{a}_y 4.73 + \vec{a}_z 7.82)$$
$$= \vec{a}_x 3.22 - \vec{a}_y 42.6 + \vec{a}_z 70.4 \text{ V/m}.$$

6.2.2 Continuous Charge Distributions

We will now obtain formulas for the electric field intensity due to volume, surface, and line charge densities. In each case, our starting point is Eq. (6.8), but with q replaced by an infinitesimal charge dQ' located at $\vec{R}'$ and with the resultant field $\vec{E}$ replaced by $d\vec{E}$, the infinitesimal field produced by dQ'. Thus we write

$$d\vec{E} = \frac{dQ'(\vec{R} - \vec{R}')}{4\pi\varepsilon_0 |\vec{R} - \vec{R}'|^3}. \tag{6.10}$$

Let us now assume that there is a volume V in which there is a volume charge density ρ (Fig. 6.4). To find the electric field intensity due to the charge distribution, we first find the field due to the charge dQ' in a small element of volume dv' located at the source point at $\vec{R}'$. The charge located in this element of volume is $dQ' = \rho(\vec{R}')\,dv'$. Since the volume is very small this charge appears to be a point charge and the electric field intensity at the field point located at $\vec{R}$ is

$$d\vec{E}(\vec{R}) = \frac{\rho(\vec{R}')\,dv'\,(\vec{R} - \vec{R}')}{4\pi\varepsilon_0 |\vec{R} - \vec{R}'|^3}.$$

To find the total electric field intensity we use the principle of superposition and sum up or, in this case, integrate over the volume charge distribution:

$$\vec{E}(\vec{R}) = \int_V \frac{\rho(\vec{R}')(\vec{R} - \vec{R}')\,dv'}{4\pi\varepsilon_0 |\vec{R} - \vec{R}'|^3}. \tag{6.11}$$

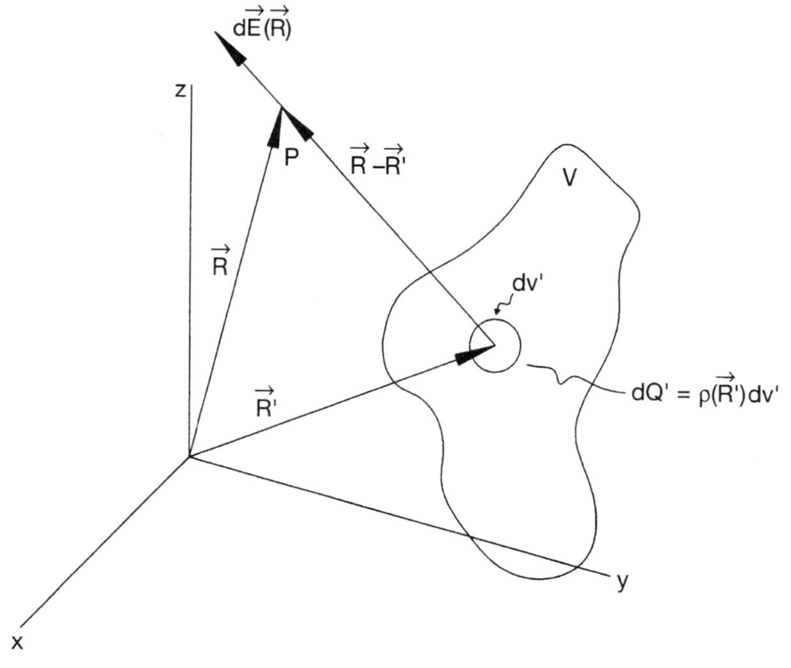

FIGURE 6.4 Electric field due to a volume charge density.

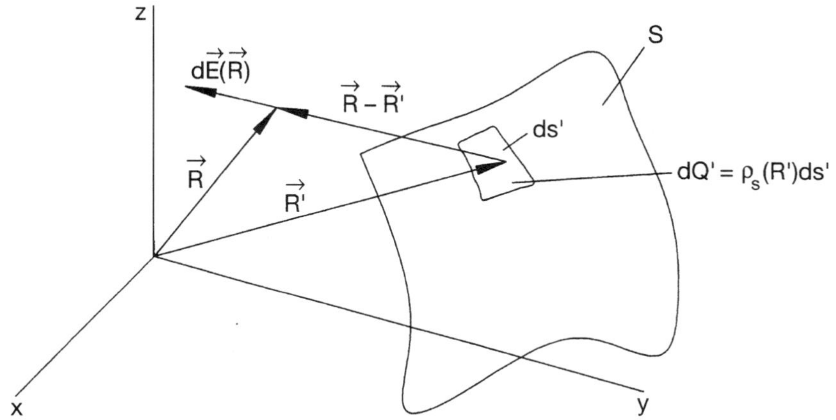

FIGURE 6.5 Electric field due to a surface charge density.

For a surface charge distribution ρ_s located on the surface S, we proceed in the same fashion. The electric field intensity at the field point $\vec{R}$ due to the element of charge $dQ' = \rho_s(\vec{R}')\,dS'$ located at the source point $\vec{R}'$ (Fig. 6.5) is

$$d\vec{E}(\vec{R}) = \frac{\rho_s(\vec{R}')\,dS'\,(\vec{R}-\vec{R}')}{4\pi\varepsilon_0|\vec{R}-\vec{R}'|^3},$$

and the total field due to the entire surface charge distribution is

$$\vec{E}(\vec{R}) = \int_S \frac{\rho_s(\vec{R}')(\vec{R}-\vec{R}')\,dS'}{4\pi\varepsilon_0|\vec{R}-\vec{R}'|^3}. \tag{6.12}$$

For a line charge distribution ρ_l located on the line L, we again proceed in the same fashion. The electric field intensity at the field point $\vec{R}$ due to the element of charge $dQ' = \rho_l(\vec{R}')\,dl'$ located at the source point $\vec{R}'$ (Fig. 6.6) is

$$d\vec{E}(\vec{R}) = \frac{\rho_l(\vec{R}')\,dl'\,(\vec{R}-\vec{R}')}{4\pi\varepsilon_0|\vec{R}-\vec{R}'|^3}, \tag{6.13}$$

and the total field due to the entire line charge distribution is

$$\vec{E}(\vec{R}) = \int_L \frac{\rho_l(\vec{R}')(\vec{R}-\vec{R}')\,dl'}{4\pi\varepsilon_0|\vec{R}-\vec{R}'|^3}. \tag{6.14}$$

Finally, if all four types of sources—point charges and volume, surface, and line charge distributions—are simultaneously present, then we simply add up the electric fields due to each source taken separately.

It will seem as if there are a lot of formulas to memorize, unless you recognize that Eqs. (6.9), (6.11), (6.12), and (6.14) are just special forms of Eq. (6.8), with the

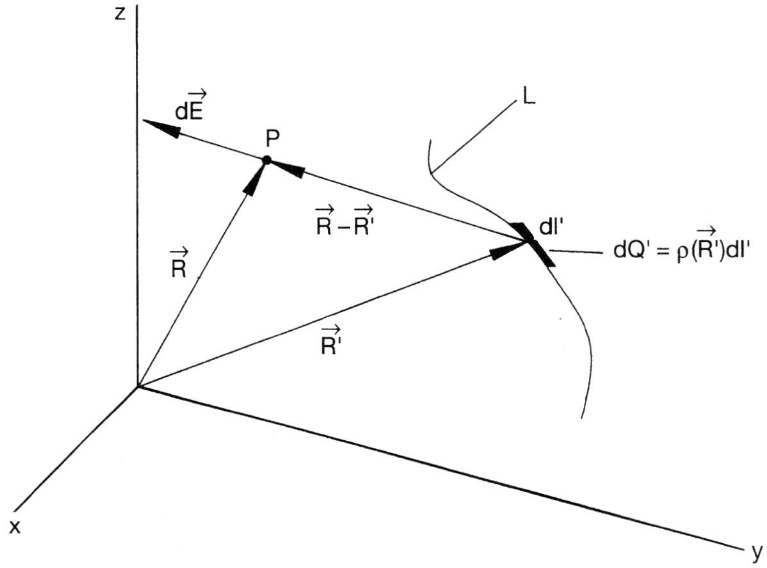

FIGURE 6.6 Electric field due to a line charge density.

equivalences (Section 3.2.5) $dQ' \Leftrightarrow q_i \Leftrightarrow \rho(\vec{R}')\,dv' \Leftrightarrow \rho_s(\vec{R}')\,dS \Leftrightarrow \rho_l(\vec{R}')\,dl'$ and a summation or integration over the location of the charge distribution.

Example 6.2 The charge density in the rectangular parallelepiped region $1 < x < 2$, $3 < y < 4, 5 < z < 6$ is given by $11x(z+1)\ \mu\text{C/m}^3$. Give an integral expression for the electric field intensity at the point $(-7, 8, -9)$ m.

Solution In this example, the source point is located at $\vec{R}' = \vec{a}_x x' + \vec{a}_y y' + \vec{a}_z z'$, the element of charge is $dQ' = 11 \times 10^{-6} x'(z'+1)\,dx'\,dy'\,dz'$, and the field point is at $\vec{R} = -\vec{a}_x 7 + \vec{a}_y 8 - \vec{a}_z 9$. Substituting in Eq. (6.11) we have

$$\vec{E} = \int_{z'=5}^{6} \int_{y'=3}^{4} \int_{x'=1}^{2} \times \frac{11 \times 10^{-6} x'(z'+1)[\vec{a}_x(-7-x') + \vec{a}_y(8-y') + \vec{a}_z(-9-z')]\,dx'\,dy'\,dz'}{4\pi\varepsilon_0 [(-7-x')^2 + (8-y')^2 + (-9-z')^2]^{3/2}}.$$

The integrals cannot be performed analytically but could be evaluated by numerical integration to whatever accuracy is desired.

Example 6.3 Repeat Example 6.2 by using MathCad.

Solution The electric field can be evaluated using MathCad by determining the individual components of the field. The software package does not evaluate the integral of a vector. The x, y, and z components of the electric field are evaluated as follows:

$$E_x(x, y, z) := \int_5^6 \int_3^4 \int_1^2 \frac{\rho_v(x', y', z)(R(x, y, z) - R'(x', y', z')) \cdot a_x}{4\pi\varepsilon_0 |R(x, y, z) - R'(x', y', z')|^3} dx'\, dy'\, dz',$$

$$E_y(x, y, z) := \int_5^6 \int_3^4 \int_1^2 \frac{\rho_v(x', y', z)(R(x, y, z) - R'(x', y', z')) \cdot a_y}{4\pi\varepsilon_0 |R(x, y, z) - R'(x', y', z')|^3} dx'\, dy'\, dz',$$

$$E_z(x, y, z) := \int_5^6 \int_3^4 \int_1^2 \frac{\rho_v(x', y', z)(R(x, y, z) - R'(x', y', z')) \cdot a_z}{4\pi\varepsilon_0 |R(x, y, z) - R'(x', y', z')|^3} dx'\, dy'\, dz',$$

where the volume charge density has to be previously defined as $\rho_v(x, y, z) := 11 \cdot 10^{-6} x(z + 1)$ C/m^3. The position vectors $R'(x', y', z') := a_x x' + a_y y' + a_z z'$ and $R(x, y, z) := a_x x + a_y y + a_z z$ will also have to be defined, as well as the three position vectors:

$$a_x := \begin{pmatrix} 1 \\ 0 \\ 0 \end{pmatrix} \qquad a_y := \begin{pmatrix} 0 \\ 1 \\ 0 \end{pmatrix} \qquad a_z := \begin{pmatrix} 0 \\ 0 \\ 1 \end{pmatrix}.$$

The results of the integration for $x = -7$, $y = 8$, and $z = -9$ can be viewed as

$$E_x(-7, 8, -9) = -1.558 \cdot 10^3,$$

$$E_y(-7, 8, -9) = 819.2,$$

$$E_z(-7, 8, -9) = -2.642 \cdot 10^3.$$

Example 6.4 A semicircular surface of charge is located as shown in Fig. 6.7 in the xy plane. The surface charge density is

$$\rho_s = A \sin\phi \text{ C/m}^2, \quad 0 \le r \le a, \quad \le \phi \le \pi.$$

Find the electric field intensity at any point on the z axis.

Solution The field point is at $\vec{R} = \vec{a}_z z$ and the source point is at $\vec{R}' = \vec{a}_{r'} r' = r'(\vec{a}_x \cos\phi' + \vec{a}_y \sin\phi')$. The element of charge is $dQ' = A(\sin\phi')r'\, d\phi'\, dr'$. Substituting in Eq. (6.12) we find

$$\vec{E}(0, 0, z) = \int_{r'=0}^{a} \int_{\phi'=0}^{\pi} \frac{A \sin\phi' [-\vec{a}_x r' \cos\phi' - \vec{a}_y r' \sin\phi' + \vec{a}_z z] r'\, d\phi'\, dr'}{4\pi\varepsilon_0 [(r')^2 + z^2]^{3/2}}.$$

When we integrate over ϕ' the x-directed term is zero because the integral of $\sin\phi' \cos\phi'\, d\phi'$ from $\phi' = 0$ to π is zero. This can be anticipated by vectorially adding the field due to the equal elements of surface charge located at $(r', \phi', 0)$ and

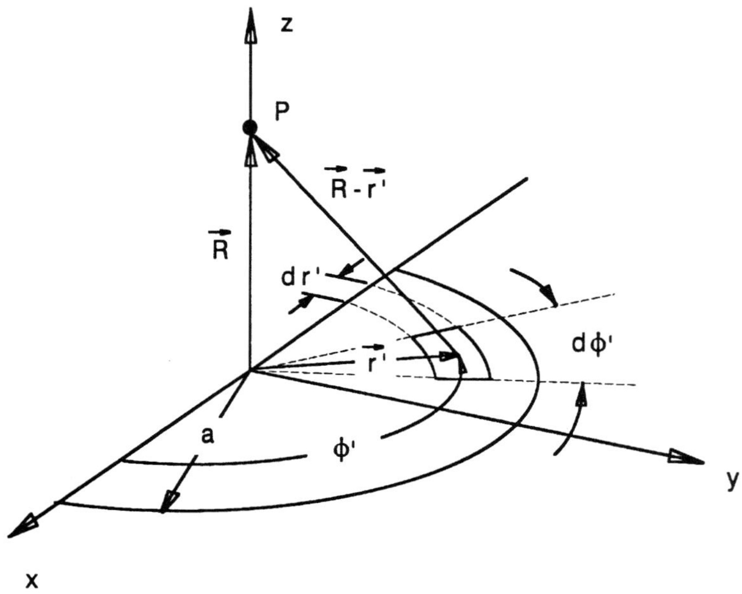

FIGURE 6.7 Semicircular surface of charge located in the xy plane.

$(r', \pi - \phi', 0)$. The remaining components are

$$\vec{E}(0, 0, z) = \frac{A}{4\pi\varepsilon_0}\left\{ -\vec{a}_y \frac{\pi}{2}\int_0^a \frac{(r')^2\, dr'}{[(r')^2 + z^2]^{3/2}} + \vec{a}_z 2z \int_0^a \frac{r'\, dr'}{[(r')^2 + z^2]^{3/2}} \right\}$$

$$= \frac{A}{4\pi\varepsilon_0}\left\{ -\vec{a}_y \frac{\pi}{2}\left[-\frac{a}{\sqrt{a^2 + z^2}} + \ln\left(\frac{a + \sqrt{a^2 + z^2}}{z}\right)\right]\right.$$

$$\left. + \vec{a}_z 2z\left[\frac{1}{z} - \frac{1}{\sqrt{a^2 + z^2}}\right]\right\}.$$

6.3 ELECTRIC POTENTIAL AND WORK IN THE ELECTROSTATIC FIELD

The electromotive force (emf) or voltage was defined in Section 3.5 as the line integral of the electric field intensity along a particular path. In the case of electrostatics, the line integral between two points is easily shown to be independent of the path chosen. Referring to Fig. 6.8 and using Eq. (6.1a), we find

$$\int_{L_1} \vec{E} \cdot d\vec{l} - \int_{L_2} \vec{E} \cdot d\vec{l} = 0$$

$$\int_{L_1} \vec{E} \cdot d\vec{l} = \int_{L_2} \vec{E} \cdot d\vec{l}. \tag{6.15}$$

Since L_1 and L_2 are any two paths, the line integral is independent of the path.

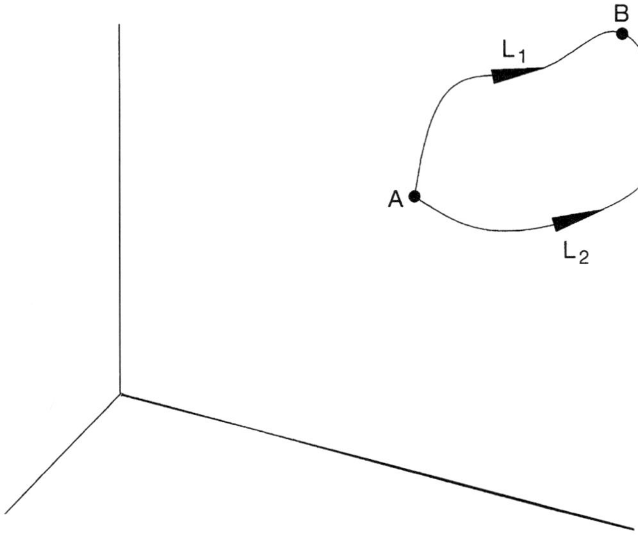

FIGURE 6.8 Two paths connecting points A and B that are used to calculate electrostatic potential and work.

Because the emf or voltage between two points is independent of the path, the voltage at point A with respect to point B is called the electric potential difference between the points and the path of integration is not specified. Hence the potential difference V_{AB} is

$$V_{AB} = \int_A^B \vec{E} \cdot d\vec{l}.$$

This may also be called the potential of point A with respect to point B. We can arbitrarily choose one point, say B, as a reference point and then the potential difference V_{AB} is called the electric potential V_A with respect to the reference point or simply the electric potential V of point A. The reference point is usually chosen at infinity when the charges are confined to a finite region of space. In that case,

$$V(\vec{R}) = \int_A^\infty \vec{E} \cdot d\vec{l} = V_A. \tag{6.16}$$

The inverse operation gives the electric field intensity in terms of the electric potential. Consider the point A in Fig. 6.9 and a point P a distance $d\vec{l}$ away. If the component of electric field intensity in the direction of $d\vec{l}$ is E_l, then the potential difference $V_{AP} = -dV = E_l\, dl$, where dV is the potential at P with respect to the potential at A. Therefore

$$E_l = -\frac{dV}{dl},$$

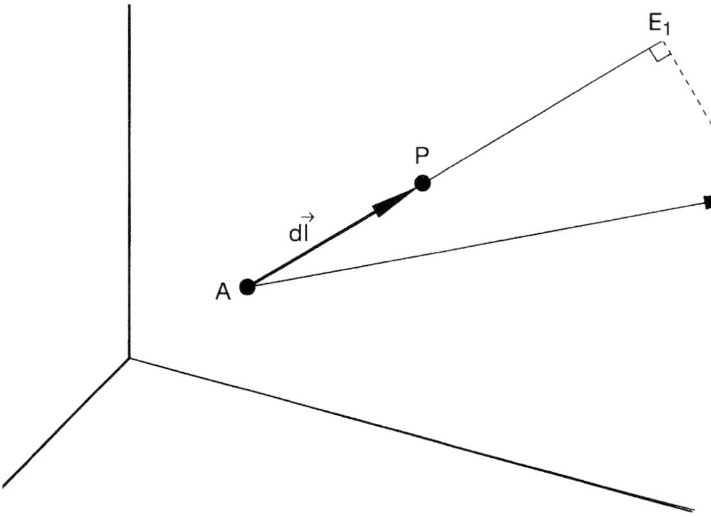

FIGURE 6.9 Calculation of electric field intensity using the gradient of the electrostatic potential.

which is the component of the negative gradient of V in the direction of $\vec{dl}$. The complete electric field intensity at A is therefore equal to the negative gradient of the electric potential,

$$\vec{E} = -\nabla V. \tag{6.17}$$

The electric potential difference has an important physical significance. The work done by the field as a small test charge q_t is moved slowly (the test charge must be small so as not to change the original electric field intensity; the motion must be slow so as to maintain the electrostatic nature of the fields) from point A to point B (Fig. 6.10) in an electrostatic field $\vec{E}$ is given by the component of the electric field force $\vec{F}_e = q_t\vec{E}$ in the direction of motion, integrated over the path:

$$W_{\mathrm{AB}} = \int_{\mathrm{A}}^{\mathrm{B}} (q_t\vec{E}) \cdot \vec{dl}.$$

We divide by the test charge q_t to get the electric potential difference V_{AB},

$$V_{\mathrm{AB}} = \frac{W_{\mathrm{AB}}}{q_t}, \tag{6.18}$$

which we now see is the work done by the field per unit test charge. The mechanical work done per unit charge is $-V_{\mathrm{AB}}$ and this is the increase in potential energy per unit charge moved from A to B. This potential energy is stored in the electric field. When the mechanical work input in moving the charge is positive, the energy stored in

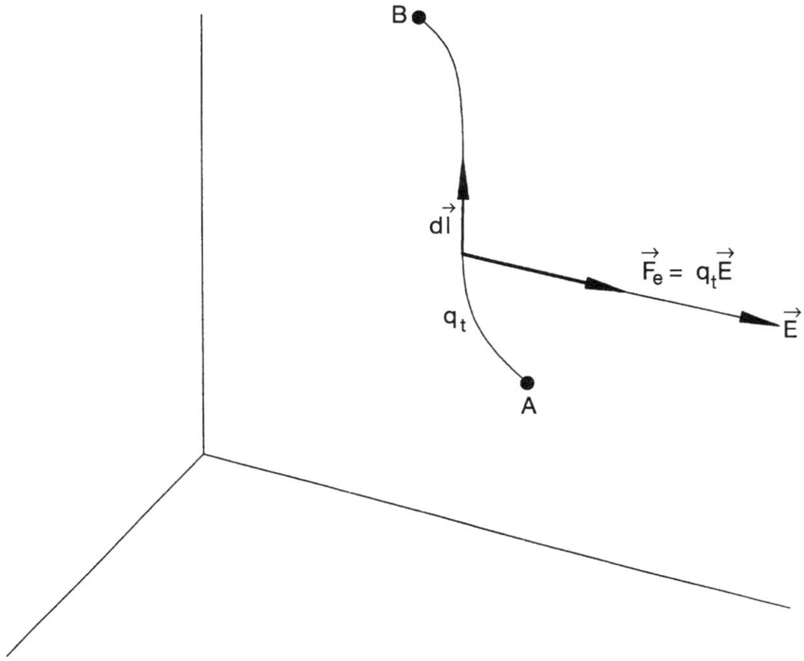

FIGURE 6.10 The force acting on a point charge, which is slowly moved from point A to point B in an electrostatic field.

the electric field is increased. When the mechanical work input in moving the charge is negative, the energy stored in the electric field is decreased.

6.3.1 Electric Potential of a Point Charge

For a point charge q located at the origin the electric potential at any field point $\vec{R}$ (Fig. 6.11a), with respect to a reference point at infinity, is easily obtained from the electric field intensity. Using Eq. (6.6) and integrating along a radial line, we obtain

$$\begin{aligned} V(\vec{R}) &= \int_R^\infty \left(\vec{a}_R \frac{q}{4\pi\varepsilon_0 R^2}\right)\cdot(\vec{a}_R\, dR) \\ &= \frac{q}{4\pi\varepsilon_0}\left(-\frac{1}{R}\right)_R^\infty \\ &= \frac{q}{4\pi\varepsilon_0 R}. \end{aligned} \tag{6.19}$$

The potential at the reference point has been taken as zero, $V(\infty) = 0$.

The electric potential at the field point $\vec{R}$ for a point charge q located at the source point $\vec{R}'$ (Fig. 6.11b) depends on the distance $R_d = |\vec{R} - \vec{R}'|$ from the point charge to

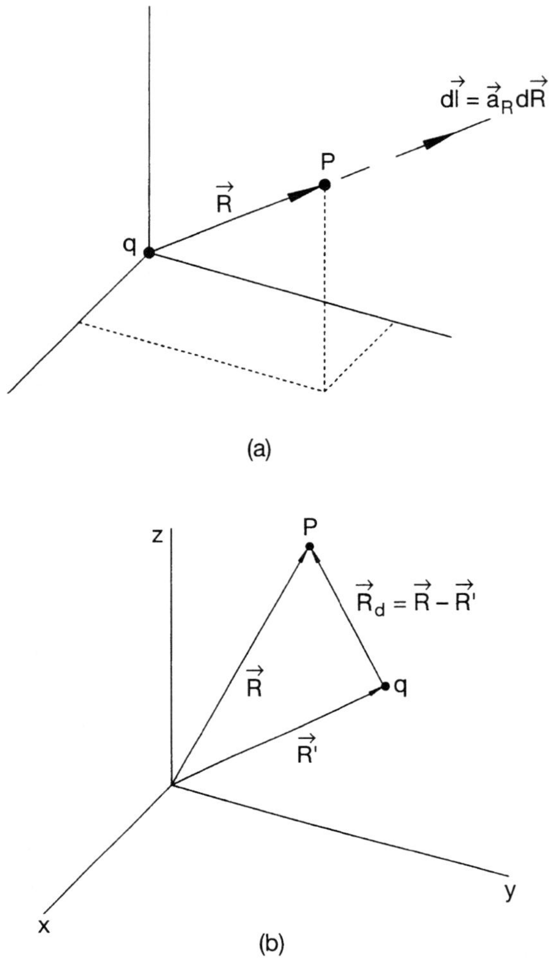

FIGURE 6.11 Electrostatic potential at a point P due to a point charge (a) located at the origin and (b) located away from the origin.

the field point:

$$V(\vec{R}) = \frac{q}{4\pi\varepsilon_0 |\vec{R} - \vec{R}'|}. \tag{6.20}$$

This equation is the Green's function for electric potential. Like Eq. (6.8), it gives the response at position $\vec{R}$ due to a point source located at $\vec{R}'$. It is the basis for the formulas for the electric potential for any charge distribution.

6.3.2 Multiple Point Charges and Continuous Charge Distributions

The electric potential of different charge distributions is found by application of the principle of superposition. Thus if we have N point sources (Fig. 6.3), $q_1, q_2, \ldots, q_N$, located at the source points $\vec{R}'_1, \vec{R}'_2, \ldots, \vec{R}'_N$, respectively, then the electric potential at the field point $\vec{R}$ is given by the sum of the fields due to each charge acting alone. Thus

$$V(\vec{R}) = \sum_{i=1}^{N} \frac{q_i}{4\pi\varepsilon_0 |\vec{R} - \vec{R}'_i|}. \tag{6.21}$$

The electric potential is a scalar quantity and the addition is therefore scalar addition.

For a continuous charge distribution the electric potential at the field point $\vec{R}$ due to the element of charge dQ' located at the source point $\vec{R}'$ is

$$dV(\vec{R}) = \frac{dQ'}{4\pi\varepsilon_0 |\vec{R} - \vec{R}'|}. \tag{6.22}$$

For a volume charge density (Fig. 6.4), with $dQ' = \rho(\vec{R}')\,dv'$,

$$V(\vec{R}) = \int_V \frac{\rho(\vec{R}')\,dv'}{4\pi\varepsilon_0 |\vec{R} - \vec{R}'|}. \tag{6.23}$$

For a surface charge density (Fig. 6.5), with $dQ' = \rho_s(\vec{R}')\,dS'$,

$$V(\vec{R}) = \int_S \frac{\rho_s(\vec{R}')\,dS'}{4\pi\varepsilon_0 |\vec{R} - \vec{R}'|}. \tag{6.24}$$

For a line charge density (Fig. 6.6), with $dQ' = \rho_l(\vec{R}')\,dl'$,

$$V(\vec{R}) = \int_L \frac{\rho_l(\vec{R}')\,dl'}{4\pi\varepsilon_0 |\vec{R} - \vec{R}'|}. \tag{6.25}$$

Note again that these formulas are all special forms of Eq. (6.20) and do not have to be memorized individually.

Example 6.5 Repeat Example 6.1, but find the electric potential. Use the result to find the electric field at (x, y, z).

Solution The source point vectors $\vec{R}'_1$ and $\vec{R}'_2$ and the field point vector $\vec{R}$ are as given in that example. Using Eq. (6.21) we obtain

$$V(x, y, z) = \frac{10^{-6}}{4\pi\varepsilon_0} \times \left[\frac{7}{\sqrt{(x-1)^2 + (y-2)^2 + (z-3)^2}} + \frac{-8}{\sqrt{(x+4)^2 + (y-5)^2 + (z-6)^2}} \right].$$

Substituting (9, − 10, 11) for (x, y, z) we calculate

$$V(9,\ -10,11)=\frac{10^{-6}}{4\pi\varepsilon_0}\left(\frac{7}{272}-\frac{8}{419}\right)=59.8\ \text{V}.$$

Use Eq. (6.17) to find the electric field intensity. Noting that

$$\frac{\partial}{\partial u}\left[\frac{1}{\sqrt{(u-u')^2+k}}\right]=-\frac{(u-u')}{[(u-u')^2+k]^{3/2}},$$

we find

$$\vec{E}(x,y,z)=-\frac{10^{-6}}{4\pi\varepsilon_0}\left\{7\left[-\vec{a}_x\frac{(x-1)}{D_1}-\vec{a}_y\frac{(y-2)}{D_1}-\vec{a}_z\frac{(z-3)}{D_1}\right]\right.$$
$$\left.-8\left[-\vec{a}_x\frac{(x+4)}{D_2}-\vec{a}_y\frac{(y-5)}{D_2}-\vec{a}_z\frac{(z-6)}{D_2}\right]\right\}$$

where

$$D_1=[(x-1)^2+(y-2)^2+(z-3)^2]^{3/2}$$

and

$$D_2=[(x+4)^2+(y-5)^2+(z-6)^2]^{3/2}.$$

When terms in the brackets are combined over their common denominator, we obtain the same result as Example 6.1.

Example 6.6 For the charge density of Example 6.3 find the electric potential at any point on the z axis and then use this result to find the z component of electric field intensity. Explain why the result cannot be used to find the y component.

Solution The field point and source point vectors and the charge dQ' are as given in that example. Using Eq. (6.24) we obtain

$$V(0,0,z)=\int_{r'=0}^{a}\int_{\phi'=0}^{\pi}\frac{A\sin\phi'\, r'\, d\phi'\, dr'}{4\pi\varepsilon_0\sqrt{(r')^2+z^2}}$$
$$=\frac{2A}{4\pi\varepsilon_0}(\sqrt{a^2+z^2}-z).$$

To find the z component of the electric field intensity, find the z component of the negative gradient of the electric potential.

$$E_z=-\frac{\partial V}{\partial z}=-\frac{2A}{4\pi\varepsilon_0}\left[\frac{z}{\sqrt{a^2+z^2}}-1\right]=\frac{A}{4\pi\varepsilon_0}2z\left(\frac{1}{z}-\frac{1}{\sqrt{a^2+z^2}}\right).$$

This result is the same as the result in Example 6.3.

To find the y component of the electric field intensity we must be able to find $\partial V/\partial y$. Since we only found an expression for V on the z axis, at $y = 0$, and not as a function of the y coordinate, we cannot take this partial derivative.

Example 6.7 Find the potential at any point (r, ϕ, z) for a uniform line charge density ρ_{l0} on the z axis from $z = -L$ to L (Fig. 6.12).

Solution The distance from an element of charge $\rho_{l0}\,dz'$ located at the source point z' on the z axis to the field point at (r, ϕ, z) is $R_d = \sqrt{r^2 + (z - z')^2}$. Hence the potential at the field point is

$$V = \int_{z'=-L}^{L} \frac{\rho_{l0}\, dz'}{4\pi\varepsilon_0 \sqrt{r^2 + (z - z')^2}}. \tag{6.26}$$

Letting $u = z - z'$ so $du = -dz'$, we obtain

$$\begin{aligned} V &= \frac{\rho_{l0}}{4\pi\varepsilon_0} \int_{u=z+L}^{z-L} \frac{-du}{\sqrt{r^2 + u^2}} = \frac{\rho_{l0}}{4\pi\varepsilon_0} \ln\left(\frac{z + L + \sqrt{z + L)^2 + r^2}}{z - L + \sqrt{(z - L)^2 + r^2}}\right) \\ &= \frac{\rho_{l0}}{4\pi\varepsilon_0} \ln\left(\frac{R_1 + z + L}{R_2 + z - L}\right), \end{aligned} \tag{6.27}$$

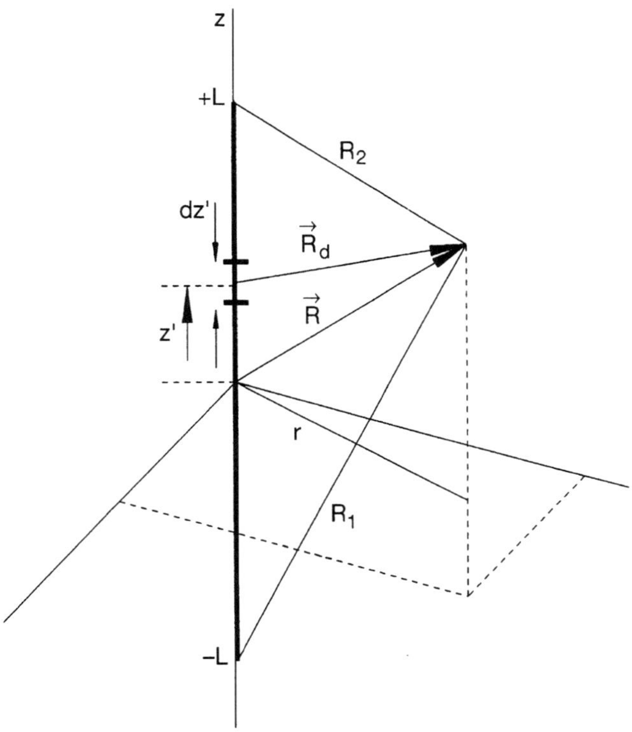

FIGURE 6.12 Uniform line charge density between $-L$ and $+L$ along the z axis.

where

$$R_1 = \sqrt{(z+L)^2 + r^2} \quad \text{and} \quad R_2 = \sqrt{(z-L)^2 + r^2}. \tag{6.28}$$

We can obtain a different form of this answer by multiplying the numerator and the denominator of the argument of the logarithm by $[R_2 - (z - L)]$ to get

$$V = \frac{\rho_{l0}}{4\pi\varepsilon_0} \ln\left(\frac{[R_1 + z + L][R_2 - (z - L)]}{r^2}\right). \tag{6.29}$$

6.3.3 Electric Dipole

Let us next calculate the electric field of two equal and opposite point charges q and $-q$ separated by a distance d. To solve for the fields, choose a coordinate system with the origin at the center point and the positive z axis (Fig. 6.13). The field point is located at (R, θ, ϕ) and the electric potential is

$$V(R, \theta, \phi) = \frac{q}{4\pi\varepsilon_0}\left(\frac{1}{R_1} - \frac{1}{R_2}\right) = \frac{q}{4\pi\varepsilon_0}\left(\frac{R_2 - R_1}{R_1 R_2}\right), \tag{6.30}$$

where

$$R_1 = \sqrt{R^2 + \left(\frac{d}{2}\right)^2 - Rd\cos\theta} = R\sqrt{1 - \left(\frac{d}{R}\right)\cos\theta + \left(\frac{d}{2R}\right)^2}$$

and

$$R_2 = \sqrt{R^2 + \left(\frac{d}{2}\right)^2 + Rd\cos\theta} = R\sqrt{1 + \left(\frac{d}{R}\right)\cos\theta + \left(\frac{d}{2R}\right)^2}.$$

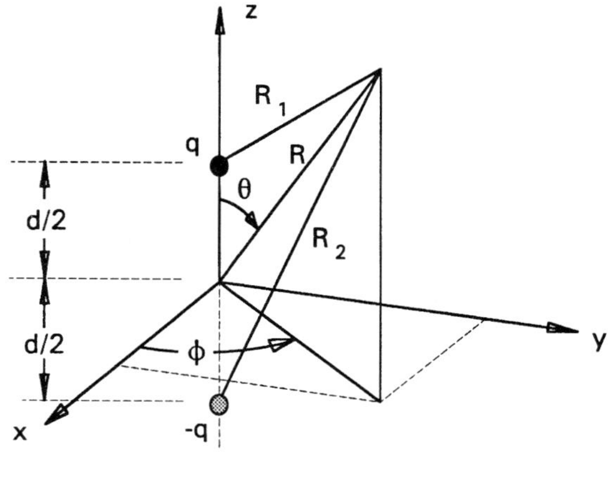

FIGURE 6.13 Electric dipole centered at the origin.

The expressions for R_1 and R_2 are from the law of cosines. The electric field intensity is found by taking the negative gradient of the electric potential.

In the special case where $(d/R) \ll 1$ it is possible to get a simpler approximate expression. In the approximations that follow we retain all terms of order (d/R) and neglect all terms of order $(d/R)^2$ and higher. We use the Maclaurin series expansion $(1 + w)^v \approx 1 + vw$, $|w| \ll 1$, to obtain

$$R_1 \approx R\sqrt{1 - \left(\frac{d}{R}\right)\cos\theta} \approx R\left[1 - \left(\frac{d}{2R}\right)\cos\theta\right],$$

$$R_2 \approx R\sqrt{1 + \left(\frac{d}{R}\right)\cos\theta} \approx R\left[1 + \left(\frac{d}{2R}\right)\cos\theta\right],$$

$$R_2 - R_1 = d\cos\theta,$$

$$R_1 R_2 = R^2\left[1 - \left(\frac{d}{2R}\right)\cos\theta\right]\left[1 + \left(\frac{d}{2R}\right)\cos\theta\right] \approx R^2, \tag{6.31}$$

and hence

$$V(R, \theta, \phi) = \frac{qd\cos\theta}{4\pi\varepsilon_0 R^2}. \tag{6.32}$$

This expression for electric potential is approximate and is valid for $(d/R) \ll 1$. If we allow d to approach zero and q to become infinite in such a way that the product qd remains finite, then $(d/R) \ll 1$ for any value of R and Eq. (6.32) gives the exact electric potential for all values of R. This charge distribution is known as an electric dipole of moment qd.

In expanding the square roots representing R_1 and R_2 we assumed that we should retain terms of order (d/R) and neglect terms of order $(d/R)^2$ and higher. This assumption was justified when the nonzero difference between the distances, Eq. (6.31), was obtained. Had this difference turned out to be zero, it would have been necessary to go back and retain the next higher-order term. In the present case, where only first-order terms were required, the difference $R_2 - R_1$ could be obtained geometrically as shown in Fig. 6.14.

The electric field intensity for the electric dipole is found by taking the negative gradient of the electric potential:

$$\begin{aligned}\vec{E}(R, \theta, \phi) &= -\left(\vec{a}_R\frac{\partial V}{\partial R} + \vec{a}_\theta\frac{1}{R}\frac{\partial V}{\partial\theta}\right)\\ &= \frac{qd}{4\pi\varepsilon_0 R^3}(\vec{a}_R 2\cos\theta + \vec{a}_\theta\sin\theta).\end{aligned} \tag{6.33}$$

A picture of the equipotential and the electric field lines surrounding the electric dipole is shown in Fig. 6.15.

We have already noted that the point charge q is a singularity of charge density: a finite charge located in zero volume. We know that a point charge is the source of an

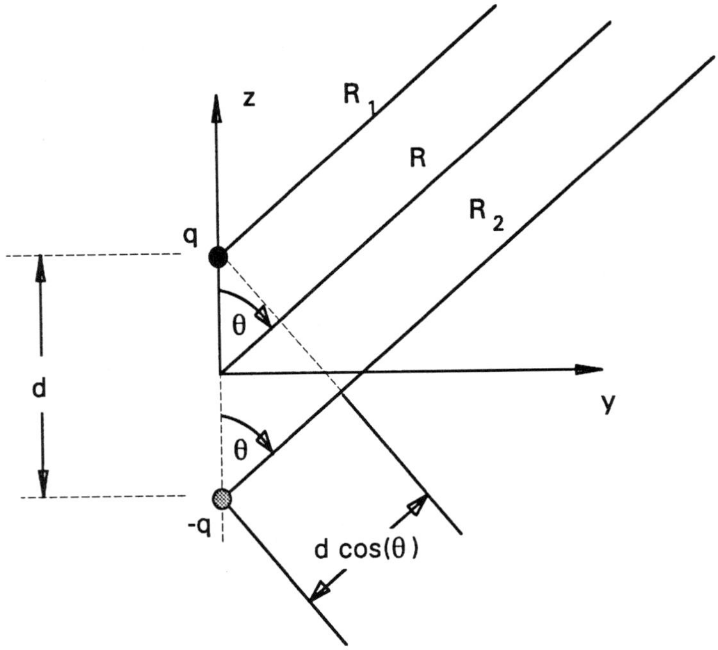

FIGURE 6.14 Electric dipole and geometric determination of the difference between R_1 and R_2.

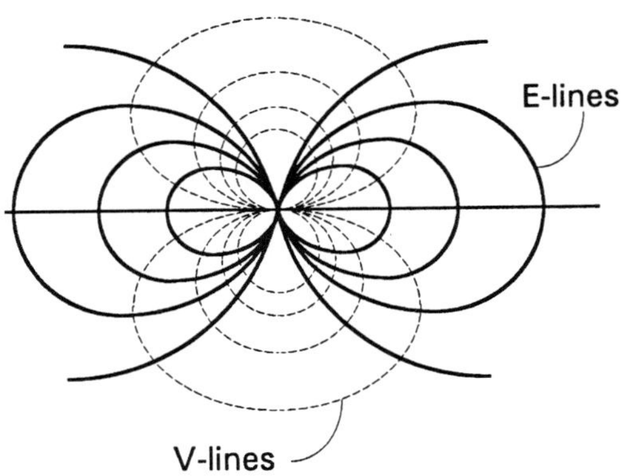

FIGURE 6.15 Equipotential and the electric field lines surrounding an electric dipole.

electric potential field, which is inversely proportional to the distance from the charge, and the source of an electric field intensity, which is inversely proportional to the square of the distance. The electric dipole is a higher-order singularity of charge density since it stems from two infinite charges of opposite sign squeezed into zero volume. The electric dipole is the source of electric potential, which is inversely proportional to the square of the distance from the charge, and of an electric field intensity, which is inversely proportional to the cube of the distance. The higher-order singularity in the source produces a higher-order singularity in the fields.

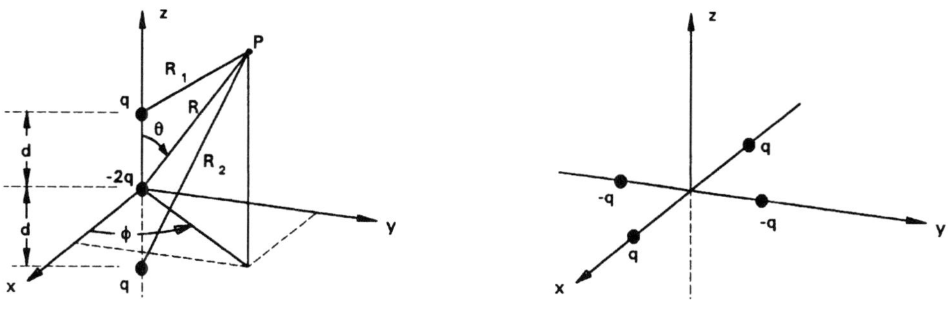

(a) Linear Quadrupole (a) Square Quadrupole

FIGURE 6.16 Quadrupole obtained by superimposing two dipoles.

By superimposing two dipoles, Figs. 6.16a and 6.16b, we can form an even higher-order charge singularity, an electric quadrupole, which will produce fields with even higher-order singularities with respect to the distance. The process of superposition can be continued indefinitely, producing a series of multipoles and their associated fields. If added together with proper weighting this series can represent the field of any charge distribution in the region $R > R_0$, where the radius R_0 includes all of the charge. We will not pursue this idea further in this text.

6.3.4 Line Charge and Paired Line Charge

Let us next look at the electric potential of a uniform line charge ρ_{l0} along the entire length of the z axis. Since the source charge is independent of the z coordinate, so are the fields. This is a two-dimensional problem, because the field lines are the same in any $z = C$ plane.

From Gauss's law the electric flux density is $D_r = \rho_{l0}/2\pi r$ and, also, $E_r = D_r/\varepsilon_0$. Then the electric potential is

$$V(r) = \int_{r'=r}^{r_0} \frac{\rho_{l0}}{2\pi\varepsilon_0 r'}\, dr'$$

$$= \frac{\rho_{l0}}{2\pi\varepsilon_0} \ln\left(\frac{r_0}{r}\right). \tag{6.34}$$

This formula gives the electric potential at a radial distance r relative to the potential at some arbitrary radius r_0. The zero potential reference cannot be taken at infinity, because the charge distribution in this problem extends to infinity. In a realistic problem with a long straight thin wire of length L with average charge density per unit length ρ_l, Eq. (6.34) gives the potential in a region $R \ll L$ from an origin at the center of the wire. Figure 6.17 shows the electric field lines (with arrowheads) and the equipotential surfaces in an end view of the line charge.

Let us next assume we have a pair of equal but opposite polarity line charges, parallel to the z axis and separated by a distance $2D$ (Fig. 6.18). Then the electric

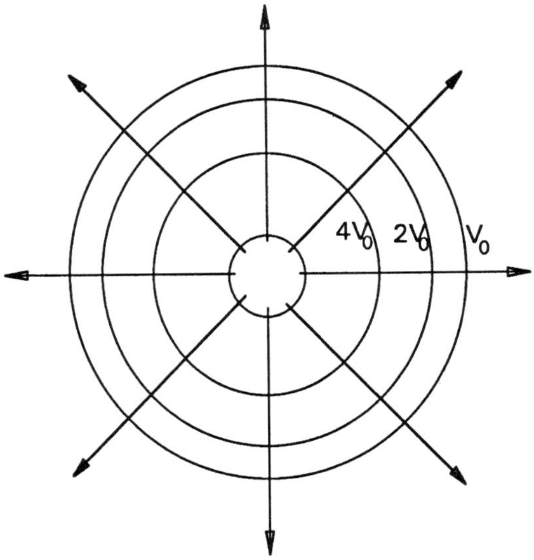

FIGURE 6.17 Electric field lines and equipotential surfaces in an end view of a line charge.

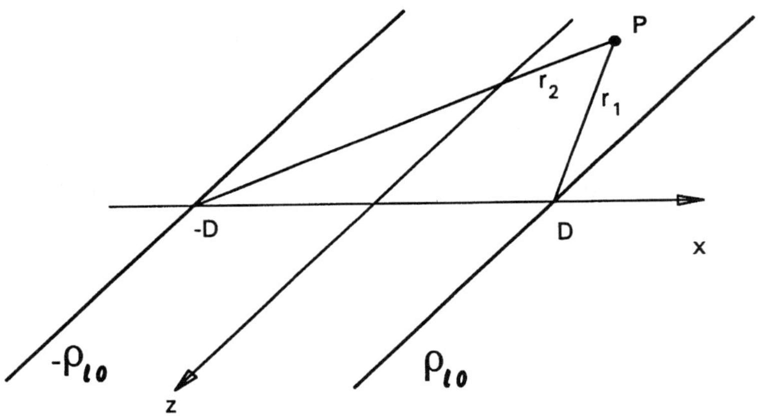

FIGURE 6.18 Pair of line charges of opposite polarity in the *xz* plane.

potential at any point (x, y) is

$$V(x, y) = \frac{\rho_{l0}}{2\pi\varepsilon_0}\left[\ln\left(\frac{r_0}{r_1}\right) - \ln\left(\frac{r_0}{r_2}\right)\right] = \frac{\rho_{l0}}{2\pi\varepsilon_0}\ln\left(\frac{r_2}{r_1}\right), \tag{6.35}$$

where

$$r_1 = \sqrt{(x - D)^2 + y^2} \quad \text{and} \quad r_2 = \sqrt{(x + D)^2 + y^2},$$

and r_0 represents some arbitrary radius at which the reference potential is defined. One equipotential surface is obvious and is consistent with our physical intuition. On the plane $x = 0$, halfway between the line charges $r_1 = r_2$, the ratio r_2/r_1 is unity, and the logarithm of the ratio is zero. This is the $V = 0$ equipotential surface. For points to the right of this plane—$x > 0$, $r_2 > r_1$—the ratio is greater than unity, the logarithm is positive, and the potential is positive. For points to the left of this plane—$x < 0$, $r_2 < r_1$—the ratio is less than unity, the logarithm is negative, and the potential is negative.

To find the equation representing other equipotential surfaces, we note that on an equipotential surface the ratio $r_2/r_1 = k$, a real positive constant. Squaring this equation, substituting for r_1 and r_2 from Eq. (6.35), multiplying by the denominator, grouping terms in powers of x and y, and completing the square, we find

$$\left(x - \frac{k^2+1}{k^2-1}D\right)^2 + y^2 = \left(\frac{2k}{k^2-1}\right)^2.$$

This is the equation of a circular cylinder with its axis parallel to the z axis passing through the point $x_0 = [(k^2+1)/(k^2-1)]D$ on the x axis and having a radius $r_0 = |2k/(k^2-1)|D$. When $k > 1$, x_0 is positive and the axis lies to the right of the positive line charge. When $k < l$, x_0 is negative and the axis lies to the left of the negative line charge. As $k \to 1$, the axis moves further and further away, with the radius becoming greater and greater. The limiting "circle" is the $x = 0$ plane.

A sketch of the equipotential surfaces is shown in Fig. 6.19.

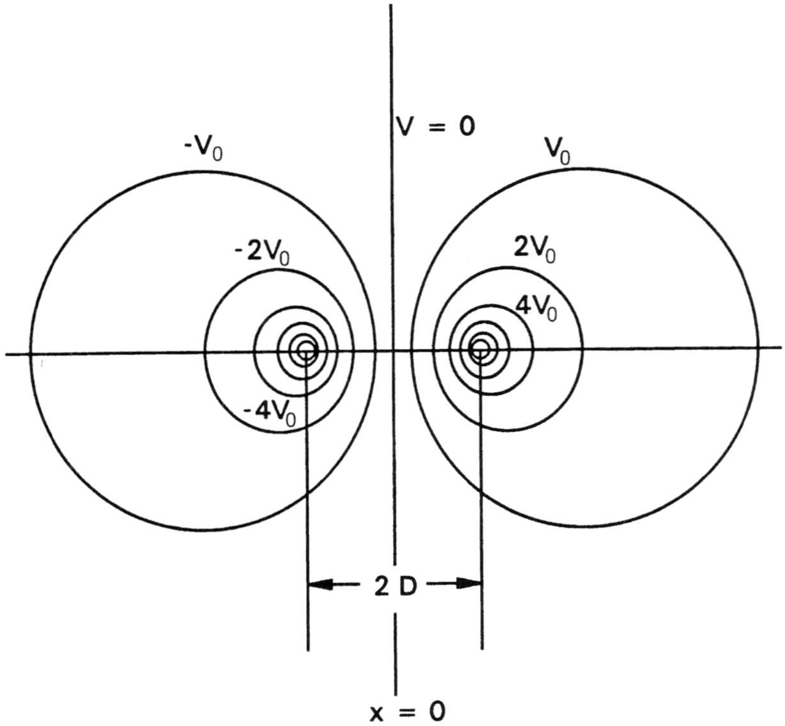

FIGURE 6.19 Equipotential surfaces for the pair of line charges in Fig. 6.18.

6.4 FIELDS FOR UNKNOWN CHARGE DISTRIBUTIONS

When the charge distribution is known or assumed as in the examples given so far, then there is no theoretical difficulty in calculating the electric field intensity or the electric potential. In practice it may be difficult to perform the integrations to obtain a closed form of solution (a formula), but a computer can always be programmed to obtain numerical results. On the other hand, if the charge distribution is not known, then the problem is much more difficult.

Consider the following example. There is a thin straight conducting wire of length $2L$ and radius $a \ll L$ with a total charge of Q and we want to find the field surrounding the wire. Assume the wire to be coaxial with the z axis and centered at the origin (Fig. 6.20).

Since the wire is a conductor it must be at some potential V_0 throughout its volume and on its surface. The charge must therefore distribute itself on the surface of the wire so as to create this equipotential volume and surface. The charge, which has never studied calculus or electrostatics, has no difficulty in doing this, but we are not in such an easy position.

As a first guess assume that the charge distributes itself uniformly over the surface of the wire so that the linear charge density is $\rho_{l0} = Q/2L$. In addition to assuming uniformity along the wire, this also neglects the charge on the flat surface at each end of the wire. Since $a \ll L$ the problem is now like Example 6.6 (Fig. 6.12), and we have already found the potential, Eq. (6.27). We can find the shape of the equipotential

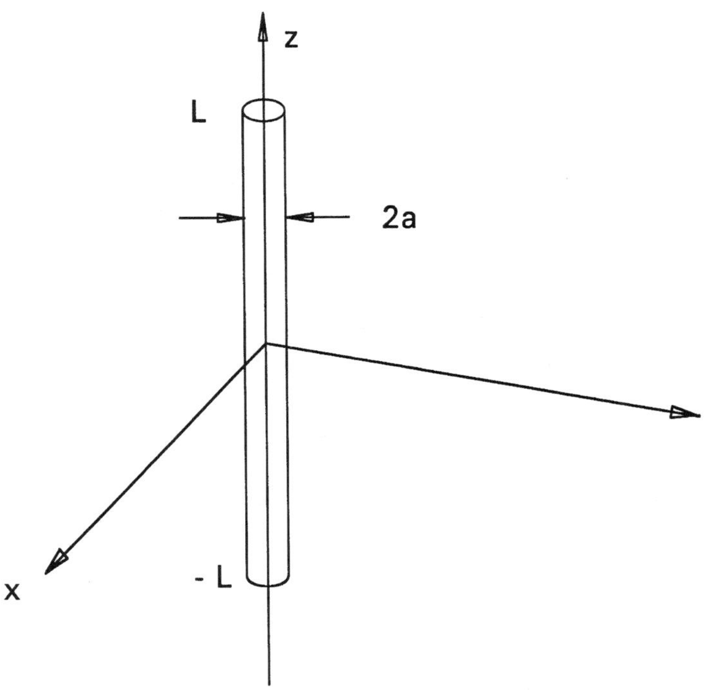

FIGURE 6.20 A thin conductor of length $2L$ and radius $a \ll L$ along the z axis and centered at the origin.

surfaces for this charge distribution as follows. We multiply the numerator and the denominator of the logarithm's argument by $[R_1 - (z + L)][R_2 - (z - L)]$. Since $R_1^2 - (z + L)^2 = R_2^2 - (z - L)^2 = r^2$, we have a different form of the result:

$$V = \frac{\rho_{l0}}{4\pi\varepsilon_0} \ln\left(\frac{R_2 - (z - L)}{R_1 - (z + L)}\right). \tag{6.36}$$

For a particular equipotential surface $V = V_0$ the argument of the logarithm in Eqs. (6.36) and (6.27) must equal the same constant k. Therefore

$$k = \frac{R_2 - (z - L)}{R_1 - (z + L)}, \quad \text{giving} \quad R_1 k - zk - Lk = R_2 - z + L, \tag{6.37}$$

and

$$k = \frac{R_1 + (z + L)}{R_2 + (z - L)}, \quad \text{giving} \quad R_2 k + zk - Lk = R_1 + z + L, \tag{6.38}$$

where

$$V_0 = \frac{\rho_{l0}}{4\pi\varepsilon_0} \ln k \quad \text{or} \quad k = \exp\left(\frac{4\pi\varepsilon_0 V_0}{\rho_{l0}}\right). \tag{6.39}$$

We add the expressions on the right in Eqs. (6.37) and (6.38) to obtain

$$R_1 + R_2 = 2L\left(\frac{k + 1}{k - 1}\right).$$

Substituting for k from Eq. (6.39) we find

$$R_1 + R_2 = 2L \coth\left(\frac{2\pi\varepsilon_0 V_0}{\rho_{l0}}\right). \tag{6.40}$$

For any V_0 the right side of this result is a constant. Equation (6.40) therefore describes an ellipse with foci at $z = L$ and $z = -L$ in any plane passing through the z axis. The surfaces of equal potential are confocal spheroids formed by rotating the ellipse about the z axis. These surfaces do not coincide with the cylindrical surface of our conducting wire and hence this cannot be an exact solution to the problem presented.

If we evaluate the potential from Eq. (6.29) on the surface of the wire $r = a$ not too near the ends of the wire, where $(z + L)^2 \gg a^2$ and $(z - L)^2 \gg a^2$, then $R_1 \approx z + L$, $R_2 \approx L - z$, and

$$\begin{aligned} V &\approx \frac{\rho_{l0}}{4\pi\varepsilon_0} \ln\left(\frac{2(z + L)2(L - z)}{a^2}\right) = \frac{\rho_l}{4\pi\varepsilon_0} \ln\left[\left(\frac{2L}{a}\right)^2 \left(1 - \frac{z^2}{L^2}\right)\right] \\ &= \frac{\rho_{l0}}{4\pi\varepsilon_0}\left[2 \ln\left(\frac{2L}{a}\right) + \ln\left(1 - \frac{z^2}{L^2}\right)\right]. \end{aligned} \tag{6.41}$$

This shows that V is a very slowly varying function of z, not too near the ends. At the ends $r = a$ and $z = \pm L$, Eq. (6.29) gives

$$V \approx \frac{\rho_{l0}}{4\pi\varepsilon_0} \ln\left(\frac{(2L + 2L)(a)}{a^2}\right) = \frac{\rho_{l0}}{4\pi\varepsilon_0} \ln\left(\frac{4L}{a}\right)$$
$$= \frac{\rho_{l0}}{4\pi\varepsilon_0}\left[\ln\left(\frac{2L}{a}\right) + \ln 2\right]. \tag{6.42}$$

For the thin wire, the ln 2 term is small compared with the first term in the brackets. Comparison of Eq. (6.42) with Eq. (6.41) shows that the potential at the end of the wire (assuming uniform charge density) is approximately half the potential at the middle. For thinner wires, the potential is more closely constant over the length of the wire, dropping to half the value more quickly at the ends. This is illustrated in Fig. 6.21,

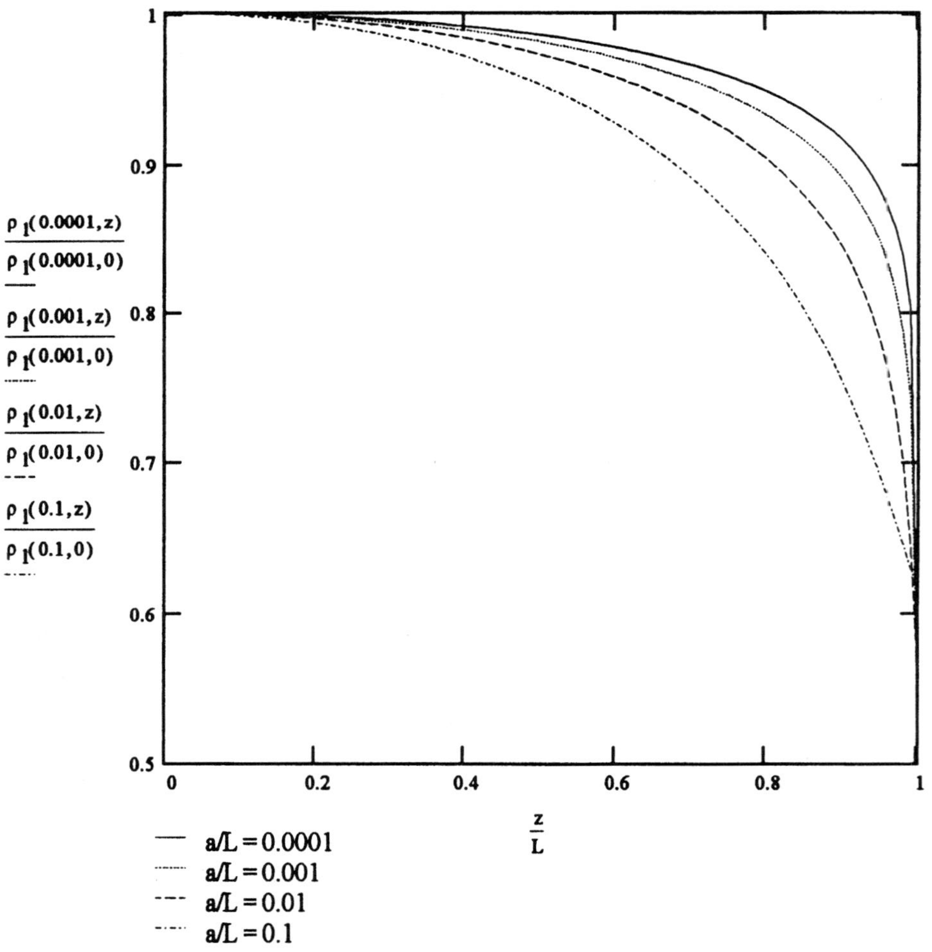

FIGURE 6.21 Normalized potential as a function of normalized distance for different a/L values.

which shows the potential along the length of the wire normalized with respect to the potential at the center of the wire, with a/L as a parameter.

Let us take a different tack and try to find an expression for the charge density along the wire. Since the entire volume of the wire is at the potential V_0, we can write the potential along the axis of the wire (the z axis) at any value of z as

$$V_0 = \int_{z'=-L}^{L} \frac{\rho_l(z')\,dz'}{4\pi\varepsilon_0\sqrt{a^2 + (z - z')^2}}, \qquad -L < z < L.$$

This is an integral equation for the unknown charge density $\rho_l(z')$ and it neglects only the charge on the flat ends of the wire. Unfortunately, there is no known closed form solution to this equation. We can find a first approximation to the solution as follows. First, we note that the integrand is largest at $z' = z$, where the denominator is smallest. At this point $\rho_l(z') = \rho_l(z)$. Second, the charge density is a slowly varying function of z', so we can approximate $\rho_l(z')$ with $\rho_l(z)$, its value at the most important point, and then take it outside the integral to obtain

$$V_0 = \frac{\rho_l(z)}{4\pi\varepsilon_0} \int_{z'=-L}^{L} \frac{dz'}{\sqrt{a^2 + (z - z')^2}}.$$

This integral is the same as that in Eq. (6.26), but with r replaced by a, so we find

$$\rho_l(z) = 4\pi\varepsilon_0 V_0 \left\{ \ln \frac{[\sqrt{(z+L)^2 + a^2} + (z+L)][\sqrt{(z-L)^2 + a^2} - (z-L)]}{a^2} \right\}^{-1}.$$

The expression within the braces is exactly the expression that we just studied and that is plotted in Fig. 6.21. Hence, for thin wires, the charge density varies slowly along the length of the wire (a confirmation of the assumption made above) and is approximately twice as large at the ends of the wire as it is in the center.

This problem is related to problems in the theory of linear antennas, which are long thin wires like a typical automobile antenna.

6.5 ENERGY IN THE ELECTRIC FIELD

It is necessary to do mechanical work to assemble a system of charges or a charge distribution in space. If the charge distribution is put into space slowly, so that no radiation takes place, then the mechanical work done becomes potential energy, which is stored in the system. This is just the same as when we lift a book off the table: the mechanical work of lifting is equal to the increased potential energy of the book. The potential energy is stored in the energy of the gravitational field of the book–Earth system. In the case of electrostatic fields, the potential energy is stored in the electric field surrounding the charge distribution. We will now find the equation relating the field to the energy.

Let us assume that there is a charge density $\rho(\vec{R})$ in a finite region of space. By a "finite region" we mean a region that could be enclosed by a sphere, centered on an arbitrary origin, of radius $R < R_0 < \infty$, where R_0 is some fixed number. The charge may be held in place in some region or may be on the surface of conductors, but, in any case, it has a particular distribution. The resulting electric potential throughout the region is given by $V(\vec{R})$. We now assume that a small additional charge $\delta\rho(\vec{R})$ is added to the charge density already present in such a way that the relative distribution of charge density is not changed. The work done in adding this charge density is equal to the increase in potential energy stored in the system or, equivalently, the increase in energy stored in the electric field, δW_e. The incremental energy required to place charge $\delta\rho(\vec{R})\,dv$ at a position where the potential is $V(\vec{R})$ is $V(\vec{R})\delta\rho(\vec{R})\,dv$. The total increment in energy for the volume is given by

$$\delta W_e = \int_V V(\vec{R})\delta\rho(\vec{R})\,dv. \tag{6.43}$$

The differential form of Gauss's law for electric fields $\rho = \nabla\cdot\vec{D}$ gives

$$\delta\rho = \delta(\nabla\cdot\vec{D}) = \nabla\cdot(\delta\vec{D}).$$

The vector identity Eq. (5.33) gives

$$V\nabla\cdot(\delta\vec{D}) = \nabla\cdot(V\delta\vec{D}) - \delta\vec{D}\cdot\nabla V$$

$$V\delta\rho = \nabla\cdot(V\delta\vec{D}) + \delta\vec{D}\cdot\vec{E},$$

where $\vec{E}$ has been substituted for $-\nabla V$. We substitute in Eq. (6.43) to obtain

$$\delta W_e = \int_V \nabla\cdot(V\delta\vec{D})\,dv + \int_V \vec{E}\cdot\delta\vec{D}\,dv.$$

Applying the divergence theorem to the first integral on the right side of this equation yields

$$\delta W_e = \int_S V\,\delta\vec{D}\cdot d\vec{S} + \int_V \vec{E}\cdot\delta\vec{D}\,dv.$$

Let the surface S expand into a sphere of radius $R\to\infty$ about an arbitrary origin. Since the charge is confined to a finite volume, at sufficiently large R it appears as a point charge, giving

$$V \propto 1/R \quad \text{and} \quad D \propto 1/R^2.$$

For the surface, $dS \propto R^2$. Multiplying these three, we find that the surface integral is proportional to $1/R$ and it therefore vanishes as $R\to\infty$. This leaves only the second integral on the right, where the volume throughout which we integrate is all of space.

$$\delta W_e = \int_V \vec{E}\cdot\delta\vec{D}\,dv.$$

The final charge distribution is obtained by starting with empty space, $\rho = 0$, and adding charge distribution $\delta\rho$ until the final charge distribution ρ is present. As this is done the electric flux density will go from $\vec{D} = 0$ to some final value $\vec{D}$. The corresponding electric energy stored in the field is

$$W_e = \int_V \int_{\vec{D}=0}^{\vec{D}} \vec{E} \cdot \delta\vec{D}\, dv.$$

Let us consider only linear isotropic material so that $\delta\vec{D} = \varepsilon\, \delta\vec{E}$, where ε may be a function of position. The total electric energy is then

$$\begin{aligned} W_e &= \int_V \left(\varepsilon \int_{\vec{E}=0}^{\vec{E}} \vec{E} \cdot \delta\vec{E} \right) dv = \int_V \left(\varepsilon \int_{\vec{E}=0}^{\vec{E}} E\, \delta E \right) dv \\ &= \tfrac{1}{2} \int_V \varepsilon |\vec{E}|^2\, dv. \end{aligned} \tag{6.44}$$

This equation gives the electric energy stored in the field. It is the result that we sought. The SI unit of energy is the joule (J).

This total energy is given as an integral throughout a volume including all of space. The energy density at each point in space is the energy per unit volume. It is therefore reasonable to assume, although not possible to prove, that the electric energy density w_e at each point in space is

$$w_e = \tfrac{1}{2} \varepsilon |\vec{E}|^2. \tag{6.45}$$

The SI units of energy density are joules per cubic meter (J/m^3).

A second expression for the energy can be obtained quite simply from the original starting point, Eq. (6.43). As the charge density is incremented by $\delta\rho$, again keeping the original relative charge distribution, the electric potential will increase proportionally to the charge density ρ. Let the space-dependent function of proportionality be $f_s(\vec{R})$ so that we can write $V(\vec{R}) = f_s(\vec{R})\rho(\vec{R})$. Then

$$\begin{aligned} W_e &= \int_V \left(\int_{\rho=0}^{\rho} V\, \delta\rho \right) dv = \int_V f_s \left(\int_{\rho=0}^{\rho} \rho\, \delta\rho \right) dv = \int_V f_s(\tfrac{1}{2}\rho^2)\, dv \\ &= \tfrac{1}{2} \int_V V\rho\, dv. \end{aligned} \tag{6.46}$$

When the volume consists only of n conductors separated by dielectric matter, then the ith conductor is at potential V_i, has surface charge density ρ_{si} on it, and has surface area S_i. Equation (6.46) can then be written as

$$\begin{aligned} W_e &= \frac{1}{2} \sum_{i=1}^{n} V_i \int_{S_i} \rho_{si}\, dS_i \\ &= \frac{1}{2} \sum_{i=1}^{n} V_i Q_i, \end{aligned} \tag{6.47}$$

where Q_i is the total charge on the surface of the ith conductor.

Example 6.8 Find the electric energy stored in the field around a spherical conductor of radius a located in free space, when a charge Q is placed on it.

Solution The potential of the conducting sphere is $Q/4\pi\varepsilon_0 a$. Using Eq. (6.47), we have

$$\begin{aligned} W_e &= \frac{1}{2}\left(\frac{Q}{4\pi\varepsilon_0 a}\right)Q \\ &= \frac{Q^2}{8\pi\varepsilon_0 a}. \end{aligned}$$

Let us use Eq. (6.44) and confirm that the same answer is obtained. The electric field intensity and the energy density are

$$\vec{E} = \begin{cases} 0 \\ \vec{a}_R \dfrac{Q}{4\pi\varepsilon_0 R^2} \end{cases} \qquad w_e = \begin{cases} 0, & R < a, \\ \dfrac{1}{2}\varepsilon_0 \left(\dfrac{Q}{4\pi\varepsilon_0 R^2}\right)^2, & R > a. \end{cases}$$

Integrating the energy density to find the total stored energy we obtain

$$\begin{aligned} W_e &= \int_{R=a}^{\infty}\int_{\phi=0}^{2\pi}\int_{\theta=0}^{\pi} \frac{1}{2}\varepsilon_0 \left(\frac{Q}{4\pi\varepsilon_0 R^2}\right)^2 R^2 \sin\theta \, d\theta \, d\phi \, dR \\ &= \frac{1}{2}\varepsilon_0 \frac{Q}{(4\pi\varepsilon_0)^2}\left[-\frac{1}{R}\right]_a^{\infty} [\phi]_0^{2\pi}[-\cos\theta]_0^{\pi} \\ &= \frac{Q^2}{8\pi\varepsilon_0 a}. \end{aligned}$$

6.6 CAPACITANCE

Consider two conducting bodies separated by a linear dielectric medium (Fig. 6.22). The bodies have equal but opposite charges Q and $-Q$, which form surface charge distributions ρ_{s1} and ρ_{s2}. Let V_1 and V_2 be the potentials of the two bodies. The

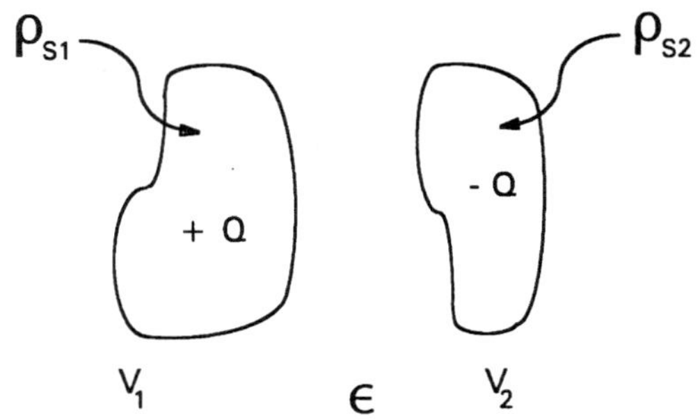

FIGURE 6.22 Two charged conductors of arbitrary shape inside a linear dielectric medium.

potential difference between them is $V = V_1 - V_2$. The electric flux density surrounding the bodies is proportional to the charge (Gauss's law) and hence so is the electric field intensity. The potential difference is the line integral of the electric field and therefore the potential difference is proportional to the electric field intensity and hence the charge. This proportionality can be expressed as

$$Q = CV, \tag{6.48}$$

where C is called the capacitance of the two-conductor system. The SI unit of capacitance is the farad (F).

The capacitance can be evaluated as

$$C = \frac{Q}{V} = \frac{\oint_S \vec{D} \cdot d\vec{S}}{\oint_L \vec{E} \cdot d\vec{l}} = \frac{\oint_S \varepsilon \vec{E} \cdot d\vec{S}}{\oint_L \vec{E} \cdot d\vec{l}}. \tag{6.49}$$

The closed surface S is taken to be any surface surrounding the positively charged body, and the line L, along which the electric field is integrated, is defined as any line that starts on the positively charged body and terminates on the negatively charged body.

The relation between the electric energy stored in the system and the capacitance can be found from Eq. (6.47):

$$W_e = \tfrac{1}{2}[V_1 Q + V_2(-Q)] = \tfrac{1}{2}(V_1 - V_2)Q = \tfrac{1}{2}VQ$$

$$= \begin{cases} \tfrac{1}{2}CV^2, & (6.50) \\ \tfrac{1}{2}Q^2/C. & (6.51) \end{cases}$$

Inverting these last expressions, we obtain formulas for the capacitance in terms of the stored electric energy

$$C = \begin{cases} \dfrac{2W_e}{V^2}, & (6.52) \\ \dfrac{Q^2}{2W_e}. & (6.53) \end{cases}$$

These are sometimes easier to use than Eq. (6.49). They also illustrate the statement that a capacitor is a device that stores electric energy.

Consider next the two bodies connected to a slowly varying voltage source (Fig. 6.23). We will clarify the meaning of "slowly" somewhat later. From the law of conservation of charge, Eq. (4.20), applied to the dashed surface, we have $I_{\text{outward}} = -i = -dQ/dt$. Taking the derivative of Eq. (6.48) with respect to time and substituting, we find

$$i = C\frac{dV}{dt}. \tag{6.54}$$

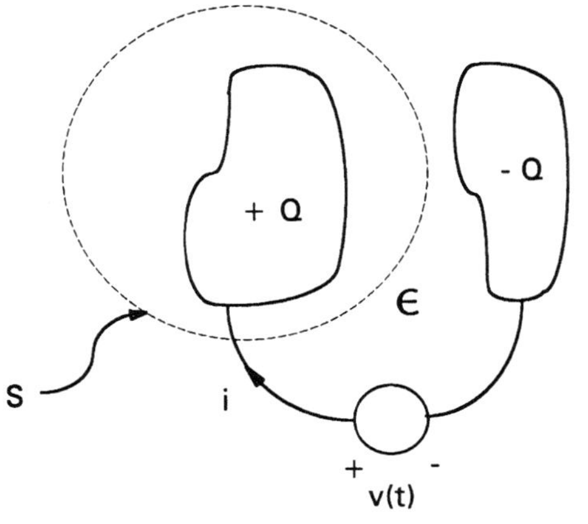

FIGURE 6.23 Two charged conductors of arbitrary shape inside a linear dielectric medium connected to a slowly varying voltage source.

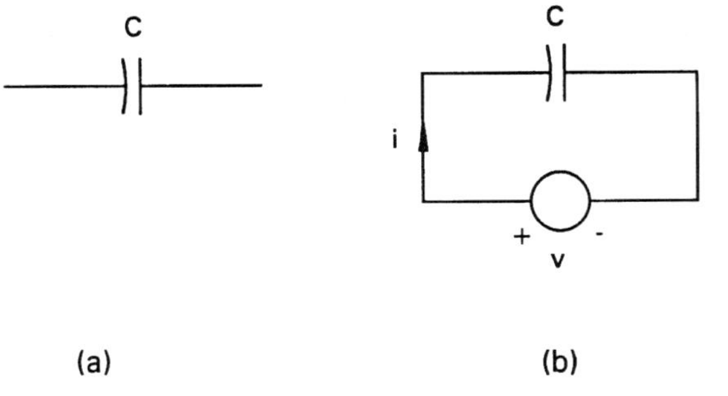

FIGURE 6.24 Electric circuit model for the two-body capacitor with a voltage source.

This is the usual voltage–current terminal relation used to define the capacitance in circuit theory courses. The symbol for the capacitor is shown in Fig. 6.24a and the electric circuit model for the two-body capacitor with a voltage source is shown in Fig. 6.24b.

6.6.1 Parallel Plate Capacitor

Consider two identically shaped parallel conducting plates with surface area S and separation between the plates d. Let the separation between the plates be small compared to the smallest linear dimension of the surface area. For example, if the plates are circular with radius a, then $d \ll a$. The region between the plates has permittivity ε.

To find the capacitance, we assume that a potential V_0 is applied across the plates (Fig. 6.25a), so that the bottom plate has charge Q distributed on its upper surface and the top plate has charge $-Q$ distributed on its lower surface. Since d is small, we can assume the electric flux density and the electric field lines are uniform in the z direction

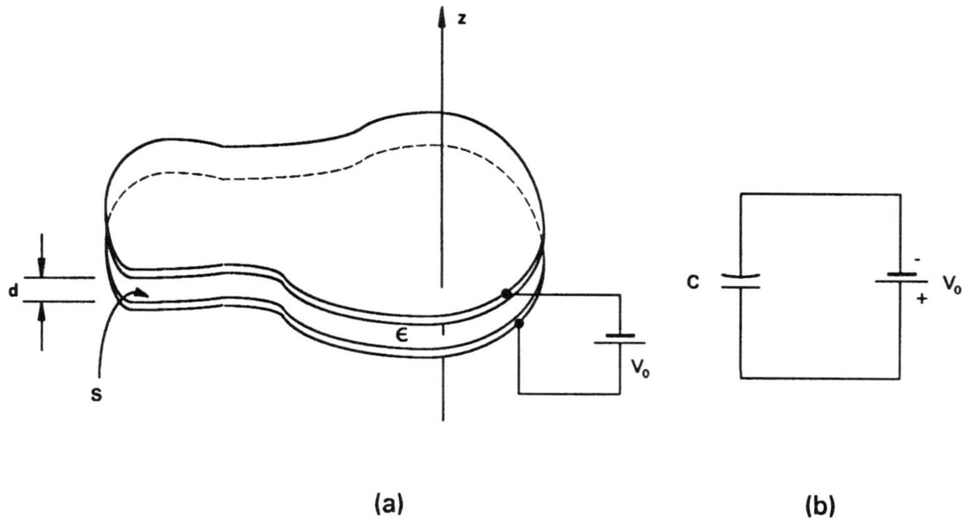

FIGURE 6.25 (a) Parallel plate capacitor connected to a battery and (b) the equivalent circuit model.

between the plates and zero outside. This is called "neglecting fringing," which is the small bulging of the true field near the edge of the plates. Fringing must occur to completely satisfy the boundary condition that $E_{\text{tangential}}$ is continuous. The surface charge density on the bottom plate is $\rho_s = Q/S$; the fields between the plates are

$$D_z = \rho_s = \frac{Q}{S} \quad \text{and} \quad E_z = \frac{D_z}{\varepsilon} = \frac{Q}{\varepsilon S}.$$

The electric energy density and the total electric energy stored between the plates are given by

$$w_e = \tfrac{1}{2}\varepsilon E_z^2 = \tfrac{1}{2}\varepsilon\left(\frac{Q}{\varepsilon S}\right)^2 = \frac{Q^2}{2\varepsilon S^2},$$

$$W_e = w_e(\text{Volume}) = w_e(Sd) = \frac{1}{2}\left(\frac{d}{\varepsilon S}\right)Q^2.$$

Equation (6.51) gives

$$C = \frac{\varepsilon S}{d}. \tag{6.55}$$

The potential across the plates is $V_0 = E_z d = (d/\varepsilon S)Q$. Comparing with Eq. (6.48), we again find $C = (\varepsilon S/d)$.

The simple electrical circuit model for this capacitor is shown in Fig. 6.25b.

Let us make things a little more interesting and assume that the region between the plates is filled with two different dielectrics. Throughout some region over the partial surface S_1 the permittivity is ε_1 and over the remainder, $S_2 = S - S_1$, the permittivity is ε_2 (Fig. 6.26a). The tangential electric field intensity at the boundary of the two dielectric regions must be continuous, so the electric field intensity in the two regions must be the same, $E_{1z} = E_{2z} = E_z$. The electric potential across the plates is $V_0 = E_z d$. The electric energy stored is

$$W_e = W_{e1} + W_{e2} = \tfrac{1}{2}\varepsilon_1 (E_z)^2 S_1 d + \tfrac{1}{2}\varepsilon_2 (E_z)^2 S_2 d = \tfrac{1}{2}\left(\frac{\varepsilon_1 S_1}{d} + \frac{\varepsilon_2 S_2}{d}\right) V_0^2.$$

Equation (6.50) gives the capacitance as

$$C = \frac{\varepsilon_1 S_1}{d} + \frac{\varepsilon_2 S_2}{d}$$
$$= C_1 + C_2, \tag{6.56}$$

where

$$C_1 = \frac{\varepsilon_1 S_1}{d} \quad \text{and} \quad C_2 = \frac{\varepsilon_2 S_2}{d}. \tag{6.57}$$

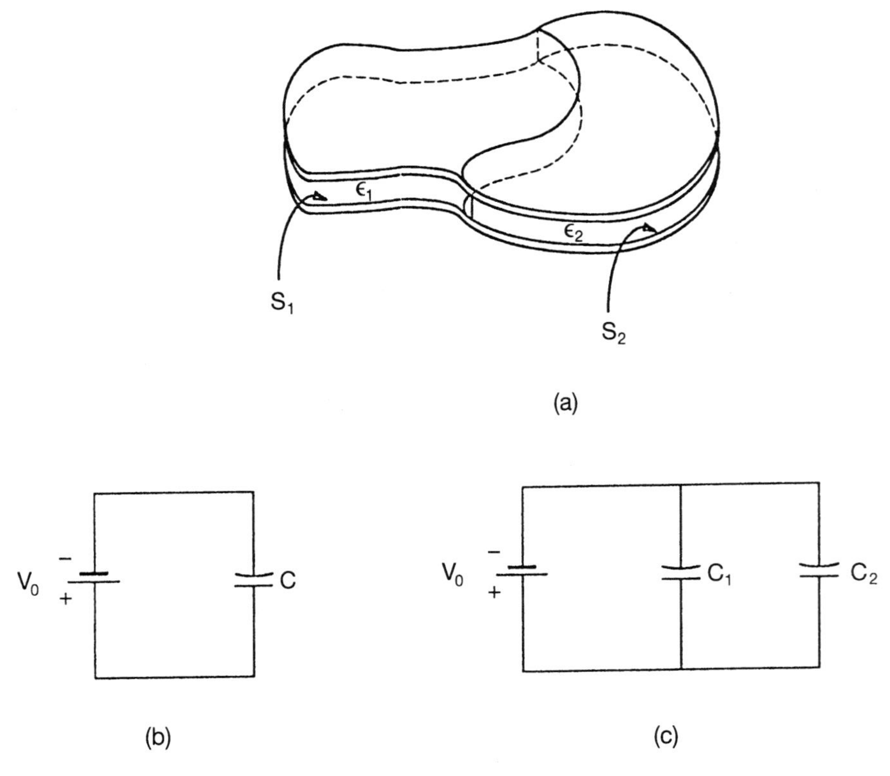

FIGURE 6.26 (a) Parallel conducting plates with two different dielectics between them and (b, c) the corresponding parallel capacitor circuit models.

The two sections of the capacitor have the same potential V_0 across them. The circuit model for this capacitance is therefore either the capacitor C (Fig. 6.26b) or the two capacitors C_1 and C_2 in parallel (Fig. 6.26c). Of course, capacitors in parallel add to give a total capacitance, as in Eq. (6.56).

Note that the electric flux densities in the two regions are different: $D_{1z} = \varepsilon_1 E_z$ and $D_{2z} = \varepsilon_2 E_z$. The surface charge densities on the two portions of the lower plate are also different: $\rho_{s1} = D_{1z} = (\varepsilon_1 V_0/d)$ and $\rho_{s2} = D_{2z} = (\varepsilon_2 V_0/d)$. The total charge is $Q = \rho_{s1} S_1 + \rho_{s2} S_2$. We substitute in Eq. (6.48), yielding the same formula for the capacitance.

Another variation of the problem is to divide the region into two dielectrics so that for $0 < z < d_1$ the permittivity is ε_1 and for $d_1 < z < d$ the permittivity is ε_2 (Fig. 6.27a). Now, at the dielectric to dielectric boundary at $z = d_1$, the normal component of electric flux density must be continuous. Hence $D_{1z} = D_{2z} = D_z = Q/S$.

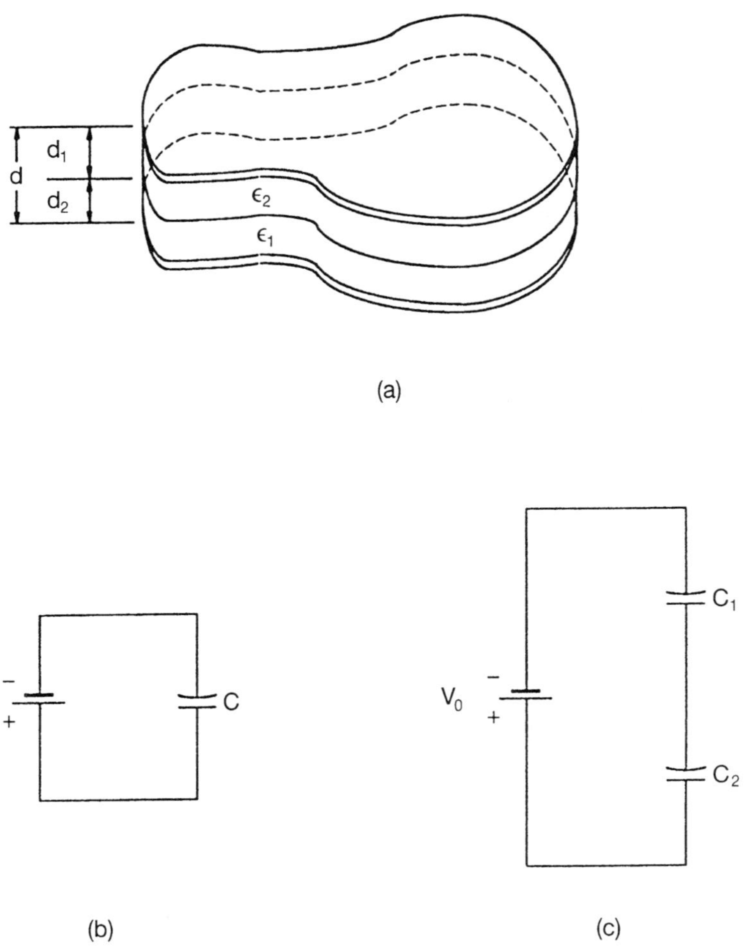

FIGURE 6.27 (a) Parallel conducting plates with two different dielectics between them and (b, c) the corresponding series capacitor circuit models.

Letting $d_2 = d - d_1$, the electric energy stored is

$$W_e = W_{e1} + W_{e2} = \frac{1}{2}\varepsilon_1\left(\frac{D_z}{\varepsilon_1}\right)^2 Sd_1 + \frac{1}{2}\varepsilon_2\left(\frac{D_z}{\varepsilon_2}\right)^2 Sd_2 = \frac{1}{2}\left(\frac{d_1}{\varepsilon_1 S} + \frac{d_2}{\varepsilon_2 S}\right)Q^2.$$

From Eq. (6.53), $1/C = 2W_e/Q^2$. Hence

$$\begin{aligned}\frac{1}{C} &= \frac{d_1}{\varepsilon_1 S} + \frac{d_2}{\varepsilon_2 S} \\ &= \frac{1}{C_1} + \frac{1}{C_2},\end{aligned} \tag{6.58}$$

where

$$C_1 = \frac{\varepsilon_1 S}{d_1} \quad \text{and} \quad C_2 = \frac{\varepsilon_2 S}{d_2}. \tag{6.59}$$

The potential across the capacitor is the sum of the potentials across each of its dielectric sections. The circuit model for this capacitor is therefore either a single capacitor C (Fig. 6.27b) or the two capacitors C_1 and C_2 in series (Fig. 6.27c). Of course, when capacitors are in series, their reciprocals add to give the reciprocal of the total capacitance, as in Eq. (6.58).

We could also divide the region into two dielectrics by an irregular boundary as shown in Fig. 6.28. In this case both the electric field intensity and the electric flux density must change from the field for the original simple parallel plate capacitor in order to keep both $E_{\text{tangential}}$ and D_{normal} continuous along the boundary. There is no longer a simple solution to this problem.

It is easy to generalize this example. Figure 6.29a shows two conductors with a uniform dielectric, ε_1, and a potential difference V_0 between them. The charge on the

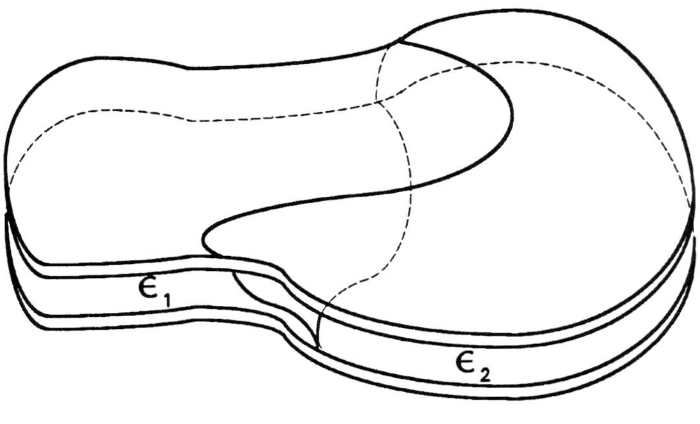

FIGURE 6.28 Parallel conducting plates with two different dielectics with an irregular boundary between them.

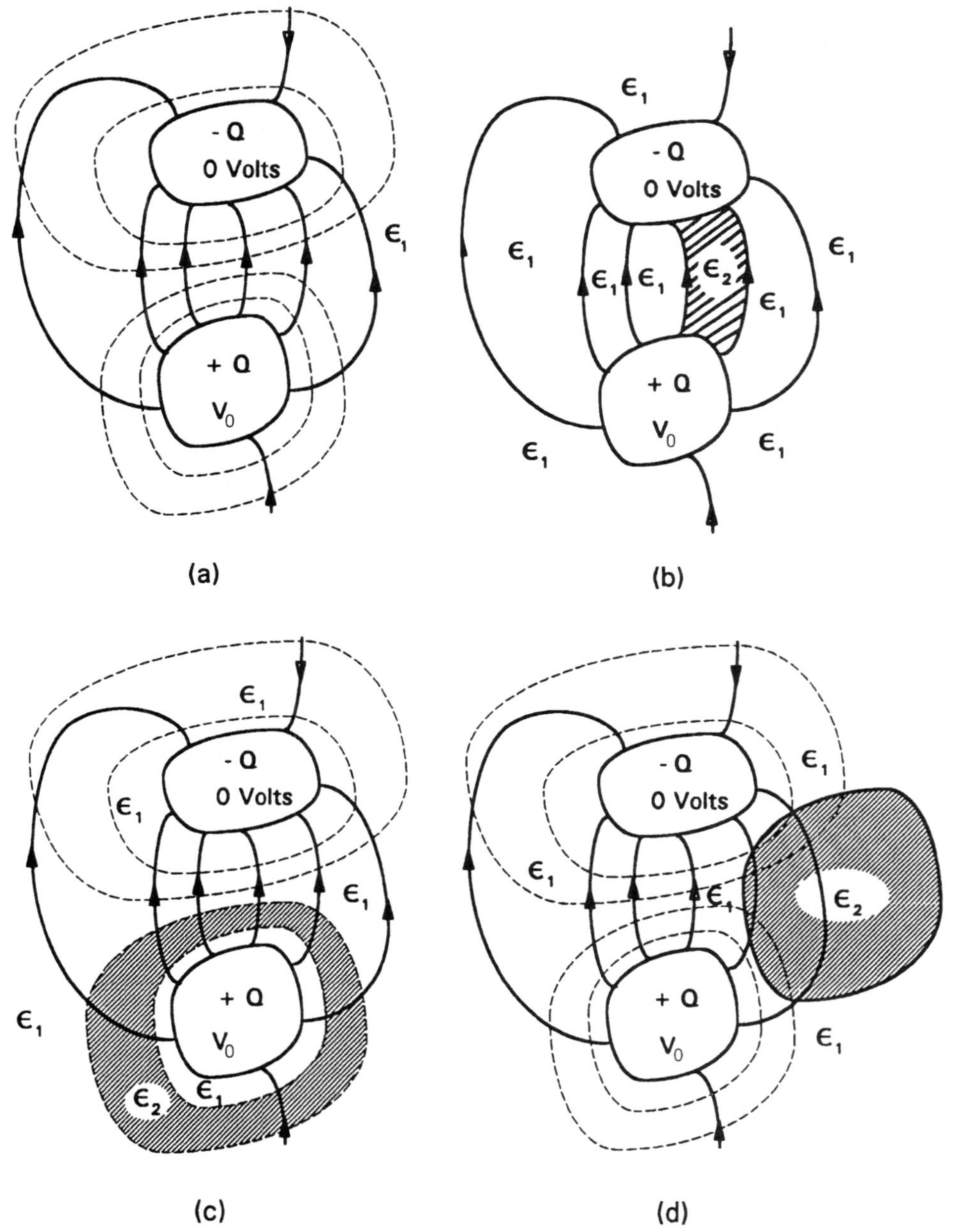

FIGURE 6.29 Generalization of Figs. 6.25–6.28 for capacitors with arbitrary conductors.

higher potential conductor is Q_0. The solid lines are the electric field lines and the dashed lines are the equipotential lines. The electric field lines form tubes connecting the two conductors. When the dielectric within an entire tube is changed to a value ε_2 (Fig. 6.29b) and the potential V_0 is maintained, then the electric field intensity is unchanged. The original $\vec{E}$ does not change because it satisfies the condition that

$E_{\text{tangential}}$ must be continuous at the boundary of the tube and because its integral along any line connecting the conductors is V_0. This capacitor can be modeled as two capacitors in parallel. Looking back to the original single dielectric capacitor, we see that the equipotential lines form shells surrounding the conductors. When the dielectric within an entire shell is changed to a value ε_2 (Fig. 6.29c) and the charge Q_0 is maintained, then the electric flux density is unchanged. The original $\vec{D}$ does not change because it still satisfies the condition that D_{normal} is continuous at the boundary of the shell and because its integral over any closed surface surrounding the higher potential conductor is Q_0. This capacitor can be modeled as two capacitors in series. When a change in dielectric is made from the original, which is neither a complete tube nor a complete shell (Fig. 6.29d), then both $\vec{E}$ and $\vec{D}$ must change and the new fields are not simply related to the original fields.

6.6.2 Coaxial Cylinders

Figure 6.30 shows two conducting cylinders coaxial with the z axis. The inner cylinder is a conductor of radius a. The outer cylinder is hollow with inner radius $b > a$. The region between the cylinders is filled with a dielectric with permittivity ε. Let us assume a potential difference between the cylinders with the $r = a$ surface at potential V_0 with respect to the $r = b$ surface. There will then be a charge per unit length ρ_{l0} distributed on the $r = a$ surface and an equal but opposite charge per unit length $-\rho_{l0}$ distributed on the $r = b$ surface. The electric flux density and the electric field intensity are, respectively, $D_r = \rho_{l0}/2\pi r$ and $E_r = \rho_{l0}/2\pi\varepsilon r$ in the region $a < r < b$ and zero elsewhere. The potential difference is

$$V_0 = \int_{r=a}^{b} \frac{\rho_{l0}}{2\pi\varepsilon r}\, dr = \frac{\rho_{l0}}{2\pi\varepsilon} \ln\left(\frac{b}{a}\right).$$

The capacitance per unit length is given by $C_l = \rho_{l0}/V_0$ and is

$$C_l = \frac{2\pi\varepsilon}{\ln(b/a)}. \tag{6.60}$$

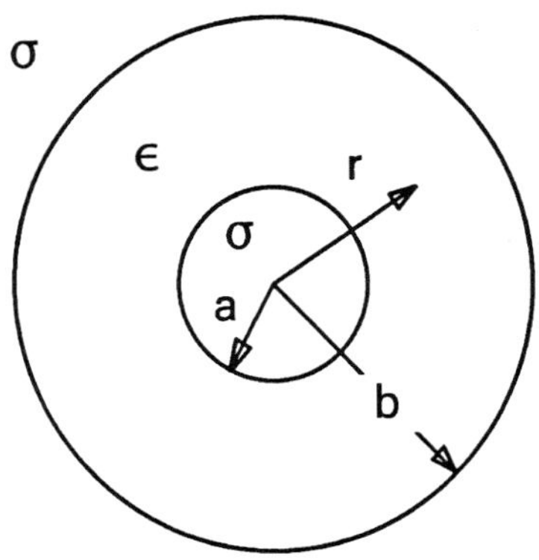

FIGURE 6.30 Coaxial conducting cylinders with a dielectric medium between them.

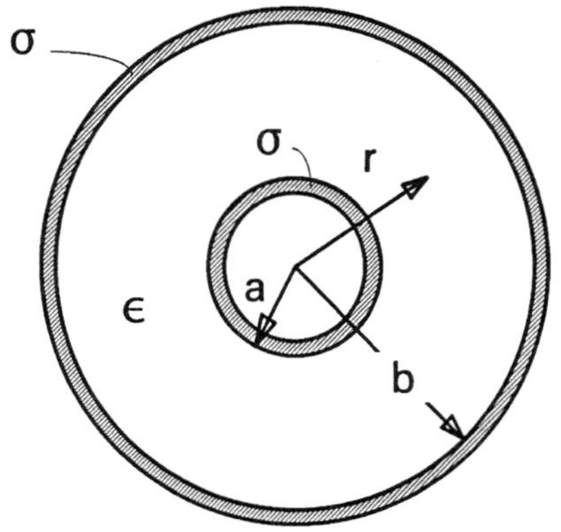

FIGURE 6.31 Coaxial conducting cylinders with a dielectric medium between them. The outer cylinder is replaced by a thin cylindrical shell and the inner cylinder is hollow.

The same result can be found in terms of the electric energy stored per unit length, W_{el}.

$$W_{el} = \int_{\substack{\text{cross}\\\text{section}}} \tfrac{1}{2}\varepsilon E^2 \, dS = \int_{r=a}^{b} \int_{\phi=0}^{2\pi} \frac{1}{2}\varepsilon\left(\frac{\rho_{l0}}{2\pi\varepsilon r}\right)^2 r \, d\phi \, dr$$

$$= \frac{1}{2}\left[\frac{1}{2\pi\varepsilon}\ln\left(\frac{b}{a}\right)\right]\rho_{l0}^2.$$

Using Eq. (6.53) in per unit length form, $C_l = \rho_{l0}^2/2W_{el}$, we obtain the same result as Eq. (6.60).

Note that the fields inside both conductors are zero. The outer cylinder could be replaced by a thin cylindrical shell and the inner cylinder could be hollow (Fig. 6.31), without affecting the fields or the capacitance.

6.7 METHOD OF IMAGES

In finding the fields from a given charge distribution we have dealt with a uniform medium occupying all of space. When more than one medium is present, more advanced methods of solution are required. In several cases, when the second medium is properly shaped, it is still possible to find the fields relatively simply. The method of images uses image charges to act as the virtual source of the fields whose true source is charge distributed on the boundary between the two media.

6.7.1 Point Charge Over a Conducting Plane

In Section 6.3.3 we obtained the field of two equal but opposite polarity point charges (see Fig. 6.13). Examination of Eq. (6.30) shows that on the plane $\theta = \pi/2$, which is the $z = 0$ plane, the distances from the two charges to the field point are equal, $R_1 = R_2$.

Consequently, the electric potential on this surface is zero, $V(R, \pi/2, \phi) = 0$. We can therefore fill the region $z < 0$ with a conductor (Fig. 6.32), without affecting the field in the region $z > 0$. Of course, the electric field in the conducting region $z < 0$ is now zero. Thus the electric potential when a point charge q is located on the z axis at $z = d/2$ over a conducting region is

$$V(R, \theta, \phi) = \begin{cases} \dfrac{q}{4\pi\varepsilon_0}\left(\dfrac{1}{R_1} - \dfrac{1}{R_2}\right), & z > 0, \qquad (6.61) \\ 0, & z < 0. \qquad (6.62) \end{cases}$$

The first term on the right side of Eq. (6.61) is the potential due to the real point charge q. The second term is the potential due to an imaginary image (as in "mirror image") charge $-q$ located below the surface of the conductor at $z = -d/2$. The true source of the field attributed to the image charge is the surface charge density ρ_s lying on the surface of the conductor at $z = 0$. To evaluate this surface charge, we find the difference between normal or z components of electric flux density just above and just below the surface. Just above the surface (Fig. 6.33),

$$\begin{aligned} D_z &= 2\left(\frac{-q}{4\pi R_1^2}\right)\cos\alpha = 2\left(\frac{-q}{4\pi R_1^2}\right)\frac{d/2}{R_1} \\ &= -\frac{qd}{4\pi[(d/2)^2 + r^2]^{3/2}}, \quad \text{at } z = 0+. \end{aligned}$$

Below the surface, $D_z = 0$, at $z = 0-$. The z component above the surface minus the z component below the surface equals the surface charge density [Eq. (6.4)]:

$$\rho_s = -\frac{qd}{4\pi[(d/2)^2 + r^2]^{3/2}}. \qquad (6.63)$$

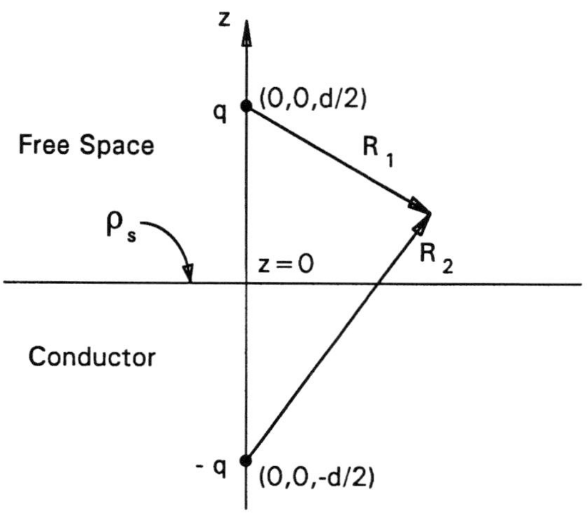

FIGURE 6.32 A point charge in free space and its image with respect to a perfect conducting boundary.

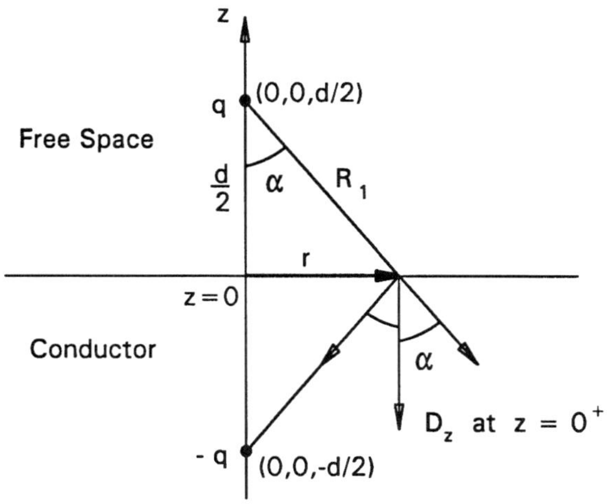

FIGURE 6.33 Evaluation of the surface charge density at the perfect conductor interface of Fig. 6.32.

Since all the electric flux lines emanating from the point charge q at $z = d/2$ end on the $z = 0$ surface, the total charge on that surface must be $Q_s = -q$. We can check that this works out correctly by integrating the surface charge density over the entire $z = 0$ surface:

$$\begin{aligned} Q_s &= \int_{r=0}^{\infty} \int_{\phi=0}^{2\pi} \rho_s r \, d\phi \, dr = \left(\frac{-qd}{4\pi}\right)(2\pi) \int_0^{\infty} \frac{r \, dr}{[(d/2)^2 + r^2]^{3/2}} \\ &= \left(\frac{-qd}{4\pi}\right)(2\pi)\left[-\frac{1}{\sqrt{(d/2)^2 + r^2}}\right]_0^{\infty} = \left(\frac{-qd}{4\pi}\right)(2\pi)\left[\frac{1}{(d/2)}\right] \\ &= -q. \end{aligned} \tag{6.64}$$

No field penetrates into the conducting region $z < 0$. The field will not be changed if the conducting region is replaced by a thin metal plate, the top surface of which is at $z = 0$.

6.7.2 Point Charge Over a Dielectric

A similar technique can be used to find the fields when a point charge is in free space above a dielectric medium (Fig. 6.34). The region $z > 0$ is free space and the point charge q is at $z = d/2$ on the z axis. The region $z < 0$ has relative permittivity ε_r. The field is due to the actual point charge and to a layer of bound surface charge q_{bs} on the $z = 0$ surface. Let us guess, from our experience with the point charge over a conducting region, that the layer of surface charge gives a field in the free-space region that is the same as the field of a point charge q_1 located on the z axis at $z = -d/2$. By symmetry, the surface charge must then give a field in the dielectric region, which is

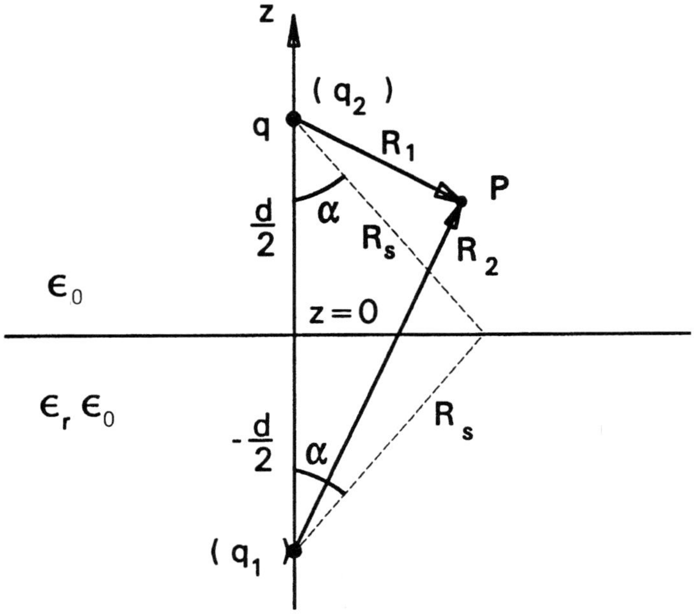

FIGURE 6.34 A point charge in free space and its image with respect to a perfect dielectric boundary.

the field of an additional point charge q_2 located on the z axis at $z = d/2$. Thus the electric potential is

$$V = \begin{cases} \dfrac{q}{4\pi\varepsilon_0 R_1} + \dfrac{q_1}{4\pi\varepsilon_0 R_2}, & z > 0, \\ \dfrac{q + q_2}{4\pi\varepsilon_0 \varepsilon_r R_1}, & z < 0. \end{cases} \tag{6.55}$$

At the interface, $z = 0$, $R_1 = R_2 = R_s$, and the electric potential is continuous.[†] Hence

$$q + q_1 = (q + q_2)/\varepsilon_r. \tag{6.66}$$

Since the boundary is a dielectric to dielectric interface, the normal or z component of electric flux density must be continuous. Hence

$$\frac{-q}{4\pi R_s^2}\cos\alpha + \frac{q_1}{4\pi R_s^2}\cos\alpha = -\frac{q + q_2}{4\pi R_s^2}\cos\alpha,$$

[†] The tangential electric field intensity along the boundary is the negative derivative in the tangential direction of the electric potential. Continuity of the tangential electric field intensity therefore requires that the electric potential on each side of the boundary be the same, except for an additive constant. The constant must be zero because a jump in V across the boundary surface would imply an infinite normal component of E. Thus the electric potential is continuous.

or

$$-q + q_1 = -q - q_2. \tag{6.67}$$

Solving Eqs. (6.66) and (6.67) simultaneously, we find

$$q_1 = -q_2 = -q\frac{\varepsilon_r - 1}{\varepsilon_r + 1}. \tag{6.68}$$

That the two image charges are equal but opposite could have been anticipated from the fact that the total free charge on the $z = 0$ surface is zero.

In solving this problem we "guessed" at the electric potential, Eq. (6.65). If our guess had been wrong, then it would not have been possible to satisfy the two boundary conditions.

6.7.3 Parallel Circular Cylindrical Conductors

Parallel wires are frequently used as transmission lines, for 60 Hz power transmission and for radio frequency transmission. An important parameter of a transmission line is the capacitance per unit length. We will now use the method of images to derive the formula for this parameter.

In Section 6.3.4 we found the field of two parallel line charges. Looking at Fig. 6.19 we see that if the region within an equipotential surface to the right of the $x = 0$ plane is replaced by a conducting circular cylinder of radius a with charge per unit length ρ_{l0} and the region within an equipotential surface to the left of the $x = 0$ plane is replaced by a conducting circular cylinder of radius b with charge per unit length $-\rho_{l0}$ (Fig. 6.35), then the fields outside the conductors will not be changed. The line charges ρ_{l0} and $-\rho_{l0}$ no longer exist, but each may be considered as an image line charge to which we attribute the fields outside the conductors, even though we know that these fields are really due to the surface charge on the cylinders.

Let l_a be the distance from the negative image line charge to the center of the circle of radius a and λ_a be the distance from the positive image line charge to the same

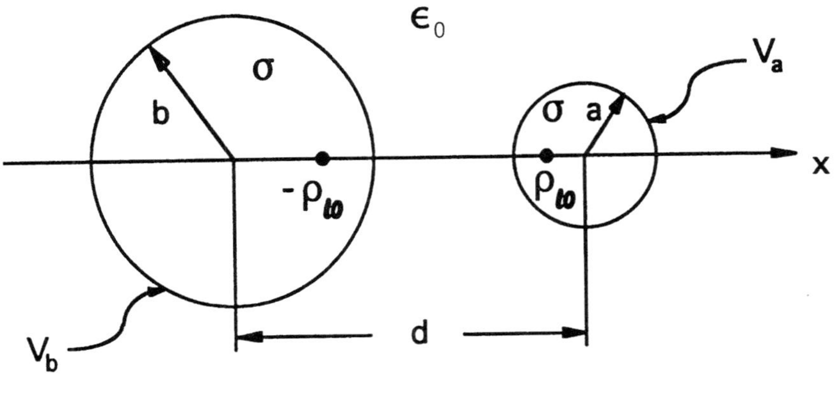

FIGURE 6.35 Parallel circular cylindrical conductors.

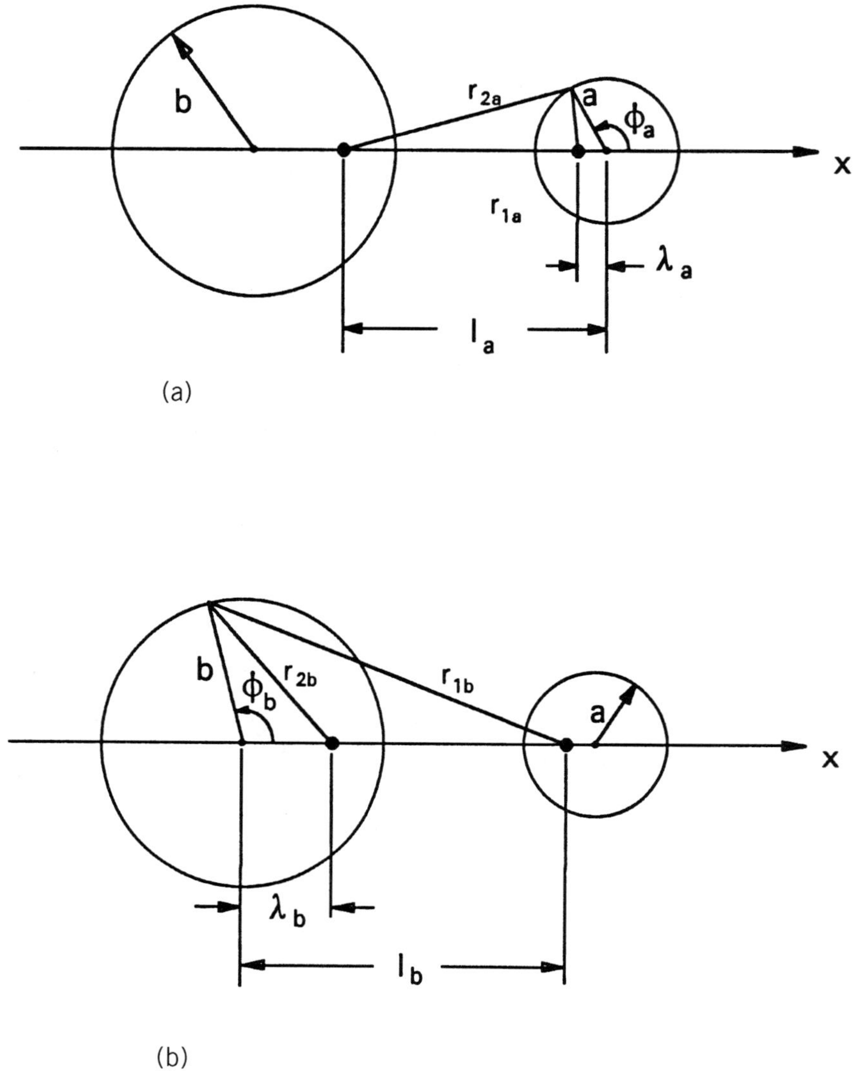

FIGURE 6.36 Parallel circular cylindrical conductors and the determination of their images.

circle's center (Fig. 6.36a). On the circle $V = V_a = (\rho_{l0}/2\pi\varepsilon_0)\ln(r_{2a}/r_{1a})$ [see Eq. (6.35)]. Using the law of cosines, we have

$$\frac{r_{2a}}{r_{1a}} = k_a = \frac{\sqrt{a^2 + l_a^2 + 2al_a\cos\phi_a}}{\sqrt{a^2 + \lambda_a^2 + 2a\lambda_a\cos\phi_a}}$$

$$= \frac{l_a\sqrt{1 + (a/l_a)^2 + 2(a/l_a)\cos\phi_a}}{a\sqrt{1 + (\lambda_a/a)^2 + 2(\lambda_a/a)\cos\phi_a}}.$$

For the ratio to be independent of ϕ_a the radicals in the numerator and the denominator must cancel. This will happen if

$$\frac{a}{l_a} = \frac{\lambda_a}{a} \quad \text{or} \quad \lambda_a = \frac{a^2}{l_a}. \tag{6.69}$$

The result is $k_a = l_a/a > 1$.

Similarly, on the equipotential circle of radius b (Fig. 6.36b), $V = V_b = (\rho_{l0}/2\pi\varepsilon_0)\ln(r_{2b}/r_{1b})$ and

$$\frac{r_{2b}}{r_{1b}} = k_b = \frac{\sqrt{b^2 + \lambda_b^2 - 2b\lambda_b \cos\phi_b}}{\sqrt{b^2 + l_b^2 - 2bl_b \cos\phi_b}}$$

$$= \frac{b\sqrt{1 + (\lambda_b/b)^2 - 2(\lambda_b/b)\cos\phi_b}}{l_b\sqrt{1 + (b/l_b)^2 - 2(b/l_b)\cos\phi_b}}.$$

Again, for the ratio to be independent of ϕ_b, the radicals in the numerator and the denominator must cancel. This requires that

$$\frac{b}{l_b} = \frac{\lambda_b}{b} \quad \text{or} \quad \lambda_b = \frac{b^2}{l_b}. \tag{6.70}$$

The result is $k_b = b/l_b < 1$.

The potential difference between the cylinders is

$$V_0 = V_a - V_b = \frac{\rho_{l0}}{2\pi\varepsilon_0} \ln\left(\frac{k_a}{k_b}\right)$$

$$= \frac{\rho_{l0}}{2\pi\varepsilon_0} \ln\left(\frac{l_a l_b}{ab}\right). \tag{6.71}$$

This expression would be more useful if we could eliminate the distances l_a and l_b and introduce the distance d between the cylinder's centers. From Figs. 6.36a and 6.36b and Eqs. (6.69) and (6.70), we have

$$d = l_a + \lambda_b = l_a + \frac{b^2}{l_b},$$

$$d = l_b + \lambda_a = l_b + \frac{a^2}{l_a},$$

which are two simultaneous equations that can be solved for l_a and l_b in terms of the dimensions a, b, and d. The second equation gives $l_b = d - (a^2/l_a)$, which is substituted in the first equation to obtain the quadratic equation

$$dl_a^2 + (b^2 - a^2 - d^2)l_a + da^2 = 0.$$

Solving for l_a gives

$$l_a = \frac{(d^2 + a^2 - b^2) \pm \sqrt{(d^2 + a^2 - b^2)^2 - 4d^2a^2}}{2d}$$

$$= \frac{(d^2 + a^2 - b^2) \pm \sqrt{(d^2 - a^2 - b^2)^2 - 4a^2b^2}}{2d}.$$

Then

$$\frac{l_a l_b}{ab} = \frac{l_a(d - a^2/l_a)}{ab} = \frac{l_a d - a^2}{ab}$$

$$= \frac{(d^2 + a^2 - b^2) \pm \sqrt{(d^2 - a^2 - b^2)^2 - 4a^2b^2} - 2a^2}{2ab}$$

$$= \frac{d^2 - a^2 - b^2}{2ab} \pm \sqrt{\left(\frac{d^2 - a^2 - b^2}{2ab}\right)^2 - 1}$$

$$= p \pm \sqrt{p^2 - 1}, \quad \text{where } p = \frac{d^2 - a^2 - b^2}{2ab}. \tag{6.72}$$

Recall the identity $\ln(p \pm \sqrt{p^2 - 1}) = \cosh^{-1}(\pm p)$ and that the argument of the inverse hyperbolic cosine must be greater than 1, $\pm p \geq 1$. Substituting Eq. (6.72) into Eq. (6.71), we obtain the result that

$$V_0 = \frac{\rho_{l0}}{2\pi\varepsilon_0} \ln(p \pm \sqrt{p^2 - 1})$$

$$= \frac{\rho_{l0}}{2\pi\varepsilon_0} \cosh^{-1}(\pm p). \tag{6.73}$$

The capacitance per unit length is $C_l = \rho_{l0}/V_0$;

$$C_l = \frac{2\pi\varepsilon_0}{\cosh^{-1}(\pm p)} = \frac{2\pi\varepsilon_0}{\cosh^{-1}\ [\pm(d^2 - a^2 - b^2)/2ab]}. \tag{6.74}$$

The physical significance of the $\pm$sign is as follows. Figure 6.35 shows the problem of two parallel wires, which we started to investigate. Clearly, $d \geq a + b$ and hence $d^2 \geq (a + b)^2 = a^2 + b^2 + 2ab$. Transposing a^2 and b^2 and dividing by $2ab$, we have $p \geq 1$. This means that we must use the positive sign for this physical configuration. Figure 6.37 also corresponds to the equipotential circles of Fig. 6.19. It represents a hollow wire of radius b enclosing an eccentrically centered wire of radius a. In this case, $b \geq a + d$ and hence $d \leq b - a$; thus $d^2 \leq (b - a)^2 = b^2 + a^2 - 2ab$. Transposing and dividing as before, we have $p \leq -1$. This means that we must use the negative sign for this physical configuration.

Note that in both situations the image line charges lie outside the region in which the fields exist. This must always be true for image solutions. If the images fell within the field region then they would cause singularities in the field, where no real charge exists. Such singularities can only be found at the position of real charges.

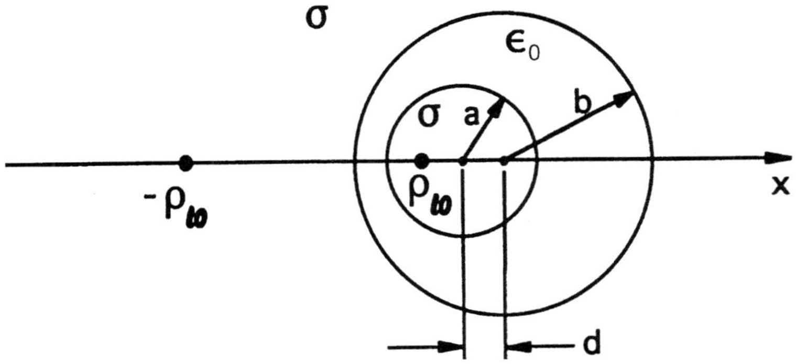

FIGURE 6.37 Parallel circular cylindrical conductors (one inside the other) and the determination of their images.

Example 6.9 (a) For the configuration of Fig. 6.35, $a = 0.5$, $b = 1.0$, and $d = 3.0$ inches. Calculate $C_l/2\pi\varepsilon_0$. (b) For the configuration of Fig. 6.37, $a = 0.5$, $b = 1.0$, and $d = 0.375$ inch. Calculate $C_l/2\pi\varepsilon_0$.

Solution

$$\text{(a)}\quad p = \frac{3.0^2 - 0.5^2 - 1.0^2}{2(0.5)(1.0)} = 7.75, \quad \frac{C_l}{2\pi\varepsilon_0} = [\cosh^{-1}(7.75)]^{-1} = 0.365.$$

$$\text{(b)}\quad p = \frac{0.375^2 - 0.5^2 - 1.0^2}{2(0.5)(1.0)} = -1.11, \quad \frac{C_l}{2\pi\varepsilon_0} = [\cosh^{-1}(1.11)]^{-1} = 2.16.$$

The second capacitance is greater than the first because the cylinders are closer together in that configuration.

Because we worked so hard to obtain this formula for capacitance, we should try to squeeze some additional results from it. First, for two parallel wires of equal radius, a common practical situation, we set $b = a$ and obtain

$$C_l = \frac{2\pi\varepsilon_0}{\cosh^{-1}[\frac{1}{2}(d/a)^2 - 1]}. \tag{6.75}$$

Second, let us look at the case of a single wire of radius a alongside a conducting plane (Fig. 6.38). The conducting plane is a circle of radius $b \to \infty$ with its center a distance d from the wire's center receding infinitely far to the left. Let $h = d - b$ so that

$$p = \frac{(d-b)(d+b) - a^2}{2ab} = \frac{h(h+2b) - a^2}{2ab}$$

$$= \frac{h}{a} + \frac{h^2 - a^2}{2ab}.$$

$$\lim_{b\to\infty} p = \frac{h}{a}.$$

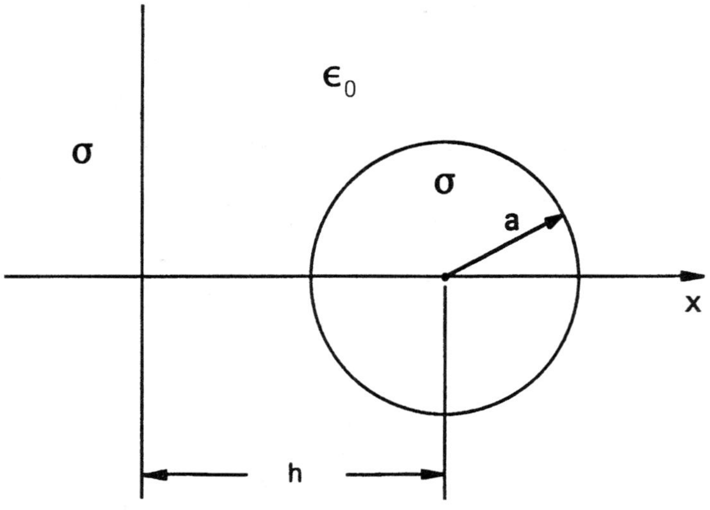

FIGURE 6.38 Conducting wire of radius a near a conducting plane.

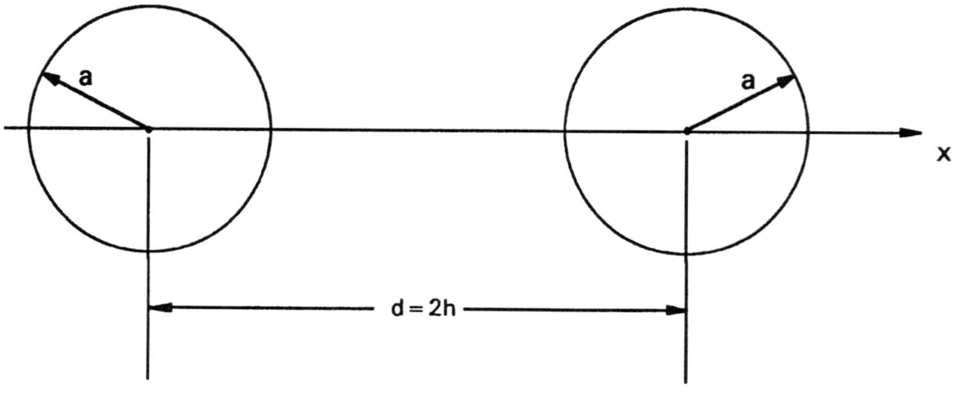

FIGURE 6.39 Two parallel wires of equal radius.

The capacitance per unit length of a wire of radius a with its axis a distance h from a conducting plane is therefore

$$C_l = \frac{2\pi\varepsilon_0}{\cosh^{-1}(h/a)}. \tag{6.76}$$

Third, returning to the first problem, the parallel wires of equal radius with axes $d = 2h$ apart will have half the capacitance per unit length as that given in Eq. (6.76) since they consist of two such capacitors in series (Fig. 6.39). Thus

$$C_l = \frac{\pi\varepsilon_0}{\cosh^{-1}(d/2a)}. \tag{6.77}$$

Equations (6.75) and (6.77) must, of course, be equal for all $d/2a > 1$.

6.8 POISSON'S AND LAPLACE'S EQUATIONS

Equation (6.23) gives an integral relation between the electric charge and the electric potential. We will next find a differential relation connecting these two quantities and examine several methods of solution.

6.8.1 Poisson's Equation

Combining Eqs. (6.2b), (6.5), and (6.17) for simple media characterized by a single constant permittivity, we have

$$\nabla\cdot\vec{D} = \nabla\cdot(\varepsilon\vec{E}) = \varepsilon\nabla\cdot(-\nabla V) = \rho$$

$$\nabla^2 V = -\frac{\rho}{\varepsilon} \tag{6.78}$$

as a direct relation between the free charge density and the electric potential at a point in space. This equation is known as Poisson's equation. The equation states that the scalar Laplacian of the potential equals the negative of the charge density divided by the permittivity. The Laplacian was discussed and formulas are given in Section 5.4.

The solution of Poisson's equation for a point charge is given by Eq. (6.20) and the solution for a known charge distribution is given by Eq. (6.23).

When the charge density is known or assumed and when the field possesses sufficient symmetry so that it is one dimensional, then it is possible to solve Poisson's equation by direct integration. This is illustrated by the following example.

Example 6.10 The charge density in the cylindrical region surrounding the z axis is $\rho = \rho_0$, $r < a$ and $\rho = 0$, $r > a$. Find the electric potential with respect to a zero potential reference at some radius $r_0 > a$.

Solution The problem has symmetry around and along the z axis, so that fields are independent of the ϕ and z coordinates. Using Eq. (5.21) we have

$$\frac{1}{r}\frac{d}{dr}\left(r\frac{dV}{dr}\right) = \begin{cases} \dfrac{-\rho_0}{\varepsilon_0}, & r < a, \\ 0, & r > a. \end{cases}$$

The partial derivatives have been replaced by ordinary derivatives because the potential is a function of the radial coordinate only. Integrating in the region $r < a$, we find

$$\frac{d}{dr}\left(r\frac{dV}{dr}\right) = -\frac{\rho_0}{\varepsilon_0}r$$

$$r\frac{dV}{dr} = -\frac{\rho_0}{2\varepsilon_0}r^2 + C_1$$

$$-E_r = \frac{dV}{dr} = -\frac{\rho_0}{2\varepsilon_0}r + C_1\frac{1}{r}$$

$$V = -\frac{\rho_0}{4\varepsilon_0}r^2 + C_1 \ln r + C_2, \quad r < a.$$

Integrating in the region $r > a$, we find

$$\frac{d}{dr}\left(r\frac{dV}{dr}\right) = 0$$

$$r\frac{dV}{dr} = C_3$$

$$-E_r = \frac{dV}{dr} = C_3\frac{1}{r}$$

$$V = C_3 \ln r + C_4, \quad a < r.$$

This gives us the solution in terms of four unknown constants of integration—C_1, C_2, C_3, and C_4. We must utilize the boundary conditions to find these constants. Let us number the conditions to be sure we have four.

1. $V(0)$ is finite, because there is no charge singularity on the z axis.
2. $V(r_0) = 0$, $r_0 > a$, as given.
3. D_r is continuous at $r = a$, because there is no surface charge there.
4. V is continuous at $r = a$.

The first condition requires that $C_1 = 0$, since $\ln r$ is singular at $r = 0$. The second condition means that $C_4 = -C_3 \ln r_0$ so that $V = C_3 \ln(r/r_0)$, $r > a$. The third condition implies that the E_r is also continuous because the permittivity is the same in each region. Thus

$$-\frac{\rho_0}{2\varepsilon_0}a = C_3\frac{1}{a}$$

$$C_3 = -\frac{\rho_0 a^2}{2\varepsilon_0}.$$

Finally, the fourth condition gives

$$-\frac{\rho_0}{4\varepsilon_0}a^2 + C_2 = -\frac{\rho_0 a^2}{2\varepsilon_0}\ln\left(\frac{a}{r_0}\right)$$

$$C_2 = \frac{\rho_0 a^2}{4\varepsilon_0} - \frac{\rho_0 a^2}{2\varepsilon_0}\ln\left(\frac{a}{r_0}\right).$$

Letting $\rho_{l0} = \rho_0 \pi a^2$, the charge per unit length around the z axis, we find

$$V = \begin{cases} \dfrac{\rho_{l0}}{4\pi\varepsilon_0}\left[1 - \left(\dfrac{r}{a}\right)^2\right] - \dfrac{\rho_{l0}}{2\pi\varepsilon_0}\ln\left(\dfrac{a}{r_0}\right), & r < a, \quad (6.79) \\ -\dfrac{\rho_{l0}}{2\pi\varepsilon_0}\ln\left(\dfrac{r}{r_0}\right), & r > a. \quad (6.80) \end{cases}$$

The result for $r > a$ is the same as the result for a line charge ρ_{l0} along the z axis, Eq. (6.34).

6.8.2 Laplace's Equation

When there is no free charge in a region, then $\rho = 0$ and

$$\nabla^2 V = 0, \tag{6.81}$$

which is known as Laplace's equation.

Laplace's equation has been studied extensively and there are several powerful methods of solution. The usefulness of these methods generally depends on the shape of the region and the boundary conditions at the edges of the region.

Many two-dimensional problems, in which the sources and the fields are independent of the z coordinate, have been solved by the method of conformal transformation, which "transforms" a complicated shaped boundary into a simple shape for which field solutions are known. The inverse transformation then transforms these known fields back to the complicated geometry. Often the capacitance can also be found by this method. This method requires a knowledge of complex variable theory and will not be covered here [1, 2].

A powerful technique for the solution of Laplace's equation is the method of separation of variables. This technique is useful when the boundaries surrounding the region coincide with the surfaces formed by holding a space variable constant in an orthogonal coordinate system. Some examples are given in Fig. 6.40. In this technique the potential function of the three orthogonal coordinates is written as a product of functions of each of the coordinates. Thus, in the Cartesian system, we would write $V(x, y, z) = X(x)Y(y)Z(z)$. When this product function is substituted in Laplace's equation, which is a second-order partial differential equation, it "separates" into three second-order ordinary differential equations. The ordinary differential equations are solvable in terms of known functions, which may be combined to form a complete solution. For example, in Cartesian coordinates, the solutions are in terms of circular functions (sine and cosine functions), hyperbolic functions (hyperbolic sine and hyperbolic cosine functions), and exponential functions. The technique will be used later on when we look for field solutions in a waveguide, but we will not discuss it further in connection with electrostatic problems.

6.8.3 Numerical Techniques

Numerical techniques can be used to find numerical values for the potential regardless of the shape of the boundary. A computer can be used to remove the tedium of finding potential values to any degree of accuracy desired.

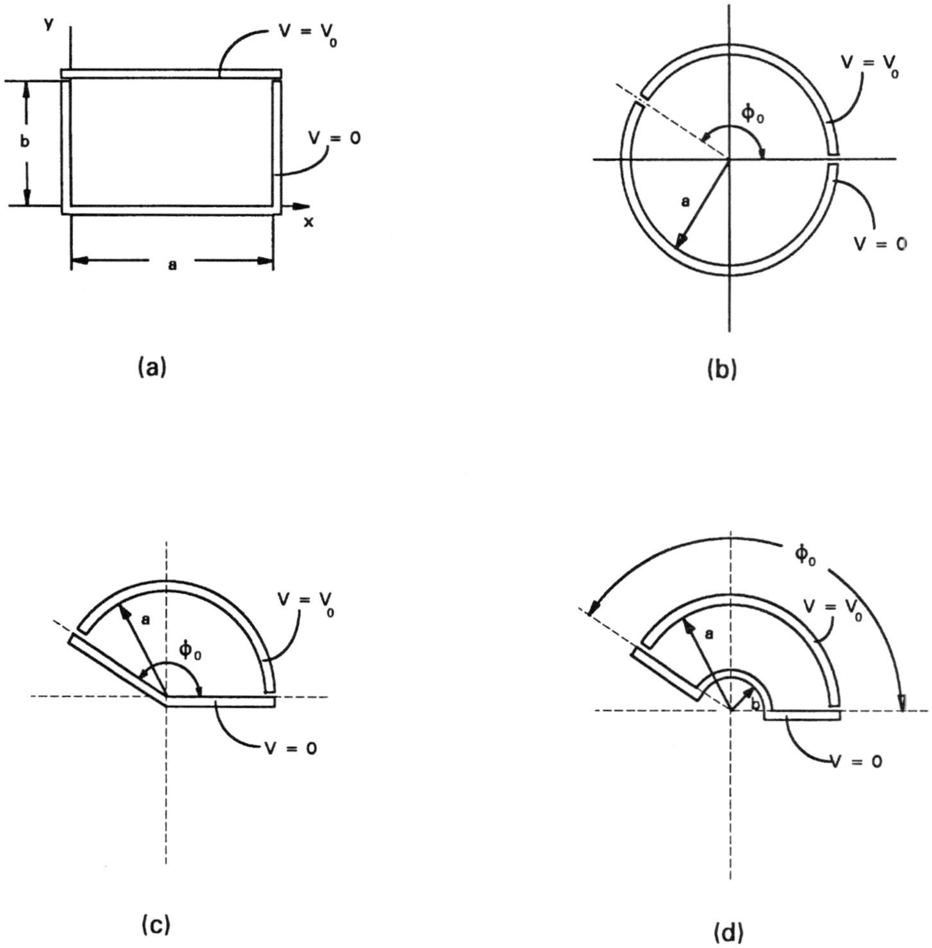

FIGURE 6.40 Example of boundary surfaces where the method of separation of variables may be used to calculate static fields.

6.8.3.1 The Finite Difference Method The finite difference method is the simplest numerical technique. This method is based on the result that the potential at any point in space is closely approximated as the average value of the potential at nearby equidistant points. In using this method the space between the region's boundaries, where the potential is given, can be divided into a grid of equally spaced points. The value of the potential at each point is first estimated and then it is calculated by systematically sweeping through the grid and averaging the potential at surrounding adjacent points.

Consider the point P_0 in the grid surrounded by the adjacent four points P_1, P_2, P_3, P_4, as shown in Fig. 6.41. The potential at point P_0, defined as V_0, is calculated in terms of the potentials at the surrounding points, V_1, V_2, V_3, and V_4, as follows:

$$V_0 = \frac{V_1 + V_2 + V_3 + V_4}{4}.$$

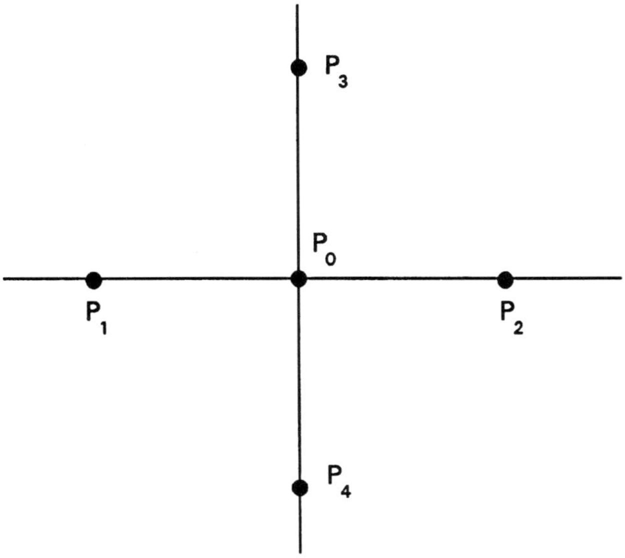

FIGURE 6.41 Grid points used in the finite difference method calculation.

At the boundary between two media of different dielectric permittivities, the potentials at the points on the boundary are calculated as follows:

$$V_0 = \frac{(\varepsilon_1 + \varepsilon_2)V_1 + (\varepsilon_1 + \varepsilon_2)V_2 + 2\varepsilon_1 V_3 + 2\varepsilon_2 V_4}{4(\varepsilon_1 + \varepsilon_2)}.$$

The potentials at the four points and the dielectric permittivities of the two media are defined in Fig. 6.42. The process is repeated until the potential values converge. The method can be applied to odd-shaped structures. The price of simplicity is numerical inefficiency [3].

A number of examples are now given to illustrate this numerical method.

Example 6.11 Write a program that uses the finite difference method to determine the potential inside the structure shown in Fig. 6.43a. Select a grid of 13 by 13 points. Try to plot equal potential lines. Use MathCad to solve the same problem.

Solution A listing of a BASIC program that uses the finite difference method to determine the potential inside the given structure, for any potential assignments on the boundaries of the structure, is given in Listing 6.1.

The following steps are taken in getting the solution to the problem:

1. An appropriate grid is selected; in our case it is a 13 by 13 grid (see Fig. 6.43b).
2. Initialize the potential array elements by setting the potential to an initial value (in this example a value of 0 was selected) except for the boundary values. (See statements 30 through 240 in Listing 6.1.)

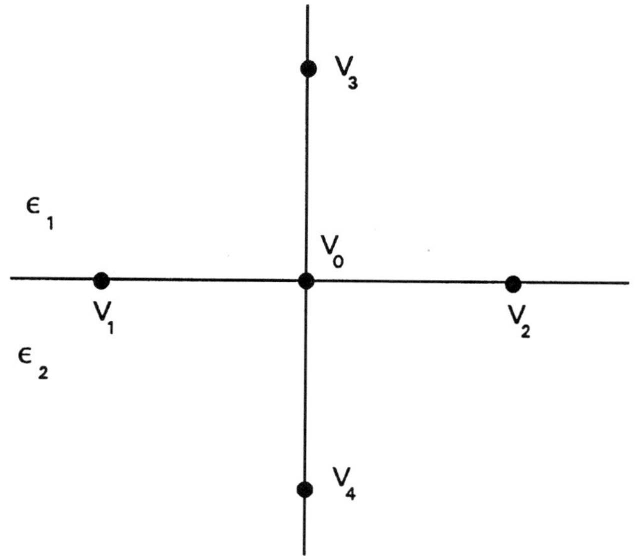

FIGURE 6.42 Grid points and the corresponding scalar potentials at the boundary between two dielectrics.

3. Replace the potential value at each grid point, except for the boundary values, by the average of the potentials at the four points around it. Update the error function (which in this case is the square root of the sum of the squares of the differences between the past and present values at all grid points). (See statements 250 through 350 in Listing 6.1.)
4. Having gone once through all the array elements, evaluate the error function and check for accuracy. If the error function is not less than the desired small value (10^{-7} volts), then go to step 3 and iterate again. If the error function is less than the desired value, then print the array elements and end the program. (See statements 360 through 440.)

A similar solution can be obtained using MathCad. The BASIC program stops the iterations when an error function is less then a chosen minimum. In MathCad, the control of the solution is done by controlling the number of iterations. The MathCad file used for the solution of the problem is given in Listing 6.2. MathCad allows iterations to be performed by using the "until" function (see Appendix B on MathCad for more information). The plot of the equipotential lines is very easily obtained by displaying a contour plot of the potential matrix, as seen in Fig. 6.44.

The capacitance per unit length of strip capacitors, microstrip lines, and other cylindrical structures can be determined using the finite difference method. The charge can be calculated for the given structure by integrating the electric flux density over a properly selected volume:

$$Q = \oint_S \varepsilon \vec{E} \cdot d\vec{S}.$$

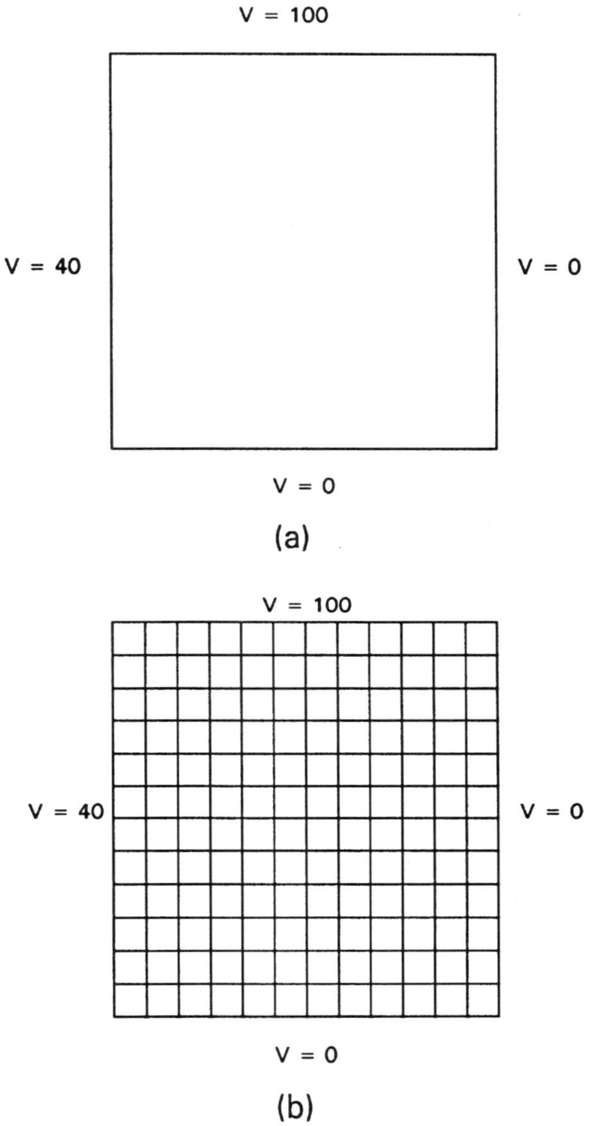

FIGURE 6.43 A rectangular structure with boundary conditions and the grid used for a finite difference calculation of the electrostatic potential.

We only need the charge per unit length to determine the capacitance per unit length. The electric field can also be approximated by the derivative of the electric potential in a direction normal to the integrating path. The calculation of the charge per unit length reduces to the closed line integral of the derivative of the electric potential in a direction normal to the integration path:

$$q = \frac{Q}{l} = \oint_L \varepsilon \vec{E} \cdot \frac{d\vec{S}}{l} = \oint_L \varepsilon(-\nabla V) \cdot d\vec{n},$$

```
10    REM FINITE ELEMENT PROGRAM
20    CLS
30    BCTOP = 100
40    BCBOTTOM = 0
50    BCLEFT = 40
60    BCRIGHT = 0
70    NI = 13
80    NJ = 13
90    DIM SHARED V(NI, NJ)
100   DIM SHARED P(NI, NJ)
110   FOR I = 1 TO NI - 1 STEP 1
120   FOR J = 1 TO NJ - 1 STEP 1
130   V(I, J) = 0
140   P(I, J) = 0
150   NEXT J
160   NEXT I
170   FOR I = 1 TO NI STEP 1
180   V(I, NJ) = BCRIGHT
190   V(I, 1) = BCLEFT
200   NEXT I
210   FOR J = 1 TO NJ STEP 1
220   V(1, NJ) = BCTOP
230   V(NI, J) = BCBOTTOM
240   NEXT J
250   FOR K = 1 TO 1000 STEP 1
260   SUM = 0
270   FOR I = 2 TO NI - 1 STEP 1
280   FOR J = 2 TO NJ - 1 STEP 1
290   P(I, J) = V(I, J)
300   V(I, J) = 1/4* (V(I-1, J) + V(I+1, J) + V(I, J-1) + V(I, J+1))
310   SUM = SUM + (V(I, J) - P(I, J))^ 2
320   NEXT J
330   NEXT I
340   SUM = SUM ^ .5
350   PRINT SUM
360   IF SUM < .0000001 THEN GOTO 380
370   NEXT K
380   FOR I = 1 TO NI STEP 1
390   FOR J = 1 TO NJ STEP 1
400   PRINT USING "###.##"; V(I, J);
410   NEXT J
420   PRINT
430   NEXT I
440   END
```

LISTING 6.1 BASIC program for finite difference method in Example 6.11.

```
i:=0..12
j:=0..12
Top boundary condition:       V_{i,12}:=100
Bottom boundary condition:    V_{i,0}:=0
Left boundary condition:      V_{0,j}:=40
Right boundary condition:     V_{12,j}:=0
k: =1..300
i: =1..11
j: =1..11
```

$$V_{i,j} := \text{until}\left(k, \frac{V_{i-1,j} - V_{i+1,j} + V_{i,j-1} + V_{i,j+1}}{4}\right)$$

LISTING 6.2 MathCad file used in Example 6.11.

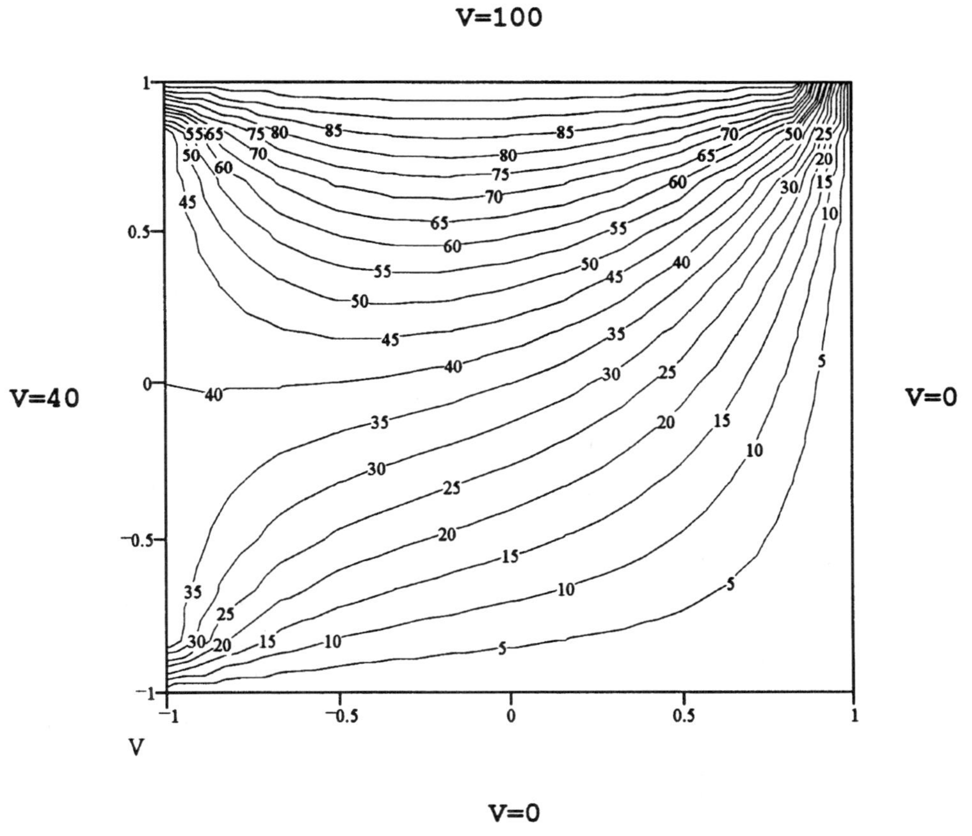

FIGURE 6.44 A display of the calculated equipotential lines for Example 6.11.

where $d\vec{n} = d\vec{S}/l$. The charge per unit length q then becomes

$$q = -\oint_L \nabla V \cdot d\vec{n} = -\sum_L \frac{\Delta V}{\Delta n} dn,$$

where dn equals the grid size in the normal direction to the integration path. Since the potentials involved in the ΔV operation are two grid spaces apart, the equation for the charge per unit length becomes

$$q = -\sum_L \frac{\Delta V}{\Delta n} dn = -\sum_L \frac{\Delta V}{2},$$

where the ΔV is taken in a direction normal to the path L.

Consider the structure shown in Fig. 6.45. The charge per unit length is calculated as

$$q = -\left(\cdots + \varepsilon_1 \cdot \frac{V_{i-2,j-1} - V_{i-2,j+1}}{2} + \varepsilon_1 \cdot \frac{V_{i-1,j-1} - V_{i-1,j+1}}{2} \right.$$
$$\left. + \frac{\varepsilon_1 + \varepsilon_2}{2} \cdot \frac{V_{i,j-1} - V_{i,j+1}}{2} + \varepsilon_2 \cdot \frac{V_{i+1,j-1} - V_{i+1,j+1}}{2} + \cdots \right)$$

where ε_1 and ε_2 are the dielectric permittivities of the two regions.

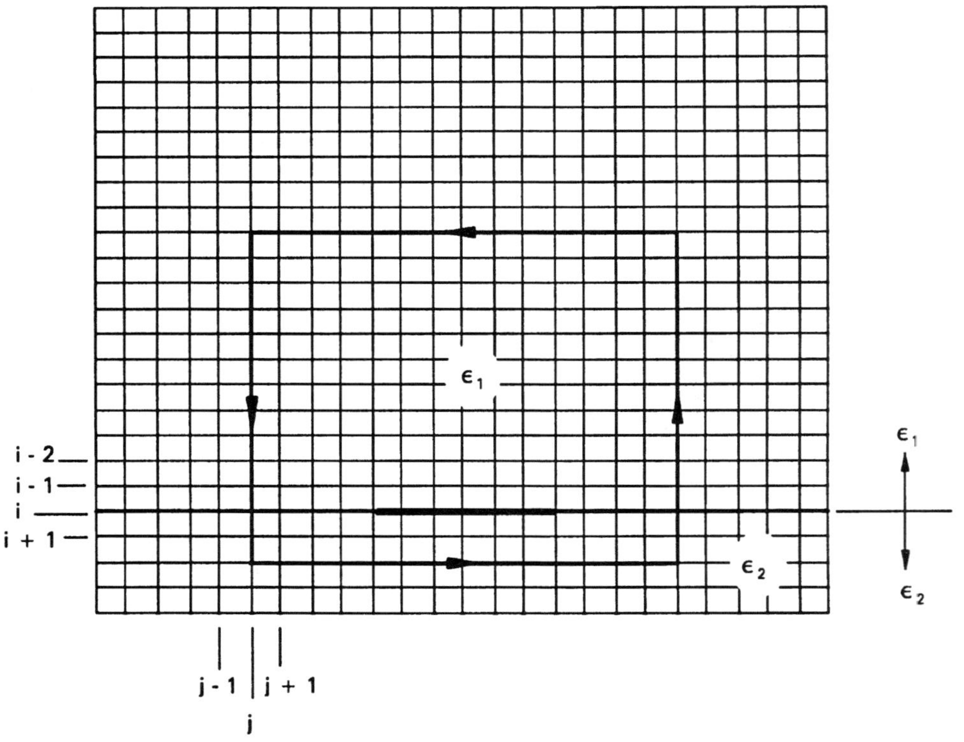

FIGURE 6.45 Determination of the capacitance per unit length using the finite difference method.

The capacitance per unit length is then calculated as

$$C = \frac{q}{V_0}.$$

Example 6.12 Determine the capacitance per unit length for the structure (the shielded microstrip line) shown in Fig. 6.46 when Teflon ($\varepsilon_r = 2.1$) is used as the dielectric.

Solution First, the potential inside the structure is determined in a similar fashion to the previous example. A grid of 50 by 50 points is selected. The grid lines are numbered from 0 to 49 in both directions. The height of the dielectric becomes 5 grid units high (from row 44 to row 49), while the width of the 10 volt strip becomes 11 grid units long (from column 19 to column 30). The potential inside the structure is determined using MathCad, and the file is given in Listing 6.3.

The charge per unit length is obtained by evaluating the closed line integral of the derivative of the electric potential in a direction normal to the integration path. The closed path surrounding the 10 volt strip is chosen midway between the strip and the boundaries, and runs along the 22nd row (from column 39 to column 10), along the 10th column (from row 22 to row 46), along the 46th row (from column 10 to

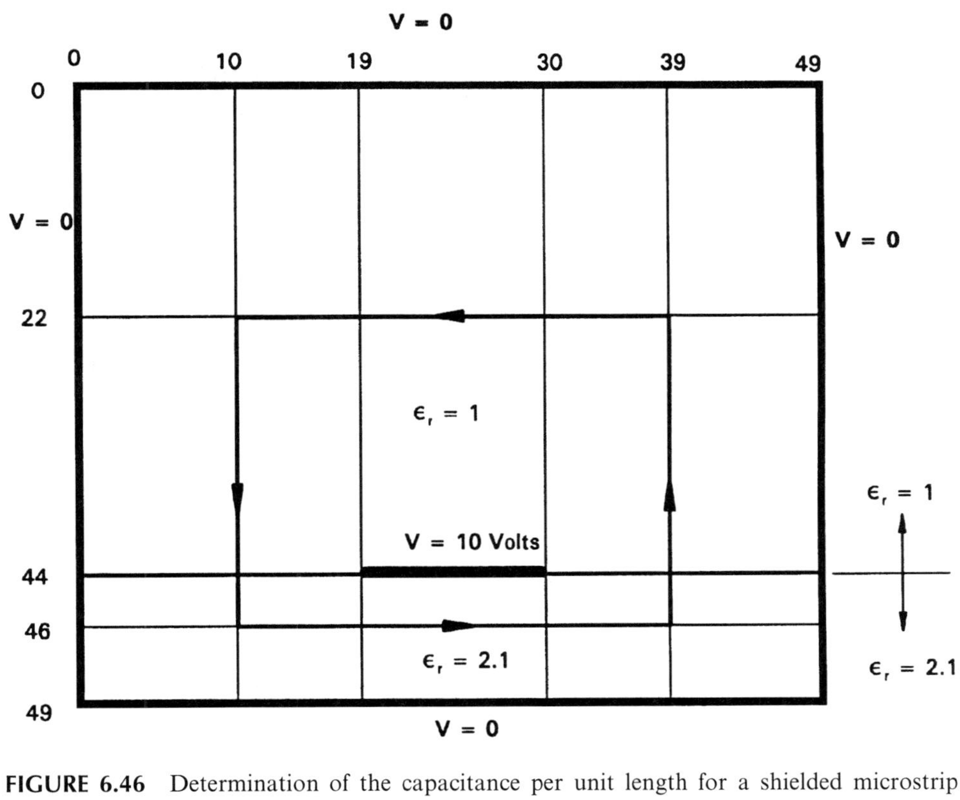

FIGURE 6.46 Determination of the capacitance per unit length for a shielded microstrip structure.

$$i := 0..49 \qquad \varepsilon_1 := 1$$
$$j := 0..49 \qquad \varepsilon_2 := 2.1$$

$$V_{i,j} := 0 \qquad \varepsilon_{e1} := \frac{2 \cdot \varepsilon_1}{\varepsilon_1 + \varepsilon} \qquad \varepsilon_{e2} := \frac{2 \cdot \varepsilon_2}{\varepsilon_1 + \varepsilon_2}$$

$$V_{i,j} := \mathrm{if}\ (i = 44,\ \mathrm{if}\ (j > 18,\ \mathrm{if}\ (j < 31, 10, 0), 0), 0)$$
$$k := 1..200$$
$$i := 1..48$$
$$j := 1..48$$

$$V_{i,j} := \mathrm{until}\left(k,\ \mathrm{if}\left(i = 44,\ \mathrm{if}\left(j > 18,\ \mathrm{if}\left(j < 31, 10,\right.\right.\right.\right.$$
$$\left.\times \frac{V_{i,j-1} + V_{i,j+1} + \varepsilon_{e1} \cdot V_{i-1,j} + \varepsilon_{e2} \cdot V_{i+1,j}}{4}\right),$$
$$\left.\times \frac{V_{i,j-1} + V_{i,j+1} + \varepsilon_{e1} \cdot V_{1-1,j} + \varepsilon_{e2} \cdot V_{i+1,j}}{4}\right),$$
$$\left.\left.\times \frac{V_{i-1,j} + V_{i+1,j} + V_{i,j-1} + V_{i,j+1}}{4}\right)\right)$$

LISTING 6.3 MathCad file used to calculate the potential for a shielded microstrip line.

column 39), and along the 39th column (from row 46 to row 22). The summation of all the contributions of the derivative of the potential along the closed path is obtained using MathCad. The capacitance per unit length is then obtained as

$$C = \frac{q}{V_0}.$$

The results are shown in Listing 6.4.

6.8.3.2 The Finite Element Method The finite element method is related. This technique is based on variational expressions that lead to greater numerical efficiency at the cost of computational complexity. The method has some advantages in dealing with odd-shaped regions [4].

PROBLEMS

6.1 A point charge of $2\,\mu$C is located at $\vec{R}'$. Evaluate the electric field intensity at $\vec{R}$ if:

a. $\vec{R}' = 0,\ \vec{R} = \vec{a}_z 3$ m.

b. $\vec{R}' = \vec{a}_y 4$ m, $\vec{R} = \vec{a}_y 4 + \vec{a}_z 3$ m.

c. $\vec{R}' = \vec{a}_x 2 + \vec{a}_y 4 - \vec{a}_z 3$ m, $\vec{R} = \vec{a}_x 2 + \vec{a}_y 4$ m.

Ans. $\vec{E} = \vec{a}_z 2000$ V/m.

$$\varepsilon_0 := \frac{10^{-9}}{36 \cdot \pi}$$

$$Q_1 := \sum_{j=38}^{10} \frac{V_{21,j} - V_{23,j}}{2} \cdot \varepsilon_1 + \sum_{i=3}^{43} \frac{V_{i,9} - V_{i,11}}{2} \cdot \varepsilon_1 + \left(\frac{V_{44,9} - V_{44,11}}{2} \cdot \frac{\varepsilon_1 + \varepsilon_2}{2} \right)$$

$$+ \sum_{i=45}^{46} \frac{V_{i,9} - V_{i,11}}{2} \cdot \varepsilon_2$$

$$Q_2 := \sum_{j=11}^{39} \frac{V_{47,j} - V_{45,j}}{2} \cdot \varepsilon_2 + \frac{V_{45,40} - V_{45,38}}{2} \cdot \varepsilon_2 + \left(\frac{V_{44,40} - V_{44,38}}{2} \cdot \frac{\varepsilon_1 + \varepsilon_2}{2} \right)$$

$$+ \sum_{i=43}^{22} \frac{V_{i,40} - V_{i,38}}{2} \cdot \varepsilon_1$$

$$Q := Q_1 + Q_2$$

$$C := \frac{Q \cdot \varepsilon_0}{10}$$

$$C = 73.177 \cdot \text{kg} \cdot \text{m}^2 \cdot \text{sec}^{-2} \cdot \text{coul}^{-2} \cdot \text{pF}$$

LISTING 6.4 MathCad partial file (to be attached to the file in Listing 6.3) used to calculate the capacitance per unit length for a shielded microstrip line.

6.2 A point charge of -6 nC is located at (2, 4, 6) m and a second point charge of 8 nC is located at (1, 3, 5) m. Evaluate the electric field intensity at (2, 3, 4) m. Evaluate its magnitude.
Ans. $\vec{E} = \vec{a}_x 25.5 - \vec{a}_y 20.6 + \vec{a}_z 9.7$ V/m, $|\vec{E}| = 34.2$ V/m.

6.3 A uniformly charged circular disk of radius a and surface charge density ρ_{s0} lies in the xy plane.

a. Find an expression for the electric field intensity at any point along the $+z$ axis.

b. Show that $\vec{E} \to \vec{a}_z \rho_{s0}/2\varepsilon_0$ as $a \to \infty$.

c. Show that $\vec{E} \to \vec{a}_z Q/4\pi\varepsilon_0 z^2$ as $a \to 0$ and find Q.

Ans. a. $\vec{E} = \vec{a}_z \dfrac{\rho_{s0} z}{2\varepsilon_0} \left(\dfrac{1}{z} - \dfrac{1}{\sqrt{a^2 + z^2}} \right)$; c. $Q = \rho_{s0} \pi a^2$.

For points along $-z$ axis the results are the negative of these.

6.4 A uniform line charge ρ_{l0} extends along the z axis from $-L$ to $+L$ (see Fig. 6.12).

a. Using Eq. (6.14), find the electric field intensity at any point (r, ϕ, z).

b. Show that as $L \to \infty$, $\vec{E} \to \vec{a}_r \rho_{l0}/2\pi\varepsilon_0 r$.

Ans. a. $\vec{E} = \dfrac{\rho_{l0}}{4\pi\varepsilon_0} \left[\vec{a}_r \dfrac{1}{r} \left(\dfrac{z+L}{R_1} - \dfrac{z-L}{R_2} \right) + \vec{a}_z \left(\dfrac{1}{R_2} - \dfrac{1}{R_1} \right) \right]$,

$$R_1 = \sqrt{r^2 + (z+L)^2}, \; R_2 = \sqrt{r^2 + (z-L)^2}.$$

6.5 A uniform line charge ρ_{l0} forms a circle of radius a centered at the origin and lying in the xy plane.

a. Find the electric field intensity at any point along the z axis.

b. The circle of charge is moved to the $z = z'$ plane. Find the electric field intensity at any point along the z axis.

c. A circular cylinder of radius a coaxial with the z axis extends from $-L$ to $+L$. The cylinder has uniform surface charge density ρ_{s0}. Find the electric field intensity along the z axis.

Ans. a. $E_z = \dfrac{\rho_{l0}\, az}{2\varepsilon_0 (a^2 + z^2)^{3/2}}$;

c. $E_z = \dfrac{\rho_{s0} a}{2\varepsilon_0}\left[\dfrac{1}{\sqrt{a^2 + (z - L)^2}} - \dfrac{1}{\sqrt{a^2 + (z + L)^2}}\right].$

6.6 Calculate the electric potential at (2, 3, 4) m for the charge distribution of Problem 6.2.

Ans. $V = 26.8$ V.

6.7 For the uniformly charged disk of Problem 6.3, find the electric potential at any point along the z axis. Show that the answer to Problem 6.3a is obtained by taking $E_z = -\partial V/\partial z$.

Ans. a. $V(0, 0, z) = \dfrac{\rho_{s0}}{2\varepsilon_0}[+\sqrt{z^2 + a^2} - |z|]$.

6.8 Repeat Problem 6.4, by finding $\vec{E} = -\nabla V$, where V is given by Eq. (6.27). Show that you get the same result.

6.9 Find the electric potential for the linear quadrupole of Fig. 6.16a. Use this result to find the electric field intensity.

Ans. $V = \dfrac{qd^2}{8\pi\varepsilon_0 R^3}(1 + 3\cos 2\theta)$,

$\vec{E} = \dfrac{3qd^2}{8\pi\varepsilon_0 R^4}[\vec{a}_R(1 + 3\cos 2\theta) + \vec{a}_\theta 2\sin 2\theta]$.

6.10 Show that Eq. (6.29) reduces to Eq. (6.34) when L is much greater than the distance to the field point at (r, ϕ, z), that is, when $(r/L) \ll 1$ and $(z/L) \ll 1$. What is r_0 in terms of L?

6.11 We can illustrate the meaning of Eq. (6.40) by drawing the cylinder of uniform charge and the ellipse that is the resultant equipotential surface.

a. Let $L = 5$ cm and $a = 5$ cm and choose V_p such that the right side of Eq. (6.40) is $2L\coth(2\pi\varepsilon_0 V_p/\rho_{l0}) = 2\sqrt{L^2 + a^2} = 14.142$ cm. Draw (full scale) the cylinder of charge and the ellipse $R_1 + R_2 = 14.142$ cm in any plane passing through the z axis. This is <u>not</u> a thin cylinder.

b. Repeat for $L = 5$ cm and $a = 0.5$ cm, with $R_1 + R_2 = 2\sqrt{L^2 + a^2} = 10.050$ cm.

Note that for the thinner (but still not very thin) cylinder the ellipse closely approximates the cylinder.

6.12 We wish to find the electric energy stored between concentric conducting spheres of radius a and $b > a$, when the inner sphere has charge Q on its surface and the region between the spheres has permittivity ε.

a. Use Eq. (6.44).

b. Use Eq. (6.46).

c. Find the capacitance C between the spheres.

d. Show that as $b \to \infty$ the results of Example 6.7 are obtained.

e. Find the capacitance of a single isolated sphere of radius a.

Ans. a,b. $W_e = \dfrac{Q^2}{8\pi\varepsilon}\left(\dfrac{1}{a} - \dfrac{1}{b}\right)$; c. $C = \dfrac{4\pi\varepsilon ab}{b-a}$; e. $C = 4\pi\varepsilon a$.

6.13 For the conducting spheres of Problem 6.12, the permittivity of the medium is inversely proportional to R, $\varepsilon = \varepsilon_1(R_0/R)$ for $a < R < b$. Find the stored electric energy and the capacitance.

Ans. $W_e = \dfrac{Q^2}{8\pi\varepsilon_1 R_0}\ln\left(\dfrac{b}{a}\right)$, $\quad C = \dfrac{4\pi\varepsilon_1 R_0}{\ln(b/a)}$.

6.14 For the spheres of Problem 6.12, $\varepsilon = \varepsilon_1$, $a < R < c$ and $\varepsilon = \varepsilon_2$, $c < R < b$. Find the capacitance and draw an equivalent circuit in terms of two capacitors. What is the effect of adding a very thin conductor at $R = c$?

Ans. $\dfrac{1}{C} = \dfrac{1}{C_1} + \dfrac{1}{C_2}$, $\quad \dfrac{1}{C_1} = \dfrac{1}{4\pi\varepsilon_1}\left(\dfrac{1}{a} - \dfrac{1}{c}\right)$, $\quad \dfrac{1}{C_2} = \dfrac{1}{4\pi\varepsilon_2}\left(\dfrac{1}{c} - \dfrac{1}{b}\right)$.

6.15 For the spheres of Problem 6.12, $\varepsilon = \varepsilon_1$, $0 < \theta < \theta_0$ and $\varepsilon = \varepsilon_2$, $\theta_0 < \theta < \pi$. Find the capacitance and draw an equivalent circuit in terms of two capacitors. Find the surface charge density on each sector of the inner sphere.

Ans. $C = C_1 + C_2$, $\quad C_1 = \dfrac{2\pi\varepsilon_1 ab(1 - \cos\theta_0)}{b-a}$, $\quad C_2 = \dfrac{2\pi\varepsilon_2 ab(\cos\theta_0 + 1)}{b-a}$;

$$\rho_{sl} = \frac{Q\varepsilon_1}{2\pi a^2 F}, \quad \rho_{s2} = \frac{Q\varepsilon_2}{2\pi a^2 F}, \quad F = \varepsilon_1(1 - \cos\theta_0) + \varepsilon_2(\cos\theta_0 + 1).$$

6.16 A conducting cylinder of radius a is coaxial with a hollow conducting cylinder of inner radius $b > a$. The region between the cylinders has permittivity ε. Assume a charge ρ_{l0} per unit length on the inner cylinder. Find the capacitance per unit length.

Ans. $C_l = \dfrac{2\pi\varepsilon}{\ln(b/a)}$.

6.17 For the conducting cylinders of Problem 6.16, let the permittivity be inversely proportional to the radius, $\varepsilon = \varepsilon_1(r_0/r)$, for $a < r < b$. Find the capacitance per unit length.

Ans. $C_l = 2\pi\varepsilon_1 r_0/(b - a)$.

6.18 For the cylinders of Problem 6.16, $\varepsilon = \varepsilon_1$, $a < r < c$ and $\varepsilon = \varepsilon_2$, $c < r < b$. Find the capacitance per unit length and draw an equivalent circuit in terms of two capacitors. What is the effect of adding a very thin conductor at $r = c$?

Ans. $\dfrac{1}{C_l} = \dfrac{1}{C_{l1}} + \dfrac{1}{C_{l2}}$, $\quad \dfrac{1}{C_{l1}} = \dfrac{1}{2\pi\varepsilon_1}\ln\left(\dfrac{c}{a}\right)$, $\quad \dfrac{1}{C_{l2}} = \dfrac{1}{2\pi\varepsilon_2}\ln\left(\dfrac{b}{c}\right)$.

6.19 For the cylinders of Problem 6.16, $\varepsilon = \varepsilon_1, 0 < \phi < \phi_0$ and $\varepsilon = \varepsilon_2, \phi_0 < \phi < 2\pi$. Find the capacitance per unit length and draw an equivalent circuit in terms of two capacitors. Find the surface charge density on each sector of the inner sphere.

Ans. $C_l = C_{l1} + C_{l2}, \quad C_{l1} = \dfrac{\varepsilon_1 \phi_0}{\ln(b/a)}, \quad C_{l2} = \dfrac{\varepsilon_2(2\pi - \phi_0)}{\ln(b/a)};$

$$\rho_{s1} = \frac{\rho_{l0}\varepsilon_1}{aF}, \quad \rho_{s2} = \frac{\rho_{l0}\varepsilon_2}{aF}, \quad F = \varepsilon_1\phi_0 + \varepsilon_2(2\pi - \phi_0).$$

6.20 Equations (6.61) and (6.62) give the electric potential for a point charge q located at $(0, 0, d/2)$ above a conducting plane at $z = 0$.

a. Find the electric potential for a point charge q located at $(x', y', z'), z' > 0$ above the conducting plane.

b. Give the integral expression for a general volume charge density ρ located over the plane.

Ans. a. $V(x, y, z) = \dfrac{q}{4\pi\varepsilon_0}\left(\dfrac{1}{R_1} - \dfrac{1}{R_2}\right), \quad 0 < z; \ V(x, y, z) = 0, \ z < 0;$

$$R_1 = \sqrt{(x - x')^2 + (y - y')^2 + (z - z')^2},$$

$$R_2 = \sqrt{(x - x')^2 + (y - y')^2 + (z + z')^2}.$$

6.21 A uniform line charge ρ_{l0} forms a circle of radius a centered on the z axis in the $z = d/2$ plane above a conducting surface in the $z = 0$ plane.

a. Find the electric potential and the electric intensity along the z axis.

b. What is the total charge on the conducting surface?

Ans. a. $V(0, 0, z) = \dfrac{a\rho_{l0}}{2\varepsilon_0}\left[\dfrac{1}{\sqrt{a^2 + (z - d/2)^2}} - \dfrac{1}{\sqrt{a^2 + (z + d/2)^2}}\right], \quad z > 0.$

6.22 A line charge ρ_{l0} is located at $x = d/2$ parallel to the z axis over a conducting surface in the $x = 0$ plane.

a. Find the electric potential $V(x, y)$.

b. Find the surface charge density on the conducting surface and show that the total charge per unit length in the z direction is $-\rho_{l0}$.

Ans. a. $V(x, y) = \dfrac{\rho_{l0}}{2\pi\varepsilon_0}\ln\left(\dfrac{r_2}{r_1}\right), \ x > 0; \quad r_1 = \sqrt{(x - d/2)^2 + y^2};$

b. $\rho_s = -\dfrac{\rho_{l0}d}{2\pi[(d/2)^2 + y^2]}; \quad r_2 = \sqrt{(x + d/2)^2 + y^2}.$

6.23 A point charge q lies halfway between two conducting surfaces. Assume the charge is at the origin, one conducting surface is at $z = d/2$ and the other is at $z = -d/2$, with free space in the region $-d/2 < z < d/2$.

a. Draw a figure showing the polarity and location of the image charges required to keep the two conducting surfaces at zero potential.

b. Find the electric potential $V(x,y)$ between the conductors.

c. What is the total surface charge on each plane?

Ans. b. $V(r, z) = \dfrac{q}{4\pi\varepsilon_0} \sum_{n=-\infty}^{\infty} \dfrac{(-1)^n}{\sqrt{r^2 + (z - nd)^2}}, \quad -\dfrac{d}{2} < z < \dfrac{d}{2}.$

6.24 Figure 6.47 shows a point charge q located in free space near three conducting surfaces. For each surface, draw a sketch showing the location and polarity of images charges required to keep the conducting surfaces at zero electric potential. Why is it not possible to use the method of images to solve the problem presented in Figure 6.47c?

6.25 Show that Eq. (6.77) can be obtained directly from Eq. (6.75).
Hint: Use the identity $\cosh^{-1}(p) = \ln(p + \sqrt{p^2 - 1})$.

6.26 Figure 6.48 shows a line charge ρ_{l0} parallel to the z axis located at $r = l_a$, $\phi = \phi_0$ in the first quadrant. Conducting surfaces cover the $x = 0$ and $y = 0$

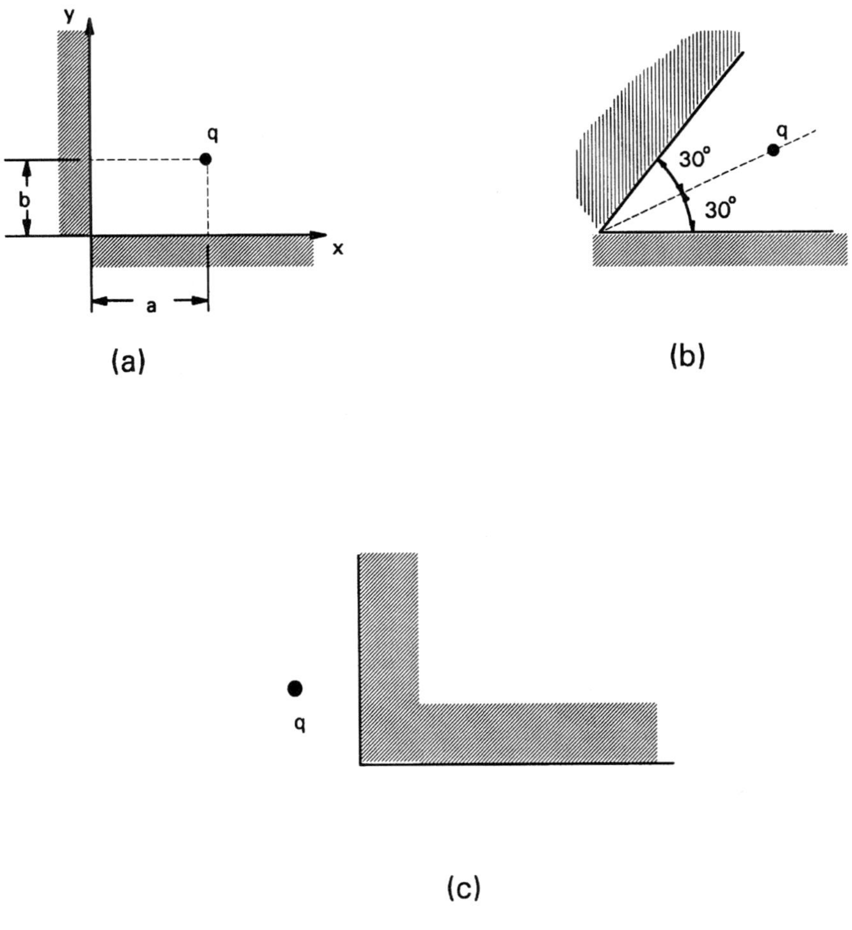

FIGURE 6.47 Problem 6.24.

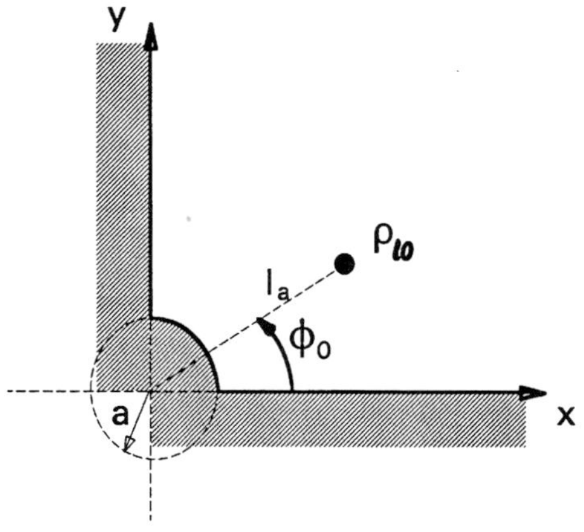

FIGURE 6.48 Problem 6.26.

planes and one-quarter of a conducting cylinder of radius $a < l_a$ fills the region $r < a$. Draw a figure showing the location and polarity of image line charges required to keep the conducting surfaces at zero potential.

Ans. Image line charges ρ_{l0} at $(a, -\pi + \phi_0)$, $(a^2/l_a, -\phi_0)$, and $(a^2/l_a, \pi - \phi_0)$. Image line charges $-\rho_{l0}$ at $(a, -\phi_0)$, $(a, \pi - \phi_0)$, $(a^2/l_a, \phi_0)$, and $(a^2/l_a, -\pi + \phi_0)$.

6.27 If the two cylinders of Fig. 6.35 are far apart compared to their radii $d \gg a$ and $d \gg b$, then the potential can be found by assuming the line charges lie along the axes of the cylinders with a spacing d between them and that the entire surface of one cylinder is approximately a distance d from the axis of the other cylinder.

a. Show that the electric potential difference between the cylinders is then $V_0 \approx (\rho_{l0}/2\pi\varepsilon_0)\ln(d^2/ab)$.

b. Find the capacitance per unit length.

c. When the cylinders have equal radii, $b = a$, then by this method $C_l \approx \pi\varepsilon_0/\ln(d/a)$. Compare this approximate value with the exact value, Eq. (6.77), by plotting $C_l/\pi\varepsilon_0$ for both formulas as (d/a) varies from 2 to 10.

The exact value differs from the approximate value because of the proximity effect. When one charged conductor approaches another the field of each charged body causes the charges on the other body to be displaced from the position they held in isolation. In this example, the proximity effect causes a 1% error in C_l when $d/a \approx 7$.

6.28 Neglecting the proximity effect (see the previous problem), find the capacitance between two spheres of radius a when their centers are d apart, $d \gg a$.

Ans. $C = \dfrac{2\pi\varepsilon_0 ad}{d - a}$.

6.29 A conducting sphere of radius a has charge Q_a on it and a conducting sphere of radius b has charge Q_b on it. The distance between the centers of the spheres is d,

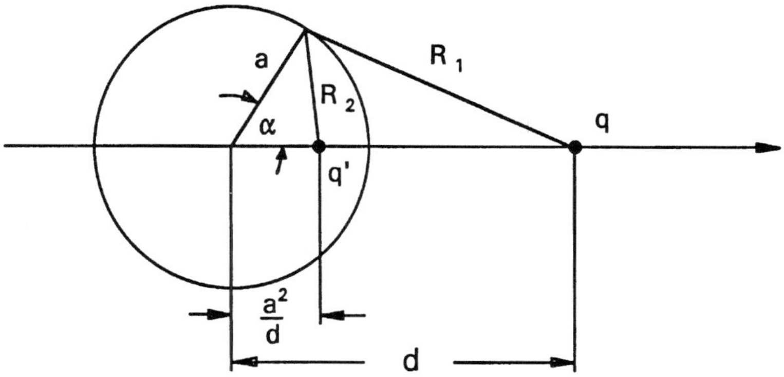

FIGURE 6.49 Problem 6.30.

where $d \gg a + b$. The total charge is Q and the two spheres are at the same potential V. Neglecting the proximity effect, find Q_a and Q_b in terms of Q.

Ans. $Q_a = Q\left(\frac{1}{b} - \frac{1}{d}\right)\left(\frac{1}{a} + \frac{1}{b} - \frac{2}{d}\right)^{-1}$,

$$Q_b = Q\left(\frac{1}{a} - \frac{1}{d}\right)\left(\frac{1}{a} + \frac{1}{b} - \frac{2}{d}\right)^{-1}.$$

6.30 A point charge q is located at distance d from the center of a conducting sphere of radius a. Show that the surface of the sphere is at zero potential if an image point charge q' is located a distance a^2/d from the center of the sphere along the line to the point charge (Fig. 6.49) and find q'. *Hint*: Use the law of cosines to write R_1 in terms of a, d, and α. Do the same for R_2 and then show that $R_2 = (a/d)R_1$.
Ans. $q' = -(a/d)q$.

6.31 Use the results of the previous problem to show that the potential of a neutral sphere (one with zero net charge) in the vicinity of a point charge q is $V = q/4\pi\varepsilon_0 d$.

6.32 A conducting sphere of radius a is centered at the origin. A point charge $-q$ is located on the z axis at $(0, 0, d)$ and a point charge $+q$ is located on the $-z$ axis at $(0, 0, -d)$; see Fig. 6.50.

a. Show that the potential at any point (R, θ, ϕ) outside the sphere is given by

$$4\pi\varepsilon_0 V = -\frac{q(R_4 - R_1)}{R_1 R_4} + q\left(\frac{a}{d}\right)\frac{(R_3 - R_2)}{R_2 R_3}.$$

b. Show that as $d \to \infty$, the potential is given by

$$4\pi\varepsilon_0 V = -\frac{2qz}{d^2} + \frac{2qa^3 \cos\theta}{d^2 R^2}.$$

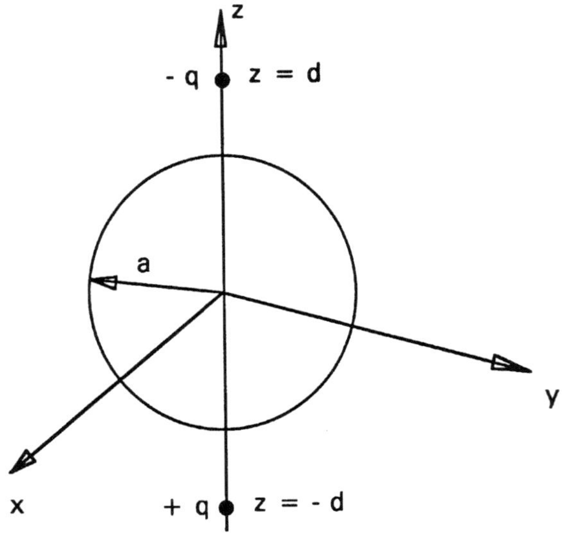

FIGURE 6.50 Problem 6.32.

c. For the potential found in part (b), find $\vec{E} = -\nabla V$, letting $E_0 = 2q/4\pi\varepsilon_0 d^2$. The resultant electric field intensity is a uniform field $\vec{a}_z E_0$ (due to the original charges, now infinitely far away) impressed on the conducting sphere. The charge distribution on the surface of the sphere gives a "reflected" field outside the sphere, which is the same as the field of an electric dipole (the image charges) of moment $4\pi\varepsilon_0 a^3 E_0$ located at the origin.

6.33 A conducting sphere of radius a is placed at the origin in a uniform dielectric medium ε. In the absence of the sphere the impressed electric field intensity is

$$\vec{E}^i = \vec{a}_z E_0 = E_0(\vec{a}_R \cos\theta - \vec{a}_\theta \sin\theta).$$

With the sphere present there is a "reflected" field outside the sphere given by

$$\vec{E}^r = \frac{A}{4\pi\varepsilon R^3}(\vec{a}_R 2\cos\theta + \vec{a}_\theta \sin\theta).$$

a. Use the boundary conditions at the surface of the sphere to show that $A = 4\pi\varepsilon a^3 E_0$ and that the surface charge density on the sphere is $\rho_s = 3\varepsilon E_0 \cos\theta$.

b. The field of ρ_s for $R > a$ is $\vec{E}^r$. What is the field of ρ_s for $R < a$?

6.34 An electric dipole of moment qd is located at the center of a hollow conducting spherical shell of inner radius a. The z axis is aligned with the dipole. In addition to the impressed dipole field

$$\vec{E}^i = \frac{qd}{4\pi\varepsilon R^3}(\vec{a}_R 2\cos\theta + \vec{a}_\theta \sin\theta)$$

there is a reflected field

$$\vec{E}^r = \vec{a}_z E_0 = E_0(\vec{a}_R \cos\theta - \vec{a}_\theta \sin\theta).$$

Find E_0 and ρ_s, the surface charge density on the inner surface of the shell. What is the field due to ρ_s for $R > a$?

Ans. $E_0 = \dfrac{qd}{4\pi\varepsilon a^2}, \quad \rho_s = -\dfrac{3qd\cos\theta}{4\pi a^3}.$

6.35 A dielectric sphere of radius a and permittivity ε_1 is placed at the origin in free space in an impressed field

$$\vec{E}^i = \vec{a}_z E_0 = E_0(\vec{a}_R \cos\theta - \vec{a}_\theta \sin\theta).$$

Due to the presence of the sphere there is an additional "reflected" field

$$\vec{E}^r = \frac{A}{R^3}(\vec{a}_R 2\cos\theta + \vec{a}_\theta \sin\theta), \quad R > a$$

outside the sphere and an additional "transmitted" field

$$\vec{E}^t = \vec{a}_z B = B(\vec{a}_R \cos\theta - \vec{a}_\theta \sin\theta), \quad R < a$$

inside the sphere. Find A and B.

Ans. $A = \left(\dfrac{\varepsilon_1 - \varepsilon_0}{\varepsilon_1 + 2\varepsilon_0}\right)a^3 E_0, \quad B = -\left(\dfrac{\varepsilon_1 - \varepsilon_0}{\varepsilon_1 + 2\varepsilon_0}\right)E_0.$

6.36 A uniform charge density ρ_0 is located in the spherical region $R < a$. The rest of space, $R > a$, is charge free. Solve Poisson's equation by direct integration to get the electric potential and the electric field intensity throughout all space.

Ans. $V = -\dfrac{\rho_0}{6\varepsilon_0}R^2 + \dfrac{\rho_0 a^2}{2\varepsilon_0}, \ R < a; \quad V = \dfrac{\rho_0 a^3}{3\varepsilon_0 R}, \quad R > a.$

The electric field intensity is given in Eqs. (4.10) and (4.11).

6.37 Write a program that uses the finite difference method to determine the potential subject to the boundary conditions in Fig. 6.51.

6.38 Write a program that uses the finite difference method to determine the potential subject to the boundary conditions in Fig. 6.52. Determine the capacitance per unit length of the structure, using the method outlined in Section 6.8.3.1.

6.39 Write a program that uses the finite difference method to determine the potential subject to the boundary conditions in Fig. 6.53. Determine the capacitance per unit length of the structure, using the method outlined in Section 6.8.3.1.

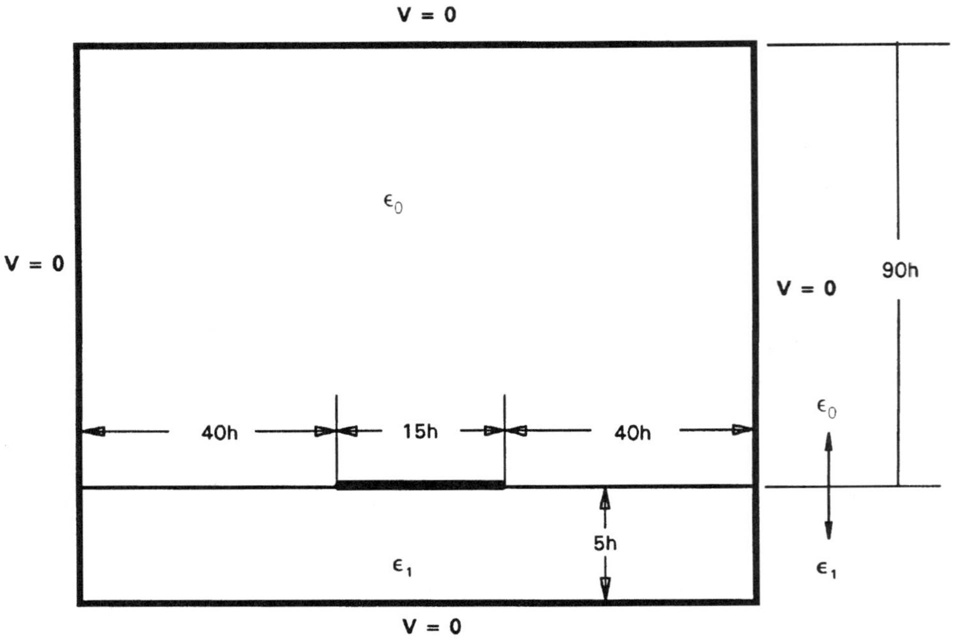

FIGURE 6.51 Problem 6.37.

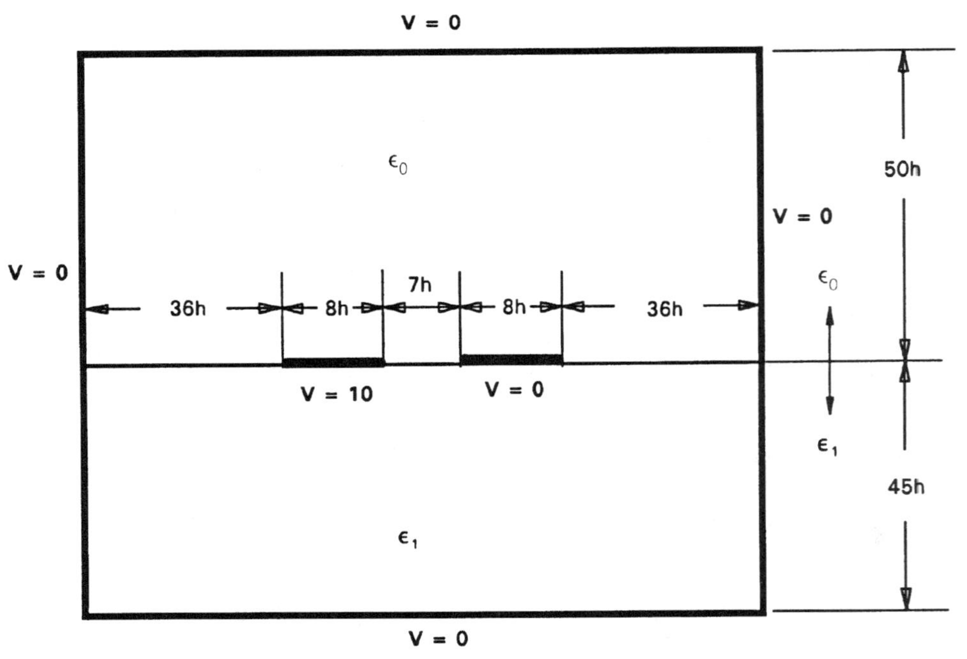

FIGURE 6.52 Problem 6.38.

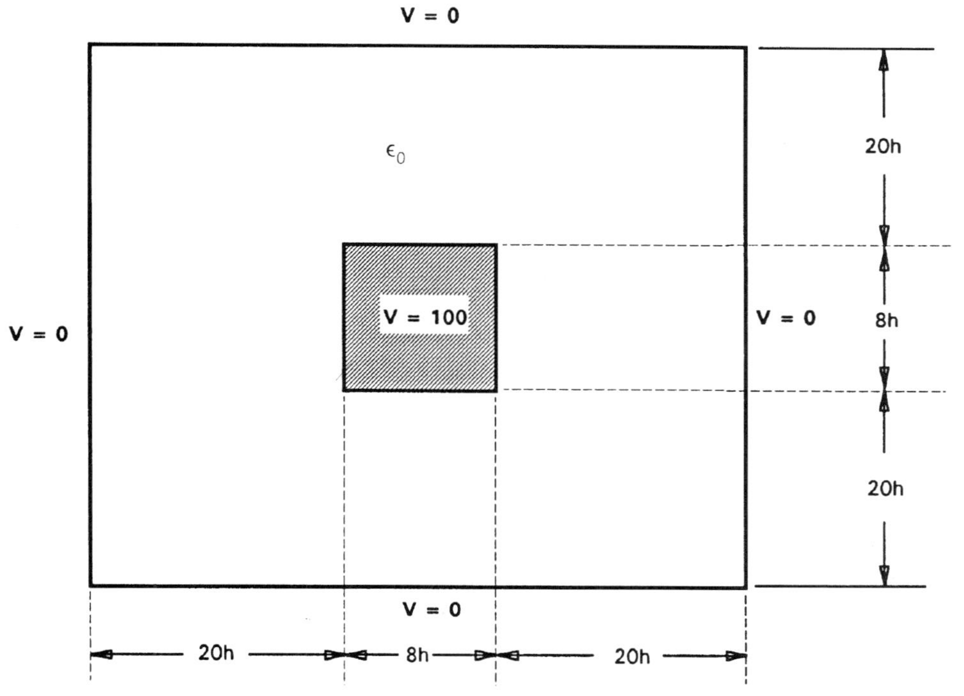

FIGURE 6.53 Problem 6.39.

REFERENCES

[1] Ruel V. Churchill, *Complex Variables and Applications*, McGraw-Hill, New York, 1960, Chaps. 9–11.

[2] Wilbur R. LePage, *Complex Variables and the Laplace Transform for Engineers*, McGraw-Hill, New York, 1961, Chap. 3.

[3] William H. Hayt, Jr., *Engineering Electromagnetics*, 5th ed., McGraw-Hill, New York, 1989, pp. 162–175.

[4] J. N. Reddy, *An Introduction to the Finite Element Method*, McGraw-Hill, New York, 1984.

CHAPTER SEVEN

Steady Currents

7.1 INTRODUCTION

Another special case of Maxwell's equations arises when there are steady currents in conducting media. Steady currents mean the current is not a function of time. Since the current depends on the electric field intensity, $\vec{J} = \sigma\vec{E}$, the electric field intensity and hence the electric charge density must also be time invariant. For steady currents the equation of continuity, Eq. (5.16), becomes

$$\nabla \cdot \vec{J} = 0 \tag{7.1}$$

and the law of conservation of charge, Eq. (4.21), becomes

$$\oint_S \vec{J} \cdot d\vec{S} = 0. \tag{7.2}$$

The boundary condition for current density becomes

$$\vec{a}_n \cdot (\vec{J}_1 - \vec{J}_2) = 0 \quad \text{or} \quad J_{1n} = J_{2n}. \tag{7.3}$$

Thus the current density normal to an interface is continuous. At a perfect dielectric to conductor boundary the current in the dielectric is zero and hence the component of the current density normal to the boundary in the conductor is also zero.

7.2 POWER LOSS

Consider a charged region with charge density ρ moving with constant velocity $\vec{u}$ in an electric field $\vec{E}$. The moving charge constitutes a current with current density $\vec{J} = \rho\vec{u}$, Eq. (3.13). The mechanical work ΔW done by the field in moving an element of charge

$\Delta Q = \rho\,\Delta v$ a distance $\Delta\vec{l} = \vec{u}\,\Delta t$ is

$$\Delta W = \Delta Q\,\Delta\vec{l}\cdot\vec{E} = \rho\,\Delta v\,\Delta t\,\vec{u}\cdot\vec{E}.$$

The power loss density p_l is the work done per unit time per unit volume, and hence

$$\begin{aligned} p_l &= \frac{\Delta W}{\Delta t\,\Delta v} = \rho\vec{u}\cdot\vec{E} \\ &= \vec{J}\cdot\vec{E}. \end{aligned} \tag{7.4}$$

The SI units of power density are watts per cubic meter, W/m^3.

In a conducting medium the current density and the electric field intensity are related by $\vec{J} = \sigma\vec{E}$. Hence the power loss density is

$$p_l = \frac{1}{\sigma}|\vec{J}|^2 = \sigma|\vec{E}|^2. \tag{7.5}$$

This power is converted into heat energy.

The total power loss in a volume is

$$P_l = \int_V p_l\,dv = \int_V \vec{J}\cdot\vec{E}\,dv. \tag{7.6}$$

The SI unit of power is the watt, W.

Next, consider a conducting body with conductivity σ between the terminal surfaces S_1 and S_2. A potential source V is connected to the terminals, giving a current I (Fig. 7.1). The potential V is the work per unit positive charge flowing from S_1 to S_2. The current I is the number of unit positive charges per second flowing from S_1 to S_2.

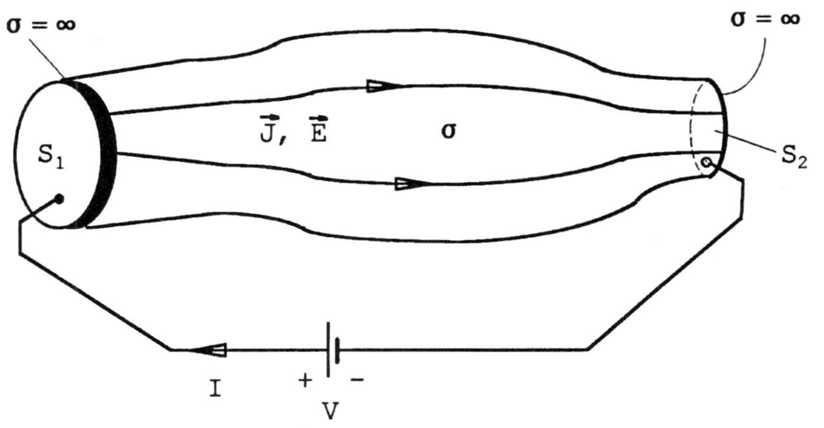

FIGURE 7.1 A conducting body with conductivity σ between the surfaces S_1 and S_2 connected to a voltage source V.

Their product is therefore the work per second or the power needed to maintain the charge flow. This is the power loss in the resistor,

$$P_l = IV. \tag{7.7}$$

The power loss in a resistor equals the product of the current and the potential.

7.3 RESISTANCE AND CONDUCTANCE

The current I equals the surface integral of the current density $\vec{J}$ through any surface cutting through the resistor. The potential V equals the line integral of the electric field $\vec{E}$ along any line from one terminal to the other. The two field quantities are simply related by a constant of proportionality, the conductivity, $\vec{J} = \sigma\vec{E}$. We may therefore anticipate that for the conducting body of Fig. 7.1, V and I are also simply related by a constant of proportionality. Thus

$$V = RI \quad \text{or} \quad I = GV, \quad \text{where } G = 1/R. \tag{7.8}$$

The constant R is the resistance of the conducting body between the surfaces S_1 and S_2. The SI unit of resistance is the ohm, Ω. The constant G is the conductance of the conducting body between the surfaces S_1 and S_2. The SI unit of conductance is the siemens (S) or mho (Ω^{-1}).

For the conducting body

$$R = \frac{V}{I} = \frac{\int_l \vec{E}\cdot d\vec{l}}{\int_l \vec{J}\cdot d\vec{S}} = \frac{\int_l (\vec{J}/\sigma)\cdot d\vec{l}}{\int_l \vec{J}\cdot d\vec{S}} = \frac{\int_l \vec{E}\cdot d\vec{l}}{\sigma\vec{E}\cdot d\vec{S}}. \tag{7.9}$$

In this equation l is any line from S_1 to S_2 and S is any surface cutting across the resistor.

The power loss in the resistor, Eq. (7.7), may now be written in terms of the resistance or the conductance as

$$P_l = VI = RI^2 = GV^2. \tag{7.10}$$

The symbol for an ideal resistor is shown in Fig. 7.2a and the electric circuit model of an ideal resistor with a potential source is shown in Fig. 7.2b.

7.3.1 Cylindrical Resistor

Consider a cylindrical conductor with conductivity σ (Fig. 7.3). The cross-sectional area of the conductor is S in any plane perpendicular to the z axis and the length of the conductor parallel to the z axis is d. With a potential difference V between the end surfaces, the electric field intensity will be uniform and in the z direction, $\vec{E} = \vec{a}_z E_0$.

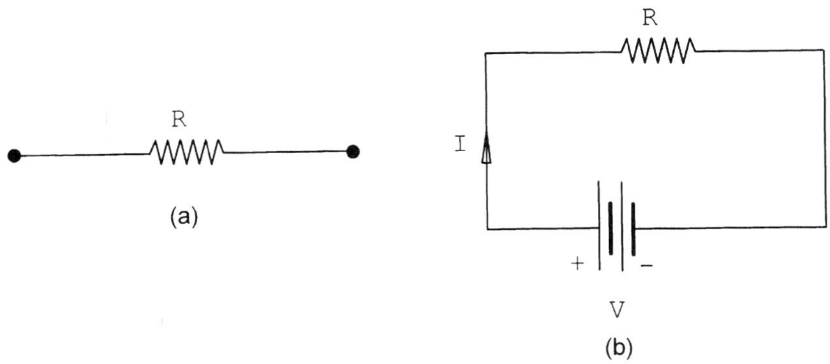

FIGURE 7.2 (a) Symbol for an ideal resistor R and (b) the electric circuit model of an ideal resistor with a potential source.

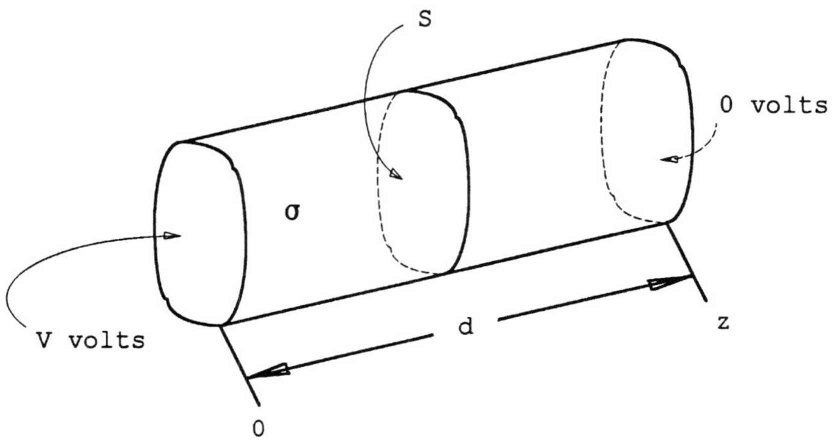

FIGURE 7.3 Cylindrical conductor with conductivity σ.

The power loss in terms of conductance and resistance is

$$P_l = \int_V \sigma E_0^2 \, dv = \sigma E_0^2 \, Sd = \frac{\sigma S}{d}(dE_0)^2 = \frac{\sigma S}{d} V^2 = GV^2, \quad G = \frac{\sigma S}{d},$$

or

$$P_l = \int_V \frac{1}{\sigma} J_0^2 \, dv = \frac{1}{\sigma} J_0^2 \, Sd = \frac{d}{\sigma S}(SJ_0)^2 = \frac{d}{\sigma S} I^2 = RI^2, \quad R = \frac{d}{\sigma S},$$

or, from Eq. (7.9),

$$R = \frac{E_0 \int_l dl}{\sigma E_0 \int_S dS} = \frac{d}{\sigma S}. \tag{7.11}$$

Of course, the three results are the same. Be careful to use this formula only for cylindrical conductors; it is *not* applicable to other shapes.

Let us now divide the conductor into two parts with different conductivities. Let the region $0 < z < d_1$ have conductivity σ_1 and the remaining region $d_1 < z < d_1 + d_2 = d$ have conductivity σ_2 (Fig. 7.4a). At the conductor to conductor interface the normal current density must be continuous, $J_{1z} = J_{2z} = I/S$. The electric field intensity in the two regions is therefore $E_{1z} = J_{1z}/\sigma_1 = I/\sigma_1 S$ and $E_{2z} = J_{2z}/\sigma_2 = I/\sigma_2 S$. The power dissipated in the resistor is the sum of the power loss in each of its parts,

$$P_l = P_{l1} + P_{l2} = \left(\frac{I}{S}\right)\left(\frac{I}{\sigma_1 S}\right)(Sd_1) + \left(\frac{I}{S}\right)\left(\frac{I}{\sigma_2 S}\right)(Sd_2)$$
$$= \left(\frac{d_1}{\sigma_1 S} + \frac{d_2}{\sigma_2 S}\right) I^2.$$

Hence the resistance is

$$R = \frac{d_1}{\sigma_1 S} + \frac{d_2}{\sigma_2 S}$$
$$= R_1 + R_2, \quad \text{where } R_1 = \frac{d_1}{\sigma_1 S} \text{ and } R_2 = \frac{d_2}{\sigma_2 S}. \tag{7.12}$$

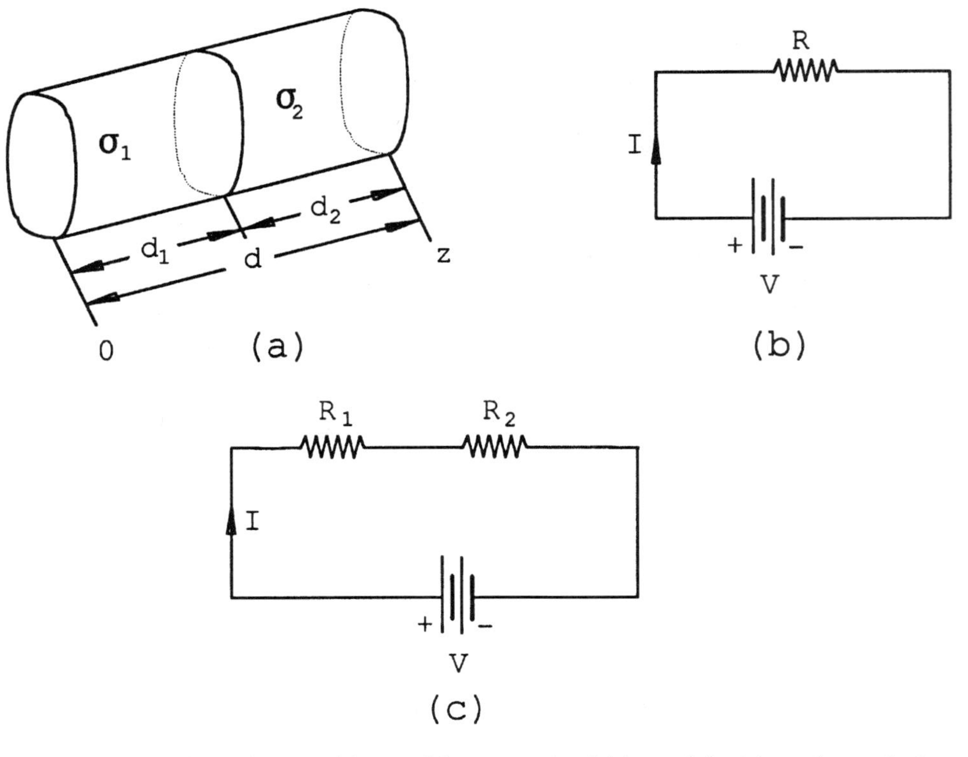

FIGURE 7.4 (a) A conductor with two different conductivities and (b, c) its series equivalent circuit models.

The two sections of the resistor have the same current I and the total potential is the sum of the potentials across each section. The resistors R_1 and R_2 are therefore connected in series. For resistors in series the total resistance is equal to the sum of the individual resistances as in Eq. (7.12). The circuit model for this physical resistor is shown as a single resistor R (Fig. 7.4b) or as two resistors R_1 and R_2 in series (Fig. 7.4c).

Next, let us divide the resistor into two conducting regions, which meet along a surface the normal to which is everywhere perpendicular to the z axis (Fig. 7.5a). Between the partial surfaces S_1 the conductivity is σ_1 and between the remaining surfaces S_2 the conductivity is σ_2. At the conductor to conductor interface the tangential (z component) of the electric field intensity must be continuous, $E_{1z} = E_{2z} = V/d$. The current density in each region is therefore $J_{1z} = \sigma_1 E_{1z} = \sigma_1 V/d$ and $J_{2z} = \sigma_2 E_{2z} = \sigma_2 V/d$. The power dissipated in the resistor is the sum of the

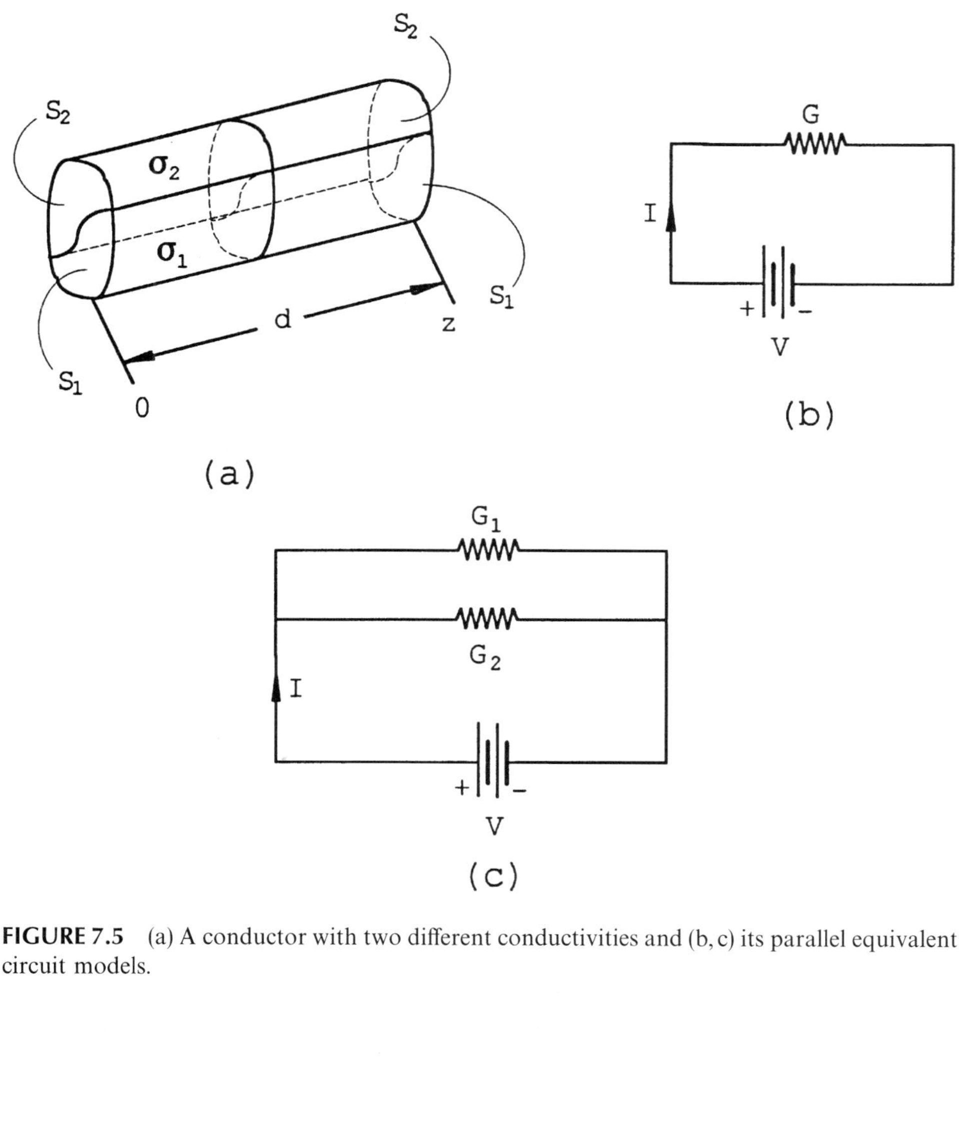

FIGURE 7.5 (a) A conductor with two different conductivities and (b, c) its parallel equivalent circuit models.

power dissipated in each of its parts,

$$P = P_1 + P_2 = \left(\frac{\sigma_1 V}{d}\right)\left(\frac{V}{d}\right)S_1 d + \left(\frac{\sigma_2 V}{d}\right)\left(\frac{V}{d}\right)S_2 d$$
$$= \left(\frac{\sigma_1 S_1}{d} + \frac{\sigma_2 S_2}{d}\right)V^2.$$

Hence the conductance is

$$G = \frac{\sigma_1 S_1}{d} + \frac{\sigma_2 S_2}{d}$$
$$= G_1 + G_2, \quad \text{where } G_1 = \frac{\sigma_1 S_1}{d} \text{ and } G_2 = \frac{\sigma_2 S_2}{d}. \tag{7.13}$$

The two sections of the resistor have the same potential V across them and the total current is the sum of the current in each section. The conductances G_1 and G_2 are therefore connected in parallel. For conductances in parallel the total conductance is equal to the sum of the individual conductances as in Eq. (7.13). The circuit model for this physical resistor is shown as a single conductance G (Fig. 7.5b) or as two conductances G_1 and G_2 in parallel (Fig. 7.5c).

We could also divide the cylindrical resistor into two regions along an irregular boundary as shown in Fig. 7.6. In this case both the electric field intensity and the current density must change from the corresponding field in the original resistor in order to make J_{normal} and $E_{\text{tangential}}$ continuous along the interface. There is no longer a simple solution to this problem.

It is easy to generalize this example to conductors of any shape. Figure 7.7a shows two perfectly conducting contact plates with a uniform conductor, σ_1, between them. A potential V is applied across the plates, driving a current I from the higher potential plate to the lower potential plate. The solid lines represent both the electric field intensity and the current density lines within the conductor. The dashed lines outline the equipotential surfaces within the conductor. The electric field lines form tubes of flow between the contact plates. If the conductivity within an entire tube is changed to σ_2 (Fig. 7.7b) and the potential V is maintained, then the electric field intensity is

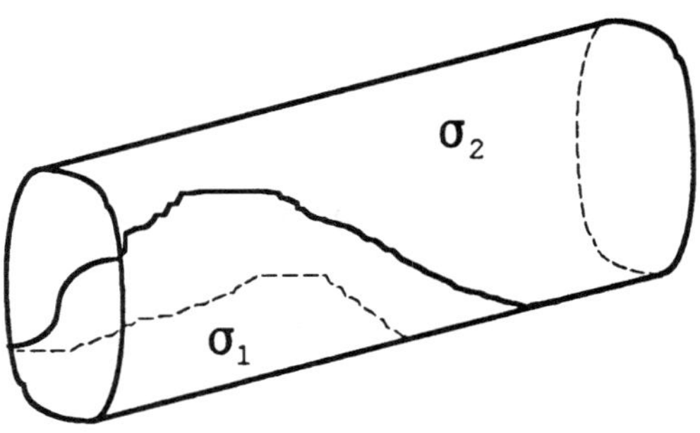

FIGURE 7.6 Cylindrical resistor divided into two regions with an irregular boundary.

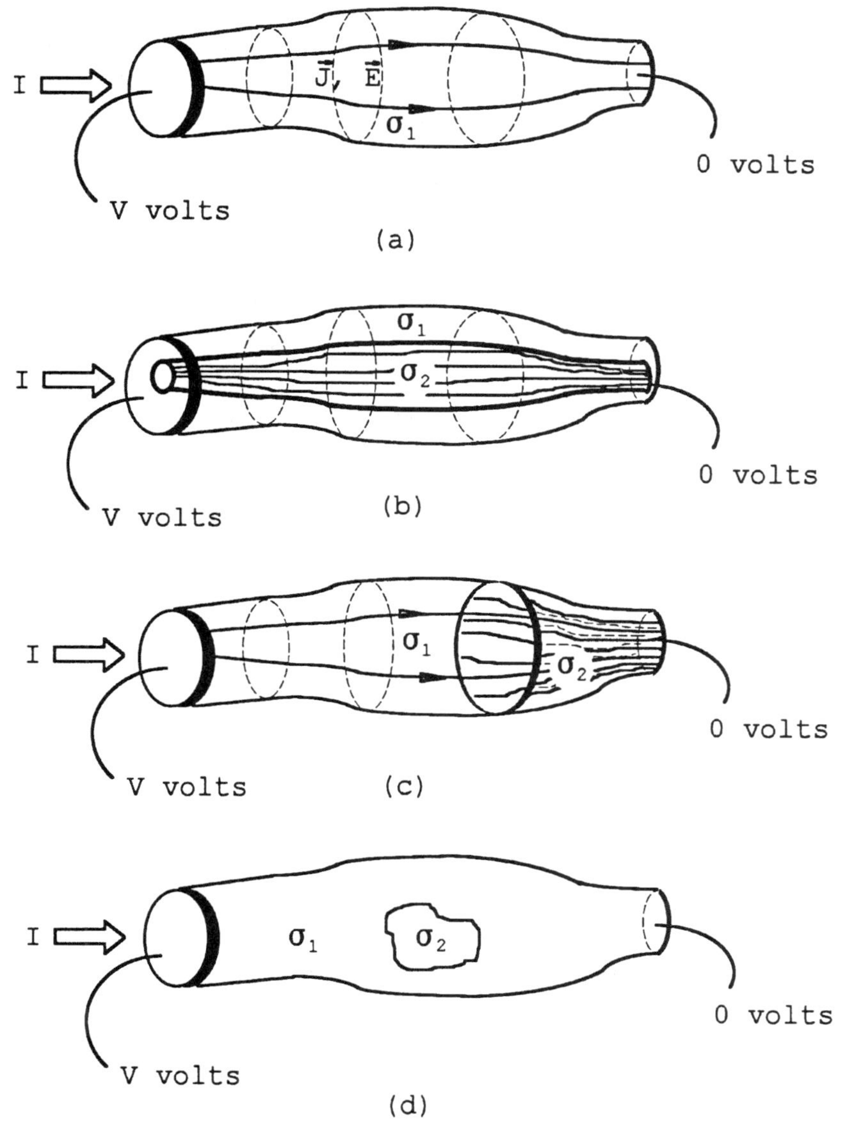

FIGURE 7.7 Generalization of Figs. 7.3–7.6 with conductors of arbitrary shape.

unchanged since the original $\vec{E}$ will satisfy the condition that $E_{\text{tangential}}$ must be continuous at the boundary of the tube and will integrate along any line connecting the plates to V. This resistor can be modeled as two resistors in parallel. Next, look to the original single conductor resistor and note that the equipotential surfaces divide the conductor along surfaces that are everywhere normal to the current density. If the conductor within a block that is an entire region between any two equipotential surfaces is changed to conductivity σ_2 (Fig. 7.7c) and the total current I is maintained, then the original current density will still satisfy the condition that J_{normal} is continuous at the boundary surfaces and will integrate to the current I over any surface between the plates. This resistor can be modeled as two resistors in series. If a change in conductivity is made that is neither a complete tube of current flow nor a complete block (Fig. 7.7d), then both the $\vec{E}$ and the $\vec{J}$ must change and the new fields are not simply related to the original fields.

Example 7.1 Find the resistance and the conductance for a circular cylinder of radius a and length d if (a) the cylinder has a single conductivity σ (Fig. 7.8a), (b) the sector $0 < \phi < \alpha$ has conductivity σ_1 and the sector $\alpha < \phi < 2\pi$ has conductivity σ_2 (Fig. 7.8b), and (c) the circular cylinder $0 < z < d_1$ has conductivity σ_1 and the circular cylinder $d_1 < z < d$ has conductivity σ_2 (Fig. 7.8c). The cylinder is excited by

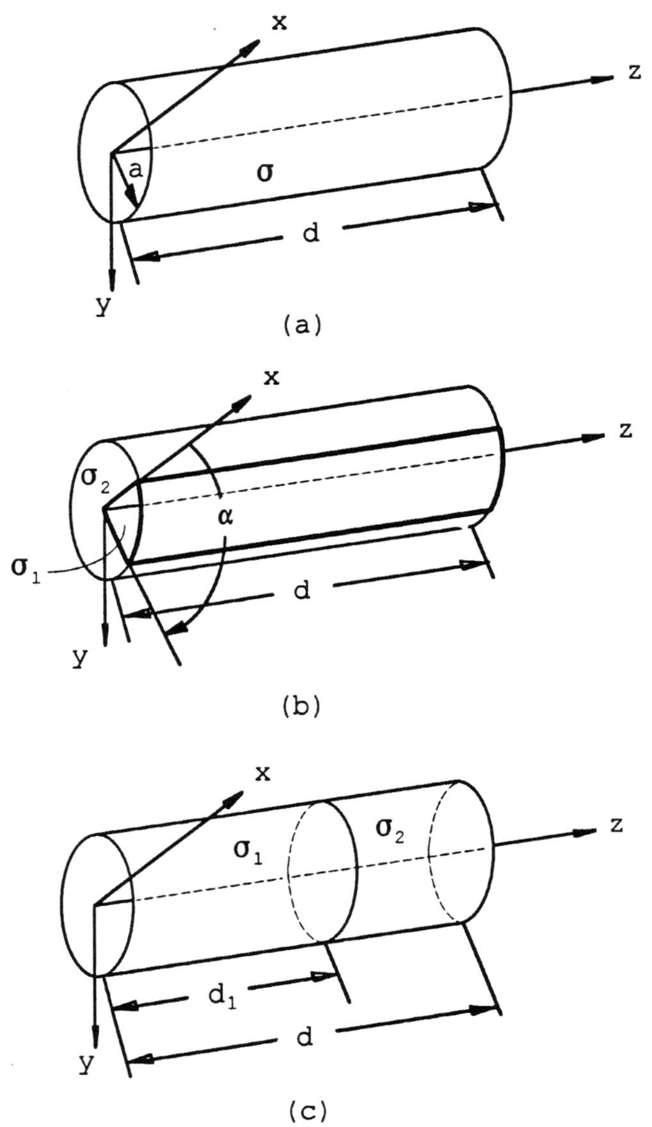

FIGURE 7.8 Circular cylindrical resistors in Example 7.1.

a potential source connected through perfectly conducting plates at $z = 0$ and $z = d$. Show that the results of (b) and (c) become the result of (a) under suitable conditions.

Solution (a) Using Eq. (7.11),

$$R = \frac{d}{\sigma S} = \frac{d}{\sigma \pi a^2}, \quad G = \frac{\sigma \pi a^2}{d}.$$

(b) The two sectors are in parallel. Hence

$$G_1 = \frac{\sigma_1 \alpha a^2}{2d}, \quad G_2 = \frac{\sigma_2 (2\pi - \alpha) a^2}{2d}.$$

$$G = G_1 + G_2 = \frac{a^2}{2d}[\sigma_1 \alpha + \sigma_2 (2\pi - \alpha)], \quad R = \frac{1}{G}.$$

When $\alpha = 2\pi$ this result is the same as part (a), with a conductivity σ_1, and when $\alpha = 0$ the result is the same as part (a), with a conductivity σ_2.

(c) The two cylinders are now in series. Hence

$$R_1 = \frac{d_1}{\sigma_1 \pi a^2}, \quad R_2 = \frac{d - d_1}{\sigma_2 \pi a^2},$$

$$R = R_1 + R_2 = \frac{1}{\pi a^2}\left(\frac{d_1}{\sigma_1} + \frac{d - d_1}{\sigma_2}\right), \quad G = \frac{1}{R}.$$

When $d_1 = d$ the result is the same as part (a), with a conductivity σ_1, and when $d_1 = 0$ the result is the same as part (a), with a conductivity σ_2.

7.3.2 Coaxial Conductor

For any shape other than the cylindrical resistor the simple formula Eq. (7.11) is not applicable, and we must use the general formula Eq. (7.9). An example of the resistance between coaxial cylinders follows.

Example 7.2 (a) Find the resistance for a coaxial resistor with inner radius a, outer radius b, length d, and conductivity σ. The potential is applied between the $r = a$ and the $r = b$ surfaces. (b) Assume the sector $0 < \phi < \alpha$ is of conductivity σ_1 and the remaining sector $\alpha < \phi < 2\pi$ is of conductivity σ_2 and calculate the resistance. (c) Assume the cylinder $a < r < c < b$ is of conductivity σ_1 and the cylinder $c < r < b$ is of conductivity σ_2 and calculate the resistance. Show that the answers to parts (b) and (c) become the answer to part (a) under suitable conditions.

Solution (a) For a current I between the cylinders, the law of conservation of charge applied to a cylinder of radius r, where $a < r < b$, gives $2\pi r d J_r - I = 0$ or

$$J_r = \frac{1}{2\pi r d} \quad \text{and} \quad E_r = \frac{1}{2\pi \sigma r d}.$$

The potential difference is

$$V = \int_a^b E_r\, dr = \frac{1}{2\pi\sigma d} \ln\left(\frac{b}{a}\right)$$

and the resistance $R = V/I$ is

$$R = \frac{1}{2\pi\sigma d} \ln\left(\frac{b}{a}\right). \tag{7.14}$$

The conductance and the conductance per unit length are

$$G = \frac{1}{R} = \frac{2\pi\sigma d}{\ln(b/a)}, \quad G_l = \frac{2\pi\sigma}{\ln(b/a)}. \tag{7.15}$$

Note that every length is in parallel with every other length and hence the conductance increases and the resistance decreases with length.

(b) With the same potential V as in (a) between the cylinders, the electric field intensity will be the same. Replacing $1/2\pi\sigma d$ with $V/\ln(b/a)$ we have

$$E_r = \frac{V}{\ln(b/a)} \frac{1}{r} \quad \text{and} \quad J_r = \begin{cases} J_{r1} = \sigma_1 E_r, & 0 < \phi < \alpha, \\ J_{r2} = \sigma_2 E_r, & \alpha < \phi < 2\pi. \end{cases}$$

The current is

$$I = d\left(\int_{\phi=0}^{\alpha} J_{r1} r\, d\phi + \int_{\phi=\alpha}^{2\pi} J_{r2} r\, d\phi\right) = \frac{Vd}{\ln(b/a)}[\sigma_1\alpha + \sigma_2(2\pi - \alpha)]$$

and the conductance $G = I/V$ is

$$G = \frac{\sigma_1 \alpha d}{\ln(b/a)} + \frac{\sigma_2(2\pi - \alpha)d}{\ln(b/a)} = G_1 + G_2.$$

To find the conductance per unit length, divide by the length d. Note that the sectors are in parallel and hence their conductances add. When $\alpha = 2\pi$ this result is the same as part (a) with conductivity σ_1. When $\alpha = 0$ the result is the same as part (a) with conductivity σ_2.

(c) With the same current I as in part (a) the current density will be the same as in part (a). The electric field intensity is then

$$E_r = \begin{cases} E_{r1} = \dfrac{1}{2\pi\sigma_1 r d}, & a < r < c, \\[2ex] E_{r2} = \dfrac{1}{2\pi\sigma_2 r d}, & c < r < b. \end{cases}$$

The potential difference between the cylinders is then

$$V = \int_{r=a}^{c} E_{r1}\, dr + \int_{r=c}^{b} E_{r2}\, dr = \frac{1}{2\pi\sigma_1 d}\ln\left(\frac{c}{a}\right) + \frac{1}{2\pi\sigma_2 d}\ln\left(\frac{b}{a}\right)$$

and the resistance $R = V/I$ is

$$R = \frac{1}{2\pi\sigma_1 d}\ln\left(\frac{c}{a}\right) + \frac{1}{2\pi\sigma_2 d}\ln\left(\frac{b}{c}\right) = R_1 + R_2 .$$

The two cylindrical shells are in series and their resistances add. With $c = b$ the result is the same as part (a), with conductivity σ_1, and with $c = a$ the result is the same as part (a), with conductivity σ_2.

7.4 EQUIVALENT CIRCUIT

Let us consider two perfectly conducting bodies embedded in a medium of infinite extent, with conductivity σ and permeability ε (Fig. 7.9a). Equation (7.9) gives the resistance R between the perfect conductors and Eq. (6.49) gives the capacitance C. An equivalent circuit is therefore as shown in (Fig. 7.9b). We can now see that every physical resistor has some capacitance associated with it and every physical capacitor has some resistance associated with it. You cannot build one device without unavoidably getting the other.

Why do we say, then, that some particular physical device is a resistor (and ignore the capacitance) or is a capacitor (and ignore the resistance)? The differential equation

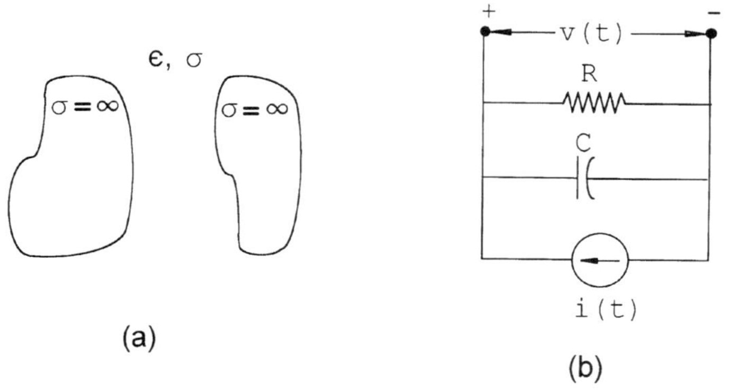

FIGURE 7.9 (a) Two perfectly conducting bodies embedded in a medium of infinite extent and (b) the equivalent circuit of the structure.

that relates the current $i(t)$ to the voltage $v(t)$ is

$$i = \frac{1}{R} v + C \frac{dv}{dt}$$

$$iR = v + RC \frac{dv}{dt}.$$

The product $RC = \tau$ is called the time constant of the circuit and has the units of time.

The term dv/dt can be approximated by $\Delta v/\Delta t$. If $\Delta v \ll v$ during any interval of time $\Delta t \approx \tau$, then the second term on the right is small compared to the first term on the right. Then $iR \approx v$, which is the i–v relation for a resistor. The physical device appears to be a pure resistor.

On the contrary, if $\Delta v \gg v$ during any interval of time $\Delta t \approx \tau$, then the second term on the right predominates over the first and $i \approx C(dv/dt)$. This is the i–v relation for a capacitor and the physical device appears to be a pure capacitor.

From a time domain viewpoint, if we are dealing with signals with a pulse width $T \ll \tau$, then the device acts as a capacitor; if $T \gg \tau$, then it acts as a resistor. From a frequency viewpoint, if the signal is band limited so that $\omega < \omega_{\max}$, then if $\tau \gg 1/\omega_{\max}$ the device appears as a capacitor; if $\tau \ll 1/\omega_{\max}$, then it acts as a resistor.

Example 7.3 (a) For a square-wave current

$$i(t) = \begin{cases} I_0/2, \ \dots, & -2T < t < -T, 0 < t < T, 2T < t < 3T, \ \dots \\ -I_0/2, \ \dots, & -T < t < 0, \ T < t < 2T, \ 3T < t < 4T, \ \dots \end{cases}$$

the voltage response is

$$v(t) = \begin{cases} \dfrac{I_0 R}{2}\left(1 - \dfrac{2}{1 + e^{-T/\tau}} e^{-t/\tau}\right), & 0 < t < T, \\ \dfrac{-I_0 R}{2}\left(1 - \dfrac{2}{1 + e^{-T/\tau}} e^{-(t-T)/\tau}\right), & T < t < 2T, \end{cases}$$

repeated every $2T$. Show that the voltage is appropriate for a capacitor if $\tau \gg T$ and for a resistor if $\tau \ll T$.

(b) For a sine-wave current

$$i(t) = I_0 \cos \omega t = I_0 \cos 2\pi f t$$

the steady-state voltage response is

$$v(t) = RI_0 \left[\frac{\cos \omega t}{1 + (\omega\tau)^2} + \frac{\omega\tau \sin \omega t}{1 + (\omega\tau)^2}\right].$$

Show that the voltage is appropriate for a capacitor if $\omega\tau \gg 1$ and for a resistor if $\omega\tau \ll 1$.

Solution (a) For $\tau \gg T$, $\exp(-T/\tau) \approx 1-(T/\tau)$ and $\exp(-t/\tau) \approx 1-(t/\tau)$ for $t < T$. Then, for $0 < t < T$,

$$\begin{aligned}
v(t) &\approx \frac{I_0 R}{2}\left[1-\frac{2(1-t/\tau)}{1+(1-T/\tau)}\right] \approx \frac{I_0 R}{2}\left[1-\left(1-\frac{t}{\tau}\right)\left(1+\frac{T}{2\tau}\right)\right] \\
&\approx \frac{I_0 R}{2}\left[1-\left(1-\frac{t}{\tau}+\frac{T}{2\tau}\right)\right] = -\frac{I_0 RT}{4\tau}\left(1-\frac{2t}{T}\right) \\
&= -\frac{I_0 T}{4C}\left(1-\frac{2t}{T}\right).
\end{aligned}$$

Similarly, for $T < t < 2T$,

$$v(t) = \frac{I_0 T}{4C}\left[1-\frac{2(t-T)}{T}\right].$$

These voltages are appropriate for an ideal capacitor.

For $\tau \ll T$, $\exp(-t/\tau) \ll 1$ during all but the first fraction of T. Thus

$$v(t) = \begin{cases} I_0 R/2, & 0 < t < T, \\ -I_0 R/2, & T < t < 2T, \end{cases}$$

which is appropriate for an ideal resistor. Figure 7.10a shows the three voltage waveforms.

(b) For $\omega\tau \gg 1$,

$$\begin{aligned}
v(t) &\approx RI_0\left[\frac{1}{(\omega\tau)^2}\cos\omega t + \frac{1}{\omega\tau}\sin\omega t\right] \approx \frac{RI_0}{\omega\tau}\sin\omega t \\
&= \frac{I_0}{\omega C}\sin\omega t,
\end{aligned}$$

which is appropriate to an ideal capacitor.

For $\omega\tau \ll 1$,

$$\begin{aligned}
v(t) &\approx RI_0(\cos\omega t + \omega\tau\sin\omega t) \\
&= RI_0\cos\omega t,
\end{aligned}$$

which is appropriate to an ideal resistor. Figure 7.10b shows the three voltage waveforms.

Multiplying Eq. (7.9) for resistance by Eq. (6.49) for capacitance we obtain

$$\tau = RC = \frac{\varepsilon}{\sigma}, \tag{7.16}$$

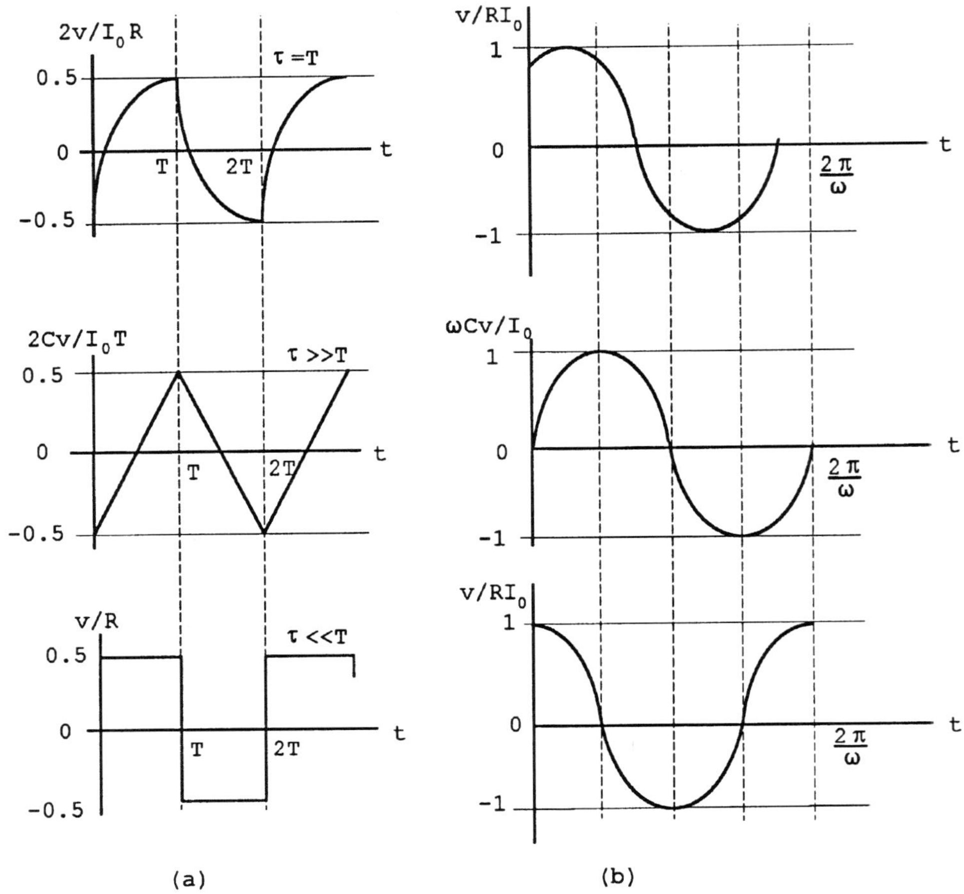

FIGURE 7.10 Voltage waveforms for Example 7.3.

showing that the time constant is independent of the shape of the conductors but depends on the permittivity and the conductivity of the medium. Table 7.1 shows some values of time constant for several materials. The time constants for copper and carbon are extremely short and these can be used to make resistors. The time constant for seawater is short, but not short compared to the pulse length in many pulse circuits. The time constants for mica and fused quartz are extremely long and these can be used to make capacitors.

One last observation about these circuits. In making a physical resistor or capacitor there will not usually be a single medium as shown in Fig. 7.9a. More typically there will be at least two media, one with constants σ and ε and the surrounding free space ($\sigma = 0$, $\varepsilon = \varepsilon_0$), as in Fig. 7.11a. The current lines are restricted to within the conducting region, but the electric field lines will extend into the free-space region as well. Thus there will be a resistance and a capacitance due to the interior fields and then an additional capacitance due to the free-space or fringing field, resulting in an equivalent circuit as shown in Fig. 7.11b. The time constant is now $\tau = R(C + C_f)$, where C_f is

TABLE 7.1 Time Constants for Materials

Material	Time Constant, τ (seconds)
Copper	0.15×10^{-18}
Carbon	0.44×10^{-15}
Seawater	0.16×10^{-9}
Transformer oil	2.0
Glass	62
Mica	4.8×10^{3}
Fused quartz	6.2×10^{6} (72 days)

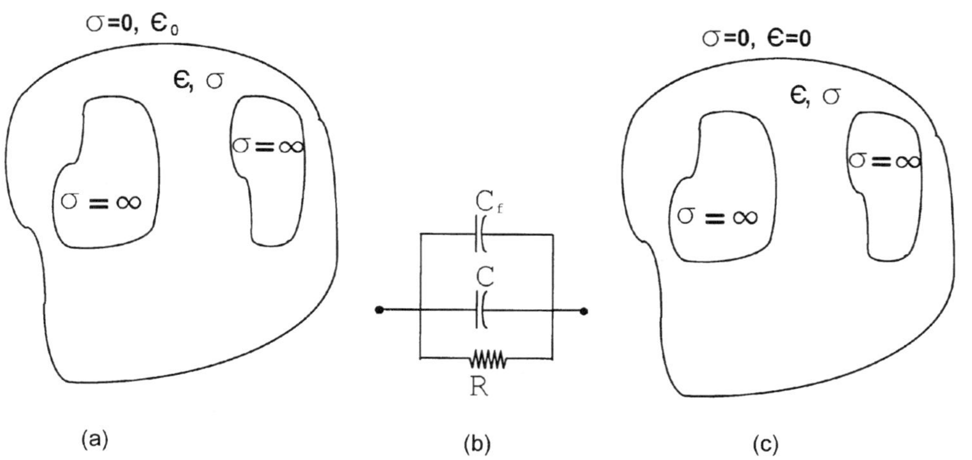

FIGURE 7.11 Illustration of fringing effects for two conductors surrounded by two different dielectric media.

the fringing capacitance. It is geometry dependent and not simply given by ε/σ. If there were a material with $\varepsilon = 0$ and $\sigma = 0$ (Fig. 7.11c), then the time constant would again be given by $\tau = \varepsilon/\sigma$.

For example, if we make a circular parallel plate physical resistor or capacitor with the distance between the plates much less than the radius of the plates (Fig. 7.12a), then the fields are largely confined within the region between the plates and the fringing field can be neglected. Then $R = d/\sigma\pi a^2$, $C = \varepsilon\pi a^2/d$, and $\tau = RC = \varepsilon/\sigma$. If the separation between the plates is not much less than the radius (Fig. 7.12b), then the fringing capacitance cannot be neglected [1,2].†

† The fringing capacitance can be approximated by $C_f = 2\varepsilon_0 a$. The time constant is then $\tau \approx (\varepsilon/\sigma)(1 + 2d/\varepsilon_r \pi a)$.

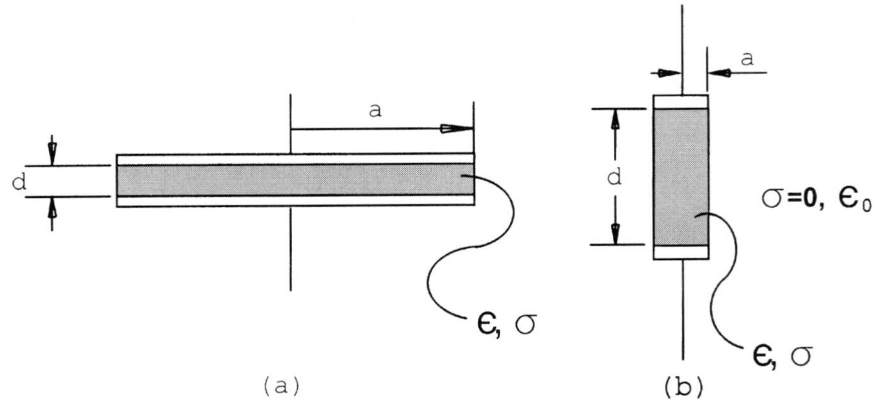

FIGURE 7.12 Circular parallel plate physical resistor or capacitor.

7.5 SKIN DEPTH AND SURFACE RESISTANCE

In discussing steady currents we noted that the current density $\vec{J}$ was uniform over the cross section of a cylindrical conductor (Fig. 7.13a). If the source is oscillatory at some frequency f, then the current is pushed toward the surface of the conductor. In fact, the current density decreases exponentially with a spatial decay constant δ_s, called the skin depth (Fig. 7.13b). Later on we will derive the formula for skin depth, but at this point we simply state that

$$\delta_s = \sqrt{\frac{2}{\omega\mu\sigma_c}}, \tag{7.17}$$

where $\omega = 2\pi f$ is the angular frequency and μ and σ_c are the permittivity and conductivity of the conductor. The SI units of the skin depth are meters, m.

Example 7.4 For copper, (a) find the formula for skin depth as a function of frequency, and evaluate the skin depth at the following frequencies: (b) 60 Hz, (c) 1 MHz, and (d) 10 GHz.

Solution (a) For copper, $\mu = \mu_0 = 4\pi \times 10^{-7}$ H/m and $\sigma_c = 5.80 \times 10^7$ S/m, so

$$(\delta_s)_{\text{copper}} = \frac{0.0661}{\sqrt{f}} \text{ m}.$$

(b) At 60 Hz, the ordinary frequency for power transmission in the United States, $\delta_s = 8.53 \times 10^{-3}$ m $= 8.53$ mm or almost 1 cm. Hence, except for extremely thick wires, the skin depth is much greater than the radius of the wire and the current density can be considered to be uniform over the cross section of a wire.

(c) At 1 MHz, in the AM radio frequency band, $\delta_s = 0.0661 \times 10^{-3}$ m $= 0.0661$ mm, which is very small compared to a typical wire radius.

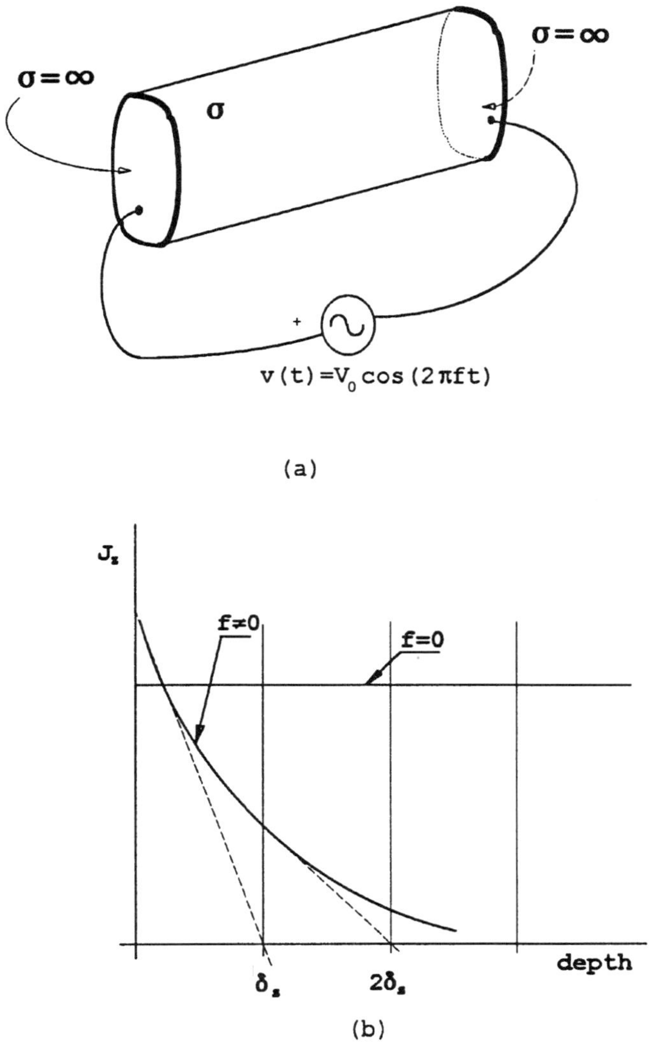

FIGURE 7.13 Illustration of the skin depth effect in conductors with a sinusoidal excitation.

(d) At 10 GHz, a typical radar frequency, $\delta_s = 0.0661 \times 10^{-5}$ m $= 0.661$ μm and, hence, the current flows in an extremely thin shell on the surface of the wire.

Let us now find the resistance of a circular wire of radius a, where

$$J_z(r) = J_0 \exp\left[-\frac{(a-r)}{\delta_s}\right], \quad a \gg \delta_s.$$

The current density and the electric field intensity at the surface of the wire $r = a$ are J_0 and J_0/σ_c, respectively. The potential difference on the wire's surface for a length

L is $V = J_0 L/\sigma$. The total current flowing in the wire is

$$I = \int_{\phi'=0}^{2\pi} \int_{r'=0}^{a} J_z(r') r' \, dr' \, d\phi' = 2\pi J_0 e^{-a/\delta_s} \int_0^a e^{r'/\delta_s} \, dr'$$

$$= 2\pi J_0 \delta_s^2 \left(\frac{a}{\delta_s} - 1 + e^{-a/\delta_s} \right).$$

Since $a/\delta_s \gg 1$ the second and third terms in the parentheses can be neglected in comparison with the first term. Thus I $\approx 2\pi a \delta_s J_0$ and the resistance of a length L of the wire is

$$(R_L)_{ac} = \frac{V}{I} = \frac{L}{\sigma 2\pi a \delta_s}$$

$$= \frac{L}{2\pi a} R_s, \quad \text{where } R_s = \frac{1}{\sigma \delta_s} = \sqrt{\frac{\omega\mu}{2\sigma_c}}. \tag{7.18}$$

The result gives the high-frequency resistance in terms of R_s the skin resistance.

We can generalize this result to a wire of any cross section by noting that the right side of the equation is the length L, divided by the perimeter of the wire $p = 2\pi a$, multiplied by the surface resistance R_s. Thus, in general, for a cylindrical wire,

$$(R_L)_{ac} = \frac{L}{p} R_s.$$

The high-frequency resistance per unit length is

$$R_{ac} = \frac{1}{p} R_s. \tag{7.19}$$

For copper, $R_s = 0.261 \times 10^{-6} \sqrt{f}\ \Omega$.

Example 7.5 Find the steady current and the high-frequency resistance per unit length of a copper coaxial cable with a solid inner wire of radius a and outer wire of inner radius b and outer radius $c > b > a$. Note that the inner and outer wires are in series. Evaluate the ratio of these resistances at 100 MHz, with $a = 0.44$ cm, $b = 1.01$ cm, and $c = 1.5$ cm.

Solution The resistance per unit length for steady current is

$$R = \frac{1}{\sigma} \left[\frac{1}{\pi a^2} + \frac{1}{\pi (c^2 - b^2)} \right].$$

For high frequencies it is

$$R_{ac} = \frac{1}{2\pi}\left(\frac{1}{a} + \frac{1}{b}\right) R_s .$$

When we evaluate for the given coax dimensions we calculate

$$R = 0.328 \times 10^{-3}\ \Omega/\text{m},$$

$$R_s = 2.61 \times 10^{-3}\ \Omega, \quad R_{ac} = 0.135\ \Omega/\text{m},$$

$$R_{ac}/R = 411 .$$

The high-frequency resistance per unit length is much greater than that for steady currents, because at high frequencies the current flows in a small depth of wire δ_s instead of uniformly over the entire wire cross section.

PROBLEMS

7.1 A spherical conductor (conductivity σ_1, permittivity ε_1) of radius a is centered at the origin within a conducting medium (σ_0, ε_0) extending throughout all space. In the absence of the sphere there is an impressed electric field

$$\vec{E}^i = \vec{a}_z E_0 = \vec{a}_R E_0 \cos\theta - \vec{a}_\theta E_0 \sin\theta$$

throughout all space. With the sphere present there is an additional field given by

$$\vec{E}^t = \vec{a}_z A = \vec{a}_r A \cos\theta - \vec{a}_\theta A \sin\theta, \quad R < a,$$

$$\vec{E}^r = \frac{B}{R^3}(\vec{a}_R 2\cos\theta + \vec{a}_\theta \sin\theta), \qquad R > a.$$

Find formulas for the constants A and B and find the surface charge density on the $R = a$ surface.

Ans. $A = -\left(\dfrac{\sigma_1 - \sigma_0}{\sigma_1 + 2\sigma_0}\right) E_0, \quad B = -a^3 A, \quad \rho_s = 3\left(\dfrac{\varepsilon_0\sigma_1 - \varepsilon_1\sigma_0}{\sigma_1 + 2\sigma_0}\right) E_0 \cos\theta .$

7.2 Find the formula for the resistance of a wire with rectangular cross section $a \times b$, length l, and conductivity σ.

Ans. $R = l/\sigma ab$.

7.3 Find the resistance of a spherical conducting shell with inner radius a, outer radius b, and conductivity σ.

Ans. $R = (b - a)/4\pi\sigma ab$.

7.4 Repeat Problem 7.3 if the conductivity is $\sigma = \sigma_0\,(R/a)$.

7.5 For the spherical shell of Problem 7.3 the region $0 \le \theta < \theta_0$ has conductivity σ_1 and the remainder, $\theta_0 < \theta \le \pi$, has conductivity σ_2.

a. Draw an equivalent circuit in terms of resistors R_1 and R_2.

b. Find formulas for the total resistance R and the resistors R_1 and R_2.

c. Show that these results match with the result of Problem 7.3 when $\theta_0 = 0$ and when $\theta_0 = \pi$.

Ans. $\dfrac{1}{R} = \dfrac{1}{R_1} + \dfrac{1}{R_2}$, $\quad R_1 = \dfrac{b-a}{2\pi ab\sigma_1(1-\cos\theta_0)}$, $\quad R_2 = \dfrac{b-a}{2\pi ab\sigma_2(\cos\theta_0 + 1)}$.

7.6 A conductor of conductivity σ consists of a sector of a cylindrical shell defined by $a \le r \le b$, $0 \le \phi \le \alpha$, and length L (Fig. 7.14).

a. Find the resistance between the faces at $\phi = 0$ and $\phi = \alpha$.

b. Take the mean radius to be $c = (a+b)/2$ and the difference in radii to be $d = b - a$. Show that for $c \gg d$ the resistance is of the same form as the result of Problem 7.2.

Ans. *a.* $R = \dfrac{\alpha}{\sigma L \ln(b/a)}$; $\quad$ *b.* $R \approx \dfrac{\alpha c}{\sigma d L}$.

7.7 Consider a pair of perfectly conducting parallel wires with circular cross section embedded in a conducting medium, σ. One wire has radius a, the other has radius b, and the distance between their axes is d (see Fig. 6.35). Find the resistance between the wires for a length L. *Hint*: Take advantage of Eq. (7.16).

Ans. $R = \dfrac{1}{2\pi\sigma L}\cosh^{-1}\left(\dfrac{d^2 - a^2 - b^2}{2ab}\right)$.

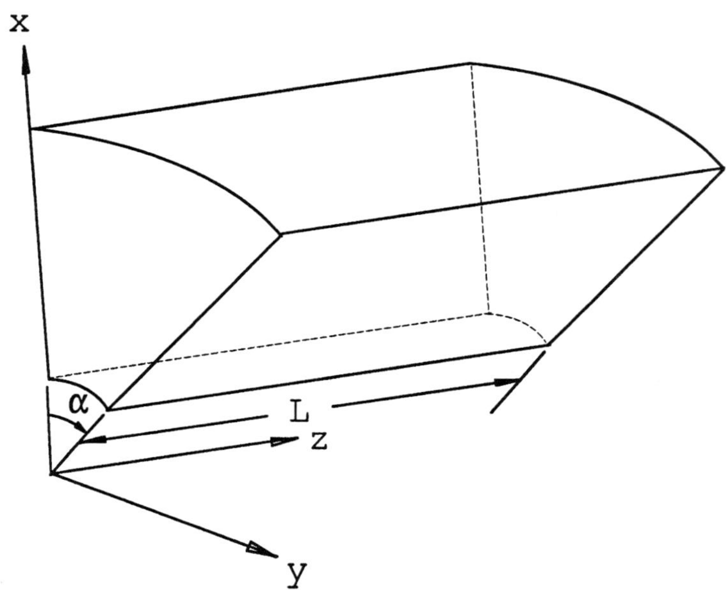

FIGURE 7.14 Problem 7.6.

7.8 Find the ratio of the high-frequency resistance to the steady-current resistance for the conductor of Problem 7.2.

Ans. $\dfrac{R_{ac}}{R} = \dfrac{ab}{2(a+b)\delta_s}$.

7.9 Find the resistance and capacitance between a perfectly conducting wire of radius a and a perfectly conducting coaxial cylindrical shell of inner radius b and outer radius c, $a < b < c$, when the medium between the wire and the shell is seawater, $\sigma = 4$ S/m, $\varepsilon = 81\,\varepsilon_0$ F/m. For what frequency range is this a good resistor? For what frequency range is this a good capacitor? For what frequency range must it be represented by a parallel RC circuit?

Ans. $\omega\tau = 0.01$ at $f = 9.95 \times 10^6$, good resistor for frequency less than 10 MHz. $\omega\tau = 100$ at $f = 99.5 \times 10^9$, good capacitor for frequency greater than 100 GHz. From 10 MHz to 100 GHz, use a parallel RC circuit to represent the physical device.

REFERENCES

[1] William R. Smythe, *Static and Dynamic Electricity*, McGraw-Hill, New York, 1950, p. 114.

[2] Sergai A. Schelkunoff and Harold T. Friis, *Antennas: Theory and Practice*, Wiley, New York, 1952, p. 438.

CHAPTER EIGHT

Magnetostatics

8.1 INTRODUCTION

In this chapter we examine the magnetic fields present when the current, which is the source of those fields, is constant with time and stationary in space. As a consequence, the magnetic fields are independent of time and all derivatives with respect to time equal zero ($\partial/\partial t = 0$). Maxwell's equations (see Table 5.1) separate into two independent sets of equations, as noted in Section 6.1, introducing electrostatic fields. The set of equations for magnetostatic fields in integral and differential form is

$$\oint_L \vec{H}\cdot d\vec{l} = \int_S \vec{J}\cdot d\vec{S} \qquad \nabla\times\vec{H} = \vec{J}, \tag{8.1a,b}$$

$$\oint_S \vec{B}\cdot d\vec{S} = 0 \qquad \nabla\cdot\vec{B} = 0 \tag{8.2a,b}$$

The boundary conditions are repeated as

$$\vec{a}_n\times(\vec{H}_1 - \vec{H}_2) = \vec{J}_s \qquad H_{1\text{tangential}} - H_{2\text{tangential}} = J_{s,\text{normal}}, \tag{8.3a,b}$$

$$\vec{a}_n\times(\vec{H}_1 - \vec{H}_2) = 0,\ \vec{J}_s = 0 \qquad H_{1\text{tangential}} = H_{2\text{tangential}}, \tag{8.3c,d}$$

$$\vec{a}_n\cdot(\vec{B}_1 - \vec{B}_2) = 0 \qquad B_{1\text{normal}} = B_{2\text{normal}}. \tag{8.4a,b}$$

Equations (8.3a,b) mean that the tangential components of magnetic field intensity are discontinuous at a surface by the current density, which lies in the surface and is normal to them. Recall that there can only be a surface current density on a perfectly conducting surface. Equations (8.3c,d) are the same equations, but with the surface current density set to zero, in which case the tangential components of magnetic field intensity are continuous. Equations (8.4a,b) state that the normal components of magnetic flux density are always continuous.

This chapter discusses the solution of magnetostatic field problems. When we study the propagation of waves along transmission lines, we will find that the magnetostatic solution is important for these time-dependent fields as well.

8.2 MAGNETIC FIELD INTENSITY OF A CURRENT DISTRIBUTION

The electric field intensity for any charge distribution was derived, in Section 6.2, from the electric field intensity for a point source. It is now our objective to find the analogous equations for the magnetic field. In order to do so, we must find the magnetic field intensity for a short current element, $\vec{a}_z I_0\, dz$, consisting of a current I_0 of length dz located at the origin and oriented along the z axis (Fig. 8.1).

8.2.1 Magnetic Field Intensity of a Current Element

In Sections 4.5.3 and 4.5.4 we found the field of current $-I_0$ flowing along the positive z axis and then spreading radially outward in a surrounding conductor (Fig. 4.17 or 4.18). The result was given in Eqs. (4.60) and (4.67) and is repeated here:

$$H_\phi = -\frac{I_0(1+\cos\theta)}{4\pi R\sin\theta}. \tag{8.5}$$

We now add a second current $+I_0$ extending along the z axis from $z = -\Delta z$ to $z = \infty$ (Fig. 8.2). The field due to the two superimposed currents is the sum of the fields due to each one separately.

$$H_\phi = -\frac{I_0(1+\cos\theta)}{4\pi R\sin\theta} + \frac{I_0(1+\cos\theta')}{4\pi R'\sin\theta'}.$$

Since the currents $+I_0$ and $-I_0$ along the positive z axis cancel each other, this is really the field due to a current $+I_0$ extending along the z axis from $z = -\Delta z$ to $z = 0$. From the figure we note that $R\sin\theta = R'\sin\theta' = r$, where r is the cylindrical radius from the z axis to the field point. Thus, putting the numerators over their common denominator and multiplying and dividing by Δz, we find

$$H_\phi = \frac{I_0\,\Delta z}{4\pi r}\left[\frac{\cos\theta' - \cos\theta}{\Delta z}\right]. \tag{8.6}$$

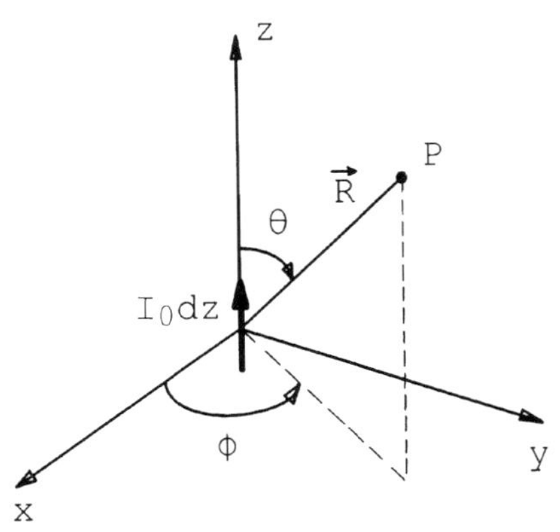

FIGURE 8.1 Short current element.

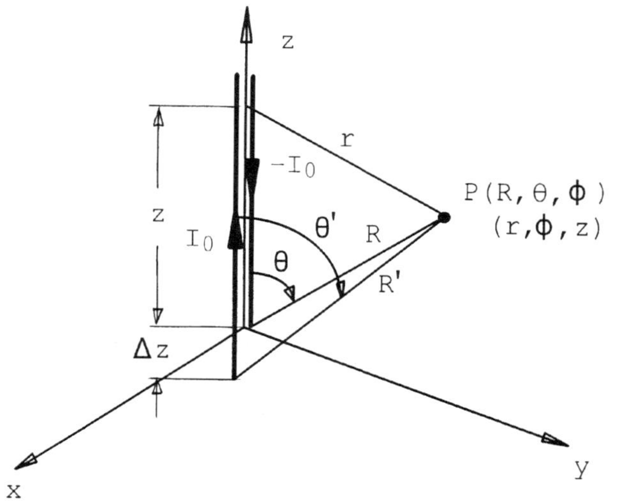

FIGURE 8.2 Superimposed filamentary currents.

As Δz approaches the differential length dz the bracketed term approaches $d(\cos\theta)/dz$. From the figure,

$$\cos\theta = \frac{z}{\sqrt{r^2 + z^2}}, \quad \sin\theta = \frac{r}{\sqrt{r^2 + z^2}}, \quad R = \sqrt{r^2 + z^2},$$

and therefore

$$\lim_{\Delta z \to 0}\left[\frac{\cos\theta' - \cos\theta}{\Delta z}\right] = \frac{d(\cos\theta)}{dz} = \frac{d}{dz}\left(\frac{z}{\sqrt{r^2 + z^2}}\right) = \frac{r^2}{(r^2 + z^2)^{3/2}} = \frac{r\sin\theta}{R^2}. \tag{8.7}$$

We write dH_ϕ for H_ϕ and dz for Δz in Eq. (8.6) and substitute from Eq. (8.7) to obtain

$$dH_\phi = \frac{I_0\, dz \sin\theta}{4\pi R^2} \tag{8.8}$$

as the magnetic field of a current element $I_0\, dz$ located at the origin along the z axis (Fig. 8.1).

The result can be generalized to find the field of a current element located at the origin but with arbitrary orientation, $I_0\, d\vec{l}$ (Fig. 8.3a). Comparing Fig. 8.1 and Fig. 8.3a we see that $d\vec{l} = \vec{a}_z\, dz$ and thus $d\vec{l} \times \vec{a}_R = (\vec{a}_z\, dz) \times \vec{a}_R = \vec{a}_\phi\, dz \sin\theta$. Therefore

$$d\vec{H} = \frac{I_0\, d\vec{l} \times \vec{a}_R}{4\pi R^2} = \frac{I_0\, d\vec{l} \times \vec{R}}{4\pi R^3}.$$

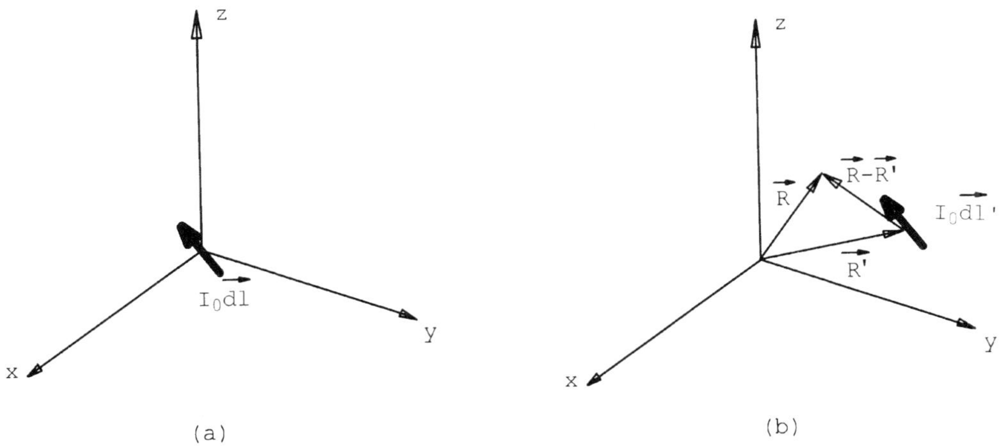

FIGURE 8.3 Current element of arbitrary orientation.

We can generalize still further by moving the current element to an arbitrary source point $\vec{R}'$ (Fig. 8.3b). Then the field at the field point $\vec{R}$ is

$$d\vec{H} = \frac{I_0\, d\vec{l}' \times (\vec{R} - \vec{R}')}{4\pi|\vec{R} - \vec{R}'|^3}. \tag{8.9}$$

This formula gives the response (magnetic field intensity $d\vec{H}$) at the field point $\vec{R}$ due to an impulsive input (current element) with arbitrary direction $\vec{a}_{dl'}$ located at $\vec{R}'$.

This formula is similar to Eq. (6.8) derived for the electrostatic electric field intensity. The magnetostatic and electrostatic formulas differ in the following ways. First, there is the obvious difference that the magnetostatic formula involves a vector product, while the electrostatic formula does not. The cross product means that the current source and the field response are in orthogonal directions. It also reflects the different nature of the field's sources. For the electrostatic field the source is electric charge, a scalar quantity. For the magnetostatic case the source is electric current, a vector quantity. Second, the electrostatic formula gives the electric field caused by a point charge, which is a physically realizable quantity. The magnetostatic formula, on the contrary, gives the contribution to the total magnetic field intensity caused by a current element, which cannot physically exist by itself. This is because currents must be continuous or there will be time-dependent charges at the points of discontinuity. The isolated current element has discontinuous current at each end.

The formula gives the field in terms of a physically unrealizable current element, but by combining current elements we can obtain continuous current distributions and thus the fields that they produce.

8.2.2 Magnetic Field Intensity of Continuous Current Distributions

If a current density $\vec{J}$ A/m^2 is present throughout some volume V, then using the equivalence, Eq. (3.17), $I_0\, d\vec{l}' \Leftrightarrow \vec{J}\, dv'$ and integrating throughout the volume, we have

$$\vec{H}(\vec{R}) = \int_V \frac{\vec{J}(\vec{R}') \times (\vec{R} - \vec{R}')\, dv'}{4\pi|\vec{R} - \vec{R}'|^3}. \tag{8.10}$$

If surface current density $\vec{J}_s$ A/m is present over a surface S, then using the equivalence $I_0\, d\vec{l}' \Leftrightarrow \vec{J}_s\, dS'$ and integrating over the surface, we have

$$\vec{H}(\vec{R}) = \int_S \frac{\vec{J}_s(\vec{R}') \times (\vec{R} - \vec{R}')\, dS'}{4\pi|\vec{R} - \vec{R}'|^3}. \tag{8.11}$$

If line current i A is present along some line L, then replacing I_0 with i and integrating along the line, we have

$$\vec{H}(\vec{R}) = \int_L \frac{i\, d\vec{l}' \times (\vec{R} - \vec{R}')}{4\pi|\vec{R} - \vec{R}'|^3}. \tag{8.12}$$

If any combination of these currents or current densities is present, then the total magnetic field intensity is found by adding the appropriate formulas.

Note that these three formulas are just different forms of the original formula, Eq. (8.9), for the current element.

Example 8.1 (a) A line current I_0 flows in the positive direction along the z axis from $z = -D$ to $z = D$ (Fig. 8.4a). Find the magnetic field intensity at any point in space. (Note that this is not a physically possible source, because the current is not continuous at the two ends.) (b) Show that as $D \to \infty$ the result approaches $H_\phi = I_0/2\pi r$, Eq. (4.54). (c) Find the magnetic field intensity at the center of a equilateral triangle with side length a and current I_0 (Fig. 8.4b). (Note that this is a physically possible current source, because the current forms a closed loop.)

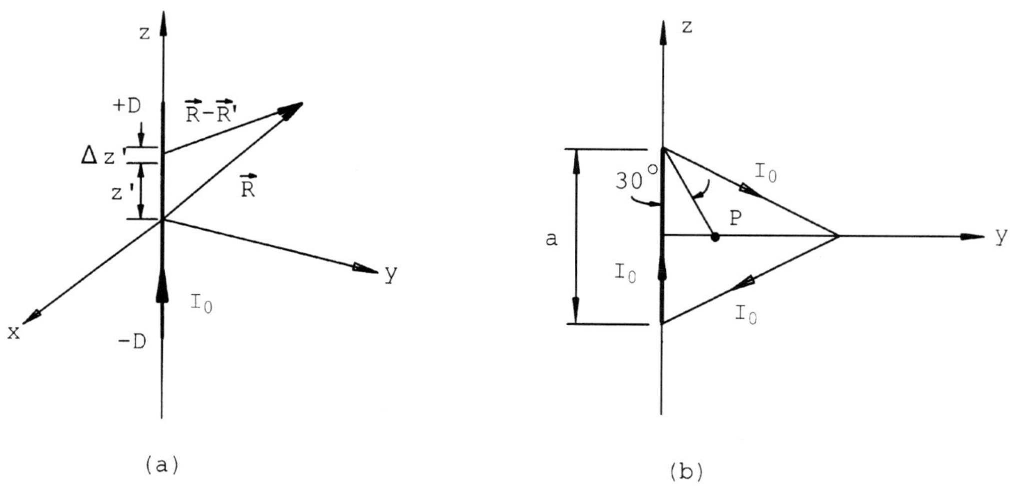

FIGURE 8.4 Example 8.1.

Solution (a) A typical source point is located at $\vec{R}' = \vec{a}_z z'$. At that point $i\,d\vec{l}' = I_0\,\vec{a}_z\,dz'$. Using cylindrical coordinates the field point is located at $\vec{R} = \vec{a}_r r + \vec{a}_z z$. Then

$$\vec{R} - \vec{R}' = \vec{a}_r r + \vec{a}_z(z - z'), \quad |\vec{R} - \vec{R}'| = [r^2 + (z - z')^2]^{1/2}.$$

Substituting in Eq. (8.12) we obtain

$$\vec{H} = \int_{z'=-D}^{D} \frac{I_0(\vec{a}_z\,dz') \times [\vec{a}_r r + \vec{a}_z(z - z')]}{4\pi[r^2 + (z - z')^2]^{3/2}} = \frac{I_0\,r\,\vec{a}_\phi}{4\pi} \int_{-D}^{D} \frac{dz'}{[r^2 + (z - z')^2]^{3/2}}.$$

Let $u = z - z'$ and $du = -dz'$. Then the upper and lower limits become $z - D$ and $z + D$, respectively. Interchanging the limits and dropping the minus sign we obtain

$$\begin{aligned} \vec{H} &= \frac{I_0\,r\,\vec{a}_\phi}{4\pi} \int_{z-D}^{z+D} \frac{du}{(r^2 + u^2)^{3/2}} \\ &= \vec{a}_\phi \frac{I_0}{4\pi r}\left[\frac{z + D}{\sqrt{r^2 + (z + D)^2}} - \frac{z - D}{\sqrt{r^2 + (z - D)^2}}\right]. \end{aligned}$$

(b) As $D \to \infty$, $z + D \to D$ and $\sqrt{r^2 + (z + D)^2} \to D$. Hence the first term in the bracket approaches $D/D = 1$. Similarly, the second term approaches -1. Thus the bracketed difference approaches 2 and

$$\lim_{D \to \infty} \vec{H} = \vec{a}_\phi \frac{I_0}{2\pi r}.$$

This result is the same as Eq. (4.54), which was derived from the Ampère–Maxwell integral law.

(c) For the side of the triangle along the z axis, we can use the result of part (a) with $z = 0$, $r = (a/2)\tan 30° = a/(2\sqrt{3})$, and $D = a/2$. This side yields $H_\phi = 3\,I_0/2\pi a$ at the center. Each of the other two sides causes the same field, so the total at the center of the triangle is

$$H_\phi = \frac{9 I_0}{2\pi a}.$$

Example 8.2 (a) A surface current $\vec{J}_s = \vec{a}_\phi J_{s0}$ flows on a cylinder of radius a coaxial with the z axis extending from $z = -D$ to $z = D$ (Fig. 8.5). Find the magnetic field intensity at any point on the z axis. (b) Find the magnetic field intensity at the center of the solenoid and at its ends.

Solution (a) A typical source point is located at $\vec{R}' = \vec{a}_{r'}\,a + \vec{a}_z z'$. The field point is at $\vec{R} = \vec{a}_z z$. Hence

$$\begin{aligned} \vec{H} &= \int_{z'=-D}^{D} \int_{\phi'=0}^{2\pi} \frac{(\vec{a}_{\phi'} J_{s0}) \times [-\vec{a}_{r'}\,a + \vec{a}_z(z - z')]\,a\,d\phi'\,dz'}{4\pi[a^2 + (z - z')]^{3/2}} \\ &= \int_{z'=-D}^{D} \int_{\phi'=0}^{2\pi} \frac{J_{s0}\,[\vec{a}t_z a + \vec{a}_{r'}(z - z')]\,a\,d\phi'\,dz'}{4\pi[a^2 + (z - z')]^{3/2}}. \end{aligned}$$

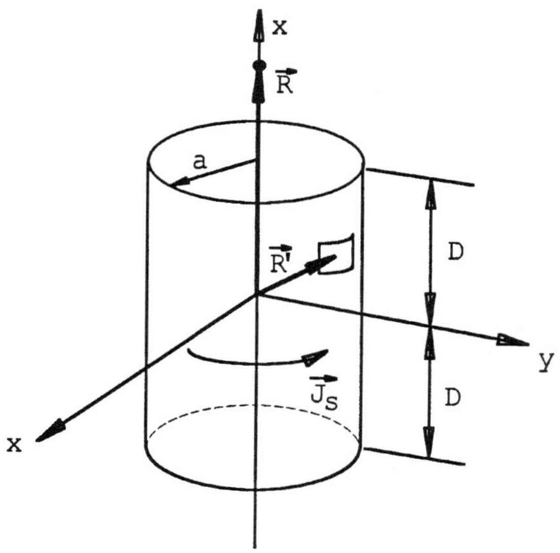

FIGURE 8.5 Surface current density on a cylinder of radius a coaxial with the z axis.

The term involving $\vec{a}_{r'}$ is zero when integrated over ϕ'. Thus

$$\vec{H} = \vec{a}_z \frac{J_{s0}\,a^2}{2} \int_{z'=-D}^{D} \frac{dz'}{[a^2 + (z - z')^2]^{3/2}}.$$

The integral is the same as the integral in the previous example. Using those results we find

$$\vec{H} = \vec{a}_z \frac{J_{s0}}{2} \left[\frac{z + D}{\sqrt{a^2 + (z + D)^2}} - \frac{z - D}{\sqrt{a^2 + (z - D)^2}} \right]. \tag{8.13}$$

(b) The magnetic field intensity at the center of the solenoid, $z = 0$, is

$$\begin{aligned} H_z &= \frac{J_{s0} D}{\sqrt{a^2 + D^2}} \\ &\approx J_{s0}, \quad D \gg a. \end{aligned}$$

At either end of the solenoid, $z = \pm D$, it is

$$\begin{aligned} H_z &= \frac{J_{s0} D}{\sqrt{a^2 + (2D)^2}} \\ &\approx J_{s0}/2, \quad D \gg a. \end{aligned}$$

For a long thin solenoid the field inside the solenoid is closely equal to the surface current density until very close to the ends, where it drops to half that value. This result is in accord with the operational definition for the magnetic field intensity given in Section 3.8.

8.3 MAGNETIC FIELD OF A CURRENT LOOP

Consider a thin wire shaped as a circular loop of radius a with current I flowing in the loop. The coordinate system is introduced with its origin at the center of the loop and with the xy plane oriented to coincide with the plane of the loop (Fig. 8.6). We will use Eq. (8.12) to find a formula for the magnetic field intensity at any point in space.

As shown in the figure,

$$d\vec{l}' = \vec{a}_{\phi'} a\, d\phi' = (-\vec{a}_x \sin\phi' + \vec{a}_y \cos\phi')\, a\, d\phi',$$

$$\vec{R}' = \vec{a}_{r'} a = (\vec{a}_x \cos\phi' + \vec{a}_y \sin\phi')\, a,$$

$$\vec{R} = \vec{a}_R R = \vec{a}_x x + \vec{a}_y y + \vec{a}_z z.$$

Therefore

$$\vec{R} - \vec{R}' = \vec{a}_x (x - a\cos\phi') + \vec{a}_y (y - a\sin\phi') + \vec{a}_z z$$

and

$$d\vec{l}' \times (\vec{R} - \vec{R}') = [\vec{a}_x z\cos\phi' + \vec{a}_y z\sin\phi' - \vec{a}_z(y\sin\phi' + x\cos\phi' - a)]\, a\, d\phi'.$$

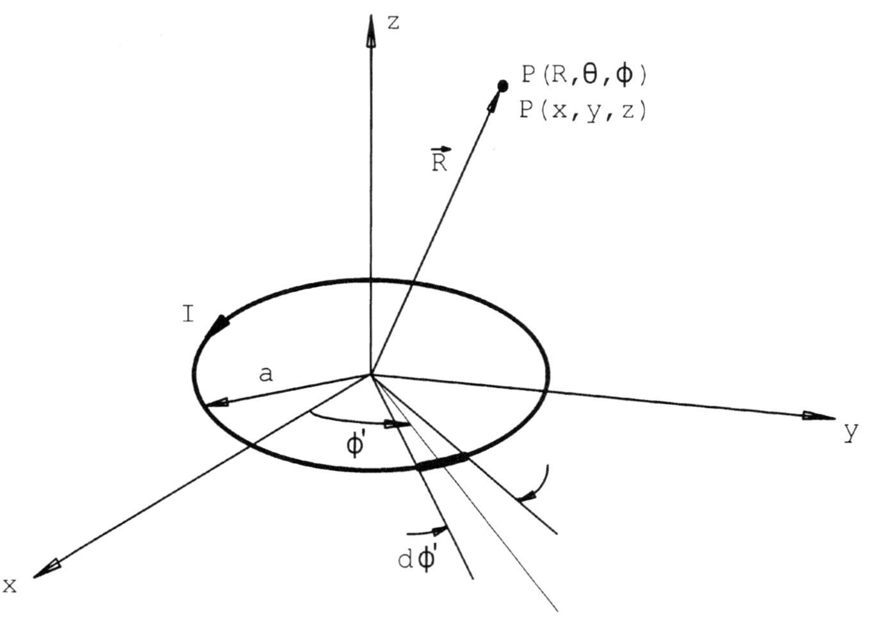

FIGURE 8.6 Circular loop carrying a current I.

In the denominator

$$
\begin{aligned}
|\vec{R} - \vec{R}'|^3 &= [(x - a\cos\phi')^2 + (y - a\sin\phi')^2 + z^2]^{3/2} \\
&= (x^2 + y^2 + z^2 - 2ax\cos\phi' - 2ay\sin\phi' + a^2)^{3/2} \\
&= R^3\left(1 - \frac{2ax}{R^2}\cos\phi' - \frac{2ay}{R^2}\sin\phi' + \frac{a^2}{R^2}\right)^{3/2}
\end{aligned}
$$

To obtain the last expression we have used $R^2 = x^2 + y^2 + z^2$. The magnetic intensity is therefore

$$
\vec{H} = \frac{Ia}{4\pi R^3}\int_{\phi'=0}^{2\pi} \frac{[\vec{a}_x z\cos\phi' + \vec{a}_y z\sin\phi' - \vec{a}_z(y\sin\phi' + x\cos\phi' - a)]\,d\phi'}{\left[1 - \dfrac{2a}{R}\left(\dfrac{x}{R}\cos\phi' + \dfrac{y}{R}\sin\phi'\right) + \left(\dfrac{a}{R}\right)^2\right]^{3/2}}. \tag{8.14}
$$

This expression could be evaluated numerically, but no closed form solution can be written for the integral.

Let us now look at the special case where the field point is on the z axis, $R = z$, $x = y = 0$. This gives

$$
\begin{aligned}
\vec{H}(0, 0, z) &= \frac{Ia}{4\pi z^3}\int_{\phi'=0}^{2\pi} \frac{(\vec{a}_x z\cos\phi' + \vec{a}_y z\sin\phi' + \vec{a}_z a)\,d\phi'}{(1 + a^2/z^2)^{3/2}} \\
&= \vec{a}_z \frac{Ia^2}{2(a^2 + z^2)^{3/2}}.
\end{aligned}
$$

The same result could be obtained more simply by starting with the field point on the z axis, instead of with the general case.

A second special case is more important. Let us assume that the field point is much further from the origin than the radius of the loop, $R \gg a$. Then we can neglect the term $(a/R)^2$ in the denominator of Eq. (8.14). Furthermore, we can use the approximation $(1 - u)^{-3/2} \approx 1 + \frac{3}{2}u$, $|u| \ll 1$, to obtain

$$
\begin{aligned}
\vec{H} = \frac{Ia}{4\pi R^3}\int_{\phi'=0}^{2\pi} &[\vec{a}_x z\cos\phi' + \vec{a}_y z\sin\phi' \\
&- \vec{a}_z(y\sin\phi' + x\cos\phi' - a)]\left(1 + \frac{3ax}{R^2}\cos\phi' + \frac{3ay}{R^2}\sin\phi'\right)d\phi'.
\end{aligned}
$$

When the terms are multiplied together and each term of the sum is integrated separately, the integration is simple. Remember that the integral of $\sin\phi'$, $\cos\phi'$, and $\sin\phi'\cos\phi'$ equals zero and the integral of $\sin^2\phi'$ and $\cos^2\phi'$ equals π. The result is

$$
\vec{H} = \frac{I\pi a^2}{4\pi R^3}\left[\vec{a}_x\frac{3xz}{R^2} + \vec{a}_y\frac{3yz}{R^2} - \vec{a}_z\left(\frac{3y^2}{R^2} + \frac{3x^2}{R^2} - 2\right)\right].
$$

Recalling that $z/R = \cos\theta$, $x/R = \sin\theta\cos\phi$, $y/R = \sin\theta\sin\phi$, and $(x^2 + y^2)/R^2 = (r/R)^2 = \sin^2\theta$, the bracketed term becomes

$$\begin{aligned}[\ \ ] &= \vec{a}_x 3\sin\theta\cos\theta\cos\phi + \vec{a}_y 3\sin\theta\cos\theta\sin\phi + \vec{a}_z(2 - 3\sin^2\theta) \\ &= 2\cos\theta(\vec{a}_x\sin\theta\cos\phi + \vec{a}_y\sin\theta\sin\phi + \vec{a}_z\cos\theta) \\ &\quad + \sin\theta(\vec{a}_x\cos\theta\cos\phi + \vec{a}_y\cos\theta\sin\phi - \vec{a}_z\sin\theta) \\ &\quad + \vec{a}_z(2 - 3\sin^2\theta - 2\cos^2\theta + \sin^2\theta) \\ &= \vec{a}_R 2\cos\theta + \vec{a}_\theta\sin\theta.\end{aligned}$$

Letting $\pi a^2 = S$, the area of the loop, we have

$$\vec{H} = \frac{IS}{4\pi R^3}(\vec{a}_R 2\cos\theta + \vec{a}_\theta\sin\theta) = \frac{m}{4\pi R^3}(\vec{a}_R 2\cos\theta + \vec{a}_\theta\sin\theta), \qquad (8.15)$$

where $m = IS$.

This is the field of an infinitesimal current loop of magnetic moment $m = IS$. Note that the form of the field is identical to the form of an electric charge dipole of electric moment qd, Eq. (6.33). The magnetic current loop and the electric charge dipole are dual elements.

Example 8.3 Find the magnetic moment of a charge q uniformly distributed throughout a sphere of radius a rotating with angular velocity $\vec{\omega} = \vec{a}_z\omega$ about the z axis.

Solution The charge density at any point within the sphere is $\rho = q/V = q/(4\pi a^3/3)$. Figure 8.7 shows the sphere of charge and a typical element of area $d\vec{S} = \vec{a}_\phi R\,d\theta\,dR$

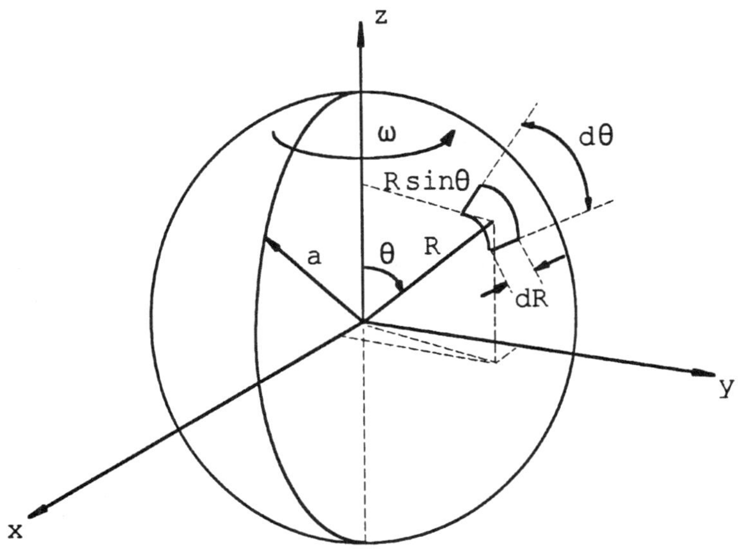

FIGURE 8.7 Sphere of charge with typical element of surface.

located at (R, θ, ϕ), $R < a$. The charge at this element of area has a velocity $\vec{u} = \vec{a}_\phi \, \omega R \sin\theta$, causing a circle of current $dI = \rho \, d\vec{S} \cdot \vec{u}$. The magnetic moment of this current about the z axis is $dm = \varepsilon\pi(R \sin\theta)^2$. Thus

$$dm = \left(\frac{3q}{4\pi a^3}\right)(R \, d\theta \, dR)(\omega R \sin\theta)(\pi R^2 \sin^2\theta)$$

$$m = \frac{3q\omega}{4a^3} \int_0^a R^4 \, dR \int_0^\pi \sin^3\theta \, d\theta$$

$$= \frac{q\omega a^2}{5}.$$

This magnetic moment is about the z axis at the origin. The magnetic field intensity is given by Eq. (8.15) at distances $R \gg a$.

8.4 MAGNETIC ENERGY

In studying the energy in the electrostatic field we recognized that in order to establish a system of charges in space, and hence an electric field due to the charges, it was necessary to do mechanical work on the charges. This expended mechanical energy was stored in the electric field as electric energy. Similarly, in order to establish a system of steady currents, and hence a magnetic field due to the currents, it is necessary to do work that is stored in the magnetic field. We will now find an expression for the energy stored in the magnetic field.

As we establish the system of currents, the fields will increase from zero to their final values. The changing magnetic flux density, $\partial\vec{B}/\partial t$, will give rise to an induced electric field intensity $\vec{E}^i$. The changing electric field will itself give rise to a displacement current density, $\vec{J}_d = \partial\vec{D}/\partial t$, but we assume that the fields change so slowly that this displacement current density is negligible compared to the real current density $\vec{J}$. The power expended in establishing the magnetic field is the integral throughout some volume of the scalar product of the induced electric field and the current density,

$$P = \frac{dW_m}{dt} = -\int_V \vec{E}^i \cdot \vec{J} \, dv.$$

The current density is related to the magnetic field intensity by the Ampère–Maxwell law $\nabla \times \vec{H} = \vec{J} + \vec{J}_d = \vec{J}$, neglecting the displacement current density. Then

$$\frac{dW_m}{dt} = -\int_V [\vec{E}^i \cdot (\nabla \times H)] \, dv$$

$$= \int_V [\nabla \cdot (\vec{E}^i \times \vec{H})] \, dv - \int_V [\vec{H} \cdot (\nabla \times \vec{E}^i)] \, dt.$$

We have used the vector identity $\nabla \cdot (\vec{E}^i \times \vec{H}) = \vec{H} \cdot (\nabla \times \vec{E}^i) - \vec{E}^i \cdot (\nabla \times \vec{H})$, Eq. (5.35). Using the divergence theorem to transform the first integral on the right into a surface integral and using the Faraday–Maxwell law, $\nabla \times \vec{E} = -\partial\vec{B}/\partial t$, in the second integral on the right, we obtain

$$\frac{dW_m}{dt} = \int_S (\vec{E}^i \times \vec{H}) \cdot d\vec{S} + \int_V \left(\vec{H} \cdot \frac{\partial \vec{B}}{\partial t}\right) dv.$$

Let the surface S expand into a sphere of radius $R \to \infty$ about an arbitrary origin. Since the source current is confined to a finite volume, at sufficiently large R all loops of current will appear to be small in size. Then $H \propto 1/R^3$ on S as shown by Eq. (8.15). Since the curl of $\vec{E}$ gives the $\vec{H}$ (really, its derivative with respect to time), and since the curl operation involves differentiation with respect to position, $E^i \propto 1/R^2$. The surface expands as $dS \propto R^2$. The surface integral is therefore proportional to $1/R^3$ and vanishes as $R \to \infty$. This leaves only the volume integral on the right, integrated throughout all space:

$$\frac{dW_m}{dt} = \int_V \vec{H} \cdot \frac{\partial \vec{B}}{\partial t}\, dv. \tag{8.16}$$

This result is true even if there are magnetic materials present.

Let us consider only linear isotropic material so that $\vec{B} = \mu\vec{H}$, where μ may be a function of position. Then

$$\begin{aligned}\frac{dW_m}{dt} &= \int_V \mu\vec{H} \cdot \frac{\partial \vec{H}}{\partial t}\, dv = \int_V \mu \frac{\partial}{\partial t}\left(\frac{1}{2}|\vec{H}|^2\right) dv \\ &= \frac{1}{2}\frac{d}{dt}\int_V \mu|\vec{H}|^2\, dv.\end{aligned}$$

Integrating both sides with respect to t and assuming $W_m = 0$ when $\vec{H} = 0$, we have

$$W_m = \frac{1}{2}\int_V \mu|\vec{H}|^2\, dv = \frac{1}{2}\int_V \frac{1}{\mu}|\vec{B}|^2\, dv = \frac{1}{2}\int_V \vec{B} \cdot \vec{H}\, dv. \tag{8.17}$$

This equation gives the magnetic energy stored in the field in terms of the magnetic field intensity throughout all space.

8.5 SELF-INDUCTANCE

Consider a current loop with constant current I in a medium with uniform permeability (Fig. 8.8). The magnetic field intensity $\vec{H}$ and hence the magnetic flux density $\vec{B}$ will be proportional to the current I. The total flux Φ linked with the loop is the integral of the flux density through any surface capping the loop. This will also be proportional to the current I. This proportionality can be expressed as

$$\Phi = LI, \tag{8.18}$$

where L is called the self-inductance or, more simply, the inductance of the loop. The SI unit of inductance is the henry, H.

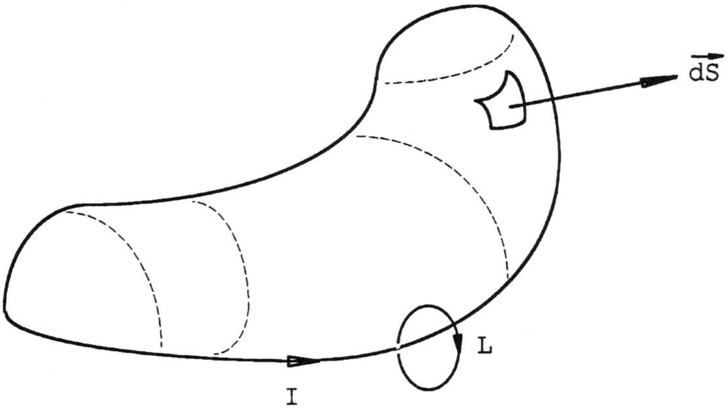

FIGURE 8.8 Current loop with constant current (I) placed in a medium with uniform permeability.

The self-inductance can therefore be written as

$$L = \frac{\Phi}{I} = \frac{\int_S \vec{B} \cdot d\vec{S}}{\oint_L \vec{H} \cdot d\vec{l}} = \frac{\int_S \mu \vec{H} \cdot d\vec{S}}{\oint_L \vec{H} \cdot d\vec{l}}. \tag{8.19}$$

In this relation the line L is any closed line surrounding or linked with the current I and the surface S is any surface capping the current loop.

To find the relation between current, inductance, and magnetic energy we must consider the process of increasing the current in the loop from zero, let us say at $t = 0$, to I at time t. In Fig. 8.9 the loop is redrawn with a current source. As the current source is increased to I, the flux through the loop will also increase. In accordance with the Faraday–Maxwell law the changing flux will induce a voltage $-d\Phi/dt$ around the loop. Thus

$$\oint_L \vec{E} \cdot d\vec{l} = -\frac{d\Phi}{dt}$$

$$V_{AB} + V_{BCA} = -\frac{d\Phi}{dt}.$$

Multiplying by I gives a term $V_{AB}\, I$, which is the power input to the current source, and a term $V_{BCA}\, I = RI^2$, which is the power dissipated in ohmic loss (heat) in the current loop. The term $V_{BA}\, I = -V_{AB}\, I$ therefore represents the power supplied by the current source. Solving for this term

$$V_{BA}\, I = \frac{d\Phi}{dt} I + RI^2.$$

The left side of the equation is the power supplied by the source. The rightmost term is the power dissipated by the resistance of the loop. The first term on the right must be

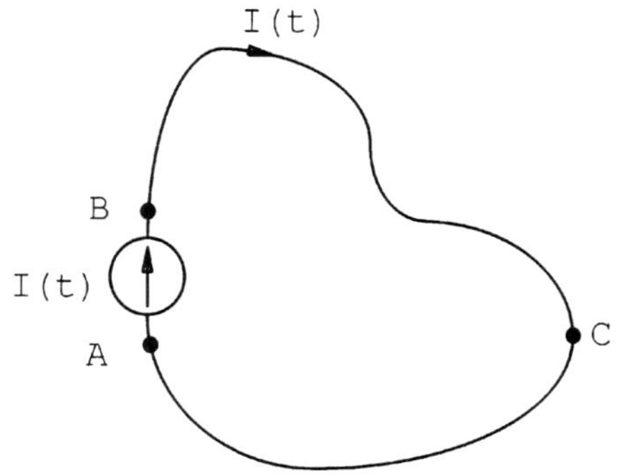

FIGURE 8.9 Loop excited by a current source.

the energy that the source supplies to the magnetic field as the current increases. Integrating with time gives

$$\int_{t'=0}^{t} V_{BA}(t')I(t')\,dt' = \int_{t'=0}^{t} \frac{d\Phi(t')}{dt'} I(t')\,dt' + \int_{t'=0}^{t} RI(t')^2\,dt'$$

$$\int_{t'=0}^{t} \frac{dW_s}{dt'}\,dt' = \int_{t'=0}^{t} L\,\frac{dI(t')}{dt'}\,I(t')\,dt' + \int_{t'=0}^{t} \frac{dW_L}{dt'}\,dt'$$

$$W_s = \tfrac{1}{2}\,L\,[I^2(t) - I^2(0)] + W_L\,.$$

The left side W_s is the energy supplied by the source. The rightmost term is the energy lost in the resistance of the loop. Since $I(0) = 0$, the remaining term is the energy supplied to the magnetic field. Thus

$$W_m = \tfrac{1}{2}\,LI^2. \tag{8.20}$$

Inverting this expression and using Eq. (8.18), we obtain formulas for the inductance in terms of the stored magnetic energy:

$$L = \begin{cases} \dfrac{2W_m}{I^2}, & (8.21) \\[2ex] \dfrac{\Phi^2}{2W_m}. & (8.22) \end{cases}$$

These are sometimes easier to use than Eq. (8.18). They also illustrate the statement that an inductor is a device that stores magnetic energy.

Combining the expression for magnetic energy storage, Eq. (8.17), with that for inductance, Eq. (8.21), gives

$$L = \frac{1}{I^2}\int_V \vec{B}\cdot\vec{H}\,dv \tag{8.23}$$

We can separate this into two regions of integration. The first is the volume V_{int}, which is internal to the conductors where the current exists. The second region is the volume V_{ext}, which is external to the conductors.

$$\begin{aligned} L &= \frac{1}{I^2} \int_{V_{\text{int}}} \vec{B} \cdot \vec{H} \, dv + \frac{1}{I^2} \int_{V_{\text{ext}}} \vec{B} \cdot \vec{H} \, dv \\ &= L_{\text{int}} + L_{\text{ext}}. \end{aligned} \tag{8.24}$$

The first term is the internal self-inductance and the second term is the external self-inductance.

Consider next the loop connected to a slowly varying current source (Fig. 8.10). We will clarify the meaning of "slowly" somewhat later. A voltage $v(t) = d\Phi/dt$ appears across the loop. Using $\Phi(t) = Li(t)$ we have

$$v(t) = L \frac{di(t)}{dt}. \tag{8.25}$$

This is the usual voltage–current terminal relation used to define the inductance in circuit theory courses. The symbol for an inductor is shown in Fig. 8.11a and the electric circuit model for the current loop with a current source is shown in Fig. 8.11b.

8.5.1 A Solenoid: External Inductance

In Example 8.2 we derived the field in a long thin solenoid. With surface current $\vec{J}_s = \vec{a}_\phi J_{s0}$ on a cylinder of radius a and length $D \gg a$ coaxial with the z axis, the magnetic field intensity within the solenoid is closely given by $\vec{H} = \vec{a}_z J_{s0}$. The magnetic flux through the solenoid is then $\Phi = \mu H_z S = \mu \pi a^2 J_{s0}$, where μ is the permeability of the solenoid's core. The total current circulating around the cylinder is $I = J_{s0} D$. The inductance of the solenoid is given by

$$L = \frac{\Phi}{I} = \frac{\mu \pi a^2}{D}. \tag{8.26}$$

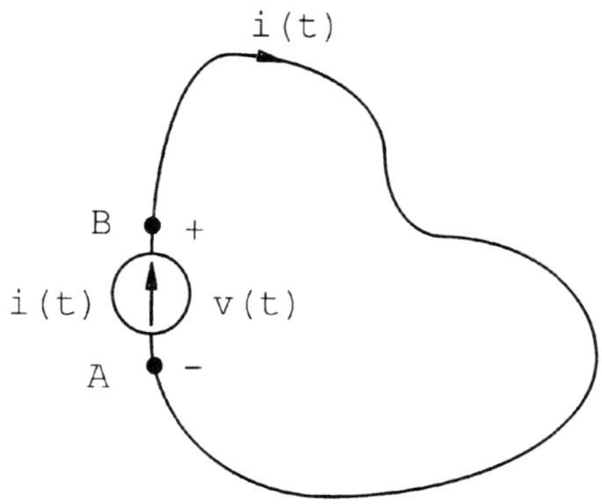

FIGURE 8.10 Loop excited by a slowly varying current source.

The condition of ϕ-directed surface current is closely approximated by winding insulated wire in a tight spiral to form a cylinder (Fig. 8.12). If there are n turns per unit length and a current I in the turns, then $\vec{J}_s \approx \vec{a}_\phi nI$. This neglects the small z-directed component of current due to the pitch of the spiral. The flux is therefore

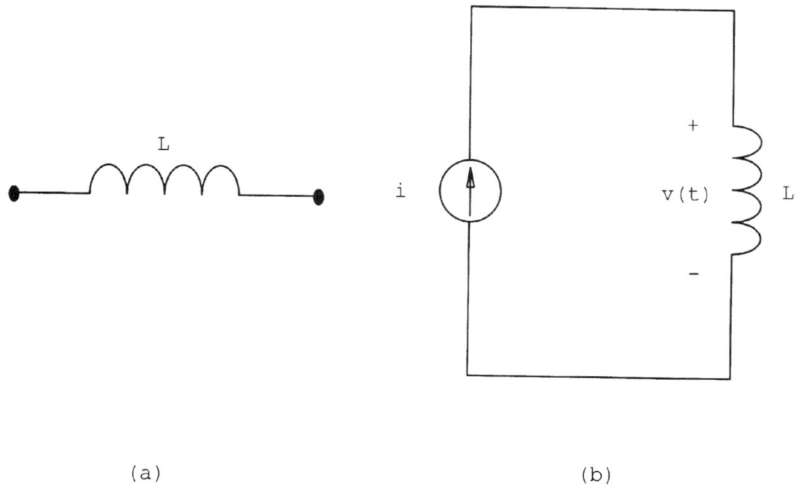

FIGURE 8.11 The inductor as a circuit element.

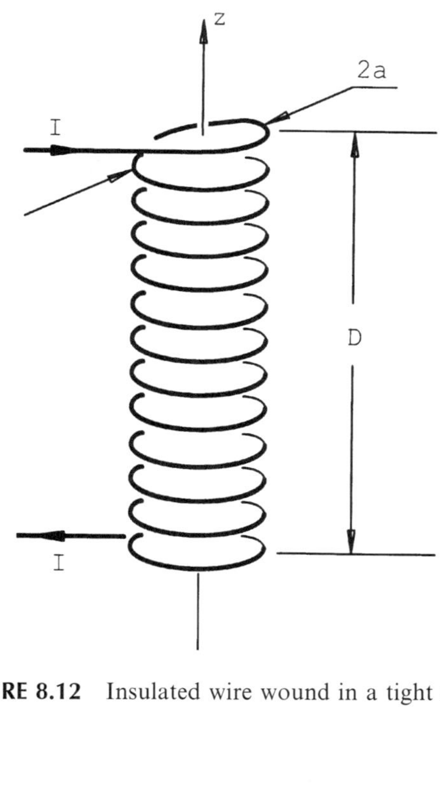

FIGURE 8.12 Insulated wire wound in a tight spiral.

$\Phi = \mu\pi a^2 nI$. This flux is linked with nD turns so that the total flux linkage is $\mu\pi a^2 n^2 DI$. Dividing by the current we get

$$L = \mu\pi a^2 n^2 D. \tag{8.27}$$

8.5.2 Coaxial Cylinders: External Inductance

Figure 8.13 shows two cylinders coaxial with the z axis. Both cylinders are made of thin conductors so that the currents flowing on them are essentially surface currents; we shall examine the effects of wire thickness later. The inner and outer cylinders have radii a and b, respectively. The region between the cylinders has uniform permeability μ. The cylinders are of length D and are shorted with a conducting plate at both ends.

The total current I_0 is in the z direction on the inner cylinder, is radially outward on the top plate, returns in the $-z$ direction on the outer cylinder, and closes the circuit radially inward on the bottom plate. The surface current density on the cylinders is thus

$$\vec{J}_s = \begin{cases} \vec{a}_z \dfrac{I_0}{2\pi a}, & r = a, \\ -\vec{a}_z \dfrac{I_0}{2\pi b}, & r = b. \end{cases}$$

On the conducting surface at $z = D$ the surface current density is $\vec{J}_s = \vec{a}_r I_0/2\pi r$, making the current continuous at $r = a$ and b. Similarly, in the $z = 0$ surface $\vec{J}_s = -\vec{a}_r I_0/2\pi r$.

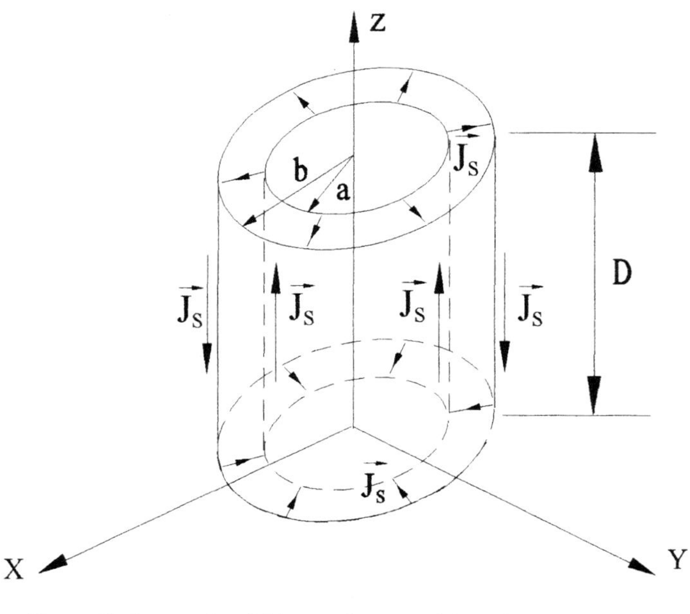

FIGURE 8.13 Two cylinders made of thin conductors with surface currents used in determining external inductance.

Using the Ampère–Maxwell law around a circle of radius r gives

$$H_\phi = \begin{cases} 0, & r < a, \\ \dfrac{I_0}{2\pi r}, & a < r < b, \\ 0, & b < r. \end{cases} \tag{8.28}$$

Note that the boundary condition that tangential magnetic field intensity be discontinuous by the surface current density, Eqs. (8.3c,d), is satisfied at the $r = a$ and b surfaces and also on the $z = 0$ and $z = D$ surfaces.

The magnetic flux through any radial plane hinged on the z axis (a $\phi = \phi_0$ surface) is

$$\Phi = \int_{z=0}^{D} \int_{r=0}^{b} \frac{\mu I_0}{2\pi r}\, dr\, dz = \frac{\mu D I_0}{2\pi} \ln\left(\frac{b}{a}\right).$$

The inductance is given by Φ / I_0:

$$L = \frac{\mu D}{2\pi} \ln\left(\frac{b}{a}\right). \tag{8.29}$$

We could have obtained the same result using the energy relations, Eqs. (8.17) and (8.21). The energy stored in the field and the inductance are

$$W_m = \frac{1}{2} \int_{z=0}^{D} \int_{\phi=0}^{2\pi} \int_{r=a}^{b} \mu \left(\frac{I_0}{2\pi r}\right)^2 r\, dr\, d\phi\, dz = \frac{1}{2} \frac{\mu D I_0^2}{2\pi} \ln\left(\frac{b}{a}\right),$$

$$L = \frac{\mu D}{2\pi} \ln\left(\frac{b}{a}\right).$$

Let us make things a little more interesting and assume that the region between the cylinders is filled with two materials with different permeabilities. Let the region $a < r < c < b$ have permeability μ_1 and the region $c < r < b$ have permeability μ_2 (Fig. 8.14a). The problem still has cylindrical symmetry (i.e., the fields are independent of ϕ) and the Ampère–Maxwell law still gives the same magnetic field intensity, Eq. (8.28). The boundary condition that H_ϕ be continuous at $r = c$ is satisfied. The magnetic flux density now depends on the radius and hence the flux through the $\phi = \phi_0$ surface is

$$\begin{aligned} \Phi &= \int_{z=0}^{D} \int_{r=a}^{c} \frac{\mu_1 I_0}{2\pi r}\, dr\, dz + \int_{z=0}^{D} \int_{r=c}^{b} \frac{\mu_2 I_0}{2\pi r}\, dr\, dz \\ &= \frac{D I_0}{2\pi} \left[\mu_1 \ln\left(\frac{c}{a}\right) + \mu_2 \ln\left(\frac{b}{c}\right)\right]. \end{aligned}$$

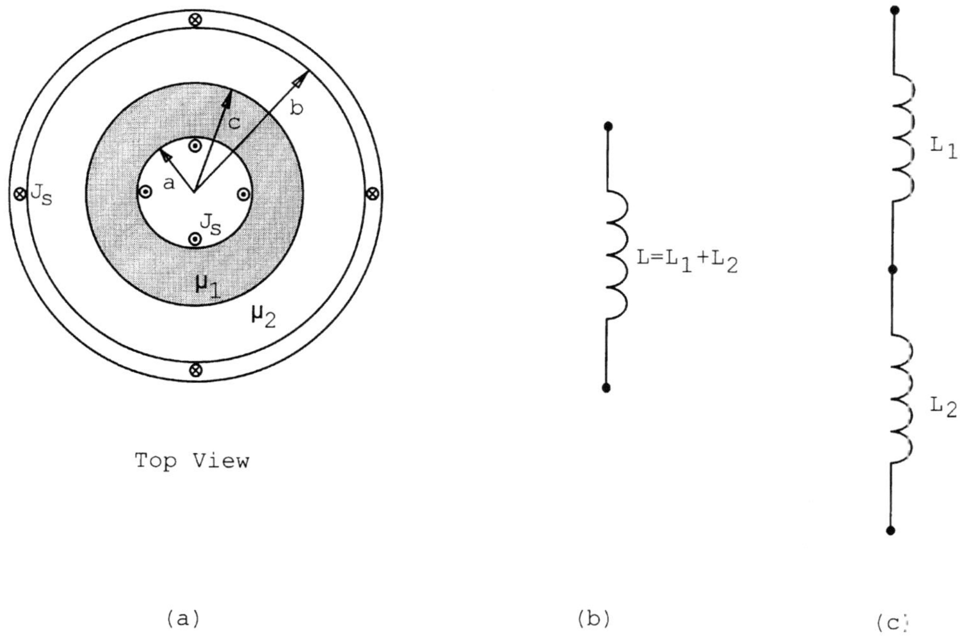

FIGURE 8.14 Coaxial cylinders filled with two materials with different magnetic permeabilities.

The inductance is now given by

$$L = \frac{\mu_1 D}{2\pi} \ln\left(\frac{c}{a}\right) + \frac{\mu_2 D}{2\pi} \ln\left(\frac{b}{c}\right)$$
$$= L_1 + L_2,$$

where

$$L_1 = \frac{\mu_1 D}{2\pi} \ln\left(\frac{c}{a}\right) \quad \text{and} \quad L_2 = \frac{\mu_2 D}{2\pi} \ln\left(\frac{b}{c}\right).$$

The current for each inductor is the same, I_0, and the circuit model for this inductor is either the single inductor L (Fig. 8.14b) or the two inductors L_1 and L_2 in series (Fig. 8.14c).

Another variation of the problem is to again assume two materials with different permeabilities, but this time let the sector $0 < \phi < \alpha$ have permeability μ_1 and the sector $\alpha < \phi < 2\pi$ have permeability μ_2 (Fig. 8.15a). In this case the flux density normal to the radial boundaries at $\phi = 0$ or 2π and at $\phi = \alpha$ must be continuous. With A an arbitrary constant, we can assume that $\vec{B} = \vec{a}_\phi A/r$ for all ϕ and it will certainly satisfy those boundary conditions. The magnetic flux through any radial plane $\phi = \phi_0$

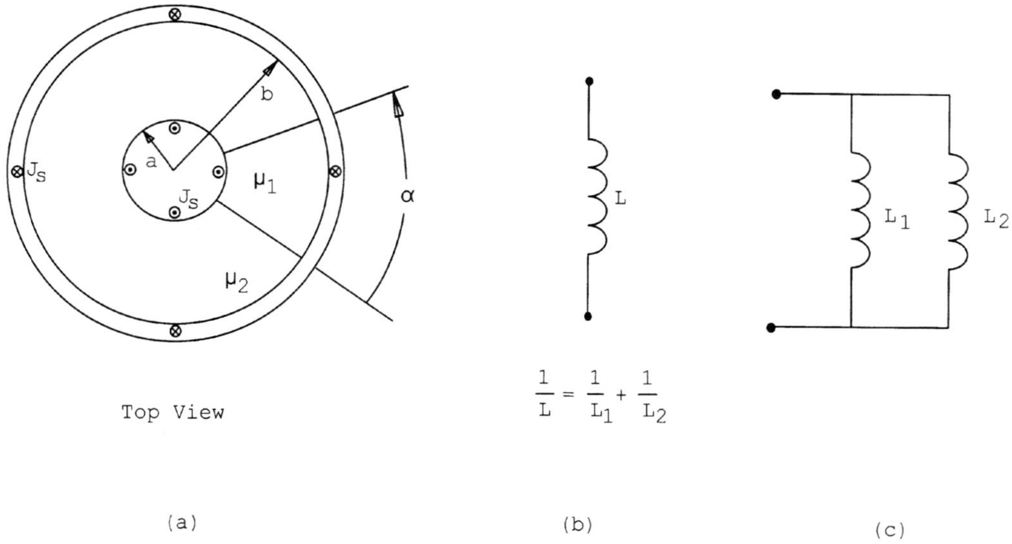

FIGURE 8.15 Inductors in parallel.

is therefore

$$\Phi = \int_{z=0}^{D} \int_{r=a}^{b} \frac{A}{r}\, dr\, dz = AD \ln\left(\frac{b}{a}\right).$$

The magnetic field intensity between the cylinders is

$$\vec{H} = \begin{cases} \vec{a}_\phi \dfrac{A}{\mu_1 r}, & 0 < \phi < \alpha, \\ \vec{a}_\phi \dfrac{A}{\mu_2 r}, & \alpha < \phi < 2\pi. \end{cases}$$

The current from the source is related to the constant A by integrating H_ϕ around a circle of radius r enclosing the inner cylinder:

$$I_0 = A\left(\frac{\alpha}{\mu_1} + \frac{2\pi - \alpha}{\mu_2}\right).$$

This is actually the sum of the currents flowing in each sector, $I_1 = A\alpha/\mu_1$ and $I_2 = A(2\pi - \alpha)/\mu_2$.

The reciprocal of the inductance is now

$$\frac{1}{L} = \frac{I_0}{\Phi} = \frac{\alpha}{\mu_1 D \ln(b/a)} + \frac{2\pi - \alpha}{\mu_2 D \ln(b/a)}$$
$$= \frac{1}{L_1} + \frac{1}{L_2},$$

where

$$L_1 = \frac{\Phi}{I_1} = \frac{\mu_1 D \ln(b/a)}{\alpha} \quad \text{and} \quad L_2 = \frac{\Phi}{I_2} = \frac{\mu_2 D \ln(b/a)}{2\pi - \alpha}.$$

The total current is the sum of the sector currents $I_0 = I_1 + I_2$ and the circuit model for this inductor is either the single inductor L (Fig. 8.15b) or the two inductors L_1 and L_2 in parallel (Fig. 8.15c).

We could also divide the region between the cylinders into two different permeability materials by an irregular boundary. In this case both the magnetic field intensity and the magnetic flux density must change from the field in the original inductor in order to make both $H_{\text{tangential}}$ and B_{normal} continuous along the irregular boundary. There is no longer a simple solution to this problem.

It is easy to generalize this example. Figure 8.16a shows the cross section of two conductors with a uniform permeability material, μ_1, and a current I_0. The magnetic flux through any surface between the conductors is Φ_0. The solid lines are both the magnetic field intensity and the magnetic flux density lines. The dashed lines, drawn everywhere normal to the solid lines, are the lines of constant magnetomotive force (Section 3.9). If the permeability of the material within an entire shell of flux lines surrounding one conductor is changed to a value μ_2 (Fig. 8.16b) and the current is unchanged, then the original $\vec{H}$ will satisfy the condition that $H_{\text{tangential}}$ must be continuous at the boundary of the shell and will integrate along a line surrounding the enclosed conductor to I_0. This inductor can be modeled as two inductors in series. On the other hand, if the permeability of the material within an entire tube of mmf lines is

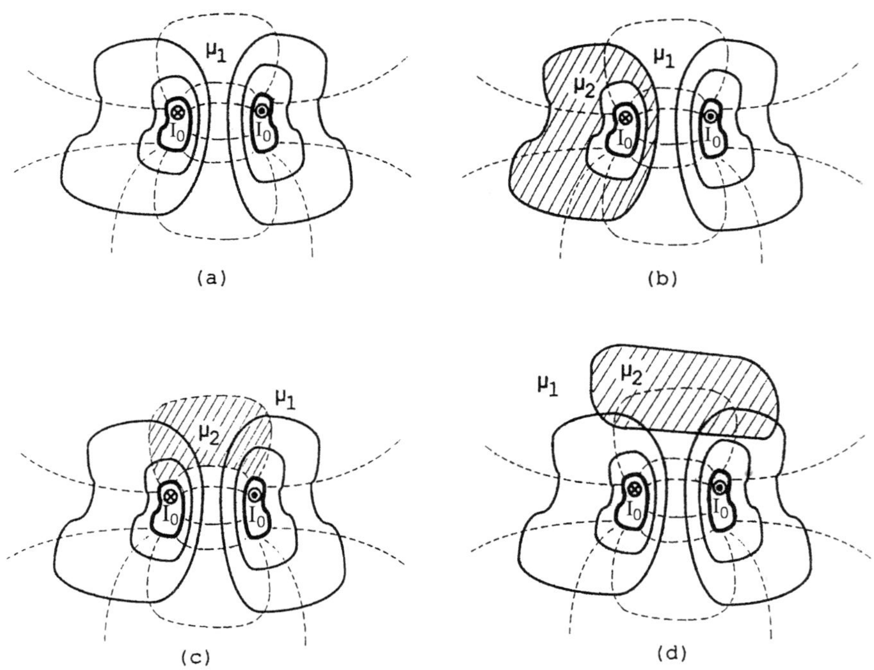

FIGURE 8.16 Generalization of inductor connections.

changed to μ_2 (Fig. 8.16c) and the magnetic flux is unchanged, then the original $\vec{B}$ will satisfy the condition that B_{normal} must be continuous at the tube boundaries and will integrate through any surface between the conductors to Φ_0. This inductor can be modeled as two inductors in parallel. If a change in permeability is made that is neither a complete shell nor a complete tube (Fig. 8.16d), then both $\vec{H}$ and $\vec{B}$ must change in form and the new fields are not simply related to the original fields.

8.5.3 A Current-Carrying Wire: Internal Inductance

Let us assume that a solid circular cylindrical wire of radius a coaxial with the z axis carries a current I_0, which is parallel to the axis of the wire and uniformly distributed over the cross section of the wire. The current density in the wire is then $J_z = I_0/\pi a^2$. The magnetic field intensity interior to the wire ($r < a$) is found by integrating around a circle of radius r.

$$2\pi r H_\phi = \frac{I_0}{\pi a^2}(\pi r^2)$$

$$H_\phi = \frac{I_0 r}{2\pi a^2}, \quad r < a.$$

The magnetic energy stored within the wire per unit length of wire is

$$\begin{aligned} W_m &= \frac{1}{2}\int_{r=0}^{a}\int_{\phi=0}^{2\pi} \mu_c \left(\frac{I_0 r}{2\pi a^2}\right)^2 r\, d\phi\, dr \\ &= \frac{1}{2}\left(\frac{\mu_c}{8\pi}\right) I_0^2. \end{aligned}$$

where μ_c is the permeability of the wire. Equating this to $L_{l,\text{int}}\ I_0^2/2$, where $L_{l,\text{int}}$ is the internal inductance per unit length, we find

$$L_{l,\text{int}} = \frac{\mu_0}{8\pi}. \tag{8.30}$$

This inductance is called internal inductance because the region in which the magnetic energy is stored coincides with the region in which the current is flowing.

Example 8.4 Find the total inductance per unit length of a coaxial wire consisting of a solid inner conductor of radius a and a coaxial cylindrical wire of inner radius b and outer radius c (Fig. 8.17). The wires have permeability μ_c and the region between the wires has permeability μ.

Solution The inductance between the cylinders (for a length D) is given by Eq. (8.29). The inductance per unit length is found by dividing by D. We have just found the internal inductance per unit length of the inner wire, Eq. (8.30). In a similar manner we can show that the internal inductance per unit length of the outer

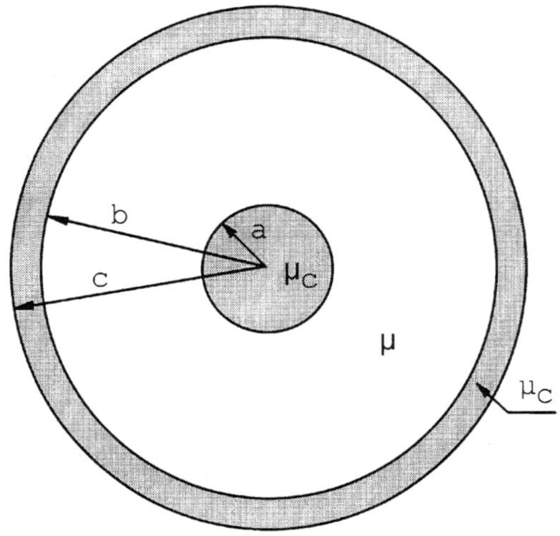

FIGURE 8.17 Coaxial wire.

cylindrical shell is

$$L_{l,\mathrm{int}} = \frac{\mu_c}{8\pi}F, \quad F = \frac{(c^4 - b^4) - 4c^2(c^2 - b^2) + 4c^4 \ln(c/b)}{(c^2 - b^2)^2}. \tag{8.31}$$

The total energy stored in the magnetic field is the sum of the energy stored in the solid wire, the region between the wires, and the outer cylinder. The total inductance per unit length is therefore the sum of the three inductances:

$$L_l = \frac{\mu_c}{8\pi} + \frac{\mu}{2\pi}\ln\left(\frac{b}{a}\right) + \frac{\mu_c}{8\pi}F.$$

This is actually the inductance for steady currents. At high frequency the internal inductance pushes the current toward the surface of the wires (the outer surface, $r = a$, of the inner wire and the inner surface, $r = b$, of the outer wire). This is the skin effect, which was discussed in Section 7.5. The inductance per unit length is then just the inductance of the region between the wires, $L_l = (\mu/2\pi)\ln(b/a)$.

Example 8.5 Find the inductance per unit length of a two-wire line. The radii of the wires is a and the separation between the wire's center lines is $d \gg a$ (Fig. 8.18).

Solution We can write the current and field of each wire with respect to a z axis drawn along the axis of the wire in the direction of the current. The wires are far apart compared to the radius and the field due to the current in one wire has negligible effect on the current distribution in the other wire. The current distribution in each wire is therefore uniform, $J_z = I/\pi a^2$. The field outside either wire due to its own current is $H_\phi = I/2\pi r$.

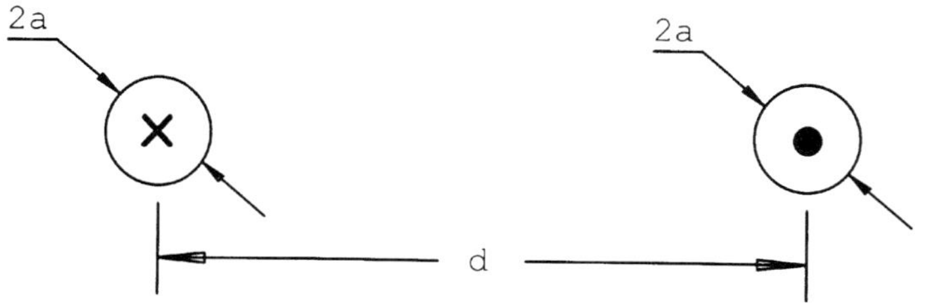

FIGURE 8.18 A two-wire line.

The flux per unit length through the surface from the edge of one wire to the axis of the other due to one of the currents is then

$$\Phi' = \int_{r=a}^{d} \frac{\mu_0 I}{2\pi r}\, dr = \frac{\mu_0 I}{2\pi} \ln\left(\frac{d}{a}\right).$$

There is an equal flux added due to the current in the other wire. The total flux between the wires is therefore $\Phi = 2\Phi'$. The external inductance per unit length is

$$L_{l,\text{ext}} = \frac{\Phi}{I} = \frac{\mu_0}{\pi} \ln\left(\frac{d}{a}\right).$$

For low frequency we could add twice the internal inductance per unit length of a single wire, so that the total inductance per unit length is

$$L_l = \frac{\mu_c}{4\pi} + \frac{\mu_0}{\pi} \ln\left(\frac{d}{a}\right).$$

In the usual case $\mu_c = \mu_0$ and the external inductance is greater than the internal inductance by a factor of at least 10 so long as $d/a > 12$.

The exact value for the external inductance per unit length, for any $d > 2a$, will be shown in Section 9.5 to be $L_{l,\text{ext}} = \mu\varepsilon/C_l$ and hence, from Eq. (6.77),

$$L_{l,\text{ext}} = \frac{\mu_0}{\pi} \cosh^{-1}\left(\frac{d}{2a}\right) = \frac{\mu_0}{\pi} \ln\left[\frac{d}{2a} + \sqrt{\left(\frac{d}{2a}\right)^2 - 1}\right].$$

where $d \gg a$ this becomes

$$L_{l,\text{ext}} \approx \frac{\mu_0}{\pi} \ln\left(\frac{d}{a}\right),$$

as we have found.

8.6 MUTUAL INDUCTANCE

Self-inductance results from magnetic flux linked with the same current that is the source of the flux. Mutual inductance results from the magnetic flux caused by a current in one circuit being linked with the current in another circuit. Consider the two current loops shown in Fig. 8.19. The total flux Φ_1 linked with loop 1 will be linearly related to the current in loop 1, I_1, and to the current in loop 2, I_2. Similarly, the total flux Φ_2 linked with loop 2 will be linearly related to the current flowing in loop 1, I_1, and to the current in loop 2, I_2. In equation form,

$$\Phi_1 = L_{11}I_1 + L_{12}I_2,$$

$$\Phi_2 = L_{21}I_1 + L_{22}I_2,$$

where L_{11} and L_{22} are the self-inductances of the loops and L_{12} and L_{21} are the mutual inductances.

We can again find the relations between the currents, the inductances, and the magnetic energy by proceeding as we did for self-inductance. Figure 8.20 shows the loops redrawn with current sources. As the currents increase slowly from zero at

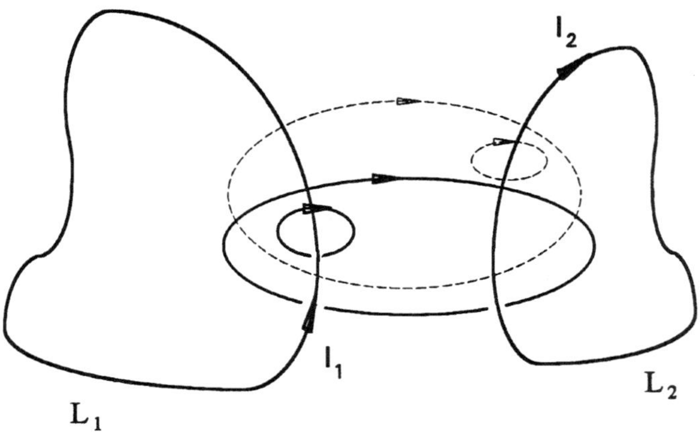

FIGURE 8.19 Two current loops linked by magnetic flux.

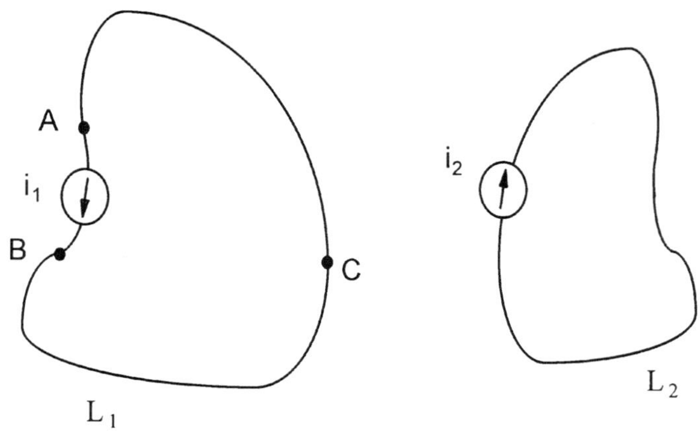

FIGURE 8.20 Two loops with current sources.

time $t=0$ to I_1 and I_2 at time t, the Faraday–Maxwell law gives

$$\oint_{L_1} \vec{E}_1 \cdot dl_1 = -\frac{d\Phi_1}{dt}$$

$$V_{\mathrm{AB}} + V_{\mathrm{BCA}} = -\left(L_{11}\frac{di_1}{dt} + L_{12}\frac{di_2}{dt}\right).$$

Multiplying by i_1 gives a term $V_{\mathrm{AB}} i_1$, which is the power input to current source 1, and a term $V_{\mathrm{BCA}} i_1 = R_1 i_1^2$, which is the power dissipated in ohmic loss in loop 1. $V_{\mathrm{BA}} i_1 = -V_{\mathrm{AB}} i_1$ is therefore the power supplied by current source 1. Rearranging the terms we have

$$V_{\mathrm{BA}} i_1 = L_{11} i_1 \frac{di_1}{dt} + L_{12} i_1 \frac{di_2}{dt} + R_1 i_1^2.$$

The left side of the equation is the power supplied by source 1, $P_{S1} = dW_{S1}/dt$. The rightmost term is the power loss in circuit 1, $P_{L1} = dW_{L1}/dt$. The remaining terms on the right are the power supplied to the magnetic field as the current increases. Integrating with time gives

$$\int_{t'=0}^{t} \frac{dW_{S1}}{dt'}\,dt' = \int_{t'=0}^{t} \left[L_{11} i_1(t')\frac{di_1(t')}{dt'} + L_{12} i_1(t')\frac{di_2(t')}{dt'}\right] dt' + \int_{t'=0}^{t} \frac{dW_{L1}}{dt'}\,dt'$$

$$W_{S1} = W_{m1} + W_{L1}, \quad W_{m1} = \int_{t'=0}^{t} \left[L_{11} i_1(t')\frac{di_1(t')}{dt'} + L_{12} i_1(t')\frac{di_2(t')}{dt'}\right] dt'.$$

Similarly, for loop 2,

$$W_{S2} = W_{m2} + W_{L2}, \quad W_{m2} = \int_{t'=0}^{t} \left[L_{21} i_2(t')\frac{di_1(t')}{dt'} + L_{22} i_2(t')\frac{di_2(t')}{dt'}\right] dt'.$$

We can increase the currents to their final values I_1 and I_2 in the following way. First, keep $i_2 = 0$ while i_1 is increased to I_1. Second, keep $i_1 = I_1$ while i_2 is increased to I_2. With this sequence

$$W_{m1} = \tfrac{1}{2} L_{11} I_1^2 + L_{12} I_1 I_2, \quad W_{m2} = \tfrac{1}{2} L_{22} I_2^2$$

$$W_m = W_{m1} + W_{m2} = \tfrac{1}{2} L_{11} I_1^2 + L_{12} I_1 I_2 + \tfrac{1}{2} L_{22} I_2^2. \tag{8.32}$$

We can also increase the currents to their final values in the opposite order. First, keep $i_1 = 0$ while i_2 is increased to I_2. Second keep $i_2 = I_2$ while i_1 is increased to I_1. With this sequence

$$W_{m1} = \tfrac{1}{2} L_{11} I_1^2, \quad W_{m2} = \tfrac{1}{2} L_{22} I_2^2 + L_{21} I_1 I_2$$

$$W_m = W_{m1} + W_{m2} = \tfrac{1}{2} L_{11} I_1^2 + L_{21} I_1 I_2 + \tfrac{1}{2} L_{22} I_2^2. \tag{8.33}$$

These two sequences must result in the same energy stored in the magnetic field. The two W_m's we have obtained must be the same and therefore

$$L_{12} = L_{21}. \tag{8.34}$$

Thus the mutual inductance from loop 1 to loop 2 equals the mutual inductance from loop 2 to loop 1.

8.6.1 Mutual Inductance of a Solenoid with Two Windings

Figure 8.21 shows two coaxial windings wrapped around a circular core of permeability μ and radius a. The two windings have n_1 and n_2 turns per unit length, respectively. The length of both coils is D.

Let a current I_1 flow in the first coil. The flux through the core will be $\Phi_1 = \mu\pi a^2 n_1 I_1$. This is linked with the second winding $n_2 D$ times so that the total flux linkage with the second coil is $\mu\pi a^2 n_1 n_2 D I_1$. Dividing by I_1 gives the mutual inductance:

$$L_{21} = \mu\pi a^2 n_1 n_2 D. \tag{8.35}$$

Interchanging the subscripts gives the same right side, $L_{21}=L_{12}$, as was expected.

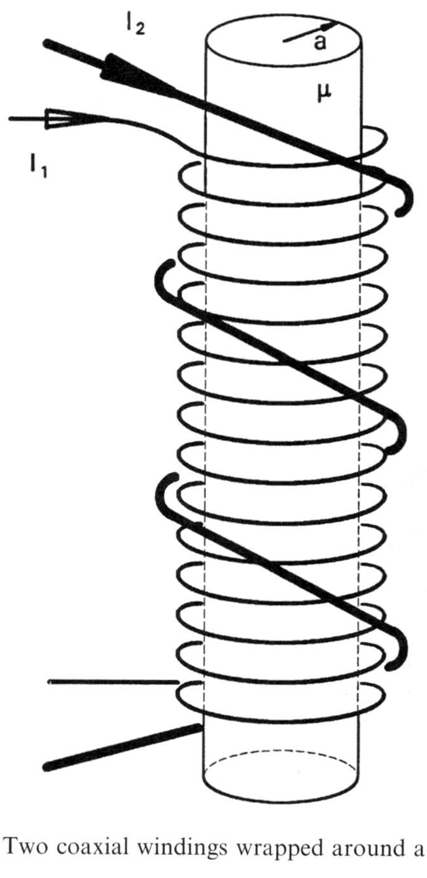

FIGURE 8.21 Two coaxial windings wrapped around a circular core.

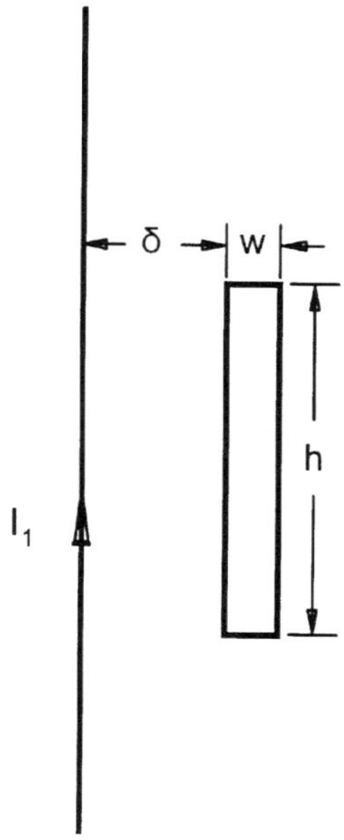

FIGURE 8.22 Infinitely long wire with a current I_1 near a rectangular loop.

8.6.2 Mutual Inductance of a Wire and a Rectangular Loop

Figure 8.22 shows an infinitely long wire extending along the z axis and a rectangular loop of height h and width w a distance d from the wire. The wire and the loop are coplanar.

With a current I_1 in the wire the magnetic field intensity is $H_\phi = I_1/2\pi r$. The flux linked with the loop is

$$\Phi_{21} = \int_{z=0}^{h} \int_{r=d}^{d+w} \frac{\mu_0 I_1}{2\pi r}\, dr\, dz = \frac{\mu_0 h I_1}{2\pi} \ln\left(\frac{d+w}{d}\right).$$

Dividing by the current we get

$$L_{21} = \frac{\mu_0 h}{2\pi} \ln\left(1 + \frac{w}{d}\right). \tag{8.36}$$

This is also the mutual inductance L_{12}.

8.7 MAGNETIC CIRCUITS

A magnetic circuit occurs when the magnetic fields are entirely or almost entirely found inside a material, which has a high permeability in comparison with the permeability of the surrounding medium. Practical examples are found in electric transformers, motors and generators, and electromagnets.

Figure 8.23 shows a high-permeability iron toroid in which a short air gap has been cut. There is a coil with N turns around the toroidal core. This is a simple example of a magnetic circuit. Let us now apply Eq. (8.1a) to the loop ABCDEA, taking the capping surface to be the area inside the dashed line in the plane of the page. Letting the subscript I refer to iron and the subscript a refer to air, we have

$$\int_S \vec{J}\cdot d\vec{S} = \oint_L \vec{H}\cdot d\vec{l}$$

$$NI = H_I l_I + H_a l_a \tag{8.37}$$

$$= \frac{B_I l_I}{\mu_I} + \frac{B_a l_a}{\mu_a}. \tag{8.38}$$

In obtaining Eq. (8.37) we have assumed that the magnetic field intensity in the iron, H_I, and in the air gap, H_a, is constant along the lines ABCDE and EA, respectively. The normal component of magnetic flux density is continuous across a boundary between two media and, therefore, $B_I = B_a = \Phi/A$, where A is the cross-sectional area of the iron core and of the air gap. Equation (8.38) becomes

$$NI = \left(\frac{l_I}{\mu_I A} + \frac{l_a}{\mu_a A}\right)\Phi$$

$$V_m = (\mathscr{R}_I + \mathscr{R}_a)\Phi, \tag{8.39}$$

where

$$V_m = NI, \quad \mathscr{R}_I = \frac{l_I}{\mu_I A}, \quad \mathscr{R}_a = \frac{l_a}{\mu_a A}. \tag{8.40}$$

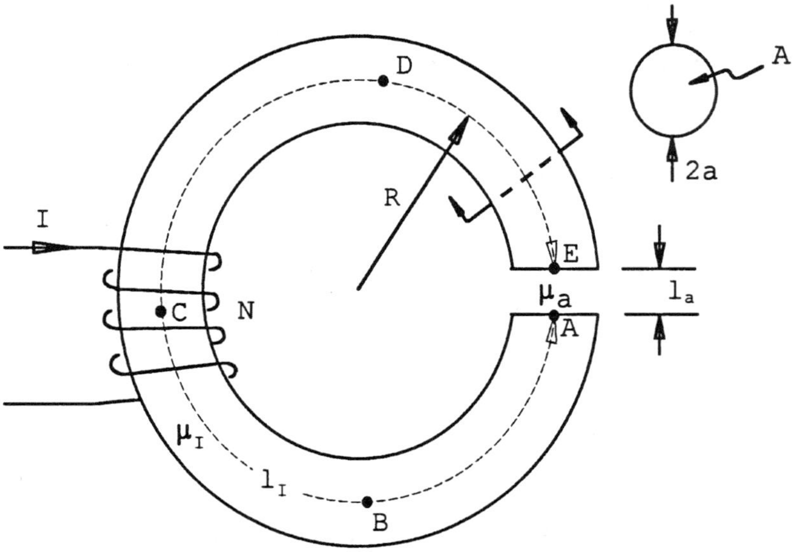

FIGURE 8.23 Toroidal magnetic circuit with air gap.

V_m is the magnetomotive force (mmf) around the loop. $\mathscr{R}_I$ and $\mathscr{R}_a$ are the reluctance, in H^{-1}, of the iron core and the air gap, respectively. Equation (8.39) shows that the mmf equals the sum of the reluctances multiplied by the magnetic flux.

There are several approximations hidden in the above derivation. We have assumed that H_I and H_a are independent of position within the iron and air gap, respectively. This is a good assumption if the mean radius of the toroid, R, is much greater than the radius of the cross-sectional area, A.

More important, we have assumed that all of the magnetic flux flows around the path ABCDEA, while, in fact, some leakage flux must be present at all points in the air surrounding the core. Integrating around the path ABCDEA shows that the mmf from D to B equals the sum of the mmf's from D to E, from E to A, and from A to B. Since the permeability of the iron is much greater than the permeability of the air, we can neglect the mmf from D to E and from A to B. This leaves $H_{\text{leakage}}\, l_{\text{DB}} \approx H_a l_a$. But $l_{\text{DB}} \gg l_a$ and therefore the leakage magnetic field intensity and magnetic flux density are much less than the magnetic field intensity and magnetic flux density in the air gap, respectively.

Example 8.6 Consider the magnetic circuit, shown in Fig. 8.24, which is part of an electromagnetic relay. The coil, which has 138 turns and a resistance of 2.6 Ω, is excited by a 24 volt dc battery. The reluctances of the two cast steel sections, with a cross-sectional area of 0.1 m^2, are $\mathscr{R}_1 = 324\ H^{-1}$ and $\mathscr{R}_2 = 122\ H^{-1}$. (a) If the two core pieces are joined together (the relay is energized), determine the flux established in the core. (b) If the two core pieces are separated by an air gap of 1 mm, as shown in Fig. 8.25, determine the new core flux, neglecting the fringing effect at the air gap, and flux leakage.

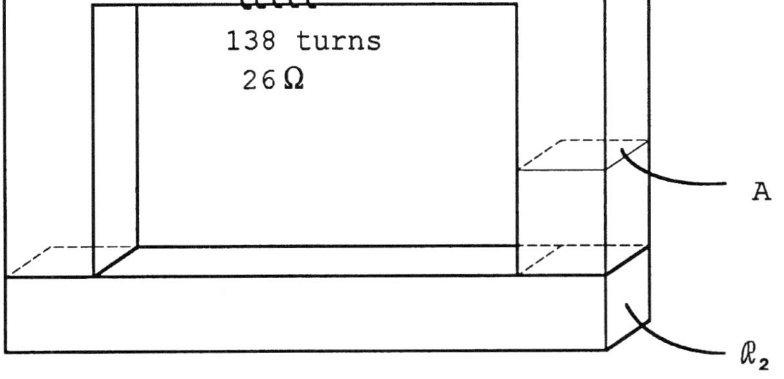

FIGURE 8.24 Magnetic circuit of an electromagnetic relay.

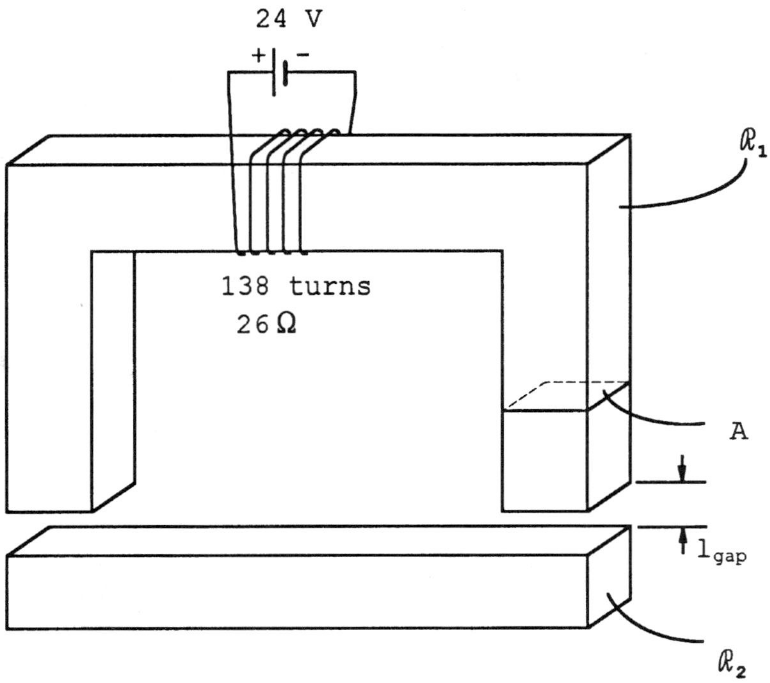

FIGURE 8.25 Magnetic circuit of an electromagnetic relay with air gap.

Solution (a) The magnetomotive force, V_m, that drives the flux through the magnetic circuit is calculated as $V_m = N \cdot I$, where the current I is the dc current obtained by applying the 24 volt battery across the coil with the 2.6 Ω resistance.

Therefore

$$I = \frac{E_{\text{bat}}}{R} = \frac{24\ \text{V}}{2.6\ \Omega} = 9.231\ \text{A}.$$

The mmf is then calculated as

$$V_m = N \cdot I = 138 \cdot 9.231 = 1273.88\ \text{A}.$$

The flux through the core is then obtained as

$$\phi = \frac{V_m}{\mathcal{R}_{\text{total}}} = \frac{V_m}{\mathcal{R}_1 + \mathcal{R}_2} = \frac{1273.88}{(324 + 122)} = 2.86\ \text{Wb}.$$

(b) The introduction of the two air gaps changes the reluctance of the magnetic circuit. The reluctance of each air gap is calculated as follows:

$$\mathcal{R}_{\text{gap}} = \frac{l_{\text{gap}}}{\mu_0 \cdot A_{\text{gap}}} = \frac{0.001}{(4\pi \times 10^{-7}) \cdot 0.1} = 7957.75\ \text{H}^{-1}.$$

The magnetomotive force (mmf) remains the same, and the new flux through the magnetic circuit is calculated as follows:

$$\phi = \frac{V_m}{\mathscr{R}_{\text{total}}} = \frac{V_m}{\mathscr{R}_1 + \mathscr{R}_2 + 2 \cdot \mathscr{R}_{\text{gap}}} = \frac{1273.88}{324 + 122 + 2 \cdot 7957.75} = \frac{1273.88}{16361.5} = 0.078 \text{ Wb}.$$

Note: Comparing the two parts of the problem shows us that the introduction of the air gaps in the magnetic circuit results in a significant reduction of the magnetic flux. In order to obtain the same flux as in part (a) through the magnetic circuit with the air gap in part (b) would require an increase of 36.7 times in the magnetomotive force, and hence the required electric current. In motors and generators, the developed torque and the generated voltage, respectively, are directly proportional to the magnetic flux of the field windings. The result obtained in the above example explains the reason why a small air gap is used in motors and generators.

There is a direct analogy between an electric and a magnetic circuit. Figure 8.26 shows an electric circuit analogy to the magnetic circuit of Fig. 8.24. Let us assume the toroidal shaped resistor (in two parts) is made of copper ($\sigma_{\text{Cu}} = 5.8 \times 10^7$) and the small gap on the right is filled with carbon ($\sigma_{\text{gap}} = 2.5 \times 10^4$). Then, for a source voltage V, there will be a current I through the copper and carbon conductors, where

$$V = (R_{\text{Cu}} + R_{\text{gap}})I, \tag{8.41}$$

where

$$R_{\text{Cu}} = \frac{l_{\text{Cu}}}{\sigma_{\text{Cu}} A}, \quad R_{\text{gap}} = \frac{l_{\text{gap}}}{\sigma_{\text{gap}} A}. \tag{8.42}$$

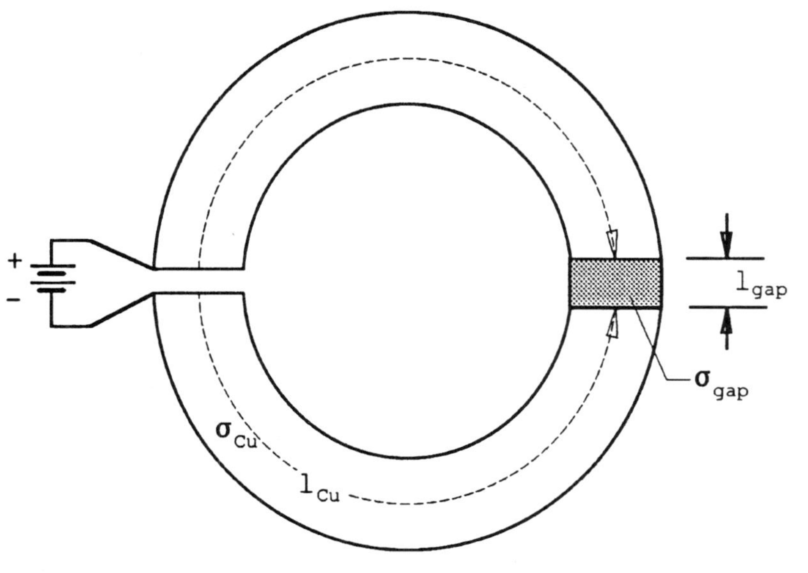

FIGURE 8.26 The equivalent electric circuit to the magnetic circuit in Fig. 8.23.

These equations are identical in form to Eqs. (8.39) and (8.40). In the case of the electric circuit the surrounding medium is air, which has zero conductivity. As a result, there is no leakage current; all of the current is confined to the toroidal conducting path. The analogy can be made complete by making the surrounding medium of the same carbon material as the gap. Then the electric circuit will also have leakage currents.

For the electric circuit we draw the circuit diagram of Fig. 8.27a. By analogy, we can draw a similar diagram for the magnetic circuit, Fig. 8.27b.

All of the techniques you have learned to solve multiple element resistive circuit problems can be used to solve the analogous magnetic circuit problems.

Example 8.7 A high-permeability iron core with two coils and an air gap in the center leg is shown in Fig. 8.28. The cross-sectional area of all branches except the center leg is A and for the center leg (and the air gap) it is A_g. Draw the equivalent circuit, give formulas for the elements, and solve for Φ_1, Φ_2, and Φ_g.

Solution The equivalent circuit is shown in Fig. 8.29. The circuit elements are

$$V_{m1} = N_1 I_1, \quad V_{m2} = N_2 I_2,$$

$$\mathscr{R}_1 = \frac{l_1}{\mu_I A}, \quad \mathscr{R}_2 = \frac{l_2}{\mu_I A}, \quad \mathscr{R}_3 = \frac{l_3}{\mu_I A_g}, \quad \mathscr{R}_a = \frac{l_a}{\mu_0 A_g}.$$

Φ_1 and Φ_2 are the mesh fluxes. The mesh equations are

$$V_{m1} = (\mathscr{R}_1 + 2\mathscr{R}_3 + \mathscr{R}_g)\Phi_1 - (2\mathscr{R}_3 + \mathscr{R}_g)\Phi_2,$$

$$-V_{m2} = -(2\mathscr{R}_3 + \mathscr{R}_g)\Phi_1 + (\mathscr{R}_2 + 2\mathscr{R}_3 + \mathscr{R}_g)\Phi_2.$$

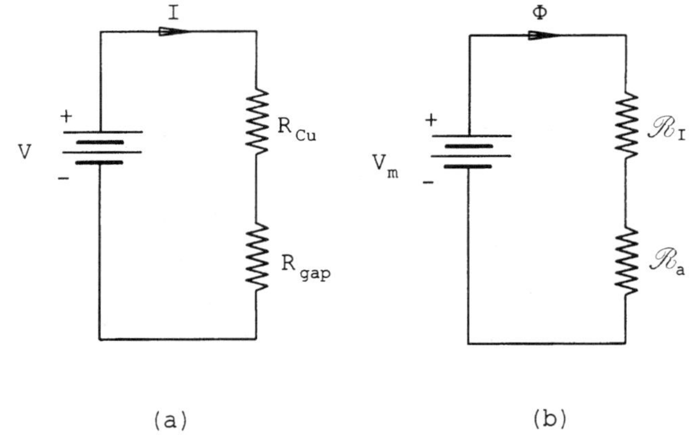

FIGURE 8.27 Circuit diagrams for both the (a) electric and (b) magnetic circuits in Fig. 8.26 and Fig. 8.23, respectively.

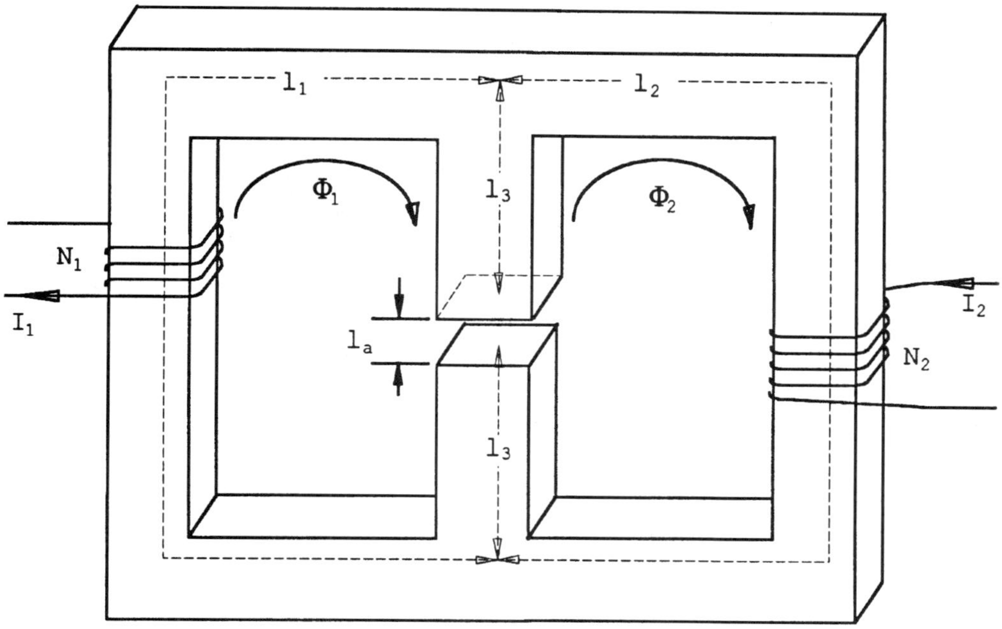

FIGURE 8.28 Magnetic circuit in Example 8.7.

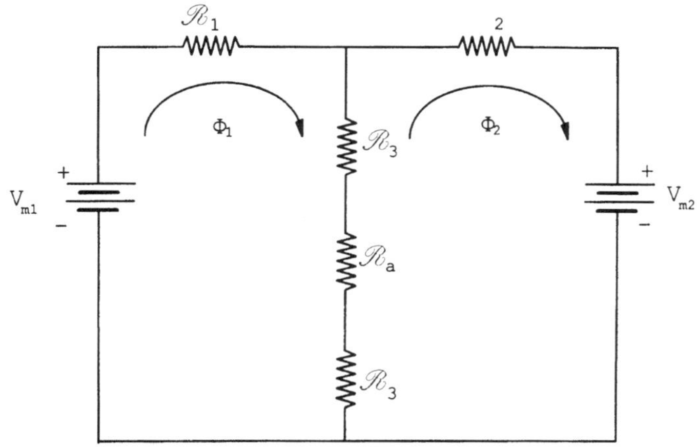

FIGURE 8.29 Equivalent electric circuit to the magnetic circuit in Fig. 8.28.

The solutions are

$$\Phi_1 = \frac{(\mathscr{R}_2 + 2\mathscr{R}_3 + \mathscr{R}_g)\,V_{m1} - (2\mathscr{R}_3 + \mathscr{R}_g)\,V_{m2}}{\mathscr{R}_1\mathscr{R}_2 + (2\mathscr{R}_3 + \mathscr{R}_g)(\mathscr{R}_1 + \mathscr{R}_2)},$$

$$\Phi_2 = \frac{(2\mathscr{R}_3 + \mathscr{R}_g)\,V_{m1} - (\mathscr{R}_1 + 2\mathscr{R}_3 + \mathscr{R}_g)\,V_{m2}}{\mathscr{R}_1\mathscr{R}_2 + (2\mathscr{R}_3 + \mathscr{R}_g)(\mathscr{R}_1 + \mathscr{R}_2)},$$

$$\Phi_g = \Phi_1 - \Phi_2 = \frac{\mathscr{R}_2 V_{m1} + \mathscr{R}_1 V_{m2}}{\mathscr{R}_1\mathscr{R}_2 + (2\mathscr{R}_3 + \mathscr{R}_g)(\mathscr{R}_1 + \mathscr{R}_2)}.$$

PROBLEMS

8.1 A rectangular loop of thin wire carries current I_0. The coordinate system is fixed with the origin in the center of the rectangle, the sides of length $2a$ parallel to the x axis, and the sides of length $2b$ parallel to the y axis. The current circulates in the right-hand sense about the z axis. See Fig. 8.30.

a. Find the magnetic field intensity at any point along the z axis.

b. Show that for $z \gg a$ and $z \gg b$ the result is $H_z \to I_0 S/2\pi z^3$, where $S = 4ab$. Show that this is the field along the axis of a magnetic dipole of moment $I_0 S$.

Ans. $H_z = \dfrac{ab}{\pi\sqrt{a^2+b^2+z^2}}\left(\dfrac{1}{b^2+z^2}+\dfrac{1}{a^2+z^2}\right)I_0.$

8.2 A surface current $\vec{a}_y J_{s0}$ lies in the $z = 0$ plane, extending from $x = x_1$ to $x = x_2$ and for all y. See Fig. 8.31.

a. Find the magnetic field intensity along the z axis. (Note that this is the equivalent to finding the field at any point in space.)

b. Specialize the result obtained to the case $x_1 = -a, x_2 = a$.

c. Show that as $a \to \infty$, $\vec{H} \to \vec{a}_x J_{s0}/2$, $z > 0$ and $\vec{H} \to -\vec{a}_x J_{s0}/2$, $z < 0$. Show that this matches the boundary condition, Eq. (8.3).

d. Returning to the answer for part (b), show that as $a \to 0$ and $2aJ_{s0} = I_0$ it matches with Eq. (4.50).

Ans. a. $\vec{H} = \dfrac{J_{s0}}{2\pi}\left\{\vec{a}_x\left[\tan^{-1}\left(\dfrac{x_2}{z}\right) - \tan^{-1}\left(\dfrac{x_1}{z}\right)\right] + \vec{a}_z\,\dfrac{1}{2}\ln\left(\dfrac{x_2^2+z^2}{x_1^2+z^2}\right)\right\};$

b. $\vec{H} = \vec{a}_x\,\dfrac{J_{s0}}{\pi}\tan^{-1}\left(\dfrac{a}{z}\right).$

8.3 In Eq. (8.13) let $D = 10a$ and plot H_z versus z/a for $0 \le z/a \le 15$.

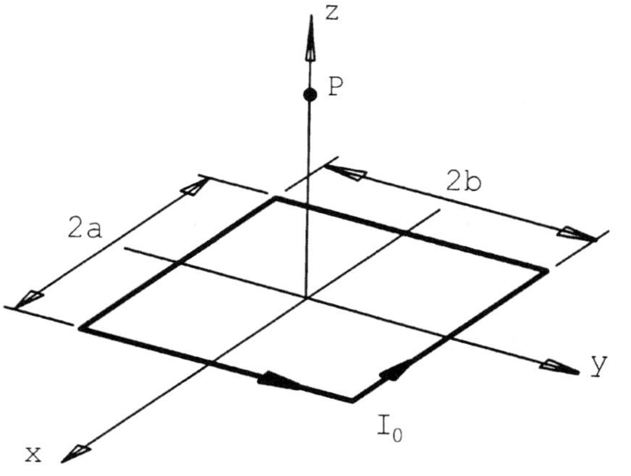

FIGURE 8.30 Rectangular loop carrying a current I_0.

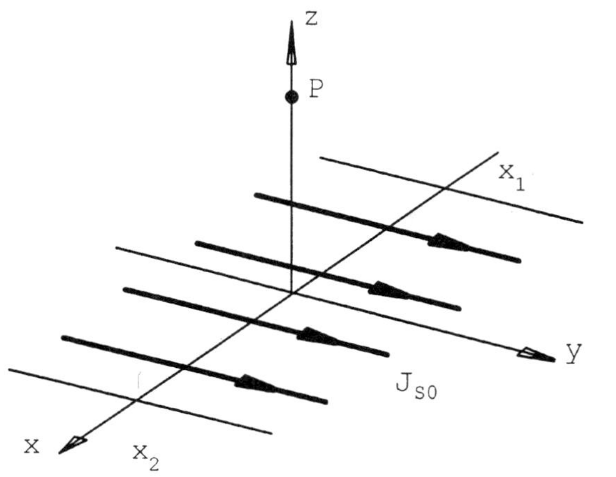

FIGURE 8.31 Surface current density in the xy plane in the y direction.

8.4 A sphere of radius a and permeability μ is placed at the origin in an impressed field:

$$\vec{H}^i = \vec{a}_z H_0 = H_0(\vec{a}_R \cos\theta - \vec{a}_\theta \sin\theta).$$

Due to the presence of the sphere there is an additional "reflected" field,

$$\vec{H}^r = \frac{A}{R^3}(\vec{a}_R 2\cos\theta + \vec{a}_\theta \sin\theta), \quad R > a,$$

outside the sphere and an additional "transmitted" field,

$$\vec{H}^t = \vec{a}_z B = B(\vec{a}_R \cos\theta - \vec{a}_\theta \sin\theta), \quad R < a,$$

inside the sphere. Find A and B. Compare the results with those of Problem 6.35.

Ans. $A = \left(\dfrac{\mu - \mu_0}{\mu + 2\mu_0}\right) a^3 H_0, \quad B = -\left(\dfrac{\mu - \mu_0}{\mu + 2\mu_0}\right) H_0.$

8.5 A spherical shell of inner radius a, outer radius b, and permeability μ is located concentric with the origin in an impressed field $\vec{H}^i = \vec{a}_z H_0$.

a. Find the magnetic field intensity inside the shell, $R < a$.
Hint: Outside the sphere there is an added dipole field, in the shell there is both a dipole and a uniform field, and in the interior there is a uniform field.

b. Show that for $\mu \gg \mu_0$, $k \approx 9(\mu_0/\mu)/2[1 - (a/b)^3]$.

Ans. $\vec{H}^t = \vec{a}_z k H_0, \quad k = \dfrac{9(\mu_0/\mu)}{(2 + \mu_0/\mu)(1 + 2\mu_0/\mu) - 2(a/b)^3(1 - \mu_0/\mu)^2}.$

The proportionality constant k is a magnetic shielding factor.

8.6 Obtain Eqs. (8.26) and (8.27) by finding the magnetic energy stored in the solenoid.

8.7 Find the mutual inductance between the straight wire and the coplanar circular wire shown in Fig. 8.32. *Hint*: Put the origin of the coordinate system at the center of the circle. Using Cartesian coordinates you will need

$$\int_{-a}^{+a} \frac{\sqrt{a^2 - x^2}}{1 + (x/d)}\, dx = \pi d^2 \left[1 - \sqrt{1 - \left(\frac{a}{d}\right)^2}\right], \quad a < d,$$

or using cylindrical coordinates you will need

$$\int_0^{\pi} \frac{d\phi}{d + r\cos\phi} = \frac{\pi}{\sqrt{d^2 - r^2}}, \quad d > r > 0.$$

Ans. $L_{21} = \mu_0 [d - \sqrt{d^2 - a^2}]$.

8.8 Find the approximate mutual inductance between the two coplanar loops shown in Fig. 8.33. Assume that h_1 is much greater than any other dimension and that the smaller loop is approximately vertically centered with respect to the larger loop. Show that as $w_1 \to \infty$ (and hence $h_1 \to \infty$ also) the mutual inductance is given by Eq. (8.36).

Ans. $L_{12} = L_{21} = \frac{\mu_0 h_2}{2\pi} \ln\left[\left(1 + \frac{w_2}{d}\right)\left(1 + \frac{w_2}{w_1 + d}\right)^{-1}\right]$.

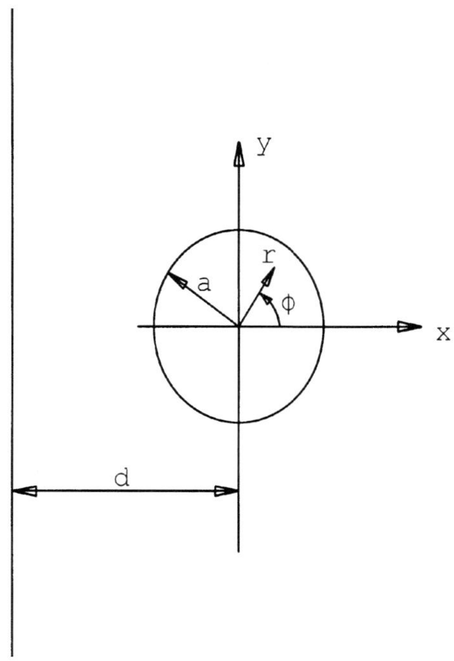

FIGURE 8.32 Straight and coplanar circular wires in Problem 8.7.

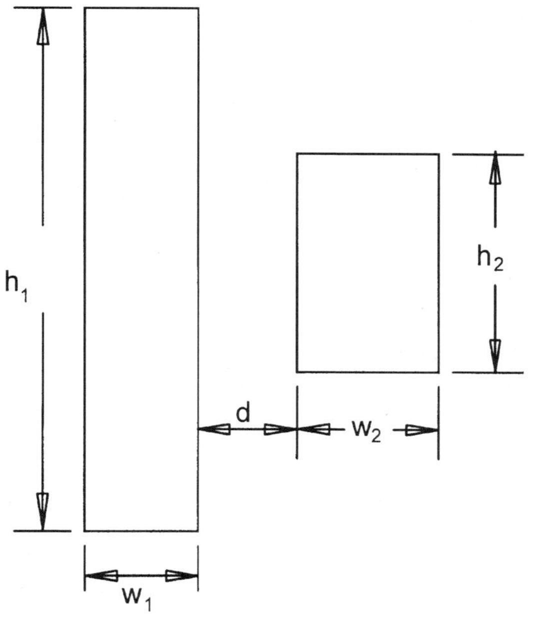

FIGURE 8.33 The two coplanar rectangular loops from Problem 8.8.

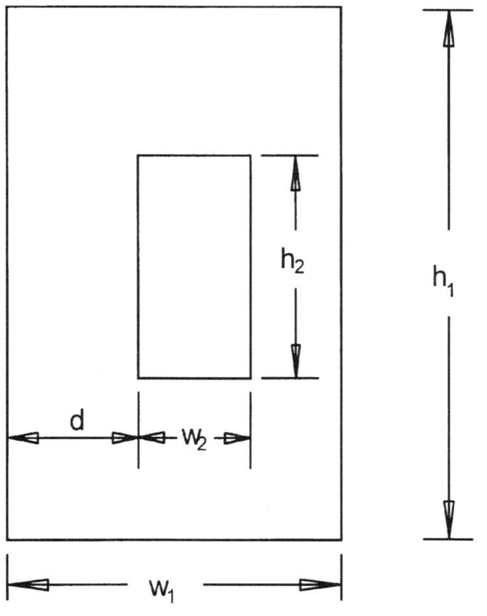

FIGURE 8.34 The two rectangular loops in Problem 8.9.

8.9 Repeat Problem 8.8 for the two loops shown in Fig. 8.34. Show that as $w_1 \to \infty$ (and hence $h_1 \to \infty$ also) the mutual inductance is given by Eq. (8.36).

Ans. $L_{12} = L_{21} = \dfrac{\mu_0 h_2}{2\pi} \ln\left[\left(1 + \dfrac{w_2}{d}\right)\left(1 + \dfrac{w_2}{w_1 - d - w_2}\right)\right]$.

8.10 A circular loop of radius a lies in the $z = 0$ plane and a second circular loop of radius $b \ll a$ lies in the $z = d$ plane, both centered on the z axis (Fig. 8.35). Find the approximate mutual inductance as follows.

a. For the current I_1, find the magnetic field intensity on the z axis at $z = d$ and assume it is constant inside the small second loop.

b. For the current I_2 assume the small loop approximates a magnetic dipole and integrate its field over the spherical surface capping the large first loop.

Ans. $L_{21} = L_{12} = \dfrac{\mu_0 \pi a^2 b^2}{2(a^2 + d^2)^{3/2}}$.

8.11 Derive Eq. (8.31). Show that as $c \to b$, $F \to 0$, meaning there is no internal inductance in the infinitesimally thin wall.

8.12 Suppose the inner conductor of the coax is made of a hollow cylinder with inner radius d and outer radius a. Show that the internal inductance is given by

$$L_{l,\text{int}} = \frac{\mu_c}{8\pi} G, \quad G = \frac{(a^4 - d^4) - 4d^2(a^2 - d^2) + 4d^4 \ln(a/d)}{(a^2 - d^2)^2}.$$

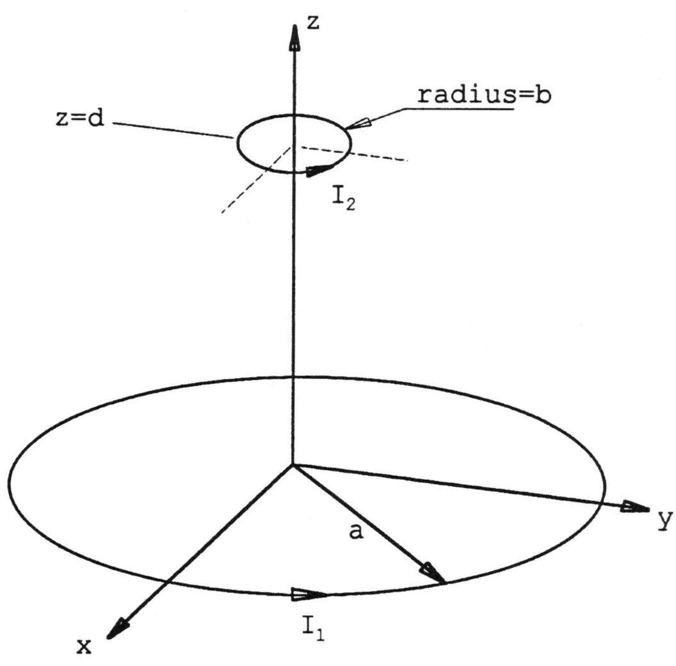

FIGURE 8.35 Two parallel circular loops with the centers on the z axis.

8.13 Figure 8.36 shows a toroid with rectangular cross section.

a. Assuming there are n turns per unit length of closely wound wire around the entire toroid (symbolically represented by the coil in the figure), show that the reluctance of the toroidal core is $\mathscr{R} = (2\pi/\mu h)[\ln(b/a)]^{-1}$.

b. Let $r_m = (a+b)/2$ be the mean radius, where $r_m \gg b - a$. Show that $\mathscr{R} \approx l/\mu A$, where $l = 2\pi r_m$ and $A = h(b-a)$.

8.14 The magnetic circuit of Fig. 8.23 has the following dimensions and values: $R = 5$ cm; $a = 1$ cm; $l_a = 0.3$ cm; $\mu_I = 1000\mu_0$ H/m; $N = 100$ turns; $I = 0.1$ A.

a. Find the reluctances of the iron core and the air gap.

b. Find the flux in the circuit.

c. Find an approximate flux in the circuit by neglecting the reluctance of the iron core. What percent error does this approximation cause?

Ans. $\mathscr{R}_I = 0.80 \times 10^6\ \text{H}^{-1}$; $\mathscr{R}_g = 7.60 \times 10^6\ \text{H}^{-1}$; $\Phi = 1.19 \times 10^{-6}$ Wb; 11%.

8.15 The magnetic circuit of Fig. 8.37 has the following dimensions and values: $l_1 = l_4 = 6$, $l_2 = 12$, $l_3 = 4$, $l_a = 0.2$, $a = 1$, $b = 1.5$, and $d = 2$ cm; $\mu = 1500\mu_0$ H/m. Evaluate the magnetic flux directed downward in the left (air gap), middle, and right branches of the circuit.

Ans. $\Phi = 0.143NI$, $-1.543NI$, $1.686NI$ μWb, respectively.

8.16 The magnetic circuit shown in Fig. 8.38 represents a rotating electric machine. The magnetic circuit sections have the following reluctances:

$$\mathscr{R}_{\text{ACB}} = \mathscr{R}_{\text{ADB}} = 30000\ \text{H/m}, \quad \mathscr{R}_{\text{AE}} = \mathscr{R}_{\text{FB}} = 4000\ \text{H/m}, \quad \mathscr{R}_{\text{GH}} = 10000\ \text{H/m}.$$

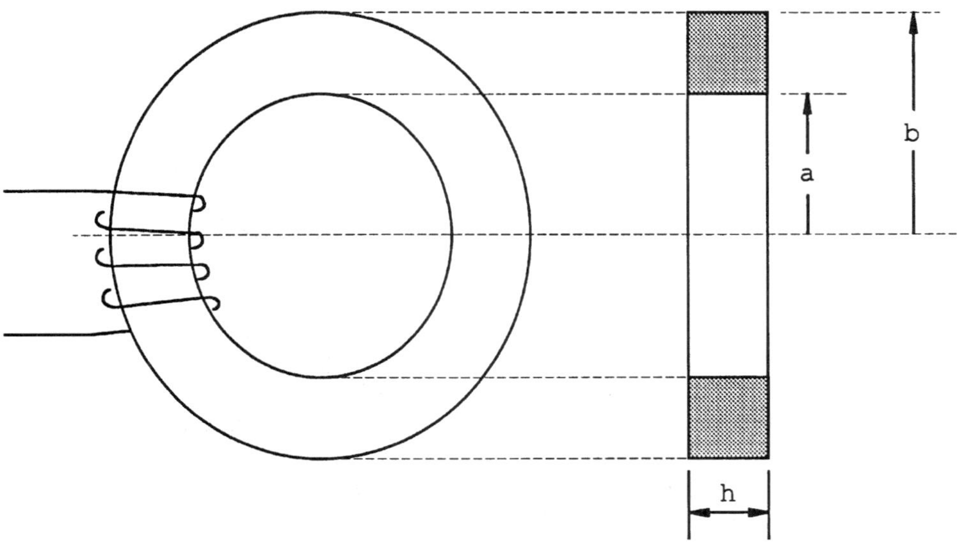

FIGURE 8.36 Toroid with rectangular cross section.

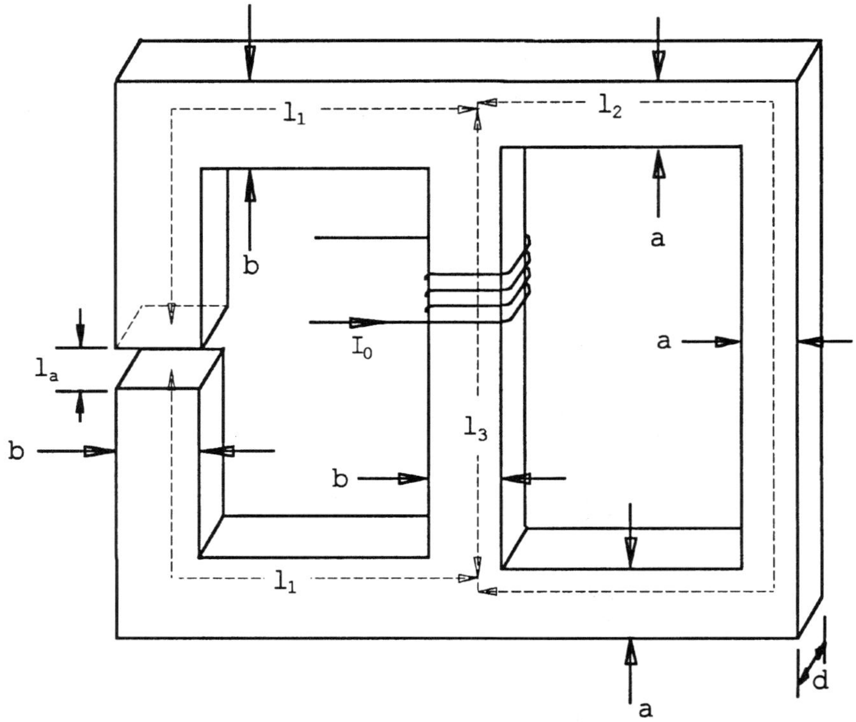

FIGURE 8.37 Magnetic circuit in Problem 8.15.

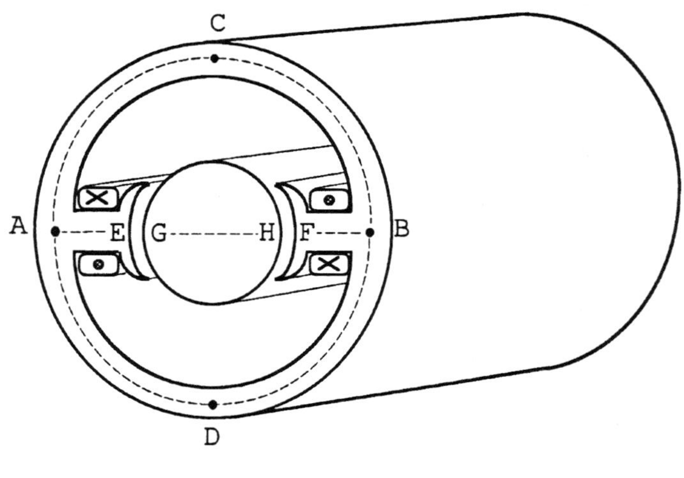

FIGURE 8.38 Magnetic circuit of a rotating machine.

The coils wrapped around the AE and FB sections are connected in series and carry a current of 2 A, each coil having 400 turns. The length of each air gap is 2 mm, and the effective area of the gap is 14.4 cm^2. Determine the flux through the air gap, assuming no flux leakage.

Ans. $\Phi = 0.0007132$ Wb.

CHAPTER NINE

Transmission Lines

9.1 INTRODUCTION

This chapter introduces the topic of waves guided by transmission lines. Transmission lines are used to carry power locally and nationally and to transmit voice, video, and data signals. We will see that when the length of a wire is comparable to or longer than one-tenth of a wavelength, $\lambda = c/f$, then we must use the theory presented here. For example, a frequency of 1 GHz, 10^9 Hz, corresponding to a digital data bit rate of 10^9 bits/s, corresponds to a wavelength of 0.3 m or 12 in., so any line longer than about 3 cm or 1 in. must be treated as a transmission line.

A further reason for studying transmission lines is that they can be used as models to represent wave transmission and reflection in free space (Chapter 11) and in waveguides (Chapter 12). The study of these problems can be simplified by applying what we will learn about transmission lines.

9.2 TIME HARMONIC FORM OF MAXWELL'S EQUATIONS

We will deal primarily with time harmonic or sinusoidal steady-state signals. In a linear time-invariant system, which includes the physical world described by Maxwell's equations in linear media, when the input or source frequency is sinusoidal at frequency f, then the output or response is also sinusoidal at frequency f. This is the same principle used in linear systems or circuit analysis courses.

If $g(u, v, w, t)$ is a typical time harmonic scalar field, then

$$g(u, v, w, t) = g_a(u, v, w)\cos[\omega t + \phi(u, v, w)],$$

where $\omega = 2\pi f$ is the angular frequency, with units of radians per second (rad/s) or, simply, s^{-1}, and f is the frequency, with units of hertz (Hz) or cycles per second (cps). We see that g is a function of the spatial coordinates and time, while the amplitude g_a and the phase angle ϕ are functions of the spatial coordinates only. Recall Euler's identity that

$$e^{j\alpha} = \cos\alpha + j\sin\alpha,$$

from which we obtain

$$\cos\alpha = \mathrm{Re}(e^{j\alpha}), \quad \sin\alpha = \mathrm{Im}(e^{j\alpha}).$$

Similarly, we can write

$$\begin{aligned} g(u, v, w, t) &= \mathrm{Re}\left[g_a(u, v, w)e^{j[\omega t + \phi(u, v, w)]}\right] \\ &= \mathrm{Re}\left[g_a(u, v, w)e^{j\phi(u, v, w)}e^{j\omega t}\right] \\ g(u, v, w, t) &= \mathrm{Re}\left[\underline{g}(u, v, w)e^{j\omega t}\right]. \end{aligned} \tag{9.1}$$

The function

$$\underline{g}(u, v, w) = g_a(u, v, w)e^{j\phi(u, v, w)}$$

is the scalar phasor field.†

The phasor field contains both the amplitude and the phase information of the original field. The original or physical field can be obtained from the phasor field by multiplying the phasor by $\exp(j\omega t)$ and taking the real part of the product.

For a time harmonic vector field we can write

$$\begin{aligned} \vec{G}(u, v, w, t) &= \vec{a}_u G_{au}(u, v, w)\cos\left[\omega t + \phi_u(u, v, w)\right] \\ &\quad + \vec{a}_v G_{av}(u, v, w)\cos\left[\omega t + \phi_v(u, v, w)\right] \\ &\quad + \vec{a}_w G_{aw}(u, v, w)\cos\left[\omega t + \phi_w(u, v, w)\right] \\ &= \mathrm{Re}\left\{\left[\vec{a}_u G_{au}(u, v, w)e^{j\phi_u(u, v, w)} + \vec{a}_v G_{av}(u, v, w)e^{j\phi_v(u, v, w)}\right.\right. \\ &\quad \left.\left. + \vec{a}_w G_{aw}(u, v, w)e^{j\phi_w(u, v, w)}\right]e^{j\omega t}\right\} \\ \vec{G}(u, v, w, t) &= \mathrm{Re}\left[\underline{\vec{G}}(u, v, w)e^{j\omega t}\right]. \end{aligned} \tag{9.2}$$

The vector phasor field $\underline{\vec{G}}(u, v, w)$, which replaces the bracketed term, contains the amplitude and phase information for each of the components of the physical vector field $\vec{G}(u, v, w, t)$. The physical field is obtained from the phasor field by multiplying the phasor by $\exp(j\omega t)$ and taking the real part of the product.

For a scalar field or a vector field, differentiation of the physical field with respect to time is the equivalent of multiplication of the phasor field by $j\omega$:

$$\frac{\partial g}{\partial t} = \mathrm{Re}(j\omega \underline{g} e^{j\omega t}) \qquad \frac{\partial \vec{G}}{\partial t} = \mathrm{Re}(j\omega \underline{\vec{G}} e^{j\omega t}).$$

† This text uses an underline to distinguish a phasor from a physical quantity. As with the vector symbol, this is an easy notation for you to use.

Conversely, integration of the physical field with respect to time is the equivalent of division of the phasor field by $j\omega$:

$$\int^t g\,dt = \mathrm{Re}\left(\frac{1}{j\omega}\underline{g}e^{j\omega t}\right) \qquad \int^t \vec{G}\,dt = \mathrm{Re}\left(\frac{1}{j\omega}\underline{\vec{G}}e^{j\omega t}\right).$$

Applying these results to Maxwell's equations written in point form, we obtain the following relations between the phasor fields:

$$\nabla \times \underline{\vec{E}} = -j\omega\mu\underline{\vec{H}}, \qquad \nabla \times \underline{\vec{H}} = \underline{\vec{J}} + j\omega\varepsilon\underline{\vec{E}},$$

$$\nabla \cdot \underline{\vec{D}} = \underline{\rho}, \qquad \nabla \cdot \underline{\vec{B}} = 0.$$

The current can be written as the sum of a source current and a conduction current $\underline{\vec{J}} = \underline{\vec{J}}_s + \sigma\underline{\vec{E}}$. In source-free space with $\underline{\vec{J}}_s = 0$ the curl equations are

$$\nabla \times \underline{\vec{E}} = -j\omega\mu\underline{\vec{H}}, \tag{9.3}$$

$$\nabla \times \underline{\vec{H}} = (\sigma + j\omega\varepsilon)\underline{\vec{E}}. \tag{9.4}$$

These are the point form of the Faraday–Maxwell and Ampère–Maxwell relations, respectively, for time harmonic fields in a source-free medium. The source-free divergence equations are not needed because they can be obtained from the curl relations.

Note that time t does not appear in these relations between the phasor fields. Time reappears in the equations when the physical fields are obtained from the phasor fields. Indeed, it is the elimination of the time variable and derivatives with respect to time that makes the use of phasor fields worthwhile.

9.2.1 Duality

The Maxwell curl relations in free space, Eqs. (9.3) and (9.4), are of exactly the same form. If we make the replacements shown in Table 9.1, then we again obtain the same curl relations. This property is called duality and we say that Maxwell's equations are dual to themselves under the replacements of Table 9.1.

The property of duality implies that if one solution is obtained to Maxwell's equations, then a second solution can be obtained by making the replacements shown.

TABLE 9.1 Duals

Replace	With
$\underline{\vec{E}}$	$-\underline{\vec{H}}$
$\underline{\vec{H}}$	$\underline{\vec{E}}$
ω	ω
$j\omega\mu$	$\sigma + j\omega\varepsilon$
$\sigma + j\omega\varepsilon$	$j\omega\mu$

9.3 TRANSVERSE ELECTROMAGNETIC WAVES

We begin by investigating the following question: Is there a time harmonic solution to Maxwell's equations that represents a wave traveling in the forward or backward z direction and that has only x and y components of the electric and magnetic fields? Mathematically we are asking if a solution of the form

$$\underline{\vec{E}}(x, y, z) = [\vec{a}_x e_x(x, y) + \vec{a}_y e_y(x, y)]\, e^{\mp\gamma z} = \vec{e}_t(x, y) e^{\mp\gamma z}, \tag{9.5}$$

$$\underline{\vec{H}}(x, y, z) = [\vec{a}_x h_x(x, y) + \vec{a}_y h_y(x, y)] e^{\mp\gamma z} = \vec{h}_t(x, y) e^{\mp\gamma z} \tag{9.6}$$

exists. If there is such a solution we would also like to know how the components are related to each other and what the value of γ is.

Before we answer these questions let us discuss the form of this pair of equations. First, in order to rationalize the $\exp(\mp\gamma z)$ term, let us for the moment assume that γ is a pure imaginary quantity, $\gamma = j\beta$. Then, when these equations are multiplied by $\exp(j\omega t)$ and the real part is taken in order to find the physical electric and magnetic field intensities, the result will have a term $\cos(\omega t \mp \beta z + \phi) = \cos\beta(ut \mp z + \phi/\beta)$, where $u = \omega/\beta$. This is recognized as being of the form $f(ut - z)$ or $h(ut + z)$, which we showed in Section 5.9 represents a wave traveling in the plus or minus z direction, respectively. The meaning of purely real or complex values of γ will be discussed soon.

Second, we note that there is no z component of the electric field intensity. The electric field therefore lies entirely in the xy plane, which is perpendicular or transverse to the direction of propagation. The wave is therefore called a transverse electric (TE) wave.

Third, we note that there is no z component of the magnetic field intensity. The magnetic field therefore lies entirely in the plane that is perpendicular or transverse to the direction of propagation. The wave is therefore called a transverse magnetic (TM) wave.

Finally, when, as in this pair of equations, the field is both transverse electric and transverse magnetic, the wave is called a transverse electromagnetic (TEM) wave.

To answer our original questions, we substitute Eqs. (9.5) and (9.6) into Maxwell's equations. In source-free space Maxwell's curl relations written in rectangular coordinates are

$$\nabla\times\underline{\vec{E}} = -j\omega\mu\underline{\vec{H}} \qquad \nabla\times\underline{\vec{H}} = (\sigma + j\omega\varepsilon)\underline{\vec{E}},$$

$$\frac{\partial \underline{E}_z}{\partial y} - \frac{\partial \underline{E}_y}{\partial z} = -j\omega\mu\underline{H}_x \qquad \frac{\partial \underline{H}_z}{\partial y} - \frac{\partial \underline{H}_y}{\partial z} = (\sigma + j\omega\varepsilon)\underline{E}_x,$$

$$\frac{\partial \underline{E}_x}{\partial z} - \frac{\partial \underline{E}_z}{\partial x} = -j\omega\mu\underline{H}_y \qquad \frac{\partial \underline{H}_x}{\partial z} - \frac{\partial \underline{H}_z}{\partial x} = (\sigma + j\omega\varepsilon)\underline{E}_y,$$

$$\frac{\partial \underline{E}_y}{\partial x} - \frac{\partial \underline{E}_x}{\partial y} = -j\omega\mu\underline{H}_z \qquad \frac{\partial \underline{H}_y}{\partial x} - \frac{\partial \underline{H}_x}{\partial y} = (\sigma + j\omega\varepsilon)\underline{E}_z,$$

and Maxwell's divergence relations are

$$\nabla\cdot(\varepsilon\underline{\vec{E}})=0 \qquad \nabla\cdot(\mu\underline{\vec{H}})=0,$$

$$\frac{\partial \underline{E}_x}{\partial x}+\frac{\partial \underline{E}_y}{\partial y}+\frac{\partial \underline{E}_z}{\partial z}=0 \qquad \frac{\partial \underline{H}_x}{\partial x}+\frac{\partial \underline{H}_y}{\partial y}+\frac{\partial \underline{H}_z}{\partial z}=0,$$

assuming ε and μ are uniform.

As we substitute we note three things. First, $\underline{E}_z$ and $\underline{H}_z$ are identically equal to zero and hence their derivatives are zero. Second, the derivative of any component with respect to z is just the component multiplied by $\mp\gamma$. Third, since all terms and their derivatives are proportional to $\exp(\mp\gamma z)$ and since the exponential is never equal to zero, the exponential can be factored out of the equations entirely. The result is therefore

$$\pm\gamma e_y=-j\omega\mu h_x \qquad \mp\gamma e_x=-j\omega\mu h_y, \tag{9.7a,b}$$

$$\frac{\partial e_y}{\partial x}-\frac{\partial e_x}{\partial y}=0 \qquad \pm\gamma h_y=(\sigma+j\omega\varepsilon)e_x, \tag{9.7c,d}$$

$$\mp\gamma h_x=(\sigma+j\omega\varepsilon)e_y \qquad \frac{\partial h_y}{\partial x}-\frac{\partial h_x}{\partial y}=0, \tag{9.7e,f}$$

and

$$\frac{\partial e_x}{\partial x}+\frac{\partial e_y}{\partial y}=0 \qquad \frac{\partial h_x}{\partial x}+\frac{\partial h_y}{\partial y}=0. \tag{9.8a,b}$$

Combining Eqs. (9.7a) and (9.7e) gives

$$e_y=\mp\frac{j\omega\mu}{\gamma}h_x=\mp\frac{j\omega\mu}{\gamma}\left(\mp\frac{\sigma+j\omega\varepsilon}{\gamma}\right)e_y=\frac{j\omega\mu(\sigma+j\omega\varepsilon)}{\gamma^2}e_y.$$

For this to be true, the term multiplying e_y on the right must be equal to unity. Hence

$$\gamma=\sqrt{j\omega\mu(\sigma+j\omega\varepsilon)}. \tag{9.9}$$

Combining Eqs. (9.7b) and (9.7d) gives the same result. This quantity, γ, is called the propagation constant of the wave. Its full meaning will be explained a little further on.

Because $\vec{e}_t(x, y)$ has no z component, as it is not a function of z, Eq. (9.7c) is equivalent to $\nabla\times\vec{e}_t=0$. This means that $\vec{e}_t$ can be written as the gradient of a scalar potential,

$$\vec{e}_t(x, y)=-\nabla V(x, y). \tag{9.10}$$

This is, of course, exactly the relation between the static electric field intensity and the static electric potential. Substitution of Eq. (9.10) into Eq. (9.8a) gives $\nabla^2 V=0$, which is Laplace's equation. Thus the electric field intensity $\underline{\vec{E}}$, Eq. (9.5), is just a two-dimensional static electric field $\vec{e}_t$ multiplied by $\exp(\mp\gamma z)$, where γ is given by Eq. (9.9).

The two-dimensional electric potential $V(x, y)$ is independent of the z coordinate and hence must represent the potential between two cylindrical conductors with sides parallel to the z axis, as shown in Fig. 9.1.

Equations (9.7b) and (9.7a) [or (9.7d) and (9.7e)] can be written

$$e_x = \pm \eta h_y \qquad e_y = \mp \eta h_x, \tag{9.11}$$

where

$$\eta = \sqrt{\frac{j\omega\mu}{\sigma + j\omega\varepsilon}} \,. \tag{9.12}$$

This quantity, η, is called the intrinsic impedance of the medium. It will be discussed further shortly.

So far we have used five of the six equations stemming from the Maxwell curl relations, Eqs. (9.7). When Eqs. (9.11) are substituted in the sixth relation, Eq. (9.7f), we get

$$\frac{\partial(\pm e_x/\eta)}{\partial x} - \frac{\partial(\mp e_y/\eta)}{\partial y} = 0$$

$$\pm\left(\frac{\partial e_x}{\partial x} + \frac{\partial e_y}{\partial y}\right) = 0,$$

which is true by virtue of Eq. (9.8a). Finally, Eq. (9.7c) is identical to Eq. (9.8b).

We summarize as follows:

1. TEM waves of the form given by Eqs. (9.5) and (9.6) are possible since they satisfy Maxwell's equations in a source-free region.

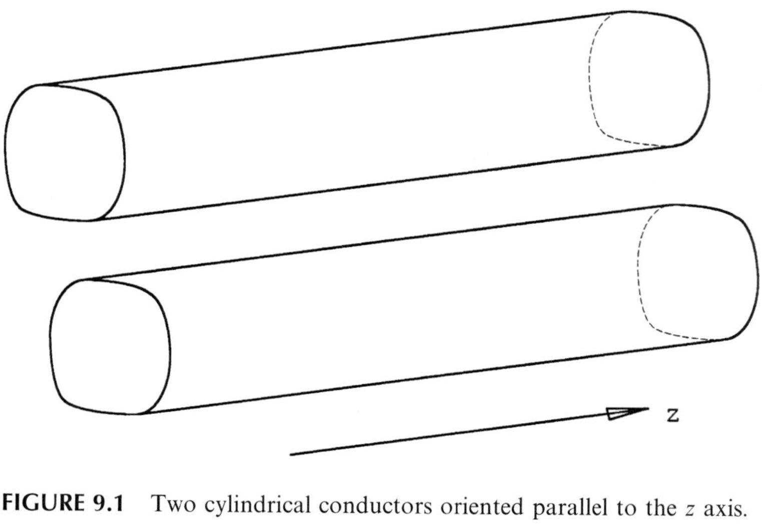

FIGURE 9.1 Two cylindrical conductors oriented parallel to the z axis.

2. The propagation constant of the waves is given by γ, Eq. (9.9).
3. The intrinsic impedance, η, of the medium in which the wave propagates, Eq. (9.12), gives the relation between the electric and magnetic field components, Eqs. (9.11).
4. The TEM wave must propagate along (at least) two cylindrical conductors.

Some of these points will now be explained more fully.

9.3.1 Cylindrical Structures

A cylindrical structure (Fig. 9.1) is any combination of conductors and dielectrics, the cross section of which is invariant in any xy or $z = C$ plane. Some examples of cylindrical structures follow.

Figure 9.2 shows cylindrical structures that consist of two conductors. In Fig. 9.2a a two-wire transmission line is shown. The conductors are parallel circular wires, each with radius a, and the spacing between their center lines is d. Power lines hung along poles (Fig. 9.2b) are an approximation to this kind of line. Another example is the line commonly used to connect a television to a roof top antenna (Fig. 9.2c). In Fig. 9.2d a coaxial line (or, simply, a coax) is shown. This has two circular coaxial conductors with radii as shown. In Fig. 9.2e a pair of parallel plates is shown. In Fig. 9.2f a microstrip line is shown. This consists of a thin conductor forming a ground plane

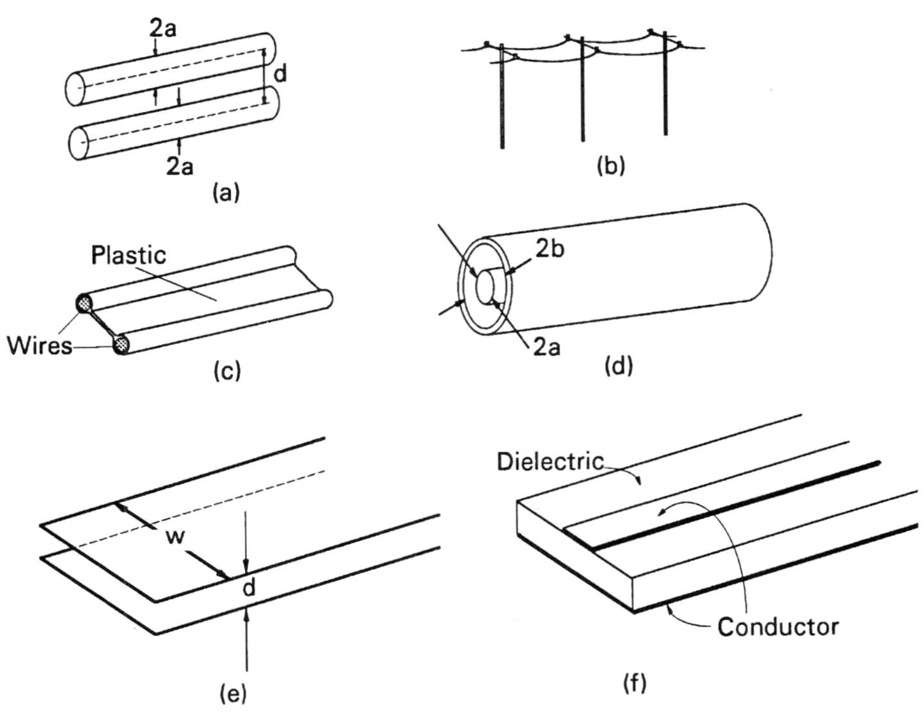

FIGURE 9.2 Cylindrical structures that consist of two conductors.

and a conducting line bonded onto a dielectric board. These lines are like printed circuits commonly used to connect or mount electronic components. In each case the dielectric is characterized by the constants μ, ε, and σ and the conductor by its conductivity $\sigma_c(\mu_c = \mu_0, \varepsilon_c = \varepsilon_0)$.

If these lines could be made with perfect conductors ($\sigma_c = \infty$) and with a single dielectric (examples in Fig. 9.2a–e), then they could support a pure TEM wave. In the practical case where the conductivity is high they support an "almost" or quasi-TEM wave and are analyzed as if the wave were purely TEM. The same is true for multiple dielectric microstrip lines.

Figure 9.3 shows some cylindrical structures that consist of a single conductor or no conductor at all. In Fig. 9.3a–c, rectangular, circular, and ridge waveguides, which are hollow cylindrical conductors, are shown. In Fig. 9.3d we have a dielectric slab, while

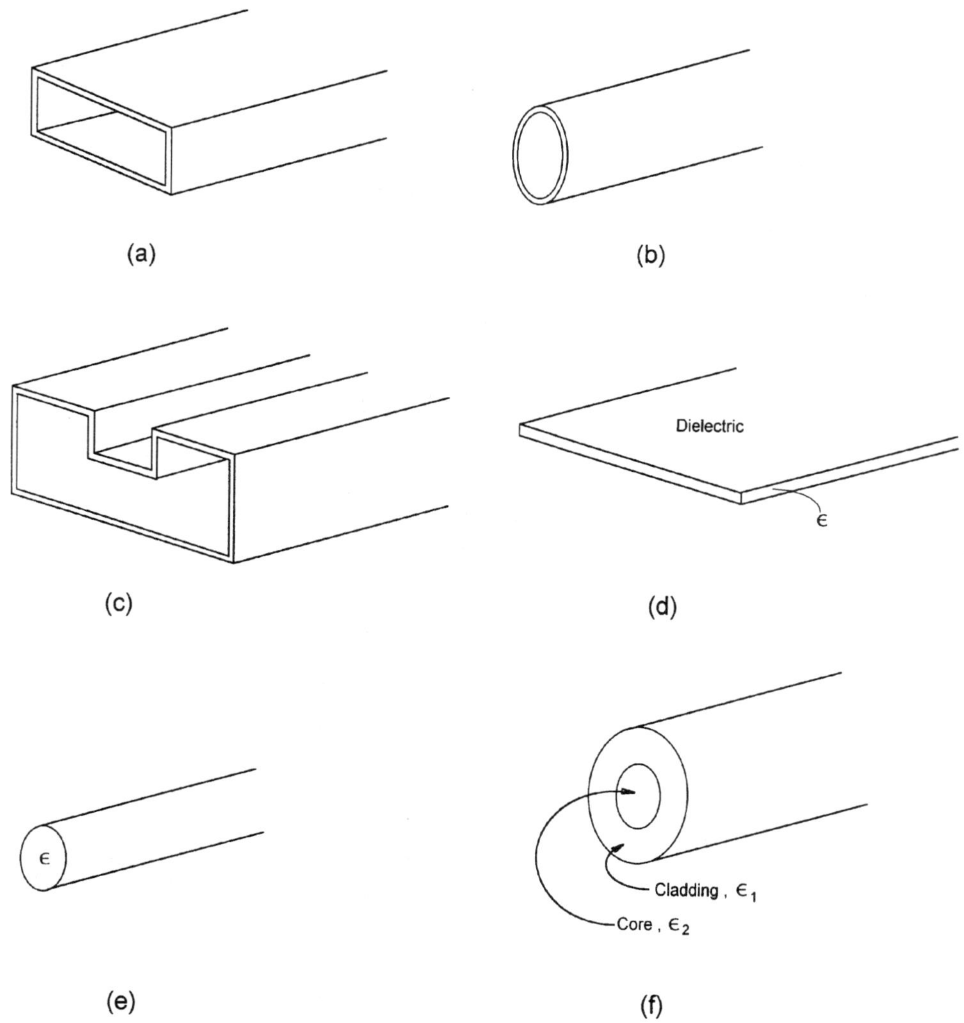

FIGURE 9.3 Cylindrical structures that consist of a single conductor or no conductor at all.

Fig. 9.3e and Fig. 9.3f show a dielectric rod and fiber optic cable, respectively. None of these lines have a pair of conductors and none can propagate a TEM wave. They will be discussed in subsequent chapters.

9.3.2 Propagation Constant

The propagation constant, Eq. (9.9), is seen to be a complex number since it is the square root of a complex number. It can therefore be written as

$$\gamma = \alpha + j\beta. \tag{9.13}$$

The quantity α is called the attenuation constant and β is called the phase constant. Since the product γz is dimensionless, the propagation constant, the attenuation constant, and the phase constant must have the dimensions of reciprocal length, m^{-1}. It is important to understand how these constants relate to the behavior of the physical wave.

Let us assume that we are dealing with any component of the electric or magnetic field. The z dependence for a forward or $+z$ traveling wave is given by

$$e^{-\gamma z} = e^{-(\alpha + j\beta)z} = e^{-\alpha z} e^{-j\beta z},$$

so that a typical phasor representing any component of the wave can be written as

$$\underline{F}(x, y, z) = F_0 f(x, y) e^{-\alpha z} e^{-j\beta z}, \quad F_0 = F_{0a} e^{j\phi_0}. \tag{9.14}$$

F_0 is a complex number, F_{0a} is its magnitude, and ϕ_0 is its angle.

The corresponding physical field is

$$\begin{aligned} F(x, y, z, t) &= \mathrm{Re}[\underline{F} e^{j\omega t}] = \mathrm{Re}[F_0 f(x, y) e^{-\alpha z} e^{-j\beta z} e^{j\omega t}] \\ &= F_{0a} f(x, y) e^{-\alpha z} \cos(\omega t - \beta z + \phi_0). \end{aligned} \tag{9.15}$$

Figure 9.4 shows F at some particular values of x and y as a function of z, with t as a parameter. The curves are drawn for $t = 0$, $t = t_1 > 0$, $t = t_2 > t_1$, and $t = t_3 > t_2$. The wave appears to be moving in the $+z$ direction as time increases, confirming that this is a forward traveling wave.

At any position the maximum value (positive or negative) of F is $F_{0a} f(x, y) \exp(-\alpha z)$ and this value is called the amplitude of the wave at that position. At this fixed position F varies sinusoidally with time at an angular frequency $\omega = 2\pi f$. The wave decays or attenuates in amplitude as it propagates in the z direction. Every $1/\alpha$ meters the wave amplitude attenuates by $\exp(-1) = 0.368$. The attenuation constant has, as already noted, the dimensions of inverse length, m^{-1}; the units are usually stated as α nepers per meter (Np/m) or as $20 \log_{10} [\exp(\alpha)] = 8.69\alpha$ decibels per meter (dB/m).

The term

$$\Phi = \omega t - \beta z + \phi_0 \tag{9.16}$$

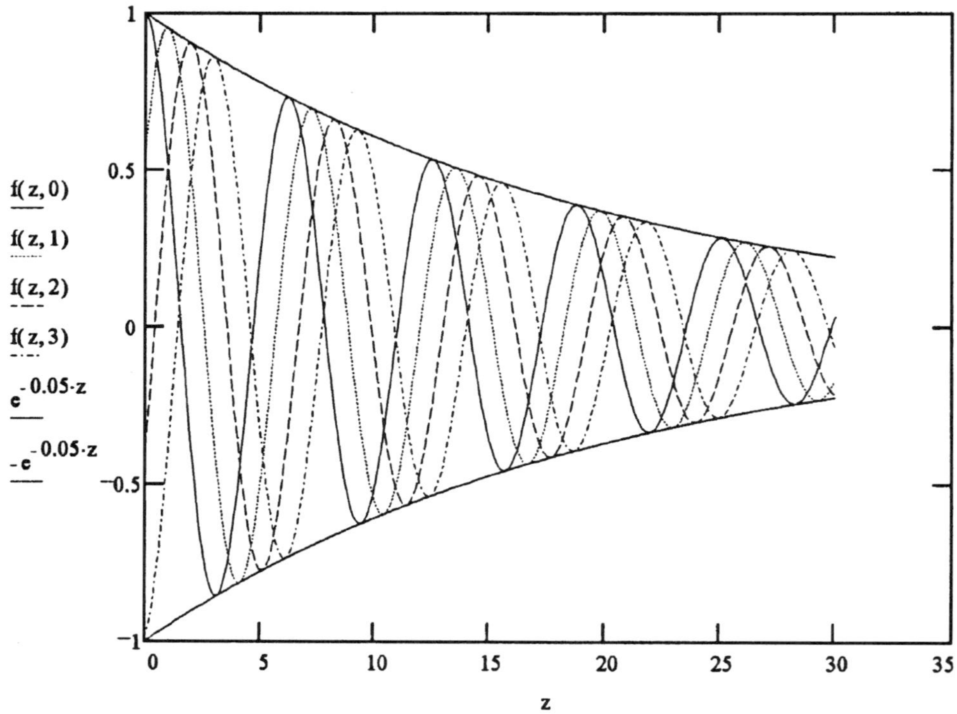

FIGURE 9.4 A wave $F(x, y, z, t)$ traveling in the positive z direction.

is called the phase of the wave. As t increases the value of z must also increase to keep the phase Φ constant. This is another way of saying that this is a forward traveling wave.

Equation (9.16) shows that at any fixed time the phase is constant in a plane that is transverse or perpendicular to the z axis. Since the surface of constant phase is a plane, this is called a plane wave. If the term βz were replaced by βr, then the surface of constant phase would be a circular cylinder around the z axis and the wave would be called a cylindrical wave. Similarly, if the term βz were replaced by βR, then the surface of constant phase would be a sphere centered at the origin and the wave would be called a spherical wave. Thus the terms plane wave, cylindrical wave, and spherical wave describe the shape of the surface of constant phase. In any case, Eq. (9.15) as it is describes a plane wave.

If the amplitude of the wave is constant on the surface of constant phase, then the wave is a uniform wave. Thus if $f(x, y) = C$, then Eq. (9.15) represents a uniform plane wave.

When we consider the positive term in the exponential, $\exp(+\gamma z)$, the physical field of a typical electric or magnetic field component is given by

$$\begin{aligned} F &= \mathrm{Re}[\underline{F}e^{j\omega t}] = \mathrm{Re}[F_0 f(x, y)e^{\alpha z}e^{j\beta z}e^{j\omega t}] \\ &= F_{0a} f(x, y)e^{\alpha z}\cos(\omega t + \beta z + \phi_0). \end{aligned}$$

Figure 9.5 shows F at some particular values of x and y as a function of z, with t as a parameter. The curves are drawn for $t = 0$, $t = t_1 > 0$, $t = t_2 > t_1$, and $t = t_3 > t_2$. The wave appears to be moving in the $-z$ direction as time increases, showing that this is a backward traveling wave. The phase of the wave is now

$$\Phi = \omega t + \beta z + \phi_0,$$

so this is still a plane wave. As t increases the value of z must now decrease to keep the phase Φ constant. This is another way of saying that this is a backward traveling wave.

In a perfect dielectric medium $\sigma = 0$ and Eq. (9.9) shows that the attenuation constant is zero,

$$\alpha = 0, \quad \beta = \omega\sqrt{\mu\varepsilon}, \quad \text{when } \sigma = 0. \tag{9.17}$$

9.3.3 Wavelength and Wave Velocity

The wavelength λ is the distance in the direction of propagation over which the phase changes by 2π radians. Thus, if $\Phi = \Phi_0$ at $(z, t) = (z_0, t_0)$, then $\Phi = \Phi_0 - 2\pi$ at $(z, t) = (z_0 + \lambda, t_0)$. Using Eq. (9.16) we obtain

$$\Phi_0 = \omega t_0 - \beta z_0 + \phi_0,$$

$$\Phi_0 - 2\pi = \omega t_0 - \beta(z_0 + \lambda) + \phi_0.$$

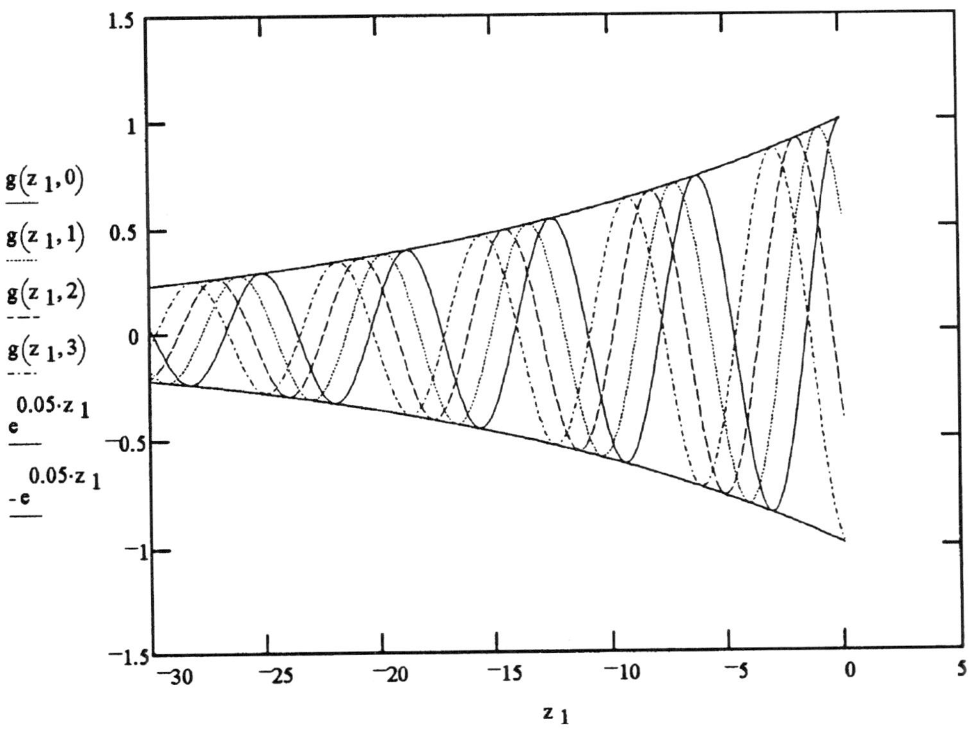

FIGURE 9.5 A wave $F(x, y, z, t)$ traveling in the negative z direction.

Subtracting the second equation from the first and dividing by β, we have

$$\lambda = \frac{2\pi}{\beta}. \tag{9.18}$$

The wavelength is shown in Figs. 9.4 and 9.5.

The phase velocity v_{ph} is the speed with which a point of constant phase moves in the direction of propagation. As z and t change by dz and dt, then Eq. (9.16) shows that Φ changes by

$$d\Phi = \omega\, dt - \beta\, dz.$$

Setting $d\Phi = 0$ we find

$$v_{\text{ph}} = \left.\frac{dz}{dt}\right|_{d\Phi=0}$$

$$v_{\text{ph}} = \frac{\omega}{\beta}. \tag{9.19}$$

In a perfect dielectric medium, substituting from Eq. (9.17),

$$v_{\text{ph}} = \frac{1}{\sqrt{\mu\varepsilon}}, \quad \text{when } \sigma = 0. \tag{9.20}$$

In free space, $v_{\text{ph}} = c = 3 \times 10^8$ m/s. Thus the phase velocity of a TEM wave in a perfect dielectric is equal to the speed of light in that medium.

When we calculate the phase velocity for TE and TM waves we will find that the phase velocity is greater than the speed of light in the medium. From physics courses we know that "nothing" can travel faster than the speed of light, so this seems to present a problem. The "nothing" refers, however, to genuine things, such as energy or information. The energy carried by a wave or the information carried by a wave cannot travel faster than the speed of light in the medium. Phase is just a mathematical quantity, which can travel at any speed.

Without proof at this point, we assert that the speed of energy or information flow (under certain generally existing conditions) is given by the group velocity, v_{g}. The group velocity is given by

$$v_{\text{g}} = \frac{\partial\omega}{\partial\beta} = \left(\frac{\partial\beta}{\partial\omega}\right)^{-1}. \tag{9.21}$$

In the case of a TEM wave propagating in a perfect dielectric $\beta = \omega\sqrt{\mu\varepsilon}$ and so $v_{\text{g}} = 1/\sqrt{\mu\varepsilon} = v_{\text{ph}}$ and the group velocity equals the phase velocity. Later on, in dealing with TE and TM waves, we will find that $v_{\text{g}} < c$.

These equations for the phase and group velocities may be shown graphically in a plot of ω versus β (Fig. 9.6). At any value of angular frequency the phase velocity is the slope of the line drawn from the origin to the point on the curve, while the group velocity is the slope of the line that is tangent to the curve.

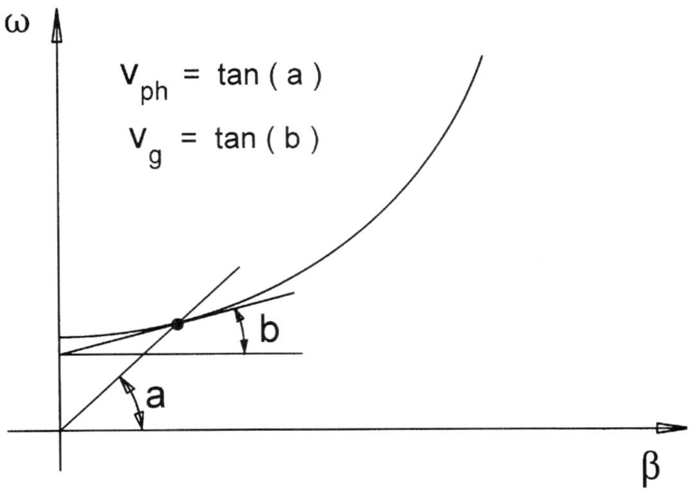

FIGURE 9.6 A plot of ω versus β for TE and TM waves and the definitions of phase and group velocities.

9.4 LINE VOLTAGE AND CURRENT

We have shown that plane TEM waves propagate on two-conductor cylindrical structures. We will now show that it is possible to define a unique voltage between the two cylindrical conductors and a unique current in the conductors for the TEM wave. These will be functions of the coordinate in the direction of propagation (z) and of time. As electrical engineers we would much rather deal with the voltage and current than with electric and magnetic fields. The reason for that is simple: the electric and magnetic fields are functions of the three spatial coordinates and time, while the voltage and current are functions of only one spatial coordinate (z) and time.

For time-varying fields it is generally not possible to define a unique voltage between two points in space. This is because

$$V_{ab} = \int_a^b \vec{E} \cdot d\vec{l}$$

is usually a function of the path connecting point a to point b. Let us, however, consider the special case of a TEM wave and a closed path lying entirely in the transverse ($z = C$) plane as shown in Fig. 9.7. Points a and d are any points on cylindrical conductor C_a and points b and c are any points lying on cylindrical conductor C_b, with all four points lying in a single transverse plane. The Faraday–Maxwell integral equation, Eq. (4.70), for time harmonic fields is

$$\oint_L \underline{\vec{E}} \cdot d\vec{l} = -j\omega\mu \int_S \underline{\vec{H}} \cdot d\vec{S}. \tag{9.22}$$

Let us apply this to the closed path abcda and the capping surface (shown shaded) in the transverse plane. The right side of the equation is equal to zero because $d\vec{S} = \vec{a}_z \, dS$

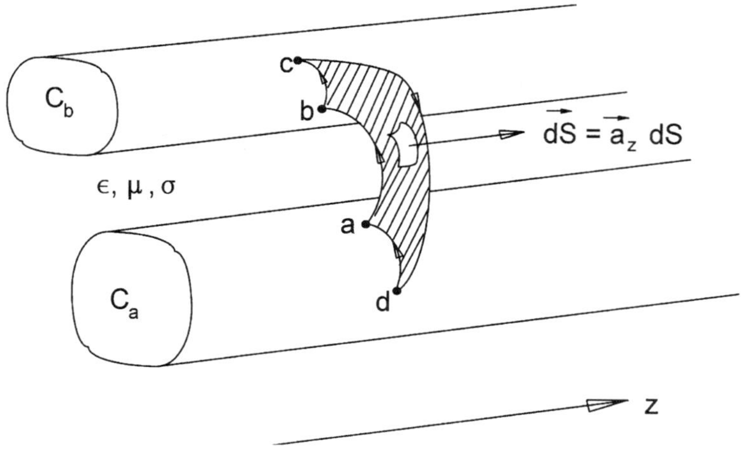

FIGURE 9.7 A closed path lying in the transverse plane of a two-conductor transmission line.

and $\underline{H}_z = 0$ for the TEM wave. Thus

$$\oint_L \underline{\vec{E}} \cdot d\vec{l} = 0$$

$$V_{ab} + V_{bc} + V_{cd} + V_{da} = 0.$$

The terms $V_{bc} = 0$ and $V_{da} = 0$ because the tangential electric field on the surface of each conductor is zero. We then have the result that $V_{ab} = -V_{cd} = V_{dc}$ independent of the paths between the conductors, as long as the paths lie in the same transverse plane. For our TEM wave it is therefore proper to speak of the voltage between the conductors at any particular $z = C$ plane. Substituting $\underline{\vec{E}}$ from Eq. (9.4), the voltage phasor as a function of z is

$$\begin{aligned} \underline{V}(z) &= \int_a^b \underline{\vec{E}} \cdot d\vec{l} = \int_a^b (\vec{e}_t e^{\mp \gamma z}) \cdot d\vec{l} = \left(\int_a^b \vec{e}_t \cdot d\vec{l} \right) e^{\mp \gamma z} \\ &= V_0 e^{\mp \gamma z}, \quad V_0 = \int_a^b \vec{e}_t \cdot d\vec{l}. \end{aligned} \tag{9.23}$$

The voltage phasor $\underline{V}(z)$ representing the voltage from C_a to C_b is therefore a voltage wave propagating in the positive or negative z direction with the same propagation constant as the electromagnetic field.

We next examine the concept of current and its relation to the magnetic field intensity. For this we use the Ampère–Maxwell integral law, Eq. (4.46), written for time harmonic fields:

$$\oint_L \underline{\vec{H}} \cdot d\vec{l} = \int_S \underline{\vec{J}} \cdot d\vec{S} + j\omega\varepsilon \int_S \underline{\vec{E}} \cdot d\vec{S}. \tag{9.24}$$

Consider an arbitrary closed loop L lying in a transverse plane and encircling the conductor C_a (Fig. 9.8). We take the capping surface to be the shaded area lying in the

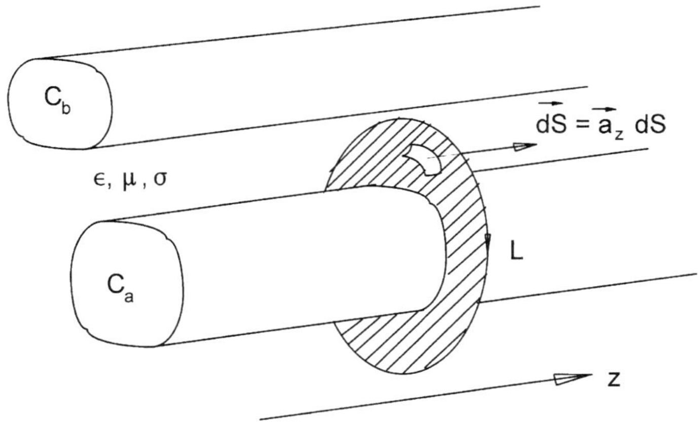

FIGURE 9.8 An arbitrary closed loop L lying in a transverse plane encircling the conductor C_a.

transverse plane. The free current represented by the first integral on the right side of the equation consists of two different currents. The first current, which is the same for any loop L, is the current in the positive z direction in the conductor C_a. This is the current phasor $\underline{I}(z)$. The second current is the current in the imperfect dielectric surrounding the wires. The current density in the dielectric is $\underline{\vec{J}} = \sigma\underline{\vec{E}}$. Hence Eq. (9.24) can be written as

$$\oint_L \underline{\vec{H}} \cdot d\vec{l} = \underline{I}(z) + (\sigma + j\omega\varepsilon)\int_S \underline{\vec{E}} \cdot d\vec{S}.$$

The integral on the right side of the equation is equal to zero because $d\vec{S} = \vec{a}_z\, dS$ and $\underline{E}_z = 0$ for the TEM wave. Substituting for $\underline{\vec{H}}$ from Eq. (9.5), the current phasor is therefore

$$\underline{I}(z) = \oint_L \underline{\vec{H}} \cdot d\vec{l} = \oint_L (\vec{h}_t e^{\mp\gamma z}) \cdot d\vec{l} = \left(\oint_L \vec{h}_t \cdot d\vec{l}\right) e^{\mp\gamma z}$$
$$= I_0 e^{\mp\gamma z}, \quad I_0 = \oint_L \vec{h}_t \cdot d\vec{l}. \tag{9.25}$$

The current phasor $\underline{I}(z)$ representing the current in the positive z direction in C_a is therefore a current wave propagating in the positive or negative z direction with the same propagation constant as the electromagnetic field. The current in the positive z direction in C_b is $-\underline{I}(z)$.

9.5 TRANSMISSION LINE PARAMETERS: INDUCTANCE, CAPACITANCE, AND CONDUCTANCE PER UNIT LENGTH

We next define the inductance L, capacitance C, and conductance G per unit length of the cylindrical line. These are constants of the line and are not functions of z because the cross section of the line is independent of z. There is no subscript l because we will only deal with per unit length quantities in this chapter.

To find a formula for the line inductance per unit length assume a steady current I_0 on the surface of C_a and returning on the surface of C_b (Fig. 9.9).[†] The field surrounding the conductors is $\vec{h}_t(x, y)$. The magnetic flux surrounding line C_a for a length l is

$$\Phi = \int_S \mu \vec{h}_t \cdot d\vec{S}$$

and the current is given by Eq. (9.25). The line inductance per length (H/m) is

$$L = \frac{\Phi/l}{I_0} = \frac{\mu \int_S \vec{h}_t(x, y) \cdot d\vec{S}}{l \oint_L \vec{h}_t(x, y) \cdot d\vec{l}}. \tag{9.26}$$

This result is the same as Eq. (8.19), specialized for a two-dimensional field.

To find a formula for the line capacitance per unit length assume that a steady voltage V_0 is connected across the wires (Fig. 9.10). The field between the conductors is $\vec{e}_t(x, y)$. The charge on conductor C_a for a length l is

$$Q = \int_S \varepsilon \vec{e}_t \cdot d\vec{S}$$

and the electric potential is given by Eq. (9.23). The line capacitance per unit length (F/m) is

$$C = \frac{Q/l}{V_0} = \frac{\varepsilon \int_S \vec{e}_t(x, y) \cdot d\vec{S}}{l \int_a^b \vec{e}_t(x, y) \cdot d\vec{l}}. \tag{9.27}$$

This result is the same as Eq. (6.49), specialized for a two-dimensional field.

To find a formula for the line conductance per unit length we again use Fig. 9.10. The current flowing from conductor C_a to conductor C_b for a length l is

$$I = \int_S \sigma \vec{e}_t(x, y) \cdot d\vec{S}$$

and the potential between the lines is again given by Eq. (9.23). The line conductance per unit length (S/m) is

$$G = \frac{I/l}{V_0} = \frac{\sigma \int_S \vec{e}_t(x, y) \cdot d\vec{S}}{l \int_a^b \vec{e}_t(x, y) \cdot d\vec{l}} = \left(\frac{\sigma}{\varepsilon}\right) C. \tag{9.28}$$

This result is the same as Eq. (7.9), specialized for a two-dimensional field.

[†] The current flows in a thin layer on the surface due to the skin effect. It is therefore only necessary to calculate the external self-inductance. The same effect can be achieved at low frequency by considering the conductors to be thin cylindrical shells.

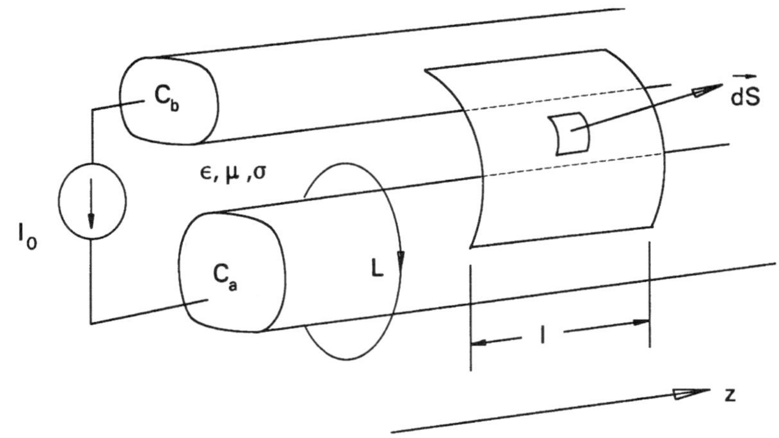

FIGURE 9.9 Determination of the line inductance per unit length for a two-wire transmission line.

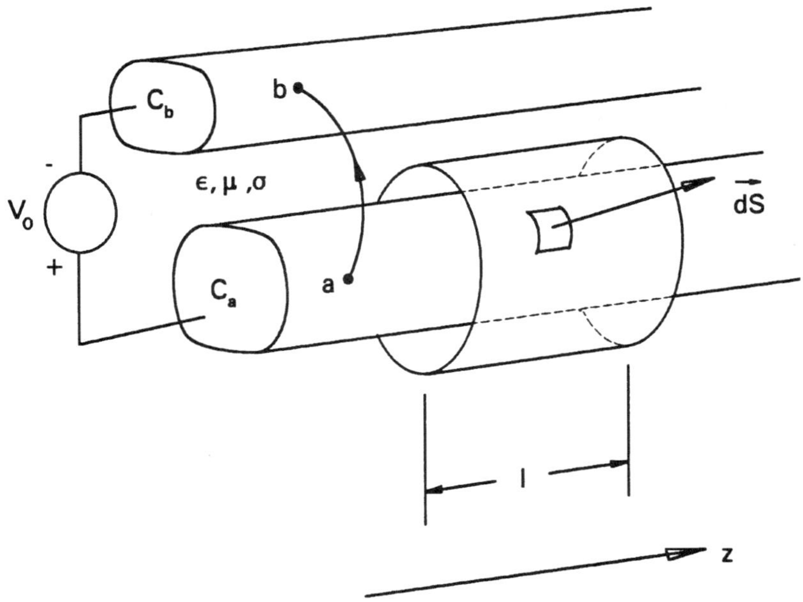

FIGURE 9.10 Determination of the line capacitance and conductance per unit length for a two-wire transmission line.

These formulas give us the inductance L, capacitance C, and conductance G per unit length for a transmission line made up of two perfectly conducting cylindrical conductors. These parameters define the transmission line just as the permeability μ, permittivity ε, and conductivity σ define the medium surrounding the conductors. A little later we will drop the requirement that the conductors be perfect and define a resistance R per unit length (Ω/m), which depends on the conductivity σ_c of the conductors.

TABLE 9.2 Transmission Line Parameters

Parameter	Parallel Plate Line	Two-wire Line	Coaxial Line	SI Unit
R	$\frac{2}{w}R_s,\ R_s=\sqrt{\frac{\omega\mu_c}{2\sigma_c}}$	$\frac{1}{\pi a}R_s$	$\frac{1}{2\pi}\left(\frac{1}{a}+\frac{1}{b}\right)R_s$	Ω/m
L	$\mu\frac{d}{w}$	$\frac{\mu}{\pi}\cosh^{-1}\left(\frac{d}{2a}\right)$	$\frac{\mu}{2\pi}\ln\left(\frac{b}{a}\right)$	H/m
G	$\sigma\frac{w}{d}$	$\frac{\pi\sigma}{\cosh^{-1}(d/2a)}$	$\frac{2\pi\sigma}{\ln(b/a)}$	S/m
C	$\varepsilon\frac{w}{d}$	$\frac{\pi\varepsilon}{\cosh^{-1}(d/2a)}$	$\frac{2\pi\varepsilon}{\ln(b/a)}$	F/m

The four transmission line parameters—R, L, G, C—depend on the shape of the conductors. For some shapes, like parallel circular wires and the coaxial cable, they are easy to calculate. Indeed, we have already done so in Chapters 6, 7, and 8. For other shapes exact formulas are not easy or even possible to obtain. In these cases numerical methods may be used or the values may be measured.

Table 9.2 gives formulas for the transmission line parameters for the parallel plate line, the two-wire line, and the coaxial cable.

Example 9.1 Determine the parameters per unit length of a two-wire power transmission line at 60 Hz, for which the conductors are made of copper and the surrounding medium is air. The radius of each wire is 1.27 cm and the distance between the wires is 75 cm.

Solution According to Table 9.2, the values of the parameters per unit length are obtained as follows:

$$R=\frac{1}{\pi a}R_s=\frac{1}{\pi a}\sqrt{\frac{\omega\mu_c}{2\sigma_c}}=\frac{1}{\pi\cdot 0.0127}\sqrt{\frac{(2\pi\cdot 0)\cdot 4\pi\times 10^{-7}}{2\cdot}(5.8\times 10^{7})}=0.05065\ \text{m}\Omega/\text{m},$$

$$L=\frac{\mu}{\pi}\cdot\cosh^{-1}\left(\frac{d}{2a}\right)=\frac{4\pi\times 10^{-7}}{\pi}\cdot\cosh^{-1}\left(\frac{0.75}{2\cdot 0.0127}\right)=1.631\ \mu\text{H/m},$$

$$C=\frac{\pi\varepsilon}{\cosh^{-1}(d/2a)}=\frac{\pi\cdot(1/36\pi)\times 10^{-9}}{\cos\text{h}^{-1}(0.75/(2\cdot 0.0127))}=6.81\ \text{pF/m},$$

$$G=0.$$

9.6 TRANSMISSION LINE EQUATIONS

We have developed definitions for line voltage $\underline{V}(z)$ and line current $\underline{I}(z)$ and we have found that each of these travel with the same propagation constant γ. The voltage and current are intimately connected to the electric field intensity and magnetic field intensity, respectively. The electric and magnetic fields are related to each other by Maxwell's equations, which implies that there must be a set of equations that relate the voltage to the current. These relations, which we will now derive, are called the transmission line equations.

9.6.1 Derivation from Maxwell's Equations

Figure 9.11 shows a closed path abcda and a surface capping the path. The line $\overline{ab}$ is in the $z = C_1$ plane and the line $\overline{cd}$ is just a short distance away in the $z + dz = C_2$ plane. The lines $\overline{bc}$ and $\overline{da}$ lie on the surface of the conductors. We now use the Faraday–Maxwell integral equation to integrate around the closed path and we replace the line integral along each path segment by its corresponding voltage.

$$\oint_L \underline{\vec{E}} \cdot d\vec{l} = -j\omega\mu \int_S \underline{\vec{H}} \cdot d\vec{S}$$

$$\underline{V}_{ab} + \underline{V}_{bc} + \underline{V}_{cd} + \underline{V}_{da} = -j\omega\underline{\Phi}_S.$$

Now $\underline{V}_{ab} = -\underline{V}_{ba} = -\underline{V}(z)$, $\underline{V}_{cd} = \underline{V}(z + \Delta z)$, and $\underline{V}_{bc} = \underline{V}_{da} = 0$ because the component of electric field along the wires is zero. The flux through the surface capping the loop $\underline{\Phi}_s$ can be given in terms of the inductance per unit length and the current as $\underline{\Phi}_s = L\,\Delta z\,\underline{I}(z + \Delta z/2)$. Substituting these terms, dividing by Δz, and taking the limit as $\Delta z \to 0$, we obtain

$$\lim_{\Delta z \to 0} \frac{\underline{V}(z + \Delta z) - \underline{V}(z)}{\Delta z} = \lim_{\Delta z \to 0} [-j\omega L\underline{I}(z + \Delta z/2)]$$

$$\frac{d\underline{V}(z)}{dz} = -j\omega L\underline{I}(z). \tag{9.29}$$

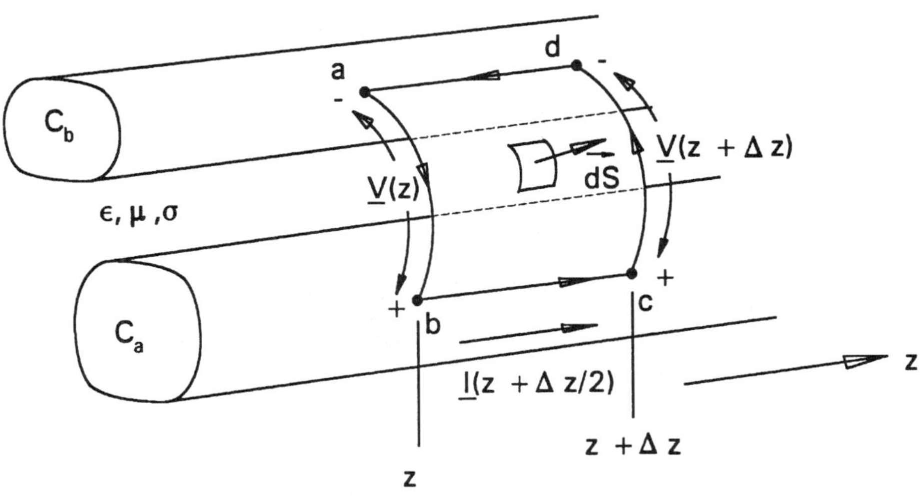

FIGURE 9.11 A closed path abcda and a surface capping the path.

This is the first of the transmission line equations relating the voltage and current phasors.

Figure 9.12 shows a closed path $\overline{abcda}$ and a capping surface S. Line $\overline{ab}$ lies in the $z = C_1$ plane. Points a and b are almost coincident so that the path $\overline{ab}$ is virtually closed. This line can be capped by the surface S_1, where $d\vec{S}_1 = \vec{a}_z\, dS_1$. Similarly, line $\overline{cd}$ lies in the $z + \Delta z = C_2$ plane, with points c and d being almost coincident. This line can be capped by the surface S_2, where $d\vec{S}_2 = -\vec{a}_z\, dS_2$. The lines $\overline{bc}$ and $\overline{da}$ are virtually coincident, but oppositely directed. To find the second transmission line equation, we use the Ampère–Maxwell integral equation:

$$\oint_L \underline{\vec{H}} \cdot d\vec{l} = (\sigma + j\omega\varepsilon)\int_S \underline{\vec{E}} \cdot d\vec{S}$$

$$\int_{L_{ab}} \underline{\vec{H}} \cdot d\vec{l} + \int_{L_{bc}} \underline{\vec{H}} \cdot d\vec{l} + \int_{L_{cd}} \underline{\vec{H}} \cdot d\vec{l} + \int_{L_{da}} \underline{\vec{H}} \cdot d\vec{l} = \underline{I}_S + (\underline{I}_d)_S.$$

For the first integral on the left, we again use the Ampère–Maxwell integral equation to obtain

$$\int_{L_{ab}} \underline{\vec{H}} \cdot d\vec{l} = \underline{I}(z) + (\sigma + j\omega\mu)\int_{S_1} \underline{\vec{E}} \cdot \vec{a}_z\, dS_1 = \underline{I}(z),$$

since $\underline{E}_z = 0$. Similarly, the third integral on the left is equal to $-\underline{I}(z + \Delta z)$. The second and fourth integrals are exactly equal but of opposite sign and cancel. On the right, $\underline{I}_S$ is the conduction current through the surface S. It is given by $G\,\Delta z\, V(z + \Delta z/2)$, where G is the conductance per unit length between the lines. Also on the right, the term $(\underline{I}_d)_S$ is the displacement current through S. It is given by

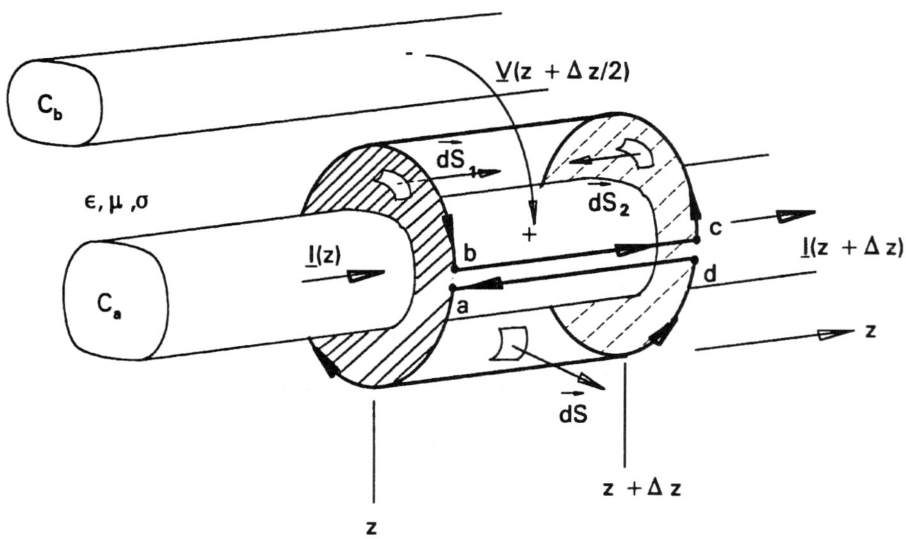

FIGURE 9.12 A closed path abcda and capping surface S.

$j\omega C\,\Delta z\,\underline{V}(z + \Delta z/2)$, where C is the capacitance per unit length between the lines. Substituting these terms, dividing by Δz, and taking the limit as $\Delta z \to 0$, we find

$$\lim_{\Delta z \to 0} \frac{\underline{I}(z) - \underline{I}(z + \Delta z)}{\Delta z} = \lim_{\Delta z \to 0} (G + j\omega C)\,\underline{V}(z + \Delta z/2)$$

$$\frac{d\underline{I}(z)}{dz} = -\,(G + j\omega C)\underline{V}(z). \tag{9.30}$$

This is the second of the transmission line equations relating the voltage and current phasors.

Equations (9.29) and (9.30) give the complete relations between the line voltage and the line current on a cylindrical two-conductor structure with infinite conductivity. We will defer finding the solution until we have introduced losses due to imperfectly conducting lines and until we have derived these equations again, from a circuit theory viewpoint.

9.6.2 Imperfectly Conducting Lines

So far we have assumed that the conductors are perfect. However, we all know that wires are typically made of copper, aluminum, or some other good, but not perfect, conductor.† This is a sticky problem. If the wires are not perfect conductors then a field $\underline{E}_z$ must be present to drive the current $\underline{I}(z)$ in the z direction. This means that our wave is not TEM, which was the basis for our entire definition of voltage and current and our derivation of the transmission line equations. Our carefully built theory is collapsing. As a practical matter, however, if the conductors are good conductors, then $\underline{E}_z$ is small $(|\underline{E}_z| \ll |\vec{e}_t \exp(\mp\gamma z)|)$ and we can simply ignore the problems with good results. Put another way, as the wire conductivity goes from perfect to some large but finite value, the field behavior cannot change abruptly or drastically.

We have "explained away" the problem, but we must still account for the resistance of the wires and the consequent power losses. We do this by adding the term $R\underline{I}(z)$ V/m to the right side of Eq. (9.29). This gives the decrease in voltage per unit length due to the resistance of the conductors. That transmission line equation becomes

$$\frac{d\underline{V}(z)}{dz} = -\,(R + j\omega L)\underline{I}(z). \tag{9.31}$$

The current is on a thin layer on the surface of the wire and the resistance per unit length (Ω/m) is found as in Eq. (7.19), repeated here:

$$R = \left(\frac{1}{p_a} + \frac{1}{p_b}\right) R_s, \quad R_s = \sqrt{\frac{\omega\mu_c}{2\sigma_c}}, \tag{9.32}$$

† With the possible introduction of room-temperature superconductors, now under intensive investigation, zero-resistance wire may be incorporated in practical devices.

where p_a and p_b are the length of the perimeters of conductors C_a and C_b and μ_c and σ_c are the permeability and conductivity of the wires. For all practical conductors the permeability is equal to the free-space permeability, $\mu_c = \mu_0$.

Table 9.2 includes formulas for resistance per unit length.

9.6.3 Complete Transmission Line Equations

The complete transmission line equations for the phasor voltage and current are therefore

$$\frac{d\underline{V}(z)}{dz} = -(R + j\omega L)\underline{I}(z), \tag{9.33}$$

$$\frac{d\underline{I}(z)}{dz} = -(G + j\omega C)\underline{V}(z). \tag{9.34}$$

We can obtain the equations for the physical voltage and current by remembering that multiplication by $j\omega$ implies differentiation with respect to time. Hence

$$\frac{\partial V(z,t)}{\partial z} = -\left[RI(z,t) + L\frac{\partial I(z,t)}{\partial t}\right], \tag{9.35}$$

$$\frac{\partial I(z,t)}{\partial z} = -\left[GV(z,t) + C\frac{\partial V(z,t)}{\partial t}\right]. \tag{9.36}$$

These equations are valid for all excitations, not only for time harmonic excitations.

We will again defer solving these equations.

9.6.4 Derivation from Circuit Theory

The transmission line equations can also be derived from the circuit theorist's viewpoint. In this case we start with many small lumped elements along the transmission line, write the Kirchhoff Voltage Law (KVL) and Kirchhoff Current Law (KCL) equations for a length Δz, and then take the limit as $\Delta z \to 0$. The resultant circuit is known as a distributed circuit, because the circuit elements are spaced along the z direction and not lumped in a region of space that is small compared to the wavelength.

Figure 9.13 shows a transmission line with the R, L, G, and C per unit length parameters in a typical length between z and $z + \Delta z$. The KVL equation around mesh **M** and KCL equation at node **N** are, respectively,

$$\underline{V}(z) = (R\,\Delta z + j\omega L\,\Delta z)\underline{I}(z) + \underline{V}(z + \Delta z),$$

$$\underline{I}(z) = (G\,\Delta z + j\omega C\,\Delta z)\underline{V}(z + \Delta z) + \underline{I}(z + \Delta z).$$

Transposing terms, dividing by Δz, and taking the limit as $\Delta z \to 0$, we obtain

$$\frac{d\underline{V}(z)}{dz} = -(R + j\omega L)\underline{I}(z)$$

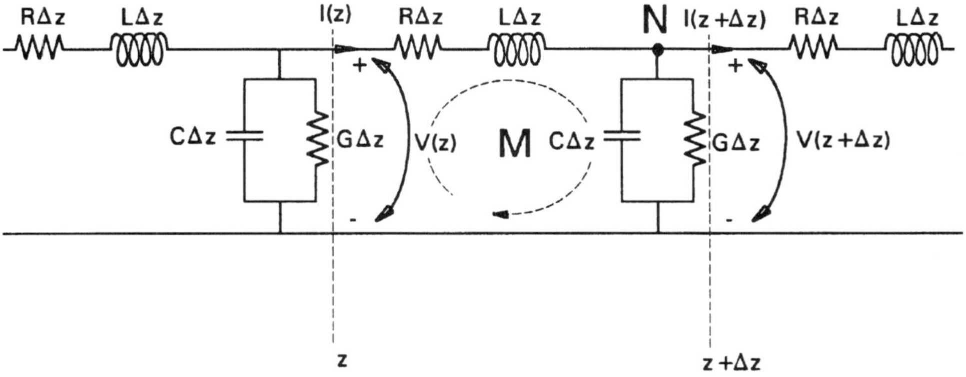

FIGURE 9.13 Lumped element equivalent circuit model for a transmission line.

and

$$\frac{d\underline{I}(z)}{dz} = -(G + j\omega C)\underline{V}(z),$$

which are the same as Eqs. (9.33) and (9.34).

This derivation of the transmission line equations is clearly simpler than the one based on Maxwell's equations. That is because it does not relate the voltage and current to the electric and magnetic field quantities.

9.7 VOLTAGE AND CURRENT WAVE EQUATIONS; PROPAGATION CONSTANT; CHARACTERISTIC IMPEDANCE

We have now derived the transmission line equations (twice). The next step is to find their solutions for time harmonic fields.

The wave equation for voltage is obtained by differentiating both sides of Eq. (9.33) with respect to z and substituting on the right from Eq. (9.34). Thus

$$\frac{d^2\underline{V}(z)}{dz^2} = -(R + j\omega L)\frac{d\underline{I}(z)}{dz} = (R + j\omega L)(G + j\omega C)\underline{V}(z)$$

$$\frac{d^2\underline{V}(z)}{dz^2} - \gamma^2\underline{V}(z) = 0, \tag{9.37}$$

where

$$\gamma = \sqrt{(R + j\omega L)(G + j\omega C)} \tag{9.38}$$

is the propagation constant. The wave equation for the current can be obtained in the same way and is of identical form:

$$\frac{d^2\underline{I}(z)}{dz^2} - \gamma^2\underline{I}(z) = 0. \tag{9.39}$$

The solutions to these two wave equations are

$$\underline{V}(z) = V_0^+ e^{-\gamma z} + V_0^- e^{+\gamma z}, \tag{9.40}$$

$$\underline{I}(z) = I_0^+ e^{-\gamma z} + I_0^- e^{+\gamma z}. \tag{9.41}$$

The propagation constant plays the same role in the voltage and current waves as it did in the electric and magnetic field waves. Indeed, except for the resistance term R, which we have now included, the values of γ given in Eqs. (9.38) and (9.9) must be identical. As before, the propagation constant is, in general, complex, $\gamma = \alpha + j\beta$, with α being the attenuation constant and β the phase constant.

The complex numbers V_0^+ and I_0^+ and V_0^- and I_0^- are not independent of each other. If we substitute Eq. (9.40) into Eq. (9.33) [or Eq. (9.41) into Eq. (9.34)] we obtain

$$\frac{V_0^+}{I_0^+} = -\frac{V_0^-}{I_0^-} = Z_0, \tag{9.42}$$

where

$$Z_0 = R_0 + jX_0 = \frac{R + j\omega L}{\gamma} = \frac{\gamma}{G + j\omega C} = \sqrt{\frac{R + j\omega L}{G + j\omega C}}. \tag{9.43}$$

The quantity Z_0 is called the characteristic impedance (Ω) of the transmission line. It is a function of the material constants, μ, ε, σ, and σ_c, and the cross-sectional geometry of the line. R_0 and X_0 are the resistive and reactive parts of Z_0.

Equation (9.42) shows that the characteristic impedance is the ratio of the forward traveling voltage wave to the forward traveling current wave and the negative of the ratio of the backward traveling voltage wave to the backward traveling current wave. We will find another meaning for the characteristic impedance later on.

Example 9.2 Design a 50 Ω coaxial cable with an outside diameter of $2b = 0.750$ in. The conductors are copper ($\sigma_c = 5.8 \times 10^7$ S/m). The dielectric material is polystyrene, characterized by $\varepsilon_r = 2.45$, $\mu_r = 1$, and an equivalent conductivity (at 1×10^9 Hz) of $\sigma = 4 \times 10^{-5}$ S/m. Calculate the transmission line parameters at 1 GHz.

Solution The design of the coax requires that we find a proper inner diameter to go with the given outer diameter and the requirement that the characteristic impedance be 50 Ω. A useful line will have low loss, so we begin by considering the line to be lossless ($R = 0$, $G = 0$). Then

$$Z_0 = \sqrt{\frac{L}{C}} = \frac{1}{2\pi}\sqrt{\frac{\mu}{\varepsilon}}\ln\left(\frac{b}{a}\right).$$

The values $\mu_0 = 4\pi \times 10^{-7}$ H/m and $\varepsilon_0 = (1/36\pi) \times 10^{-9}$ F/m give $\sqrt{\mu_0/\varepsilon_0} = 120\pi = 377\ \Omega$ and therefore

$$50 = \frac{120\pi}{2\pi} \frac{1}{\sqrt{2.45}} \ln\left(\frac{0.750}{2a}\right)$$

$$\ln\left(\frac{0.750}{2a}\right) = \left(\frac{50}{60}\right)\sqrt{2.45} = 1.30, \quad \frac{0.750}{2a} = \exp(1.30) = 3.69$$

$$2a = 0.204 \text{ in.}$$

We can now calculate the line parameters.

$$C = \frac{2\pi\varepsilon}{\ln(b/a)} = (2\pi)\left(\frac{1}{36\pi} \times 10^{-9}\right)(2.45)\left(\frac{1}{1.30}\right)$$
$$= 0.104 \times 10^{-9} \text{ F/m} = 0.104 \text{ nF/m}.$$

$$L = \frac{\mu}{2\pi} \ln\left(\frac{b}{a}\right) = \left(\frac{1}{2\pi}\right)(4\pi \times 10^{-7})(1.30)$$
$$= 0.260 \times 10^{-6} \text{ H/m} = 0.260\ \mu\text{H/m}.$$

(As a check, $\sqrt{L/C} = \sqrt{(0.261 \times 10^{-6})/(0.104 \times 10^{-9})} = 50.$)

$$G = \frac{2\pi\sigma}{\ln(b/a)} = 2\pi(4 \times 10^{-5})\left(\frac{1}{1.30}\right)$$
$$= 0.193 \times 10^{-3} \text{ S/m} = 0.193 \text{ mS/m}.$$

$$R = \frac{1}{2\pi}\left(\frac{1}{a} + \frac{1}{b}\right)R_s, \quad R_s = \sqrt{\frac{\omega\mu}{2\sigma_c}}$$

$$R_s = \sqrt{\frac{(2\pi)(10^9)(4\pi \times 10^{-7})}{2 \times 5.8 \times 10^7}} = 8.25 \times 10^{-3}\ \Omega$$

$$R = \frac{1}{\pi}\left(\frac{1}{0.204} + \frac{1}{0.75}\right)(39.4)(8.25 \times 10^{-3}) = 0.645\ \Omega/\text{m}.$$

The factor of 39.4 converts from per inch to per meter.

9.7.1 Circuit Representation of the Transmission Line

Physical resistors, inductors, and capacitors are represented by symbols in circuit diagrams. These symbols are unrelated to the physical construction of the particular device. Similarly, the physical transmission line is represented by a symbol as shown in Fig. 9.14. Although this symbol looks like a pair of parallel wires it can be used to represent any transmission line. The parameters of the line are the resistance R, inductance L, conductance G, and capacitance C per unit length and the length of the line d. Alternatively, for time harmonic problems, we can specify the propagation

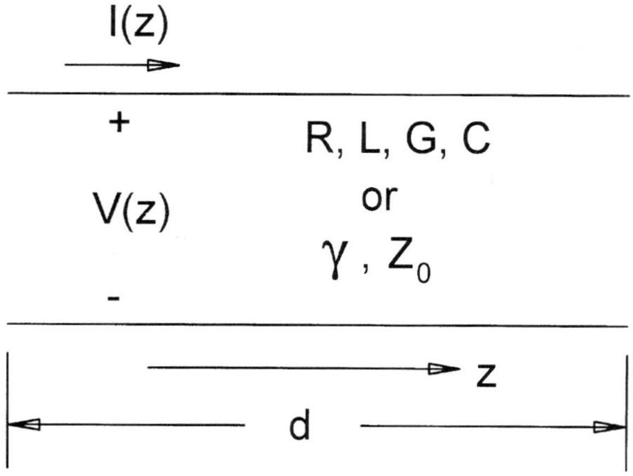

FIGURE 9.14 The transmission line circuit representation and its defining parameters.

constant $\gamma = \alpha + j\beta$ and the characteristic impedance $Z_0 = R_0 + jX_0$ (perhaps as functions of angular frequency ω) and the length of the line d.

9.7.2 Distortion, Dispersion, and Distortionless Lines

When the power company transmits power to our homes, schools, or offices it transmits at a single frequency—60 Hz in the United States. A single-frequency signal carries no information. When we transmit information, whether it is voice communications, a TV channel, or digital data, we must transmit a band of frequencies. We would like the signal to arrive at the receiver exactly as it was sent at the transmitter—in other words, without distortion. Our desire to transmit a distortion-free signal places strict requirements on the channel, transmission medium, or line.

Distortion in transmission lines comes from two sources. First, if different frequency components within the signal are attenuated differently, then the received signal will be distorted. Second, if different frequency components travel through the line with different velocities, then the received signal will also be distorted. A simple example will illustrate these types of distortion.

Example 9.3 A transmitted signal consists of three frequency terms:

$$f_t(t) = 15\sin(600\pi t) + 5\sin(1800\pi t) + 3\sin(3000\pi t).$$

The signal is transmitted over a 100 km long line. Calculate the received signal if: (a) the attenuation constant is 0.060 Np/km and the phase velocity is 8.0×10^4 km/s for all frequencies; (b) the attenuation constant is 0.060, 0.065, and 0.070 Np/km at 300, 900, and 1500 Hz, respectively, and the phase velocity is constant at 8.0×10^4 km/s; (c) the attenuation constant is fixed at 0.060 Np/km and the phase velocity is 3.6×10^4, 8.0×10^4, and 12.9×10^4 km/s at 300, 900, and 1500 Hz, respectively; and (d) the attenuation constants are as in (b) and the phase velocities are as in (c).

Solution (a) Each term will be attenuated by $\exp(-6.0) = 2.48 \times 10^{-3}$ and delayed by $d/v_{\text{ph}} = 1.25 \times 10^{-3}$ s. The received signal is

$$f_{ra}(t) = 2.48 \times 10^{-3}\{15\sin[600\pi(t - 0.00125)] + 5\sin[1800\pi(t - 0.00125)] + 3\sin[3000\pi(t - 0.00125)]\}.$$

(b) The three terms in the signal will be attenuated by $\exp(-6.0) = 2.48 \times 10^{-3}$, $\exp(-6.5) = 1.50 \times 10^{-3}$, and $\exp(-7.0) = 0.91 \times 10^{-3}$, respectively, and they will be delayed by 1.25×10^{-3} s. The result is

$$f_{rb}(t) = 2.48 \times 10^{-3}\{15\sin[600\pi(t - 0.00125)] + 3.03\sin[1800\pi(t - 0.00125)] + 1.10\sin[3000\pi(t - 0.00125)]\}.$$

(c) Each term will be attenuated by $\exp(-6.0) = 2.48 \times 10^{-3}$ and delayed by $d/v_{\text{ph}} = 2.78 \times 10^{-3}$, 1.25×10^{-3}, and 0.78×10^{-3}, respectively. The received signal is

$$f_{rc}(t) = 2.48 \times 10^{-3}\{15\sin[600\pi(t - 0.00278)] + 5\sin[1800\pi(t - 0.00125)] + 3\sin[3000\pi(t - 0.00078)]\}.$$

(d) The received signal is

$$f_{rd}(t) = 2.48 \times 10^{-3}\{15\sin[600\pi(t - 0.00278)] + 3.03\sin[1800\pi(t - 0.00125)] + 1.10\sin[3000\pi(t - 0.00078)]\}.$$

The signals are shown in Fig. 9.15.

We see that in (a) the wave is attenuated and delayed, but transmitted and received signals have the same waveform. Thus when the attenuation constant and the phase (signal) velocity are independent of frequency, transmission is distortionless. In (b), the second two terms are attenuated more severely than the first and the received signal is a different waveform than the transmitted signal, meaning it is distorted. In (c), the relative amplitudes of the terms are the same in the transmitted and received signals, but the received terms suffer different delays. This leads to distortion as well. When the channel causes the signal velocity to vary with frequency it is said to be a dispersive channel. Dispersive channels distort the transmitted signal. In (d), both effects lead to distortion.

For a distortionless line we must have

$$\alpha = \text{constant}, \quad \beta = \frac{\omega}{v_{\text{ph}}} = (\text{constant})f; \tag{9.44}$$

that is, the attenuation constant must be independent of frequency and the phase constant must be proportional to frequency over the bandwidth of the signal.

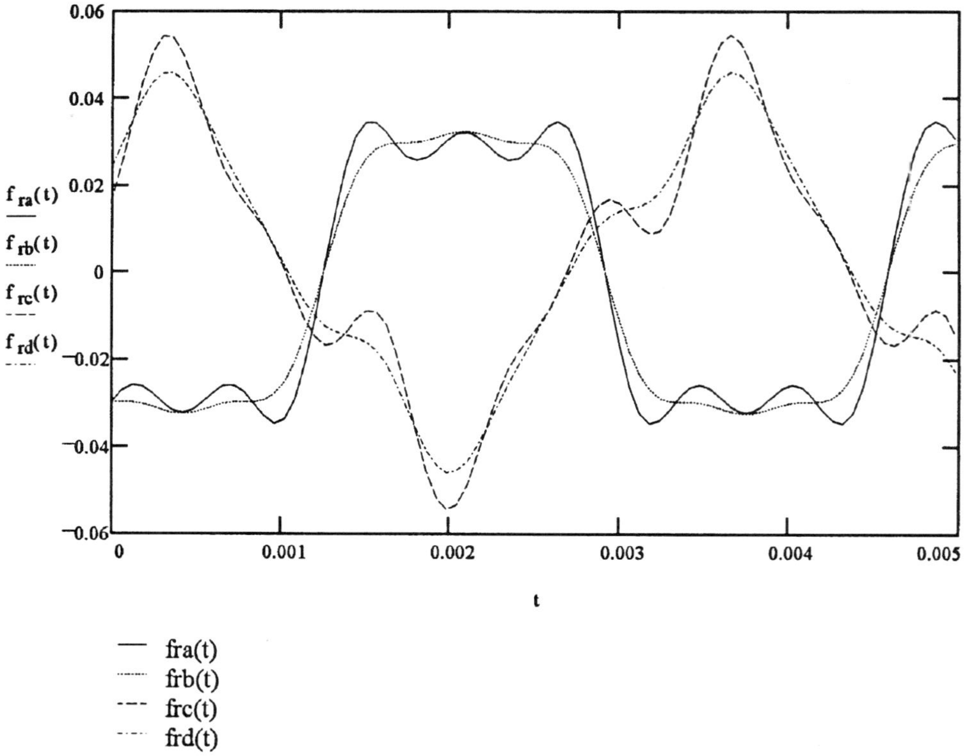

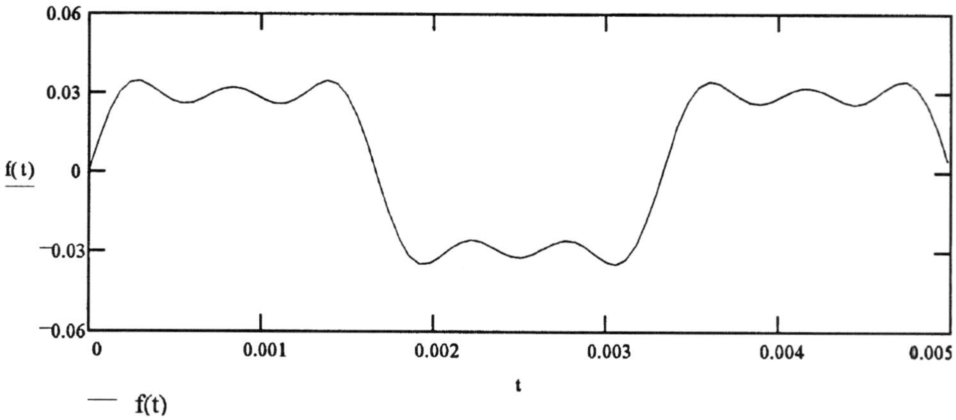

FIGURE 9.15 A plot of transmitted and received signals on a lossy and dispersive transmission line. The transmitted signal consists of three sinusoidal terms.

If we could be lucky enough to find lossless materials, we could build lossless lines, $R = 0$, $G = 0$. In that case,

$$\gamma = j\omega\sqrt{LC}, \quad \alpha = 0, \quad \beta = \omega\sqrt{LC},$$

$$Z_0 = \sqrt{\frac{L}{C}}, \quad v_{ph} = \frac{1}{\sqrt{LC}}. \tag{9.45}$$

The lossless line has no dispersion and no distortion.

It is possible to have a distortionless line even when the line has losses. By factoring the propagation constant

$$\gamma = \sqrt{RG\left(1 + \frac{j\omega L}{R}\right)\left(1 + \frac{j\omega C}{G}\right)},$$

we can see that if

$$\frac{L}{R} = \frac{C}{G} \tag{9.46}$$

then the two terms in the parentheses are identical, making a perfect square. This gives

$$\gamma = \sqrt{RG\left(1 + \frac{j\omega L}{R}\right)^2} = \sqrt{RG}\left(1 + \frac{j\omega L}{R}\right),$$

$$\alpha = \sqrt{RG}, \quad \beta = \omega L\sqrt{\frac{G}{R}} = \omega L\sqrt{\frac{C}{L}} = \omega\sqrt{LC}. \tag{9.47}$$

Thus, when the condition of Eq. (9.46) holds, the line has an attenuation constant that is independent of frequency and a phase constant that is proportional to frequency, yielding a distortionless line.

The characteristic impedance for the distortionless line is

$$Z_0 = \sqrt{\left(\frac{R}{G}\right)\frac{(1 + j\omega L/R)}{(1 + j\omega C/G)}} = \sqrt{\frac{R}{G}} = \sqrt{\frac{L}{C}},$$

just as it is for the lossless line.

In most practical situations, $G/C \ll R/L$ and the line is not distortionless. In many cases, it is possible to increase the value of inductance per unit length to make the desired equality hold, or almost hold. This can be done by substituting a higher permeability material along the line or by adding lumped inductances at regular intervals, which are small compared to the wavelength. These lumped inductors are called loading coils and may be treated as additional inductance per unit length if placed closely enough together.

Example 9.4 A voice-frequency circuit on a long-distance phone cable has parameters $R = 86.0\ \Omega$/mile, $L = 1$ mH/mile, $G = 1.4\ \mu$S/mile, and $C = 0.062\ \mu$F/mile, which may be considered to be constant with frequency. The voice-frequency band has low $f_l = 200$ Hz, middle $f_m = 1000$ Hz, and high $f_h = 3300$ Hz frequencies. (a) Calculate the attenuation constant and phase velocity at each frequency. (b) Find the value of a loading coil to be inserted every 0.1 mile to produce a distortionless line. Check that the loading coil spacing is a fraction of the shortest wavelength. (c) Assume a loading coil of 75 mH is added every $\frac{1}{2}$ mile and that each loading coil has a 6 Ω resistance. Repeat part (a).

Solution (a) At $f_l = 200$,

$$\begin{aligned}\gamma &= \sqrt{[86 + j2\pi(200)(0.001)][1.4 + j2\pi(200)(0.062)] \times 10^{-6}} \\ &= \sqrt{(86.0\angle 0.837^\circ)(77.9\angle 88.97^\circ) \times 10^{-6}} = 0.0819\angle 44.9^\circ \\ &= 0.0580 + j0.0578, \\ \alpha &= 0.0580 \text{ Np/mile}, \quad v_{\text{ph}} = \omega/\beta = 2.17 \times 10^4 \text{ miles/s}.\end{aligned}$$

Similarly, at the middle frequency of 1000 Hz, $\alpha = 0.124$ Np/mile and $v_{\text{ph}} = 4.69 \times 10^4$ miles/s, and at the high frequency of 3300 MHz, $\alpha = 0.209$ Np/mile and $v_{\text{ph}} = 7.83 \times 10^4$ miles/s. From low to high frequency the attenuation constant and the phase velocity have increased by a factor of 3.6.

(b) Using Eq. (9.46), the total inductance per unit length required to give a distortionless line is

$$L_T = \frac{RC}{G} = 3.8 \text{ H/mile}.$$

The line inductance is very small compared to this, so the loading coils with 0.1 mile spacing should have an inductance of 0.38 H. With this inductance the propagation constant is

$$\gamma = \sqrt{RG} + j\omega\sqrt{L_T C} = 0.011 + j0.485 \times 10^{-3}\,\omega,$$

so the attenuation constant is 0.011 Np/mile and the phase velocity is 2060 miles/s for all frequencies. The shortest wavelength occurs at the highest frequency and is $\lambda = v_{\text{ph}}/f = 0.62$ mile. The between-coil spacing of 0.1 mile is sufficiently smaller than the wavelength.

(c) With these values, the inductance per unit length is 0.151 H/mile and the resistance per unit length is 98 Ω/mile. Then, at 200 Hz,

$$\begin{aligned}\gamma &= \sqrt{[98 + j2\pi(200)(0.151)][1.4 + j2\pi(200)(0.062)] \times 10^{-6}} \\ &= \sqrt{(214.\angle 62.7^\circ)(77.9\angle 88.97^\circ) \times 10^{-6}} = 0.129\angle 75.8^\circ \\ &= 0.0316 + j0.125, \\ \alpha &= 0.0316 \text{ Np/mile}, \quad v_{\text{ph}} = \omega/\beta = 1.01 \times 10^4 \text{ miles/s}.\end{aligned}$$

Similarly, at the middle frequency of 1000 Hz, $\alpha = 0.0317$ Np/mile and $v_{\text{ph}} = 1.04 \times 10^4$ miles/s and at the high frequency of 3300 MHz, $\alpha = 0.0317$ Np/mile and

$v_{ph} = 1.04 \times 10^4$ miles/s. These values are sufficiently constant to give very little distortion. The shortest wavelength is $\lambda = v_{ph}/f = 3.15$ miles, so the $\frac{1}{2}$ mile spacing is sufficiently small. The larger spacing between loading coils makes this solution more practical and makes the attenuation constant and phase velocity sufficiently constant to give very little distortion.

9.7.3 Low-Loss Formulas for γ and Z_0

The full formula for the propagation constant γ, Eq. (9.38), and the characteristic impedance Z_0, Eq. (9.43), can always be used to calculate these quantities. In most practical situations, however, we deal with transmission lines that are designed to produce low losses by keeping the conductance and resistance terms as small as possible. In these cases it is possible to simplify the calculation and eliminate the need to multiply, divide, and find the square roots of complex numbers.

Whenever we say something is "small," we must compare it to another quantity expressed in the same units and say "a is small compared to b" or $a \ll b$. In the equations for γ and Z_0, R is added to $j\omega L$ (units of Ω/m for both) and G is added to $j\omega C$ (units of S/m for both). Thus if $R \ll \omega L$ and $G \ll \omega C$, then

$$\begin{aligned} Z_0 &= \sqrt{\frac{j\omega L}{j\omega C}\left(\frac{1 + R/j\omega L}{1 + G/j\omega C}\right)} = \sqrt{\frac{L}{C}}\left(1 + \frac{R}{j\omega L}\right)^{1/2}\left(1 + \frac{G}{j\omega C}\right)^{-1/2} \\ &\approx \sqrt{\frac{L}{C}}\left(1 + \frac{R}{2j\omega L}\right)\left(1 - \frac{G}{2j\omega c}\right) \approx \sqrt{\frac{L}{C}}\left[1 + \frac{1}{2j\omega}\left(\frac{R}{L} - \frac{G}{C}\right)\right] \\ &= \sqrt{\frac{L}{C}} - j\frac{1}{2\omega}\sqrt{\frac{L}{C}}\left(\frac{R}{L} - \frac{G}{C}\right) \end{aligned}$$

$$Z_0 = R_0 + jX_0, \quad R_0 = \sqrt{\frac{L}{C}}, \quad X_0 = -\frac{1}{2\omega}\sqrt{\frac{L}{C}}\left(\frac{R}{L} - \frac{G}{C}\right) \tag{9.48}$$

and

$$\begin{aligned} \gamma &= \sqrt{(j\omega L)(j\omega C)\left(1 + \frac{R}{j\omega L}\right)\left(1 + \frac{G}{j\omega C}\right)} \\ &\approx j\omega\sqrt{LC}\left(1 + \frac{R}{2j\omega L}\right)\left(1 + \frac{G}{2j\omega C}\right) \\ &\approx j\omega\sqrt{LC}\left[1 + \frac{1}{2j\omega}\left(\frac{R}{L} + \frac{G}{C}\right)\right] \\ &= \frac{1}{2}\left(R\sqrt{\frac{C}{L}} + G\sqrt{\frac{L}{C}}\right) + j\omega\sqrt{LC} \end{aligned}$$

$$\alpha = \frac{1}{2}\left(R\sqrt{\frac{C}{L}} + G\sqrt{\frac{L}{C}}\right) = \frac{1}{2}\left(\frac{R}{R_0} + GR_0\right), \quad \beta = \omega\sqrt{LC}. \tag{9.49}$$

Generally, the characteristic resistance R_0 is much greater than the characteristic reactance X_0 and $Z_0 \approx R_0 = \sqrt{L/C}$. The terms characteristic impedance and characteristic resistance are then used interchangeably.

These formulas are easier to use than the general formulas.

Example 9.5 For the coaxial cable designed in Example 9.1, show that the low-loss condition holds and calculate the characteristic impedance and the propagation constant at 1 GHz.

Solution The low-loss formulas require

$$R = 8.25 \times 10^{-3} \ll \omega L = 2\pi(10^9)(0.260 \times 10^{-6}) = 1.63 \times 10^3$$

and

$$G = 0.193 \times 10^{-3} \ll \omega C = 2\pi(10^9)(0.104 \times 10^{-9}) = 0.653$$

both of which are clearly satisfied. We can now calculate

$$R_0 = \sqrt{\frac{L}{C}} = 50\ \Omega,$$

$$X_0 = -\frac{1}{2\omega}\sqrt{\frac{L}{C}}\left(\frac{R}{L} - \frac{G}{C}\right) = -\frac{50}{4\pi(10^9)}\left(\frac{0.645}{0.260 \times 10^{-6}} - \frac{0.193 \times 10^{-3}}{0.104 \times 10^{-9}}\right)$$

$$= -2.49 \times 10^{-3}\ \Omega,$$

$$\alpha = \frac{1}{2}\left(\frac{R}{R_0} + GR_0\right) = \frac{1}{2}\left[\frac{0.645}{50} + (0.193 \times 10^{-3})(50)\right]$$

$$= 0.0113\ \text{Np/m},$$

$$\beta = \omega\sqrt{LC} = 2\pi(10^9)\sqrt{(0.260 \times 10^{-6})(0.104 \times 10^{-9})}$$

$$= 32.7\ \text{rad/m}.$$

Clearly, $|X_0| \ll R_0$ and we can take $Z_0 = R_0 = 50$. The phase velocity is $v_{\text{ph}} = 2\pi f/\beta = 1.92 \times 10^8$ m/s and the wavelength is $\lambda = v_{\text{ph}}/f = 0.192$ m.

9.8 PRINTED CIRCUIT TRANSMISSION LINES

The need for batch fabricated, low-cost, small size, and highly integrated microwave assemblies has driven the continuous development of alternate forms of microwave transmission lines. Some of the more important ones, in approximate chronological order, are illustrated in Fig. 9.16. These include rectangular waveguide, circular waveguide, coaxial transmission lines, triplate, microstrip, and coplanar waveguide. Of these, the microstrip and the coplanar waveguide are most important in applications where active and passive devices must be integrated together. Printed circuits in

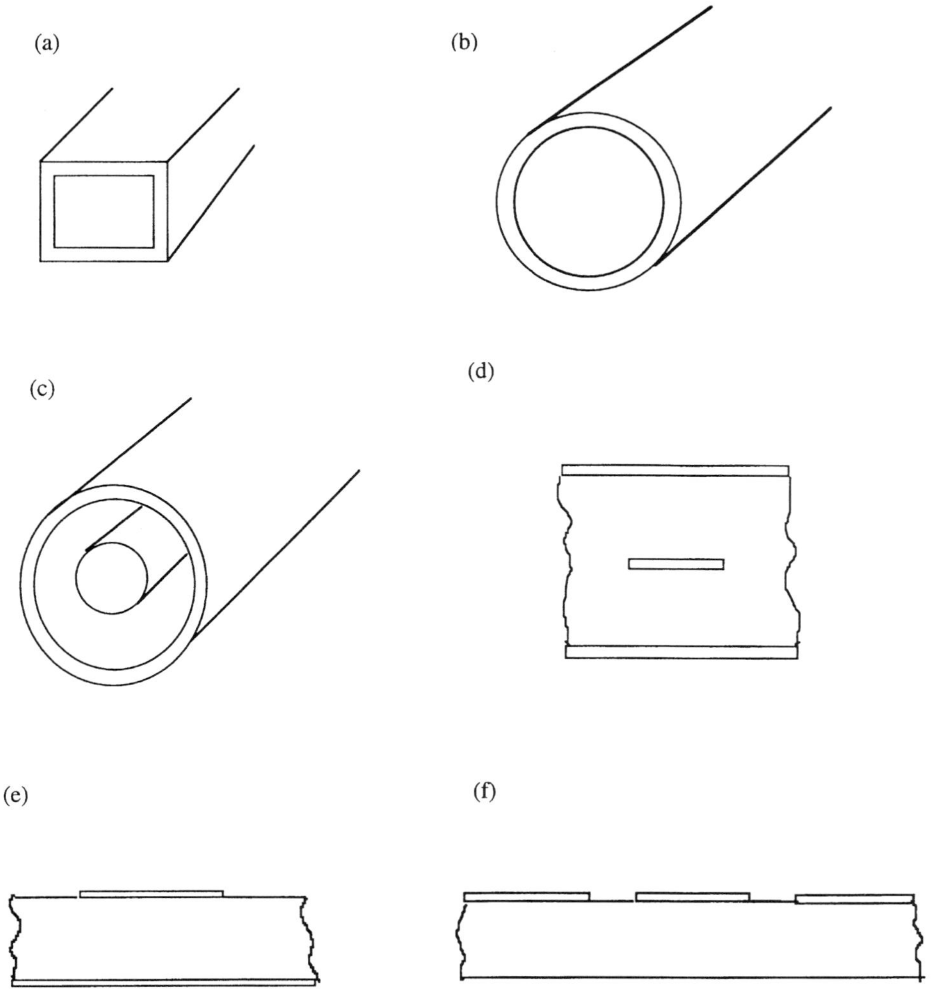

FIGURE 9.16 Microwave transmission lines: (a) rectangular waveguide, (b) circular waveguide, (c) coaxial line, (d) triplate, (e) microstrip, and (f) coplanar waveguide.

general are important because they may be batch fabricated using semiconductor photolithographic techniques. In this introductory section the evolution of these transmission lines and their qualitative properties will be described. In other sections more detailed characteristics will be discussed.

9.8.1 Conventional Waveguides (Rectangular, Circular, and Coaxial)

Rectangular and circular waveguides propagate waves in TE and TM modes, and all the others guide waves in either a TEM or quasi-TEM mode. Both rectangular and circular waveguides are very low loss, can handle high power levels, and, as a result, are heavily used in applications such as antenna feeds and feed-networks (e.g., power combiners/dividers, couplers, low-loss ferrite phase shifters). Coaxial transmission lines have higher loss but are used to interconnect passive and active components that have coaxial connectors.

9.8.2 Strip Transmission Lines

Conventional waveguides are basically formed using machine shop processes. They are either precision machined or extruded, which is a very costly, time-consuming process. Strip transmission lines, in contrast, are made using semiconductor-like processes. The starting material is a low-loss dielectric substrate such as alumina (Al_2O_3), quartz (SiO_2), or sapphire. The substrate may also be woven glasses, which are soft and easily cut or drilled. They also may be semiconducting or semi-insulating substrates made from high-resistivity silicon or semi-insulating GaAs.

The substrates are lapped to the correct thickness, polished, and coated with metals. The transmission lines are formed by coating the substrate with photoresist (a spin-on polymer material that is sensitive to certain wavelengths of light), exposing sections of photoresist to light, removing either the exposed or unexposed resist, and then etching away the unwanted metal, leaving the desired transmission line pattern. In this photolithographic process, many transmission lines or circuit patterns may be processed simultaneously and accurately.

Triplate was one of the first printed circuit transmission lines to gain prominence and acceptance by the microwave community. It is formed by etching the desired strip width on one substrate (using the photolithographic process described above) and joining it with another substrate with metal on one side. The properties of this transmission line are very similar to coaxial lines in that it propagates a pure TEM mode, is low loss, and can carry dc current, but it also has the inherent advantage that it is made with photolithographic processes. A disadvantage of triplate is that active or passive components cannot easily be connected in series or in parallel without drilling holes or creating cavities in the dielectric media.

9.9 MICROSTRIP LINE PARAMETERS

Microstrip line (Fig. 9.17a) has become the transmission line of choice for many applications. This is because of the ease of fabrication using photoetching techniques and because lumped elements such as resistors, capacitors, diodes, and transistors can simply be mounted on the exposed upper surface. It is also possible to fabricate fully integrated circuits on semiconductor substrates. To prevent losses due to radiation, the circuits are frequently enclosed in a conducting box Fig. (9.17b).

The dielectric board that holds the conducting ground plane and strip is most frequently made of alumina, quartz, or Duroid, having relative permittivities ε_r of about 10, 3.8, and 2.3, respectively. For full integration of semiconductor devices, intrinsic silicon ($\varepsilon_r = 11.8$) is frequently used. The phase velocity for a TEM wave in the free-space region above the board is $v_{\text{ph}} = c = 1/\sqrt{\mu_0 \varepsilon_0}$ and the phase velocity for the dielectric region is $v_{\text{ph}} = c/\sqrt{\varepsilon_r}$. The wave cannot travel at both velocities and hence there is no truly TEM wave solution. Because the strip conductor is close to the ground plane and because the permittivity of the dielectric is greater than that of free space, most of the energy is concentrated in the dielectric region between the line and

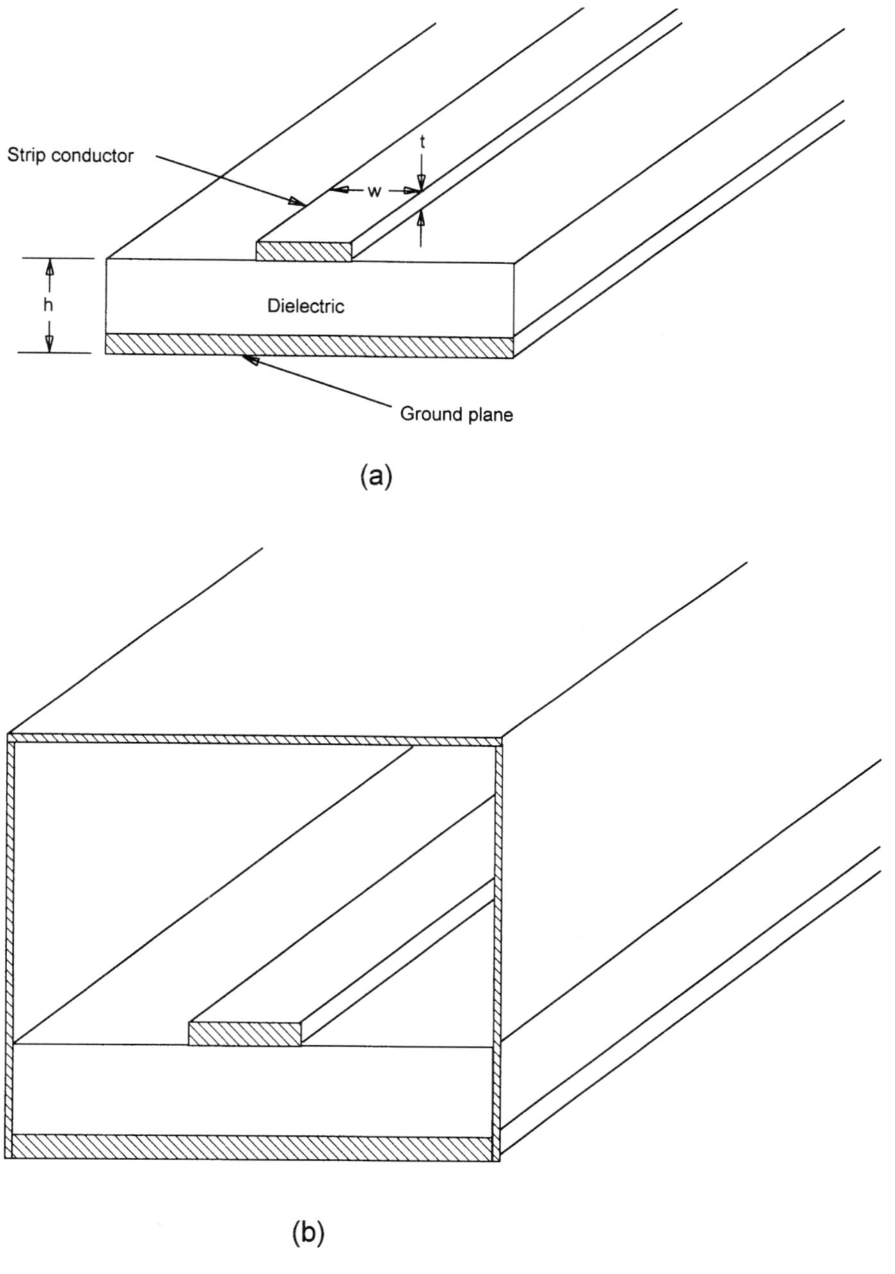

FIGURE 9.17 Unshielded and shielded microstrip transmission lines.

the ground plane. The wave approximates that which would exist if all the space were dielectric and the analysis therefore assumes the wave is "quasi-TEM." We may expect an actual phase velocity somewhere between the free-space and the dielectric value, but closer to the dielectric value.

Various formulas have been developed using different mathematical and numerical techniques for calculating the important transmission line quantities Z_0, λ or v_{ph}, and α, and there is an extensive literature. We will get some simple results and then give suitable formulas for analysis and design.

9.9.1 Simple Approximations to Characteristic Impedance

Assume that $h \ll w$, so that the strip is close to the ground plane in comparison with its width. The strip and ground plane then approximate a parallel plate transmission line, with parameters

$$C = \varepsilon \frac{w}{h}, \quad L = \mu_0 \frac{h}{w}, \quad Z_0 = \sqrt{\frac{L}{C}} = \frac{\eta_0}{\sqrt{\varepsilon_r}}\left(\frac{h}{w}\right), \quad \frac{h}{w} \ll 1.$$

The field between the strip and the ground plane actually spreads out (fringes) and we can account for this by writing

$$Z_0 = \frac{\eta_0}{\sqrt{\varepsilon_r}}\left(\frac{h}{w + 2h}\right), \quad \frac{h}{w} \ll 1. \tag{9.50}$$

With this improvement the approximation is useful for $w/h > 1$, but the approximation deteriorates as the relative permittivity deviates from unity, except if $w/h \gg 1$.

9.9.2 Design Formulas†

As noted above, the quasi-TEM wave will propagate as if the relative permittivity were somewhere between unity for free space and the value for the dielectric region, ε_r. An approximate value for the effective relative permittivity is

$$\varepsilon_{\text{eff}} = \frac{(\varepsilon_r + 1)}{2} + \frac{(\varepsilon_r - 1)}{2\sqrt{1 + 12h/w}}.$$

The phase velocity is $v_{\text{ph}} = c/\sqrt{\varepsilon_{\text{eff}}}$ and the phase constant is $\beta = \omega\sqrt{\varepsilon_{\text{eff}}}/c$, $c = 3 \times 10^8$ m/s.

For analysis purposes we must be able to calculate the characteristic impedance from the dimensions of the line. Suitable approximate formulas are

$$Z_0 = \begin{cases} \dfrac{\eta_0}{2\pi\sqrt{\varepsilon_{\text{eff}}}} \ln\left(\dfrac{8h}{w} + \dfrac{w}{4h}\right), & w/h \le 1, \\[2ex] \dfrac{\eta_0}{\sqrt{\varepsilon_{\text{eff}}}\,[w/h + 1.393 + 0.667\ln(w/h + 1.444)]}, & w/h \ge 1. \end{cases} \tag{9.51}$$

† All these formulas are from Ref. [1].

For design purposes we need to find the value of w/h for a desired characteristic impedance. Suitable approximate formulas are

$$\frac{w}{h} = \begin{cases} \dfrac{8e^A}{e^{2A} - 2}, & \dfrac{w}{h} < 2, \\[2ex] \dfrac{2}{\pi}\left\{B - 1 - \ln(2B - 1) + \dfrac{\varepsilon_r - 1}{2\varepsilon_r}\left[\ln(B - 1) + 0.39 - \dfrac{0.61}{\varepsilon_r}\right]\right\}, & \dfrac{w}{h} > 2, \end{cases} \tag{9.52}$$

where

$$A = \frac{Z_0}{60}\sqrt{\frac{\varepsilon_r + 1}{2}} + \frac{\varepsilon_r - 1}{\varepsilon_r + 1}\left(0.23 + \frac{0.11}{\varepsilon_r}\right),$$

$$B = \frac{\pi\eta_0}{2Z_0\sqrt{\varepsilon_r}}.$$

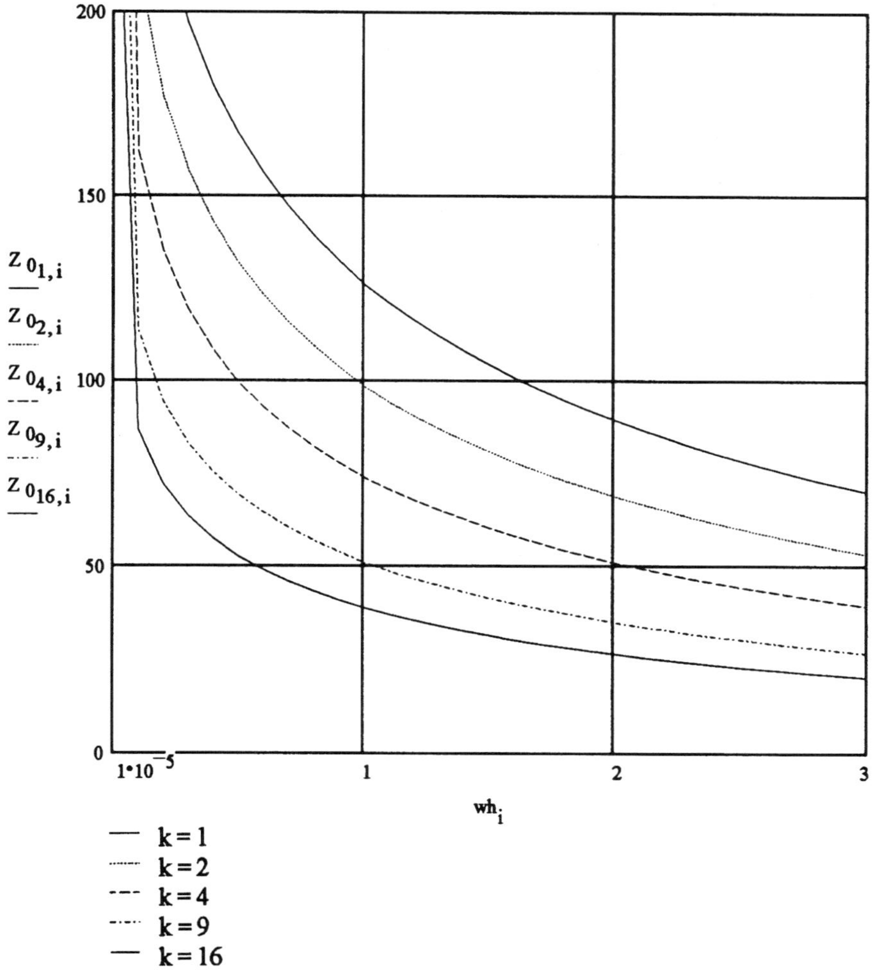

FIGURE 9.18 Characteristic impedance versus width-to-height ratio for a single strip microstrip ($k = \varepsilon_r$).

Figure 9.18 [2] shows characteristic impedance and Fig. 9.19 [2] shows the velocity and wavelength versus the ratio of strip width to dielectric substrate height with relative permittivity as a parameter.

Example 9.6 Design a 50 Ω line on a 0.015 in. thick quartz ($\varepsilon_r = 3.8$) substrate. Find the corresponding phase velocity.

Solution Figure 9.18 shows $w/h \approx 2.1$ and hence $w = 0.032$ in. for the given $h = 0.015$ in. From the Fig. 9.19 $v_{\text{ph}} \approx 1.7 \times 10^8$ m/s.

Figure 9.18 has indicated that $w/h > 2$ so we use the second design equation. Performing the calculations we find $B = 6.09$, $w/h = 2.14$, and $w = 0.032$ in., confirming the value found from the figure. [Using the first design equation gives $A = 1.44$, $w/h = 2.14$, and $w = 0.032$ in. The two equations give the same result because we are

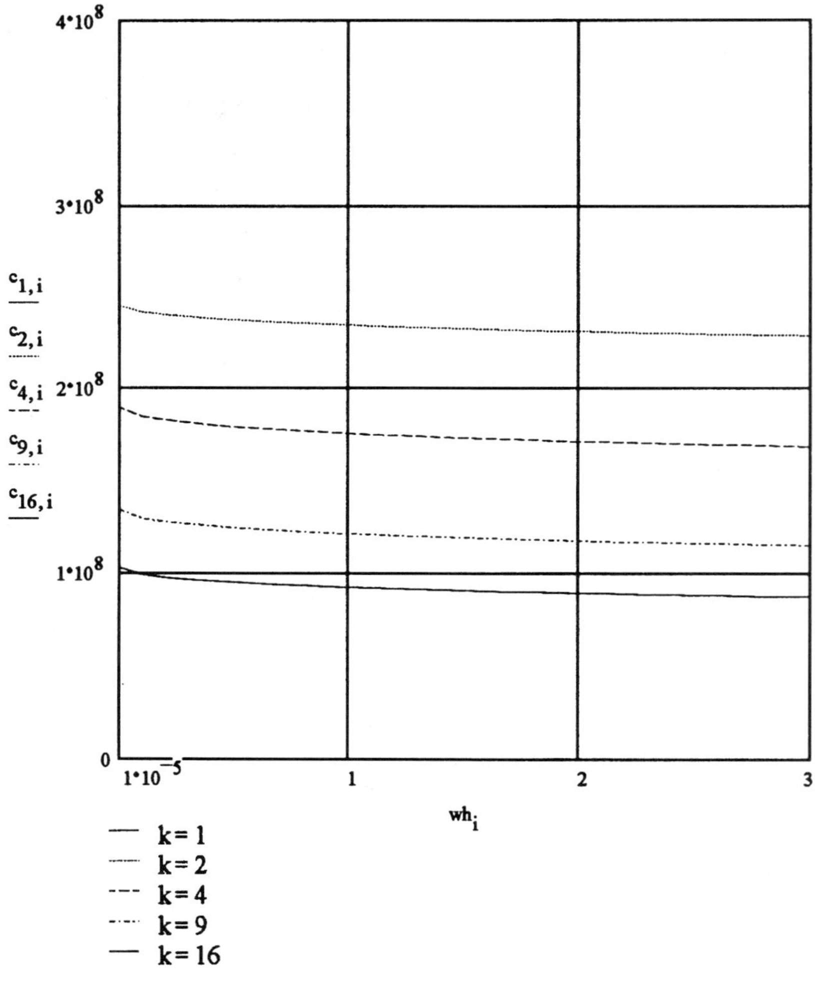

FIGURE 9.19 Velocity of propagation versus width-to-height ratio for single strip microstrip ($k = \varepsilon_r$).

close to the point ($w/h = 2$) where they meet.] The effective dielectric constant is calculated from Eq. (9.51) to be $\varepsilon_e = 2.94$, so $v_{ph} = 1.75 \times 10^8$ m/s.

For the sake of interest, our approximate result, Eq. (9.50), gives a value of $w/h = 1.87$, which is so far off as to be useless.

Example 9.7 Construct a MathCad file that accepts as input variables the dimensions and the relative dielectric permittivity of a microstrip line, and calculate its characteristic impedance and the corresponding phase velocity.

Solution The MathCad file is given in Listing 9.1. First, the given variables are specified. The ratio of width to height of the microstrip line is then calculated. The effective relative dielectric permittivity is calculated according to the equation in Section 9.8.2. The characteristic impedance is then calculated using the MathCad "if" statement. If the logic expression ($wh < 1$) is true, then the value obtained from the first expression is assigned to Z_0; if the logic expression is false, the value obtained from the second expression is assigned to Z_0. The phase velocity is calculated as

$$v_{ph} := \frac{c}{\sqrt{\varepsilon_{eff}}},$$

where $c = 3 \times 10^8$ m/s.

9.10 COPLANAR WAVEGUIDES

The coplanar waveguide (CPW), shown in Fig. 9.16f, has found extensive use in hybrid microwave integrated circuits (HMICs) and monolithic microwave integrated

$$\varepsilon_r := 2.1 \qquad w := 204 \quad \text{mils} \qquad \eta_0 := 377$$
$$h := 64 \quad \text{mils}$$

$$wh := \frac{w}{h}$$

$$e_{eff} := \frac{\varepsilon_r + 1}{2} + \frac{\varepsilon_r - 1}{2 \cdot \sqrt{1 + \frac{12}{wh}}}$$

$$Z_0 := \text{if}\left[wh < 1, \frac{\eta_0}{2 \cdot \pi \cdot \sqrt{e_{eff}} \cdot \ln\left(\frac{8}{wh} + \frac{wh}{4}\right)}, \times \frac{\eta_0}{\sqrt{e_{eff}} \cdot ((wh + 1.393 + 0.667 \cdot \ln(wh + 1.444))}\right]$$

$$Z_0 = 50.125\ \Omega$$

LISTING 9.1 MathCad file used to calculate the impedance of a microstrip line.

circuits (MMICs). Its utility and popularity have grown due to its geometry, where all conductors (center conductor and ground planes) are formed on the same surface of a dielectric or semi-insulating surface. This is important because active and passive components may easily be attached in parallel between the center conductor and ground planes, or in series in a gap etched in the center conductor. This geometric advantage avoids the need to drill holes or create cavities.

As described in earlier sections, any transmission line may be fully specified by three quantities: characteristic impedance (Z_0), propagation constant (β), and the physical length of the line (l). An expression for the characteristic impedance of the line has been derived by C. P. Wen [3] by mapping the coplanar waveguide strips onto that of an equivalent parallel plate capacitor using the Schwartz–Christofel transformation and by using the expression

$$Z_0 = \frac{\sqrt{\varepsilon_{re}}}{cC},$$

where c, C, and ε_{re} are the speed of light in free space, capacitance per unit length, and effective dielectric constant of the dielectric medium between the two conductors, respectively. It can be shown that the characteristic impedance for a coplanar waveguide may be expressed as

$$Z_0 = \frac{30\pi}{\sqrt{\varepsilon_{re}}} \frac{K'\ (k)}{K\ (k)},$$

where K and K' are complete elliptic functions of the first and second kind and

$$k = \frac{S}{S + 2W}.$$

Note that S is the spacing between the center conductor and ground planes and W is the width of the center conductor. A closed form expression for ε_{re} has been obtained by Davis et al. [4] and may be expressed as

$$\varepsilon_{re} = \frac{\varepsilon_r + 1}{2} \left\{ \tan\left(0.775 \ln\left(\frac{h}{w}\right) + 1.75\right) + \frac{kw}{h}(0.04 - 0.7k + 0.01)(1 - 0.1\varepsilon_r)(0.25 + k) \right\}$$

Note that h is defined as the thickness of the dielectric material by which the thin conductors are supported. Also note that to first order the above expression for effective dielectric constant may be expressed as the mean between ε_r and air for large h:

$$\varepsilon_{re} \simeq \frac{\varepsilon_r + 1}{2}.$$

9.11 TRIPLATE

Triplate, shown in Fig. 9.16d, was one of the first forms of printed circuit transmission lines to be used at microwave frequencies. The dominant mode of operation is a transverse electromagnetic (TEM) wave and may be analyzed in the same manner as a coplanar waveguide.

The expression for the characteristic impedance is shown to be

$$Z_0 = \frac{30\pi}{\sqrt{\varepsilon_r}} \frac{K'(k)}{K(k)}$$

where again K and K' are complete elliptic functions of the first and second kind. The function k may be expressed as

$$k = \tanh\left(\frac{\pi w}{2b}\right).$$

Note that w is the width of the center conductor and b is the distance between ground planes.

PROBLEMS

9.1 The electric field intensity in a source-free lossless medium is

$$\vec{E} = \vec{a}_y E_0 \sin\left(\frac{\pi x}{a}\right) \cos(\omega t - \beta z).$$

a. Give the corresponding phasor $\underline{\vec{E}}$.
b. Find the magnetic field intensity phasor $\underline{\vec{H}}$.
c. Find the expression for the phase constant β.
d. Give the physical magnetic field intensity $\vec{H}$.

Ans. $\beta = \sqrt{\omega^2 \mu\varepsilon - (\pi/a)^2}$,

$$\vec{H} = -\vec{a}_x \frac{\beta}{\omega\mu} E_0 \sin\left(\frac{\pi x}{a}\right) \cos(\omega t - \beta z) - \vec{a}_z \frac{(\pi/a)}{\omega\mu} E_0 \cos\left(\frac{\pi x}{a}\right) \sin(\omega t - \beta z).$$

9.2 Repeat Problem 9.1 if the electric field intensity is

$$\vec{E} = \vec{a}_y E_0 \sin\left(\frac{\pi x}{a}\right) \cos(\omega t + \beta z).$$

Ans. $\beta = \sqrt{\omega^2 \mu\varepsilon - (\pi/a)^2}$,

$$\vec{H} = \vec{a}_x \frac{\beta}{\omega\mu} E_0 \sin\left(\frac{\pi x}{a}\right) \cos(\omega t + \beta z) - \vec{a}_z \frac{(\pi/a)}{\omega\mu} E_0 \cos\left(\frac{\pi x}{a}\right) \sin(\omega t + \beta z).$$

9.3 The magnetic field intensity in a source-free lossless medium is

$$\vec{H} = \vec{a}_y H_0 \sin\left(\frac{\pi x}{a}\right)\cos(\omega t - \beta z).$$

a. Give the corresponding phasor $\underline{\vec{H}}$.

b. Find the electric field intensity phasor $\underline{\vec{E}}$.

c. Find the expression for the phase constant β.

d. Give the physical electric field intensity $\vec{E}$.

Ans. $\beta = \sqrt{\omega^2 \mu\varepsilon - (\pi/a)^2}$,

$$\vec{E} = \vec{a}_x \frac{\beta}{\omega\varepsilon} H_0 \sin\left(\frac{\pi x}{a}\right)\cos(\omega t - \beta z) + \vec{a}_z \frac{(\pi/a)}{\omega\varepsilon} H_0 \cos\left(\frac{\pi x}{a}\right)\sin(\omega t - \beta z).$$

9.4 Show that the physical and phasor fields of Problems 9.1 and 9.3 are dual sets of fields.

9.5 The equation of continuity is $\nabla \cdot \vec{J} = -(\partial\rho/\partial t)$.

a. Give the phasor form of this equation.

b. Obtain the phasor form from Maxwell's equations.

9.6 Confirm that the substitutions of Table 9.1 transform Eqs. (9.3) and (9.4) into themselves.

9.7 The waves of Problems 9.1, 9.2, and 9.3 have $\beta = \sqrt{\omega^2 \mu\varepsilon - (\pi/a)^2}$.

a. Find the phase velocity. Is it greater or less than c?

b. Find the group velocity. Is it greater or less than c?

c. Show that $v_{\text{ph}} v_{\text{g}} = c^2$.

9.8 Design a two-wire transmission line with a characteristic resistance of 300 Ω. Assume the spacing between the center lines is 1 cm, that the medium is air, and that the wires are copper. Calculate the resistance, inductance, and capacitance per unit length at 100 MHz. Show that $R \ll \omega L$ and calculate the phase and attenuation constants.

Ans. $2a = 0.163$ cm, $R = 1.02\ \Omega/\text{m}$, $L = 1.0\ \mu\text{H/m}$, $C = 11.1$ pF/m, $\beta = 2.1$ rad/m, $\alpha = 0.0017$ Np/m $= 0.015$ dB/m.

9.9 Design 100 and 150 Ω lines for the substrate of Example 9.5. Use the figures and the design equations. Find the phase velocity.

Ans. For 100 Ω, $w = 0.0079$ in., $v_{\text{ph}} = 1.83 \times 10^8$ m/s.
For 150 Ω, $w = 0.0022$ in., $v_{\text{ph}} = 1.88 \times 10^8$ m/s.

9.10 Design 150, 100, 75, and 50 Ω microstrip lines on a Duroid ($\varepsilon_r = 2.3$) substrate that is 0.030 in. thick. Find the width of the conducting strip and the phase velocity.

Ans. For 150 Ω, $w = 0.0087$ in., $v_{\text{ph}} = 2.27 \times 10^8$ m/s.
For 100 Ω, $w = 0.0259$ in., $v_{\text{ph}} = 2.22 \times 10^8$ m/s.
For 75 Ω, $w = 0.0462$ in., $v_{\text{ph}} = 2.28 \times 10^8$ m/s.
For 50 Ω, $w = 0.0899$ in., $v_{\text{ph}} = 2.15 \times 10^8$ m/s.

REFERENCES

[1] I. J. Bahl and D. K. Trivedi, "A designer's guide to microstrip line," *Microwaves*, May 1977, pp. 90–96.

[2] T. G. Bryant and J. A. Weiss, "Parameters of microstrip transmission lines and of coupled pairs of microstrip lines," *IEEE Trans. Microwave Theory Techniques*, Vol. MTT-16, Dec. 1968, pp. 1021–1027.

[3] C. P. Wen, "Coplanar Waveguide: A Surface Strip Transmission Line Suitable for Non-Reciprocal Gyromagnetic Device Application," *IEEE Trans. Microwave Theory Tech.*, Vol. MTT-17, Dec. 1969, pp. 1087–1090.

[4] M. E. Davis et al. "Finite-Boundary Correction to the Coplanar Waveguide Analysis," *IEEE Trans. Microwave Theory Tech.*, Vol. MTT-21, Sept-1973, pp. 594–596.

CHAPTER TEN

The Terminated Transmission Line

10.1 INTRODUCTION

In Chapter 9 we developed transmission line concepts, defining voltage and current and finding the transmission line equations that relate each to the other. We also found and discussed the transmission line parameters, resistance, inductance, conductance, and capacitance per unit length and the characteristic impedance and the propagation constant.

In this chapter we will find solutions to the transmission line equations for time harmonic and transient excitations when the line is terminated in any load. We will study the need for and methods of matching loads to the line and we will look briefly at the problem of broad-band matching.

10.2 TIME HARMONIC SOLUTIONS

So far we have considered the properties of the transmission line itself but have not considered it as part of a complete circuit. Figure 10.1 shows a transmission line with propagation constant γ, characteristic impedance Z_0, and length d terminated by a load impedance $Z_L = R_L + jX_L$ and connected to a voltage generator with open circuit voltage $\underline{V}_g$ and internal generator impedance Z_g. The angular frequency of the source is ω. For convenience we take the load to be located at $z = 0$ and the source to be located at $z = -d$.

The voltage and current phasors on the transmission line are

$$\underline{V}(z) = V_0^+ e^{-\gamma z} + V_0^- e^{+\gamma z}, \tag{10.1}$$

$$\underline{I}(z) = I_0^+ e^{-\gamma z} + I_0^- e^{+\gamma z}. \tag{10.2}$$

In both equations, the first term on the right is the incident wave traveling from the generator toward the load and the second term is the reflected wave traveling from the load toward the generator.

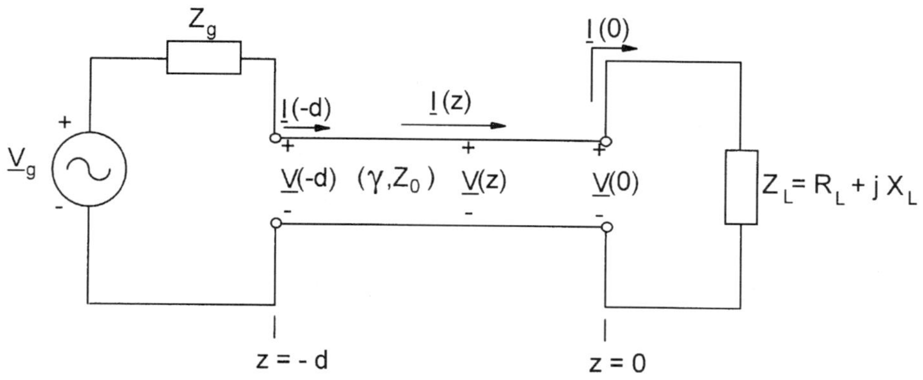

FIGURE 10.1 Terminated transmission line.

The current amplitudes are related to the voltage amplitudes by the characteristic impedance, Eq. (9.42), so

$$\underline{I}(z) = \frac{1}{Z_0}(V_0^+ e^{-\gamma z} - V_0^- e^{+\gamma z}). \tag{10.3}$$

We are now left with two unknown constants, V_0^+ and V_0^-. These are determined by the terminal conditions at the two ends of the transmission line.

At the load, $z = 0$, the ratio of the voltage phasor to the current phasor equals the load impedance. Thus

$$Z_L = \frac{\underline{V}(0)}{\underline{I}(0)} = Z_0 \frac{V_0^+ + V_0^-}{V_0^+ - V_0^-}.$$

This equation can be solved for V_0^-, giving

$$V_0^- = \frac{Z_L - Z_0}{Z_L + Z_0} V_0^+. \tag{10.4}$$

It is easy to see that if the load impedance equals the characteristic impedance, $Z_L = Z_0$, then there is no reflected or backward traveling wave on the line, $V_0^- = 0$. Indeed, an alternate definition of the characteristic impedance is that the characteristic impedance is that load impedance that causes no reflected wave on the line.

When a line is terminated in its characteristic impedance it is matched. There is no reflected wave on a matched line.

At the input, $z = -d$, the ratio of the voltage phasor to the current phasor gives the input impedance to the transmission line:

$$Z_{\text{in}} = \frac{\underline{V}(-d)}{\underline{I}(-d)} = Z_0 \frac{V_0^+ e^{\gamma d} + V_0^- e^{-\gamma d}}{V_0^+ e^{\gamma d} - V_0^- e^{-\gamma d}}.$$

Dividing the numerator and the denominator by V_0^+ and substituting from Eq. (10.4), we obtain

$$Z_{\text{in}} = Z_0 \frac{e^{\gamma d} + \left(\dfrac{Z_L - Z_0}{Z_L + Z_0}\right) e^{-\gamma d}}{e^{\gamma d} - \left(\dfrac{Z_L - Z_0}{Z_L + Z_0}\right) e^{-\gamma d}} = Z_0 \frac{Z_L(e^{\gamma d} + e^{-\gamma d}) + Z_0(e^{\gamma d} - e^{-\gamma d})}{Z_0(e^{\gamma d} + e^{-\gamma d}) + Z_L(e^{\gamma d} - e^{-\gamma d})}$$

$$Z_{\text{in}} = Z_0 \frac{Z_L + Z_0 \tanh \gamma d}{Z_0 + Z_L \tanh \gamma d}. \tag{10.5}$$

This is a very important result and we will return to it and discuss it further. First, let us finish the problem we started.

We now know the input impedance to the loaded transmission line. Using the equivalent circuit of Fig. 10.2, we obtain

$$\underline{V}_{\text{in}} = \underline{V}(-d) = \frac{Z_{\text{in}}}{Z_g + Z_{\text{in}}} \underline{V}_g,$$

$$\underline{I}_{\text{in}} = \underline{I}(-d) = \frac{1}{Z_g + Z_{\text{in}}} \underline{V}_g.$$

Substituting for $\underline{V}(-d)$ and $\underline{I}(-d)$ from Eqs. (10.1) and (10.3), respectively, we can solve for V_0^+:

$$V_0^+ = \frac{1}{2}\left(\frac{Z_{\text{in}} + Z_0}{Z_{\text{in}} + Z_g}\right) e^{-\gamma d} \underline{V}_g. \tag{10.6}$$

We have now completely solved the problem, except for some computational difficulties due to complex numbers. Equation (10.5) gives Z_{in}, Eq. (10.6) gives V_0^+ and Eq. (10.4) gives V_0^-. Knowing these, we know the voltage and current at any point on the line.

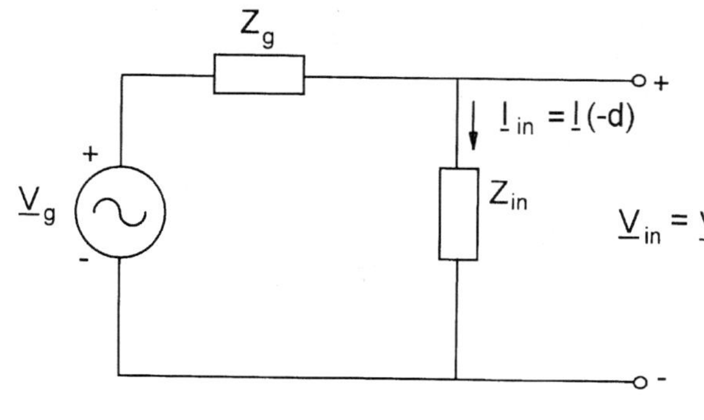

FIGURE 10.2 Input impedance of terminated transmission line.

With a little luck, you will never have to do the calculations that these formulas require. Just for completeness, however, you should know that

$$\tanh \gamma d = \tanh\,[(\alpha + j\beta)d] = \frac{\tanh \alpha d + j \tan \beta d}{1 + j \tanh \alpha d \tan \beta d}.$$

10.2.1 Lossless Transmission Line

The equations derived in the last section can be used to calculate the voltage and current at any point on the transmission line. When the transmission line is assumed to be lossless the equations are simplified. In this case $Z_0 = R_0$, $\gamma = j\beta$, and $\tanh \gamma d = j \tan \beta d$. We then have

$$Z_{\text{in}} = R_0 \frac{Z_L + jR_0 \tan \beta d}{R_0 + jZ_L \tan \beta d}, \tag{10.7}$$

$$V_0^+ = \frac{1}{2}\left(\frac{Z_{\text{in}} + R_0}{Z_{\text{in}} + Z_g}\right) e^{-j\beta d} \underline{V}_g, \tag{10.8}$$

and

$$V_0^- = \frac{Z_L - R_0}{Z_L + R_0} V_0^+ . \tag{10.9}$$

Example 10.1 A lossless line is $3\lambda/8$ long and is terminated with an impedance $Z_L = R_0(1 + j)$. The generator impedance is matched to the line $Z_g = R_0$. Find the input impedance and the forward and backward voltage wave amplitudes.

Solution The argument $\beta d = (2\pi/\lambda)(3\lambda/8) = 3\pi/4 = 135^\circ$ so $\tan \beta d = -1$ and $\exp(j\beta d) = (1\angle 135^\circ)$. Then

$$\begin{aligned}
Z_{\text{in}} &= R_0\left[\frac{(1+j) + j(-1)}{1 + j(1+j)(-1)}\right] = R_0\left(\frac{1}{2.24\angle -26.6^\circ}\right) = (0.447\angle 26.6^\circ)R_0 \\
&= (0.4 + j0.2)R_0 \\
V_0^+ &= \frac{1}{2}\left[\frac{0.4 + j0.2) + 1}{(0.4 + j0.2) + 1}\right](1\angle -135^\circ)\,\underline{V}_g = (0.500\angle -135.0^\circ)\,\underline{V}_g, \\
V_0^- &= \left[\frac{(1+j) - 1}{(1+j) + 1}\right](0.500\angle -135.0^\circ)\underline{V}_g = \left(\frac{1\angle 90^\circ}{2.24\angle 26.6^\circ}\right)(0.500\angle -135.0^\circ)\underline{V}_g \\
&= (0.224\angle -71.6^\circ)\,\underline{V}_g.
\end{aligned}$$

10.2.2 Power on the Transmission Line

The instantaneous power for a voltage $v(t)$ and a current $i(t)$ is $p(t) = v(t)i(t)$. In the case of time harmonic waves we can write this in terms of the phasors. Let the voltage phasor be $\underline{V} = |V|\exp(j\phi_V)$ and the current phasor be $\underline{I} = |I|\exp(j\phi_I)$; then

$$\begin{aligned}
p(t) &= \mathrm{Re}(\underline{V}e^{j\omega t})\,\mathrm{Re}(\underline{I}e^{j\omega t}) \\
&= [\tfrac{1}{2}(\underline{V}e^{j\omega t} + \underline{V}^*e^{-j\omega t})]\,[\tfrac{1}{2}(\underline{I}e^{j\omega t} + \underline{I}^*e^{-j\omega t})] \\
&= \tfrac{1}{4}[\underline{V}\underline{I}^* + (\underline{V}\underline{I}^*)^*] + \tfrac{1}{4}[\underline{V}\underline{I}e^{2j\omega t} + (\underline{V}\underline{I}e^{2j\omega t})^*] \\
&= \tfrac{1}{2}\mathrm{Re}(\underline{V}\underline{I}^*) + \tfrac{1}{2}\mathrm{Re}(\underline{V}\underline{I}e^{2j\omega t})
\end{aligned}$$

$$p(t) = \tfrac{1}{2}|V|\,|I|\cos(\phi_V - \phi_I) + \tfrac{1}{2}|V|\,|I|\cos(2\omega t + \phi_V + \phi_I). \tag{10.10}$$

We have denoted the complex conjugate by a superscript asterisk and used the result that if $w = u + jv$ then $u = \mathrm{Re}(w) = (w + w^*)/2$.

Equation (10.10) shows that the instantaneous power consists of two terms. The first is a constant term, which is independent of time, and the second is a term at twice the frequency of the voltage and current.

The time average power is the instantaneous power averaged over one period of time, $T = 1/f$. From our result, we see that

$$P_{\mathrm{Avg}} = \frac{1}{T}\int_0^T p(\tau)\,d\tau, \quad T = \frac{1}{f}$$

$$P_{\mathrm{Avg}} = \tfrac{1}{2}\mathrm{Re}(\underline{V}\underline{I}^*) = \tfrac{1}{2}|V|\,|I|\cos(\phi_V - \phi_I). \tag{10.11}$$

The term $\cos(\phi_V - \phi_I)$ is called the power factor. Equation (10.11) gives a convenient form for calculating the time average power in terms of the voltage and current phasors.

Other forms of this equation are also convenient. Since the impedance $Z = R + jX = \underline{V}/\underline{I}$ or the admittance $Y = G + jB = \underline{I}/\underline{V} = I/Z$, then

$$P_{\mathrm{Avg}} = \tfrac{1}{2}R|\underline{I}|^2 = \tfrac{1}{2}G|\underline{V}|^2. \tag{10.12}$$

Example 10.2 For the lossless loaded line of Example 10.1, find the time average power delivered to (a) the load, (b) the line input, and (c) the internal impedance of the generator and (d) find the time average power produced by the generator.

Solution (a) At the load,

$$\begin{aligned}
\underline{I}(0) &= \frac{1}{R_0}(V_0^+ - V_0^-) = \frac{1}{R_0}[(0.500\angle -135.0^\circ) - (0.224\angle -71.6^\circ)]\,\underline{V}_g \\
&= \frac{1}{R_0}(-0.424 - j0.141)\,\underline{V}_g = \frac{1}{R_0}(0.447\angle -161.5^\circ)\,\underline{V}_g,
\end{aligned}$$

$$\begin{aligned}
P_{L,\mathrm{Avg}} &= \tfrac{1}{2}R_L|I(0)|^2 = \tfrac{1}{2}R_0\left[\frac{1}{R_0}(0.447)|\underline{V}_g|\right]^2 \\
&= 0.100\frac{|\underline{V}_g|^2}{R_0}.
\end{aligned}$$

(b) At the line input,

$$\underline{I}(-d) = \frac{1}{R_0}(V_0^+ e^{j\beta d} - V_0^- e^{-j\beta d})$$

$$= \frac{1}{R_0}[(0.500\angle -135.0^\circ)(1\angle 135^\circ) - (0.224\angle -71.6^\circ)(1\angle -135^\circ)]\underline{V}_g$$

$$= \frac{1}{R_0}(0.700 - j0.100)\underline{V}_g = \frac{1}{R_0}(0.707\angle -8.1^\circ)\underline{V}_g,$$

$$P_{\text{in,Avg}} = \tfrac{1}{2}R_{\text{in}}|I(-d)|^2 = \tfrac{1}{2}(0.4R_0)\left[\frac{1}{R_0}(0.707)|\underline{V}_g|\right]^2$$

$$= 0.100\,\frac{|\underline{V}_g|^2}{R_0}.$$

(c) Delivered to the generator impedance,

$$P_{Z_g,\text{Avg}} = \tfrac{1}{2}R_g|\underline{I}(-d)|^2 = \tfrac{1}{2}R_0\left[\frac{1}{R_0}(0.707)|\underline{V}_g|\right]^2$$

$$= 0.250\,\frac{|\underline{V}_g|^2}{R_0}.$$

(d) Produced by the generator,

$$P_{\text{gen,Avg}} = \tfrac{1}{2}\text{Re}\left[\underline{V}_g\underline{I}(-d)^*\right] = \tfrac{1}{2}\text{Re}\left[(\underline{V}_g)\frac{1}{R_0}(0.700 - j0.100)\underline{V}_g^*\right]$$

$$= 0.350\,\frac{|\underline{V}_g|^2}{R_0}.$$

Note that the power input to the line is the same as the power input to the load. This is because the line itself is lossless. The power produced by the generator equals the power delivered to the generator internal impedance plus the power delivered to the load.

10.2.3 Impedance Transformation

Equation (10.5) for a line with losses and Eq. (10.7) for the lossless line are impedance transformation formulas. They take impedance seen looking into the load, transform it through the line of length d, and give the input impedance. More generally, in Fig. 10.3, if $Z(z')$ is the impedance seen looking toward the load at some plane z', then the impedance $Z(z)$ seen looking toward the load at the plane z is

$$Z(z) = Z_0\,\frac{Z(z') + Z_0 \tanh \gamma(z' - z)}{Z_0 + Z(z') \tanh \gamma(z' - z)} \tag{10.13}$$

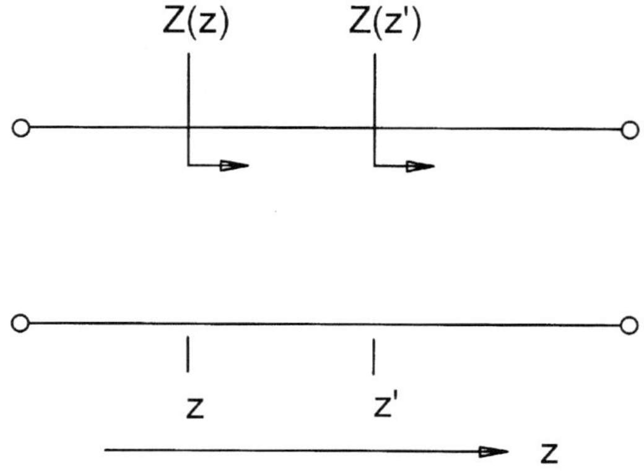

FIGURE 10.3 Impedance transformation along a lossless transmission line.

for the line with losses and

$$Z(z) = R_0 \frac{Z(z') + jR_0 \tan \beta(z' - z)}{R_0 + jZ(z') \tan \beta(z' - z)} \tag{10.14}$$

for the lossless line. A length of transmission line $z' - z$ transforms the impedance $Z(z')$ into the impedance $Z(z)$.

10.2.4 Impedance Transformation on a Lossless Line

Most lines are made as close to lossless as is possible. For distances d or $z' - z$ that are short compared to the reciprocal of the attenuation constant, $1/\alpha$, they may be considered to be lossless. It is therefore important to examine the lossless case carefully.

First, we recall that $\beta = 2\pi/\lambda$ and that $\tan n\pi = 0$, $n = \pm 0, 1, 2, \ldots$. Thus, if $z' - z = n(\lambda/2)$, then $Z(z) = Z(z')$. In words, over an integral number of half-wavelengths, an impedance is transformed into itself. The impedance seen looking toward the load repeats for every half-wavelength of travel along the lossless line, $Z(z) = Z(z \pm \lambda/2)$.

Second, if the distance $z' - z = \pm \lambda/4$, then $\tan \beta(z' - z) = \pm \infty$ and $Z(z) = R_0^2/Z(z')$. This can also be written as $Z(z)/R_0 = [Z(z')/R_0]^{-1}$. In words, over a quarter- wavelength, a normalized impedance is transformed into its reciprocal. We will use this property when we discuss impedance matching (Section 10.7).

Now let us use Eq. (10.14) to look at different loads Z_L connected to the end of the line.

We have already noted that if the load impedance equals the line's characteristic impedance $Z_L = Z_0$, then there is no reflected wave. As a consequence, $Z_{in} = Z_0$ or $Z_{in} = R_0$ for the lossy or lossless line, respectively, independent of the line length d. We could define the characteristic impedance as the impedance that when connected as the load makes the impedance seen looking toward the load equal to itself for all points on the line.

Now let us suppose we connect a pure reactance, $Z_L = jX_L$, as a load on the lossless line. Then

$$Z_{\text{in}} = jX_{\text{in}} = jR_0 \frac{X_L + R_0 \tan \beta d}{R_0 - X_L \tan \beta d},$$

showing that the input impedance is also purely reactive for any line length. This result matches with physical considerations regarding power input. The time average power input to a pure reactance load is zero. There are no losses in the line. The time average power input to the line, which equals the sum of the time average power loss in the line and the time average power into the load, is zero and, hence, the input impedance must be reactive.

Next, connect a short circuit, $Z_L = 0$, as the load. Then the input impedance for any length d is

$$Z_{\text{in}} = jX_{\text{in}} = jR_0 \tan \beta d.$$

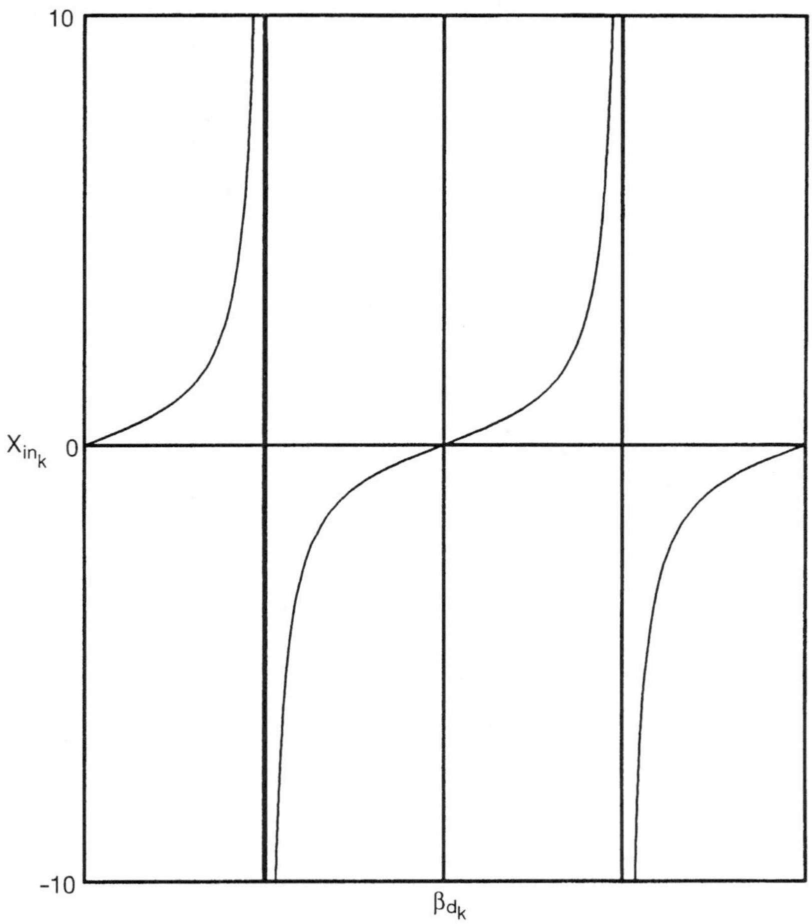

FIGURE 10.4 Normalized input reactance of a short circuited transmission line.

The impedance is purely reactive. Since no power is dissipated in the short circuit and no power is lost in the line, no power can go into the line. When $\beta d < \pi/2$ the reactance is inductive. This is because the current close to the short is high, leading to large magnetic energy storage, but the voltage close to the short is low, leading to small electric energy storage. Predominant magnetic energy storage means the device appears inductive. As d increases the input reactance goes repetitively from zero to plus infinity, switches to minus infinity, and returns to zero. This behavior is shown in Fig. 10.4. A shorted line can be used to obtain any desired value of input reactance.

Finally, connect an open circuit, $Z_L = \infty$, as the load. Then the input impedance for any length d is

$$Z_{\text{in}} = jX_{\text{in}} = -jR_0 \cot \beta d.$$

The impedance is again purely reactive. Since no power is dissipated in the open circuit and no power is lost in the line, no power can go into the line. When $\beta d < \pi/2$ the reactance is capacitive. This is because the voltage close to the open is high,

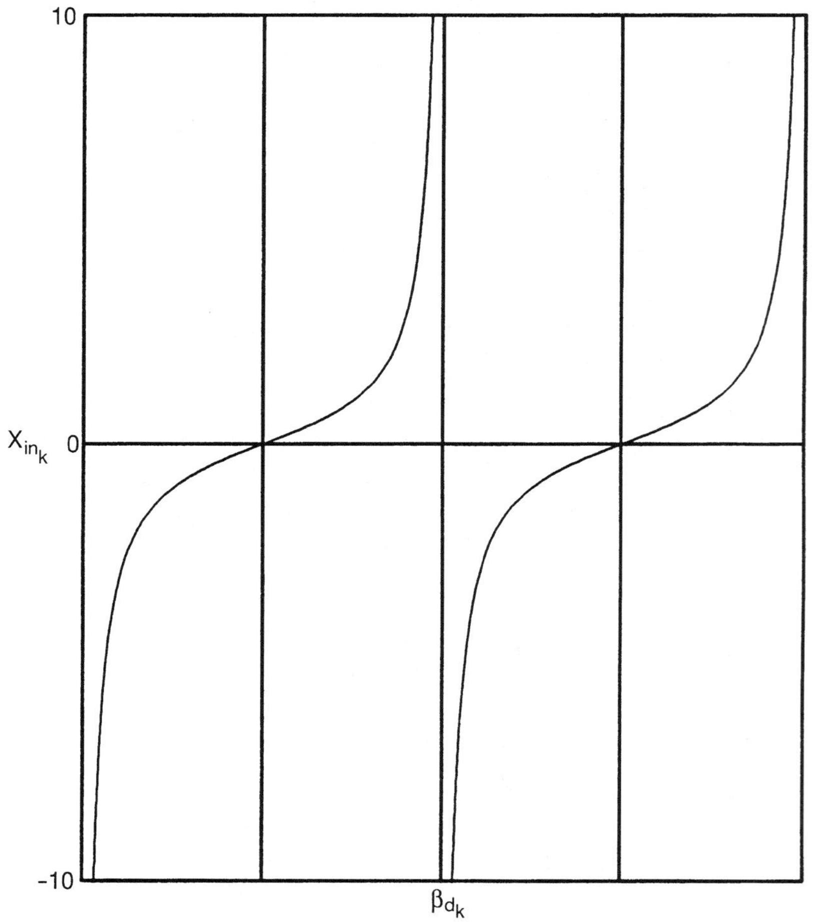

FIGURE 10.5 Normalized input reactance of an open circuited transmission line.

leading to large electric energy storage, but the current close to the open is low, leading to small magnetic energy storage. Predominant electric energy storage means the device appears capacitive. As d increases the input reactance goes repetitively from minus infinity to zero, to plus infinity, and switches back to minus infinity. This behavior is shown in Fig. 10.5. An open line can also be used to obtain any desired value of input reactance.

10.3 REFLECTION COEFFICIENT

We have seen that when the transmission line is terminated in any impedance, except its characteristic impedance, then there is a backward reflected or negative z traveling voltage and current wave on the line. This is because the forward incident or positive z going wave alone cannot satisfy the terminal relation that $\underline{V}(0)/\underline{I}(0) = Z_L$.

The voltage reflection coefficient $\Gamma(z)$ is defined at any plane z as the ratio of the reflected voltage wave phasor to the incident voltage wave phasor at that value of z. Thus

$$\Gamma(z) = \frac{V_0^- e^{\gamma z}}{V_0^+ e^{-\gamma z}} = \frac{V_0^-}{V_0^+} e^{2\gamma z}. \tag{10.15}$$

Generally, the word "voltage" is omitted and Γ is simply called the reflection coefficient. As the ratio of two voltages it is dimensionless.

If we let z' and z be any two planes along the transmission line (Fig. 10.3), then dividing $\Gamma(z)$ by $\Gamma(z')$ gives the reflection coefficient at one plane in terms of the reflection coefficient at the other:

$$\Gamma(z) = \Gamma(z') e^{2\gamma(z - z')}. \tag{10.16}$$

We will study this important transformation of the reflection coefficient a little further on.

Let us now obtain a formula for the reflection coefficient in terms of the impedance. We repeat Eqs. (10.1) and (10.3):

$$\underline{V}(z) = V_0^+ e^{-\gamma z} + V_0^- e^{\gamma z},$$

$$Z_0 \underline{I}(z) = V_0^+ e^{-\gamma z} - V_0^- e^{\gamma z}.$$

To solve for V_0^+ we add the equations and then divide by $2\exp(-\gamma z)$. To solve for V_0^- we subtract the equations and then divide by $2\exp(\gamma z)$. These manipulations give

$$V_0^+ = [\underline{V}(z) + Z_0 \underline{I}(z)]\, e^{\gamma z}/2,$$

$$V_0^- = [\underline{V}(z) - Z_0 \underline{I}(z)]\, e^{-\gamma z}/2.$$

We substitute in Eq. (10.15), divide numerator and denominator by $I(z)$, and use $V(z)/I(z) = Z(z)$ to obtain

$$\Gamma(z) = \frac{Z(z) - Z_0}{Z(z) + Z_0}. \tag{10.17}$$

This equation gives a relation between the impedance looking toward the load at any plane and the reflection coefficient at that plane.

The relation is easily inverted to give

$$Z(z) = Z_0 \frac{1 + \Gamma(z)}{1 - \Gamma(z)}. \tag{10.18}$$

Using these equations we have an alternate method for finding the input impedance. At the load $z = 0$ and $Z(0) = Z_L$. The reflection coefficient at the load is

$$\Gamma_L = \Gamma(0) = \frac{Z_L - Z_0}{Z_L + Z_0}. \tag{10.19}$$

Equation (10.16) can be used to find the reflection coefficient at the line input, $z = -d$, in terms of the reflection coefficient at the load, $z' = 0$:

$$\Gamma_{\text{in}} = \Gamma(-d) = \Gamma_L e^{-2\gamma d}. \tag{10.20}$$

Finally, Eq. (10.18) can be used to find the input impedance,

$$Z_{\text{in}} = Z_0 \frac{1 + \Gamma_{\text{in}}}{1 - \Gamma_{\text{in}}}. \tag{10.21}$$

These three equations break the calculation of input impedance into three relatively simple steps instead of the more complicated single step of Eq. (10.5). More important, they set the stage for an important graphical technique.

For the lossless transmission line—$Z_0 = R_0$, $\gamma = j\beta$—the three equations become

$$\Gamma_L = \frac{Z_L - R_0}{Z_L + R_0}, \tag{10.22}$$

$$\Gamma_{\text{in}} = \Gamma_L e^{-2j\beta d}, \tag{10.23}$$

$$Z_{\text{in}} = R_0 \frac{1 + \Gamma_{\text{in}}}{1 - \Gamma_{\text{in}}}. \tag{10.24}$$

Example 10.3 A satellite transmitter preamplifier is connected to the transmitting antenna via a 1.125λ long, 50 Ω lossless transmission line. The antenna input impedance is measured to be 25 Ω. Determine the reflection coefficient at the antenna and at the output of the preamplifier.

Solution Using Eq. (10.22), the reflection coefficient at the load is calculated as

$$\Gamma_L = \frac{Z_L - R_0}{Z_L + R_0} = \frac{25 - 50}{25 + 50} = -\frac{1}{3}.$$

The reflection coefficient at the input of the line is then calculated using Eq. (10.23):

$$\Gamma_{\text{in}} = \Gamma_L e^{-2j\beta d} = -\tfrac{1}{3} e^{-2j(2\pi/\lambda)10.125\lambda} = -\tfrac{1}{3} e^{-j40.5\pi}.$$

Recognizing that $e^{\pm j2k\pi} = 1$ and that $-1 = e^{\pm j\pi}$, the expression of the input reflection coefficient becomes

$$\Gamma_{\text{in}} = \tfrac{1}{3} e^{(j\pi - j40.5\pi)} = \tfrac{1}{3} e^{(j\pi - j0.5\pi - j40\pi)} = \tfrac{1}{3} e^{j\pi/2} = j\tfrac{1}{3}.$$

Whether for the lossy or the lossless line, if the line is terminated in its characteristic impedance, $Z_L = Z_0$, then $\Gamma_L = \Gamma_{\text{in}} = 0$ and $Z_{\text{in}} = Z_0$.

Also, if the line is terminated in a pure reactance, $Z_L = jX_L$, then the magnitude of the reflection coefficient at the load is equal to unity, $|\Gamma_L| = 1$. For the lossless line, the magnitude of the reflection coefficient at the input must also be unity, $|\Gamma_{\text{in}}| = 1$, and we can show (Problem 10.5) that the input impedance is also a pure reactance, $Z_{\text{in}} = jX_{\text{in}}$. This confirms the result obtained in Section 10.2.4.

10.3.1 The Gamma Plane

By graphically displaying the transformation of the reflection coefficient along the transmission line, Eq. (10.16), we will get a clear understanding of the transformation's meaning. We introduce the complex gamma plane or Γ plane for this purpose.

Figure 10.6 shows this plane. The horizontal axis (the abscissa) represents the real part of the reflection coefficient, $\text{Re}(\Gamma)$, and the vertical axis (the ordinate) represents the imaginary part, $\text{Im}(\Gamma)$, to the same scale. A circle concentric with the origin is the locus of the points with constant magnitude of the reflection coefficient, $|\Gamma| = C_M$. A radial line outward from the origin is the locus of the points with constant angle of the reflection coefficient, $\angle\Gamma = C_A$.

As long as we are dealing with passive loads, $\text{Re}(Z_L) = R_L \geq 0$, then the reflected power cannot be greater than the incident power and the reflected wave magnitude must not be greater than the incident wave magnitude. It follows that $|\Gamma| \leq 1$ and we therefore only draw the part of the Γ plane inside the unit circle.

Let us first look at the lossless line form of Eq. (10.16):

$$\Gamma(z) = \Gamma(z') e^{2j\beta(z-z')}. \tag{10.25}$$

We see that

$$|\Gamma(z)| = |\Gamma(z')| \quad \text{and} \quad \angle\Gamma(z) = \angle\Gamma(z') + 2\beta(z - z'). \tag{10.26}$$

This means that the magnitude of the reflection coefficient is constant as we move along the lossless transmission line; as $(z - z')$ changes we move along a circle

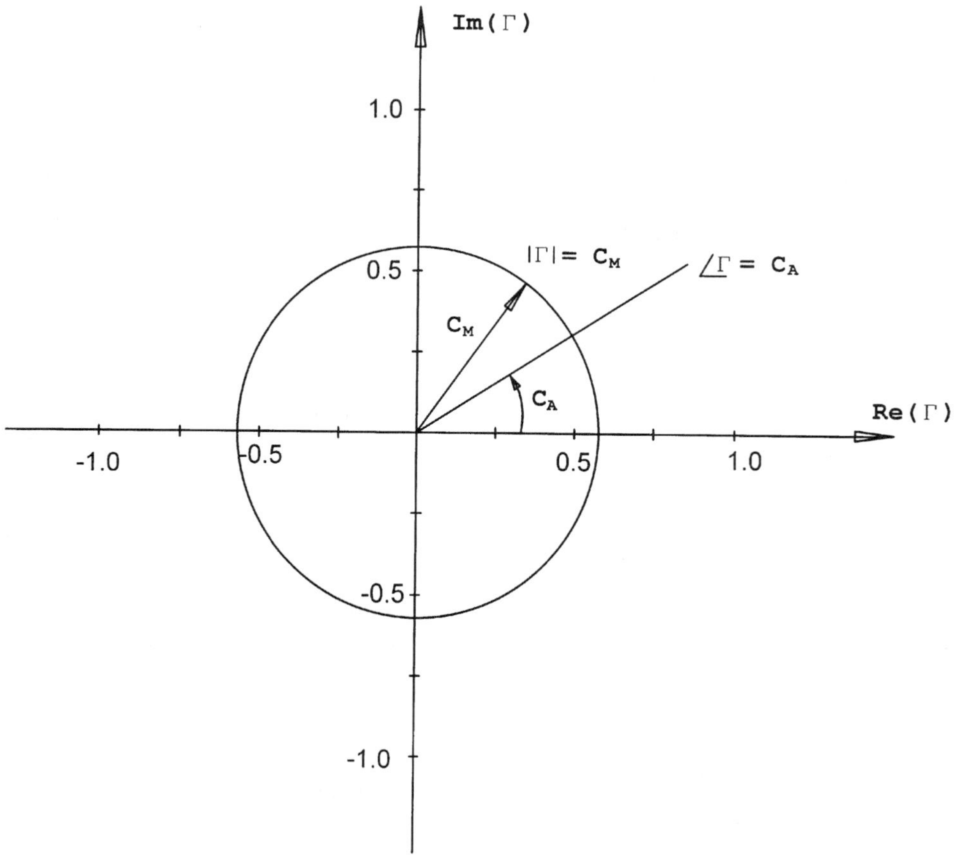

FIGURE 10.6 Plot of the reflection coefficient (Γ plane) for a lossless transmission line.

centered at the origin in the Γ plane. The angle of the reflection coefficient changes by $2\beta(z - z') = 4\pi(z - z')/\lambda$ radians or $720(z - z')/\lambda$ degrees. Every half-wavelength, $|z - z'| = \lambda/2$, the angle changes by 2π radians or 360°, which brings it back to the same point in the Γ plane.

Figure 10.7 shows a lossless transmission line with increasing z toward the load and decreasing z toward the generator or source. The plane of z' is a typical plane along the line and z_1 is any plane toward the load from z'. The difference $z_1 - z'$ is positive and the points $\Gamma(z')$ and $\Gamma(z_1)$ are as shown in Fig. 10.8. Rotation is in the positive or counterclockwise direction in the Γ plane as we move toward the load. On the other hand, z_2 is any plane toward the generator from z'. The difference $z_2 - z'$ is negative and the point $\Gamma(z_2)$ is as shown in Fig. 10.8. Rotation is in the negative or clockwise direction in the Γ plane as we move toward the generator.

To facilitate this graphical transformation along the line, we can add a linear scale along the bottom, ranging from zero to one, for the magnitude of Γ, and an angular scale outside the rim of the $|\Gamma| = 1$ circle for the angle of Γ. Also, as a reminder, we add arrows along the outside showing which way to rotate for a transformation toward the load or toward the generator (Fig. 10.9).

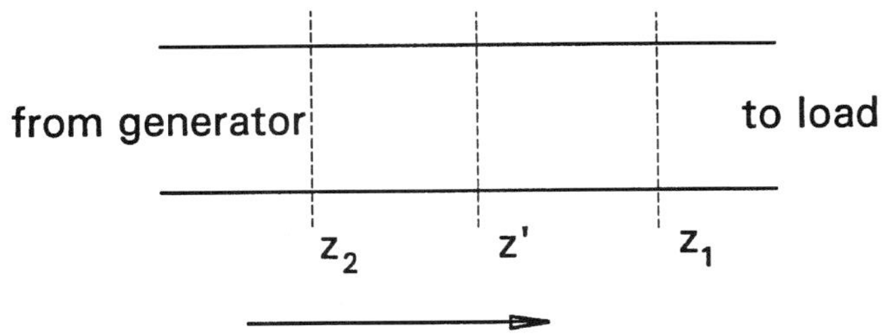

FIGURE 10.7 Lossless transmission line with increasing z direction from generator toward load.

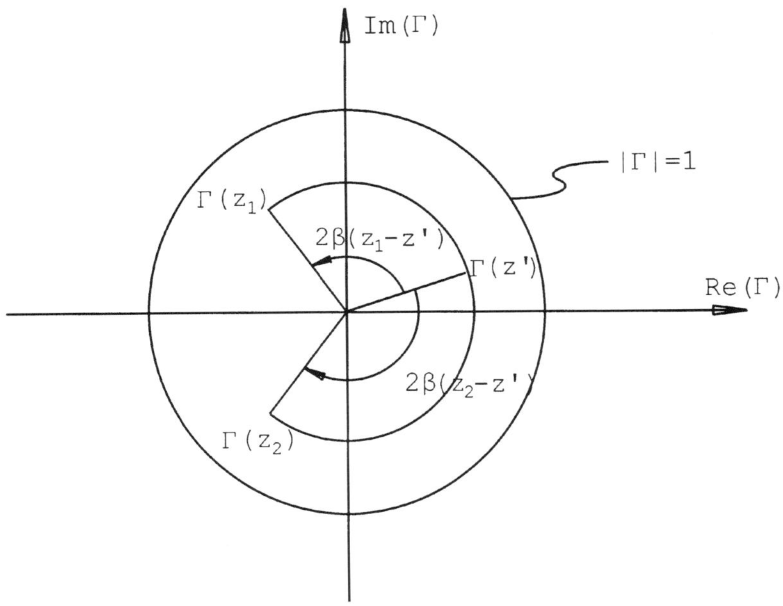

FIGURE 10.8 Reflection coefficients for the points along the lossless transmission line illustrated in Fig. 10.7.

Example 10.4 The reflection coefficient on a lossless transmission line is measured as $0.551\angle 25^\circ$ at 4.93 m along the line. Find the reflection coefficient at 6.89 m and 3.85 m along the line. The distances are measured on a ruler placed along the line, with increasing values toward the load. The wavelength is 10.00 m.

Solution Since the line is lossless the magnitude of the reflection coefficient is 0.551 all along the line. To go from 4.93 to 6.89 meters we must rotate $720(z - z')/\lambda = 720(6.89 - 4.93)/10 = 141^\circ$ in the counterclockwise direction. This changes the angle from 25° to $(25^\circ + 141^\circ) = 166^\circ$. To go from 4.93 to 3.85 meters we must rotate $720(z - z')/\lambda = 720(3.85 - 4.93)/10 = -78^\circ$ in the clockwise direction. This changes the angle from 25° to $(25^\circ - 78^\circ) = -53^\circ$. These points are shown in Fig. 10.10.

We can reduce the calculation a little by adding to the angle scale a pair of scales marked in fractions of a wavelength. Since a movement by a half-wavelength along

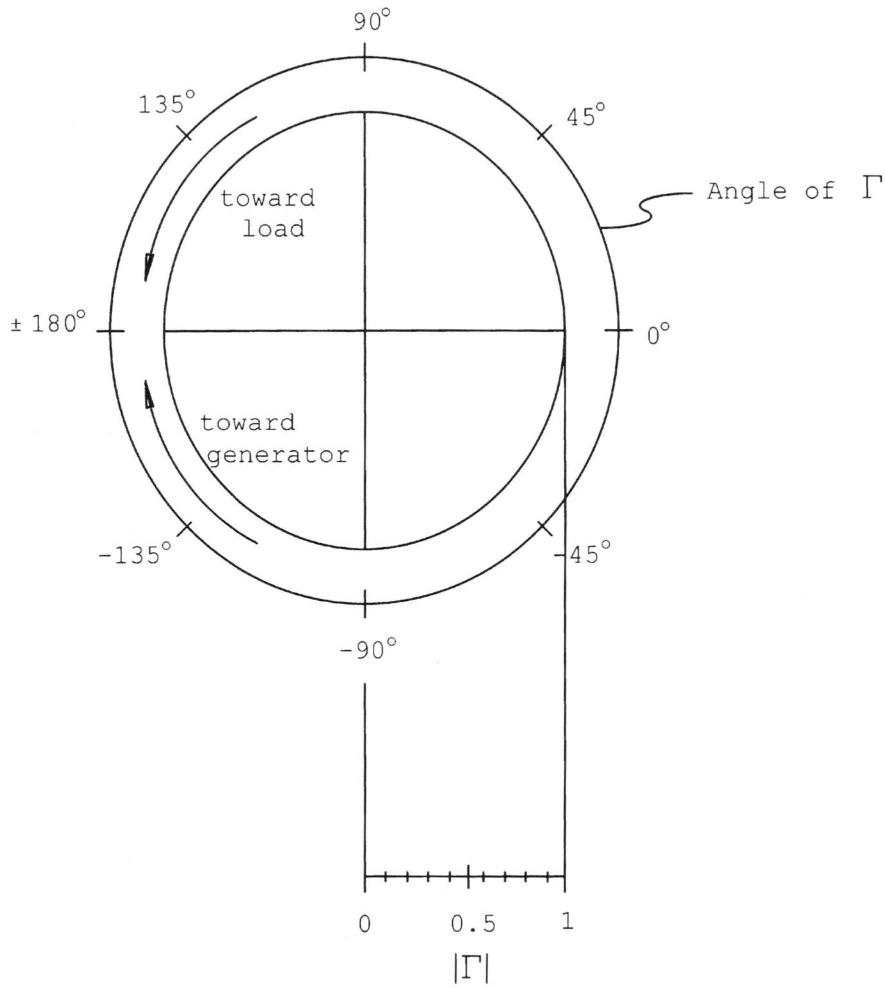

FIGURE 10.9 Reflection coefficient plane with linear scale for the magnitude ($|\Gamma|$) and angular scale for the phase ($\angle\Gamma$).

the line represents a complete rotation of 360°, these scales range from zero to 0.500 wavelengths (Fig. 10.11). We illustrate their use by repeating the last example.

Example 10.5 Repeat the previous example using the "wavelengths toward generator" and "wavelengths toward load" scales.

Solution We start by finding $\Gamma = 0.551\angle 25°$ on the Γ plane (Fig. 10.12). Draw a circle of radius 0.551 and a radial line at 25° to locate point a. The radial line shows a starting or tare value† of 0.285 wavelength toward the load and 0.215 wavelength toward the generator.

† The tare value is the value at the starting point. Rotations are always relative to the tare value. A scale shows 22 pounds before you step on it. When you are standing on the scale it shows 169 pounds. Your actual weight is $169 - 22 = 147$ pounds. The starting or tare value is 22 pounds and you must subtract this from the scale reading to get your weight.

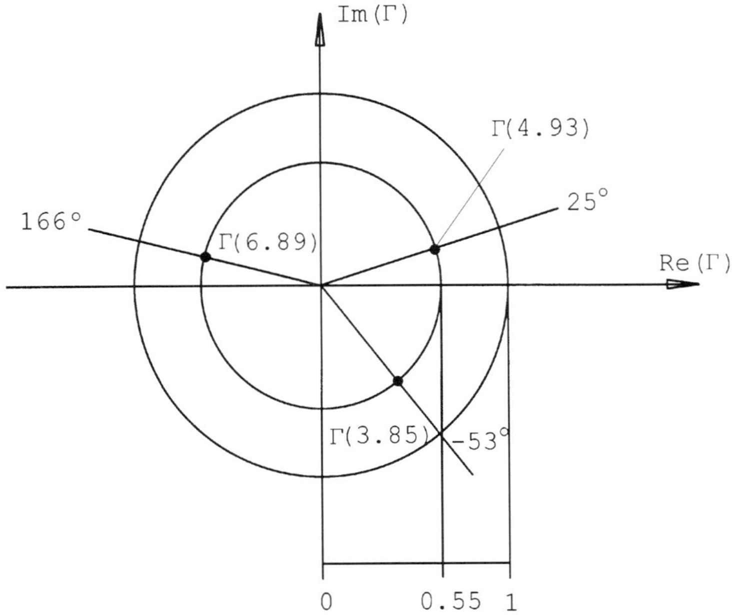

FIGURE 10.10 Reflection coefficients in Example 10.4.

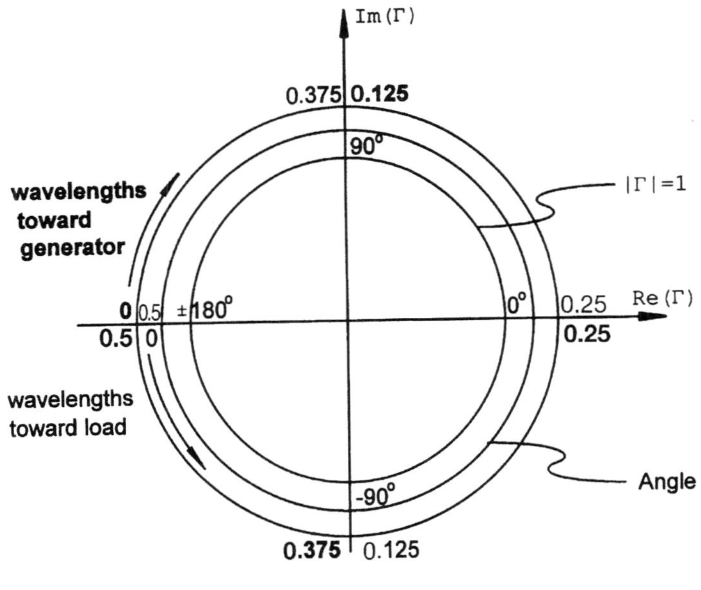

FIGURE 10.11 Reflection coefficient plane with added scales marked in fractions of a wavelength.

To go from 4.93 to 6.89 meters we must rotate $(6.89 - 4.93)/10 = 0.196$ wavelength toward the load. This moves the reflection coefficient from 0.285 to $0.285 + 0.196 = 0.481$ wavelength toward the load. Draw a radial line through this value and read the angle of the reflection coefficient as 166°. Point b shows

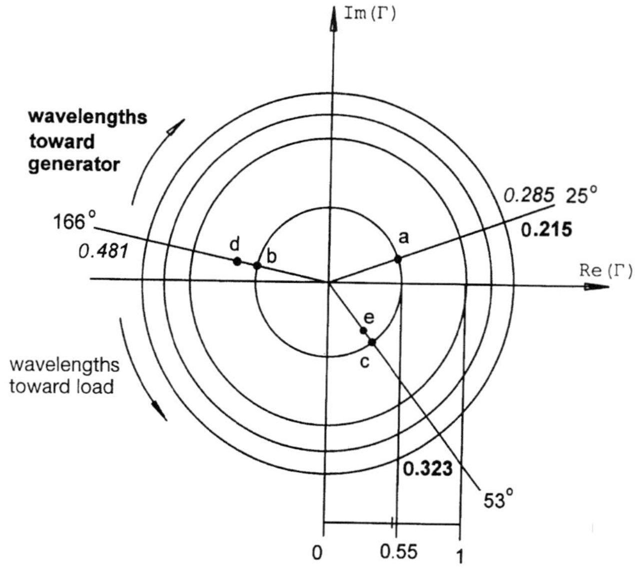

FIGURE 10.12 Reflection coefficients in Example 10.5.

the reflection coefficient. To go from 4.93 to 3.85 meters we must rotate $(4.93-3.85)/10 = 0.108$ wavelength toward the generator. This moves the reflection coefficient from 0.215 to $0.215 + 0.108 = 0.323$ wavelength toward the generator. Draw a radial line through this value and read the angle of the reflection coefficient as $-53°$. Point c shows the reflection coefficient.

When the line is not lossless the magnitude of the reflection coefficient changes as the plane along the line changes. Equation (10.26) becomes

$$|\Gamma(z)| = |\Gamma(z')|e^{2\alpha(z-z')} \quad \text{and} \quad \angle\Gamma(z) = \angle\Gamma(z') + 2\beta(z-z'). \tag{10.27}$$

Moving toward the load ($z > z'$), the magnitude increases; moving toward the generator ($z < z'$), it decreases.

Example 10.6 Repeat Example 10.4 if the line's attenuation constant is 0.025 Np/m.

Solution The angle of the attenuation constant is unchanged from that in the previous example. Going toward the load from 4.93 to 6.89 m the magnitude increases to $|\Gamma| = 0.551 \exp[0.05(6.89-4.93)] = 0.608$. Going toward the generator from 4.93 to 3.85 m the magnitude decreases to $|\Gamma| = 0.551 \exp[0.05(3.85-4.93)] = 0.522$. These values of Γ are also shown on Fig. 10.12 as points d and e, respectively. Note that as we go toward the load the value of reflection coefficient spirals outward.

10.3.2 The Normalized Impedance Plane

Just as the reflection coefficient can be pictured in the Γ plane, the impedance can be represented in an impedance plane. In order to make a single graphical representation for all transmission lines, we deal with a dimensionless quantity, the normalized impedance. The normalized impedance is the actual impedance divided by the characteristic impedance of the line, $\hat{Z}(z) = \hat{R} + j\hat{X} = Z(z)/Z_0$.[†]

The normalized impedance plane is shown in Fig. 10.13. The horizontal axis represents the real or resistive part of the normalized impedance, $\hat{R} = \text{Re}(\hat{Z}) = \text{Re}(Z/Z_0)$, and vertical lines are lines of constant normalized resistance. The vertical axis represents the imaginary or reactive part of the normalized impedance, $\hat{X} = \text{Im}(\hat{Z}) = \text{Im}(Z/Z_0)$, and horizontal lines are lines of constant normalized reactance.

In terms of normalized impedance, Eqs. (0.17) and (10.18) become

$$\Gamma(z) = \frac{\hat{Z}(z) - 1}{\hat{Z}(z) + 1} \tag{10.28}$$

and

$$\hat{Z}(z) = \frac{1 + \Gamma(z)}{1 - \Gamma(z)}. \tag{10.29}$$

Example 10.7 The load impedance of a 50 Ω transmission line is $Z_L = 150 + j150\ \Omega$. Determine the normalized load impedance and the load reflection coefficient. If the line has a length of 0.277λ, determine the input impedance to the line.

Solution The normalized load impedance is calculated as follows:

$$\hat{Z}_L = \frac{Z_L}{Z_0} = \frac{150 + j150}{50} = 3 + j3.$$

The load reflection coefficient will then be given by

$$\Gamma_L = \frac{\hat{Z}_L - 1}{\hat{Z}_L + 1} = \frac{(3 + j3) - 1}{(3 + j3) + 1} = \frac{2 + j3}{4 + j3} = \frac{3.61\, e^{j56.31^\circ}}{5e^{j36.87^\circ}} = 0.72e^{j19.44^\circ}.$$

In order to determine the input impedance to the line, a graphical approach is used. The load reflection coefficient is plotted on the Γ plane in Fig. 10.14. The line connecting the center of the circle and the point that represents the location of the reflection coefficient is extended until the scale indicating fractions of wavelength is intersected. The reading obtained is 0.223λ. Since the length of the transmission line is 0.277λ, then the input reflection coefficient will be located at the intersection of the circle of constant reflection coefficient with the horizontal line corresponding to $0.223\lambda + 0.277\lambda = 0.5\lambda$:

$$\Gamma_{\text{in}} = 0.72e^{j180^\circ} = -0.72.$$

[†] In this text we use a circumflex or "hat" ˆ to indicate a normalized value of impedance or admittance. This is a notation that you can easily use as well.

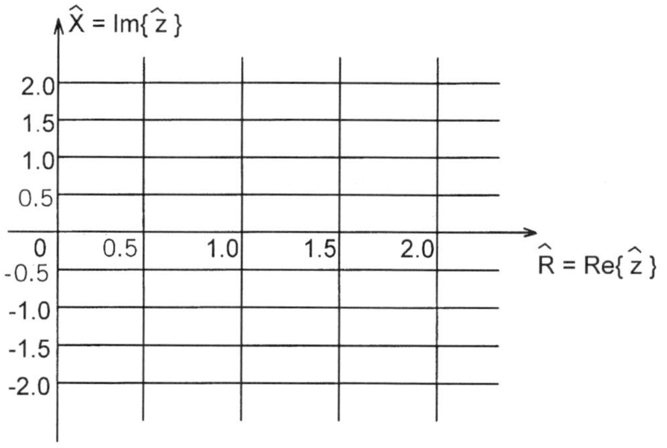

FIGURE 10.13 Normalized impedance plane showing lines of constant resistance and reactance.

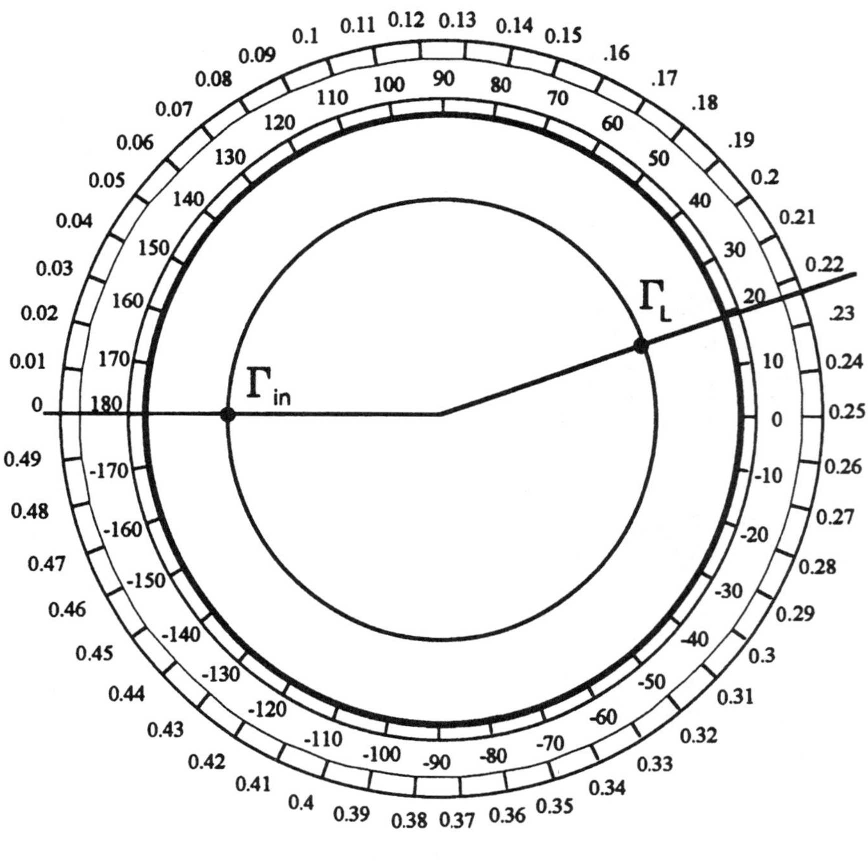

FIGURE 10.14 Reflection coefficients in Example 10.7.

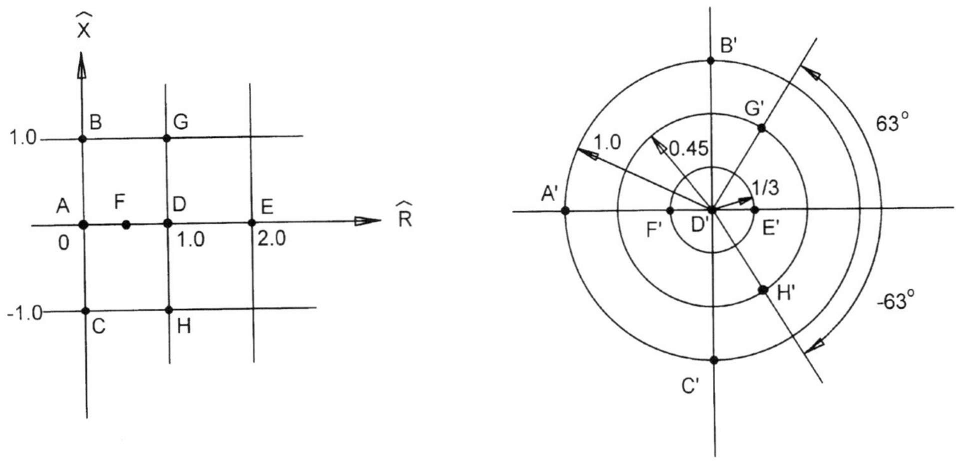

FIGURE 10.15 Illustration of the correspondence between points in the normalized impedance plane and reflection coefficient plane.

The normalized input impedance is then determined as

$$\hat{Z}_{in} = \frac{1 + \Gamma_{in}}{1 - \Gamma_{in}} = \frac{1 + (-0.72)}{1 - (-0.72)} = 0.1628.$$

The actual input impedance will be

$$Z_{in} = \hat{Z}_{in} \cdot Z_0 = 0.1628 \cdot 50\ \Omega = 8.14\ \Omega.$$

Equation (10.28) means that each point in the normalized impedance plane corresponds to a point in the Γ plane. Similarly, Eq. (10.29) means that each point on the Γ plane corresponds to a point in the normalized impedance plane. Figure 10.15 shows the correspondence between points A, B, C, ... and points A', B', C', ..., respectively. You should verify each pair of points.

If we could plot the vertical and horizontal lines of the impedance plane onto the Γ plane, then we would not need to do the computations required by Eqs. (0.28) and (10.29) in order to convert back and forth between values of normalized impedance and reflection coefficient. This plotting results in the Smith chart.

10.4 SMITH CHART

To plot the lines of constant $\hat{R}$ on the Γ plane we need an equation that involves Γ with $\hat{R}$ as a parameter, but without $\hat{X}$. Similarly, to plot the lines of constant $\hat{X}$ on the Γ plane we need an equation that involves Γ with $\hat{X}$ as a parameter, but without $\hat{R}$. This separation can be accomplished by setting $\hat{Z} = \hat{R} + j\hat{X}$ in Eq. (10.28) and taking the real and imaginary parts of both sides of the equation. Thus, letting $\Gamma = \Gamma_R + j\Gamma_i$,

$$
\begin{aligned}
\hat{Z} &= \frac{1+\Gamma}{1-\Gamma} \\
\hat{R} + j\hat{X} &= \frac{(1+\Gamma_r) + j\Gamma_i}{(1-\Gamma_r) - j\Gamma_i} = \frac{[(1+\Gamma_r) + j\Gamma_i][(1-\Gamma_r) + j\Gamma_i]}{[(1-\Gamma_r) - j\Gamma_i][(1-\Gamma_r) + j\Gamma_i]} \\
&= \frac{(1 - \Gamma_r^2 - \Gamma_i^2) + j2\Gamma_i}{(1-\Gamma_r)^2 + \Gamma_i^2}.
\end{aligned} \tag{10.30}
$$

We take the real part of both sides of this equation, multiply by the denominator of the right side, and group terms in powers of Γ_r and Γ_i to obtain

$$
\begin{aligned}
(1+\hat{R})\Gamma_r^2 - 2\hat{R}\Gamma_r + (1+\hat{R})\Gamma_i^2 &= 1 - \hat{R} \\
\Gamma_r^2 - 2\frac{\hat{R}}{1+\hat{R}}\Gamma_r + \Gamma_i^2 &= \frac{1-\hat{R}}{1+\hat{R}}.
\end{aligned}
$$

By adding $[\hat{R}/(1+\hat{R})]^2$ to both sides of the equation we complete the square on the left side and obtain the result

$$
\left(\Gamma_r - \frac{\hat{R}}{1+\hat{R}}\right)^2 + \Gamma_i^2 = \left(\frac{1}{1+\hat{R}}\right)^2. \tag{10.31}
$$

This is in the form of $(x - X_0)^2 + (y - Y_0)^2 = r^2$, which is the equation of a circle in the xy plane with its center at (X_0, Y_0) and with the radius r. Thus Eq. (10.31) represents a circle in the Γ plane with its center at $[\Gamma_{r0} = \hat{R}/(1+\hat{R}),\ \Gamma_{i0} = 0]$ and with a radius $1/(1+\hat{R})$. Figure 10.16a shows a table for several values of $\hat{R}$ and Fig. 10.16b shows corresponding circles in the Γ plane. These circles are the vertical lines of Fig. 10.13 mapped into the Γ plane.

Next, we take the imaginary part of both sides of Eq. (10.30), multiply by the denominator of the right side, and group terms in powers of Γ_r and Γ_i to obtain

$$
\begin{aligned}
\hat{X}\Gamma_r^2 - 2\hat{X}\Gamma_r + \hat{X}\Gamma_i^2 - 2\Gamma_i &= -\hat{X}, \\
(\Gamma_r^2 - 2\Gamma_r) + \left(\Gamma_i^2 = \frac{2}{\hat{X}}\Gamma_i\right) &= -1.
\end{aligned}
$$

By adding 1 to the first parentheses and $1/\hat{X}^2$ to the second and the same terms to the right side of the equation, we complete the squares on the left side and obtain the result

$$
(\Gamma_r - 1)^2 + \left(\Gamma_i - \frac{1}{\hat{X}}\right)^2 = \left(\frac{1}{\hat{X}}\right)^2. \tag{10.32}
$$

This represents a circle in the Γ plane with center at $[\Gamma_{r0} = 1,\ \Gamma_{i0} = 1/\hat{X}]$ and with a radius $1/|\hat{X}|$. Figure 10.17a shows a table for several values of $\hat{X}$ and Fig. 10.17b shows the corresponding arcs of circles in the Γ plane. These arcs are the horizontal lines of Fig. 10.13 mapped into the Γ plane.

$\hat{R}$	Γ_{r0}	Γ_{i0}	Radius
0	0	0	1
1/2	1/3	0	2/3
1	1/2	0	1/2
2	2/3	0	1/3
∞	1	0	0

(a)

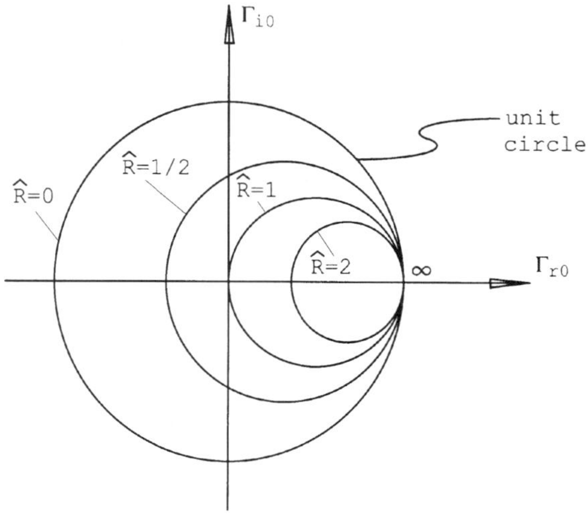

(b)

FIGURE 10.16 Mapping of constant resistance lines in the normalized impedance plane into circles of constant resistance in the reflection coefficient plane.

$\hat{X}$	Γ_{r0}	Γ_{i0}	Radius
0	1	∞	∞
1/2	1	2	2
1	1	1	1
2	1	1/2	1/2
∞	1	0	0

(a)

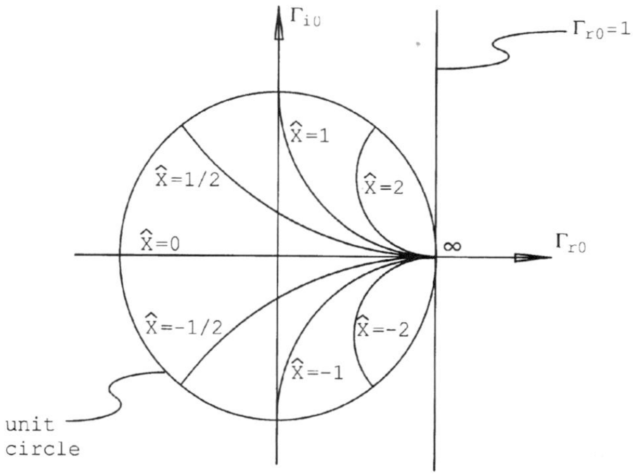

(b)

FIGURE 10.17 Mapping of constant reactance lines in the normalized impedance plane into circles of constant reactance in the reflection coefficient plane.

In Fig. 10.18 we superimpose the two sets of circles to represent completely the transformation of the normalized impedance plane onto the Γ plane. At the same time we suppress the circles and radial lines representing the lines of constant magnitude and angle of reflection coefficient. These lines would clutter the picture and are easy to reproduce with the aid of a compass and a straight edge. The constant normalized

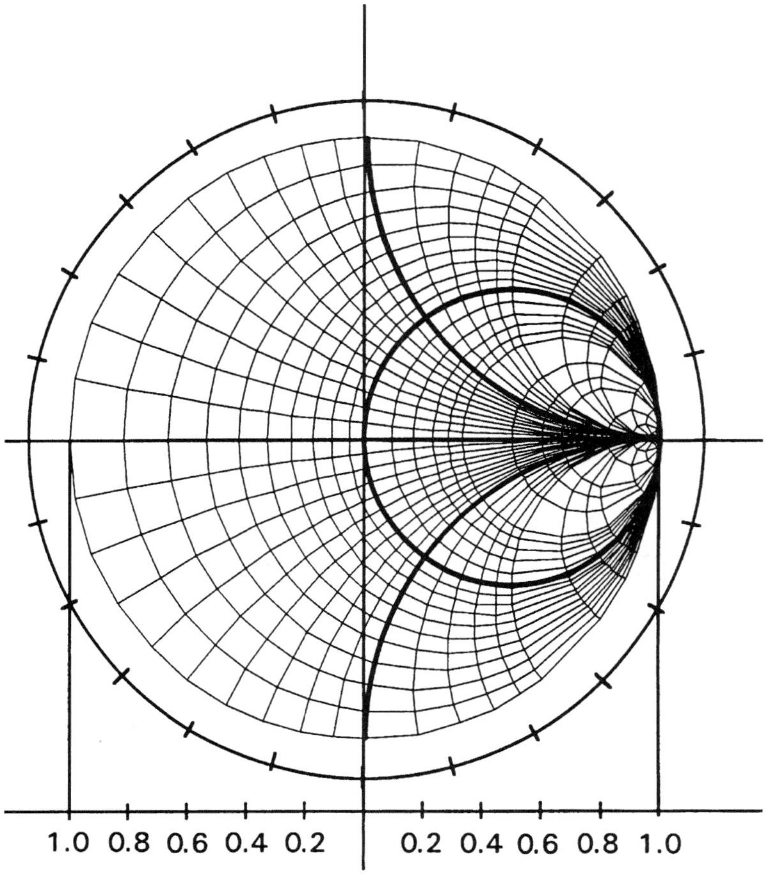

FIGURE 10.18 Smith chart.

resistance circles are clearly marked along the horizontal axis, starting from $R/Z_0 = 0$ on the left, through $R/Z_0 = 1.0$ at the origin, to $R/Z_0 = \infty$ on the right. The constant normalized reactance circles are clearly marked along the rim of the chart. Positive or inductive reactance is above the horizontal axis, going from $X/Z_0 = 0$ at the left, through $X/Z_0 = 1.0$ at the top, to $X/Z_0 = \infty$ at the right. Negative or capacitive reactance is below the horizontal axis, going from $X/Z_0 = 0$ at the left, through $X/Z_0 = -1.0$ at the bottom, to $X/Z_0 = -\infty$ at the right.

We have now derived the complete Smith chart. Let us do some examples to illustrate its use.

Example 10.8 Assume the line of Example 10.4 has a characteristic resistance of 50 Ω. Use the Smith chart to find the reflection coefficient at 6.89 m and at 3.95 m and find the impedance looking toward the load at each of these planes.

Solution $|\Gamma| = 0.551$ all along the line and we start by drawing a circle with this radius, measured on the "REFL COEF" scale at the bottom of the chart in Fig. 10.19. Draw a radial line through 25°. The intersection is marked $\Gamma(0.493)$ and we

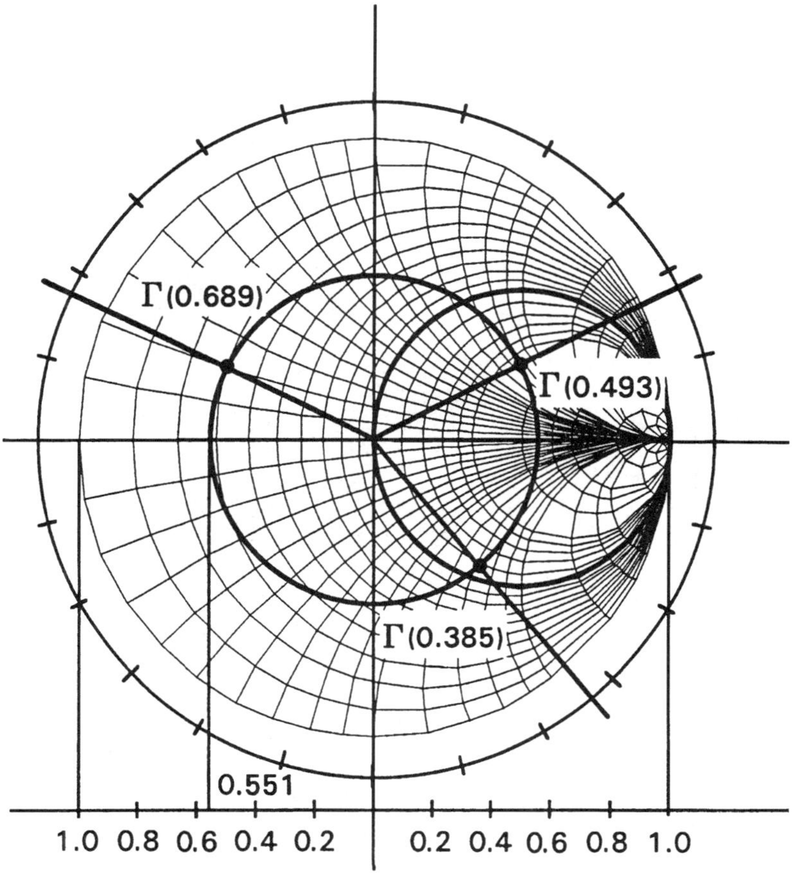

FIGURE 10.19 Smith chart solution to Example 10.8.

read the normalized impedance $\hat{Z}(0.493) = 2.3 + j1.60$. The unnormalized value is $Z(0.493) = 115 + j80$. Draw a second radial line through 0.481 wavelength toward the load (see Example 10.5) to get the point marked $\Gamma(0.689)$. The angle of the reflection coefficient is 166°, the normalized impedance is $\hat{Z}(0.689) = 0.30 + j0.11$, and the unnormalized impedance is $Z(0.689) = 15.0 + j5.50$. Draw a third radial line through 0.323 wavelength toward the generator to get the point marked $\Gamma(0.385)$. The angle of the reflection coefficient is $-53°$, the normalized impedance is $\hat{Z}(0.385) = 1.10 - j1.40$, and the unnormalized impedance is $Z(0.385) = 55.0 - j70.0$.

Example 10.9 For the line and load of Example 10.1 use the Smith Chart to find Γ_L, Γ_{in}, and Z_{in}.

Solution We enter the chart in Fig. 10.20 at $\hat{Z}_L = Z_L/Z_0 = 1 + j$ and draw the circle of constant $|\Gamma|$ and a radial line through the point. Using the compass on the scale below we read $|\Gamma| = 0.46$ and from the radial line we read $\angle\Gamma = 64°$. The radial line gives a tare reading of 0.1615 wavelength toward the generator for the load position.

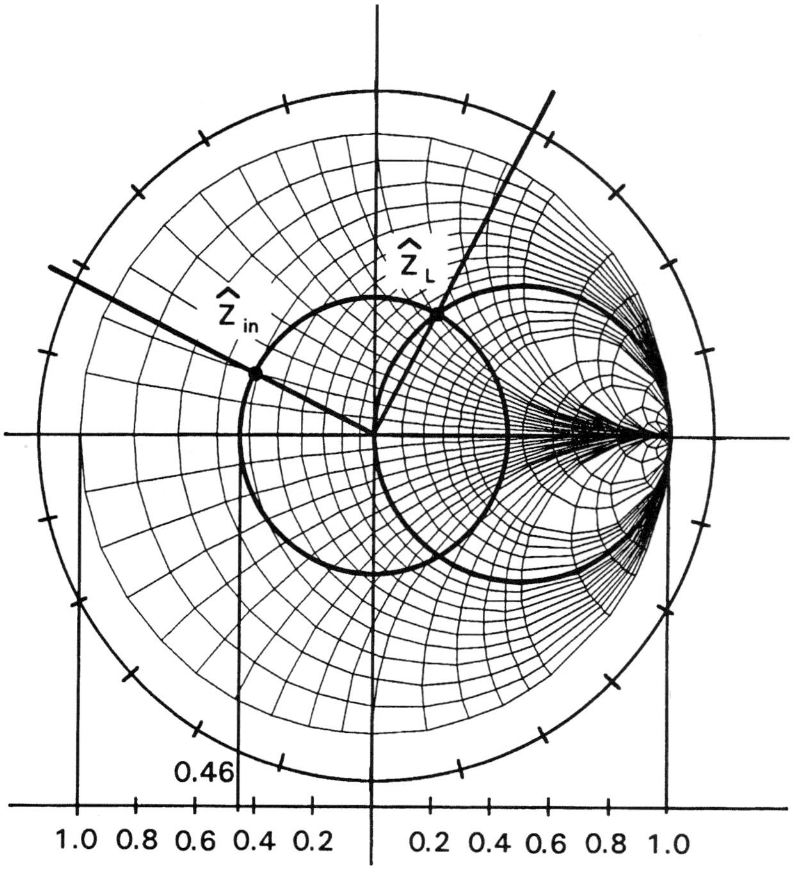

FIGURE 10.20 Smith chart solution to Example 10.9.

We add $3/8 = 0.3750$ wavelength to get the input point at 0.5365 wavelength toward the generator. The scale goes up to 0.5 and then starts over, so we draw a radial line at 0.0365 wavelength toward the generator. On the lossless line the $|\Gamma|$ is constant so the point where this radial line crosses our circle gives $\Gamma_{in} = 0.46 \angle 154^\circ$. The same point shows $\hat{Z}_{in} = 0.400 + j0.200$.

10.4.1 Uses for the Smith Chart

The previous two examples illustrate the use of the Smith chart as a graphical substitute for the complex number calculations of Eqs. (10.28) and (10.29) and Eq. (10.25). As with any graphical technique, the accuracy is limited. When a calculator is used to get greater accuracy, the Smith chart should still be used as an approximate check. Currently, digital computers and tested software make use of the chart for computation of small importance.

The Smith chart is still used in microwave circuit problems as a means of graphically portraying the results of calculations or measurements, generally with frequency as a parameter. Most microwave software will actually display a Smith chart on the monitor with calculated points plotted on the chart. We will also do this to illustrate future results.

10.4.2 Smith Chart with Admittance Coordinates

So far we have dealt with the transformation between the impedance plane and the Γ plane. Sometimes it is desirable to use admittance rather than impedance. In that case, we will need a transformation between the admittance plane and the Γ plane. Fortunately we can use the previous results.

The admittance is $Y(z) = G + jB = 1/Z(z) = I(z)/V(z)$ and the characteristic admittance is $Y_0 = 1/Z_0$. The normalized admittance is $\hat{Y} = \hat{G} + j\hat{B} = Y/Y_0 = YZ_0$. In terms of admittance, Eq. (10.28) becomes

$$\Gamma(z) = \frac{1 - Y(z)Z_0}{1 + Y(z)Z_0}$$

$$-\Gamma(z) = \frac{\hat{Y}(z) - 1}{\hat{Y}(z) + 1}, \tag{10.33}$$

and its inverse [corresponding to Eq. (10.29)] becomes

$$\hat{Y}(z) = \frac{1 + [-\Gamma(z)]}{1 - [-\Gamma(z)]}. \tag{10.34}$$

The relations between normalized impedance $\hat{Z}$ and reflection coefficient Γ are identical to the relations between normalized admittance $\hat{Y}$ and the negative of the reflection coefficient $-\Gamma$.

The normalized admittance plane is shown in Fig. 10.21a and the $-\Gamma$ plane is shown in Fig. 10.21b. The vertical lines in the $\hat{Y}$ plane are lines of constant normalized conductance and the horizontal lines are lines of constant normalized susceptance. These transform into circles (Fig. 10.21c) in the $-\Gamma$ plane. Thus exactly the same chart can be used for the admittance to reflection coefficient transformation, but the angle of the reflection coefficient must be corrected by adding or subtracting 180°. Alternately, we can read the angle of the reflection coefficient from the line drawn from the point through the origin out to the diametrically opposite point on the outer scale.

Example 10.10 For $\hat{Y} = 0.5 - j0.5$, find the reflection coefficient.

Solution The point is located on the Smith chart (Fig. 10.22). Drawing a circle through the point and using the compass to measure the distance along the scale gives $|\Gamma| = 0.45$. Drawing the (heavy) radial line outward shows $-116°$ and adding 180° gives $\angle\Gamma = 64°$. Alternately, drawing the (thin) radial line in the opposite direction gives 64° immediately.

To switch from the Γ plane to the $-\Gamma$ plane, or the reverse, we move from the point to the diametrically opposite point. Hence to switch from the normalized impedance

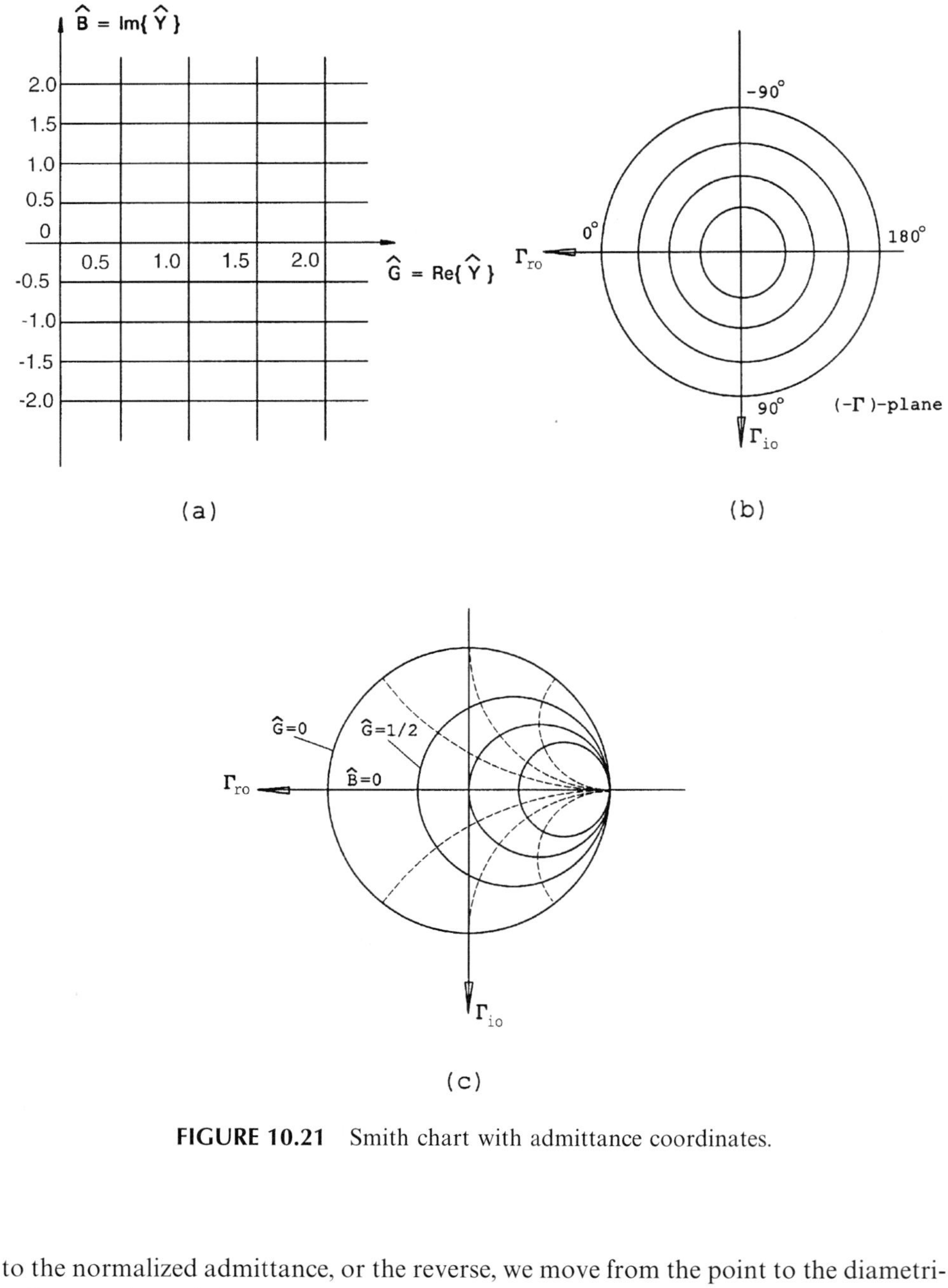

FIGURE 10.21 Smith chart with admittance coordinates.

to the normalized admittance, or the reverse, we move from the point to the diametrically opposite point.

Example 10.11 Find the normalized admittance for a normalized impedance $\hat{Z} = 1 + j$.

Solution Figure 10.23 shows $\hat{Z}$. The diametrically opposite point is found to be $\hat{Y} = 0.5 - j0.5$.

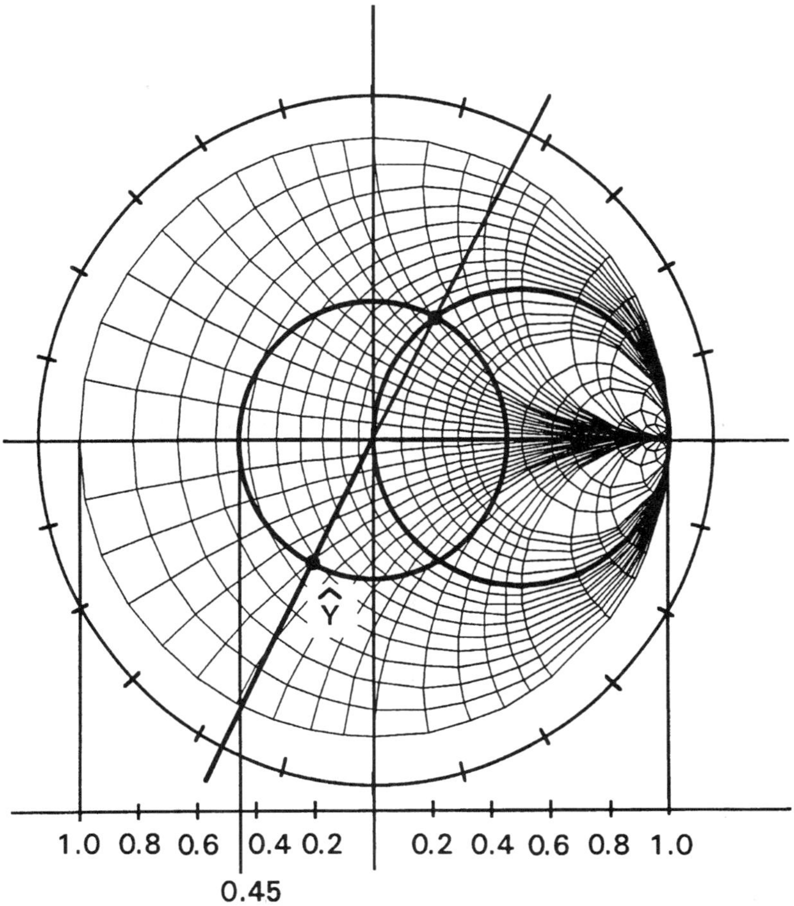

FIGURE 10.22 Smith chart solution to Example 10.10.

10.5 STANDING WAVES AND STANDING WAVE RATIO

It is important to look at the physical voltage and current wave on the terminated transmission line.

The voltage phasor can be written as

$$\underline{V}(z) = V_0^+ e^{-\alpha z} e^{-j\beta z}\left[1 + \frac{V_0^- e^{\alpha z} e^{j\beta z}}{V_0^+ e^{-\alpha z} e^{-j\beta z}}\right]$$

$$= V_0^+ e^{-\alpha z} e^{-j\beta z}[1 + \Gamma_L e^{2\alpha z} e^{j2\beta z}]$$

$$= \{V_0^+ [1 + |\Gamma_L| e^{2\alpha z} e^{j(2\beta z + \phi_L)}] e^{-\alpha z}\} e^{-j\beta z}, \qquad (10.35)$$

where $\Gamma_L = |\Gamma_L| \exp(j\phi_L)$ is the reflection coefficient at the load, which is located at $z = 0$. In this form the voltage is written as a forward travelling wave with amplitude given by the term within braces. The term within the braces is a complex function of z and can be replaced by the complex function $F(z) = |F(z)| \exp[j\Psi(z)]$. The physical

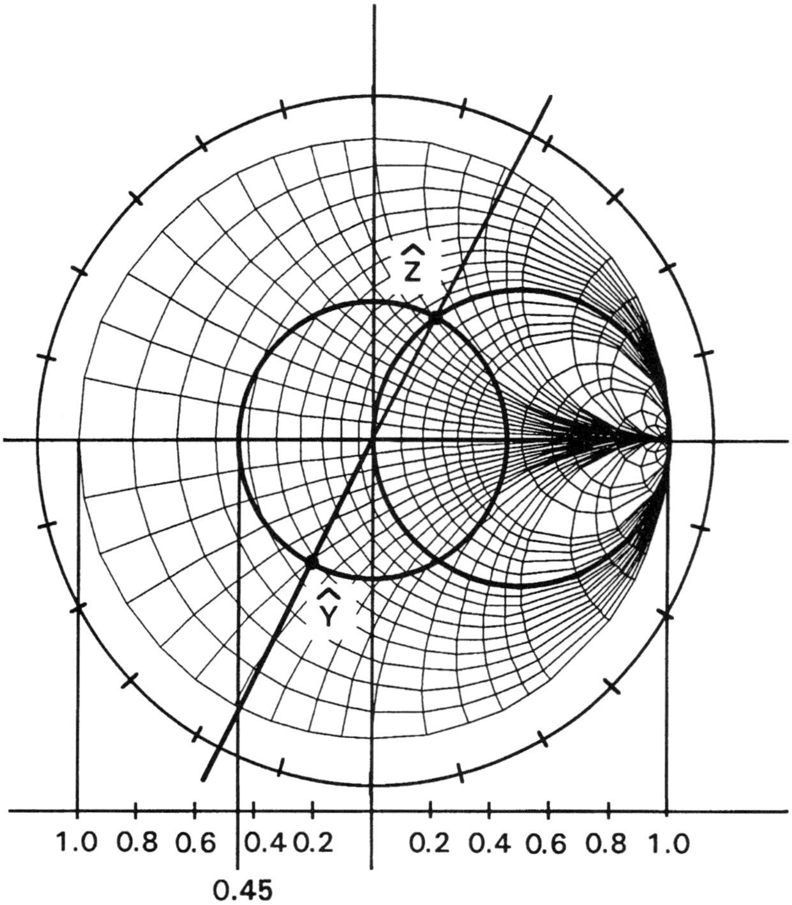

FIGURE 10.23 Smith chart solution to Example 10.11.

voltage wave along the line is found by multiplying by exp($j\omega t$) and taking the real part:

$$V(z,t) = |F(z)| \cos[\omega t - \beta z + \Psi(z)].$$

At any value of z along the line the voltage varies sinusoidally with time with amplitude

$$|F(z)| = |V_0^+||1 + |\Gamma_L| e^{2\alpha z} e^{j(2\beta z + \phi_L)}| e^{-\alpha z}. \quad (10.36)$$

To evaluate the magnitude of F, multiply the function by its complex conjugate, giving

$$|F(z)|^2 = |V_0^+|^2 [1 + 2|\Gamma_L| e^{2\alpha z} \cos(2\beta z + \phi_L) + |\Gamma_L|^2 e^{4\alpha z}] e^{-2\alpha z}. \quad (10.37)$$

The term within the brackets will have a maximum value when the cosine function equals $+1$. This occurs at half-wavelength intervals along the line, at points that we will call z_{max}. At these points the amplitude of the voltage wave is

$$V_{max} = |F(z_{max})| = (V_0^+|(1 + |\Gamma_L|e^{2\alpha z})e^{-\alpha z}.$$

The term within the brackets in Eq. (10.37) will have a minimum value when the cosine function equals -1. This also occurs at half-wavelength intervals, at points that we will call $z_{min} = z_{max} \pm \lambda/4$. At these points the amplitude of the voltage wave is

$$V_{min} = |F(z_{min})| = |V_0^+|(1 - |\Gamma_L|e^{2\alpha z})e^{-\alpha z}.$$

The standing wave ratio (SWR) S is defined as the ratio of a maximum voltage amplitude to an adjacent minimum voltage amplitude, so

$$S(z) = \frac{V_{max}}{V_{min}} = \frac{1 + |\Gamma_L|e^{2\alpha z}}{1 - |\Gamma_L|e^{2\alpha z}} = \frac{1 + |\Gamma(z)|}{1 - |\Gamma(z)|}. \tag{10.38}$$

The standing wave ratio is a positive real dimensionless number that is always greater than or equal to unity. For a lossless line, $\alpha = 0$, it is constant along the line. For a lossy line, $S \to 1$ as we go toward the source and z becomes more negative. To be exact, for a lossy line V_{max} in Eq. (10.38) should be evaluated at z_{max} and V_{min} at $z_{min} = z_{max} \pm \lambda/4$. For a low-loss line $\exp(2\alpha\lambda/4) \approx 1$ and Eq. (10.38) is correct. When the line is matched, $Z_L = Z_0$, $\Gamma_L = 0$, then $S = 1$. When the load is an open or short circuit or a pure reactance, then $|\Gamma_L| = 1$ and the SWR is infinite at the load and much greater than unity all along a low-loss line.

A similar analysis for the current wave gives

$$\underline{I}(z) = \left\{\frac{V_0^+}{Z_0}[1 - |\Gamma_L|e^{2\alpha z}e^{j(2\beta z + \phi_L)}]e^{-\alpha z}\right\}e^{-j\beta z} \tag{10.39}$$

and

$$I(z, t) = |G(z)|\cos[\omega t - \beta z + \zeta(z)],$$

where

$$|G(z)| = \frac{|V_0^+|}{|Z_0|}\left|1 - |\Gamma_L|e^{2\alpha z}e^{j(2\beta z + \phi_L)}\right|e^{-\alpha z}. \tag{10.40}$$

To evaluate the magnitude, we again multiply the function by its complex conjugate:

$$|G(z)|^2 = \frac{|V_0^+|^2}{|Z_0|^2}[1 - 2|\Gamma_L|e^{2\alpha z}\cos(2\beta z + \phi_L) + |\Gamma_L|^2e^{4\alpha z}]e^{-2\alpha z}. \tag{10.41}$$

Comparing with Eq. (10.37) for voltage we see the important difference of the change in sign for the term with the cosine function. For the current, the term within the brackets will have a maximum value when the cosine function equals -1. This occurs at the points that we called $z_{\min}$. At these points the amplitude of the current wave is

$$I_{\max} = |G(z_{\min})| = \frac{|V_0^+|}{|Z_0|}(1 + |\Gamma_L|e^{2\alpha z})e^{-\alpha z}.$$

The term within the brackets of Eq. (10.41) will have a minimum value when the cosine function equals $+1$. This occurs at points that we called $z_{\max}$. At these points the amplitude of the current wave is

$$I_{\min} = |G(z_{\max})| = \frac{|V_0^+|}{|Z_0|}(1 - |\Gamma_L|e^{2\alpha z})e^{-\alpha z}.$$

The ratio of the maximum current amplitude to the minimum current amplitude is seen to be the same S as for voltage. It is therefore not necessary to distinguish between a voltage and a current standing wave ratio.

Figure 10.24 shows the envelopes within which the voltage and the current oscillate sinusoidally at the angular frequency ω. The envelopes are stationary with time and the waves are called standing waves. The figure illustrates that the voltage maximum and the current minimum are located together at $z = z_{\max}$ and, similarly, the voltage minimum and the current maximum are located together at $z = z_{\min}$.

From Eq. (10.37) we see that

$$2\beta z_{\max} + \phi_L = 2n\pi, \quad z_{\max} = \frac{n\pi}{\beta} - \frac{\phi_L}{2\beta}. \tag{10.42}$$

Substituting this into Eqs. (10.35) and (10.39) we have

$$Z(z_{\max}) = \frac{\underline{V}(z_{\max})}{\underline{I}(z_{\max})} = Z_0\left(\frac{1 + |\Gamma_L|e^{2\alpha z}}{1 - |\Gamma_L|e^{2\alpha z}}\right) = Z_0 S.$$

Hence the normalized impedance looking toward the load at the plane of the voltage maximum and the current minimum is equal to the standing wave ratio:

$$\hat{Z}(z_{\max}) = \frac{Z(z_{\max})}{Z_0} = S. \tag{10.43}$$

On the Smith chart, used as an impedance chart, this corresponds to the points along the positive real axis of the Γ plane, shown circled with a solid line in Fig. 10.25.

Similarly, since

$$2\beta z_{\min} + \phi_L = (2n + 1)\pi, \quad z_{\max} = \frac{(n + 1/2)\pi}{\beta} - \frac{\phi_L}{2\beta}, \tag{10.44}$$

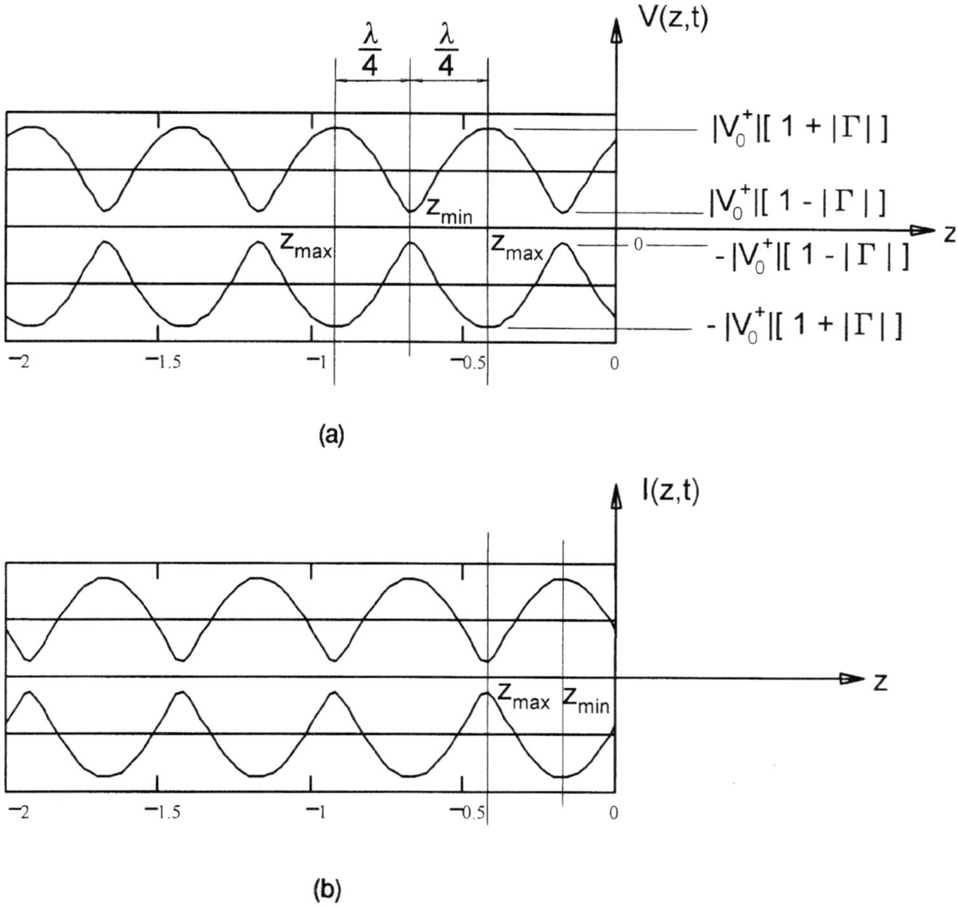

FIGURE 10.24 Plots of voltage and current amplitude envelopes.

then

$$Z(z_{\min}) = \frac{\underline{V}(z_{\min})}{\underline{I}(z_{\min})} = Z_0\left(\frac{1 - |\Gamma_L|e^{2\alpha z}}{1 + |\Gamma_L|e^{2\alpha z}}\right) = Z_0\left(\frac{1}{S}\right).$$

Hence the normalized impedance looking toward the load at the point of the voltage minimum and the current maximum is equal to the reciprocal of the standing wave ratio:

$$\hat{Z}(z_{\min}) = \frac{Z(z_{\min})}{Z_0} = \frac{1}{S}. \tag{10.45}$$

On the Smith chart, used as an impedance chart, this corresponds to the points along the negative real axis of the Γ plane, shown circled with a dashed line in Fig. 10.25.

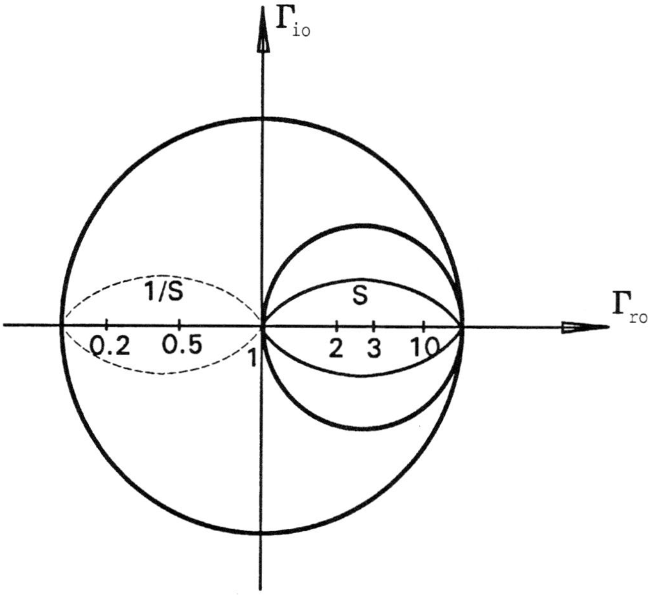

FIGURE 10.25 Plots of normalized impedances at the minimum and maximum voltage points for a standing wave on a transmission line.

The standing wave ratio is useful because it is an easy quantity to measure in the laboratory. When a probe that produces a voltage proportional to electric field intensity amplitude (or its square) is moved along the line and its output is displayed on a voltmeter, it is easy to measure a voltage proportional to V_{max} and V_{min} and hence the SWR. Knowing S we can calculate

$$|\Gamma| = \frac{S-1}{S+1}. \tag{10.46}$$

By measuring the distance between successive z_{min} we can calculate the wavelength. Furthermore, by measuring the distance from z_{min} to the load, we can also find the angle of the reflection coefficient and then the load impedance.

Example 10.12 Figure 10.26 shows a 75 Ω lossless air line with a ruler and a voltage-sensitive probe. With a short circuit as a load voltage, minima (zero volts) are detected at 1.15, 3.25, and 5.35 cm. With a load Z_L connected, a voltage ratio of $V_{max}/V_{min} = 2.80$ is measured and voltage minima are detected at 2.48, 4.58, and 6.68 cm. (a) Find the wavelength and the frequency. (b) Find the reflection coefficient at the load. (c) Find the load impedance and admittance.

Solution (a) Successive voltage minima are a half-wavelength apart. Thus $\lambda/2 = 3.25 - 1.15 = 5.35 - 3.25 = 2.10$ cm and $\lambda = 4.20$ cm. For a transmission line with air dielectric, $v_{ph} = c = 3 \times 10^{10}$ cm/s and hence $f = v_{ph}/\lambda = 7.14 \times 10^9$ Hz or 7.14 GHz.

(b) When the short circuit is connected as a load, a voltage minimum (zero volts) occurs at the load. The voltage minima detected along the line must be an integral

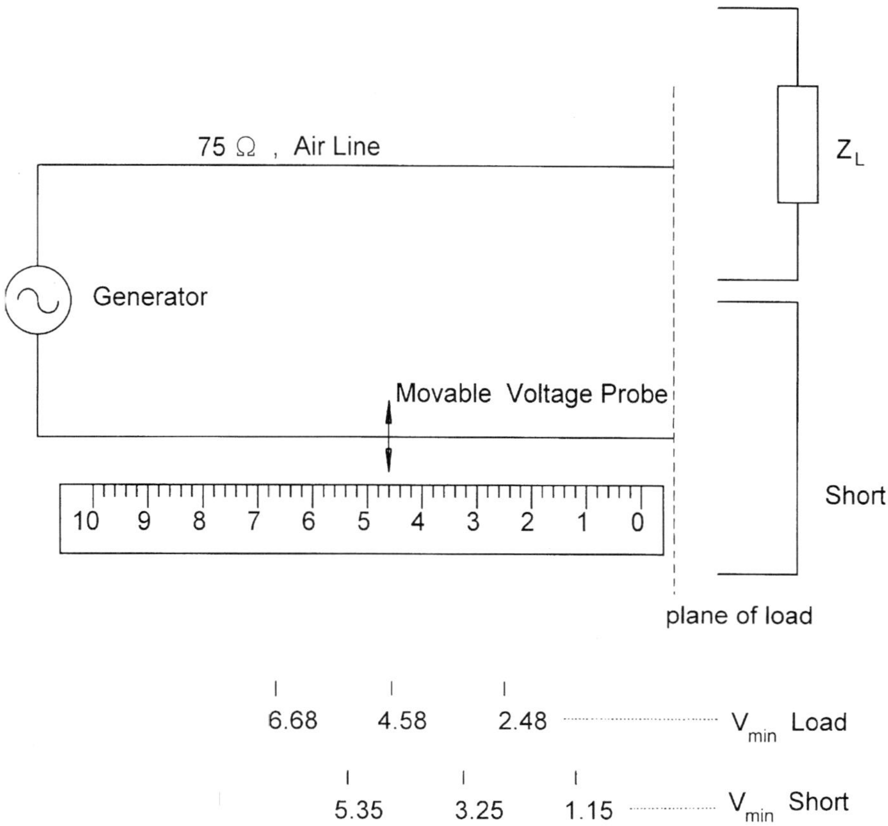

FIGURE 10.26 A 75 Ω lossless transmission line with a ruler and a movable voltage probe.

number of half-wavelengths away from the load. The positions of these minima are the electrical equivalent of the load position. With Z_L connected as a load, the minima have all shifted toward the generator by $d = 2.48 - 1.15 = 4.58 - 3.25 = 6.68 - 5.35 = 1.33$ cm or $d/\lambda = 0.317$ wavelength. The voltage ratio measured tells us that $S = 2.80$ and hence the reflection coefficient lies on the circle going through $R/Z_0 = 2.80$ on the Smith chart (point A, Fig. 10.27). Point B corresponds to the position of the voltage minimum with Z_L attached. To move from the voltage minimum with the load connected (point B) to the position of the load, we rotate 0.317 wavelength toward the load to point C. At point C we find $\Gamma_L = 0.475\angle 48.5°$.

(c) From point C we read $\hat{Z}_L = 1.30 + j1.20$ so $Z_L = 97.5 + j90.0\ \Omega$. From the diametrically opposite point D we read $\hat{Y}_L = 0.42 - j0.38$ so $Y_L = (5.60 - j5.07) \times 10^{-3}$ s.

10.6 RESONANT CIRCUITS

The subject of natural frequencies and resonance is important in basic lumped circuit theory and we begin with a short review. For a more complete discussion see any text on basic circuit theory. References [1–3] are examples of such texts.

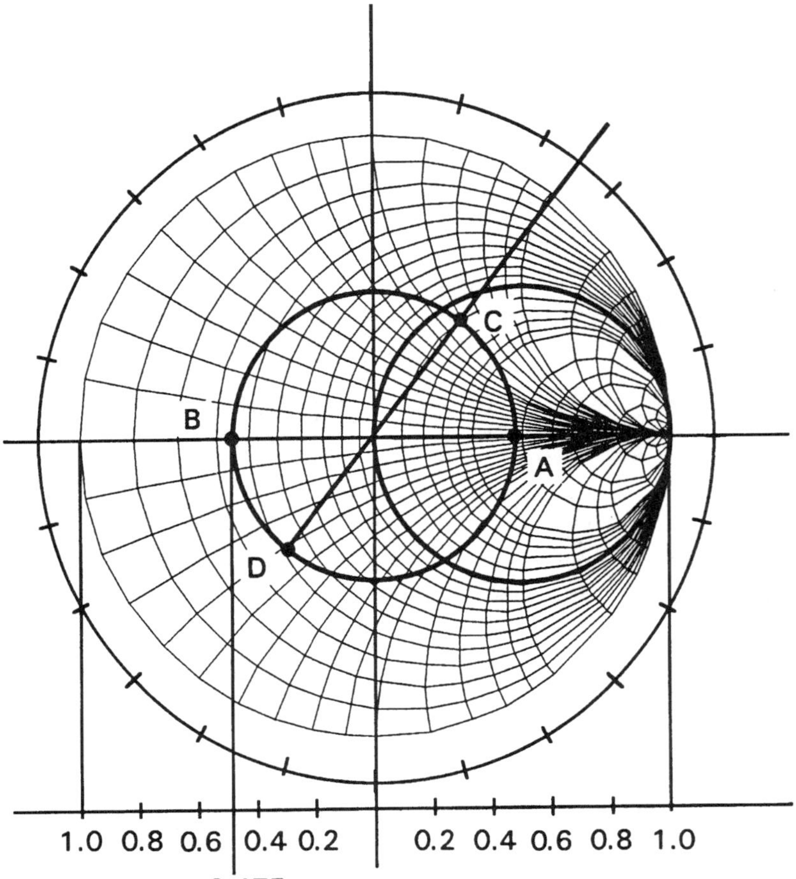

FIGURE 10.27 Smith chart solution to Example 10.12.

Figure 10.28 shows a series *RLC* circuit with a sinusoidal voltage source $v(t) = V_0 \cos \omega t$. The impedance seen by the source is

$$Z(j\omega) = R + j\left(\omega L - \frac{1}{\omega C}\right) = R + jX.$$

When the source angular frequency is $\omega = \omega_0 = 2\pi f_0 = 1/\sqrt{LC}$ the reactance X is zero, the magnitude of the impedance is a minimum, and the magnitude of the admittance $Y = 1/Z$ is a maximum. This frequency is called the resonant frequency of the circuit and the circuit is said to be "in resonance." When the circuit is in resonance the impedance is pure real, $Z = R$, the current amplitude is a maximum, and the current and source voltage are in phase.

Let us look at the power and energy relations in the resonant circuit. At resonance the current is $i(t) = (V_0/R)\cos \omega_0 t$. The time average power dissipated in the resistor is

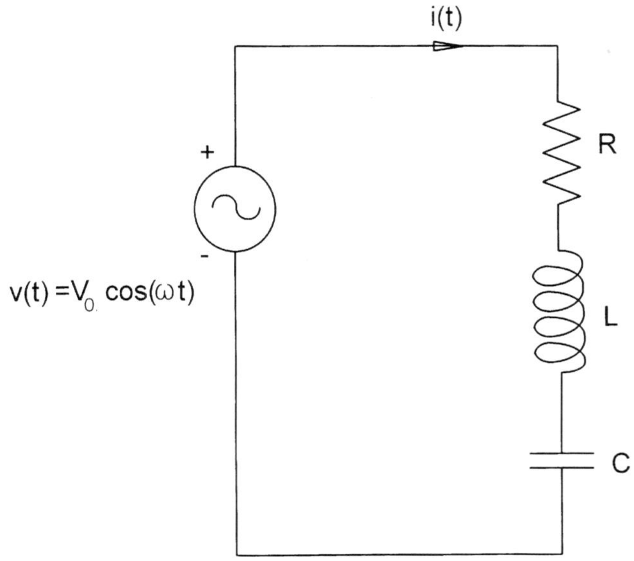

FIGURE 10.28 Series resonant *RLC* circuit with a sinusoidal source.

$$P_{\text{Avg}} = \tfrac{1}{2} R \left(\frac{V_0}{R} \right)^2 = \frac{V_0^2}{2R}, \quad \omega = \omega_0. \tag{10.47}$$

The instantaneous magnetic energy stored in the inductor is

$$W_m(t) = \tfrac{1}{2} L \left(\frac{V_0}{R} \right)^2 \cos^2 \omega_0 t, \quad \omega = \omega_0,$$

and the time average magnetic energy is

$$W_{m,\text{Avg}} = \tfrac{1}{4} L \left(\frac{V_0}{R} \right)^2, \quad \omega = \omega_0.$$

The voltage phasor across the capacitor is $\underline{V}_c = V_0/(j\omega_0 \text{RC})$ and the instantaneous voltage is $v_c(t) = V_0/\omega_0 RC) \sin \omega_0 t$. The instantaneous electric energy stored in the capacitor is

$$W_e(t) = \tfrac{1}{2} C \left(\frac{V_0}{\omega_0 RC} \right)^2 \sin^2 \omega_0 t = \tfrac{1}{2} L \left(\frac{V_0}{R} \right)^2 \sin^2 \omega_0 t, \quad \omega = \omega_0,$$

and the time average electric energy is

$$W_{e,\text{Avg}} = \tfrac{1}{4} L \left(\frac{V_0}{\omega_0 RC} \right)^2 = \tfrac{1}{4} L \left(\frac{V_0}{R} \right)^2, \quad \omega = \omega_0.$$

We see that the amplitudes of the two instantaneous energy terms are identical (at resonance) and that when one energy is a maximum the other is zero. The total instantaneous electromagnetic energy stored at resonance is

$$W_T = W_T(t) = W_m(t) + W_e(t) = \tfrac{1}{2}L\left(\frac{V_0}{R}\right)^2 = W_{m,\text{Avg}} + W_{e,\text{Avg}}. \qquad (10.48)$$

This is a constant, independent of time. The energy is exchanged between the inductor and the capacitor.

We can define the Q (pronounced "cue," not to be confused with charge) or quality factor of the circuit as

$$Q = \omega_0 \frac{\text{Time average energy stored at } \omega_0}{\text{Time average power loss at } \omega_0} = \frac{\omega_0 W_T}{P_{\text{Avg}}}, \qquad (10.49)$$

which is dimensionless. Substituting from Eqs. (10.48) and (10.47), we obtain

$$Q = \frac{\omega_0 L}{R}, \quad \text{for series } RLC \text{ circuit.} \qquad (10.50)$$

Since the impedance amplitude is a minimum and the current amplitude is a maximum at the resonant frequency, the power transferred to the circuit is a maximum. The half-power frequencies are the frequencies f_1 and $f_2 > f_1$ at which the power transferred is one-half the maximum value. The power is reduced by one-half when the current amplitude is reduced by $1/\sqrt{2}$. This occurs when the magnitude of the reactance equals the resistance, $R = \pm X = \pm(\omega L - 1/\omega C)$, or

$$R = -\left(\omega_1 L - \frac{1}{\omega_1 C}\right).$$

$$\omega_1 = -\frac{R}{2L} + \sqrt{\left(\frac{R}{2L}\right)^2 + \frac{1}{LC}}$$

and

$$R = +\left(\omega_2 L - \frac{1}{\omega_2 C}\right),$$

$$\omega_2 = +\frac{R}{2L} + \sqrt{\left(\frac{R}{2L}\right)^2 + \frac{1}{LC}}.$$

In solving each quadratic equation we have kept only the solution for positive frequency. The difference $\omega_2 - \omega_1 = R/L$ rad/s gives the angular bandwidth. The fractional bandwidth is the bandwidth divided by the resonant frequency:

$$\frac{\text{Bandwidth}}{\text{Resonant frequency}} = \frac{f_2 - f_1}{f_0} = \frac{\omega_2 - \omega_1}{\omega_0} = \frac{R}{\omega_0 L} = \frac{1}{Q}. \qquad (10.51)$$

A high-Q circuit has a small fractional bandwidth and is very selective in terms of the frequencies for which power is delivered to the circuit. We also see, from Eq. (10.49), that a high-Q circuit has a large ratio of energy stored to energy lost per cycle. These properties mean that high-Q resonators can be used to realize both low-loss narrow-bandwidth filters and, by suitably cascading resonant circuits, low-loss shaped broad-band filters.

The relation between Q and bandwidth has been derived only for the series RLC circuit, but it is true for all high-Q circuits.

The parallel RLC circuit (Fig. 10.29) is the dual of the series circuit. At $\omega_0 = 1/\sqrt{LC}$ the admittance $Y(j\omega) = G + j(\omega C - 1/\omega L)$ is a minimum and the impedance $Z = 1/Y$ is a maximum. At ω_0 the circuit is in parallel resonance or antiresonance:

$$Q = \frac{\omega_0 C}{G} = \frac{R}{\omega_0 L}, \quad \text{for parallel } RLC \text{ circuit.} \tag{10.52}$$

Now let us find the natural frequencies of a series RLC circuit. These are the frequencies at which the circuit will oscillate if it is excited only by initial conditions, voltage on the capacitors and current in the inductors, with no forcing function. No forcing function means all voltage sources are replaced by their internal impedance (a short circuit for an ideal voltage source) and all current sources are replaced by their internal admittance (an open circuit for an ideal current source). Figure 10.30 shows the series RLC circuit of Fig. 10.27 with the ideal voltage source set to zero.

Replacing $j\omega$ by complex frequency p, the voltage across the three components must be zero: $\underline{V} = [R + (pL + 1/pC)]\underline{I} = 0$. This requires that the impedance function in the brackets be zero and therefore that p be a solution to the quadratic equation $LCp^2 + RCp + 1 = 0$, which is the characteristic equation of the circuit. The solutions are

$$p_1 = -\frac{R}{2L} + \sqrt{\left(\frac{R}{2L}\right)^2 - \frac{1}{LC}}, \quad p_2 = -\frac{R}{2L} - \sqrt{\left(\frac{R}{2L}\right)^2 - \frac{1}{LC}}.$$

If $(R/2L)^2 > (1/LC)$ the two roots are real and the natural "oscillations" are exponential decays. If $(R/2L)^2 < (1/LC)$ the two roots are complex conjugates and we have

$$p_1 = -\hat{\xi} + j\hat{\omega}, \quad p_2 = -\hat{\xi} - j\hat{\omega},$$

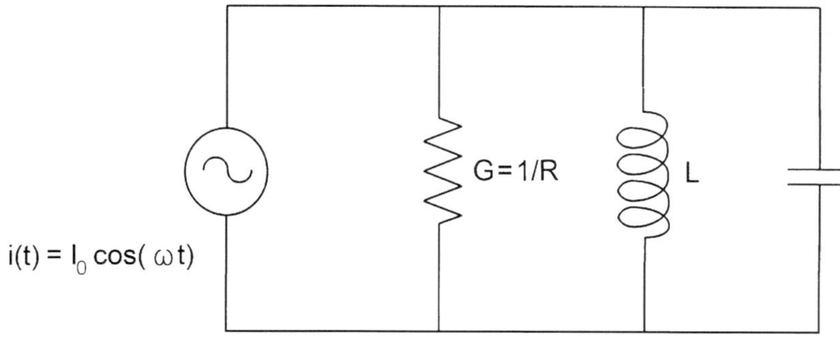

FIGURE 10.29 Parallel resonant RLC circuit with a sinusoidal source.

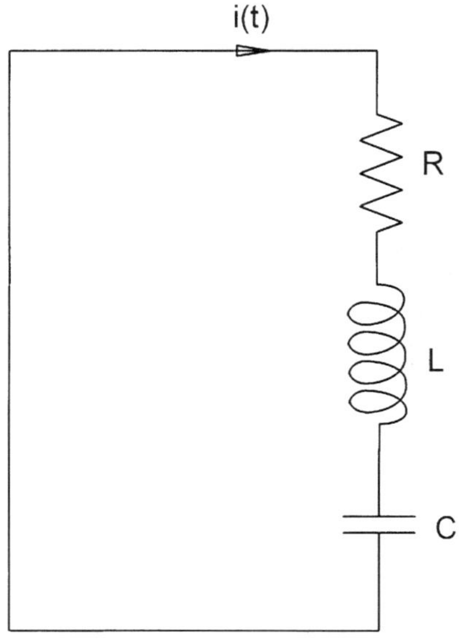

FIGURE 10.30 Source-free series *RLC* circuit.

where

$$\hat{\xi} = \frac{R}{2L}, \quad \hat{\omega} = \sqrt{\omega_0^2 - \hat{\xi}^2}, \quad \omega_0 = \frac{1}{\sqrt{LC}}.$$

The current in the circuit (and the voltage for each element) oscillates at a damped or natural angular frequency $\hat{\omega}$ rad/s and decays at the damping rate $\hat{\xi}$ Np/s.

Note that the complex numbers p_1 and p_2 are the natural frequencies of the circuit and that these are different from the resonant frequency ω_0.

The Q can be written in a third form:

$$Q = \frac{\omega_0 L}{R} = \frac{\omega_0}{2\hat{\xi}}. \tag{10.53}$$

It is important to understand the difference between the natural frequency and the resonant frequency. The natural frequencies are the frequencies at which the unforced or source-free circuit oscillates. They are the solution to the characteristic equation and may be real, imaginary, or complex. On the other hand, the resonant frequency is a real frequency. It is the frequency at which, for a constant-amplitude time harmonic voltage source, the input current is maximized, because the reactive part of the input impedance is zero. Similarly, the antiresonant frequency is the frequency at which, for a constant-amplitude time harmonic current source, the input voltage is maximized, because the susceptive part of the input admittance is zero.

For a lossless circuit $R = 0$ and the natural frequencies are equal to the resonant frequencies. In particular, for the simple LC circuit the natural frequency and the resonant angular frequency are given by $\hat{\omega} = \omega_0 = 1/\sqrt{LC}$. For a low-loss circuit,

$Q \gg 1$ or $\hat{\zeta} \ll \omega_0$, the natural frequencies are almost equal to the resonant frequencies: $\hat{\zeta} \approx R/2L$ and $\hat{\omega} \approx \omega_0$.

10.6.1 Natural Frequency: Shorted and Open Lossless Lines

The voltage and current phasors on a lossless transmission line are given by

$$\underline{V}(z) = V_0^+ e^{-j\beta z} + V_0^- e^{+j\beta z}, \tag{10.54}$$

$$\underline{I}(z) = (V_0^+ e^{-j\beta z} - V_0^- e^{+j\beta z})/R_0. \tag{10.55}$$

Figure 10.31 shows a lossless transmission line of length d, which is short-circuited at both ends. Since $\underline{V}(0) = 0$ we must have $V_0^- = -V_0^+ (\Gamma_L = -1)$. This means the voltage phasor is

$$\underline{V}(z) = -2jV_0^+ \sin \beta z, \tag{10.56}$$

and the corresponding current phasor is

$$\underline{I}(z) = \frac{2V_0^+}{R_0} \cos \beta z. \tag{10.57}$$

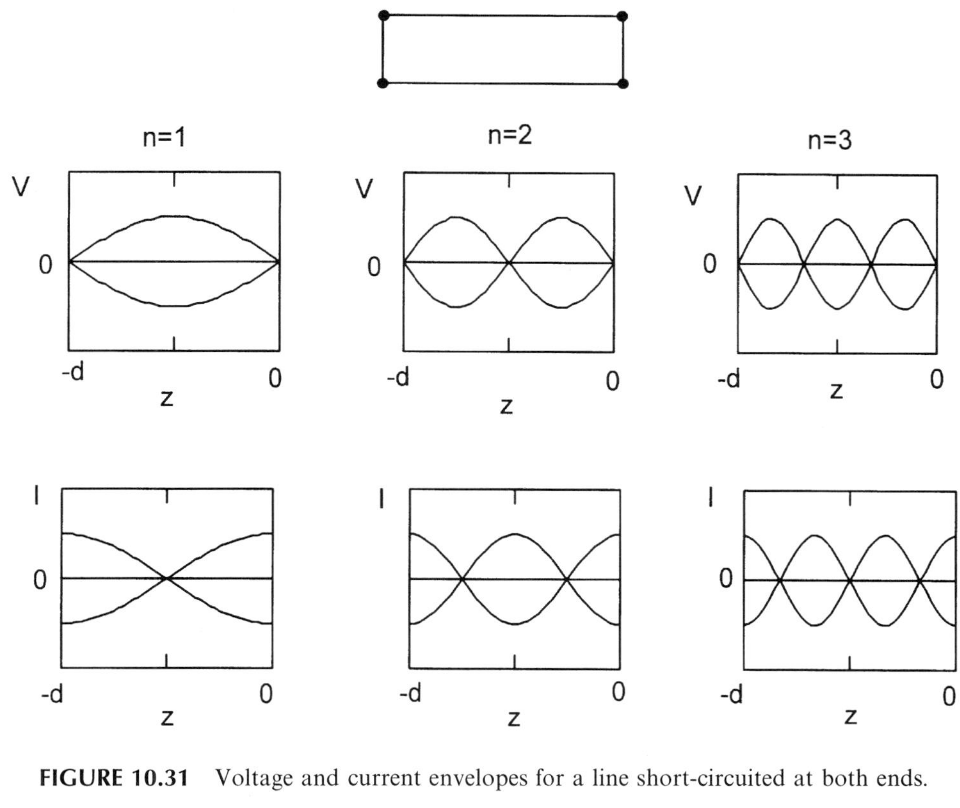

FIGURE 10.31 Voltage and current envelopes for a line short-circuited at both ends.

These are pure standing waves ($S = \infty$). The voltage at $z = -d$ must also be zero, $\underline{V}(-d) = 0$, and hence in Eq. (10.56) either $V_0^+ = 0$ or $\sin \beta d = 0$. The first possibility leads to the trivial solution that the voltage and current are everywhere equal to zero. The second possibility requires that $\beta d = \beta_n d = n\pi$ or

$$\lambda_n = \frac{2d}{n}, \qquad n = 1, 2, 3, \ldots,$$
$$f_n = \frac{v_{\text{ph}}}{\lambda_n} = \frac{n}{2d\sqrt{LC}}, \qquad \omega_n = 2\pi f_n = \frac{n\pi}{d\sqrt{LC}}. \tag{10.58}$$

Thus, in order to satisfy the boundary condition of zero voltage at the ends, only certain natural frequencies of oscillation f_n are possible. These natural frequencies are analogous to the natural frequency of the LC circuit, except that there are now an infinite number of them. At these frequencies we have

$$\underline{V}_n(z) = -2jV_0^+ \sin\left(\frac{n\pi z}{d}\right), \tag{10.59}$$

$$\underline{I}_n(z) = \frac{2V_0^+}{R_0} \cos\left(\frac{n\pi z}{d}\right), \tag{10.60}$$

which is sketched in Fig. 10.31 for $n = 1, 2, 3$. Taking V_0^+ to be real, the corresponding physical voltage and current are

$$V_n(z, t) = 2V_0^+ \sin\left(\frac{n\pi z}{d}\right) \sin \omega_n t, \tag{10.61}$$

$$I_n(z, t) = \frac{2V_0^+}{R_0} \cos\left(\frac{n\pi z}{d}\right) \cos \omega_n t. \tag{10.62}$$

Figure 10.32 shows a lossless line of length d terminated at each end in an open circuit. These terminations require that $\underline{I}(0) = 0$ and $\underline{I}(-d) = 0$. From Eq. (10.55), the first condition gives $V_0^- = V_0^+ (\Gamma_L = +1, S = \infty)$, and hence $\underline{I}(z) = -2j(V_0^+/R_0) \sin \beta z$. The second condition again requires that $\sin \beta d = 0$, which leads to the same natural frequencies, Eq. (10.58), and to the phasors

$$\underline{V}_n(z) = 2V_0^+ \cos\left(\frac{n\pi z}{d}\right),$$

$$\underline{I}_n(z) = -j\left(\frac{2V_0^+}{R_0}\right) \sin\left(\frac{n\pi z}{d}\right),$$

and the physical voltage and current

$$V_n(z, t) = 2V_0^+ \cos\left(\frac{n\pi z}{d}\right) \cos \omega_n t,$$

$$I_n(z, t) = \left(\frac{2V_0^+}{R_0}\right) \sin\left(\frac{n\pi z}{d}\right) \sin \omega_n t.$$

The voltage and current are again sketched in Fig. 10.32 for $n = 1, 2, 3$.

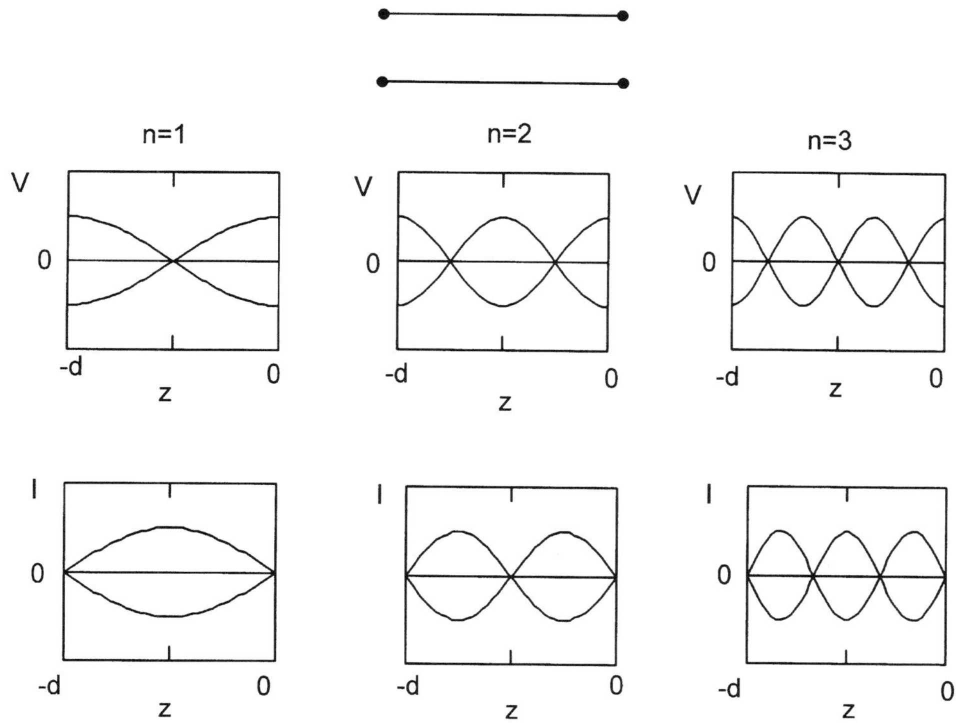

FIGURE 10.32 Voltage and current envelopes for a line open-circuited at both ends.

The third possibility is that the line is shorted at one end, say, $z = 0$, and open at the other, $z = -d$ (Fig. 10.33), requiring that $\underline{V}(0) = 0$ and $\underline{I}(-d) = 0$. The first condition requires that $V_0^- = -V_0^+$, again giving Eqs. (10.56) and (10.57). The second condition requires that $\cos \beta d = 0$ and therefore that $\beta_n d = (2n-1)\pi/2$. The natural frequencies are

$$\begin{aligned} \lambda_n &= \frac{4d}{2n-1}, \qquad n = 1, 2, 3, \ldots, \\ f_n &= \frac{v_{\text{ph}}}{\lambda_n} = \frac{2n-1}{4d\sqrt{LC}}, \quad \omega_n = 2\pi f_n = \frac{(2n-1)\pi}{2d\sqrt{LC}}. \end{aligned} \tag{10.63}$$

The voltage and current phasors are

$$\underline{V}_n(z) = -2jV_0^+ \sin\left[\frac{(2n-1)\pi z}{2d}\right],$$

$$\underline{I}_n(z) = \frac{2V_0^+}{R_0} \cos\left[\frac{(2n-1)\pi z}{2d}\right].$$

The voltage and current are again sketched in Fig. 10.33 for $n = 1, 2, 3$.

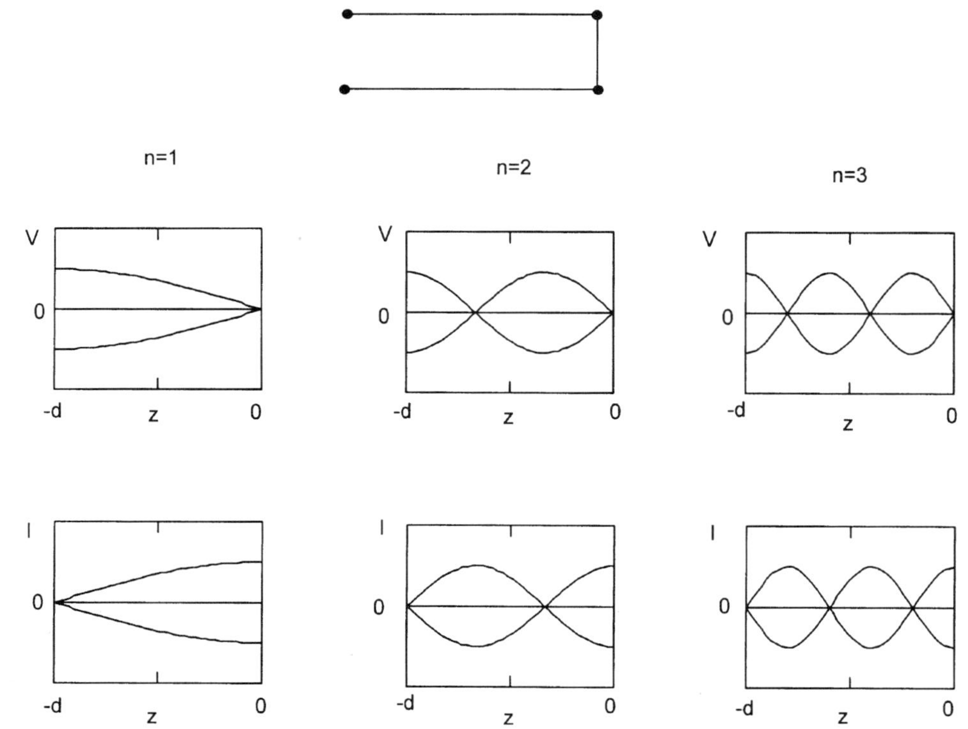

FIGURE 10.33 Voltage and current envelopes for a line open-circuited at one end and short-circuited at the other.

10.6.2 Resonant Frequency

Let us now consider an ideal voltage generator $\underline{V}_g$ with zero internal impedance, $Z_g = 0$, connected to a short-circuited low-loss (but not lossless) transmission line of length d (Fig. 10.34). The natural frequencies of the lossless line, $\omega_n = n\pi/d\sqrt{LC}$, are the resonant frequencies of the low-loss line. When the source frequency equals the nth resonant frequency, the voltage and current on the low-loss line will be practically the same as the voltage and current on the lossless line. Using Eqs. (10.59) and (10.60), we have

$$\underline{V}(-d) = \underline{V}_g, \quad \underline{V}(z) = -2jV_0^+ \sin\left(\frac{n\pi z}{d}\right), \qquad -d < z \le 0, \tag{10.64}$$

$$\underline{I}(z) = \frac{2V_0^+}{R_0} \cos\left(\frac{n\pi z}{d}\right), \qquad -d \le z \le 0. \tag{10.65}$$

Note that Eq. (10.64) includes the voltage source at $z = -d$ and is not the same as Eq. (10.59).

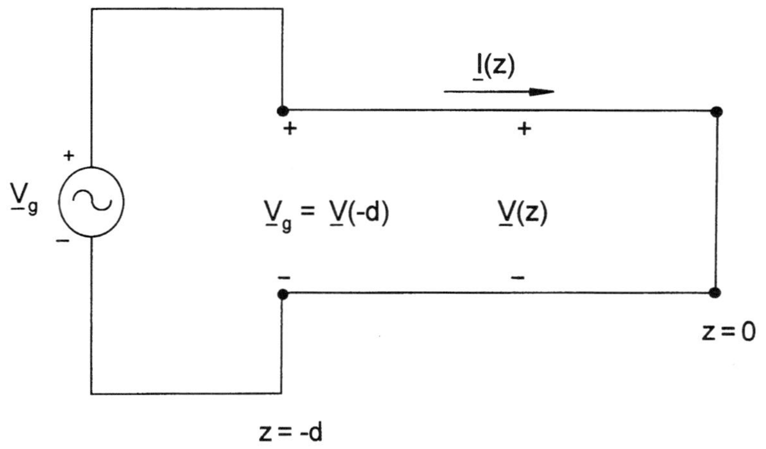

FIGURE 10.34 Ideal voltage generator connected to a short circuited low-loss transmission line.

The time average magnetic energy stored in a length of line dz located at z is $dW_{m,\text{Avg}} = \frac{1}{4}L|\underline{I}(z)|^2\,dz$ so the total time average magnetic energy stored along the line is

$$W_{m,\text{Avg}} = \frac{1}{4}\int_{-d}^{0} L\left(\frac{2|V_0^+|}{R_0}\right)^2 \cos^2\left(\frac{n\pi z}{d}\right)dz$$

$$= \tfrac{1}{8}\,Ld\left(\frac{2|V_0^+|}{R_0}\right)^2. \tag{10.66}$$

The time average electric energy stored in a length of line dz located at z is $dW_{e,\text{Avg}} = \frac{1}{4}C|\underline{V}(z)|^2\,dz$ so the total time average electric energy stored along the line is

$$W_{e,\text{Avg}} = \frac{1}{4}\int_{-d}^{0} C(2|V_0^+|)^2 \sin^2\left(\frac{n\pi z}{d}\right)dz$$

$$= \tfrac{1}{8}\,Cd(2|V_0^+|)^2. \tag{10.67}$$

The time average power loss in a length of line dz located at z is $dP_{\text{Avg}} = \frac{1}{2}[R|\underline{I}(z)|^2 + G|\underline{V}(z)|^2]\,dz$, so the total time average power loss along the line is

$$P_{\text{Avg}} = \frac{1}{2}\int_{-d}^{0}\left[R\left(\frac{2|V_0^+|}{R_0}\right)^2 \cos^2\left(\frac{n\pi z}{d}\right) + G(2|V_0^+|)^2 \sin^2\left(\frac{n\pi z}{d}\right)\right]dz$$

$$= \frac{1}{4}\left(\frac{Rd}{R_0^2} + Gd\right)(2|V_0^+|)^2. \tag{10.68}$$

The Q of the line at nth resonant frequency is

$$Q_n = \frac{\omega_n(W_{m,\text{Avg}} + W_{e,\text{Avg}})}{P_{\text{Avg}}} = \frac{\omega_n \frac{1}{8} d(L/R_0^2 + C)}{\frac{1}{4} d(R/R_0^2 + G)}$$

$$= \omega_n \left(\frac{R}{L} + \frac{G}{C}\right)^{-1}, \quad \omega_n = \frac{n\pi}{d\sqrt{LC}}, \tag{10.69}$$

where we have used $R_0 = \sqrt{L/C}$.

In the vicinity of ω_n the input impedance of the short-circuited line can be modeled as a series circuit of R_n, L_n, and C_n (Fig. 10.35a). The time average magnetic energy $W_{m,\text{Avg}} = \frac{1}{4} L_n |I(0)|^2$, where $I(0)$ is obtained from Eq. (10.65), and must equal the energy given by Eq. (10.66). Therefore

$$\frac{1}{4} L_n \left(\frac{2|V_0^+|}{R_0}\right)^2 = \frac{1}{8} Ld \left(\frac{2|V_0^+|}{R_0}\right)^2$$

$$L_n = \frac{1}{2} Ld. \tag{10.70}$$

For the series RLC circuit $Q_n = \omega_n L_n / R_n$ or $R_n = \omega_n L_n / Q_n$. Using Eqs. (10.69) and (10.70) gives

$$R_n = \frac{1}{2}\left(\frac{R}{L} + \frac{G}{C}\right) Ld. \tag{10.71}$$

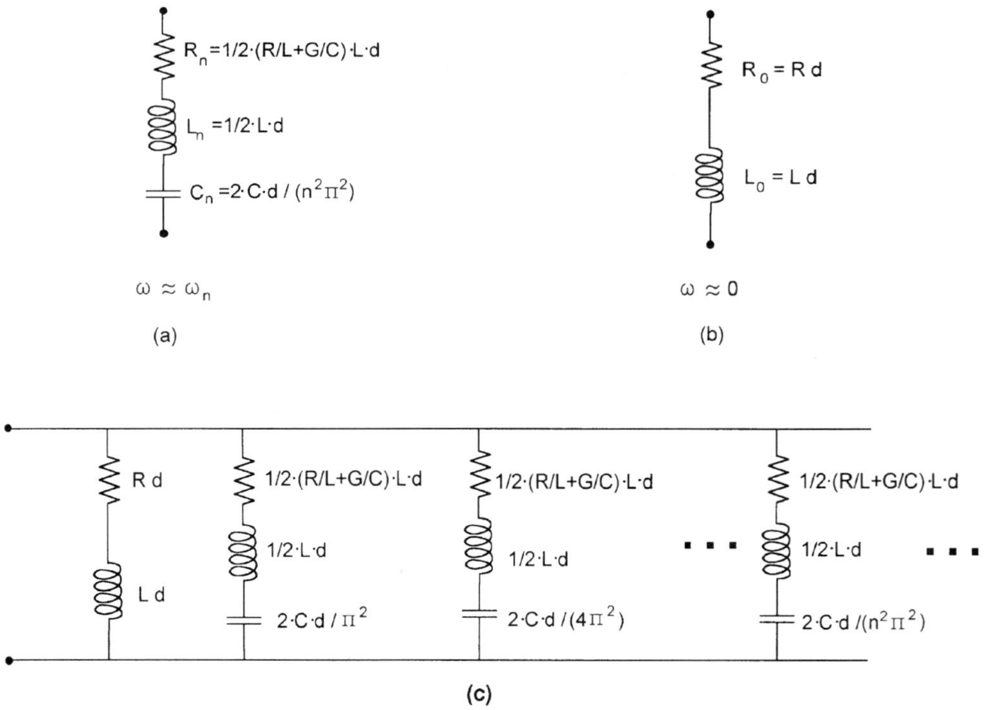

FIGURE 10.35 Lumped element model for the short circuited transmission line.

Finally, using $\omega_n = 1/\sqrt{L_n C_n} = n\pi/d\sqrt{LC}$ or $C_n = d^2LC/n^2\pi^2L_n$ gives

$$C_n = \frac{2Cd}{n^2\pi^2}. \tag{10.72}$$

For a direct current $\omega = 0$ input to the shorted line, the source will see the total inductance of the line Ld in series with the total resistance of the line Rd. The model for the shorted transmission line at $\omega \approx 0$ is then as shown in Fig. 10.35b.

The line can be modeled at all frequencies by placing these circuits in parallel (Fig. 10.35c). This has the correct behavior at zero frequency and at each resonant frequency and hence it is a complete equivalent circuit representation for the line. The input admittance is the sum of the branch admittances:

$$Y_{\text{in}} = \frac{1}{Rd + j\omega Ld} + \sum_{n=1}^{\infty} \frac{1}{R_n + j(\omega L_n - 1/\omega C_n)}. \tag{10.73}$$

This result can also be obtained from Eq. (10.5), for $Z_L = 0$ and a low-loss line; see Problem 10.14.

A similar derivation gives the equivalent circuit for an open circuited transmission line; see Problem 10.15.

It is important to understand how this low-loss problem was solved. First, we treated the transmission line as if it were lossless, obtaining the voltage and the current along the line. Second, we used these lossless results to calculate the losses in the low-loss line. This approximation is valid because the low-loss and the lossless voltages and currents must be almost the same. Finally, we used the losses calculated in this way to "correct" or perturb the lossless solution so it would represent the low-loss solution. This is a technique that we will use repeatedly.

Example 10.13 Use the lossless solution for the transmission line to calculate the low-loss attenuation constant.

Solution The forward voltage and current waves on a lossless transmission line are $\underline{V}(z) = V_0^+ \exp(-j\omega z)$ and $\underline{I}(z) = (V_0^+/R_0)\exp(-j\omega z)$. The time average power carried by the waves is

$$P_{\text{Avg}} = \tfrac{1}{2}\underline{V}\underline{I}^* = \frac{1}{2}\frac{|V_0^+|^2}{R_0}. \tag{10.74}$$

On a lossy line the power carried by a forward wave is $P_{\text{Avg}}(z) = P_0 \exp(-2\alpha z)$, where P_0 is the power at $z = 0$. The derivative of this with respect to z is the rate of increase with distance of time average power carried and its negative is the rate of decrease with distance. The rate of decrease equals the time average power loss per unit length, $P_{L,\text{Avg}}$. Hence

$$-\frac{dP_{\text{Avg}}}{dz} = P_{L,\text{Avg}} = 2\alpha P_{\text{Avg}}(z)$$

$$\alpha = \frac{P_{L,\text{Avg}}}{2P_{\text{Avg}}}. \tag{10.75}$$

The time average power loss per unit length is

$$P_{L,\mathrm{Avg}} = \tfrac{1}{2}(R|I|^2 + G|V|^2) = \frac{1}{2}\left(\frac{R}{R_0^2} + G\right)|V_0^+|^2. \tag{10.76}$$

Substituting Eqs. (10.76) and (10.74) in Eq. (10.75), we obtain

$$\alpha = \frac{\tfrac{1}{2}(R/R_0^2 + G)|V_0^+|^2}{2(\tfrac{1}{2}|V_0^+|^2/R_0)} = \frac{1}{2}\left(\frac{R}{R_0} + GR_0\right),$$

which is the same result as Eq. (9.49).

10.7 MATCHING NETWORKS

It is generally desirable to terminate a transmission line in a matched load. By this we mean that the input impedance to the load should be equal to the characteristic impedance of the transmission line. As we have seen, when the load impedance equals the line impedance there is no reflection at the load. Reflections at the load are undesirable for several reasons.

First, the reflected wave carries away some of the power that is carried to the load by the incident wave. For voltage and current waves at the load,

$$\underline{V}_L = V_0^+(1 + \Gamma_L), \quad \underline{I}_L = \frac{V_0^+}{R_0}(1 - \Gamma_L),$$

the time average power delivered to the load is

$$P_{L,\mathrm{Avg}} = \tfrac{1}{2}\,\mathrm{Re}(\underline{V}_L \underline{I}_L^*) = \frac{|V_0^+|^2}{2R_0}(1 - |\Gamma_L|^2). \tag{10.77}$$

since $\mathrm{Re}(\Gamma_L - \Gamma_L^*) = 0$. Thus $|\Gamma_L|^2$ is the fraction of the incident power reflected from the load. This fraction must usually be minimized.

Second, a microwave circuit frequently consists of several cascaded devices. For example, Fig. 10.36 shows a source feeding a pair of antennas. The line from the source must be split into two lines, and one line must contain a phase shifter and an attenuator so that the phase and power of the signal delivered to one antenna can be adjusted relative to the phase and power delivered to the other antenna. If all the devices are matched, the phase and the power to the antennas will differ by the forward phase shift of the phase shifter and by the attenuation of the attenuator, as desired. If there are multiple reflections due to mismatches, then the net phase and power difference is difficult to predict.

Third, the reflection coefficient is generally a function of frequency, making the reflected signal a function of frequency. This will cause distortion of signal waveforms.

In this text we will introduce methods of matching. A complete discussion of matching is closely related to filter theory and is covered in more advanced texts.

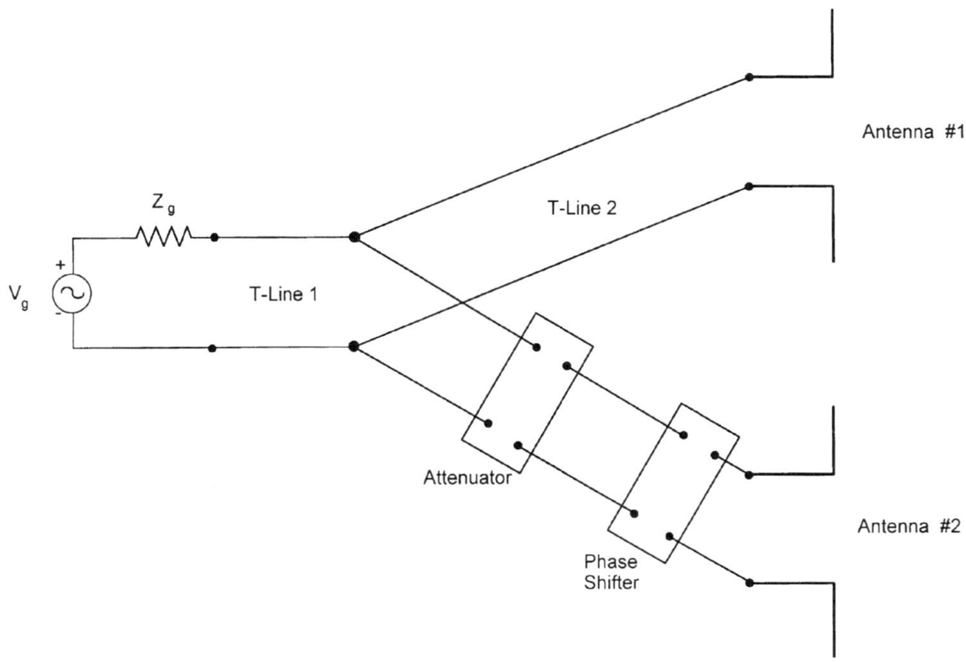

FIGURE 10.36 One source feeding two antennas. One antenna is connected to the source with a transmission line. The second antenna contains an attenuator and phase shifter in series.

10.7.1 Quarter-Wave Matching Section

A quarter-wavelength long line can be used to match a resistive load R_L to a transmission line with characteristic resistance R_0 (Fig. 10.37). Let R_Q be the characteristic resistance and let λ_Q be the wavelength for the line, which has length $d = \lambda_Q/4$. We have seen (Section 10.2.4) that a quarter-wavelength line transforms a normalized load admittance into its normalized impedance, $Z_{\text{in}}/R_Q = (Z_L/R_Q)^{-1}$ or, for a resistive load, $Z_{\text{in}} = R_Q^2/R_L$. We obtain a match by making $Z_{\text{in}} = R_0$. This requires that we design the value of R_Q to be

$$R_Q = \sqrt{R_L R_0}. \tag{10.78}$$

Figure 10.38 shows this transformation through the quarter-wavelength line on the Smith chart. Note that normalization is with respect to R_Q.

Example 10.14 Design a quarter-wave matching section to match a 150 Ω load to a 50 Ω line at 150 MHz. Assume the matching section is an air line.

Solution The characteristic resistance of the matching section is designed to be $R_Q = \sqrt{150 \times 50} = 86.6\ \Omega$. The wavelength is $\lambda_Q = c/f = 3 \times 10^8/150 \times 10^6 = 2.00$ m and the line length is $d = \lambda_Q/4 = 0.500$ m.

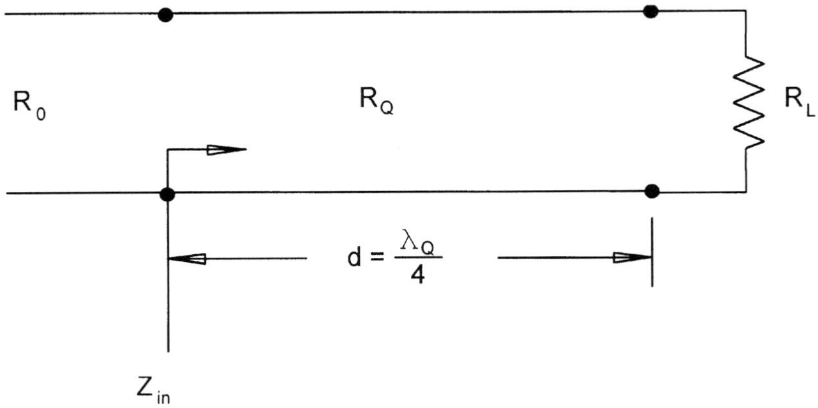

FIGURE 10.37 Single-section quarter-wave transformer matching a resistive load to a resistive source.

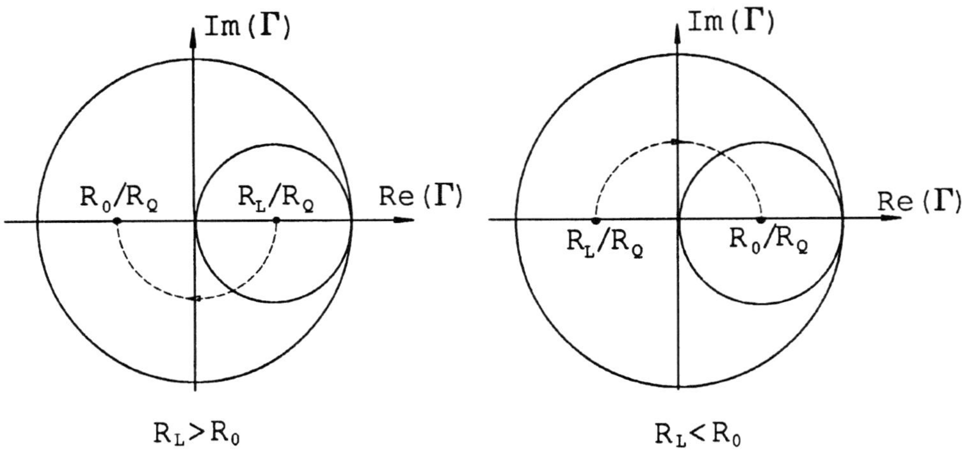

FIGURE 10.38 Smith chart illustration of the impedance transformation using the quarter-wavelength line.

The quarter-wavelength section can also be used to match a nonresistive load to a line. To do this (Fig. 10.39), we must add a length of line l with characteristic resistance R_A between the load and the matching section. This added line is designed to transform the load $Z_{L'} = R_{L'} + jX_{L'}$ into a pure resistance R_L. This is best explained by an example.

Example 10.15 Design a quarter-wave matching section and added line, to match a $150 + j50\ \Omega$ load to a $50\ \Omega$ line at 150 MHz. Assume the matching section and the added line are air lines and the added line has $R_A = 100\ \Omega$.

Solution The wavelength on both lines is 2.00 m. See the Smith chart, Fig. 10.40, to follow the solution. Normalizing the load with respect to $100\ \Omega$ gives $\hat{Z}_{L'} = 1.5 + j0.5$, which is point A on the chart. Rotating clockwise toward the generator we reach point

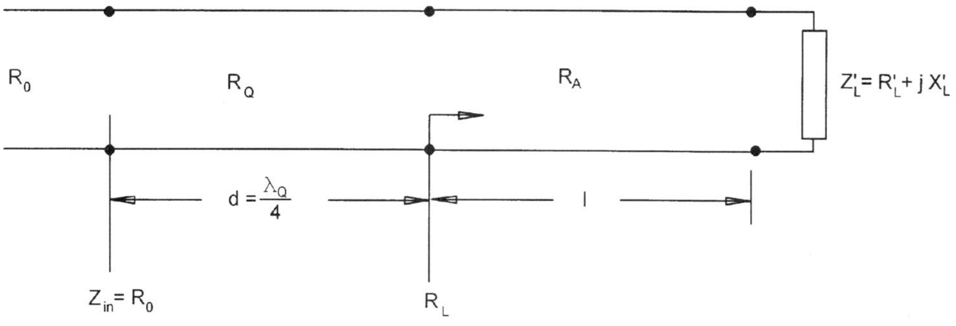

FIGURE 10.39 Transmission line matching of a complex impedance using a length of line and a quarter-wave section.

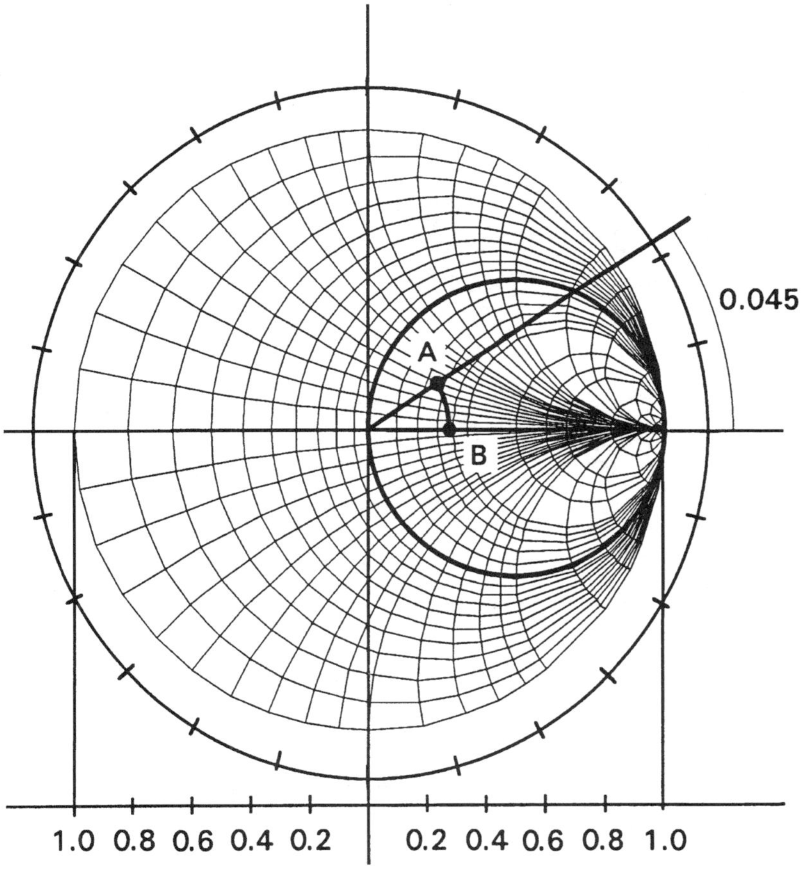

FIGURE 10.40 Smith chart solution to Example 10.15.

B, which is purely resistive. Reading $R_L/R_A = 1.75$ from the chart and unnormalizing with respect to 100 Ω, we get $R_L = 175\ \Omega$. The chart (outermost scale) shows a line length of $l = (0.250 - 0.202)\lambda = 0.096$ m. The quarter-wave line should have $R_Q = \sqrt{175 \times 50} = 93.5\ \Omega$ and length $d = 0.500$ m.

10.7.2 Single Stub Matching

Figure 10.41 shows a transmission line with characteristic conductance $G_{0l} = 1/R_{0l}$ loaded with an admittance $Y_L = G_L + j\beta_L$. A distance d_c toward the generator there is a shorted line with characteristic conductance G_{0s} connected in parallel with the line. This shorted stub has length d_s. The length d_c from the load to the stub is chosen to make the real part of the admittance seen looking toward the load equal to the line's characteristic conductance, so the loaded line has input admittance $Y_c = G_{0l} + jB_c$. We recall (Section 10.2.4) that by varying the length of a shorted line from zero to a half-wavelength we can get any input susceptance we want from minus to plus infinity. We therefore choose the length of the stub d_s to make the admittance seen looking into the stub be $Y_s = -jB_c$. Admittances in parallel add and the total admittance looking into the loaded line with the stub in parallel is $Y_t = Y_s + Y_c = (-jB_c) + (G_{0l} + jB_c) = G_{0l}$. Thus the line from the source to the stub connection sees a total load G_{0l} and is therefore matched.

It is possible to calculate d_c, B_c, and d_s analytically, but it is easier and more instructive to use the Smith chart to find these values. We enter the chart, Fig. 10.42, at

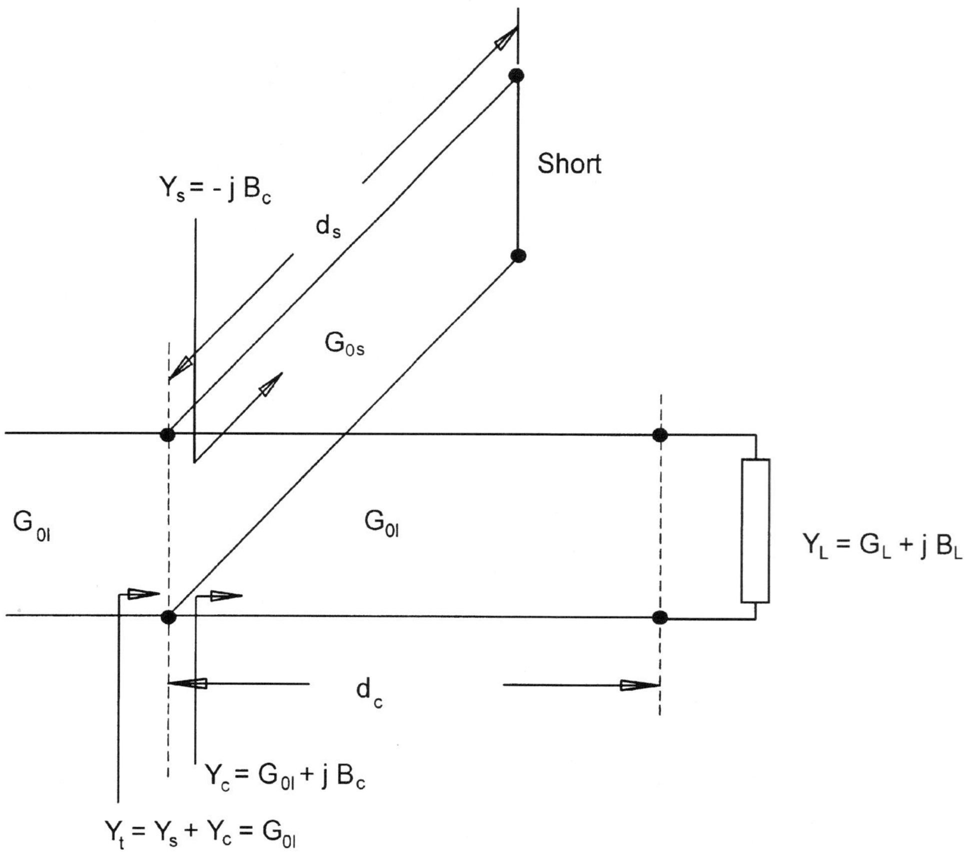

FIGURE 10.41 Single stub matching with a short parallel transmission line.

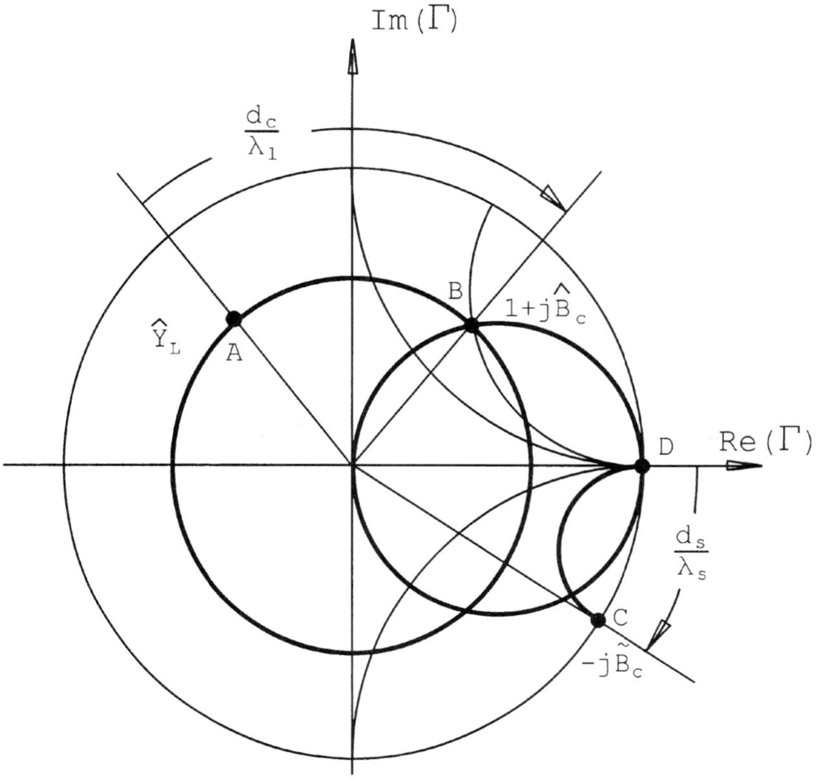

FIGURE 10.42 Normalized admittance chart illustration of single stub matching.

the load admittance normalized with respect to the line's characteristic conductance $\hat{Y}_L = Y_L/G_{0l}$, point A. We rotate clockwise toward the generator on a circle of constant $|\Gamma|$ until we reach the $G/G_0 = 1$ circle, point B. The difference between the outer scale readings ("wavelengths toward the generator") at the radial lines drawn through points A and B gives d_c/λ_l. The circle of constant susceptance through point B gives the normalized value $\hat{B}_c = B_c/G_{0l}$. Unnormalizing with respect to the line's characteristic conductance and renormalizing with respect to the stub's characteristic conductance gives $\tilde{Y}_s = -j\tilde{B}_c = -j\hat{B}_c(G_{0l}/G_{0s})$, point C. The circumflex $\hat{}$ shows normalization with respect to G_{0l} and the tilde $\sim$ shows normalization with respect to G_{0s}. Rotating counterclockwise toward the load we reach the short, point D. The difference between the penultimate outer scale readings ("wavelengths toward the load") at the radial lines drawn through points D and C gives d_s/λ_s.

Example 10.16 Design a single stub match for a 50 Ω line with an $80 + j25$ Ω load. The frequency is 5 GHz, the lines are air lines, and the stub characteristic resistance is 60 Ω.

Solution The wavelength on all lines is $\lambda = (3 \times 10^8)/(5 \times 10^9) = 0.0600$ m or 6.00 cm. The normalized load impedance is $\hat{Z}_L = Z_L/R_{0l} = 1.6 + j0.5$, point A′ in Fig. 10.43, and $\hat{Y}_L$ is diametrically opposite, point A. Rotating clockwise toward the generator

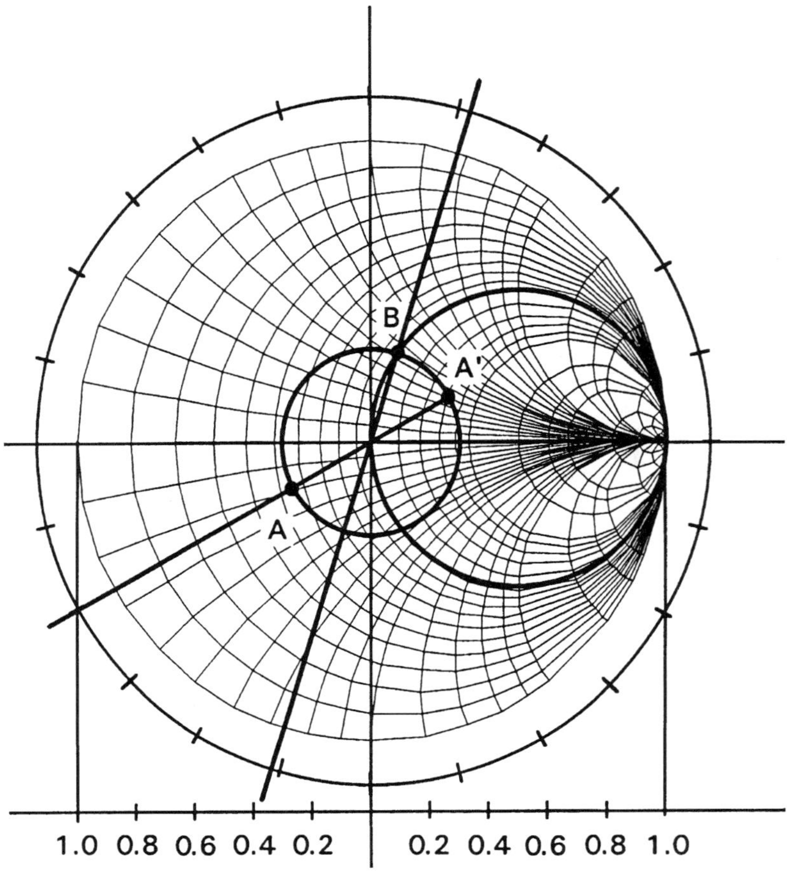

FIGURE 10.43 Smith chart solution to Example 10.16.

from point A to point B located on the $G/G_0 = 1$ circle gives $d_c/\lambda = (0.500 - 0.459) + 0.150 = 0.191$ or $d_c = 1.146$ cm. At point B, $\hat{B}_c = 0.65$, so the input admittance to the stub must be $Y_s = -j0.65/50 = -j0.013$ S. Normalizing this with respect to the stub characteristic conductance gives $\tilde{Y}_s = -j0.013 \times 60 = -j0.78$, point C. Rotating from point C counterclockwise toward the load to the short at point D we get $d_s/\lambda = 0.250 - 0.106 = 0.144$ or $d_s = 0.864$ cm.

10.7.3 Broad-Band Matching

We have seen two methods for designing matching circuits at a particular frequency. All communications signals, however, consist of a band of frequencies and we are really required to match as well as possible over the entire band. For example, the system design specification might be that the reflection coefficient magnitude must be less than 0.1 from a low to a high frequency. Each design must be tested over the bandwidth of interest and the result compared with the design criterion.

Let us look at the simple quarter-wave matching design of Example 10.14. Suppose your boss told you to match the load of 150 Ω to the 50 Ω line over the signal range from a low frequency of 120 MHz to a high frequency of 180 MHz, with $|\Gamma| < 0.10$. If you do the design at the center frequency of 150 MHz, as is reasonable, you get the results of that example, $R_Q = 86.6\ \Omega$ and $d = 0.500$ m.

Now you check at the low-frequency end, 120 MHz. There, the wavelength is $\lambda = 2.50$ m and $d/\lambda = 0.200$. Since the frequency has decreased, the wavelength has increased and the fixed physical length d has gotten smaller as a fraction of a wavelength. What is the reflection coefficient at 120 MHz? On the Smith chart, Fig. 10.44, we start at $Z_L/R_Q = 1.73$, point A, and rotate 0.200 wavelength clockwise toward the generator to point B. Reading the normalized impedance at point B, $0.62 - j0.20$, unnormalizing with respect to 86.6 Ω, and normalizing with respect to 50 Ω gives point C, $1.07 - j0.35$, as the normalized load impedance on the 50 Ω line at 120 MHz. This corresponds to a $|\Gamma| = 0.18$, which is "out of spec." Similarly, at the high-frequency end, $d/\lambda = 0.300$ and we get points D and E, with the same result. Plotting intermediate frequencies gives the curve connecting points C and E, with frequency a parameter along the curve. The boss is waiting and things are not going well. What should you do?

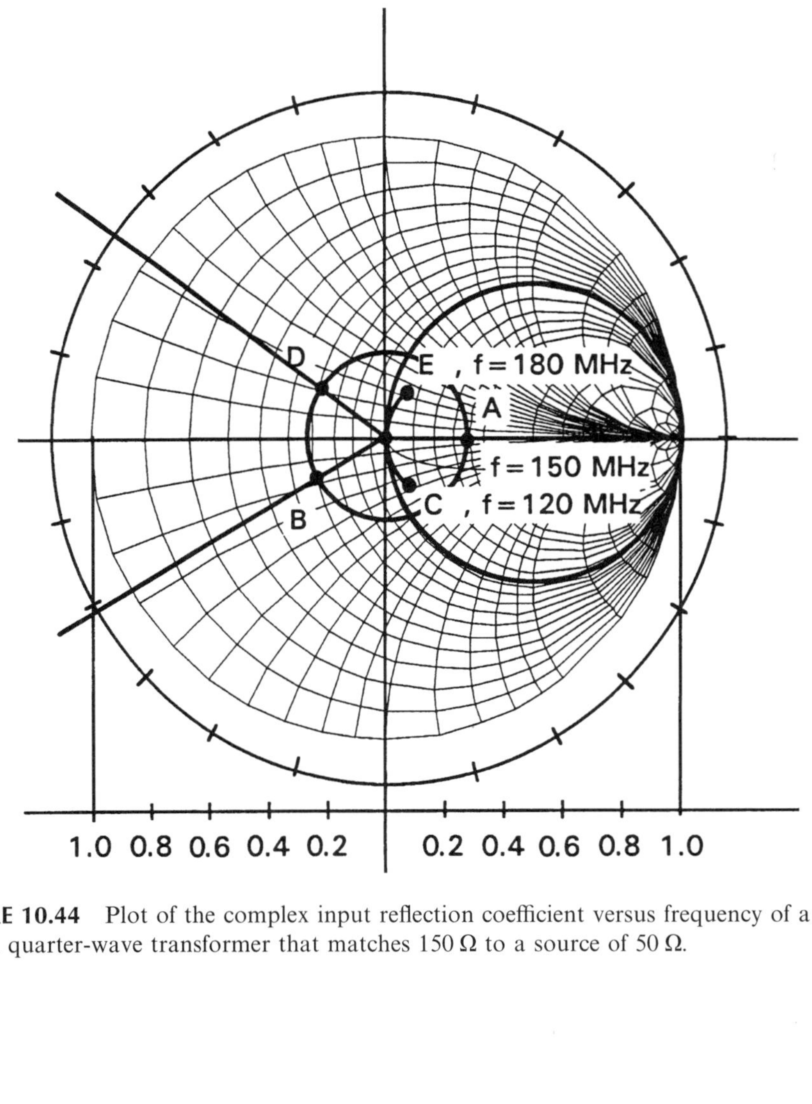

FIGURE 10.44 Plot of the complex input reflection coefficient versus frequency of a single-section quarter-wave transformer that matches 150 Ω to a source of 50 Ω.

If you can make the transition from the load resistance ($R_L = 150\ \Omega$) to the line resistance ($R_0 = 50\ \Omega$) more gradual, you can reduce the reflection coefficient at any frequency and hence increase the bandwidth over which $|\Gamma| < 0.10$. Let us try two quarter-wavelength steps (Fig. 10.45). Arbitrarily, let the first step transform the 150 Ω load to 100 Ω and the second step transform from 100 Ω to 50 Ω. Then $R_{Q1} = \sqrt{150 \times 100} = 122.5\ \Omega$ and $R_{Q2} = \sqrt{100 \times 50} = 70.7\ \Omega$, with $d_1 = d_2 = d = 0.500$ m, so each step is a quarter-wavelength long at the center frequency, 150 MHz. Follow the solution on Fig. 10.46. The figure is drawn on an expanded Smith chart, which shows only the portion of the Γ plane with $|\Gamma| < 0.22$. At 120 MHz, for which $d/\lambda = 0.200$, the first step takes us from point A, where $R_L/R_{Q1} = 1.224$, rotating clockwise by 0.200 to point B, $0.850 - j0.098$. Multiply by R_{01} and divide by R_{Q2} to get point C, $1.472 - j0.170$. Rotating clockwise by 0.200 ($0.272 + 0.200 = 0.472$) gives point D, $0.675 - j0.100$. Multiply by R_{Q2} and divide by R_0 to get point E, $0.954 - j0.141$. This corresponds to $|\Gamma| = 0{,}075$, well within "spec." At 180 MHz, the first step takes us from point A rotating clockwise by 0.300 to point F, $0.850 + j0.098$. Multiply by R_{Q1} and divide by R_{Q2} to get point G, $1.472 + j0.170$. Rotating clockwise by 0.300 ($0.228 + 0.300 = 0.528$ or 0.028) gives point H, $0.675 + j0.100$. Multiply by R_{Q2} and divide by R_0 to get point I, $0.954 + j0.141$. This again corresponds to $|\Gamma| = 0.075$. The curve from E to I shows the reflection coefficient from the low to the high frequency. This design should make your boss happy.

If a wider bandwidth or a smaller reflection coefficient were needed, more steps could be inserted, making the adjustment required for matching even more gradual. The steps are generally not chosen arbitrarily, as in our example. They can be designed for maximum flatness of the reflection coefficient magnitude over the bandwidth, for equal ripple in the band, for maximum steepness of drop-off outside the band, or by some other criterion. These design methods are generally discussed in ourses, books, or articles on filter design and are beyond the scope of this text. The important point of this discussion is that a practical system requires that specifications be met across a band of frequencies because all communications systems include such a band.

10.7.4 PSPICE Simulation of Transmission Line Circuits

Transmission line circuits can be simulated using PSPICE, an electric circuit simulation software program. A library of device subcircuits and models is included with PSPICE. Models for lossless and lossy transmission lines are available, thus making the simulation of these circuits possible.

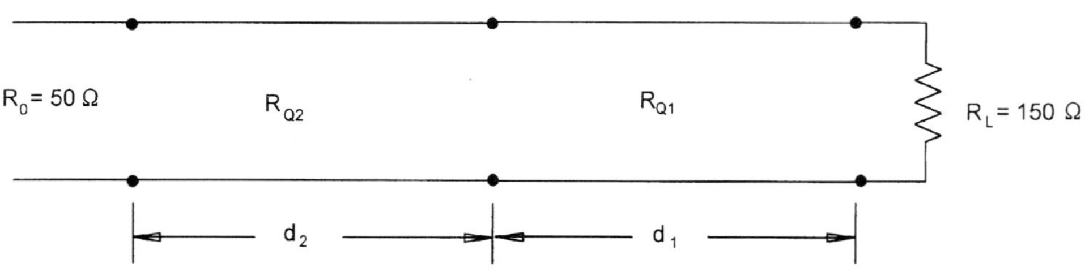

FIGURE 10.45 Two-section quarter-wave transformer.

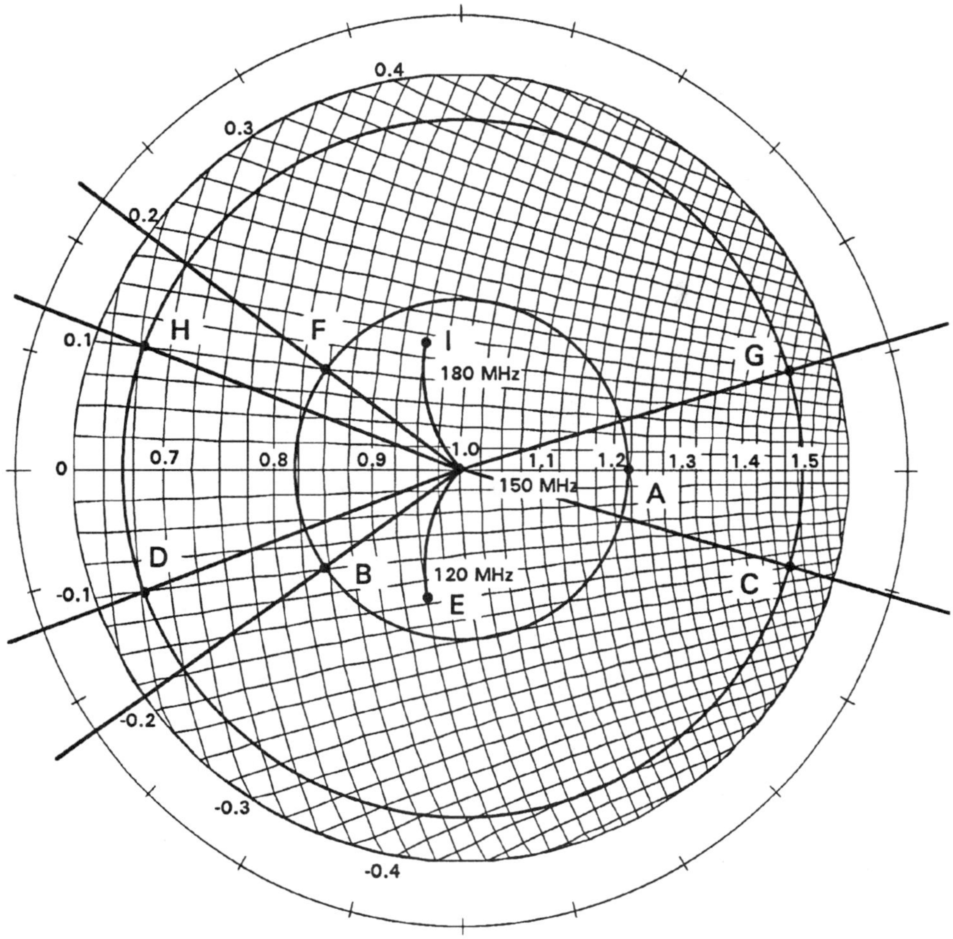

FIGURE 10.46 Plot of the complex input reflection coefficient versus frequency of a two-section quarter-wave transformer that matches 150 Ω to a source of 50 Ω.

PSPICE will allow the display of voltages and currents at any node and through any element in the circuit versus time or frequency but will not display reflection coefficients. A simple device [4] is used in conjunction with the rest of the simulated circuit in order to obtain voltages or currents that are proportional to the reflection coefficient. The description of the device is introduced in Listing 10.1, and for the user at this point, the device will have to be a part of the simulated circuit. The device will briefly be described in a later chapter as a directional coupler.

10.7.4.1 PSPICE Quarter-Wave Transformers The listing of the circuit used in PSPICE to simulate the quarter-wave transformer in Example 10.14 is shown in Listing 10.2. The subcircuit used to determine the input reflection coefficient is located in a library called TLLIB.LIB, and a copy of the listing of the model is provided in Listing 10.1. The magnitude of the reflection coefficient is then calculated as the ratio of the magnitudes of the voltages at nodes 4 and 5, respectively. The phase angle on

```
************ TLLIB.LIB ************
*DIRECTIONAL COUPLER TO BE USED FOR REFLECTION
COEFFICIENT MEASUREMENT*
.SUBCKT DIRCOUP A B C D G PARAMS: FO = 150MEGHZ
TA A G B G ZO = 55.27708 F = {FO} NL = 0.25
TB A E B D ZO = 497.49373 F = {FO} NL = 0.25
TC E G D G ZO = 55.27708 F = {FO} NL = 0.25
TEQ F G C G ZO = 50 F = {FO} NL = 0.25
REE E G 50
EFE F G E G 1
RNE C G 50
RFE D G 50
. ENDS
```

LISTING 10.1 Directional coupler used to evaluate *S* parameters in PSPICE.

```
*** QUARTER WAVE TRANSFORMER EXAMPLE ***
VS 1 0 AC 1 0
RS 1 2 50
X1 2 3 4 5 0 DIRCOUP
TL1 3 0 6 0 ZO = 86.60254038 F = 150MEG NL = 0.25
RL 6 0 150
. LIB TLLIB.LIB
. AC LIN 81 1MEG 300MEG
. PROBE
. END
```

LISTING 10.2 PSPICE file used to simulate a single-section quarter-wave transformer.

the reflection coefficient can be obtained as the difference of the phases of the same voltages. The plot of the magnitude of the reflection coefficient, as a result of the ac analysis performed by PSPICE, is shown in Fig. 10.47.

Example 10.17 Simulate the broad-band matching network from Section 10.7.3. Determine the bandwidth over which the reflection coefficient stays below 0.1.

Solution The listing of the circuit that simulates the broad-band matching network is given in Listing 10.3. The circuit diagram is shown in Fig. 10.48. The plot of the reflection coefficient for frequencies ranging from 10 to 290 MHz is shown in Fig. 10.49. Using the cursor feature of the PROBE program, we locate the low frequency at which the reflection coefficient magnitude is equal to 0.1. The result shows a frequency of 112.036 MHz. Using the same procedure, the high frequency at which the reflection coefficient magnitude is 0.1 is found to be 187.986 MHz. The bandwidth over which the reflection coefficient magnitude stays below 0.1 is then calculated to be BW = 187.986 MHz − 112.036 MHz = 75.95 MHz. The values of the reflection coefficient magnitude at 120 MHz and 180 MHz can be found using the same technique:

$$|\Gamma(120\text{ MHz})| = 0.0701,$$

$$|\Gamma(180\text{ MHz})| = 0.0701.$$

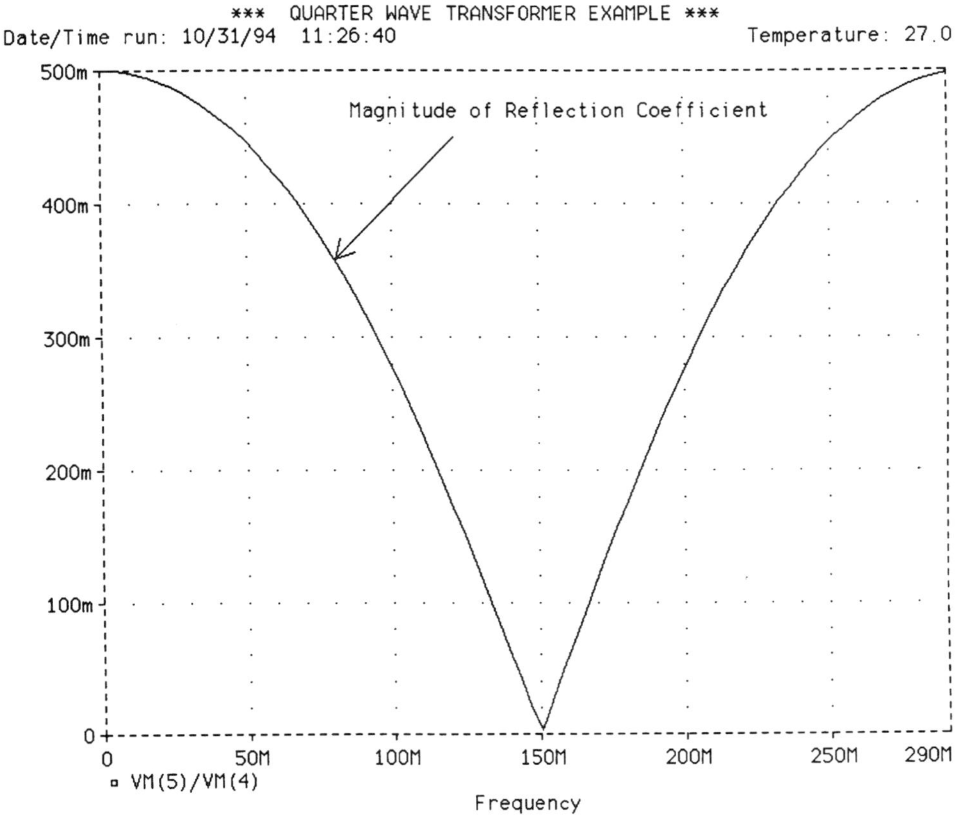

FIGURE 10.47 PSPICE plot of the magnitude of the reflection coefficient versus frequency for Example 10.14.

```
*** 2 SECTION QUARTER WAVE TRANSFORMER EXAMPLE ***
VS 1 0 AC 1 0
RS 1 2 50
X1 2 3 4 5 0 DIRCOUP
TL1 3 0 7 0 Z0 = 70.7107 F = 150MEG NL = 0.25
TL2 7 0 6 0 Z0 = 122.4745 F = 150MEG NL = 0.25
RL 6 0 150
.LIB  TLLIB.LIB
.AC LIN 81 1MEG 300MEG
.PROBE
.END
```

LISTING 10.3 PSPICE file used to simulate a two-section quarter-wave transformer.

The difference between these results and the one obtained in Section 10.7.3 is due to the limited accuracy in using the Smith chart.

10.7.4.2 Stub Matching Using PSPICE The PSPICE circuit used to simulate the single stub matching network in Example 10.16 is shown in Fig. 10.50. The same

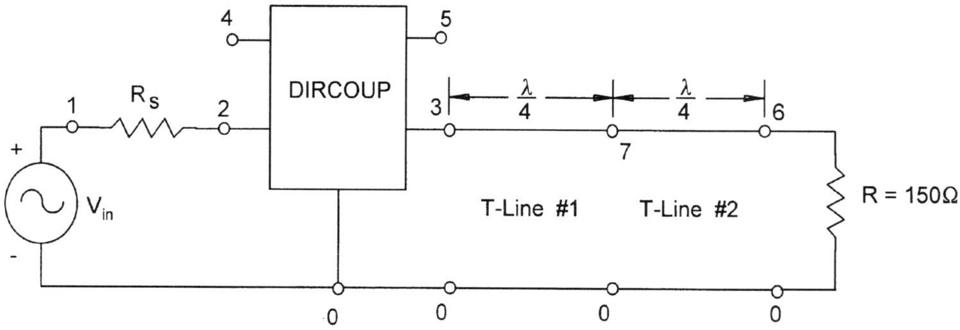

FIGURE 10.48 Circuit diagram used in PSPICE simulation of the two-section quarter-wave transformer.

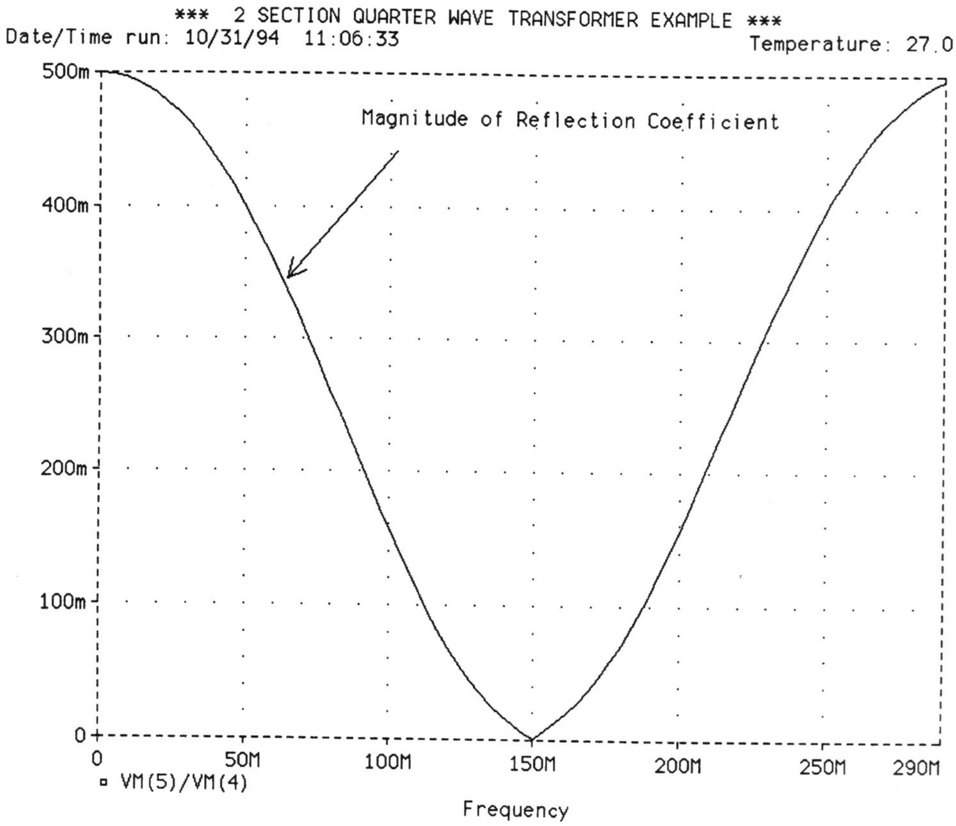

FIGURE 10.49 PSPICE plot of the magnitude of the reflection coefficient versus frequency for Example 10.17.

device, DIRCOUP, is used to determine the reflection coefficient at the input of the matching section. The reflection coefficient magnitude is calculated as the ratio of the magnitudes of the voltages at nodes 5 and 4. The listing of the PSPICE file is shown in Listing 10.4. The amplitude of the input reflection coefficient is shown in Fig. 10.51.

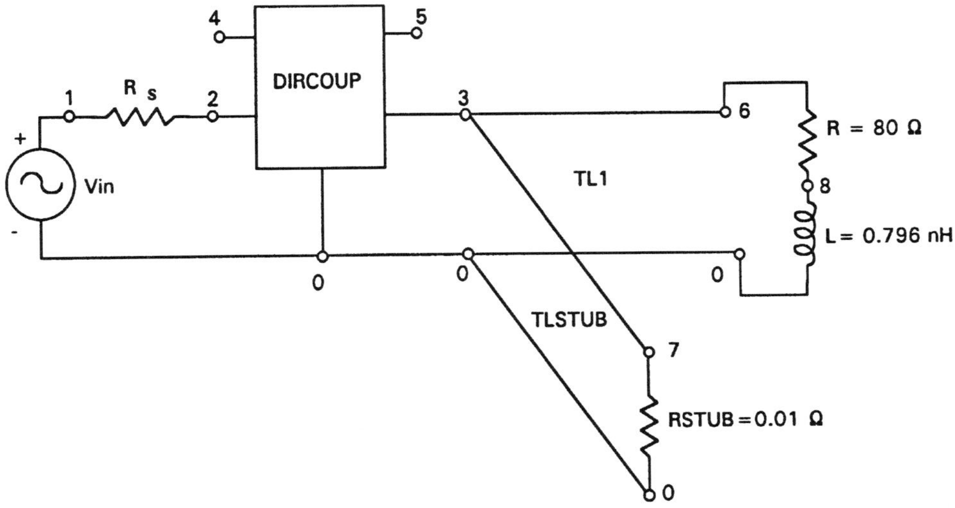

FIGURE 10.50 Single stub matching circuit used in the PSPICE simulation of Example 10.16.

```
*** SINGLE STUB MATCHING ZL = 80 + j25 ***
VIN 1 0 AC 1 0
RS 1 2 50
X1 2 3 4 5 0 DIRCOUP
TL1 3 0 6 0 Z0 = 50 F = 150MEG NL = 0.191
TLSTUB 3 0 7 0 Z0 = 60 F = 150MEG NL = .144
RSTUB 7 0 0.01
RL 6 8 80
LL 8 0 26.525824nH
.LIB  TLLIB.LIB
.AC LIN 81 1MEG 300MEG
.PROBE
.END
```

LISTING 10.4 PSPICE file used to simulate single stub matching ($Z_L = 80 + j25$).

The plot of the amplitude also gives a clear perspective of the limited bandwidth over which this matching technique would work.

Example 10.18 Design a single stub matching network that would match a load impedance of $Z_L = 100 - j100\ \Omega$ to a 50 Ω line. Use a 50 Ω shorted stub. Simulate the circuit using PSPICE.

Solution The Smith chart is used to determine the values of d_c and d_s. The normalized load impedance is

$$\hat{Z}_L = \frac{Z_L}{Z_0} = \frac{100 - \ j100}{50} = 2 - j2.$$

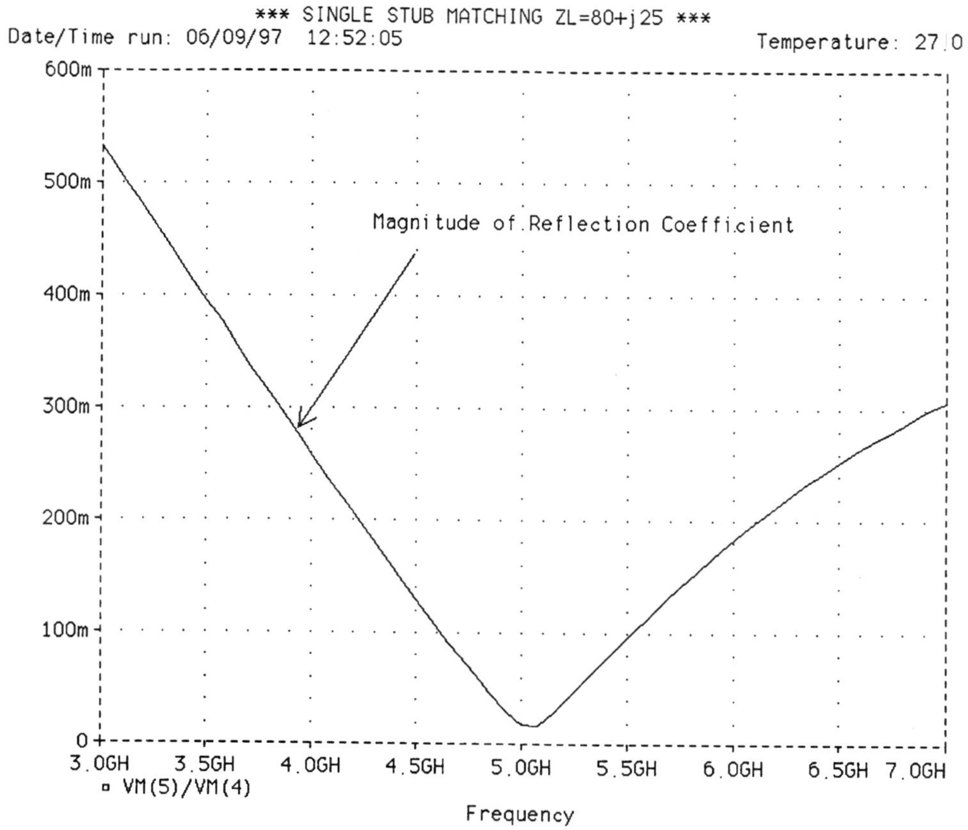

FIGURE 10.51 PSPICE plot of the magnitude of the reflection coefficient versus frequency for Example 10.16.

The location of the normalized load impedance is marked by point A on the Smith chart in Fig. 10.52. The circle of constant absolute value of the reflection coefficient is then drawn, and the diametrically opposed point is marked as B, which represents the location of the load admittance. The distance d_c from the load to where the stub is to be connected is then found to be $d_c = 0.178\lambda - 0.04\lambda = 0.138\lambda$. The normalized susceptance of the stub is found to be $\hat{B}_{\text{stub}} = -1.6$. The length of the shorted stub is then found to be $d_s = (0.339\lambda - 0.25\lambda) = 0.089\lambda$.

The PSPICE listing of the circuit file that simulates the above single stub matching problem is given in Listing 10.5. The load impedance is realized as the series connection of a resistor with $R_L = 100\ \Omega$ and a capacitor with $C_L = 0.31831$ pF that has a capacitive reactance of $X_c = -100\ \Omega$ at a frequency of 5 GHz (the selected design frequency for the simulation). The reflection coefficient magnitude is plotted in Fig. 10.53. Using the PSPICE feature that allows the analysis to be repeated for different values of one parameter in the circuit, the lengths of the transmission lines TL2 and TL3 can be "tuned." The Smith chart design can be used to give rough starting values for d_c and d_s and then "tuned" until desired values of the reflection coefficient are reached.

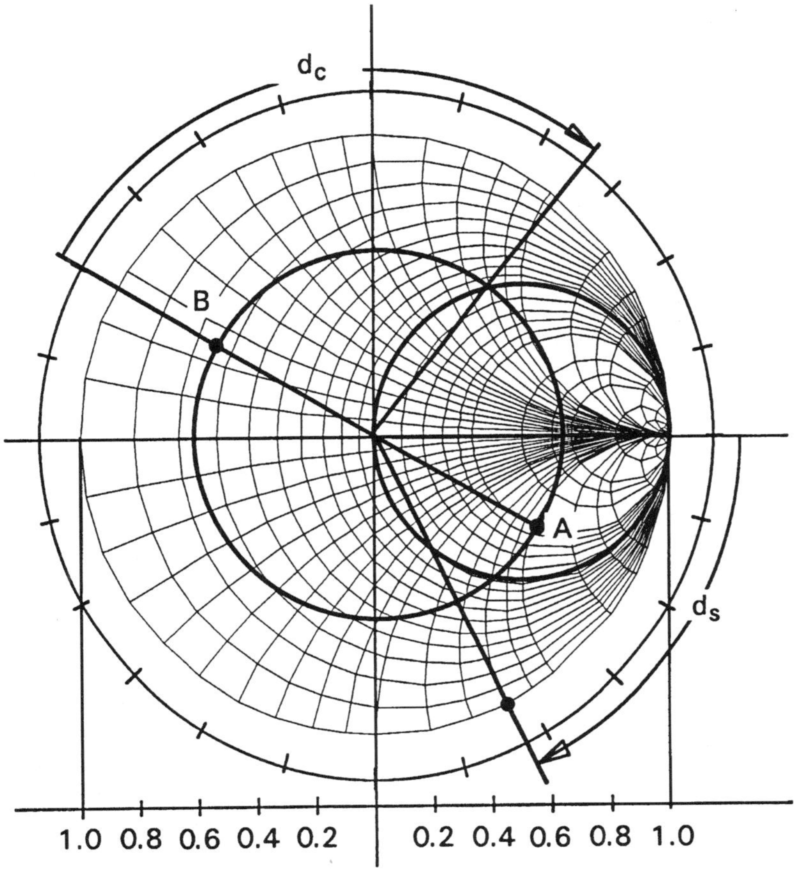

FIGURE 10.52 Smith chart solution to Example 10.18.

```
*** SINGLE STUB MATCHING ZL = 100 - j100 ***
.PARAM LDC = 0.137 LDS = 0.089
VIN 1 0 AC 1 0
RS 1 2 50
X1 2 3 4 5 0 DIRCOUP
TL1 3 0 6 0 Z0 = 50 F = 5GHZ NL = {LDC}
TLSTUB 3 0 7 0 Z0 = 50 F = 5GHZ NL = {LDS}
RSTUB 7 0 0.0000000001
RL 6 8 100
CL 8 0 .31831pF
.LIB TLLIB.LIB
.AC LIN 81 1GHZ 10GHZ
.PROBE
.END
```

LISTING 10.5 PSPICE file used to simulate single stub matching ($Z_L = 100 - j100$).

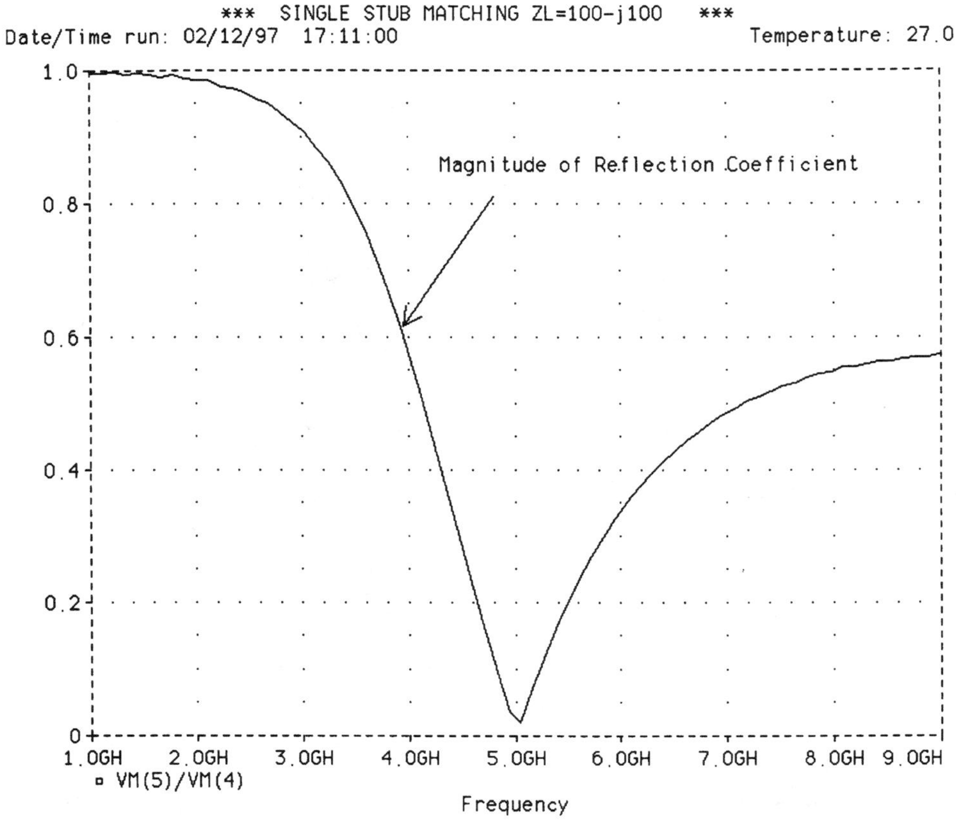

FIGURE 10.53 PSPICE plot of the magnitude of the reflection coefficient versus frequency for Example 10.18.

10.8 TRANSIENTS ON TRANSMISSION LINES

So far we have considered time harmonic sources exciting transmission lines. As a last topic we will look at transmission lines excited by other than sinusoidal sources.

Two notes of introduction are required here. First, we will deal with the Laplace transform of voltages and currents. The student who has not yet studied Laplace transforms can think of these terms as voltage and current phasors with complex frequency s. Second, we will assume that all initial conditions are zero. For a more general discussion, see Ref. [5].

Consider a transmission line (Fig. 10.54a,b) excited by a source or generator voltage $v_g(t)$ with Laplace transform $V_g(s)$ and internal impedance $Z_g(s)$. The load impedance is $Z_L(s)$. The transmission line extends from the input to the line at $x = 0$ to the load at $x = d$ and is characterized by its characteristic impedance $Z_0(s) = \sqrt{(R + sL)/(G + sC)}$ and propagation constant $\gamma(s) = \sqrt{(R + sL)(G + sC)}$. The distance coordinate is changed from z to x to emphasize the shift in origin to the input at the generator-to-line connection. The voltage and current at any point x along the line are $v(x, t)$ and $i(x, t)$, with Laplace transforms $V(x, s)$ and $I(x, s)$.

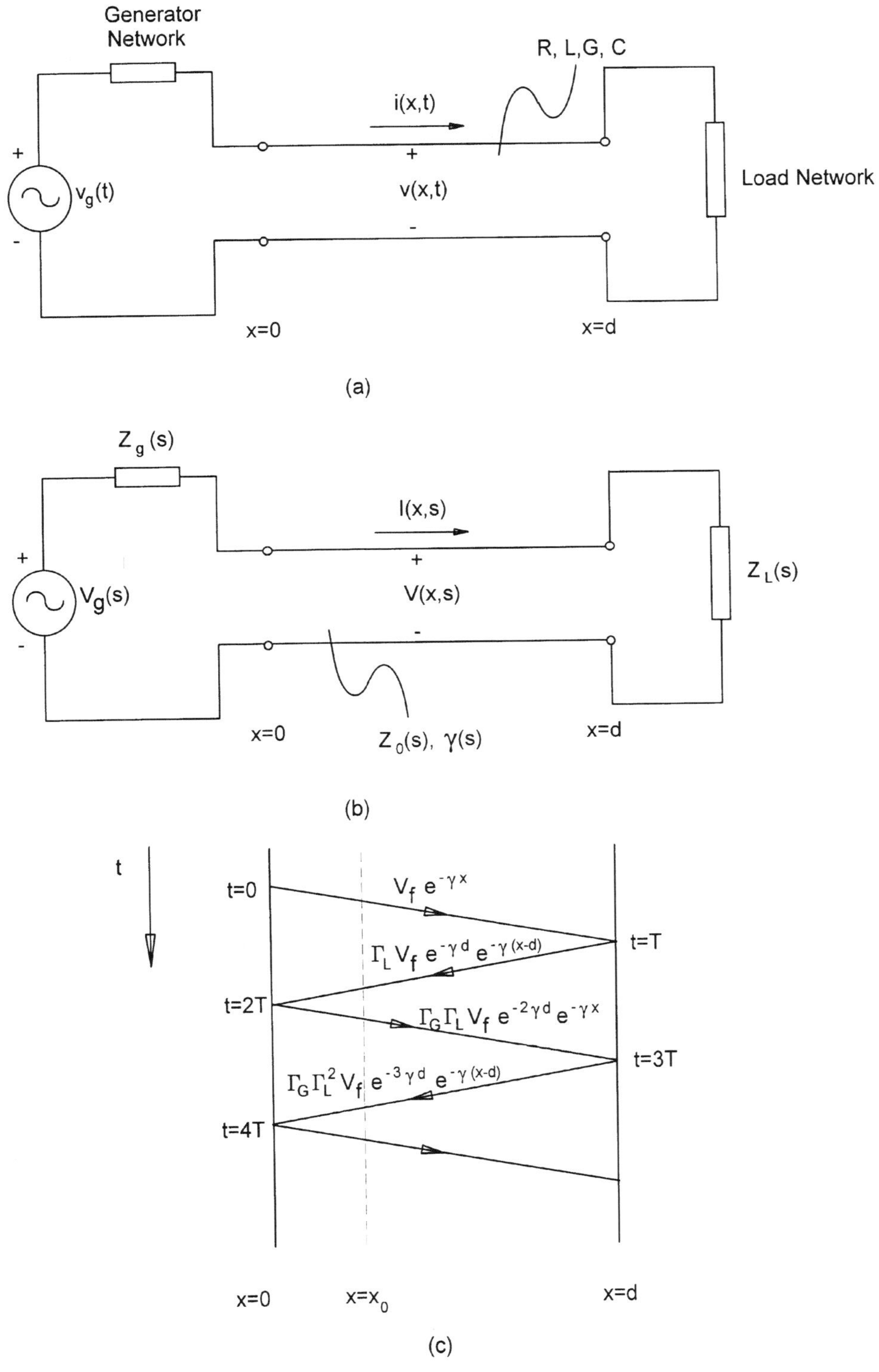

FIGURE 10.54 Distance–time diagram for transients on transmission lines.

Starting at time $t = 0$ a forward voltage wave $v_f(t)$, with transform $V_f(s)$, is launched at the line's input toward the load. The wave $V_f \exp(-\gamma x)$ travels toward the load. At the load the wave is partially reflected so that a transformed voltage wave $\Gamma_L V_f \exp(-\gamma d)$, where

$$\Gamma_L(s) = \frac{Z_L(s) - Z_0(s)}{Z_L(s) + Z_0(s)},$$

is launched back toward the generator. The transformed voltage wave $\Gamma_L V_f \exp(-\gamma d) \exp[\gamma(x - d)]$ reaches the generator ($x = 0$) and is in turn partially reflected so that a transformed voltage wave $\Gamma_G \Gamma_L V_f \exp(-2\gamma d)$, where

$$\Gamma_G(s) = \frac{Z_G(s) - Z_0(s)}{Z_G(s) + Z_0(s)},$$

is launched forward toward the load. The process is repeated indefinitely. The transformed current is almost the same. Simply replace each reflection coefficient by its negative and divide the total by Z_0.

In order to keep track of the multiple reflections we can draw a voltage distance–time diagram as in Fig. 10.54c. In this diagram distance is shown horizontally and time vertically downward. The time for a wave to travel from one end of the line to the other is T. At any time and position the transformed voltage is found by summing all the waves that have reached that point on the line by that time. For example, the figure shows a dotted line at x_0 and t_0. The transformed voltage $V(x_0, s)$ is the sum of the five waves above that cross-point:

$$V(x_0, s)|_{t=t_0} = V_f[e^{-\gamma x_0} + \Gamma_L e^{-\gamma d} e^{\gamma(x_0 - d)} + \Gamma_G \Gamma_L e^{-2\gamma d} e^{-\gamma x_0} + \Gamma_G \Gamma_L^2 e^{-3\gamma d} e^{\gamma(x_0 - d)} + \Gamma_G^2 \Gamma_L^2 e^{-4\gamma d} e^{-\gamma x_0}].$$

The corresponding transformed current is

$$I(x_0, s)|_{t=t_0} = \frac{V_f}{Z_0}[e^{-\gamma x_0} - \Gamma_L e^{-\gamma d} e^{\gamma(x_0 - d)} + \Gamma_G \Gamma_L e^{-2\gamma d} e^{-\gamma x_0} - \Gamma_G \Gamma_L^2 e^{-3\gamma d} e^{\gamma(x_0 - d)} + \Gamma_G^2 \Gamma_L^2 e^{-4\gamma d} e^{-\gamma x_0}].$$

The distance–time diagram is conceptually useful but somewhat misleading. It implies that there is a single velocity associated with the wave so that T is the time it takes the wave to travel a distance d. This condition is only met when the line has no dispersion, either because it is lossless—$R = 0$ and $G = 0$—or because $L/R = C/G$; see Section 9.7.2.

The distance–time diagram is actually a mixing of the Laplace transform or s domain and the time domain. The Laplace transform of the voltage at any position x_0 can be taken as the sum of all the waves passing through the vertical dotted line. When the inverse transform is taken to obtain $v(x_0, t)$ and then the result is evaluated at t_0, all of the terms resulting from the waves that have not yet reached x_0 will be zeroed. Let us see how this works.

Summing all the waves passing through x,

$$V(x,s) = V_f[1 + \Gamma_L\Gamma_G e^{-2\gamma d} + (\Gamma_L\Gamma_G e^{-2\gamma d})^2 + (\Gamma_L\Gamma_G e^{-2\gamma d})^3 + \cdots]e^{-\gamma x}$$
$$+ V_f[1 + \Gamma_L\Gamma_G e^{-2\gamma d} + (\Gamma_L\Gamma_G e^{-2\gamma d})^2$$
$$+ (\Gamma_L\Gamma_G e^{-2\gamma d})^3 + \cdots]\Gamma_L e^{-\gamma d} e^{\gamma(x-d)}. \tag{10.79}$$

The first bracketed term represents all the forward traveling waves and the second represents all the backward traveling waves.

Two examples will illustrate what this equation means.

Example 10.19 Assume a lossless line, so $Z_0(s) = \sqrt{L/C} = R_0$, $\gamma(s) = s\sqrt{LC}$, and $\gamma d = sT$, where $T = d\sqrt{LC} = d/u$. Assume that $R_L = 2R_0$, so $\Gamma_L = \frac{1}{3}$ and that $Z_G = 0$ so that $\Gamma_G = -1$. Let the voltage at the input to the line be a step voltage, $v_f(t) = V_0 u(t)$, so its Laplace transform is $V_f(s) = V_0/s$. (a) Find the voltage at any position on the line. (b) Evaluate the voltage at $x = d/3$. (c) Evaluate the voltage at the load $x = d$.

Solution (a) Substituting the given values, Eq. (10.79) becomes

$$V(x,s) = V_0\left[\frac{1}{s} - \frac{1}{3s}e^{-2sT} + \frac{1}{9s}e^{-4sT} - \cdots\right]e^{-s(x/d)T}$$
$$+ V_0\left[\frac{1}{s} - \frac{1}{3s}e^{-2sT} + \frac{1}{9s}e^{-4sT} - \cdots\right]\left(\frac{1}{3}\right)e^{-sT}e^{s(x/d-1)T}. \tag{10.80}$$

If the inverse transform of $F(s)$ is $f(t)u(t)$, then the inverse transform of $F(s)\exp(-st_1)$ is $f(t-t_1)u(t-t_1)$. We take the inverse transform of Eq. (10.80) term by term to get

$$v(x,t) = V_0\left[u\left(t - \frac{x}{d}T\right) - \frac{1}{3}u\left(t - \frac{x}{d}T - 2T\right) + \frac{1}{9}u\left(t - \frac{x}{d}T - 4T\right) - \cdots\right]$$
$$+ \left(\frac{1}{3}\right)V_0\left[u\left(t + \frac{x}{d}T - 2T\right) - \frac{1}{3}u\left(t + \frac{x}{d}T - 4T\right)\right.$$
$$\left. + \frac{1}{9}u\left(t + \frac{x}{d}T - 6T\right) - \cdots\right].$$

The distance–time diagram is shown in Fig. 10.55a.

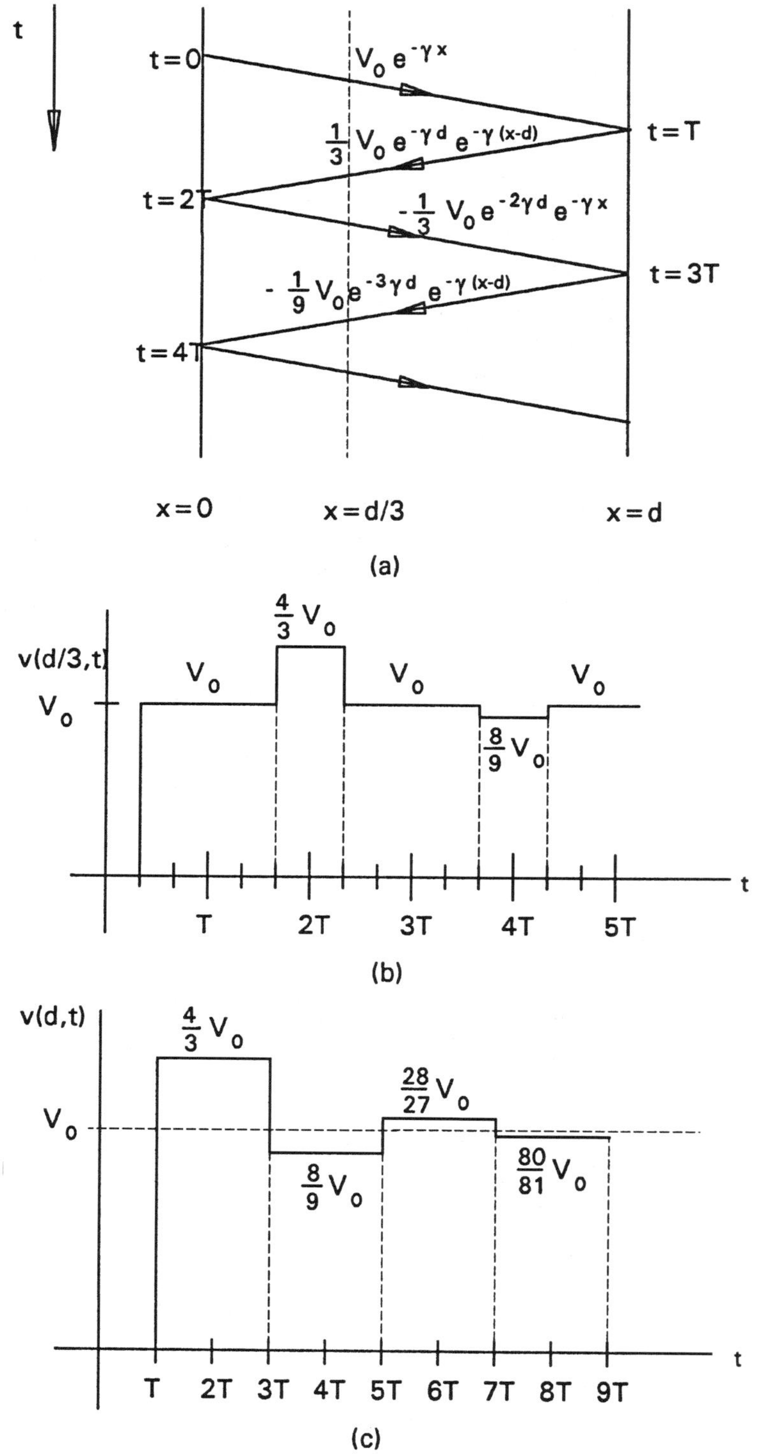

FIGURE 10.55 Distance–time diagram and transient response for a voltage step input in Example 10.19.

(b) Evaluating at $x = d/3$ we find

$$v\left(\frac{d}{3}, t\right) = 0, \qquad t < \frac{T}{3},$$

$$v\left(\frac{d}{3}, t\right) = V_0, \qquad \frac{T}{3} < t < \frac{5T}{3},$$

$$v\left(\frac{d}{3}, t\right) = V_0\left(1 + \frac{1}{3}\right) = \frac{4}{3}V_0, \qquad \frac{5T}{3} < t < \frac{7T}{3},$$

$$v\left(\frac{d}{3}, t\right) = V_0\left(1 + \frac{1}{3} - \frac{1}{3}\right) = V_0, \qquad \frac{7T}{3} < t < \frac{11T}{3},$$

$$v\left(\frac{d}{3}, t\right) = V_0\left(1 + \frac{1}{3} - \frac{1}{3} - \frac{1}{9}\right) = \frac{8}{9}V_0, \qquad \frac{11T}{3} < t < \frac{13T}{3},$$

$$v\left(\frac{d}{3}, t\right) = V_0\left(1 + \frac{1}{3} - \frac{1}{3} - \frac{1}{9} + \frac{1}{9}\right) = V_0, \qquad \frac{13T}{3} < t < \frac{17T}{3},$$

$$\vdots \qquad\qquad \vdots$$

Figure 10.55b shows $v(d/3, t)$.

(c) At the load $x = d$ the two bracketed terms are combined to give

$$v(d,t) = \frac{4}{3}V_0\left[u(t-T) - \frac{1}{3}u(t-3T) + \frac{1}{9}u(t-5T) - \cdots\right].$$

Then

$$v(d,t) = 0, \qquad t < T,$$

$$v(d,t) = \frac{4}{3}V_0, \qquad T < t < 3T,$$

$$v(d,t) = \frac{4}{3}V_0\left(1 - \frac{1}{3}\right) = \frac{8}{9}V_0, \qquad 3T < t < 5T,$$

$$v(d,t) = \frac{4}{3}V_0\left(1 - \frac{1}{3} + \frac{1}{9}\right) = \frac{28}{27}V_0, \qquad 5T < t < 7T,$$

$$v(d,t) = \frac{4}{3}V_0\left(1 - \frac{1}{3} + \frac{1}{9} - \frac{1}{27}\right) = \frac{80}{81}V_0, \qquad 7T < t < 9T,$$

$$\vdots \qquad\qquad \vdots$$

Figure 10.55c shows $v(d, t)$.

Example 10.20 Find $v(d, t)$ for the previous example if the load is a capacitor $Z_L = 1/sC$.

Solution The load reflection coefficient is now

$$\Gamma_L = \frac{1/sC - R_0}{1/sC + R_0} = \frac{-s + 1/\tau}{s + 1/\tau},$$

where $\tau = R_0 C$. Either by substituting in Eq. (10.79) or from the distance–time diagram, Fig. 10.56a, we get

$$V(d,s) = \frac{V_0}{s}[(1+\Gamma_L)e^{-sT} - (1+\Gamma_L)\Gamma_L e^{-3sT} + (1+\Gamma_l)\gamma_L^2 e^{-5sT} - \cdots].$$

Noting that $(1+\Gamma_L) = (2/\tau)/(s+1/\tau)$, this gives

$$\begin{aligned} V(d,s) &= 2V_0\left\{\left[\frac{1/\tau}{s(s+1/\tau)}\right]e^{-sT} - \left[\frac{(1/\tau)(-s+1/\tau)}{s(s+1/\tau)^2}\right]e^{-3sT}\right. \\ &\quad \left. + \left[\frac{(1/\tau)(-s+1/\tau)^2}{s(s+1/\tau)^3}\right]e^{-5sT} - \cdots\right\} \\ &= 2V_0\left\{\left[\frac{1}{s} - \frac{1}{(s+1/\tau)}\right]e^{-sT} - \left[\frac{1}{s} - \frac{2/\tau}{(s+1/\tau)^2} - \frac{1}{(s+1/\tau)}\right]e^{-3sT}\right. \\ &\quad \left. + \left[\frac{1}{s} - \frac{4/\tau^2}{(s+1/\tau)^3} - \frac{1}{(s+1/\tau)}\right]e^{-5sT} - \cdots\right\}. \end{aligned}$$

Taking the inverse transform we obtain

$$\begin{aligned} v(d,t) &= 2V_0[1 - e^{-(t-T)/\tau}]u(t-T) - 2V_0\left\{1 - \left[\frac{2}{\tau}(t-3T)+1\right]e^{-(t-3T)/\tau}\right\}u(t-3T) \\ &\quad + 2V_0\left\{1 - \left[\frac{2}{\tau^2}(t-5T)^2+1\right]e^{-(t-5T)/\tau}\right\}u(t-5T) - \cdots \end{aligned}$$

Figure 10.56b shows this result for the case $\tau = T$.

The voltage $V_f(s)$ launched into the transmission line sees the characteristic impedance of the line. This voltage is therefore given in terms of the generator voltage by

$$V_f(s) = \frac{Z_0(s)}{Z_g(s) + Z_0(s)} V_g(s). \tag{10.81}$$

Substituting this result into Eq. (10.74) gives the transform of the voltage along the line in terms of the generator voltage.

10.8.1 PSPICE Transient Analysis of Transmission Lines

PSPICE allows the display of voltage versus time for a transmission line excited with pulsed signals via the transient analysis. This feature allows the investigation of the voltage associated with the transmission line as a function of time. If the voltage at

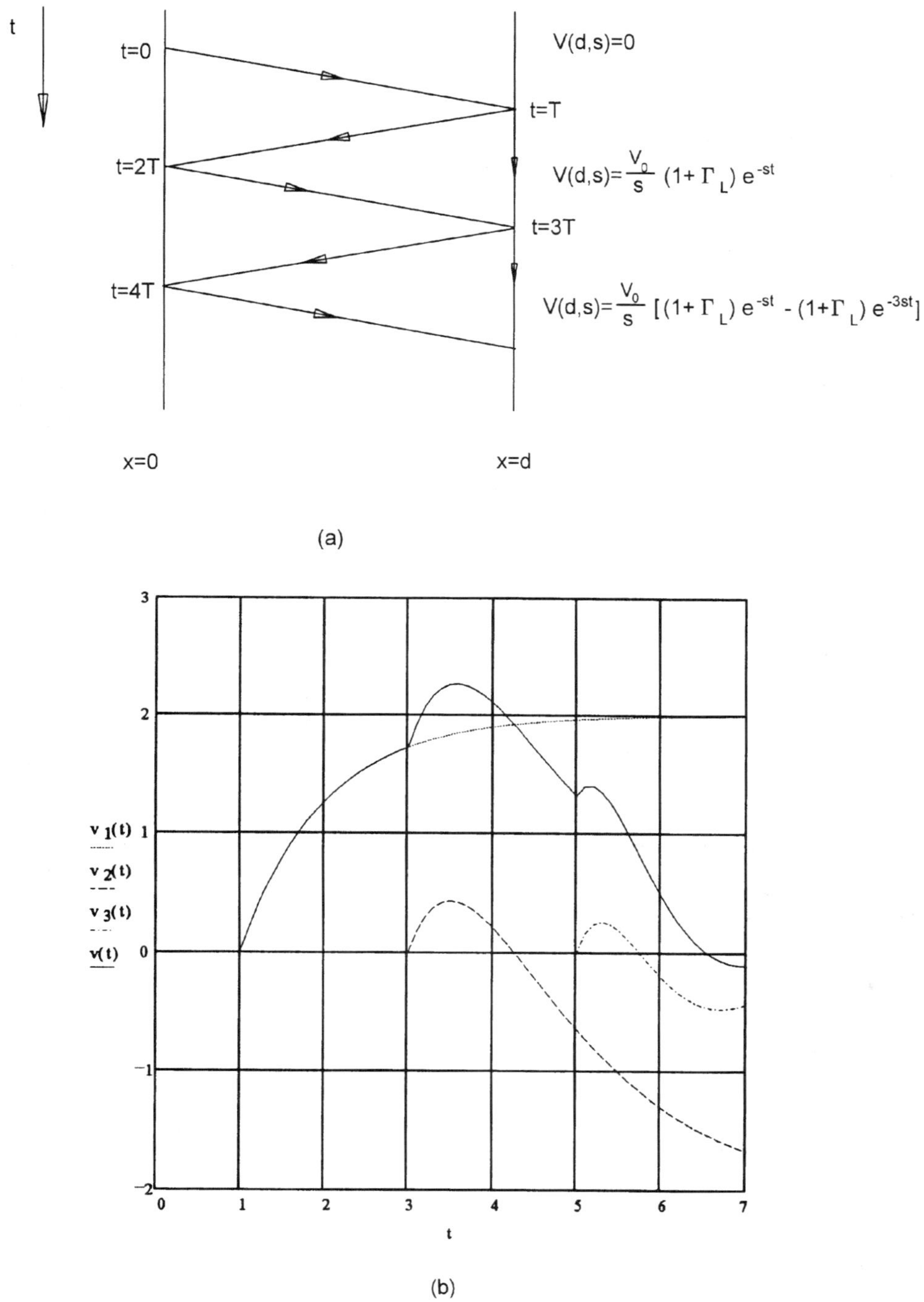

FIGURE 10.56 Distance–time diagram and transient response for a voltage step input in Example 10.20.

some distance d from the load is required, the simulation circuit will be made of two transmission lines with lengths $L - d$ and d, where L is the original length of the transmission line.

Example 10.21 Display the voltage $v(d/3)$ versus time for a loaded 50 Ω transmission line with length d, when the load is 100 Ω (see Example 10.19).

Solution The circuit used to obtain the required voltage is shown in Fig. 10.57. The listing of the PSPICE file is given in Listing 10.6. The input signal is a pulse source with a pulse duration of 20 ms and a period of 40 ms. The rise and fall times (parameters required by PSPICE) are set at 1 μs each. The initial transmission line was "broken" into two sections in order to probe the voltage at the required distance. The total time delay (the time it would take a wave to travel from one end of the transmission line to the other) introduced by the transmission line is chosen to be $T = 1$ ms; thus, in order to display the voltage $v(d/3)$, the two transmission lines TL1 and TL2 have time delays of $\frac{1}{3}$ ms and $\frac{2}{3}$ ms, respectively.

The plot of the voltages at the input of each section of transmission line and the load voltage are displayed in Fig. 10.58 for a time interval from 0 to 10 ms, which corresponds to $10T$. The results of the simulation duplicate the results obtained by hand in Fig. 10.55b,c.

Example 10.22 Simulate the circuit in Example 10.20 using PSPICE.

Solution The circuit used to simulate Example 10.20 is similar to the one used in Example 10.21 with the exception that the load resistor is replaced by a capacitor.

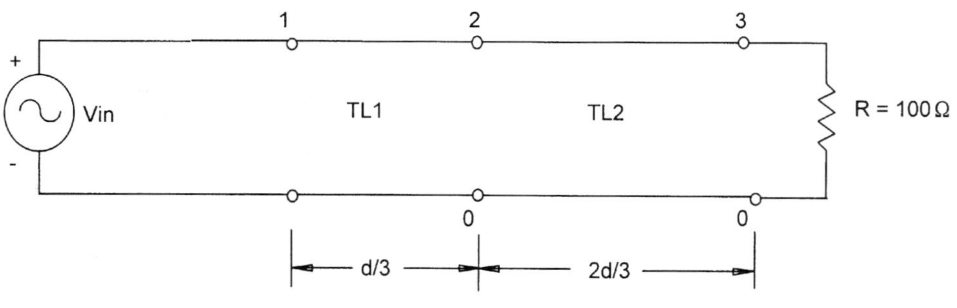

FIGURE 10.57 PSPICE circuit used for the transient response in Example 10.21.

```
*** TRANSIENT RESPONSE OF TRANSMISSION LINE ***
VIN 1 0 PULSE (0 1 0 1US 1US 20MS 40MS)
TL1 1 0 2 0 ZO = 50 TD = {1/3000}
TL2 2 0 3 0 ZO = 50 TD = {2/3000}
RL 3 0 100
.TRAN 1US 40MS
.PROBE
.END
```

LISTING 10.6 PSPICE file used to simulate the transient response on transmission lines.

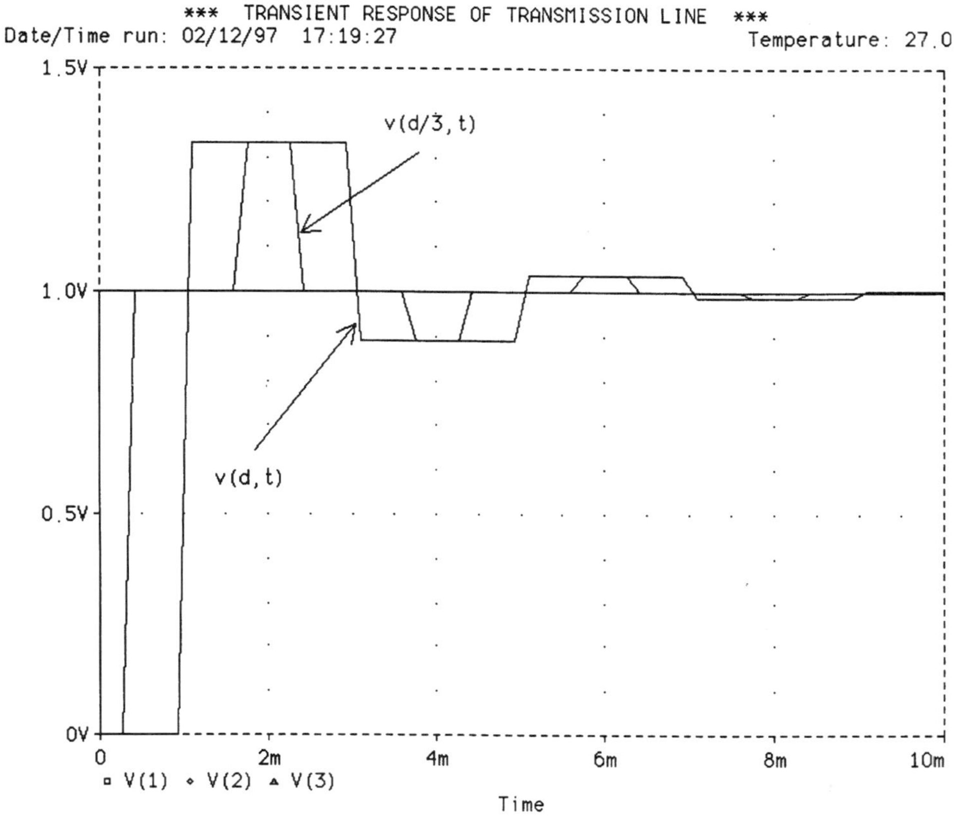

FIGURE 10.58 PSPICE simulation of the circuit in Fig. 10.57.

The value of the capacitor is chosen such that $\tau = T$. Therefore $C = \tau/R_0 = T/R_0 = 1\ \text{ms}/50\ \Omega = 20\ \mu F$. The voltage $v(d,t)$ and the input voltage are shown in Fig. 10.59 for a time interval from 0 to 10 ms. The results of the simulation duplicate the results generated in Fig. 10.56b.

10.8.2 Distortion Revisited

Equation (10.79) can be written in closed form by noting that the two bracketed series are identical and that each is an infinite geometric series that is summed using $1/(1-a) = [1 + a + a^2 + \cdots]$, $|a| < 1$. The result is

$$V(x,s) = V_f \frac{e^{-\gamma x} + \Gamma_L e^{-\gamma d} e^{\gamma(x-d)}}{1 - \Gamma_L \Gamma_G e^{-2\gamma d}}.$$

Evaluating this at the load and using $V_f = [Z_0/(Z_g + Z_0)]\, V_g$, we obtain

$$\begin{aligned} V(d,s) &= \left\{ \left[\frac{Z_0(s)}{Z_g(s) + Z_0(s)} \right] \left[\frac{1 + \Gamma_L(s)}{1 - \Gamma_L(s)\Gamma_G(s) e^{-2\gamma(s)d}} \right] e^{-\gamma(s)d} \right\} V_g(s) \\ &= H(s)\, V_g(s). \end{aligned} \tag{10.82}$$

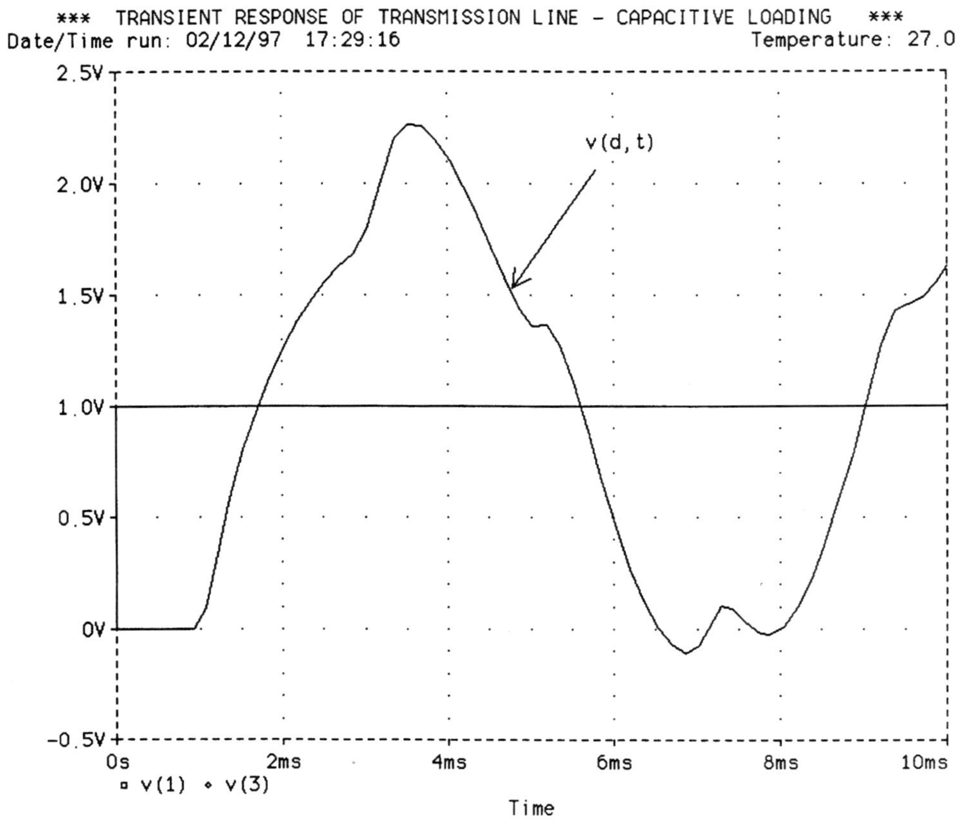

FIGURE 10.59 PSPICE simulation of the transient response of a transmission line with capacitive loading.

where the dependence of each factor on s has been shown explicitly. $H(s)$, which replaces the term within the braces, is the transfer function from the generator or source voltage to the receiver or load voltage.

For distortion-free transmission we must have

$$H(s) = h_0 e^{-sT_d},$$

which represents attenuation by the factor h_0 and a time delay by T_d. To obtain distortion-free transmission the following conditions must be satisfied.

1. The load must be matched to the line, $Z_L = Z_0$. This will make the load reflection coefficient $\Gamma_L = 0$ and yield

$$H(s) = \left[\frac{Z_0(s)}{Z_g(s) + Z_0(s)} \right] e^{-\gamma(s)d}.$$

2. The generator impedance must be proportional to the characteristic impedance, $Z_g(s) = \alpha Z_0(s)$, to give

$$H(s) = \left[\frac{1}{1+\alpha}\right] e^{-\gamma(s)d} = h_0 e^{-\gamma(s)d}.$$

3. The transmission line itself must be distortionless (Section 9.7.2). This requires a lossless line—$R = 0$ and $G = 0$—or that $L/R = C/G$, giving $\gamma(s) = s\sqrt{LC}$ and

$$H(s) = h_0 e^{-s\sqrt{LC}d} = h_0 e^{-sT}, \quad T = d\sqrt{LC} = d/u.$$

When conditions 1 and 3 are satisfied, but not condition 2, the load voltage will reproduce the line input voltage waveform, but this will be different from the generator voltage waveform. Generally, the generator impedance is matched as closely as possible to the line impedance, $Z_g \approx Z_0$, so that the generator reflection coefficient will be small. As seen already, when condition 3 is satisfied, $Z_0 = \sqrt{L/C} = R_0$, a pure resistance.

The requirement that the load and the generator be matched to the line emphasizes the importance of designing compatible complete systems rather than stringing together unmatched parts.

PROBLEMS

10.1 Repeat Examples 10.1 and 10.2 if the lossless line is $\lambda/8$ long.
Ans. $Z_{\text{in}} = (2-j)R_0, P_{L,\text{Avg}} = P_{\text{in,Avg}} = 0.100|\underline{V}_g|^2/R_0$,
$P_{Z_g,\text{Avg}} = 0.050|\underline{V}_g|^2/R_0, P_{\text{gen,Avg}} = 0.150|\underline{V}_g|^2/R_0$.

10.2 Repeat Examples 10.1 and 10.2 if $Z_g = (0.4 - j0.2)R_0$.
Ans. $Z_{\text{in}} = (0.4 + j0.2)R_0$, $P_{L,\text{Avg}} = P_{\text{in,Avg}} = P_{Z_g,\text{Avg}} = 0.3125|\underline{V}_g|^2/R_0$,
$P_{\text{gen,Avg}} = 0.625|\underline{V}_g|^2/R_0$.

Note that the generator impedance is the complex conjugate of Z_{in}. For a given load and line length, the maximum possible power is transferred to the line (and therefore to the load) when $Z_g = Z_{\text{in}}^*$.

10.3 The impedance looking toward the load at the plane $z = z_0$ is $R_0(1+j)$ on a lossless line.

a. Find the impedance looking toward the load at $z = z_0 - 3\lambda/8$.

b. Find the impedance looking toward the load at $z = z_0 + \lambda/8$.

Ans. $(0.4 + j0.2)R_0$.

10.4 Find the shortest distances in fractions of a wavelength from the load to the plane where Z_{in} equals $jR_0/2, jR_0, j2R_0, -j2R_0, -jR_0$, and $-jR_0/2$, if:

a. The load is a short circuit.

b. The load is an open circuit.

Ans. a. $0.074\lambda, 0.125\lambda, 0.176\lambda, 0.324\lambda, 0.375\lambda, 0.426\lambda$.

10.5 Show that if $\Gamma(z) = 1\angle\phi$, then $Z(z) = jZ_0 \cot(\phi/2)$. On a lossless line show that $Z(z) = jR_0 \cot(\phi/2)$, which is purely reactive.

10.6 Use Eqs. (10.22), (10.23), and (10.24) to find Z_{in} for Example 10.1. Is the same result obtained?

10.7 An air line is excited at 100 MHz and the reflection coefficient at z_0 is measured as $\Gamma(z_0) = 0.4\angle -60°$. Find the reflection coefficient a distance 0.6 m toward the generator and 0.8 m toward the load.

a. Assume the line is lossless.

b. Assume the line has an attenuation constant of 0.12 Np/m.

c. Plot the five reflection coefficients in the Γ plane.

Ans. a. $\Gamma(z_0 - 0.6) = 0.4\angle 156°$; $\Gamma(z_0 + 0.8) = 0.4\angle 132°$;
b. $\Gamma(z_0 - 0.6) = 0.346\angle 156°$; $\Gamma(z_0 + 0.8) = 0.485\angle 132°$.

10.8 A 50 ohm lossless air line is terminated with an impedance $Z_L = 75 + j25\ \Omega$ at a frequency of 1.5 GHz. The line is 0.628 m long. Use a Smith chart to find:

a. The load admittance, the reflection coefficient at the load, and the standing wave ratio on the line.

b. The distance to the three closest voltage maxima and the three closest voltage minima.

c. The input impedance and admittance.

Ans. a. $Y_L = 0.012 - j0.004$ S, $\Gamma_L = 0.28\angle 34°$, $S = 1.78$;
b. $d_{max} = 0.0094, 0.1094, 0.2094$ m;
c. $Z_{in} = 54 - j30\ \Omega$.

10.9 A 300 Ω lossless air line is terminated in a device with an input impedance $Z_L = 300 + j5(f - 150)$, where f is expressed in megahertz. The frequency band of excitation extends from $f_l = 120$, through $f_0 = 150$ to $f_u = 180$ MHz.

a. Using a Smith chart, plot the normalized load impedance with frequency as a parameter over the frequency band. Find the load reflection coefficient at f_l, f_0, and f_u.

b. Using a Smith chart, find λ_0—the wavelength at f_0—and the impedance looking toward the load at f_l, f_0, and f_u a distance $\lambda_0/2$ from the load.

Ans. a. $\Gamma_L = 0.24\angle -76°$, 0, $0.24\angle 76°$ at f_l, f_0, f_u;
b. $Z = 491 - j17$, 300, $491 + j17\ \Omega$ at f_l, f_0, f_u.

10.10 A lossless line consists of three cascaded transmission lines (Fig. 10.60). The first line, connected to a matched voltage source, has characteristic resistance $R_{01} = 50\ \Omega$. The second and third lines have characteristic resistances $R_{02} = 100\ \Omega$ and $R_{03} = 300\ \Omega$, respectively. The line lengths are $\lambda_1/4$, $3\lambda_2/8$, and $3\lambda_3/8$, where λ_i is the wavelength in the respective line. The load on the third line is its characteristic resistance R_{03}.

a. Use a Smith chart to find the input reflection coefficient and the input impedance.

b. The voltage source is 100 V (zero to peak). Calculate the power delivered to the load.

Ans. a. $Z_{in} = 15 - j20\ \Omega$; b. $P_L = 16.2$ W.

10.11 Figure 10.61 shows two secondary lines with characteristic resistances R_{01} and R_{02} connected in parallel as a load on the primary line, with characteristic resistance R_0. All three lines are lossless. The parallel connection means that the load admittance on the primary line is the sum of the input admittances to the secondary lines, $Y_L = Y_{\text{in},1} + Y_{\text{in},2}$.

a. Show that if $R_{01} = R_{02} = 2R_0$ and the two secondary lines have matched loads, then the primary line is also matched.

b. Find a more general relation between the characteristic resistances such that the primary line is matched when the secondary lines are matched. If $R_0 = 50\ \Omega$ and $R_{01} = 75\ \Omega$, what value of R_{02} satisfies your relation?

What fraction of the input power to the primary line is delivered to each of the matched loads on the secondary lines?

Ans. b. $R_{02} = 150\ \Omega$.

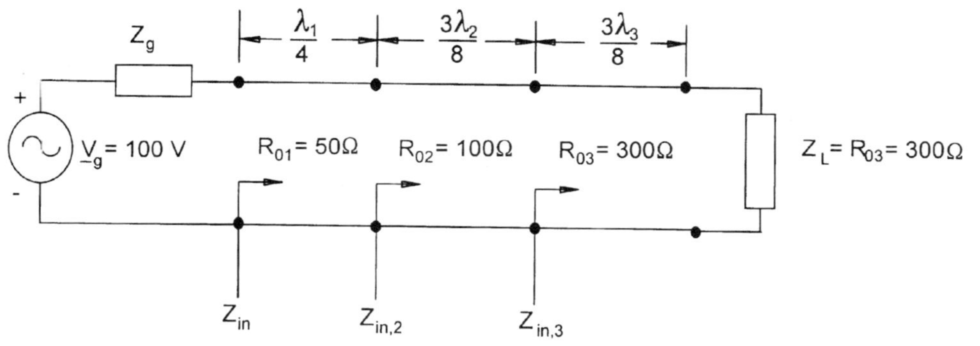

FIGURE 10.60 Circuit diagram for Problem 10.10.

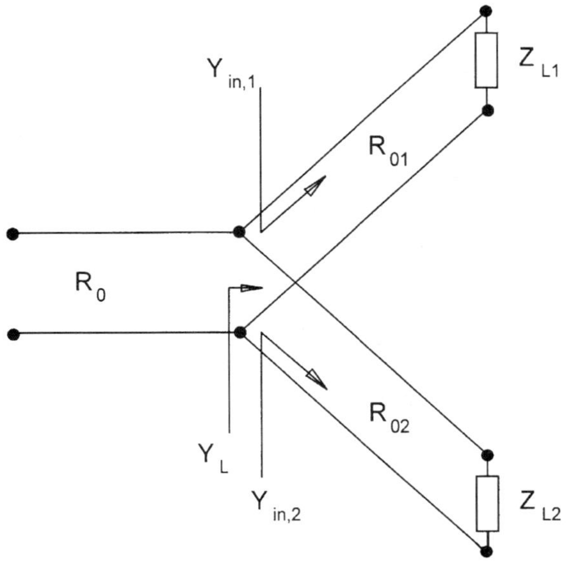

FIGURE 10.61 Circuit diagram for Problem 10.11.

10.12 For the lines of Fig. 10.61, let $R_0 = 50\,\Omega$, $R_{01} = 100\,\Omega$, and $Z_{L1} = 200 + j100\,\Omega$.

a. Find the two shortest lengths d_{max} and d_{min} (in wave lengths) for which the input impedance $Z_{in,1}$ is purely resistive (real). Do this analytically and using the Smith chart. Find the corresponding values of $R_{in,1\,max}$ and $R_{in,1\,min}$.

b. For which of the lengths is it possible to choose a matched secondary line and load R_{02} so that the primary line will be matched? Find this value of R_{02}.

Ans. a. $d_{max} = 0.0369\lambda$, $R_{in,1\,max} = 261.7\,\Omega$; $d_{min} = 0.2869\lambda$, $R_{in,1\,min} = 38.2\,\Omega$.

10.13 Repeat Example 10.12 with the following measurements. With the short connected there are adjacent voltage minima at 1.10 and 4.20 cm. With Z_L connected the voltage standing wave ratio is 3.20 and there is a voltage minimum at 2.00 cm.

Ans. $Z_L = 53 + j75\,\Omega$.

10.14 Equation (10.73) and hence the equivalent circuit of Fig. 10.35c can be obtained directly from the general expression for input impedance, Eq. (10.5).

a. Show that for a shorted line, $Y_{in} = (1/Z_0)\coth \gamma d$.

b. The cotangent can be expanded as

$$\coth u = \frac{1}{u} + \sum_{n=1}^{\infty} \frac{2u}{n^2\pi^2 + u^2}.$$

Show that $Z_0\gamma d = (R + j\omega L)d$ and $Z_0/\gamma d = [(G + j\omega C)d]^{-1} \approx (1/j\omega Cd) + (G/\omega^2 C^2 d)$, for a low-loss line, and hence

$$Y_{in} = \frac{1}{(R + j\omega L)d} + \sum_{n=1}^{\infty} \frac{1}{\frac{1}{2}\left(\frac{R}{L} + \frac{G\omega_n^2}{C\omega^2}\right)Ld + j\left(\omega L_n - \frac{1}{\omega C_n}\right)}.$$

The real term in the denominator is negligible compared to the imaginary term, except when $\omega \approx \omega_n$. We can therefore let the ratio $\omega/\omega_n = 1$. Show that Eq. (10.73) is obtained.

10.15 The circuit of Fig. 10.62 can represent a transmission line that is open at both ends. This is the dual of Fig. 10.35c. Consider an ideal current source $\underline{I}_g$ with infinite internal impedance exciting the open line. For the lossless line the voltage and current in the vicinity of the resonant frequencies $\omega_n = n\pi/d\sqrt{LC}$ are

$$\underline{I}(-d) = \underline{I}_g, \quad \underline{I}(z) = -j\left(\frac{2V_0^+}{R_0}\right)\sin\left(\frac{n\pi z}{d}\right), \qquad -d < z \le 0,$$

$$\underline{V}(z) = 2V_0^+ \cos\left(\frac{n\pi z}{d}\right), \qquad -d \le z \le 0.$$

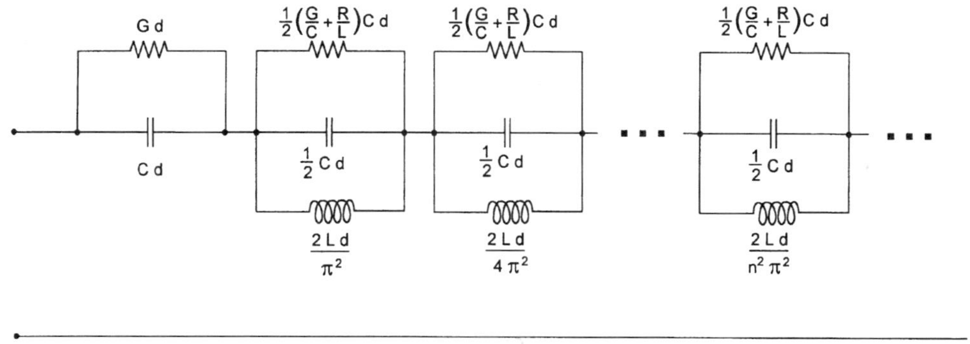

FIGURE 10.62 Model of a transmission line that is open-circuited at both ends used in Problem 10.15.

a. Show that the time average electric energy and the time average magnetic energy stored along the line are

$$W_{e,\text{Avg}} = \tfrac{1}{8}\,Cd(2|V_0^+|)^2, \quad W_{m,\text{Avg}} = \tfrac{1}{8}\,Ld\left(\frac{2|V_0^+|}{R_0}\right)^2.$$

b. Show that the time average power loss along the line is given by Eq. (10.68).

c. Show that Q_n is given by Eq. (10.69).

d. Equating $W_{e,\text{Avg}}$ with $\frac{1}{4}C_n|V(0)|^2$, show that $C_n = Cd/2$. Using Eq. (10.52), $Q_n = (\omega_n C_n/G_n)$, show that $G_n = \frac{1}{2}(G/C + R/L)Cd$. Using $\omega_n = 1/\sqrt{L_n C_n}$, show that $L_n = 2Ld/n^2\pi^2$.

e. When a direct voltage is applied to the open circuited line, the source will see the total capacitance of the line Cd in parallel with the total conductance Gd. Show that the circuit model of Fig. 10.62 gives

$$Z_{\text{in}} = \frac{1}{Gd + j\omega Cd} + \sum_{n=1}^{\infty} \frac{1}{G_n + j(\omega C_N - 1/\omega L_n)}.$$

10.16 Design a microstrip line and microstrip quarter-wave matching section for the lines of Example 10.14. The lines should be on 10 mil quartz substrate ($h = 0.01$ in., $\varepsilon_r = 3.8$).

10.17 A quarter-wave monopole antenna above a ground plane has an input resistance of 36.5 Ω. The antenna is to be fed by a 50 Ω coaxial cable with an appropriate matching quarter-wave coaxial section.

a. Design the feed coax with an outside diameter of 1 cm and a Teflon dielectric ($\varepsilon_r = 2.1$). Design the matching coax with the same outside diameter and dielectric.

b. Design the matching coax to have the same inner and outer diameters as the feed cable but a different dielectric. What listed dielectric has the relative permittivity closest to your design?

Ans. a. For feed coax, $b = 1.000$, $a = 0.299$ cm. For the matching section, $b = 1.000$, $a = 0.356$ cm.

b. $\varepsilon_r = 2.88$. Polystyrene has $\varepsilon_r = 2.7$.

10.18 A quarter-wave matching section is designed to match a load R_L to a line R_0 at a frequency f_0.

a. Show that the reflection coefficient at the input to the quarter-wave matching section is

$$\Gamma_{\text{in}} = \frac{R_L - R_0}{(R_L + R_0) + j2\sqrt{R_0 R_L}\,\tan\beta d} = \frac{\hat{R}_L - 1}{(\hat{R}_L + 1) + j2\sqrt{\hat{R}_L}\,\tan\beta d}.$$

where $\beta d = (\pi/2)\,(f/f_0)$, $\hat{R}_L = R_L/R_0$.

b. Plot $|\Gamma_{\text{in}}|$ versus f/f_0 for $0 \leq (f/f_0) \leq 2$, with $\hat{R}_L = 2, 4, 10$. Show that the same curves are obtained for $\hat{R}_L = 0.5, 0.25, 0.1$, respectively.

10.19 A 50 Ω lossless air line is excited at a frequency of 500 MHz and is terminated in a load impedance of $Z_L = 100 - j70\ \Omega$.

a. Use a 50 Ω lossless air line as a shorted stub and design a single stub match with the shortest possible length between the stub, d_{ca}, and the load and the shortest possible stub length, d_{sa}.

b. There is a second solution with a greater length d_{cb} between the load and the stub. This length is still less than a half-wavelength at 500 MHz. Design a second single stub match with this distance and a corresponding length d_{sb}.

c. The actual signal occupies a frequency band from 400 to 600 MHz. For each of the designs find the reflection coefficient on the generator side of the stub at 400 MHz and at 600 MHz, assuming the load impedance remains fixed. Which design gives a better match across the band?

In this problem $d_{ca} < d_{cb}$ and $d_{sa} < d_{sb}$, so that both values of the part (a) solution have a smaller change in length, measured in wavelengths, than the values in part (b).

Ans. a. $d_{ca} = 0.128\lambda$, $d_{sa} = 0.110\lambda$.

b. $d_{cb} = 0.291\lambda$, $d_{sb} = 0.390\lambda$.

c. For design (a), $\Gamma = 0.393\angle 86^\circ$ and $0.279\angle -148^\circ$ at 400 and 600 MHz. For design (b), $\Gamma = 0.481\angle 168^\circ$ and $0.951\angle -154^\circ$ at 400 and 600 MHz.

10.20 Repeat Problem 10.19 if $Z_L = 16.8 + j11.7\ \Omega$. In this problem $d_{ca} < d_{cb}$ and $d_{sa} > d_{sb}$. We must actually do the calculation to decide which design gives a broader bandwidth.

Ans. a. $d_{ca} = 0.041\lambda$, $d_{sa} = 0.390\lambda$.

b. $d_{cb} = 0.378\lambda$, $d_{sb} = 0.110\lambda$.

c. For design (a), $\Gamma = 0.384\angle 121^\circ$ and $0.898\angle -152^\circ$ at 400 and 600 MHz. For design (b), $\Gamma = 0.672\angle 98^\circ$ and $0.528\angle -172^\circ$ at 400 and 600 MHz.

10.21 Repeat Problem 10.20 using an open circuited stub. The solutions now have $d_{ca} < d_{cb}$ and $d_{sa} < d_{sb}$, so the part (a) solution can easily be recognized as a broader bandwidth design. The design choice between shorted and open stubs cannot be made solely on the basis of bandwidth. In many physical situations, such as a coax, an "open" may be difficult to obtain because of radiation from the end. On the other hand, in microstrip a short requires a physical conducting connection through the dielectric substrate, which may present fabrication problems. Each problem requires its own solution.

Ans. a. $d_{ca} = 0.041\lambda$, $d_{sa} = 0.140\lambda$.

b. $d_{cb} = 0.378\lambda$, $d_{sb} = 0.360\lambda$.

c. For design (a), $\Gamma = 0.210\angle 119^\circ$ and $0.326\angle -99^\circ$ at 400 and 600 MHz. For design (b), $\Gamma = 0.939\angle 149^\circ$ and $0.515\angle -178^\circ$ at 400 and 600 MHz.

10.22 For Example 10.19 plot $i(d/3, t)$ and $i(d, t)$ for the same time intervals shown in Figs. 10.55b and 10.55c.

10.23 For the line, source, and load of Example 10.19 the input signal is a pulse given by $v_f(t) = V_0[u(t) - u(t - T/2)]$. Plot $v(d/3, t)$ and $v(d, t)$ for the same time intervals shown in Figs. 10.55b and 10.55c.

10.24 The voltage source $v_g(t)$ is matched to a lossless transmission line, $Z_g = R_0$, which is short circuited, $Z_L = 0$, at its termination. Plot the voltage and current along the line $(0 \le x \le d)$ at $t = 0.8T$, $1.2T$, $1.8T$, and $2.2T$ when:

a. $v_g(t) = V_0 u(t)$.

b. $v_g(t) = V_0[u(t) - u(t - 0.1T)]$.

10.25 Repeat Example 10.20 if the generator is matched to the line, $Z_g = R_0$, when:

a. $v_g(t) = V_0 u(t)$.

b. $v_g(t) = V_0[u(t) - u(t - 0.1T)]$.

10.26 Simulate the solution of Problem 10.17 using PSPICE. Compare the results by displaying the reflection coefficient.

10.27 Simulate the two solutions of Problem 10.19 using PSPICE, and specify which design gives a better match across the defined bandwidth, by displaying the reflection coefficient over the bandwidth.

10.28 Simulate the two solutions of Problem 10.20 using PSPICE, and specify which design gives a better match across the defined bandwidth, by displaying the reflection coefficient over the bandwidth.

10.29 Simulate the two solutions of Problem 10.21 using PSPICE, and specify which design gives a better match across the defined bandwidth, by displaying the reflection coefficient over the bandwidth. In the PSPICE simulation, the open circuit is modeled by a very high resistance, say, $10^9\ \Omega$.

10.30 Use PSPICE to display the currents $i(d/3, t)$ and $i(d, t)$ as required in Problem 10.22.

10.31 Use PSPICE to display the voltages $v(d/3, t)$ and $v(d, t)$ for the input signal specified in Problem 10.23, using the same value for T as in Example 10.21.

10.32 Solve Problem 10.24 using PSPICE. Use the same value of T as in Example 10.21.

10.33 Solve Problem 10.25 using PSPICE, by using the same values of T and load capacitance as in Example 10.22.

REFERENCES

[1] James W. Nilsson, *Electric Circuits*, 3rd ed., Addison-Wesley, Reading, MA, 1990, pp. 266–273, 492–510.

[2] Lawrence P. Huelsman, *Basic Circuit Theory*, 2nd ed., Prentice-Hall, Englewood Cliffs, NJ, 1991, pp. 210, 319.

[3] Egon Brenner and Mansour Javid, *Analysis of Electric Circuits*, McGraw-Hill, New York, 1959, pp. 380–390.

[4] Charles E. Smith, "Frequency domain analysis of RF and microwave circuits using SPICE," *IEEE Trans. Microwave Theory Techniques*, Vol. 42, No. 10, Oct. 1994, pp. 1904–1909.

[5] Mansour Javid and Egon Brenner, *Analysis, Transmission, and Filtering of Signals*, McGraw-Hill, New York, 1963, Chap. 11.

CHAPTER ELEVEN

Reflection and Transmission at Plane Boundaries and Two-Dimensional Waveguides

11.1 INTRODUCTION

In this chapter we will first study the propagation of plane waves in unbounded media. We will see how to represent the direction of propagation and the orientation of the electric field. We will then obtain expressions for the power carried by the wave. After that we will study the reflection and transmission of waves when they are incident at a plane boundary between two media. We will see how to model these results in terms of transmission lines and that will allow us to use the techniques of analysis that we studied in the previous chapter. Finally, we will study the transmission of waves in two-dimensional waveguides, such as the parallel plate waveguide and the slab dielectric waveguide. These serve as an introduction to closed waveguides and fiber optic lines.

11.2 PROPAGATION VECTOR

In Section 9.3 we assumed a solution [Eqs. (9.5) and (9.6)] to Maxwell's equations, which was TEM to the direction of propagation along the z axis. The wave components were found to be proportional to $\exp(\mp \gamma z)$, where

$$\gamma = \alpha + j\beta = \sqrt{j\omega\mu(\sigma + j\omega\varepsilon)}.$$

We noted that γ is the propagation constant, α is the attenuation constant, and β is the phase constant.

Let us now suppose that we wish to represent the same wave, but we want to consider it as propagating along the z' axis as shown in Fig. 11.1. The electric and magnetic field components are now proportional to $\exp(\mp \gamma z')$. By taking projections

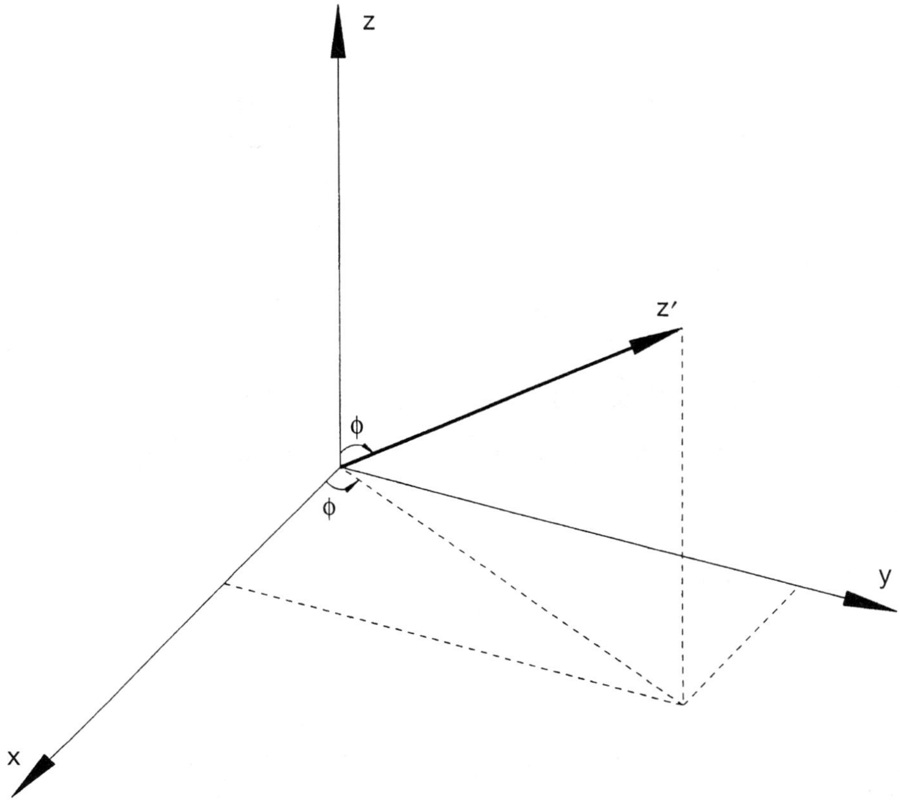

FIGURE 11.1 Coordinate system for propagating waves.

along the x, y, and z axes we can write

$$\begin{aligned}\gamma z' &= (\gamma \sin\theta \cos\phi)x + (\gamma \sin\theta \sin\phi)y + (\gamma \cos\theta)z \\ &= \gamma_x x + \gamma_y y + \gamma_z z,\end{aligned}$$

where

$$\gamma_x^2 + \gamma_y^2 + \gamma_z^2 = \gamma^2.$$

We can generalize this concept as follows. The propagation constant can be represented as a propagation or wave vector

$$\vec{\gamma} = \vec{a}_x\gamma_x + \vec{a}_y\gamma_y + \vec{a}_z\gamma_z \tag{11.1}$$

with the condition that

$$\vec{\gamma}\cdot\vec{\gamma} = \gamma^2 = j\omega\mu(\sigma + j\omega\varepsilon). \tag{11.2}$$

The propagation vector is the sum of its real and imaginary parts, $\vec{\gamma} = \vec{\alpha} + j\vec{\beta}$, giving the attenuation vector and the phase vector, respectively. Substituting in Eq. (11.2) we get

$$\begin{aligned}\gamma^2 &= \vec{\gamma}\cdot\vec{\gamma} = (\vec{\alpha} + j\vec{\beta})\cdot(\vec{\alpha} + j\vec{\beta}) \\ &= (\vec{\alpha}\cdot\vec{\alpha} - \vec{\beta}\cdot\vec{\beta}) + j2\vec{\alpha}\cdot\vec{\beta}\end{aligned}$$

or

$$-\omega^2\mu\varepsilon = \mathrm{Re}(\gamma^2) = (\alpha_x^2 + \alpha_y^2 + \alpha_z^2) - (\beta_x^2 + \beta_y^2 + \beta_z^2), \tag{11.3}$$

$$\omega\mu\sigma = \mathrm{Im}(\gamma^2) = 2(\alpha_x\beta_x + \alpha_y\beta_y + \alpha_z\beta_z). \tag{11.4}$$

The position vector $\vec{R} = \vec{a}_x x + \vec{a}_y y + \vec{a}_z z$ is the vector from the origin to any point in space. The wave components are now proportional to

$$e^{\mp\vec{\gamma}\cdot\vec{R}} = e^{\mp\vec{\alpha}\cdot\vec{R}}e^{\mp j\vec{\beta}\cdot\vec{R}}.$$

The plane surface $\vec{\alpha}\cdot\vec{R} = C_1$ gives the surface of constant amplitude and the plane surface $\vec{\beta}\cdot\vec{R} = C_2$ gives the surface of constant phase. These surfaces are shown in Fig. 11.2 for a two-dimensional case where $\gamma_y = 0$. The wave is a plane wave because the surface of constant phase is a plane. Note that the planes are normal to the vectors

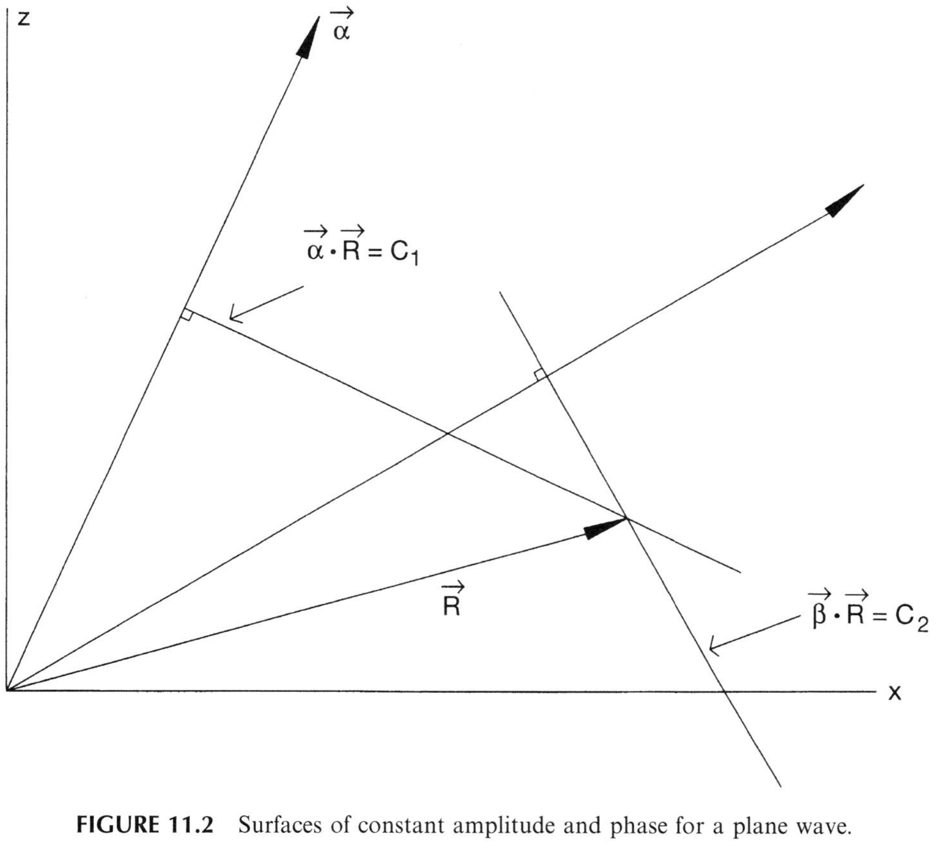

FIGURE 11.2 Surfaces of constant amplitude and phase for a plane wave.

$\vec{\alpha}$ and $\vec{\beta}$, which are not necessarily collinear. As time increases the plane of constant phase moves at the phase velocity $v_{\text{ph}} = \omega/\beta$ in the direction of the phase vector for a forward going wave or in the opposite direction for a backward going wave.

A wave is a uniform wave if the amplitude of the wave is constant in a surface of constant phase. If $\vec{\alpha}$ and $\vec{\beta}$ are not collinear, the wave is nonuniform because points on a surface of constant phase lie on different surfaces of constant amplitude. If the two vectors are collinear and if $\vec{e}_t$ and $\vec{h}_t$ are constant vectors, then the wave is uniform.

Example 11.1 A 100 kHz wave in dry soil ($\varepsilon_r = 3.5$, $\mu_r = 1$, $\sigma = 10^{-5}$ S/m) has an electric field given by

$$\underline{\vec{E}} = \vec{a}_y E_0 \exp[-\alpha_x x] \exp[-j(2.50 \times 10^{-3} x + \beta_z z)].$$

Evaluate the propagation vector.

Solution The propagation constant is given by

$$\begin{aligned}\gamma &= \sqrt{j(200\pi \times 10^3)(4\pi \times 10^{-7})\left[10^{-5} + j(200\pi \times 10^3)\left(\frac{3.5}{36\pi} \times 10^{-9}\right)\right]} \\ &= \sqrt{-1.54 \times 10^{-5} + j7.90 \times 10^{-6}} = 4.06 \times 10^{-6} + j1.68 \times 10^{-5}\ \text{m}^{-1}.\end{aligned}$$

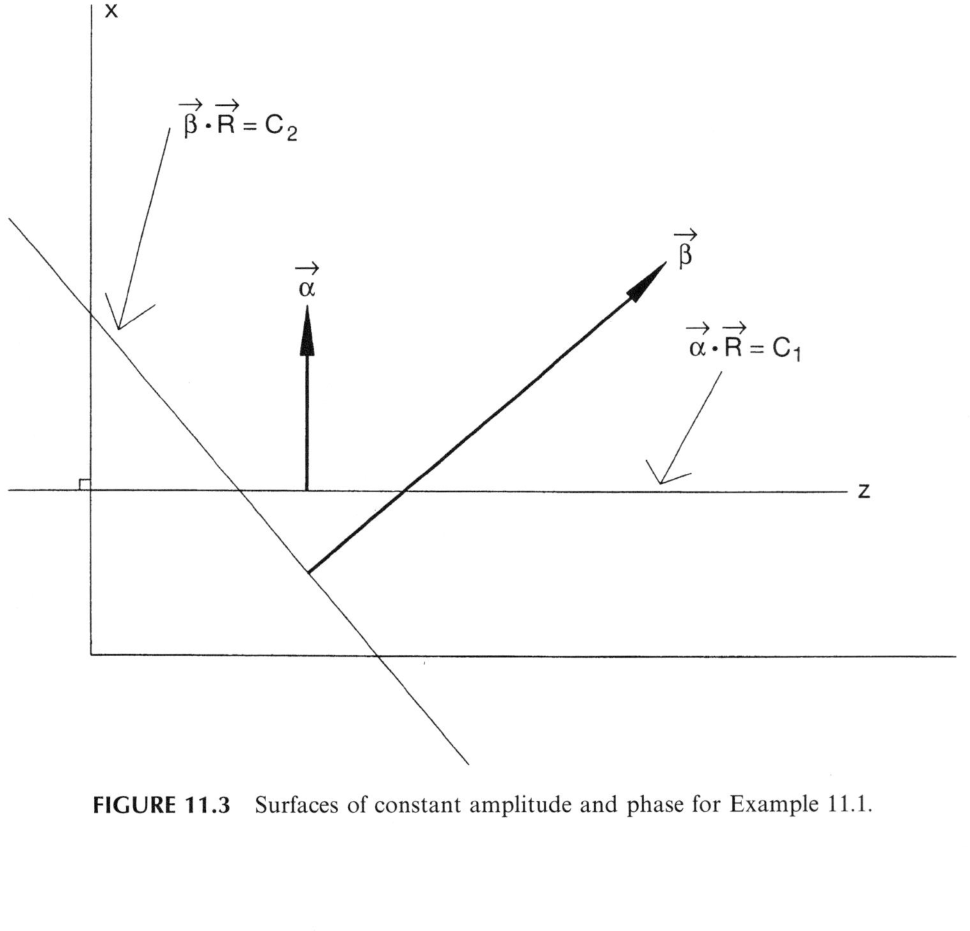

FIGURE 11.3 Surfaces of constant amplitude and phase for Example 11.1.

Using Eqs. (11.3) and (11.4) we calculate

$$\alpha_x = \frac{\mathrm{Im}(\gamma^2)}{2\beta_x} = \frac{7.90 \times 10^{-6}}{2(2.50 \times 10^{-3})} = 1.58 \times 10^{-3} \text{ Np/m},$$

$$\beta_z^2 = -\mathrm{Re}(\gamma^2) - \beta_x^2 + \alpha_x^2 = 1.54 \times 10^{-5} - (2.50 \times 10^{-3})^2 + (1.58 \times 10^{-3})^2$$

$$\beta_z = 3.41 \times 10^{-3} \text{ rad/m}.$$

The electric field intensity is therefore

$$\underline{\vec{E}} = \vec{a}_y E_0 \exp[-1.58 \times 10^{-3}x] \exp[-j(2.50 \times 10^{-3}x + 3.41 \times 10^{-3}z)].$$

and the magnetic field intensity can be found from $\nabla \times \underline{\vec{E}} = -j\omega\mu\underline{\vec{H}}$. Substituting the resultant $\underline{\vec{H}}$ in $\nabla \times \underline{\vec{H}} = (\sigma + j\omega\varepsilon)\underline{\vec{E}}$ must give the original $\underline{\vec{E}}$.

Figure 11.3 shows $\vec{\alpha}$ and the plane of constant attenuation and $\vec{\beta}$ and the plane of constant phase for this example.

11.3 PROPAGATION CONSTANT AND INTRINSIC IMPEDANCE FORMULAS

The propagation constant and the intrinsic impedance are given by

$$\gamma = \alpha + j\beta = \sqrt{j\omega\mu(\sigma + j\omega\varepsilon)}, \tag{11.5}$$

$$\eta = R + jX = \sqrt{\frac{j\omega\mu}{\sigma + j\omega\varepsilon}}. \tag{11.6}$$

Given the constants of the medium it is always possible to calculate values of α and β directly from these formulas. In several special cases it is helpful to develop special formulas that are simpler to evaluate and, more important, clearly show the relation between α and β and $\omega\mu$, $\omega\varepsilon$, and σ.

11.3.1 Perfect Dielectric Medium

In a perfect or lossless dielectric medium the conductivity is zero, $\sigma = 0$. In that case

$$\gamma = j\omega\sqrt{\mu\varepsilon}, \quad \alpha = 0, \quad \beta = \omega\sqrt{\mu\varepsilon} = \omega\sqrt{\mu_0\varepsilon_0}\sqrt{\mu_r\varepsilon_r} = \frac{\omega\sqrt{\mu_r\varepsilon_r}}{3 \times 10^8}. \tag{11.7}$$

$$\eta = \sqrt{\frac{\mu}{\varepsilon}} = \sqrt{\frac{\mu_0}{\varepsilon_0}}\sqrt{\frac{\mu_r}{\varepsilon_r}} = 120\pi\sqrt{\frac{\mu_r}{\varepsilon_r}}. \tag{11.8}$$

The phase velocity in free space $c = 1/\sqrt{\mu_0\varepsilon_0} = 3 \times 10^8$ m/s and the intrinsic impedance of free space $\eta_0 = \sqrt{\mu_C/\varepsilon_0} = 120\pi$ are easy values to remember and together they can be used to calculate the constants ε_0 and μ_0. A nice number to remember also is that the phase velocity of light in free space is approximately 1 foot per nanosecond, $c \approx 1$ ft/ns. More accurately, $c = 11.8$ in./ns.

11.3.2 Good or Low-Loss Dielectric

For a material and frequency for which $\sigma/\omega\varepsilon \ll 1$ the dielectric is said to be good or low loss. In this case,

$$\gamma = j\omega\sqrt{\mu\varepsilon}\left(1+\frac{\sigma}{j\omega\varepsilon}\right)^{1/2} \approx j\omega\sqrt{\mu\varepsilon}\left(1+\frac{\sigma}{2j\omega\varepsilon}\right) = \frac{\sigma}{2}\sqrt{\frac{\mu}{\varepsilon}} + j\omega\sqrt{\mu\varepsilon},$$

$$\alpha = \frac{\sigma}{2}\sqrt{\frac{\mu}{\varepsilon}} = \frac{1}{2}\left(\frac{\sigma}{\omega\varepsilon}\right)\beta, \quad \beta = \omega\sqrt{\mu\varepsilon}. \tag{11.9}$$

$$\eta = \sqrt{\frac{\mu}{\varepsilon}}\left(1+\frac{\sigma}{j\omega\varepsilon}\right)^{-1/2} \approx \sqrt{\frac{\mu}{\varepsilon}}\left(1-\frac{\sigma}{2j\omega\varepsilon}\right) = \sqrt{\frac{\mu}{\varepsilon}} + j\frac{1}{2}\left(\frac{\sigma}{\omega\varepsilon}\right)\sqrt{\frac{\mu}{\varepsilon}},$$

$$R = \sqrt{\frac{\mu}{\varepsilon}}, \quad X = \frac{1}{2}\left(\frac{\sigma}{\omega\varepsilon}\right)\sqrt{\frac{\mu}{\varepsilon}} = \frac{1}{2}\left(\frac{\sigma}{\omega\varepsilon}\right)R. \tag{11.10}$$

Since $X \ll R$ the reactive part of the intrinsic impedance is generally neglected.

The wave amplitude attenuates exponentially at a rate of α Np/m. The distance $\delta_s = 1/\alpha$ in which the amplitude attenuates by $\exp(-1) \approx 0.37$ is called the skin depth. For a good dielectric, $\alpha = 1/\delta_s \ll \beta = 2\pi/\lambda$ and hence $\delta_s \gg j$. In a good dielectric the wave travels many wavelengths before attenuating by 0.37.

The physical mechanism for losses in a good dielectric are not generally related to conductivity. It is represented better by attributing a complex permittivity to the medium. Thus we let

$$\varepsilon = \varepsilon' - j\varepsilon'', \quad \varepsilon'' \ll \varepsilon'. \tag{11.11}$$

The Ampère–Maxwell equation then becomes

$$\nabla \times \underline{\vec{H}} = j\omega\varepsilon\underline{\vec{E}} = (j\omega\varepsilon' + \omega\varepsilon'')\underline{\vec{E}}$$

instead of $\nabla \times \underline{\vec{H}} = (j\omega\varepsilon + \sigma)\underline{\vec{E}}$. Comparing these two forms of the Ampère–Maxwell equation we can rewrite Eqs. (11.9) and (11.10) by replacing $\omega\varepsilon$ with $\omega\varepsilon'$ and by replacing σ with $\omega\varepsilon''$. The results are

$$\alpha = \frac{\omega\varepsilon''}{2}\sqrt{\frac{\mu}{\varepsilon'}} = \frac{1}{2}\left(\frac{\varepsilon''}{\varepsilon'}\right)\beta, \quad \beta = \omega\sqrt{\mu\varepsilon'}. \tag{11.12}$$

$$R = \sqrt{\frac{\mu}{\varepsilon'}}, \quad X = \frac{1}{2}\left(\frac{\varepsilon''}{\varepsilon'}\right)\sqrt{\frac{\mu}{\varepsilon'}} = \frac{1}{2}\left(\frac{\varepsilon''}{\varepsilon'}\right)R. \tag{11.13}$$

Tables of dielectric properties of materials often list the loss tangent of the dielectric, which is defined as

$$\tan \delta_l = \frac{\varepsilon''}{\varepsilon'}. \tag{11.14}$$

The physical phenomena in the material are complicated and the loss tangent generally varies slowly with frequency. For example, the loss tangent of Teflon is between 0.0003 and 0.0004 up to 40 GHz and then rises to 0.0014 to 80 GHz. Within a typical bandwidth, $\varepsilon''/\varepsilon'$ is constant and the attenuation constant is therefore proportional to frequency, giving attenuation distortion.

Example 11.2 For Teflon at 10 GHz the dielectric constant is $\varepsilon'/\varepsilon_0 = 2.08$ and the loss tangent is $\varepsilon''/\varepsilon' = 0.00037$. Calculate the phase constant, wavelength, attenuation constant, skin depth, and intrinsic impedance.

Solution

$$\beta = \frac{\omega\sqrt{\varepsilon'/\varepsilon_0}}{c} = \frac{2\pi \times 10 \times 10^9 \sqrt{2.08}}{3 \times 10^8} = 302 \text{ rad/m},$$

$$\lambda = 2\pi/\beta = 0.0208 \text{ m},$$

$$\alpha = \tfrac{1}{2}(302)(0.00037) = 0.0559 \text{ Np/m}, \quad \delta_s = 1/\alpha = 17.9 \text{ m},$$

$$\eta = R + jX, \quad R = \frac{120\pi}{\sqrt{2.08}} = 261\,\Omega, \quad X = \tfrac{1}{2}(0.00037)(261) = 0.048\,\Omega.$$

Note that $\delta_s \approx 860\lambda$ and $R \gg X$, so that $\eta \approx R$.

11.3.3 Good Conductors

When the material and frequency are such that $\omega\varepsilon/\sigma \ll 1$ the material is called a good conductor. In this case

$$\gamma = \sqrt{j\omega\mu\sigma}\left(1 + \frac{j\omega\varepsilon}{\sigma}\right)^{1/2} \approx \sqrt{j\omega\mu\sigma} = \sqrt{\frac{\omega\mu\sigma}{2}}(1 + j),$$

$$\alpha = \beta = \sqrt{\frac{\omega\mu\sigma}{2}}. \tag{11.15}$$

$$\eta = \sqrt{\frac{j\omega\mu}{\sigma}}\left(1 + \frac{j\omega\varepsilon}{\sigma}\right)^{-1/2} \approx \sqrt{\frac{j\omega\mu}{\sigma}} = \sqrt{\frac{\omega\mu}{2\sigma}}(1 + j),$$

$$R = X = \sqrt{\frac{\omega\mu}{2\sigma}}. \tag{11.16}$$

Example 11.3 For copper, $\mu = \mu_0$, $\varepsilon = \varepsilon_0$, $\sigma = 5.8 \times 10^7$ S/m; calculate the attenuation and phase constants, the skin depth, the wavelength, the phase velocity, and the intrinsic impedance, at 10 GHz.

Solution We have $\omega\varepsilon/\sigma = 9.61 \times 10^{-9} \ll 1$. Therefore

$$\alpha = \beta = \sqrt{\frac{(2\pi \times 10 \times 10^9)(4\pi \times 10^{-7})(5.8 \times 10^7)}{2}}$$

$$= 1.51 \times 10^6 \text{ Np/m or rad/m},$$

$$\delta_s = 1/\alpha = 0.661 \times 10^{-6} \text{ m}, \quad \lambda = 2\pi/\beta = 4.15 \times 10^{-6} \text{ m},$$

$$v_{\text{ph}} = f\lambda = 41.5 \times 10^3 \text{ m/s},$$

$$\eta = \sqrt{\frac{\omega\mu}{2\sigma}}(1 + j) = \frac{1.51 \times 10^6}{5.8 \times 10^7}(1 + j) = 0.0261(1 + j)\,\Omega.$$

Note that the skin depth and the wavelength differ by a factor of only 2π and that the phase velocity is very slow compared to the phase velocity in free space. Signals travel at the speed of light in the medium surrounding the conductor because the energy is in the wave in the medium, not in the conductor. The conductor serves as a directional guide to the wave.

11.3.4 General Formulas

The following formulas for γ and η are easily derived from the general formulas, Eqs. (11.5) and (11.6). They allow the computation of the propagation constant and the intrinsic impedance without having to take the square root of a complex number.

$$\alpha = \sqrt{\omega\mu(g - \omega\varepsilon)/2}, \qquad \beta = \sqrt{\omega\mu(g + \omega\varepsilon)/2}, \tag{11.17}$$

$$\eta = R + jX = (\beta + j\alpha)/g, \quad g = \sqrt{\sigma^2 + (\omega\varepsilon)^2}. \tag{11.18}$$

This form of the formulas clearly shows that when $\alpha \ll \beta$ then it is also true that $X \ll R$.

11.3.5 Distortion

In Section 9.7.2 we found that for the transmission channel to be distortionless we must have the attenuation constant independent of frequency and the phase constant proportional to frequency. Equation (11.7) shows that these conditions are satisfied for the perfect dielectric. Equation (11.12) shows that the low-loss dielectric exhibits attenuation distortion, but not phase distortion. Equation (11.15) shows that both the attenuation constant and the phase constant are proportional to the square root of the frequency for a good conductor, so that transmitting in a good conductor results in distortion due to dispersion and attenuation.

11.4 VECTOR FIELD POLARIZATION

The polarization of a vector field describes the variation with time of the direction of the field at a point in space.

When the direction of the field at a point lies along a fixed line in space independent of time, then the field is linearly polarized. For example, the electric field phasor

$$\underline{\vec{E}}(z) = |E_0|(\vec{a}_x + \vec{a}_y\kappa)e^{-(\alpha+j\beta)z}, \quad \kappa = \text{real constant},$$

corresponds to a physical field

$$\vec{E}(z, t) = |E_0|e^{-\alpha z}(\vec{a}_x + \vec{a}_y\kappa)\cos(\omega t - \beta z), \quad \kappa = \text{real}. \tag{11.19}$$

At $z = 0$ the electric field intensity lies in the xy plane along a line at an angle $\tan^{-1}\kappa$ with the x axis (Fig. 11.4). As time varies, the amplitude of the field varies sinusoidally along the line up to a maximum value of $|E_0|\sqrt{1+\kappa^2}$. The figure shows time as a parameter along the vector. This field is linearly polarized at $z = 0$. For a different value of z, say, $z = z_0 > 0$, the amplitude is reduced by $\exp(-\alpha z_0)$ and the phase is retarded by βz_0, but this simultaneous attenuation and phase shift of both components of the field leaves the field linearly polarized along the same line.

The wave is elliptically polarized if the direction and amplitude of the field at a point vary with time along an ellipse. For example, the electric field phasor,

$$\underline{\vec{E}}(z) = |E_0|(\vec{a}_x + \vec{a}_y\kappa)e^{-(\alpha+j\beta)z}, \quad \kappa = |\kappa|e^{j\psi}, \tag{11.20}$$

corresponds to a physical field

$$\vec{E}(z, t) = |E_0|e^{-\alpha z}[\vec{a}_x\cos(\omega t - \beta z) + \vec{a}_y|\kappa|\cos(\omega t - \beta z + \psi)]. \tag{11.21}$$

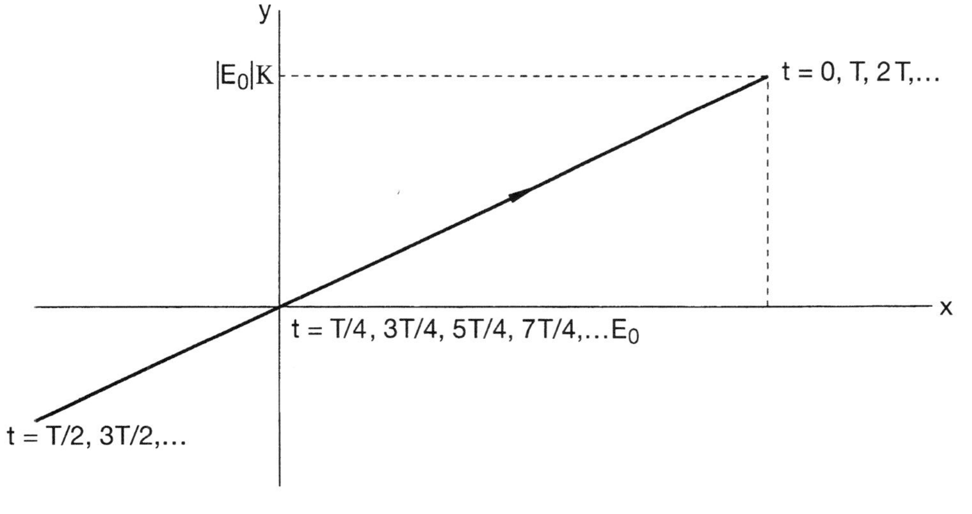

FIGURE 11.4 Electric field intensity plot in the xy plane versus time for a linearly polarized wave.

Unless $\psi = 0$ or π the field intensity will rotate as time changes.

Consider the case $\psi = -\pi/2$. Then at $z = 0$ the electric field is

$$\underline{\vec{E}}(0) = |E_0|(\vec{a}_x - j\vec{a}_y|\kappa|), \quad (11.22)$$

$$\vec{E}(0, t) = |E_0|(\vec{a}_x \cos \omega t + \vec{a}_y|\kappa| \sin \omega t). \quad (11.23)$$

When $t = \ldots, -T, 0, T, 2T, \ldots$, where $T = 2\pi/\omega = 1/f$ is the period of oscillation, then the field is along the positive x axis with amplitude $|E_0|$ (Fig. 11.5). At $t = \ldots, T/4 - T, T/4, T/4 + T, \ldots$, the field has rotated to be along the positive y axis with amplitude $|E_0||\kappa|$. At $t = \ldots, T/2 - T, T/2, T/2 + T, \ldots$, the field is along the negative x axis; and so on. The tip of the field vector traces an ellipse, the axes of which are along the x and y axes. This wave is elliptically polarized.

When the amplitude of an elliptically polarized wave is constant, then the wave is circularly polarized. In the preceding case, the wave is circularly polarized at $z = 0$ when $|\kappa| = 1$, giving

$$\underline{\vec{E}}(0) = |E_0|(\vec{a}_x - j\vec{a}_y), \quad (11.24)$$

$$\vec{E}(0, t) = |E_0|(\vec{a}_x \cos \omega t + \vec{a}_y \sin \omega t). \quad (11.25)$$

Still for the case $\psi = -\pi/2$, putting the fingers of your right hand in the direction in which the electric field vector is rotating (counterclockwise), points your thumb in the direction of propagation (positive z). This field is therefore right-handed elliptically or circularly polarized.

For a different value of z, say, $z = z_0 > 0$, the amplitude is reduced by $\exp(-\alpha z_0)$ and the phase is retarded by βz_0, but this simultaneous attenuation and phase shift of both components of the field leaves the field right-handed elliptically or circularly polarized along the same axes.

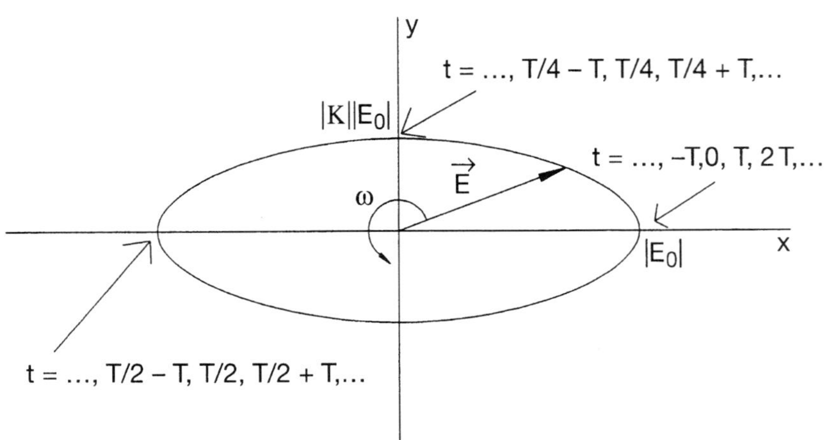

FIGURE 11.5 Electric field intensity plot in the xy plane versus time for a right-handed elliptically polarized wave.

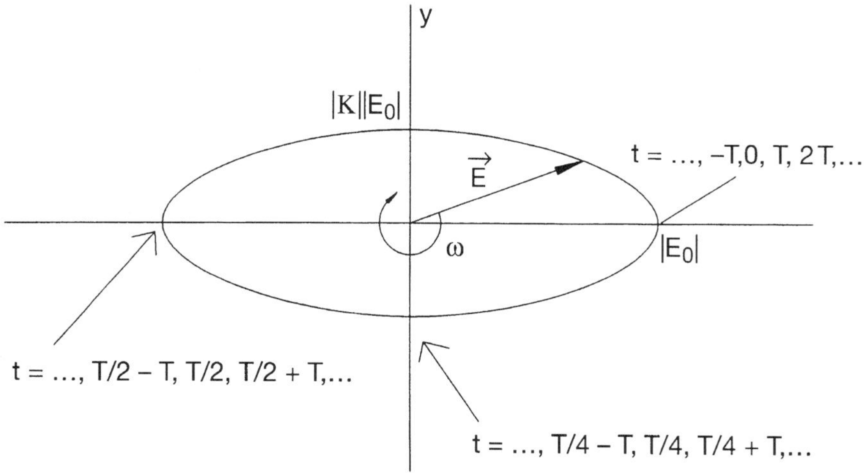

FIGURE 11.6 Electric field intensity plot in the xy plane versus time for a left-handed elliptically polarized wave.

Return now to Eqs. (11.20) and (11.21) and let $\psi = +\pi/2$, resulting at $z = 0$ in

$$\vec{E}(0, t) = |E_0|(\vec{a}_x \cos \omega t - \vec{a}_y |\kappa| \sin \omega t). \tag{11.26}$$

Figure 11.6 shows this field with time as a parameter and it appears just as in the previous example except that the field rotates in the opposite direction. This field is left-handed elliptically (or left-handed circularly when $|\kappa| = 1$) polarized.

When $\psi \neq -\pi/2, 0, \pi/2, \pi$ then the wave is still elliptically polarized, but the axes of the ellipse are rotated from the xy axes.

The magnetic field intensity is also linearly, elliptically, or circularly polarized. In a lossless medium ($\sigma = 0$, $\alpha = 0$) the magnetic field is always perpendicular to the electric field, but this is not true in a lossy medium.

Polarization is important in several different contexts. For one, it allows reuse of the frequency spectrum. Figure 11.7 shows two transmitters and two receivers. If one transmitter–receiver pair will only transmit and receive vertically polarized waves and the other will only transmit and receive horizontally polarized waves, then the two pairs can operate in the same frequency bands without interference. The same is true if the pairing is for right-handed and left-handed polarization; this frequency reuse is common in communications satellite transmissions. Another use of polarization is that some materials, such as ferrites, have different propagation constants for right-handed and left-handed polarized waves. This is used to design microwave devices such as isolators, circulators, and phase shifters [1, 2].

11.5 POWER AND ENERGY DENSITY: THE POYNTING VECTOR

In Section 10.2.2 we discussed the power carried by the voltage and current waves on a transmission line. We now examine the more general question of the power carried

FIGURE 11.7 Two pairs of transmitters and receivers for transmission and reception of plane waves.

by the electromagnetic field. We do this first for the physical time-dependent fields and second for the phasor fields.

11.5.1 Instantaneous Power and Energy

Our starting point is Maxwell's instantaneous equations

$$\nabla \times \vec{E} = -\mu \frac{\partial \vec{H}}{\partial t}, \quad \nabla \times \vec{H} = \vec{J} + \varepsilon \frac{\partial \vec{E}}{\partial t}.$$

In the second equation the current density is the sum of the current from sources and the conduction current, $\vec{J} = \vec{J}_s + \sigma \vec{E}$. We apply the vector identity $\nabla \cdot (\vec{A} \times \vec{B}) \equiv \vec{B} \cdot (\nabla \times \vec{A}) - \vec{A} \cdot (\nabla \times \vec{B})$ to the electric and magnetic field intensities and then substitute from these equations:

$$\begin{aligned}\nabla \cdot (\vec{E} \times \vec{H}) &= \vec{H} \cdot (\nabla \times \vec{E}) - \vec{E} \cdot (\nabla \times \vec{H}) \\ &= \vec{H} \cdot \left(-\mu \frac{\partial \vec{H}}{\partial t} \right) - \vec{E} \cdot \left(\vec{J}_s + \sigma \vec{E} + \varepsilon \frac{\partial \vec{E}}{\partial t} \right) \\ &= -\mu \vec{H} \cdot \frac{\partial \vec{H}}{\partial t} - \vec{J}_s \cdot \vec{E} - \sigma |\vec{E}|^2 - \varepsilon \vec{E} \cdot \frac{\partial \vec{E}}{\partial t}. \end{aligned} \tag{11.27}$$

Using the chain rule for differentiation we have for any vector field

$$\frac{\partial}{\partial t} \left(\frac{1}{2} |\vec{A}|^2 \right) = \frac{1}{2} \frac{\partial}{\partial t} (\vec{A} \cdot \vec{A}) = \frac{1}{2} \left(\frac{\partial \vec{A}}{\partial t} \cdot \vec{A} + \vec{A} \cdot \frac{\partial \vec{A}}{\partial t} \right) = \vec{A} \cdot \frac{\partial \vec{A}}{\partial t},$$

which we use for the first and last terms on the right side of Eq. (11.27). Finally, we integrate both sides throughout a volume V, which is surrounded by a closed surface S (Fig. 11.8), to obtain

$$\int_V \nabla \cdot (\vec{E} \times \vec{H})\, dv = -\int_V \left[\frac{\partial}{\partial t} (\tfrac{1}{2} \mu |\vec{H}|^2) + \frac{\partial}{\partial t} (\tfrac{1}{2} \varepsilon |\vec{E}|^2) + \sigma |\vec{E}|^2 + \vec{J}_s \cdot \vec{E} \right] dv.$$

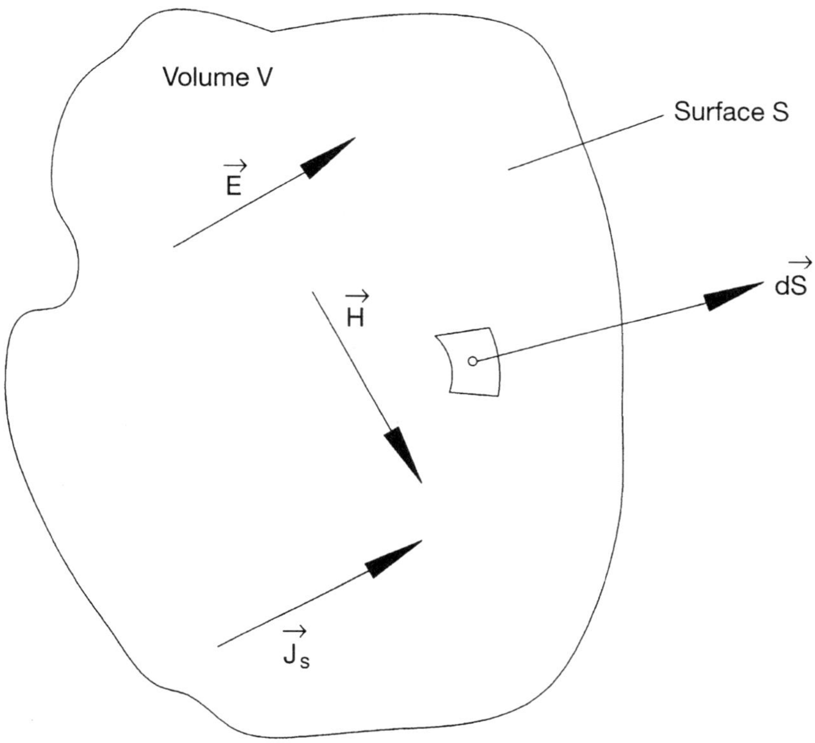

FIGURE 11.8 A volume (V) enclosed by a closed surface (S) used in the derivation of mathematical expression for the power and energy in a wave.

We apply the divergence theorem to the term on the left and rearrange the terms to get

$$\int_V (-\vec{J}_s \cdot \vec{E})\, dv = \int_V \left[\frac{\partial}{\partial t}(\tfrac{1}{2}\mu|\vec{H}|^2) + \frac{\partial}{\partial t}(\tfrac{1}{2}\varepsilon|\vec{E}|^2) + \sigma|\vec{E}|^2 \right] dv + \oint_S \vec{E} \times \vec{H} \cdot d\vec{S}. \tag{11.28}$$

This finishes the vector manipulation and yields an important result, which is known as the instantaneous Poynting theorem. In order to understand the physical meaning of the equation we will examine each term.

Let us explain each term, starting on the left.

$-\vec{J}_s \cdot \vec{E} = p_s(\vec{R}, t)$	$p_s(\vec{R}, t)$ is the instantaneous power
$\int_V p_s(\vec{R}, t)\, dv = P_s(t)$	density (W/m^3) input from sources at each point $\vec{R}$ in space. $P_s(t)$ is the total instantaneous power (W) input from sources within the volume.
$\frac{1}{2}\mu|\vec{H}|^2 = w_m(\vec{R}, t)$	$w_m(\vec{R}, t)$ is the instantaneous magnetic energy density (J/m^3) stored in the magnetic field at each point in space.

$$\frac{\partial}{\partial t}(\tfrac{1}{2}\mu|\vec{H}|^2) = \frac{\partial w_m}{\partial t} = p_m(\vec{R}, t)$$

$p_m(\vec{R}, t)$ is the time rate of increase of magnetic energy density or the instantaneous power input per unit volume (W/m^3) into magnetic energy at each point in space.

$$\tfrac{1}{2}\varepsilon|\vec{E}|^2 = w_e(\vec{R}, t)$$

$w_e(\vec{R}, t)$ is the instantaneous electric energy density (J/m^3) stored in the electric field at each point in space.

$$\frac{\partial}{\partial t}(\tfrac{1}{2}\varepsilon|\vec{E}|^2) = \frac{\partial w_e}{\partial t} = p_e(\vec{R}, t)$$

$p_e(\vec{R}, t)$ is the time rate of increase of electric energy density or the instantaneous power input per unit volume (W/m^3) into electric energy at each point in space.

$$\sigma|\vec{E}|^2 = p_l(\vec{R}, t)$$

$p_l(\vec{R}, t)$ is the instantaneous power density (W/m^3) dissipated due to conduction current in the imperfectly conducting medium at each point in space.

$$\int_V [p_m + p_e + p_l]\, dv = P_V(t)$$

$$P_m + P_e + P_l = P_V$$

$P_V(t)$ is the total instantaneous power into magnetic and electric energy storage, $P_m(t)$ and $P_e(t)$, plus the total instantaneous power dissipated, $P_l(t)$, within the volume.

Equation (11.28) is evidently a power balance equation in which total power input from sources equals total power output to energy storage and dissipation within the volume. On the power input (left) side we have accounted for all sources in the volume. On the power output (right) side we have accounted for the power stored in magnetic and electric fields and the power dissipated as heat. Still not accounted for is the power that leaves the volume through its surface. We therefore interpret the last term on the right as the instantaneous power flowing outward through the surface S surrounding the volume. We interpret the integrand

$$\vec{S}(\vec{R}, t) = \vec{E}(\vec{R}, t) \times \vec{H}(\vec{R}, t) \tag{11.29}$$

as the instantaneous power density or the instantaneous power per unit area carried by the electromagnetic field.†

The vector $\vec{S} = \vec{E} \times \vec{H}$ is called the instantaneous Poynting vector. Its direction gives the direction of power flow, its magnitude gives the amount of power flow, and its units are power per unit area (W/m^2).

We complete the listing of terms in Eq. (11.28).

$$\vec{S}(\vec{R}, t) = \vec{E}(\vec{R}, t) \times \vec{H}(\vec{R}, t)$$

$\vec{S}(\vec{R}, t)$ is the instantaneous Poynting vector, which gives the instantaneous direction and magnitude of the power flow (W/m^2) at each point in space.

† Care must be taken not to confuse the instantaneous Poynting vector $\vec{S}$ with the surface S or the differential surface $d\vec{S}$.

$\int_S \vec{S}(\vec{R}, t)\cdot d\vec{S} = P_f(t)$ — $P_f(t)$ is the instantaneous power (W) flowing outward through the surface S.

Equation (11.28) can be rewritten to show the power balance as

$$P_s(t) = P_m(t) + P_e(t) + P_l(t) + P_f(t). \tag{11.30}$$

Example 11.4 A plane wave in free space is given by

$$\vec{E}(z, t) = \vec{a}_x E_0 \cos(\omega t - \beta z), \quad \beta = \omega\sqrt{\mu_0 \varepsilon_0}$$

$$\vec{H}(z, t) = \vec{a}_y \frac{E_0}{\eta_0} \cos(\omega t - \beta z), \quad \eta_0 = \sqrt{\frac{\mu}{\varepsilon_0}}.$$

(a) Find formulas for the instantaneous Poynting vector and the energy density at any point in space. (b) Find formulas for the power transferred into energy storage within the volume and the power flow outward through the surface of a rectangular parallelepiped $0 < x < a$, $0 < y < b$, $0 < z < d$. Show that these add to zero.

Solution There are no sources and no power is dissipated.

(a) The desired terms are

$$\vec{S}(z, t) = \vec{E} \times \vec{H} = \vec{a}_z \frac{E_0^2}{\eta_0} \cos^2(\omega t - \beta z).$$

$$w_m(z, t) = \tfrac{1}{2}\mu|\vec{H}|^2 = \tfrac{1}{2}\mu_0 \frac{E_0^2}{\eta_0^2} \cos^2(\omega t - \beta z) = \tfrac{1}{2}\varepsilon_0 E_0^2 \cos^2(\omega t - \beta z).$$

$$w_e(z, t) = \tfrac{1}{2}\varepsilon|\vec{E}|^2 = \tfrac{1}{2}\varepsilon_0 E_0^2 \cos^2(\omega t - \beta z) = w_m(z, t).$$

(b) Since $w_m = w_e$ the instantaneous magnetic and electric power density into energy storage at any point is

$$\begin{aligned} p_m(z, t) = p_e(z, t) &= \frac{\partial}{\partial t}[\tfrac{1}{2}\varepsilon_0 E_0^2 \cos^2(\omega t - \beta z)] \\ &= -\omega\varepsilon_0 E_0^2 \sin(\omega t - \beta z)\cos(\omega t - \beta z) \\ &= -\tfrac{1}{2}\omega\varepsilon_0 E_0^2 \sin 2(\omega t - \beta z) \end{aligned}$$

and the power into instantaneous magnetic and electric energy in the volume is

$$\begin{aligned} P_m(t) = P_e(t) &= \int_{x=0}^{a}\int_{y=0}^{b}\int_{z=0}^{d} [-\tfrac{1}{2}\omega\varepsilon_0 E_0^2 \sin 2(\omega t - \beta z)]\, dz\, dy\, dx \\ &= -\frac{abE_0^2}{4\eta_0}[\cos 2(\omega t - \beta d) - \cos 2\omega t]. \end{aligned} \tag{11.31}$$

The instantaneous power flow outward through the surrounding surface is found by integrating $\vec{S} \cdot d\vec{S}$ over the surface of the rectangular parallelepiped. Because $\vec{S}$ is in the z direction, $\vec{S} \cdot d\vec{S} = 0$ on the four surfaces, which are not perpendicular to the z axis. Over the remaining two surfaces we find

$$\begin{aligned} P_f(t) &= \int_{x=0}^{a} \int_{y=0}^{b} \vec{S}(0,t) \cdot (-\vec{a}_z\, dy\, dx) + \int_{x=0}^{a} \int_{y=0}^{b} \vec{S}(d,t) \cdot (\vec{a}_z\, dy\, dx) \\ &= \frac{abE_0^2}{\eta_0} [\cos^2(\omega t - \beta d) - \cos^2 \omega t] \\ &= \frac{abE_0^2}{2\eta_0} [\cos 2(\omega t - \beta d) - \cos 2\omega t]. \end{aligned} \tag{11.32}$$

The sum $P_m(t) + P_e(t) + P_f(t) = 0$, as required by Eq. (11.30).

11.5.2 Complex Power and Energy

For time harmonic fields we would like a form of the power balance equations, Eqs. (11.28) and (11.30), that does not require us to shift from phasor quantities to physical or instantaneous quantities. Also, we are not generally as interested in instantaneous power or energy as we are in time average power or energy. To deal with these needs, we develop a complex form of the same results.

We start with Maxwell's equations in phasor form

$$\nabla \times \underline{\vec{E}} = -j\omega\mu\underline{\vec{H}}, \quad \nabla \times \underline{\vec{H}} = \underline{\vec{J}} + j\omega\varepsilon\underline{\vec{E}},$$

where $\underline{\vec{J}} = \underline{\vec{J}}_s + \sigma\underline{\vec{E}}$. Proceeding as in the last section,

$$\begin{aligned} \nabla \cdot (\underline{\vec{E}} \times \underline{\vec{H}}^*) &= \underline{\vec{H}}^* \cdot (\nabla \times \underline{\vec{E}}) - \underline{\vec{E}} \cdot (\nabla \times \underline{\vec{H}}^*) \\ &= \underline{\vec{H}}^* \cdot (-j\omega\mu\underline{\vec{H}}) - \underline{\vec{E}} \cdot (\underline{\vec{J}}_s^* + \sigma\underline{\vec{E}}^* - j\omega\varepsilon\underline{\vec{E}}^*) \\ &= -j\omega\mu|\underline{\vec{H}}|^2 - \underline{\vec{J}}_s^* \cdot \underline{\vec{E}} - \sigma|\underline{\vec{E}}|^2 + j\omega\varepsilon|\underline{\vec{E}}|^2. \end{aligned}$$

In using these results remember that $|\underline{\vec{A}}|^2 = \underline{\vec{A}} \cdot \underline{\vec{A}}^*$, which means the magnitude with respect to both the real and imaginary parts and with respect to the coordinate directions. For example, if $\underline{\vec{A}} = \vec{a}_x(a + jb) + \vec{a}_y(c + jd)$, then $|\underline{\vec{A}}|^2 = a^2 + b^2 + c^2 + d^2$.

As before, we integrate both sides throughout a volume V surrounded by a surface S, apply the divergence theorem to the left side of the equation, and rearrange terms to obtain

$$\int_V \tfrac{1}{2}(-\underline{\vec{J}}_s^* \cdot \underline{\vec{E}})\, dv = 2j\omega \int_V [\tfrac{1}{4}\mu|\underline{\vec{H}}|^2 - \tfrac{1}{4}\varepsilon|\underline{\vec{E}}|^2]\, dv + \int_V \tfrac{1}{2}\sigma|\underline{\vec{E}}|^2\, dv + \oint_S \tfrac{1}{2}(\underline{\vec{E}} \times \underline{\vec{H}}^*) \cdot d\vec{S}. \tag{11.33}$$

This is the complex Poynting theorem, which is a statement of conservation of complex power.

The complex Poynting vector is defined as

$$\vec{S}_c(\vec{R}) = \underline{\vec{E}}(\vec{R}) \times \underline{\vec{H}}^*(\vec{R}). \tag{11.34}$$

Equating the real and imaginary parts of both sides of Eq. (11.33), we obtain

$$\int_V \tfrac{1}{2}\text{Re}(-\underline{\vec{J}}_s^* \cdot \underline{\vec{E}})\, dv = \int_V \tfrac{1}{2}\sigma|\underline{\vec{E}}|^2\, dv + \oint_V \tfrac{1}{2}\text{Re}(\underline{\vec{E}} \times \underline{\vec{H}}^*) \cdot d\vec{S}, \tag{11.35}$$

$$\int_V \tfrac{1}{2}\text{Im}(-\underline{\vec{J}}_s^* \cdot \underline{\vec{E}})\, dv = 2\omega \int_V [\tfrac{1}{4}\mu|\underline{\vec{H}}|^2 - \tfrac{1}{4}\varepsilon|\underline{\vec{E}}|^2]\, dv + \oint_S \tfrac{1}{2}\text{Im}(\underline{\vec{E}} \times \underline{\vec{H}}^*) \cdot d\vec{S}. \tag{11.36}$$

Let us interpret these terms.

$\tfrac{1}{2}(-\underline{\vec{J}}_s^* \cdot \underline{\vec{E}}) = p_{sc}$	p_{sc}(W/m^3) is the complex power density input from sources.
$\int_V \tfrac{1}{2}(-\underline{\vec{J}}_s^* \cdot \underline{\vec{E}})\, dv = P_{sc}$	P_{sc}(W) is the complex power input from sources within the volume.
$\int_V \tfrac{1}{2}\text{Re}(-\underline{\vec{J}}_s^* \cdot \underline{\vec{E}})\, dv = P_{s,\text{Avg}}$	$P_{s,\text{Avg}}$ is the real or time average power input from sources in the volume.
$\int_V \tfrac{1}{2}\text{Im}(-\underline{\vec{J}}_s^* \cdot \underline{\vec{E}})\, dv = P_{s,\text{reac}}$	$P_{s,\text{reac}}$ is the reactive power input from sources in the volume.
$\tfrac{1}{2}\sigma|\underline{\vec{E}}|^2 = p_{l,\text{Avg}}(\vec{R})$	$p_{l,\text{Avg}}$ (W/m^3) is the real or time average power density dissipated at any point in the volume.
$\int_V \tfrac{1}{2}\sigma|\underline{\vec{E}}|^2\, dv = P_{l,\text{Avg}}$	$P_{l,\text{Avg}}$ (W) is the total time average power dissipated in the volume.
$\tfrac{1}{4}\mu|\underline{\vec{H}}|^2 = w_{m,\text{Avg}}(\vec{R})$	$w_{m,\text{Avg}}$ (J/m^3) is the time average magnetic energy density stored at any point in the volume.
$\int_V \tfrac{1}{4}\mu|\underline{\vec{H}}|^2\, dv = W_{m,\text{Avg}}$	$W_{m,\text{Avg}}$ (J) is the total time average magnetic energy stored in the volume.
$\tfrac{1}{4}\varepsilon|\underline{\vec{E}}|^2 = w_{e,\text{Avg}}(\vec{R})$	$w_{e,\text{Avg}}$ (J/m^3) is the time average electric energy density stored at any point in the volume.
$\int_V \tfrac{1}{4}\varepsilon|\underline{\vec{E}}|^2\, dv = W_{e,\text{Avg}}$	$W_{e,\text{Avg}}$ (J) is the total time average electric energy stored in the volume.
$\tfrac{1}{2}\text{Re}(\vec{S}_c) = \tfrac{1}{2}\text{Re}(\underline{\vec{E}} \times \underline{\vec{H}}^*) = \vec{p}_{f,\text{Avg}}$	$\vec{p}_{f,\text{Avg}}$ (W/m^2) is the time average power density flowing at any point in space.

$\oint_S \frac{1}{2}\mathrm{Re}(\underline{\vec{E}} \times \underline{\vec{H}}^*) \cdot d\vec{S} = P_{f,\,\mathrm{Avg}}$ — $P_{f,\,\mathrm{Avg}}$ (W) is the total real or time average power flow outward through the surface S.

$\oint_S \frac{1}{2}\mathrm{Im}(\underline{\vec{E}} \times \underline{\vec{H}}^*) \cdot d\vec{S} = P_{f,\,\mathrm{reac}}$ — $P_{f,\,\mathrm{reac}}$ (W) is the reactive power flow outward through the surface S.

Equations (11.35) and (11.36) are real and reactive power balance equations:

$$P_{s,\,\mathrm{Avg}} = P_{l,\,\mathrm{Avg}} + P_{f,\,\mathrm{Avg}}, \tag{11.37}$$

$$P_{s,\,\mathrm{reac}} = 2\omega(W_{m,\,\mathrm{Avg}} - W_{m,\,\mathrm{Avg}}) + P_{f,\,\mathrm{reac}}. \tag{11.38}$$

The most important term, which we will use frequently, is

$$\vec{p}_{f,\,\mathrm{Avg}} = \tfrac{1}{2}\mathrm{Re}(\underline{\vec{E}} \times \underline{\vec{H}}^*). \tag{11.39}$$

This gives the time average power density carried by the wave.

Example 11.5 A plane wave in a lossy medium is given by

$$\underline{E}_x = E_0 e^{-\alpha z} e^{-j\beta z}, \quad \underline{H}_y = \frac{E_0}{\eta} e^{-\alpha z} e^{-j\beta z}.$$

Find the terms in the complex Poynting theorem for the volume of Example 11.4.

Solution

$$\tfrac{1}{4}\mu|\underline{\vec{H}}|^2 = \frac{\mu|E_0|^2}{4\eta\eta^*} e^{-2\alpha z}, \qquad \int_V \tfrac{1}{4}\mu|\underline{\vec{H}}|^2\, dv = \frac{ab\mu|E_0|^2}{8\eta\eta^*\alpha}(1 - e^{-2\alpha d}).$$

$$\tfrac{1}{4}\varepsilon|\underline{\vec{E}}|^2 = \tfrac{1}{4}\varepsilon|E_0|^2 e^{-2\alpha z}, \qquad \int_V \tfrac{1}{4}\varepsilon|\underline{\vec{E}}|^2\, dv = \frac{ab\varepsilon|E_0|^2}{8\alpha}(1 - e^{-2\alpha d}).$$

$$\tfrac{1}{2}\sigma|\underline{\vec{E}}|^2 = \tfrac{1}{2}\alpha|E_0|^2 e^{-2\alpha z}, \qquad \int_V \tfrac{1}{2}\sigma|\underline{\vec{E}}|^2\, dv = \frac{ab\sigma|E_0|^2}{4\alpha}(1 - e^{-2\alpha d}).$$

$$\vec{S}_c = \underline{\vec{E}} \times \underline{\vec{H}}^* = \vec{a}_z \frac{|E_0|^2}{\eta^*} e^{-2\alpha z}, \qquad \oint_S \tfrac{1}{2}\vec{S}_c \cdot d\vec{S} = -\frac{ab|E_0|^2}{2\eta^*}(1 - e^{-2\alpha d}).$$

Substituting in Eq. (11.33) and dividing both sides by $ab|E_0|^2[1 - \exp(-2\alpha d)]$ gives

$$0 = 2j\omega\left(\frac{\mu}{8\eta\eta^*\alpha} - \frac{\varepsilon}{8\alpha}\right) + \frac{\alpha}{4\alpha} - \frac{1}{2\eta^*}.$$

Multiplying by $4\eta\eta^*\alpha$ gives

$$0 = j\omega\mu - j\omega\varepsilon\eta\eta^* + \sigma\eta\eta^* - 2\eta\alpha. \tag{11.40}$$

The first two terms on the right are pure imaginary, the third term is pure real, and the final term is complex. If the medium is lossless, $\sigma = 0$, then $\alpha = 0$ and $\eta\eta^* = \mu/\varepsilon$, and the equality clearly holds.

11.6 REFLECTION AND TRANSMISSION AT PLANE BOUNDARIES

So far we have been discussing a plane wave propagating in unbounded space. We now look at the problem of a plane wave that is incident on a plane boundary dividing the space into two regions characterized by different conductivity, permittivity, and permeability. We all know from the common experience of looking in a mirror or through a piece of glass that when an incident wave comes to the boundary between the two regions a reflected wave and a transmitted wave are produced. From our study of electromagnetic fields we know that at the boundary both the tangential electric and magnetic field intensities must be continuous. These boundary conditions require that there be reflected and transmitted waves and allow us to determine their characteristics.

In the following subsections we have several objectives. We will first obtain expressions for the propagation vectors in each medium and then find the relation that must hold between their directions. We will then apply these results to several different physical situations, such as the two media being perfect dielectrics and one being a perfect dielectric and the other a perfect conductor. In order to find complete field solutions in the two regions we will show that we can break up our incoming or incident wave into a transverse electric and a transverse magnetic wave and then deal with these separately. We will develop reflection and transmission formulas and transmission line models for the wave behavior. In subsequent sections we will apply these results to the parallel plate waveguide and the dielectric slab waveguide.

11.6.1 Relations Between Propagation Vectors and Snell's Law

We must start by assuming a geometry for our problem. We let the yz plane be the plane boundary between the two materials, so that for $x < 0$ we have a material characterized by μ_1, ε_1, and σ_1 and for $x > 0$ we have a material characterized by μ_2, ε_2, and σ_2 (Fig. 11.9). The unit vector $\vec{a}_x$ is therefore normal to the boundary.

We assume that somewhere outside our picture at $x \ll 0$ there is a source that produces a plane wave at angular frequency ω and that this wave travels toward the boundary. This wave is called the incident wave and its propagation vector $\vec{\gamma}_i$ is at an angle θ_i, called the angle of incidence, with the negative x axis. The normal to the boundary, $\vec{a}_x$, and the incident propagation vector, $\vec{\gamma}_i$, determine a plane, called the plane of incidence. We take the y axis to be normal to that plane, $\vec{a}_y = (\vec{\gamma}_i \times \vec{a}_x)/|\vec{\gamma}_i \times \vec{a}_x|$. $\vec{a}_y$ must lie in the boundary plane since it is normal to $\vec{a}_x$. The z axis is chosen to form a right-hand system, $\vec{a}_z = \vec{a}_x \times \vec{a}_y$, and it also lies in the boundary plane. The propagation vector is two dimensional since it has no y component.

At the boundary a reflected wave is produced. The reflected wave propagates away from the boundary back into the same region as the incident wave with a propagation vector $\vec{\gamma}_r$, making an angle θ_r, called the angle of reflection, with the negative x axis. Also at the boundary, a transmitted wave is produced (in optics the transmitted wave

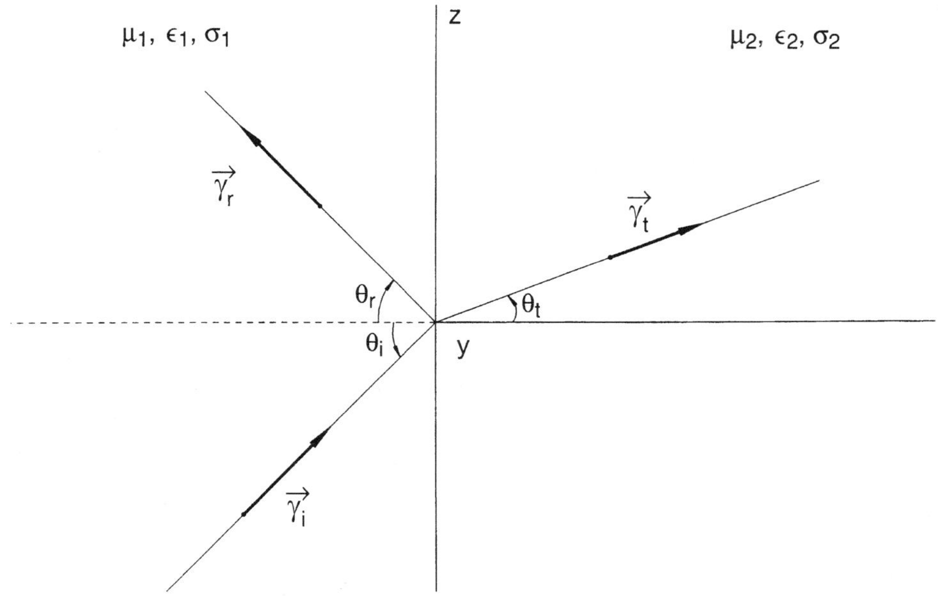

FIGURE 11.9 Illustration of the incident, reflected, and transmitted waves at boundary of two arbitrary media characterized by μ, ε, σ.

is generally called the refracted wave). The transmitted wave propagates away from the boundary into the second medium with a propagation vector $\vec{\gamma}_t$, making an angle θ_t, called the angle of transmission, with the positive x axis.[†]

The three propagation vectors are written in rectangular component form as

$$\vec{\gamma}_i = \vec{a}_x\gamma_{ix} + \vec{a}_z\gamma_{iz} = \gamma_1(\vec{a}_x \cos\theta_i + \vec{a}_z \sin\theta_i), \tag{11.41}$$

$$\vec{\gamma}_r = \vec{a}_x\gamma_{rx} + \vec{a}_z\gamma_{rz} = \gamma_1(-\vec{a}_x \cos\theta_r + \vec{a}_z \sin\theta_r), \tag{11.42}$$

$$\gamma_1 = \sqrt{j\omega\mu_1(\sigma_1 + j\omega\varepsilon_1)} = \sqrt{\gamma_{ix}^2 + \gamma_{iz}^2} = \sqrt{\gamma_{rx}^2 + \gamma_{rz}^2}, \tag{11.43}$$

$$\vec{\gamma}_t = \vec{a}_x\gamma_{tx} + \vec{a}_z\gamma_{tz} = \gamma_2(\vec{a}_x \cos\theta_t + \vec{a}_z \sin\theta_t), \tag{11.44}$$

$$\gamma_2 = \sqrt{j\omega\mu_2(\sigma_2 + j\omega\varepsilon_2)} = \sqrt{\gamma_{tx}^2 + \gamma_{tz}^2}. \tag{11.45}$$

All electric and magnetic field components of the incident wave are proportional to $\exp(-\vec{\gamma}_i \cdot \vec{R})$. Similarly, reflected wave and transmitted wave field components are proportional to $\exp(-\vec{\gamma}_r \cdot \vec{R})$ and $\exp(-\vec{\gamma}_t \cdot \vec{R})$, respectively. Along the boundary at $x = 0$ the tangential components of the field must be continuous for all values of the z coordinate and for all values of time. By assuming the angular frequency of the reflected and the transmitted waves to be the same as the angular frequency of the incident wave, we have already assured that if there is continuity at one instant

[†] We will see as we study these angles that they may be complex rather than real. It is still possible to use the term "angle" in a formal sense and to proceed without chance of error.

of time, then there will be continuity for all time. For continuity to hold for all values of the z coordinate each field component must have the same functional relation to z at $x = 0$. Thus

$$\exp(-\vec{\gamma}_i \cdot \vec{R})|_{x=0} = \exp(-\vec{\gamma}_r \cdot \vec{R})|_{x=0} = \exp(-\vec{\gamma}_t \cdot \vec{R})|_{x=0}$$

or

$$\gamma_{iz} = \gamma_{rz} = \gamma_{tz}$$

$$\gamma_1 \sin\theta_i = \gamma_1 \sin\theta_r = \gamma_2 \sin\theta_t.$$

The first pair of equalities gives

$$\theta_r = \theta_i, \tag{11.46}$$

meaning that the angle of reflection is equal to the angle of incidence. Equality of the first and last term gives

$$\sin\theta_t = \frac{\gamma_1}{\gamma_2}\sin\theta_i. \tag{11.47}$$

This is a general form of Snell's law. This equation requires discussion of its meaning for different media.

11.6.1.1 Snell's Law for Perfect Dielectrics Let us first look at the case where both media are perfect dielectrics, $\sigma_1 = \sigma_2 = 0$. Then $\gamma_1 = j\beta_1 = j\omega\sqrt{\mu_1\varepsilon_1}$, $\gamma_2 = j\beta_2 = j\omega\sqrt{\mu_2\varepsilon_2}$, and Eq. (11.47) becomes

$$\sqrt{\mu_1\varepsilon_1}\sin\theta_i = \sqrt{\mu_2\varepsilon_2}\sin\theta_t. \tag{11.48}$$

There are two cases that we must consider. For Case 1 $\mu_1\varepsilon_1 < \mu_2\varepsilon_2$, and for Case 2 $\mu_1\varepsilon_1 > \mu_2\varepsilon_2$. For simplicity of writing, let

$$r = \sqrt{\frac{\mu_1\varepsilon_1}{\mu_2\varepsilon_2}}, \tag{11.49}$$

so that Case 1 has $r < 1$ and Case 2 has $r > 1$.

Case 1: $\mu_1\varepsilon_1 < \mu_2\varepsilon_2$ or $r < 1$ In this case the wave is incident from an optically less dense onto an optically more dense medium, say, from air to glass or water. For all values of θ_i, the value of $\sin\theta_t \le 1$ and the angle of transmission is calculated as

$$\theta_t = \sin^{-1}(r\sin\theta_i). \tag{11.50}$$

This is illustrated graphically in Fig. 11.10, which shows the two sides of Eq. (11.50) as the equal z components of their respective phase vectors. As the angle of incidence varies over its full range $0 \leq \theta_i \leq \pi/2$, the angle of transmission varies over the range $0 \leq \theta_t \leq \sin^{-1} r < \pi/2$. The figure also shows the x component of the phase vectors. β_{tx} has a real value for the full range of θ_i. The transmitted wave is a plane uniform unattenuated wave propagating in the region $x > 0$. Figure 11.11 shows the planes of constant phase for the incident, reflected, and transmitted wave.

Case 2: $\mu_1\varepsilon_1 > \mu_2\varepsilon_2$ *or* $r > 1$ In this case the wave is incident from an optically more dense onto an optically less dense medium, say, from glass or water to air. Figure 11.12 now illustrates the two phase vectors, this time with $|\vec{\beta}_i| > |\vec{\beta}_t|$. We now see a change in behavior for the transmitted wave at the critical angle of incidence θ_c. As the angle of incidence varies over the range $0 \leq \theta_i \leq \theta_c < \pi/2$, where

$$\theta_c = \sin^{-1}\left(\frac{1}{r}\right) = \sin^{-1}\left(\sqrt{\frac{\mu_2\varepsilon_2}{\mu_1\varepsilon_1}}\right), \tag{11.51}$$

then the angle of transmission has a real value in the range $0 \leq \theta_t \leq \pi/2$. When the angle of incidence increases further so that $\theta_c < \theta_i \leq \pi/2$, we can no longer interpret θ_t as a real angle. For this range of angle of incidence we still have

$$\beta_{tz} = \beta_{iz} = \omega\sqrt{\mu_1\varepsilon_1}\sin\theta_i, \tag{11.52}$$

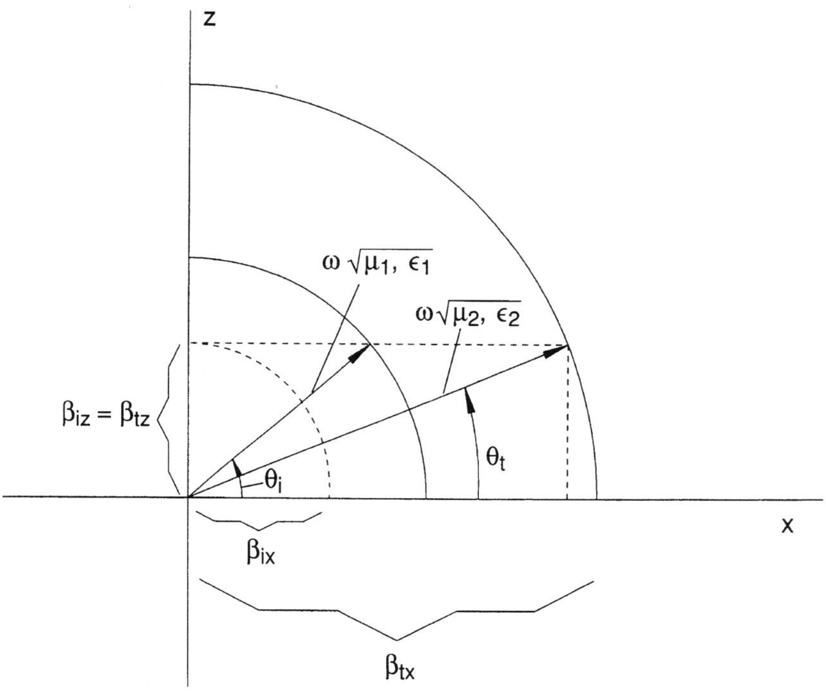

FIGURE 11.10 Graphical illustration of Snell's law (Case 1).

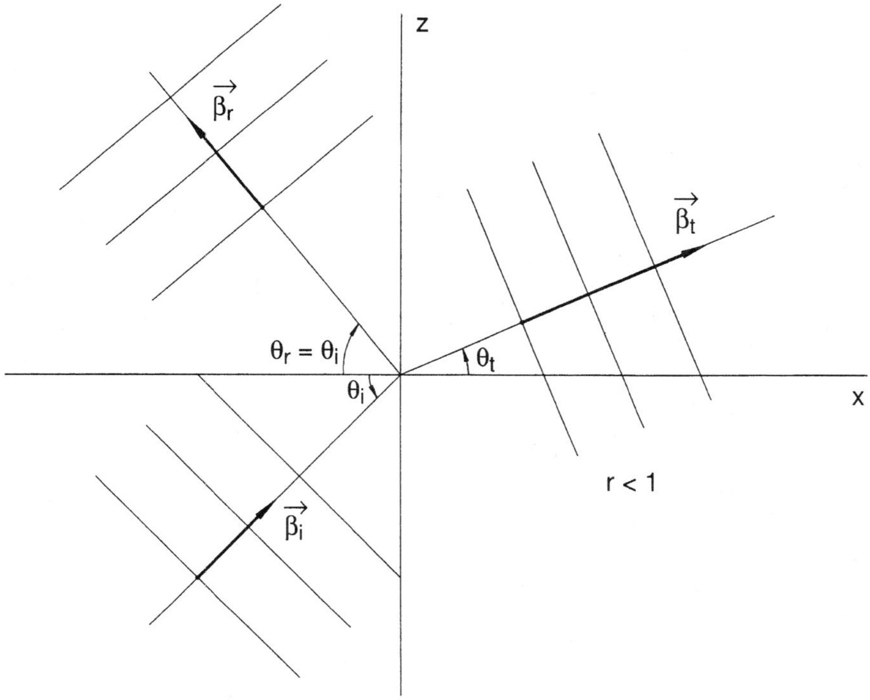

FIGURE 11.11 Planes of constant phase for the incident, reflected, and transmitted waves (Case 1).

but we must look carefully at the x component of the propagation constant, $\gamma_{tx} = j\beta_{tx} = j\omega\sqrt{\mu_2\varepsilon_2}\cos\theta_t$. Now,

$$\cos\theta_t = \sqrt{1-\sin^2\theta_t} = \sqrt{1-r^2\sin^2\theta_i}$$

$$= \begin{cases} \sqrt{1-r^2\sin^2\theta_i}, & 0 \le \theta_i \le \theta_c, \qquad (11.53) \\ -j\sqrt{r^2\sin^2\theta_i - 1}, & \theta_c \le \theta_i \le \pi/2, \qquad (11.54) \end{cases}$$

so that

$$\gamma_{tx} = \begin{cases} j\beta_{tx} = j\omega\sqrt{\mu_2\varepsilon_2}\sqrt{1-r^2\sin^2\theta_i}, & 0 \le \theta_i \le \theta_c \qquad (11.55) \\ \alpha_{tx} = \omega\sqrt{\mu_2\varepsilon_2}\sqrt{r^2\sin^2\theta_i - 1}, & \theta_c \le \theta_i \le \pi/2 \qquad (11.56) \end{cases}$$

For the transmitted wave we see that

$$e^{-\vec{\gamma}_t\cdot\vec{R}} = \begin{cases} e^{-j\beta_{tx}x}e^{-j\beta_{tz}z}, & 0 \le \theta_i \le \theta_c, \\ e^{-\alpha_{tx}x}e^{-j\beta_{tz}z}, & \theta_c \le \theta_i \le \pi/2. \end{cases}$$

For $\theta_i < \theta_c$ this represents a plane uniform unattenuated wave traveling at a real angle θ_t from the x axis (Fig. 11.13a). For $\theta_t > \theta_c$ this represents a nonuniform plane wave

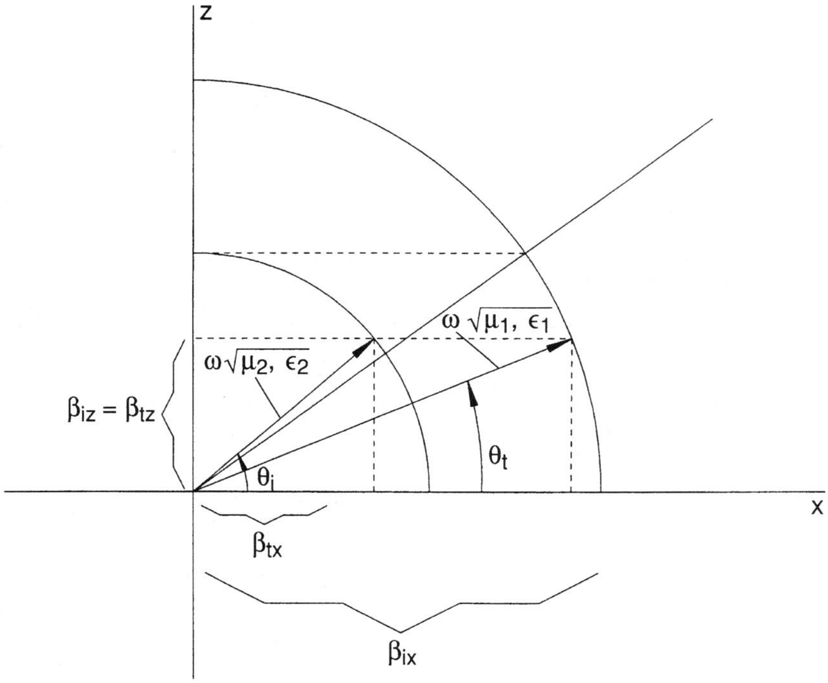

FIGURE 11.12 Graphical illustration of Snell's law (Case 2).

with planes of constant phase normal to the z axis and planes of constant amplitude normal to the x axis (Fig. 11.13b).

There is a little tricky mathematical point with regard to the square root in Eqs. (11.53) and (11.54) and in Eqs. (11.55) and (11.56). Whenever we take a square root we have a choice of positive or negative sign. The choice must be made on the basis of physical reasoning. In Eq. (11.53) the sign is chosen so that β_{tx} in Eq. (11.55) will be positive and the wave will travel away from the boundary. In Eq. (11.54) the sign is chosen so that α_{tx} in Eq. (11.56) will be positive and the wave's amplitude will decrease as x increases away from the boundary.

For the incident angle greater than the critical angle the wave attenuates in the x direction even though there are no losses in the medium. Clearly the wave cannot be carrying any time average power in the x direction. As $x \to \infty$ the amplitude of each field component and hence any power carried by the wave approaches zero. It cannot be going to zero because of power losses along the way, since the medium is lossless. Therefore the wave cannot be carrying time average power in the x direction at any $x > 0$. A wave that attenuates in a lossless medium is called an evanescent wave.

Because the wave is not carrying any time average power in the x direction for $x > 0$ it also cannot be carrying any time average power in the x direction for $x < 0$. Thus the x component of incident time average power, $(|E_0|^2/2\eta_1)\cos\theta_i$, must be reflected and carried away from the boundary by the reflected wave. For Case 2 ($r > 1$) and $\theta_i > \theta_c$ the wave is said to suffer total reflection at the boundary. By total reflection we mean that all of the power directed toward the boundary is reflected from it; none of the power is transmitted.

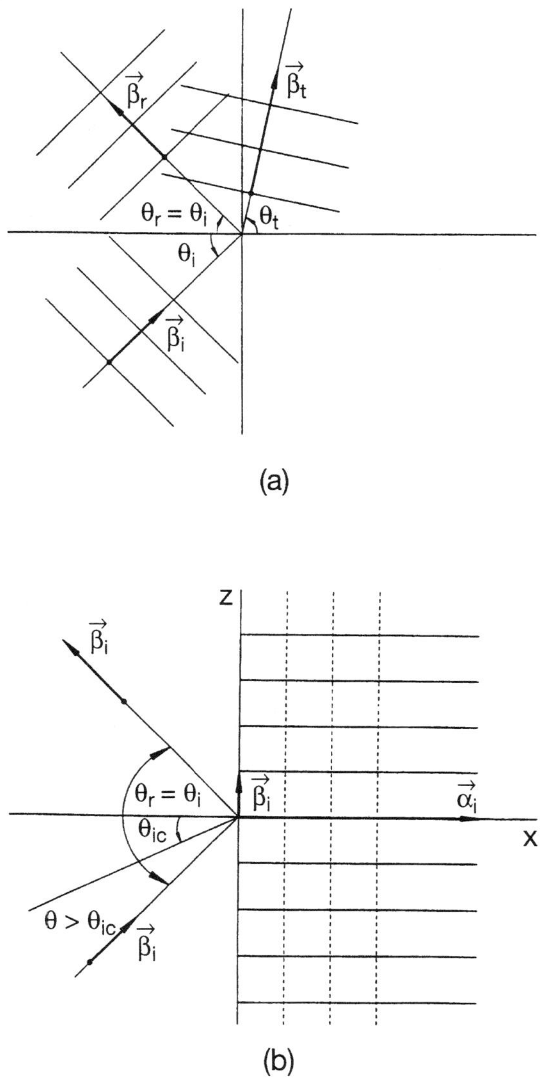

FIGURE 11.13 Planes of constant phase for a uniform and nonuniform plane wave.

11.6.1.2 Snell's Law for Perfect Dielectric to Conductor Boundary Let us now look at the situation when the wave is incident from a perfect dielectric medium and the transmitted wave is in a conducting medium, as, for example, at an air–copper boundary. In this case $\sigma_1 = 0$ and $\gamma_1 = j\omega\sqrt{\mu_1\varepsilon_1}$. From Eqs. (11.45) and (11.47) we have

$$\begin{aligned}\gamma_{tx} &= \sqrt{\gamma_2^2 - \gamma_{tz}^2} = \sqrt{j\omega\mu_2(\sigma_2 + j\omega\varepsilon_2) - (j\omega\sqrt{\mu_1\varepsilon_1}\sin\theta_i)^2} \\ &= \sqrt{j\omega\mu_2\sigma_2 - \omega^2\mu_2\varepsilon_2 + \omega^2\mu_1\varepsilon_1\sin^2\theta_i} \\ &= \sqrt{j\omega\mu_2\sigma_2}\sqrt{1 - \frac{\omega\varepsilon_2}{\sigma_2}\left(\frac{\mu_1\varepsilon_1}{\mu_2\varepsilon_2}\sin^2\theta_i - 1\right)}.\end{aligned} \tag{11.57}$$

In the usual case the conductor is a good conductor, $\sigma_2 \gg \omega\varepsilon_2$, and the imaginary term in the second square root can be neglected. We then have

$$\gamma_{tx} = \alpha_{tx} + j\beta_{tx} \approx \sqrt{j\omega\mu_2\sigma_2} = \sqrt{\frac{\omega\mu_2\sigma_2}{2}}(1+j)$$

$$= \frac{1}{\delta_s}(1+j), \quad \delta_s = \sqrt{\frac{2}{\omega\mu_2\sigma_2}}. \tag{11.58}$$

δ_s is the skin depth, which we have already encountered.

For the transmitted wave we now see that

$$e^{-\vec{\gamma}_t \cdot \vec{R}} = e^{-\alpha_{tx}x}e^{-j\beta_{tx}x}e^{-j\beta_{tz}z}$$

$$= e^{-x/\delta_s}e^{-j[x/\delta_s + (\omega\sqrt{\mu_1\varepsilon_1}\sin\theta_i)z]}.$$

The first exponential on the right means that the transmitted wave is attenuating as it propagates in the x direction, due to the losses in the conductor. The second exponential means that the planes of equal phase are tilted from the x direction toward the z direction by an angle ψ, where

$$\tan\psi = \frac{\beta_{tz}}{\beta_{tx}} = \frac{\omega\sqrt{\mu_1\varepsilon_1}\sin\theta_i}{\sqrt{\omega\mu_2\sigma_2/2}}$$

$$= \sqrt{2\left(\frac{\omega\varepsilon_2}{\sigma_2}\right)\left(\frac{\mu_1\varepsilon_1}{\mu_2\varepsilon_2}\right)}\sin\theta_i. \tag{11.59}$$

The permeabilities are generally equal to the permeability of free space and, in any case, their ratio is never far from unity. The ratio of the permittivities varies more but is never too large. For the good conductor that we are considering, $\sigma_2 \gg \omega\varepsilon_2$. All these factors mean that $\tan\psi \ll 1$ and that $\psi \approx 0$ regardless of the angle of incidence.

The result is that the transmitted wave in a good conductor propagates normally into the conductor, decaying by $\exp(-1)$ for every skin depth of distance from the surface. In equation form,

$$e^{-\vec{\gamma}_t \cdot \vec{R}} \approx e^{-x/\delta_s}e^{-jx/\delta_s}, \quad \delta_s = \sqrt{2/\omega\mu_2\sigma_2}, \quad 0 \le \theta_i \le \pi/2. \tag{11.60}$$

11.6.2 Resolution of Incident Field into TE and TM Fields

The incident wave assumed at the start of the last section is a transverse electromagnetic (TEM) wave. This means that the electric and magnetic field intensities of the incident wave lie in the plane that is transverse or normal to its propagation vector, $\vec{\gamma}_i$. We now show that this incident field can be split into two sets of fields, one transverse electric (TE) and one transverse magnetic (TM) to the x and z directions. This split is convenient for the analysis that follows.

Figure 11.14 shows the incident propagation vector and the x, y, z coordinate system, which we have defined. In the figure an x' axis has been added along the

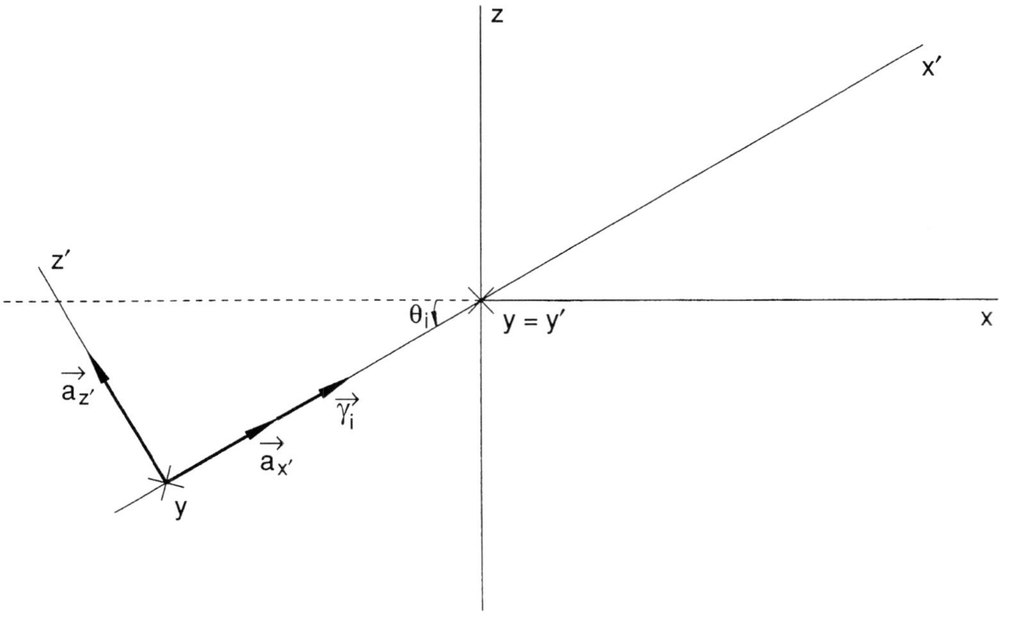

FIGURE 11.14 Incident propagation vector in the xyz coordinate system.

propagation vector and a z' axis has been added to complete the x', y, z' coordinates. The y axis is common to both sets of coordinates. The transformation between the two coordinate systems is

$$\vec{a}_{x'} = \vec{a}_x \cos\theta_i + \vec{a}_z \sin\theta_i,$$

$$\vec{a}_{z'} = -\vec{a}_x \sin\theta_i + \vec{a}_z \cos\theta_i.$$

To check this transformation for yourself, visualize the figure at the two extremes, $\theta_i = 0$ and $\theta_i = \pi/2$.

We can assume the incident electric field intensity is linearly polarized at an angle ξ with respect to the y axis (Fig. 11.15). The incident field is then

$$\begin{aligned}\underline{\vec{E}}^i_T &= E_a(\vec{a}_y \cos\xi + \vec{a}_{z'} \sin\xi)e^{-\vec{\gamma}_i\cdot\vec{R}} \\ &= E_a(-\vec{a}_x \sin\xi\sin\theta_i + \vec{a}_y\cos\xi + \vec{a}_z\sin\xi\cos\theta_i)e^{-\vec{\gamma}_i\cdot\vec{R}},\end{aligned} \tag{11.61}$$

$$\begin{aligned}\underline{\vec{H}}^i_T &= (E_a/\eta_1)(-\vec{a}_y\sin\xi + \vec{a}_{z'}\cos\xi)e^{-\vec{\gamma}_i\cdot\vec{R}} \\ &= (E_a/\eta_1)(-\vec{a}_x\cos\xi\sin\theta_i - \vec{a}_y\sin\xi + \vec{a}_z\cos\xi\cos\theta_i)e^{-\vec{\gamma}_i\cdot\vec{R}}.\end{aligned} \tag{11.62}$$

With $\vec{\gamma}_i = \vec{a}_{x'}\gamma_i$ we have

$$\vec{\gamma}_i\cdot\vec{R} = \gamma_1[(\cos\theta_i)x + (\sin\theta_i)z].$$

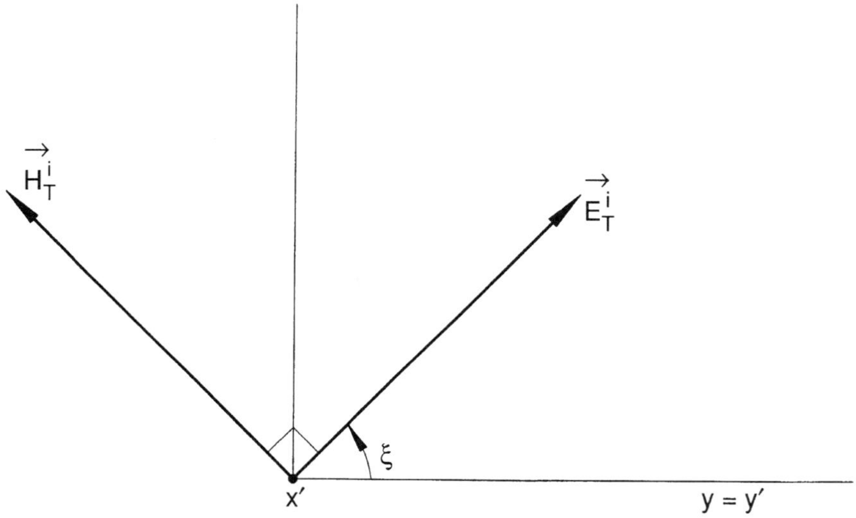

FIGURE 11.15 Incident linearly polarized electric field intensity.

These fields can be separated into two sets of fields. One has the total y-directed electric field and the other has the total y-directed magnetic field. This separation gives

$$\underline{\vec{E}}^i_{\text{TE}} = \vec{a}_y E_0 e^{-(\gamma_1 \cos\theta_i)x} e^{-(\gamma_1 \sin\theta_i)z}, \quad E_0 = E_a \cos\xi, \tag{11.63}$$

$$\underline{\vec{H}}^i_{\text{TE}} = \frac{E_0}{\eta_1}(-\vec{a}_x \sin\theta_i + \vec{a}_z \cos\theta_i) e^{-(\gamma_1 \cos\theta_i)x} e^{-(\gamma_1 \sin\theta_i)z}, \tag{11.64}$$

and

$$\underline{\vec{H}}^i_{\text{TM}} = -\vec{a}_y H_0 e^{-(\gamma_1 \cos\theta_i)x} e^{-(\gamma_1 \sin\theta_i)z}, \quad H_0 = (E_a/\eta_1)\sin\xi, \tag{11.65}$$

$$\underline{\vec{E}}^i_{\text{TM}} = \eta_1 H_0(-\vec{a}_x \sin\theta_i + \vec{a}_z \cos\theta_i) e^{-(\gamma_1 \cos\theta_i)x} e^{-(\gamma_1 \sin\theta_i)z}. \tag{11.66}$$

Adding the two sets gives the original field.

Equations (11.63) and (11.64) represent a wave that is a solution to Maxwell's equations and that is transverse electric (TE) to both the x direction and the z direction. The wave can also be described as having a magnetic field that lies in the plane of incidence (the xz plane).

Equations (11.65) and (11.66) represent a wave that is a solution to Maxwell's equations and that is transverse magnetic (TM) to both the x direction and the z direction. The wave can also be described as having an electric field that lies in the plane of incidence (the xz plane).

In the next two sections we will consider the reflection of each of these field pairs. Since any incident field can be split into these two field pairs, the solutions obtained for each pair alone can be superimposed to give the complete solution.

While the principle of superposition tells us that the response to the total wave can be found as the sum of the responses to the two sets that sum to the total wave, it does not tell us that quantities involving products, such as power and energy, can be summed to give the correct totals. Problem 11.13 shows that the time average power carried by the original wave is correctly given as the sum of the time average power carried by each wave. Problem 11.14 gives a similar result for the time average energy stored in the electric and magnetic fields.

11.6.3 TE Wave or Perpendicular Polarization

We have learned a great deal about the transmitted wave by studying its propagation vector for different angles of incidence and properties of the transmission medium. We have not, however, written out the field components of the reflected and transmitted waves or calculated their amplitudes relative to the incident wave. We will now do that for the TE wave.

Figure 11.16 shows the geometry for the incident, reflected, and transmitted TE waves. For each wave the electric field intensity is in the y direction and the magnetic field intensity is drawn so that $\underline{\vec{E}} \times \underline{\vec{H}}$ is directed along the wave's propagation vector. This is frequently called perpendicular polarization, because the electric field is perpendicular to the plane of incidence. We have already written out the incident field components, Eqs. (11.63) and (11.64), and we repeat them here, dropping the TE

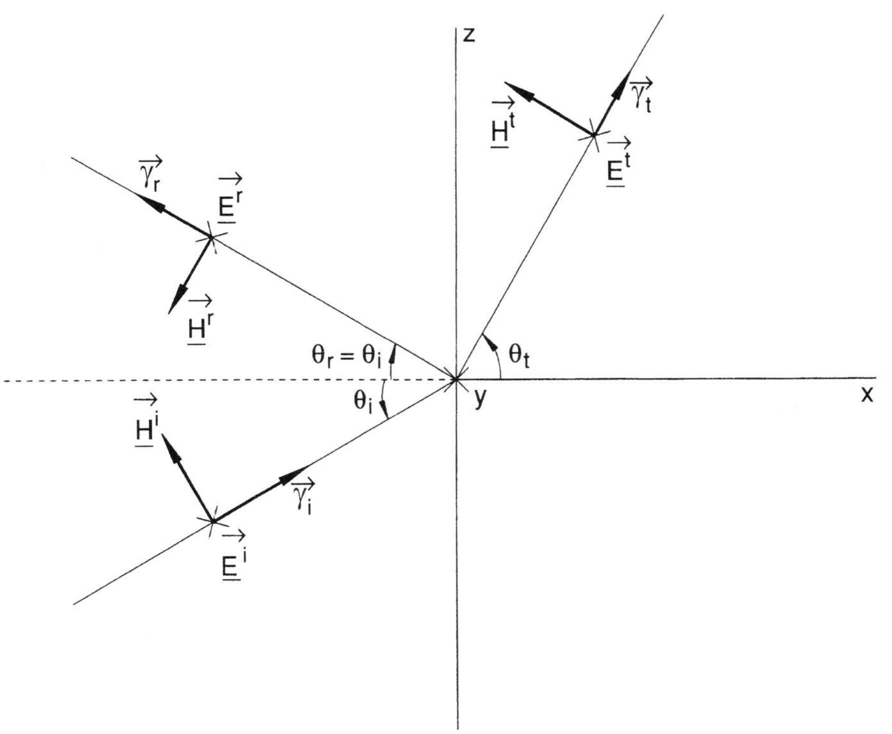

FIGURE 11.16 Geometry for the incident, reflected, and transmitted TE waves.

subscript.

$$\underline{\vec{E}}^i = \vec{a}_y E_0 e^{-(\gamma_1 \cos\theta_i)x} e^{-(\gamma_1 \sin\theta_i)z}, \tag{11.67}$$

$$\underline{\vec{H}}^i = \frac{E_0}{\eta_1}(-\vec{a}_x \sin\theta_i + \vec{a}_z \cos\theta_i) e^{-(\gamma_1 \cos\theta_i)x} e^{-(\gamma_1 \sin\theta_i)z}. \tag{11.68}$$

The complete reflected field components are

$$\underline{\vec{E}}^r = \vec{a}_y \Gamma_{TE} E_0 e^{(\gamma_1 \cos\theta_i)x} e^{-(\gamma_1 \sin\theta_i)z}, \tag{11.69}$$

$$\underline{\vec{H}}^r = \frac{\Gamma_{TE} E_0}{\eta_1}(-\vec{a}_x \sin\theta_i - \vec{a}_z \cos\theta_i) e^{(\gamma_1 \cos\theta_i)x} e^{-(\gamma_1 \sin\theta_i)z}. \tag{11.70}$$

The reflected electric field intensity has an amplitude $\Gamma_{TE} E_0$, where Γ_{TE} is the TE field reflection coefficient at the boundary,

$$\Gamma_{TE} = \left.\frac{\underline{E}_y^r}{\underline{E}_y^i}\right|_{x=0}. \tag{11.71}$$

The complete transmitted wave field components are

$$\underline{\vec{E}}^t = \vec{a}_y \tau_{TE} E_0 e^{-(\gamma_2 \cos\theta_t)x} e^{-(\gamma_1 \sin\theta_i)z}, \tag{11.72}$$

$$\underline{\vec{H}}^t = \frac{\tau_{TE} E_0}{\eta_2}(-\vec{a}_x \sin\theta_t + \vec{a}_z \cos\theta_t) e^{-(\gamma_2 \cos\theta_t)x} e^{-(\gamma_1 \sin\theta_i)z}, \tag{11.73}$$

$$\sin\theta_t = \frac{\gamma_1}{\gamma_2}\sin\theta_i, \quad \cos\theta_t = \sqrt{1 - \left(\frac{\gamma_1}{\gamma_2}\right)^2 \sin^2\theta_i}.$$

The transmitted electric field intensity has an amplitude $\tau_{TE} E_0$, where τ_{TE} is the TE field transmission coefficient at the boundary

$$\tau_{TE} = \left.\frac{\underline{E}_y^t}{\underline{E}_y^i}\right|_{x=0}. \tag{11.74}$$

There are two unknowns, the reflection and the transmission coefficients. These are determined by satisfying the boundary conditions that the tangential electric and magnetic field intensities must be continuous at $x = 0$. In equation form this is

$$\underline{\vec{E}}_y^i|_{x=0} + \underline{E}_y^r|_{x=0} = \underline{E}_y^t|_{x=0}$$

$$1 + \Gamma_{TE} = \tau_{TE},$$

and

$$\underline{H}_z^i|_{x=0} + \underline{H}_z^r|_{x=0} = \underline{H}_z^t|_{x=0}$$

$$\frac{\cos\theta_i}{\eta_1} + \Gamma_{TE}\left[-\frac{\cos\theta_i}{\eta_1}\right] = \tau_{TE}\frac{\cos\theta_t}{\eta_2}.$$

The two equations are simply solved to give

$$\Gamma_{\mathrm{TE}} = \frac{(\eta_2/\cos\theta_t) - (\eta_1/\cos\theta_i)}{(\eta_2/\cos\theta_t) + (\eta_1/\cos\theta_i)}, \tag{11.75}$$

$$\Gamma_{\mathrm{TE}} = \frac{2(\eta_2/\cos\theta_t)}{(\eta_2/\cos\theta_t) + (\eta_1/\cos\theta_i)}. \tag{11.76}$$

It is instructive to obtain these same results from a different viewpoint. The tangential electric and magnetic field intensities must be continuous at the boundary. It follows that their ratio, $\underline{E}_y/\underline{H}_z$, which is an impedance, must be continuous. But this was exactly the condition required at the boundary between two transmission lines: $\underline{V}$ and $\underline{I}$ must be continuous and hence the impedance $\underline{V}/\underline{I}$ must be continuous. We can therefore model our reflection problem as two transmission lines (Fig. 11.17). The line on the left, $x < 0$, has characteristic impedance given by the ratio of the transverse (to the x axis) electric and magnetic field intensities and a propagation constant equal to the x component of the propagation constant, both for the forward going incident wave. Hence let

$$Z_{01} = Z_{1,\mathrm{TE}} = \frac{\underline{E}_y^i}{\underline{H}_z^i} = \frac{\eta_1}{\cos\theta_i}, \quad \gamma_{1,\mathrm{TE}} = \vec{a}_x \cdot \vec{\gamma}_i = \gamma_1 \cos\theta_i. \tag{11.77}$$

Similarly, the line on the right, $x > 0$, has characteristic impedance given by the ratio of the transverse (to the x axis) electric and magnetic field intensities and a propagation constant equal to the x component of the propagation constant, both for the forward going transmitted wave. Hence let

$$Z_{02} = Z_{2,\mathrm{TE}} = \frac{\underline{E}_y^t}{\underline{H}_z^t} = \frac{\eta_2}{\cos\theta_t}, \quad \gamma_{2,\mathrm{TE}} = \vec{a}_x \cdot \vec{\gamma}_t = \gamma_2 \cos\theta_t. \tag{11.78}$$

The line to the right extends indefinitely and hence this line carries only a forward $(+x)$ traveling wave and its input impedance at $x = 0$ equals its characteristic impedance Z_{02}. This is the load impedance at $x = 0$ on the line to the left. At the

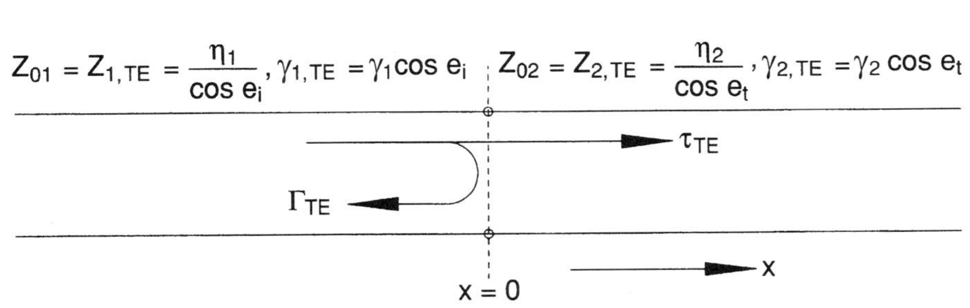

FIGURE 11.17 Two-transmission-line model for the reflection problem for TE waves.

junction the voltage reflection and transmission coefficients are

$$\Gamma = \frac{Z_{02} - Z_{01}}{Z_{02} + Z_{01}}, \quad \tau = \frac{2Z_{02}}{Z_{02} + Z_{01}}. \tag{11.79}$$

Substitution of the characteristic impedances gives the results of Eqs. (11.75) and (11.76).

We will find modeling with transmission lines to be helpful to our understanding of the reflection problem.

11.6.3.1 Perfect Dielectric to Perfect Dielectric Boundary Reflection when both media are perfect dielectrics was studied in Section 11.6.1.1. In this situation

$$Z_{1,\mathrm{TE}} = \sqrt{\frac{\mu_1}{\varepsilon_1}} \frac{1}{\cos\theta_i}, \quad Z_{2,\mathrm{TE}} = \sqrt{\frac{\mu_2}{\varepsilon_2}} \frac{1}{\cos\theta_t}$$

$$\cos\theta_t = \sqrt{1 - r^2 \sin^2\theta_i}, \quad r = \sqrt{\frac{\mu_1 \varepsilon_1}{\mu_2 \varepsilon_2}}.$$

As we saw in Case 1 of that section, when $r < 1$, so that we have transmission from the less dense medium into the more dense medium, then $\cos\theta_t$ is real for all values of θ_i. The impedance $Z_{2,\mathrm{TE}}$ is therefore real or resistive and it follows that time average power can be transmitted in the x direction.

On the other hand, for Case 2, when $r > 1$, so that we have transmission from the more dense to the less dense medium, then $\cos\theta_t$ is real for $0 < \theta_i < \theta_c$ but is negative imaginary for $\theta_c < \theta_i < \pi/2$—Eqs. (11.53) and (11.54). The impedance is therefore

$$Z_{2,\mathrm{TE}} = \begin{cases} \sqrt{\mu_2/\varepsilon_2}/\sqrt{1 - r^2 \sin^2\theta_i}, & 0 \le \theta_i \le \theta_c, \quad (11.80) \\ j\sqrt{\mu_2/\varepsilon_2}/\sqrt{r^2 \sin^2\theta_i - 1}, & \theta_c \le \theta_i \le \pi/2. \quad (11.81) \end{cases}$$

For the angle of incidence less than the critical angle, $Z_{2,\mathrm{TE}}$ is real or resistive and hence time average power can be transmitted in the x direction. For the angle of incidence greater than the critical angle, $Z_{2,\mathrm{TE}}$ is inductive reactive and hence no time average power can be transmitted in the x direction.

Let us now calculate the time average power on both sides of the boundary. For $x < 0$ the incident and reflected waves add to give a total field

$$\underline{\vec{E}}_1 = \underline{\vec{E}}^i + \underline{\vec{E}}^r = \vec{a}_y E_0 [e^{-j(\beta_1 \cos\theta_i)x} + \Gamma_{\mathrm{TE}} e^{j(\beta_1 \cos\theta_i)x}] e^{-j(\beta_1 \sin\theta_i)z}, \tag{11.82}$$

$$\underline{\vec{H}}_1 = \underline{\vec{H}}^i + \underline{\vec{H}}^r = \frac{E_0}{\eta_1}[-\vec{a}_x \sin\theta_i (e^{-j(\beta_1 \cos\theta_i)x} + \Gamma_{\mathrm{TE}} e^{j(\beta_1 \cos\theta_i)x})$$

$$+ \vec{a}_z \cos\theta_i (e^{-j(\beta_1 \cos\theta_i)x} - \Gamma_{\mathrm{TE}} e^{j(\beta_1 \cos\theta_i)x})] e^{-j(\beta_1 \sin\theta_i)z}. \tag{11.83}$$

Using Eq. (11.39), the time average power density for $x < 0$ is

$$\vec{p}_{f1,\text{Avg}} = \tfrac{1}{2}\text{Re}(\underline{\vec{E}}_1 \times \underline{\vec{H}}_1^*)$$

$$= \frac{|E_0|^2}{2\eta_1}\{\vec{a}_x \cos\theta_i (1 - |\Gamma_{\text{TE}}|^2)$$

$$+ \vec{a}_z \sin\theta_i [1 + |\Gamma_{\text{TE}}|^2 + 2|\Gamma_{\text{TE}}|\cos(2\beta_1(\cos\theta_i)x + \phi)]\},$$

where $\Gamma_{\text{TE}} = |\Gamma_{\text{TE}}|\exp(j\phi)$.

The factor $|E_0|^2/2\eta_1$ is the time average power density carried by the incident wave. Multiplying by $\cos\theta_i$ gives the component of time average power density carried by the incident wave in the x direction. For Case 1 ($r < 1$) and for Case 2 ($r > 1$) with $\theta_i < \theta_c$, the reflection coefficient is real and the reflected wave carries back a fraction $|\Gamma_{\text{TE}}|^2$ of the incident x-directed power. For Case 2 with $\theta_i > \theta_c$, $|\Gamma_{\text{TE}}| = 1$ and therefore all the incident x-directed power is reflected. This goes along with our previous observations that no real time average power is transmitted in the x direction above the critical angle and that power is totally reflected at the boundary.

For the transmitted wave $x > 0$, using Eqs. (11.72) and (11.73) with $\gamma_1 = j\beta_1$ and $\gamma_2 = j\beta_2$, the time average power density is

$$\vec{p}_{f2,\text{Avg}} = \frac{|\tau_{\text{TE}}|^2 |E_0|^2}{2\eta_2}(\vec{a}_x \cos\theta_t + \vec{a}_z \sin\theta_t)$$

for Case 1 and for Case 2 with $\theta_i < \theta_c$. For Case 2 with $\theta_i > \theta_c$ it is

$$\vec{p}_{f2,\text{Avg}} = \frac{|\tau_{\text{TE}}|^2 |E_0|^2}{2\eta_2}\vec{a}_z r \sin\theta_i \, e^{-2(\omega\sqrt{\mu_2\varepsilon_2}\sqrt{r^2\sin^2\theta_i - 1})x},$$

which shows there is no x-directed time average power, as we have already noted.

11.6.3.2 *Perfect Dielectric to Boundary* When the second medium is a perfect conductor, then $\eta_2 = 0$ and $Z_{2,\text{TE}} = 0$ and therefore $\Gamma_{\text{TE}} = -1$ and $\tau_{\text{TE}} = 0$. The total wave in region 1 is

$$\underline{\vec{E}}_1 = \vec{a}_y E_0 [e^{-j(\beta_1 \cos\theta_i)x} - e^{j(\beta_1 \cos\theta_i)x}] e^{-j(\beta_1 \sin\theta_i)z}$$

$$= \vec{a}_y(-2jE_0)\sin[(\beta_1\cos\theta_i)x]\, e^{-j(\beta_1 \sin\theta_i)z}, \quad (11.84)$$

$$\underline{\vec{H}}_1 = \frac{E_0}{\eta_1}[-\vec{a}_x \sin\theta_i (e^{-j(\beta_1\cos\theta_i)x} - e^{j(\beta_1\cos\theta_i)x})$$

$$+ \vec{a}_z \cos\theta_i (e^{-j(\beta_1\cos\theta_i)x} + e^{j(\beta_1\cos\theta_i)x})]\, e^{-j(\beta_1\sin\theta_i)z}$$

$$= \frac{2E_0}{\eta_1}\{\vec{a}_x j \sin\theta_i \sin[(\beta_1\cos\theta_i)x]$$

$$+ \vec{a}_z \cos\theta_i \cos[(\beta_1\cos\theta_i)x]\}\, e^{-j(\beta_1\sin\theta_i)z}. \quad (11.85)$$

The wave is a standing wave in the x direction and a traveling wave in the z direction. The electric field not only goes to zero at the surface of the perfect conductor but also periodically at $x = -n\pi/\beta_1 \cos\theta_i$ back from the surface. We will discuss this further when we discuss parallel plate waveguides in Section 11.8.

The time average power density is

$$\vec{p}_{f1,\,\mathrm{Avg}} = \vec{a}_z \frac{2|E_0|^2}{\eta_1} \sin\theta_i \sin^2[(\beta_1 \cos\theta_i)x], \tag{11.86}$$

which is totally z directed.

When the second medium is a good, but not perfect, conductor, the wave does penetrate into it and, as we have shown, attenuate rapidly in the x direction. We have already found that the transmitted wave is virtually normal to the boundary. Furthermore, the field in the dielectric region will be almost the same as it is when the second medium is a perfect conductor. Using Eq. (11.85), the tangential magnetic field intensity at $x = 0-$ just to the left of the boundary is

$$H_{\mathrm{tang}} = \underline{H}_z|_{x=0} = (2E_0/\eta_1)\cos\theta_i\, e^{-j\beta_1(\sin\theta_i)z}.$$

The tangential magnetic field intensity is continuous and so this is also its value at $x = 0+$ just to the right of the boundary. To obtain the magnetic field intensity for $x > 0$ we multiply this by $\exp(-\gamma_2 x)$ to obtain

$$\begin{aligned}\underline{\vec{H}}^t &= \vec{a}_z H_{\mathrm{tang}} e^{-x/\delta_s} e^{-jx/\delta_s} \\ &= \vec{a}_z \frac{2E_0}{\eta_1} \cos\theta_i\, e^{-j(\beta_1 \sin\theta_i)z} e^{-x/\delta_s} e^{-jx/\delta_s}.\end{aligned}$$

The electric field is $\underline{E}_y = \eta_2 \underline{H}_z$,

$$\underline{\vec{E}}^t = \vec{a}_y E_0 \frac{2\eta_2}{\eta_1} \cos\theta_i\, e^{-j(\beta_1 \sin\theta_i)z} e^{-x/\delta_s} e^{-jx/\delta_s}.$$

The factor $(2\eta_2/\eta_1)\cos\theta_i$ is equal to the transmission coefficient τ_{TE}, Eq. (11.76), when $|\eta_2| \ll \eta_1$ and $\theta_t \approx 0$, as they are for the good conductor. This coefficient multiplies $E_0 \exp[(-j\beta_1 \sin\theta_i)z]$, which is the tangential electric field at the boundary.

The time average power density is

$$\begin{aligned}\vec{p}_{f2,\,\mathrm{Avg}} &= \tfrac{1}{2}\mathrm{Re}(\underline{\vec{E}}^t \times \underline{\vec{H}}^{t*}) = \vec{a}_x \frac{1}{2}\left(\frac{2E_0}{\eta_1}\cos\theta_i\right)^2 \mathrm{Re}(\eta_2) e^{-2x/\delta_s} \\ &= \vec{a}_x \frac{1}{2}\left(\frac{2E_0}{\eta_1}\cos\theta_i\right)^2 \sqrt{\frac{\omega\mu_2}{2\sigma_2}}\, e^{-2x/\delta_s}.\end{aligned}$$

Evaluating at $x = 0$ gives the time average power density into the good conductor:

$$p_{f2x,\,\mathrm{Avg}}\big|_{x=0} = \frac{1}{2}\left(\frac{2E_0}{\eta_1}\cos\theta_i\right)^2 \sqrt{\frac{\omega\mu_2}{2\sigma_2}}$$

$$= \tfrac{1}{2}|H_{\mathrm{tang}}|^2 R_s, \quad R_s = \sqrt{\frac{\omega\mu_2}{2\sigma_2}}. \tag{11.87}$$

Generally we are not interested in the fields within the conducting region but are only concerned with the time average power density flowing into the conductor. This is given simply by Eq. (11.87), again demonstrating that for a good conducting boundary we can evaluate the loss from the adjacent dielectric by solving the problem as if the conductor were perfect and then using the tangential magnetic field magnitude and the surface resistance to give the power loss.

11.6.4 TM Wave or Parallel Polarization

Figure 11.18 shows the geometry for the incident, reflected, and transmitted TM waves. For each wave the z component of the electric field intensity is in the $+z$ direction and the magnetic field intensity is drawn so that $\vec{E} \times \vec{H}$ is directed along the wave's propagation vector. This is frequently called parallel polarization because the electric field lies in the plane of incidence. We have already written out the incident

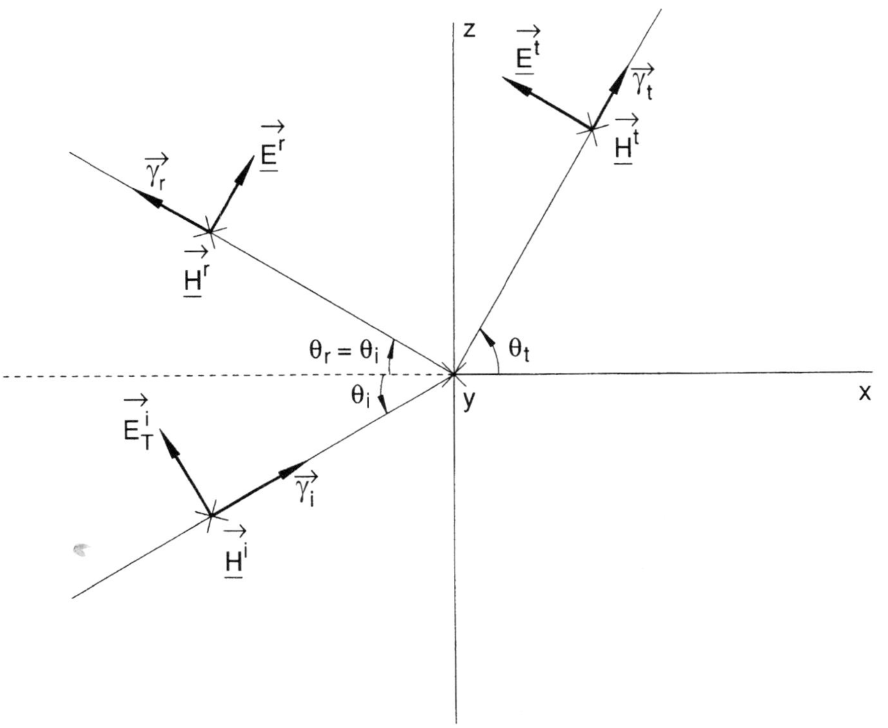

FIGURE 11.18 Geometry for the incident, reflected, and transmitted TM waves.

field components, Eqs. (11.65) and (11.66), and we repeat them here, dropping the TM subscript.

$$\underline{\vec{H}}^i = -\vec{a}_y H_0 e^{-(\gamma_1 \cos\theta_i)x} e^{-(\gamma_1 \sin\theta_i)z}, \tag{11.88}$$

$$\underline{\vec{E}}^i = \eta_1 H_0(-\vec{a}_x \sin\theta_i + \vec{a}_z \cos\theta_i) e^{-(\gamma_1 \cos\theta_i)x} e^{-(\gamma_1 \sin\theta_i)z}. \tag{11.89}$$

The complete reflected field components are

$$\underline{\vec{H}}^r = \vec{a}_y \Gamma_{\mathrm{TM}} H_0 e^{(\gamma_1 \cos\theta_i)x} e^{-(\gamma_1 \sin\theta_i)z}, \tag{11.90}$$

$$\underline{\vec{E}}^r = \eta_1 \Gamma_{\mathrm{TM}} H_0(\vec{a}_x \sin\theta_i + \vec{a}_z \cos\theta_i) e^{(\gamma_1 \cos\theta_i)x} e^{-(\gamma_1 \sin\theta_i)z}. \tag{11.91}$$

The TM reflection coefficient is Γ_{TM}, defined as the ratio of the reflected to incident tangential component of the electric field intensity at the boundary,

$$\Gamma_{\mathrm{TM}} = \left.\frac{\underline{E}_z^r}{\underline{E}_z^i}\right|_{x=0}.$$

This definition is consistent with the definition for the TE case, Eq. (11.71), since for the TE fields $\underline{E}_y$ is the component of the electric field that is tangential to the boundary.

The complete transmitted field components are

$$\underline{\vec{H}}^t = -\vec{a}_y \tau_{\mathrm{TM}} H_0 \frac{\eta_1 \cos\theta_i}{\eta_2 \cos\theta_t} e^{-(\gamma_2 \cos\theta_t)x} e^{-(\gamma_1 \sin\theta_i)z}, \tag{11.92}$$

$$\underline{\vec{E}}^t = \eta_2 \tau_{\mathrm{TM}} H_0 \frac{\eta_1 \cos\theta_i}{\eta_2 \cos\theta_t}(-\vec{a}_x \sin\theta_t + \vec{a}_z \cos\theta_t) e^{-(\gamma_2 \cos\theta_t)x} e^{-(\gamma_1 \sin\theta_i)z}, \tag{11.93}$$

$$\sin\theta_t = \frac{\gamma_1}{\gamma_2}\sin\theta_i, \quad \cos\theta_t = \sqrt{1 - \left(\frac{\gamma_1}{\gamma_2}\right)^2 \sin^2\theta_i}.$$

The TM transmission coefficient is τ_{TM}, defined as the ratio of the transmitted to incident tangential component of the electric field intensity at the boundary,

$$\tau_{\mathrm{TM}} = \left.\frac{\underline{E}_z^t}{\underline{E}_z^i}\right|_{x=0}.$$

The seemingly arbitrary factor $(\eta_1 \cos\theta_i)/(\eta_2 \cos\theta_t)$ was included in the transmitted wave amplitude to make this come out correctly.

The reflection and transmission coefficients are obtained by matching the tangential electric and magnetic field intensity components at the boundary. This gives

$$\underline{E}_z^i|_{x=0} + \underline{E}_z^r|_{x=0} = \underline{E}_z^t|_{x=0}$$

$$1 + \Gamma_{\mathrm{TM}} = \tau_{\mathrm{TM}}, \tag{11.94}$$

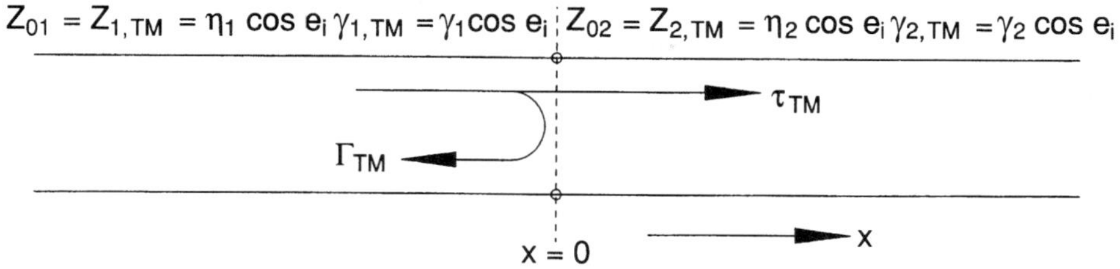

FIGURE 11.19 Two-transmission-line model for the reflection problem for TE waves.

and

$$\underline{H}^i_y|_{x=0} + \underline{H}^r_y|_{x=0} = \underline{H}^t_y|_{x=0}$$

$$-1 + \Gamma_{TM} = -\tau_{TM}\frac{\eta_1 \cos\theta_i}{\eta_2 \cos\theta_t}. \tag{11.95}$$

The two equations are solved to give

$$\Gamma_{TM} = \frac{\eta_2 \cos\theta_t - \eta_1 \cos\theta_i}{\eta_2 \cos\theta_t + \eta_1 \cos\theta_i}, \tag{11.96}$$

$$\tau_{TM} = \frac{2\eta_2 \cos\theta_t}{\eta_2 \cos\theta_t + \eta_1 \cos\theta_i}. \tag{11.97}$$

We can again obtain the same results from a transmission line model (Fig. 11.19). Using the same reasoning as in the TE wave, define a characteristic impedance and propagation constant as

$$Z_{01} = Z_{1,TM} = -\frac{\underline{E}^i_z}{\underline{H}^i_y} = \eta_1 \cos\theta_i, \quad \gamma_{1,TM} = \vec{a}_x \cdot \vec{\gamma}_i = \gamma_1 \cos\theta_i, \tag{11.98}$$

$$Z_{02} = Z_{2,TM} = -\frac{\underline{E}^t_z}{\underline{H}^t_y} = \eta_2 \cos\theta_t, \quad \gamma_{2,TM} = \vec{a}_x \cdot \vec{\gamma}_t = \gamma_2 \cos\theta_t. \tag{11.99}$$

Substituting in Eq. (11.79) gives the results of Eqs. (11.96) and (11.97).

11.6.4.1 Perfect Dielectric to Perfect Dielectric Boundary

We again note that this situation was studied in Section 11.6.1.1. When both media are perfect dielectrics we have

$$Z_{1,TM} = \sqrt{\frac{\mu_1}{\varepsilon_1}}\cos\theta_i, \quad Z_{2,TM} = \sqrt{\frac{\mu_2}{\varepsilon_2}}\cos\theta_t, \tag{11.100}$$

$$\cos\theta_t = \sqrt{1 - r^2 \sin^2\theta_i}, \quad r = \sqrt{\frac{\mu_1 \varepsilon_1}{\mu_2 \varepsilon_2}}.$$

For $r < 1$, Case 1, $\cos\theta_t$ is real for all values of θ_i. The impedance $Z_{2,\mathrm{TM}}$ is therefore real or resistive and time average power can be transmitted through both media in the x direction.

For $r > 1$, Case 2, $\cos\theta_t$ is real for angles of incidence less than the critical angle, but it is negative imaginary for incidence angles greater than the critical angle. The impedance is therefore

$$Z_{2,\mathrm{TM}} = \begin{cases} \sqrt{\mu_2/\varepsilon_2}\sqrt{1 - r^2\sin^2\theta_i}, & 0 \le \theta_i \le \theta_c, \quad (11.101) \\ -j\sqrt{\mu_2/\varepsilon_2}\sqrt{r^2\sin^2\theta_i - 1}, & \theta_c \le \theta_i \le \pi/2. \quad (11.102) \end{cases}$$

For the angle of incidence less than the critical angle $Z_{2,\mathrm{TM}}$ is real or resistive and hence time average power can be transmitted in the x direction. For the angle of incidence greater than the critical angle $Z_{2,\mathrm{TM}}$ is capacitive reactive and hence no time average power can be transmitted in the x direction.

Let us now calculate the time average power on both sides of the boundary. For $x < 0$ the incident and reflected waves add to give a total field

$$\underline{\vec{H}}_1 = \underline{\vec{H}}^i + \underline{\vec{H}}^r = -\vec{a}_y H_0[e^{-j(\beta_1\cos\theta_i)x} - \Gamma_{\mathrm{TM}}e^{j(\beta_1\cos\theta_i)x}]e^{-j(\beta_1\sin\theta_1)z}, \quad (11.103)$$

$$\underline{\vec{E}}_1 = \underline{\vec{E}}^i + \underline{\vec{E}}^r = \eta_1 H_0[-\vec{a}_x\sin\theta_i(e^{-j(\beta_1\cos\theta_i)x} - \Gamma_{\mathrm{TM}}e^{j(\beta_1\cos\theta_i)x})$$
$$+ \vec{a}_z\cos\theta_i(e^{-j(\beta_1\cos\theta_i)x} + \Gamma_{\mathrm{TM}}e^{j(\beta_1\cos\theta_i)x})]e^{-j(\beta_1\sin\theta_i)z}. \quad (11.104)$$

The time average power density for $x < 0$ is

$$\vec{p}_{f1,\mathrm{Avg}} = \tfrac{1}{2}\eta_1|H_0|^2\{\vec{a}_x\cos\theta_i(1 - |\Gamma_{\mathrm{TM}}|^2)$$
$$+ \vec{a}_z\sin\theta_i[1 + |\Gamma_{\mathrm{TM}}|^2 - 2|\Gamma_{\mathrm{TM}}|\cos(2\beta_1(\cos\theta_i)x + \phi)]\},$$

where $\Gamma_{\mathrm{TM}} = |\Gamma_{\mathrm{TM}}|\exp(j\phi)$. This expression is almost identical to the expression for the TE wave and its terms have the same significance.

For the transmitted wave $x > 0$, using Eqs. (11.92) and (11.93) with $\gamma_1 = j\beta_1$ and $\gamma_2 = j\beta_2$, the time average power density is

$$\vec{p}_{f2,\mathrm{Avg}} = \tfrac{1}{2}\eta_2|\tau_{\mathrm{TM}}|^2|H_0|^2\left(\frac{\eta_1\cos\theta_i}{\eta_2\cos\theta_t}\right)^2(\vec{a}_x\cos\theta_t + \vec{a}_z\sin\theta_t)$$

for Case 1 and for Case 2 with $\theta_i < \theta_c$. For Case 2 with $\theta_i > \theta_c$, it is

$$\vec{p}_{f2,\mathrm{Avg}} = \tfrac{1}{2}\eta_2|\tau_{\mathrm{TM}}|^2|H_0|^2\left(\frac{\eta_1\cos\theta_i}{\eta_2\cos\theta_t}\right)^2\vec{a}_z r\sin\theta_i\, e^{-2\omega(\sqrt{\mu_2\varepsilon_2}\sqrt{r^2\sin^2\theta_i - 1})x},$$

which shows there is no x-directed time average power, as we have already noted.

There is one other important result to obtain. The reflection coefficient will be zero when the two impedances in the transmission line model are equal, $Z_{1,\mathrm{TM}} = Z_{2,\mathrm{TM}}$ or $\eta_1\cos\theta_i = \eta_2\cos\theta_t$. The angle of incidence for which this occurs is known as the polarizing angle or the Brewster angle and we will denote this angle as $\theta_{iP,\,TM}$. The polarizing angle is obtained by squaring both sides of this equality, substituting for cosine in terms of the sine, and using Snell's law $\sin\theta_t = \sqrt{\mu_1\varepsilon_1/\mu_2\varepsilon_2}\sin\theta_i$:

$$\frac{\mu_1}{\varepsilon_1}(1-\sin^2\theta_{iP,\,TM}) = \frac{\mu_2}{\varepsilon_2}(1-\sin^2\theta_t)$$
$$= \frac{\mu_2}{\varepsilon_2}\left(1-\frac{\mu_1\varepsilon_1}{\mu_2\varepsilon_2}\sin^2\theta_{iP,\,TM}\right).$$

Solving for $\theta_{iP,\,TM}$ gives

$$\sin\theta_{iP,\,TM} = \sqrt{\frac{1-\mu_2\varepsilon_1/\mu_1\varepsilon_2}{1-(\varepsilon_1/\varepsilon_2)^2}} \tag{11.105}$$

for the polarizing angle. As long as the permittivities are unequal, the denominator is unequal to zero and the polarizing angle exists. In the usual situation the permeabilities are equal and then the polarizing angle is

$$\sin\theta_{iP,TM} = \sqrt{\frac{\varepsilon_2}{\varepsilon_1+\varepsilon_2}} \quad \text{or} \quad \tan\theta_{iP,TM} = \sqrt{\frac{\varepsilon_2}{\varepsilon_1}}. \tag{11.106}$$

At the polarizing angle $\tau_{TM} = 1$.

For the TE case we would get a polarizing angle when $Z_{1,TE} = Z_{2,TE}$ or $\eta_1/\cos\theta_i = \eta_2/\cos\theta_t$. Solving for the polarizing angle gives

$$\sin\theta_{iP,\,TE} = \sqrt{\frac{1-\mu_1\varepsilon_2/\mu_2\varepsilon_1}{1-(\mu_1/\mu_2)^2}}, \tag{11.107}$$

which is the dual of the formula for $\sin\theta_{iP,\,TM}$. In the usual case the permeabilities are equal, the denominator is zero, and therefore no polarizing angle can exist.

When a wave with mixed TE and TM polarization is incident on a boundary at the polarizing angle $\theta_{iP,\,TM}$, the reflected wave will consist only of a TE polarized wave: that is, the reflected wave will only have an electric field in the y direction. The reflected wave has therefore been polarized.

Example 11.6 (a) Calculate the polarizing angle for light in air striking dry soil ($\varepsilon_r = 4$). (b) Assume an incident wave at the polarizing angle containing equal powers of TE and TM waves. Calculate the relative powers in the reflected and transmitted waves.

Solution (a) The polarizing angle will be $\theta_{iP,\,TM} = \tan^{-1}\sqrt{(4\varepsilon_0/\varepsilon_0)} = 63.4°$.

(b) For the TM wave the reflection coefficient and the reflected power will be zero. The transmission coefficient will be unity and all the TM power will be transmitted at an angle of $\theta_t = \sin^{-1}(\sqrt{\tfrac{1}{4}}\sin 63.4°) = 26.6°$.

For the TE wave, $Z_{1,\text{TE}} = 377/\cos 63.4^\circ = 843\ \Omega$ and $Z_{2,\text{TE}} = (377/\sqrt{4})/\cos 26.6^\circ = 211\ \Omega$. The reflection and transmission coefficients are

$$\Gamma_{\text{TE}} = \frac{211 - 843}{211 + 843} = -0.6 \quad \text{and} \quad \tau_{\text{TE}} = \frac{2(211)}{211 + 843} = 0.4.$$

For an incident TE wave amplitude of E_0 the incident power is $\frac{1}{2}E_0^2/377$. The reflected power is $(0.6)^2 = 0.36$ times as much. The transmitted power is $\frac{1}{2}(0.4E_0)^2/(377/\sqrt{4})$, which is $(0.4)^2\sqrt{4} = 0.32$ times the incident power.

The reflected wave power is totally TE. Of the original incident power, half was TE and half was TM. The total transmitted power is therefore $0.5(0.32 + 1.0) = 0.66$ times the total incident power. The fraction that is TE is $0.5 \times 0.32/0.66 = 0.242$ and the fraction that is TM is $0.5 \times 1.0/0.66 = 0.758$.

Note that for the TE wave the sum of the reflected and transmitted powers does not equal the incident power. This equality holds only for the x component.

11.6.4.2 Perfect Dielectric to Conductor Boundary With the second medium a perfect conductor we have $Z_{2,\text{TM}} = 0$ and therefore $\Gamma_{\text{TM}} = -1$ and $\tau_{\text{TM}} = 0$. This gives

$$\underline{\vec{H}}_1 = -2\vec{a}_y H_0 \cos[(\beta_1 \cos\theta_i)x]\, e^{-j(\beta_1 \sin\theta_i)z}, \tag{11.108}$$

$$\begin{aligned}\underline{\vec{E}}_1 = 2\eta_1 H_0\{&-\vec{a}_x \sin\theta_i \cos[(\beta_1\cos\theta_i)x] \\ &- j\vec{a}_z \cos\theta_i \sin[(\beta_1 \cos\theta_i)x]\}\, e^{-j(\beta_1 \sin\theta_i)z},\end{aligned} \tag{11.109}$$

for the field in the dielectric region. The time average power density is

$$\vec{p}_{f1,\text{Avg}} = \vec{a}_z 2\eta_1 |H_0|^2 \sin\theta_i \cos^2[(\beta_1 \cos\theta_i)x]. \tag{11.110}$$

If the second medium is a good conductor the time average power density flowing into the conductor at the boundary is

$$\begin{aligned} p_{f2x,\text{Avg}}\big|_{x=0} &= \tfrac{1}{2}|H_{\text{tang}}|^2 R_s = \tfrac{1}{2}|-2H_0|^2\sqrt{\frac{\omega\mu_2}{2\sigma_2}} \\ &= 2|H_0|^2\sqrt{\frac{\omega\mu_2}{2\sigma_2}}. \end{aligned} \tag{11.111}$$

11.7 STRATIFIED MEDIA

Let us now look at the problem of N multiple dielectrics with plane boundaries normal to the x axis (Fig. 11.20). For simplicity, we will assume each of the dielectrics is lossless.

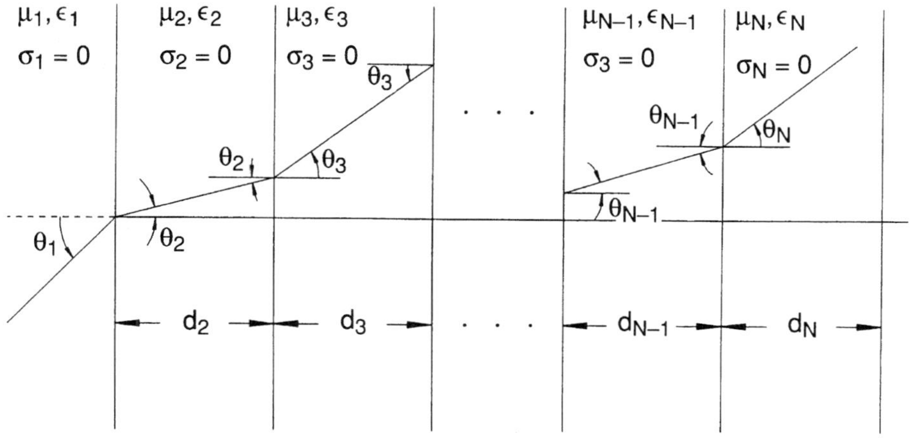

FIGURE 11.20 N multiple dielectrics with plane boundaries normal to the x axis.

The angles are found from Snell's law,

$$\sin\theta_2 = \sqrt{\frac{\mu_1\varepsilon_1}{\mu_2\varepsilon_2}}\sin\theta_1, \quad \sin\theta_3 = \sqrt{\frac{\mu_2\varepsilon_2}{\mu_3\varepsilon_3}}\sin\theta_2, \ldots,$$

$$\sin\theta_N = \sqrt{\frac{\mu_{N-1}\varepsilon_{N-1}}{\mu_N\varepsilon_N}}\sin\theta_{N-1}.$$

The overall relation between the incident angle and the finally transmitted angle is easily found to be

$$\sin\theta_N = \sqrt{\frac{\mu_1\varepsilon_1}{\mu_N\varepsilon_N}}\sin\theta_1.$$

Let us further assume that each angle is less than the critical angle at its boundary, so that all of the angles are real. That means that the x component of the phase constant in each medium is real,

$$\beta_{xn} = \beta_n\cos\theta_n = \omega\sqrt{\mu_n\varepsilon_n}\cos\theta_n,$$

and that the wave impedance in each medium is also real,

$$R_n = \begin{cases} \dfrac{\underline{E}_y^+}{\underline{H}_z^+} = \dfrac{\eta_n}{\cos\theta_n} = \sqrt{\dfrac{\mu_n}{\varepsilon_n}}\dfrac{1}{\cos\theta_n}, & \text{TE,} \\ -\dfrac{\underline{E}_z^+}{\underline{H}_y^+} = \eta_n\cos\theta n = \sqrt{\dfrac{\mu_n}{\varepsilon_n}}\cos\theta_n, & \text{TM.} \end{cases}$$

We can draw a transmission line model as shown in Fig. 11.21. At each point on the line the voltage and current waves are analogous to or model the transverse electric

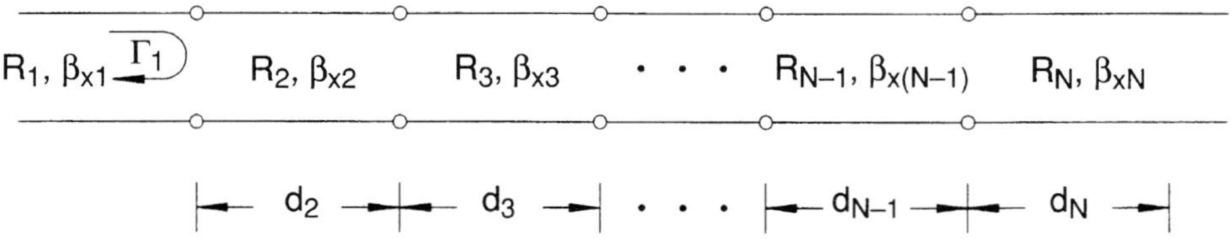

FIGURE 11.21 Transmission line model for the structure in Fig. 11.20.

and magnetic field components, respectively:

$$\left.\begin{array}{l}\underline{V}(x) \text{ models } \underline{E}_y(x)\\ \underline{I}(x) \text{ models } \underline{H}_z(x)\end{array}\right\}\text{TE,}\qquad \left.\begin{array}{l}\underline{V}(x) \text{ models } \underline{E}_z(x)\\ \underline{I}(x) \text{ models } \underline{H}_y(x)\end{array}\right\}\text{TM.}$$

We can now use the transmission line techniques of Chapter 10 to solve our problem. The impedance seen looking toward the right at each boundary, Z_n, is calculated by starting at the right end where $Z_N = R_N$ and using the impedance transformation Eq. (10.7) successively:

$$Z_{n-1} = R_{n-1}\frac{Z_n + jR_{n-1}\tan(\beta_{x,n-1}d_{n-1})}{R_{n-1} + jZ_n\tan(\beta_{x,n-1}d_{n-1})}, \quad n = N, N-1, \ldots, 3. \tag{11.112}$$

The calculation can also be done using the Smith chart. After obtaining Z_2 the reflection coefficient at the first boundary is found from

$$\Gamma_1 = \frac{Z_2 - R_1}{Z_2 + R_1}.$$

The time average power density flow in the x direction is

$$p_{fx,\text{Avg}} = \frac{|E_0|^2}{2R_1}(1 - |\Gamma_1|^2) = \frac{|E_0|^2}{2\eta_1}\cos\theta_1\,(1 - |\Gamma_1|^2), \quad \text{for TE waves}$$

and

$$p_{fx,\text{Avg}} = \tfrac{1}{2}R_1|H_0|^2(1 - |\Gamma_1|^2) = \tfrac{1}{2}\eta_1|H_0|^2\cos\theta_1\,(1 - |\Gamma_1|^2), \quad \text{for TM waves.}$$

Since the media are lossless, this is also the transmitted power in the output or Nth medium. Frequently this is all that we want to know and we can consider our problem as solved.

Example 11.7 A plane wave at frequency f_0 is incident on a plastic dielectric sheet with relative permittivity $\varepsilon_r = 9$ and with thickness of one-half wavelength (Fig. 11.22). Assume the incident wave is TE and calculate the reflection coefficient and fraction of transmitted power as a function of the angle of incidence θ_1.

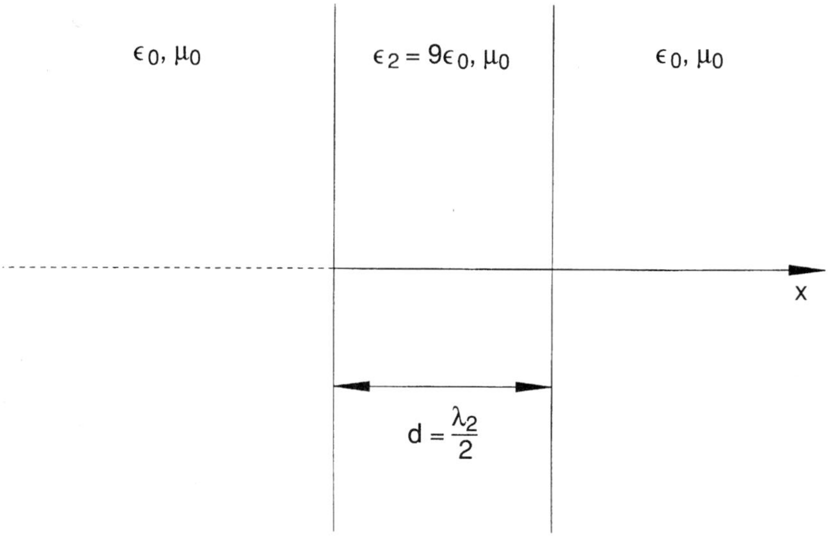

FIGURE 11.22 Structure in Example 11.7.

Solution The thickness of the plastic is $d = \lambda_0/2 = c/2\sqrt{\varepsilon_r} f_0$. The angle of transmission through the dielectric is $\sin\theta_2 = (\sin\theta_1)/3$ and the angle of transmission out the other side of the plastic is equal to the angle of incidence $\theta_3 = \theta_1$. For TE waves we have $R_1 = \eta_0/\cos\theta_1$, $R_2 = \eta_0/3\cos\theta_2$, and $R_3 = R_1$. The x-directed phase constant in the plastic sheet is $\beta_{x2} = 2\pi f_0\sqrt{\varepsilon_r}(\cos\theta_2)/c$ and hence $\beta_{x2}d = \pi\cos\theta_2$. The impedance seen at the first air–plastic boundary is

$$\begin{aligned} Z_2 &= R_2\frac{R_3 + jR_2\tan(\beta_{x2}d)}{R_2 + jR_3\tan(\beta_{x2}d)} \\ &= \left(\frac{\eta_0}{3\cos\theta_2}\right)\frac{(1/\cos\theta_1) + j(1/3\cos\theta_2)\tan(\pi\cos\theta_2)}{(1/3\cos\theta_2) + j(1/\cos\theta_1)\tan(\pi\cos\theta_2)}. \end{aligned}$$

We can use Z_2 to calculate Γ and then find the fractional transmitted power from $(1 - |\Gamma|^2)$. The fractional transmitted power is shown in Fig. 11.23.

Example 11.8 For the same geometry, assume normal incidence ($\theta_1 = 0$) and find the fraction of power transmitted as a function of frequency for the range $f_0/2 < f < 3f_0/2$.

Solution For normal incidence $\theta_1 = \theta_2 = \theta_3 = 0$. Then $R_1 = R_3 = \eta_0$ and $R_2 = \eta_0/3$. The phase constant in the plastic sheet is $\beta_{x2} = \beta_2 = 2\pi f\sqrt{\varepsilon_r}/c$ and $\beta_2 d = \pi(f/f_0)$. The impedance seen at the first air–plastic boundary is therefore

$$Z_2 = R_2\frac{R_3 + jR_2\tan(\beta_2 d)}{R_2 + jR_3\tan(\beta_2 d)} = \left(\frac{\eta_0}{3}\right)\frac{1 + j\frac{1}{3}\tan(\pi f/f_0)}{\frac{1}{3} + j\tan(\pi f/f_0)}.$$

The fractional transmitted power is shown in Fig. 11.24.

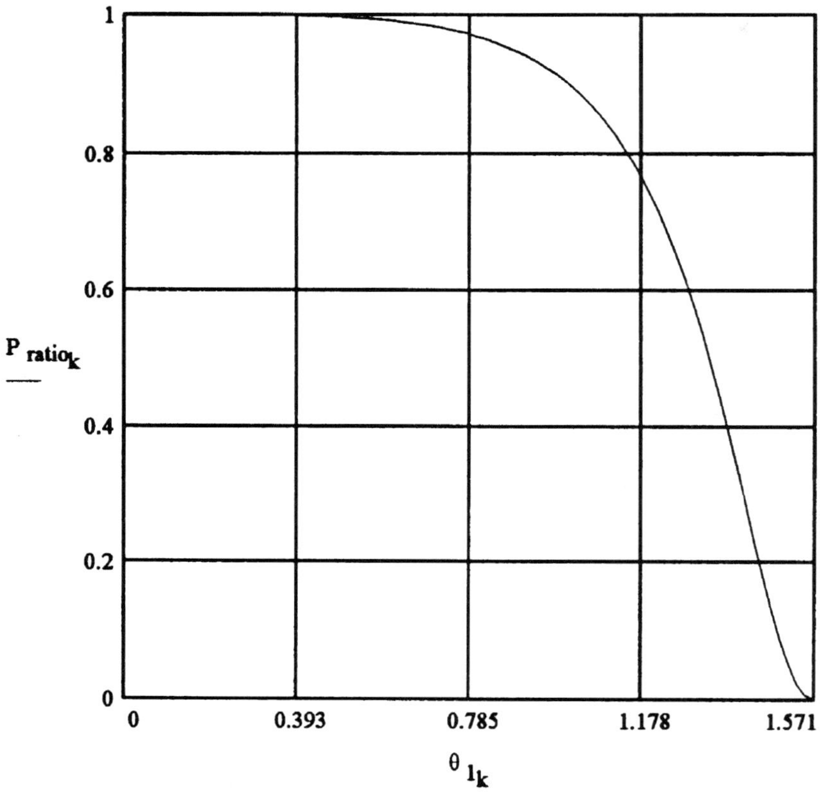

FIGURE 11.23 Fractional transmitted power in Example 11.7.

If we should want to know the forward and backward wave amplitudes in each section of the dielectric, we must do a little more work. Let us consider a typical section of the transmission line as shown in Fig. 11.25. The characteristic resistance is R_n, the phase constant is β_{xn}, and the length is d_n. $\underline{V}_{n+1}$ and $\underline{I}_{n+1}$ and $\underline{V}_n$ and $\underline{I}_n$ are the output ($x_n = 0$) end and the input ($x_n = -d_n$) end voltage and current, respectively. On the line we have

$$\underline{V}(x_n) = V_n^+ e^{-j\beta_{xn}x_n} + V_n^- e^{j\beta_{xn}x_n},$$

$$\underline{I}(x_n) = (V_n^+ e^{-j\beta_{xn}x_n} - V_n^- e^{j\beta_{xn}x_n})/R_n,$$

so that

$$\underline{V}_{n+1} = V_n^+ + V_n^-,$$

$$\underline{I}_{n+1} = (V_n^+ - V_n^-)/R_n.$$

Solving these gives

$$V_n^+ = (\underline{V}_{n+1} + R_n\underline{I}_{n+1})/2, \tag{11.113}$$

$$V_n^- = (\underline{V}_{n+1} - R_n\underline{I}_{n+1})/2, \tag{11.114}$$

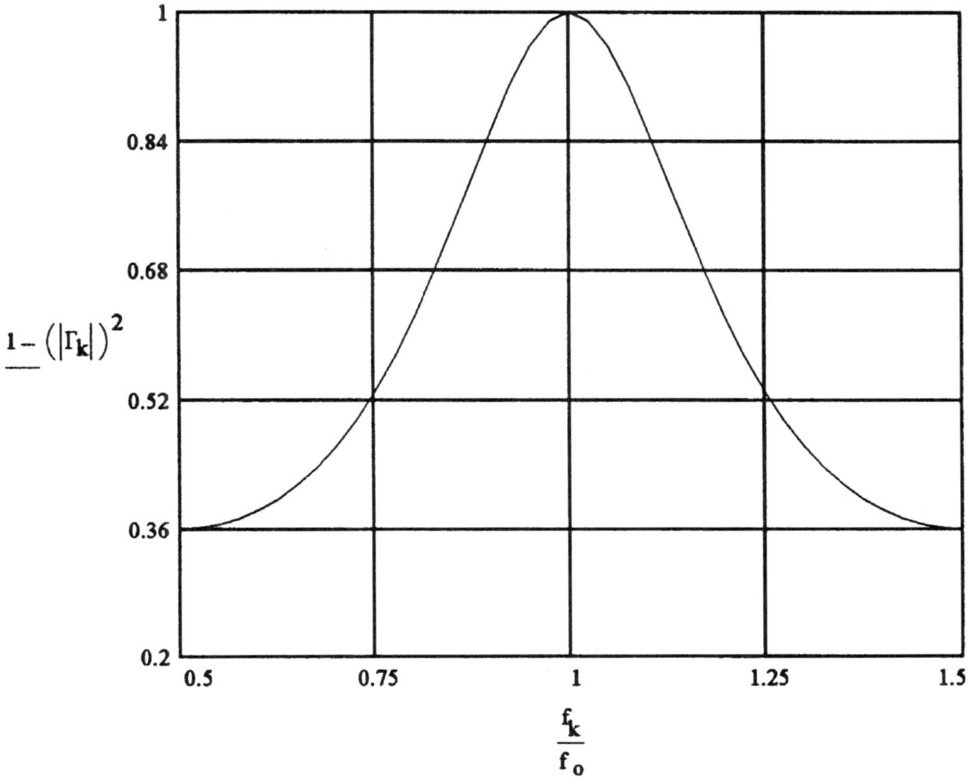

FIGURE 11.24 Fractional transmitted power in Example 11.8.

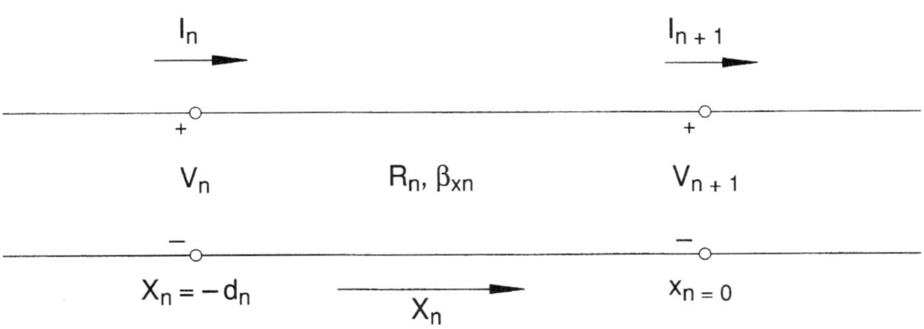

FIGURE 11.25 Typical section of transmission line.

from which we can obtain the input voltage and current

$$\underline{V}_n = V_n^+ e^{j\beta_{xn}d_n} + V_n^- e^{-j\beta_{xn}d_n}, \tag{11.115}$$

$$\underline{I}_n = (V_n^+ e^{j\beta_{xn}d_n} - V_n^- e^{-j\beta_{xn}d_n})/R_n. \tag{11.116}$$

We now solve by assuming we know the final output voltage, $\underline{V}_N = V_{N_0}$. Then $\underline{I}_N = V_{N_0}/R_N$. Using Eqs. (11.113) and (11.114) and then Eqs. (11.115) and (11.116), with $n = N - 1$, gives $\underline{V}_{N-1}$ and $\underline{I}_{N-1}$. The procedure is repeated until we obtain V_1^+ and I_1^+, which correspond to the incident transverse electric and magnetic fields.

Example 11.9 For the same geometry as in Example 11.7, assume an angle of incidence of 45° and an incident electric field 10 V/m at the first air–plastic boundary. Solve for the complete fields everywhere at $f = 1.2f_0$ with $f_0 = 300$ MHz.

Solution For the TE wave, $\theta_2 = \sin^{-1}[(1/\sqrt{9})\sin(\pi/4)] = 0.238$ rad or 13.6°. We then have $R_1 = R_3 = 377/\cos(\pi/4) = 533\,\Omega$, $R_2 = 377/\sqrt{9}\cos(0.238) = 129\,\Omega$, $\beta_{x1} = \beta_{x3} = (1.2f_0/c)\cos(\pi/4) = 0.849$ and $\beta_{x2} = (1.2f_0\sqrt{9}/c)\cos(0.238) = 3.50$. We also have $\beta_{x2}d = \pi(1.2)\cos(0.238) = 3.66$ rad, so that $\exp(j\beta_{2x}d) = -0.867 - j0.499$. Finally, the phase constant in the z direction is $\beta_z = (1.2f_0/c)\sin\pi/4) = (1.2\sqrt{9}f_0/c)\sin(0.238) = 0.849$ in all three media.

We start the problem at the output end by assuming $\underline{V}_3 = V_{N_0}$.
We have $\underline{I}_3 = V_{N_0}/R_3 = V_{N_0}/533$ and then Eqs. (11.113) and (11.114) give

$$V_2^+ = V_{N_0}(1 + 129/533)/2 = 0.621V_{N_0},$$

$$V_2^- = V_{N_0}(1 - 129/533)/2 = 0.379V_{N_0},$$

and Eqs. (11.115) and (11.116) give

$$\underline{V}_2 = V_{N_0}[0.621(-0.867 - j0.498) + 0.379(-0.867 + j0.498)]$$

$$= (-0.867 - j0.121)V_{N_0},$$

$$\underline{I}_2 = V_{N_0}[0.621(-0.867 - j0.498) - 0.379(-0.867 + j0.498)]/R_2$$

$$= (-0.210 - j0.499)V_{N_0}/129.$$

Again using Eqs. (1.113) and (11.114) gives

$$V_1^+ = [(-0.867 - j0.121) + (533/129)(-0.210 - j0.499)]V_{N_0}/2$$

$$= (-0.869 - j1.084)V_{N_0},$$

$$V_1^- = [(-0.867 - j0.121) - (533/129)(-0.210 - j0.499)]V_{N_0}/2$$

$$= (j0.964)V_{N_0}.$$

V_1^+ should correspond to the incident amplitude of 10 V/m. We can achieve this result by letting

$$V_{N_0} = 10/(-0.869 - j1.084) = -4.504 + j5.618.$$

This gives $V_1^+ = 10$, $V_1^- = -5.414 - j4.34$, $V_2^+ = -2.797 + j3.489$, and $V_2^- = -1.707 + j2.129$ V/m. The electric and magnetic field intensities in each region are therefore

$$\underline{E}_{y1} = [10e^{-j0.849x_1} + (-5.414 - j4.34)e^{j0.849x_1}]e^{-j0.849z},$$

$$\underline{H}_{z1} = [10e^{-j0.849x_1} - (-5.414 - j4.34)e^{j0.849x_1}]e^{-j0.849z}/R_1,$$

$$\underline{E}_{y2} = [(-2.797 + j3.489)e^{-j3.50x_2} + (-1.707 + j2.129)e^{j3.50x_2}]e^{-j0.849z},$$

$$\underline{H}_{z2} = [(-2.797 + j3.489)e^{-j3.50x_2} - (-1.707 + j2.129)e^{j3.50x_2}]e^{-j0.849z}/R_2,$$

$$\underline{E}_{y3} = (-4.504 + j5.618)e^{-0.849x_3}e^{-j0.849z},$$

$$\underline{H}_{z3} = (-4.504 + j5.618)e^{-\beta_{3x}x_3}e^{-j0.849z}/R_3.$$

11.8 PARALLEL PLATE WAVEGUIDE

Let us look carefully at the field in the dielectric region ($x < 0$) when the second medium ($x > 0$) is a perfect conductor. The electric and magnetic field intensities are given by Eqs. (11.84) and (11.85) for the TE wave and by Eqs. (11.108) and (11.109) for the TM wave. In each case we see that the component of the electric field that is tangential to the boundary, E_y for the TE wave and E_z for the TM wave, is proportional to $\sin[(\beta_1 \cos\theta_i)x]$, where $\beta_1 = \omega\sqrt{\mu_1\varepsilon_1}$. These field components are, of course, equal to zero at the boundary, $x = 0$. Due to the periodicity of the sine function they are also equal to zero at distances d_n back from the boundary, where

$$d_n = \frac{n\pi}{\beta_1 \cos\theta_i}, \quad n = 1, 2, 3, \dots .$$

If a second perfectly conducting plate were added in the region $x < -d_n$ then the same TE and TM waves would satisfy the boundary condition that $E_{\text{tang}} = 0$ at $x = 0$ and at $x = -d_n$. These TE and TM waves are each solutions to Maxwell's equations, which satisfy the boundary conditions at both conducting surfaces, and hence these waves will propagate between the two plates undisturbed by the addition of the second conducting surface.

Let us now start with two perfectly conducting plates separated by a fixed distance d of perfect dielectric medium characterized by μ and ε, as shown in Fig. 11.26. This is the geometry of a parallel plate waveguide. The waves we have been discussing must have

$$(\beta\cos\theta_i)d = n\pi, \quad \beta = \omega\sqrt{\mu\varepsilon}, \quad n = 1, 2, 3, \dots, \tag{11.117}$$

in order to exist between the plates. This means that $\beta\cos\theta_i = n\pi/d$ and we can define a new z-directed phase constant β_n such that

$$\begin{aligned} \beta_n &= \beta\sin\theta_i = \beta\sqrt{1 - \cos^2\theta_i} \\ &= \sqrt{\omega^2\mu\varepsilon - \left(\frac{n\pi}{d}\right)^2}, \quad n = 1, 2, 3, \dots . \end{aligned} \tag{11.118}$$

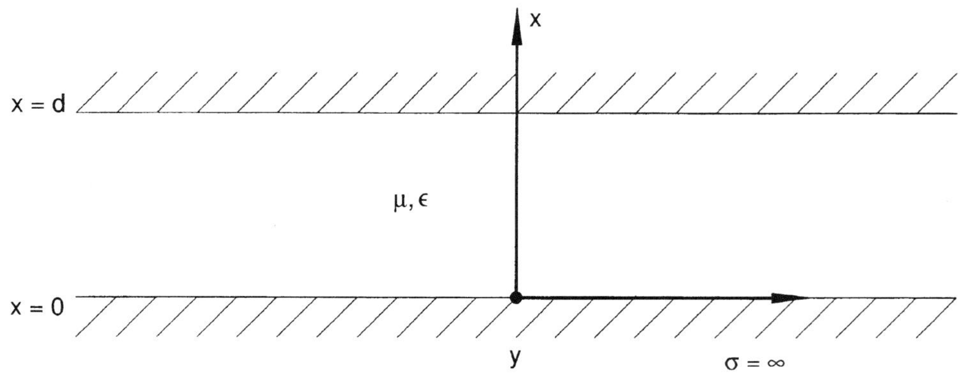

FIGURE 11.26 Parallel plate waveguide.

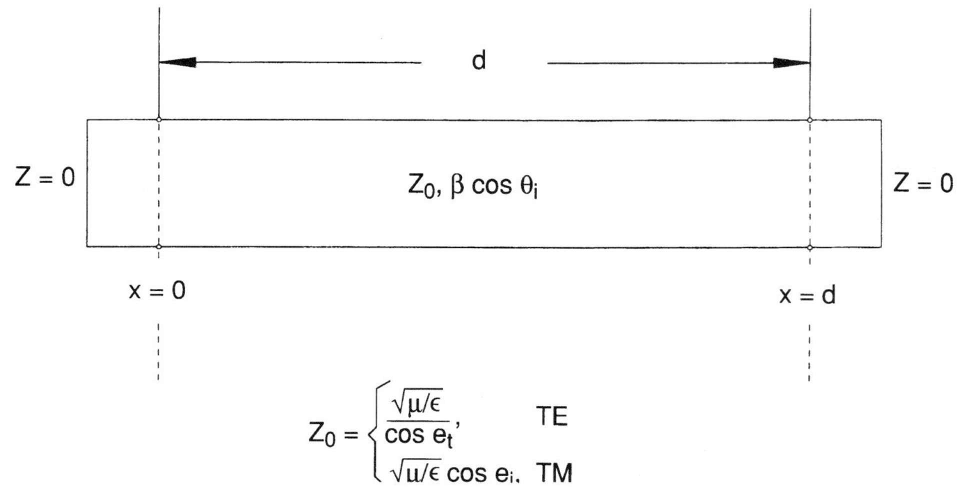

FIGURE 11.27 Transmission line of length d shorted at both ends representing perfectly conducting planes of a parallel plate wave guide.

The same results can be obtained from a different viewpoint. What we have found is a resonance condition in the x direction. The results can also be obtained from the transmission line models. Figure 11.27 shows a transmission line of length d shorted at both ends to represent the perfectly conducting plates. The transmission line's x-directed phase constant is $\beta \cos \theta_i$ for the TE and TM waves and the line impedance Z_0 is given by $\eta/\cos \theta_i$, Eq. (11.77), for the TE wave and by $\eta \cos \theta_i$, Eq. (11.100), for the TM wave. For x direction resonance, the sum of the impedance (or the admittance) looking to the right plus the impedance (or the admittance) to the left at any point on the line must be zero.†

† This condition actually gives the natural frequencies. In the lossess case, which we are studying, the natural frequencies equal the resonant frequencies.

Most simply, if we add the impedances looking to the right and left at the left short (see the dashed line in the figure) we have

$$Z_{\text{left}} + Z_{\text{right}} = 0$$

$$0 + jZ_0 \tan\left[(\beta \cos\theta_i)d\right] = 0.$$

The zeros of the tangent occur when the argument equals $n\pi$. Hence we get $(\beta \cos\theta_i)d = n\pi$, which is the same result as Eq. (11.117).

The electric field for the TE wave, Eq. (11.84), becomes

$$\vec{\underline{E}}_n = \vec{a}_y E_{0n} \sin\left(\frac{n\pi x}{d}\right) e^{-j\beta_n z}, \quad n = 1, 2, 3, \ldots . \tag{11.119}$$

The corresponding magnetic field can either be found by similarly modifying Eq. (11.85) or, more simply, from $-j\omega\mu\underline{\vec{H}} = \nabla \times \underline{\vec{E}}$, giving

$$\vec{\underline{H}}_n = \frac{E_{0n}}{\omega\mu}\left[-\vec{a}_x \beta_n \sin\left(\frac{n\pi x}{d}\right) + \vec{a}_z j\left(\frac{n\pi}{d}\right)\cos\left(\frac{n\pi x}{d}\right)\right] e^{-j\beta_n z}. \tag{11.120}$$

We see that there are an infinite number of transverse electric wave solutions to Maxwell's equations and the boundary conditions. Each solution is called a mode and these are called the TE_n modes for a parallel plate waveguide.

We can similarly substitute for $\beta \cos\theta_i$ and $\beta \sin\theta_i$ in Eq. (11.108) to obtain the magnetic field for the TM wave:

$$\vec{\underline{H}}_n = \vec{a}_y H_{0n} \cos\left(\frac{n\pi x}{d}\right) e^{-j\beta_n z}, \quad n = 1, 2, 3, \ldots . \tag{11.121}$$

The corresponding electric field can either be obtained by similarly modifying Eq. (11.109) or, more simply, from $j\omega\varepsilon\underline{\vec{E}} = \nabla \times \underline{\vec{H}}$, giving

$$\vec{\underline{E}}_n = \frac{H_{0n}}{\omega\varepsilon}\left[\vec{a}_x \beta_n \cos\left(\frac{n\pi x}{d}\right) + \vec{a}_z j\left(\frac{n\pi}{d}\right)\sin\left(\frac{n\pi x}{d}\right)\right] e^{-j\beta_n z}. \tag{11.122}$$

We see that there are an infinite number of transverse magnetic wave solutions to Maxwell's equations and the boundary conditions. These are called the TM_n modes for a parallel plate waveguide.

The physical fields are obtained from these phasor fields, as always, by multiplying by $\exp(j\omega t)$ and taking the real part of the product. Every field component will then have time appear as $\cos(\omega t - \beta_n z + \phi)$. The phase of the wave is $\Phi = \omega t - \beta_n z + \phi$ and β_n, Eq. (11.118), is the phase constant of the nth mode. This equation shows that

β_n is zero at an angular frequency ω_{cn} such that $\omega_{cn}\sqrt{\mu\varepsilon} = n\pi/d$. Thus

$$\omega_{cn} = \frac{n\pi}{d\sqrt{\mu\varepsilon}} = \frac{n\pi c}{d}, \quad f_{cn} = \frac{nc}{2d}, \quad c = \frac{1}{\sqrt{\mu\varepsilon}}, \tag{11.123}$$

where c is the velocity of light in the dielectric medium. When $\omega > \omega_{cn}$ then β_n is real and the TE_n or TM_n wave is a plane unattenuated wave traveling in the z direction.

Conversely, when $\omega < \omega_{cn}$ then β_n is imaginary. The form of the TE and TM solutions changes and we can obtain the new form by replacing $j\beta_n$ with α_n, where

$$\alpha_n = \sqrt{\left(\frac{n\pi}{d}\right)^2 - \omega^2\mu\varepsilon} \tag{11.124}$$

is the attenuation constant of the nth mode. The TE_n solutions are then

$$\underline{\vec{E}}_n = \vec{a}_y E_{0n} \sin\left(\frac{n\pi x}{d}\right) e^{-\alpha_n z}, \tag{11.125}$$

$$\underline{\vec{H}}_n = \frac{E_{0n}}{j\omega\mu}\left[-\vec{a}_x\alpha_n \sin\left(\frac{n\pi x}{d}\right) - \vec{a}_z\left(\frac{n\pi}{d}\right)\cos\left(\frac{n\pi x}{d}\right)\right]e^{-\alpha_n z}. \tag{11.126}$$

The TM_n solutions are

$$\underline{\vec{H}}_n = \vec{a}_y H_{0n} \cos\left(\frac{n\pi x}{d}\right) e^{-\alpha_n z}, \tag{11.127}$$

$$\underline{\vec{E}}_n = \frac{H_{0n}}{j\omega\varepsilon}\left[\vec{a}_x\alpha_n \cos\left(\frac{n\pi x}{d}\right) - \vec{a}_z\left(\frac{n\pi}{d}\right)\sin\left(\frac{n\pi x}{d}\right)\right]e^{-\alpha_n z}. \tag{11.128}$$

These waves attenuate in the z direction. With perfectly conducting boundaries and a perfect dielectric medium, the attenuation cannot be due to power loss. It therefore follows that these waves carry no time average power. They are evanescent waves.

We have come to behavior that is new and different from anything we have seen before. The TEM wave on the two-conductor transmission lines, which we studied in Chapter 9, are traveling waves at any frequency. They are not evanescent waves at any frequency. For the TE and TM waves, which we are now studying, when the frequency is below f_{cn} then the nth mode is evanescent or cutoff, meaning that it carries no time average power. When the frequency is greater than f_{cn} then the wave is a traveling wave and, we shall see, carries time average power in the z direction. The frequency at which the transition occurs, f_{cn}, is called the cutoff frequency of the nth mode.

The mode with the lowest cutoff frequency is called the dominant mode. The cutoff frequency for the TE_1 and the TM_1 modes is lower than the cutoff frequency for any other TE or TM modes, but the TEM mode, which we studied in the previous chapter, is the dominant mode for the parallel plate waveguide because it propagates at all frequencies down to zero. When there are two or more modes with the same cutoff frequency, those modes are called degenerate modes. All of the modes in the parallel plate waveguide are degenerate because the TE_n and the TM_n modes have the same cutoff frequency.

Let us think for a moment about the currents that flow on the surfaces of the plates. For the TE modes the tangential magnetic field intensity is in the z direction at the plates' surfaces. The surface current is normal to the magnetic field intensity and hence flows in the y direction. For the TM modes (and for the TEM mode) the tangential magnetic field intensity is in the y direction at the plates' surfaces and the surface current is in the z direction. We could construct our plates with thin laminated conducting sheets separated by insulating varnish so that current can flow along the sheets but not perpendicular to them. In this way we make a directional conductor, with perfect (or very high) conductivity along the sheets and zero (or very low) conductivity perpendicular to the sheets. If the sheets were in the y direction (Fig. 11.28) then the surface currents required by the TE modes could flow freely, but the surface currents required by the TEM and TM modes could not flow. As a consequence, the TEM and TM waves would be suppressed, the TE waves would not be changed, there would be no degenerate modes, and the TE_1 mode would be dominant mode on this structure.

The phase velocity of the nth mode is

$$v_{\text{ph},n} = \frac{\omega}{\beta_n} = \frac{\omega}{\sqrt{\omega^2\mu\varepsilon - (n\pi/d)^2}}$$

$$= \frac{c}{\sqrt{1-(f_{cn}/f)^2}}, \quad f > f_{cn}. \tag{11.129}$$

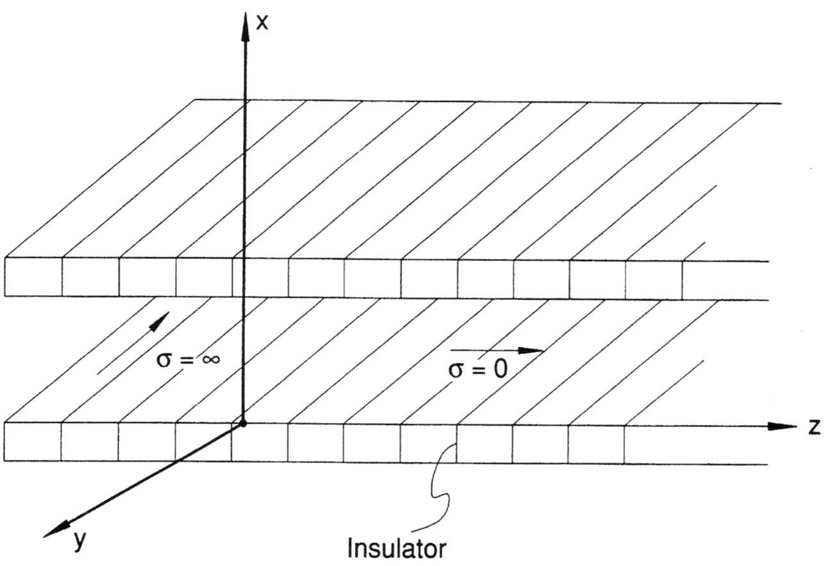

FIGURE 11.28 Parallel plate waveguide with thin laminated conductive sheets.

The phase velocity is infinite at the cutoff frequency and approaches the speed of light in the medium as frequency increases (Fig. 11.29). Recall that this velocity is only the velocity of a mathematical quantity, the phase, and not the velocity of information or energy flow.

The group velocity of the nth mode is

$$v_{g,n} = \left(\frac{\partial \beta_n}{\partial \omega}\right)^{-1} = \left(\frac{\omega\mu\varepsilon}{\sqrt{\omega^2\mu\varepsilon - (n\pi/d)^2}}\right)^{-1}$$

$$= c\sqrt{1-(f_{cn}/f)^2}, \quad f > f_{cn}. \tag{11.130}$$

The group velocity is zero at the cutoff frequency and approaches the speed of light in the medium as frequency increases (Fig. 11.29). Recall that the group velocity,

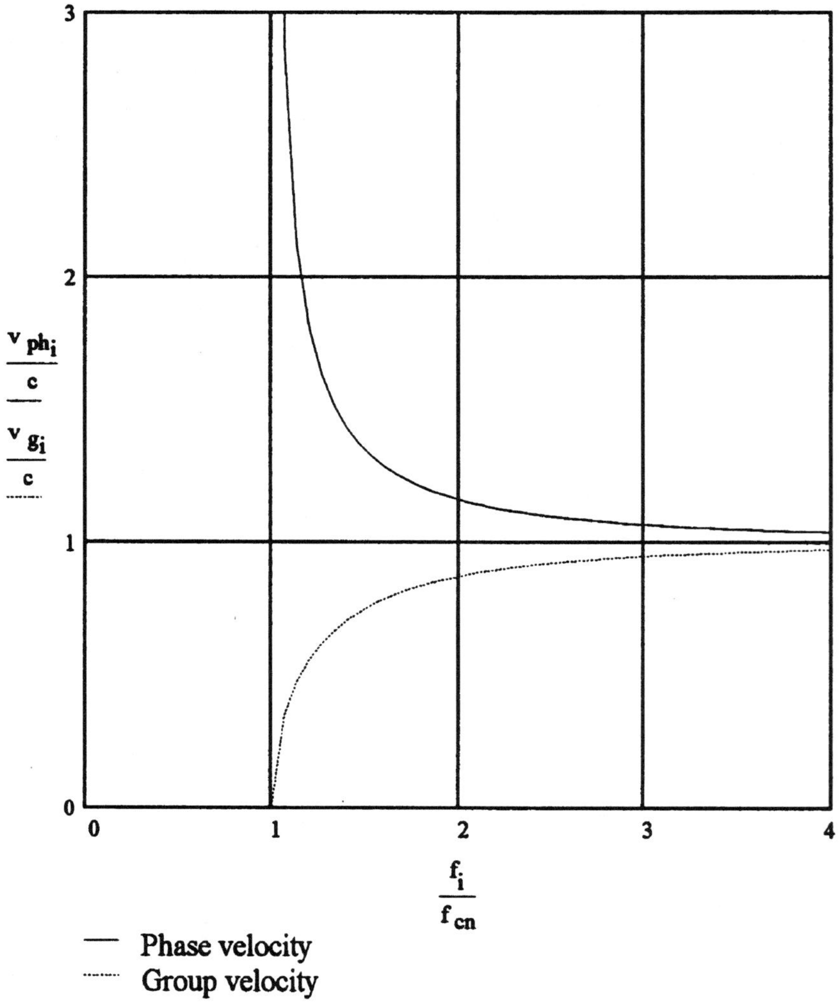

FIGURE 11.29 Plot of phase and group velocities.

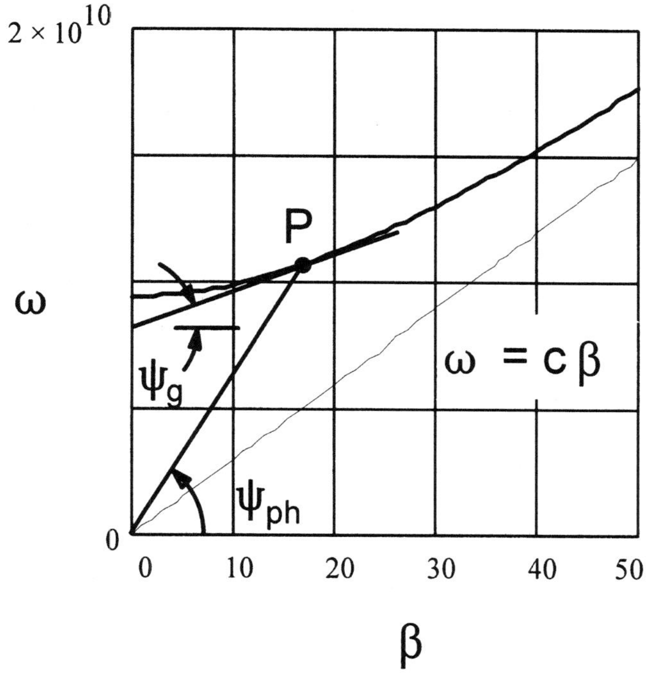

FIGURE 11.30 Parallel plate waveguide dispersion diagram.

subject to certain generally satisfied conditions, is the velocity of information flow. Equations (11.129) and (11.130) show that in the parallel plate waveguide $v_{g,n}v_{ph,n} = c^2$.

These relations are frequently illustrated graphically with an ω–β or dispersion diagram, which plots angular frequency versus the phase constant. Figure 11.30 is the ω–β diagram for the parallel plate waveguide. If we choose any point on a curve, say, point P in the figure, then the slope of the line drawn from the origin to the point is the phase velocity at P: $\tan\psi_{ph} = v_{ph}$. Also, the slope of the tangent to the curve is the group velocity at P: $\tan\psi_g = v_g$.

There will be no phase dispersion if the group velocity is constant over the frequency band of the signal. Figure 11.30 shows that for a bandwidth $\Delta\omega$ well above the cutoff frequency the slope of the ω–β diagram is approximately constant, giving little dispersive distortion.

We can define a mode impedance for each mode as the ratio of the transverse electric field intensity to the transverse magnetic intensity. For the TE_n mode this gives

$$Z_{\text{TE}_n} = -\frac{\underline{E}_y}{\underline{H}_x} = \begin{cases} \dfrac{j\omega\mu}{\alpha_n}, & f < f_{cn}, \\ \dfrac{\omega\mu}{\beta_n}, & f > f_{cn}, \end{cases} \tag{11.131}$$

and for the TM_n mode this gives

$$Z_{\mathrm{TM}_n} = \frac{E_x}{H_y} = \begin{cases} \dfrac{\alpha_n}{j\omega\varepsilon}, & f < f_{cn}, \\ \dfrac{\beta_n}{\omega\varepsilon}, & f > f_{cn}. \end{cases} \tag{11.132}$$

The mode impedance is purely reactive (inductive for TE and capacitive for TM modes) when the wave is cutoff, consistent with not carrying time average power. The mode impedance is resistive when it is propagating, which is consistent with carrying time average power.

11.8.1 Power and Attenuation

Let us now calculate time average power carried by the propagating modes and their attenuation due to imperfectly conducting walls.

We start with a TE_n mode with $f > f_{cn}$. The time average power density (W/m^2) is

$$\vec{p}_{f,\,\mathrm{Avg,\,TE}_n} = \tfrac{1}{2}\mathrm{Re}(\underline{\vec{E}} \times \underline{\vec{H}}^*) = \vec{a}_z \tfrac{1}{2}|E_{0n}|^2 \left(\frac{\beta_n}{\omega\mu}\right) \sin^2\left(\frac{n\pi x}{d}\right).$$

The time average power through the transverse cross section of the guide, for a width w, is

$$p_{f,\,\mathrm{Avg,\,TE}_n} = \int_{y=0}^{w} \int_{x=0}^{d} \vec{p}_{f,\,\mathrm{Avg,\,TE}_n} \cdot \vec{a}_z \, dx\, dy = \tfrac{1}{2}|E_{0n}|^2 \left(\frac{\beta_n}{\omega\mu}\right)\left(\frac{dw}{2}\right). \tag{11.133}$$

The time average power loss density (W/m^2) in the imperfectly conducting walls is

$$\begin{aligned} p_{l,\,\mathrm{Avg,\,TE}_n} &= \tfrac{1}{2}[|H_{\mathrm{tang}}|_{x=0}^2 + |H_{\mathrm{tang}}|_{x=d}^2] R_s \\ &= \frac{1}{2}\left(\frac{|E_{0n}|}{\omega\mu}\right)^2 \left\{ \left[\left(\frac{n\pi}{d}\right)^2 \cos^2\left(\frac{n\pi x}{d}\right)\right]_{x=0} + \left[\left(\frac{n\pi}{d}\right)^2 \cos^2\left(\frac{n\pi x}{d}\right)\right]_{x=d} \right\} R_s \\ &= \frac{1}{2}\left(\frac{|E_{0n}|}{\omega\mu}\right)^2 \left\{ 2\left(\frac{n\pi}{d}\right)^2 \right\} R_s. \end{aligned}$$

For a width w of the wall on each side of the guide the time average power loss per unit length (W/m) in the z direction is

$$\begin{aligned} P_{l,\,\mathrm{Avg,\,TE}_n} &= \int_{y=0}^{w} p_{l,\,\mathrm{Avg,TE}_n}\, dy \\ &= \frac{1}{2}\left(\frac{|E_{0n}|}{\omega\mu}\right)^2 \left[2\left(\frac{n\pi}{d}\right)^2\right] w R_s. \end{aligned} \tag{11.134}$$

If we let P_0 be the time average power through a width w of the guide at $z = 0$, then the time average power through a width w of the guide at any other z is

$P(z) = P_0 \exp(-2\alpha_{cn} z)$, where α_{cn} is the attenuation of the nth mode field component due to conduction losses in the walls. The derivative of this with respect to z is the rate of increase of power with distance, which is the negative of the power loss per unit length

$$\frac{dP(z)}{dz} = -2\alpha_{cn} P(z) = -P_l(z)$$

and therefore

$$\alpha_{cn} = \frac{P_l(z)}{2P(z)}. \tag{11.135}$$

Substituting from the above results gives

$$\alpha_{cn} = \frac{P_{l,\,\text{Avg, TE}_n}}{2P_{f,\,\text{Avg, TE}_n}} = \frac{2}{\omega\mu d\beta_n}\left(\frac{n\pi}{d}\right)^2 R_s, \quad R_s\sqrt{\frac{\omega\mu}{2\sigma_c}} \quad (\text{TE}_n). \tag{11.136}$$

This equation is plotted for the $n = 1$ mode in Fig. 11.31. The mode is highly attenuated at frequency just greater than the cutoff frequency. The attenuation decreases rapidly for increasing frequency and falls off as $f^{-3/2}$ for high frequency.

We will now find the corresponding formula for the TM_n mode with $f > f_{cn}$. The time average power density (W/m^2) is

$$\vec{p}_{f,\,\text{Avg, TM}_n} = \vec{a}_z \tfrac{1}{2}|H_{0n}|^2\left(\frac{\beta_n}{\omega\varepsilon}\right)\cos^2\left(\frac{n\pi x}{d}\right).$$

The time average power through the transverse cross section of the guide, for a width w, is

$$P_{f,\,\text{Avg, TM}_n} = \tfrac{1}{2}|H_{0n}|^2\left(\frac{\beta_n}{\omega\varepsilon}\right)\left(\frac{dw}{2}\right). \tag{11.137}$$

The time average power loss density (W/m^2) in the imperfectly conducting walls is

$$\begin{aligned} p_{l,\,\text{Avg,TE}_n} &= \tfrac{1}{2}|H_{0n}|^2\left[\cos^2\left(\frac{n\pi x}{d}\right)_{x=0} + \cos^2\left(\frac{n\pi x}{d}\right)_{x=d}\right]R_s \\ &= \tfrac{1}{2}|H_{0n}|^2(2)R_s. \end{aligned}$$

For a width w of the wall on each side of the guide the time average power loss per unit length (W/m) in the z direction is

$$P_{l,\,\text{Avg, TM}_n} = \tfrac{1}{2}|H_{0n}|^2(2w)R_s.$$

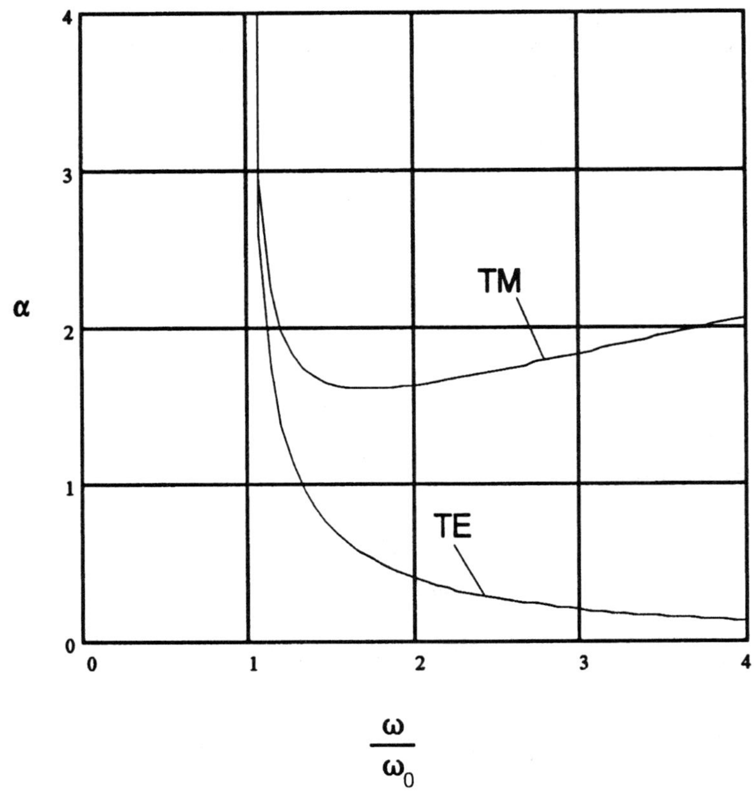

FIGURE 11.31 Parallel plate waveguide normalized attenuation.

Substituting these in Eq. (11.135) we obtain the attenuation constant due to imperfectly conducting walls for the TM_n mode,

$$\alpha_{cn} = \frac{P_{l,\,\text{Avg},\,\text{TM}_n}}{2P_{f,\,\text{Avg},\,\text{TM}_n}} = \frac{2\omega\varepsilon}{d\beta_n} R_s, \quad R_s = \sqrt{\frac{\omega\mu}{2\sigma_c}} \quad (\text{TM}_n). \tag{11.138}$$

This equation is also plotted for $n = 1$ in Fig. 11.31, which shows that this mode is also highly attenuated at frequency just greater than the cutoff frequency. The attenuation decreases rapidly for increasing frequency but then rises as $f^{1/2}$ for high frequency. Clearly, at high frequency it is desirable to use the TE mode rather than the TM mode, particularly for long-distance transmission, because of its lower attenuation.

11.9 SYMMETRIC DIELECTRIC SLAB WAVEGUIDE

In Section 11.6.1.1 we found that when a wave was incident from a more dense dielectric medium, characterized by μ_1 and ε_1, onto a less dense dielectric medium, characterized by μ_2 and ε_2, and when the angle of incidence was greater than the critical angle, then all of the incident power was reflected at the boundary. Let us add a third dielectric medium characterized by ε_3 and μ_3, with $\mu_3\varepsilon_3 < \mu_1\varepsilon_1$, in the region $x < -d$ (Fig. 11.32). The reflected wave from the boundary at $x = 0$ is incident on the added boundary at $x = -d$. If its angle of incidence is greater than the critical angle at this boundary, then it will be re-reflected without loss of power. Since no power escapes from the core region, the power must be guided in the z direction. This process is like the one we have just studied for a wave between parallel conducting plates.

Figure 11.32 shows an asymmetric dielectric slab waveguide. We will solve the somewhat simpler problem of the symmetric dielectric slab waveguide for which the second and third dielectric media are identical. We will simplify the problem a little further by examining the usual situation for which all the permeabilities are equal to the permeability of free space. Hence for the core we have μ_0, ε_1 and for the two external dielectrics we have μ_0, ε_2.

Before we derive the modal equation let us recapitulate what we already know about the propagation constants in each region. In the dielectric slab and in the two external regions the z-directed phase constant is the same, so we can use $\beta_z \equiv \beta_{1z} = \beta_{2z}$. In the slab the propagation constant is

$$\vec{\gamma}_1 = \vec{a}_x j\beta_{1x} + \vec{a}_z j\beta_z, \quad \vec{\gamma}_1 \cdot \vec{\gamma}_1 = -\omega^2\mu_0\varepsilon_1 = -\beta_{1x}^2 - \beta_z^2, \tag{11.139}$$

and for the two external regions it is

$$\vec{\gamma}_2 = \vec{a}_x \alpha_{2x} + \vec{a}_z j\beta_z, \quad \vec{\gamma}_2 \cdot \vec{\gamma}_2 = -\omega^2\mu_0\varepsilon_2 = \alpha_{2x}^2 - \beta_z^2. \tag{11.140}$$

From the first equation we get

$$\beta_z = \sqrt{\omega^2\mu_0\varepsilon_1 - \beta_{1x}^2} \tag{11.141}$$

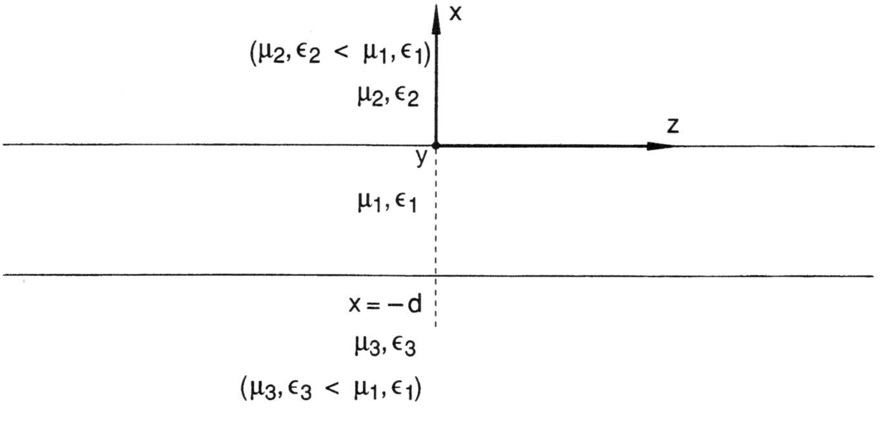

FIGURE 11.32 Asymmetric dielectric slab waveguide.

and from the second we get

$$\alpha \equiv \alpha_{2x} = \sqrt{\beta_z^2 - \omega^2 \mu_0 \varepsilon_2} = \sqrt{\omega^2 \mu_0 (\varepsilon_1 - \varepsilon_2) - \beta_{1x}^2}. \qquad (11.142)$$

We have written β_z and $\alpha \equiv \alpha_{2x}$ in terms of the parameters of the media and β_{1x}. We must now find an equation for β_{1x}.

Figure 11.33 shows the transmission line model for the x-directed propagation. We have found that for angles of incidence greater than the critical angle, the characteristic impedance of the lines to the right of $x = 0$ and to the left of $x = -d$ are pure imaginary, while the characteristic impedance of the center section or core is real. Thus $Z_{01} = R_{01}$ and $Z_{02} = jX_{02}$. For x direction resonance, the sum of the impedance (or the admittance) looking to the right plus the impedance (or the admittance) to the left at any point on the line must be zero. Let us choose the midpoint of the line $x = -d/2$ (shown dashed in the figure) as the place to apply this formula. Then, letting $Z_{\text{in, right}}$ be the impedance looking to the right, we have

$$Z_{\text{in, right}} = Z_{01} \frac{Z_{02} + jZ_{01} \tan(\beta_{1x} d/2)}{Z_{01} + jZ_{02} \tan(\beta_{1x} d/2)}$$

$$Z_{\text{in, right}} = jR_{01} \frac{X_{02} + R_{01} \tan(\beta_{1x} d/2)}{R_{01} - X_{02} \tan(\beta_{1x} d/2)}. \qquad (11.143)$$

Because of symmetry, the impedance to the left is identical to this—$Z_{\text{in, left}} = Z_{\text{in, right}}$—and the sum of the two impedances is therefore $2Z_{\text{in, right}}$. For $Z_{\text{in, right}} = 0$, we set its numerator to zero and obtain

$$-\cot\left(\frac{\beta_{1x} d}{2}\right) = \frac{R_{01}}{X_{02}}. \qquad (11.144)$$

The admittance seen looking to the right (and left) is $Y_{\text{in, right}} = 1/Z_{\text{in, right}}$ and the sum of the two admittances is $2Y_{\text{in, right}}$. For this to be zero, we set the denominator of

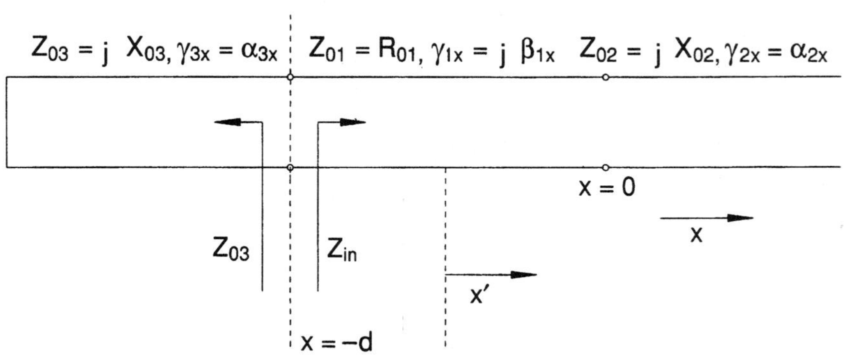

FIGURE 11.33 Transmission line model for the x-directed propagation.

Eq. (11.143) to zero and obtain

$$\tan\left(\frac{\beta_{1x}d}{2}\right) = \frac{R_{01}}{X_{02}}. \tag{11.145}$$

Equations (11.144) and (11.145) are the modal equations for the symmetric dielectric slab waveguide. We must solve these equations for β_{1x}. Unfortunately this term is hidden in the right side as well as being shown explicitly in the left. In order to solve these equations we must distinguish between TE and TM Waves.

11.9.1 Transverse Electric (TE) Wave Solutions

From Eqs. (11.77) and (11.78) for TE waves we have

$$R_{01} = R_{1,\mathrm{TE}} = \frac{\eta_1}{\cos\theta_i} = \frac{\sqrt{\mu_0/\varepsilon_1}}{\beta_{1x}/\omega\sqrt{\mu_0\varepsilon_1}} = \frac{\omega\mu_0}{\beta_{1x}}, \tag{11.146}$$

$$X_{02} = X_{2,\mathrm{TE}} = \frac{\sqrt{\mu_0/\varepsilon_2}}{\alpha_{2x}/\omega\sqrt{\mu_0\varepsilon_2}} = \frac{\omega\mu_0}{\alpha_{2x}}, \tag{11.147}$$

which, substituted in Eqs. (11.144) and (11.145), gives

$$-\cot\left(\frac{\beta_{1x}d}{2}\right) = \frac{\alpha}{\beta_{1x}} = \frac{\sqrt{\omega^2\mu_0(\varepsilon_1-\varepsilon_2)d^2-(\beta_{1x}d)^2}}{\beta_{1x}d} \tag{11.148}$$

$$-\cot\left(\frac{\beta_{1x}d}{2}\right) = \frac{\sqrt{C^2-(\beta_{1x}d)^2}}{\beta_{1x}d}, \quad C = \omega\sqrt{\mu_0(\varepsilon_1-\varepsilon_2)}\,d \tag{11.149}$$

and

$$\tan\left(\frac{\beta_{1x}d}{2}\right) = \frac{\alpha}{\beta_{1x}} \tag{11.150}$$

$$\tan\left(\frac{\beta_{1x}d}{2}\right) = \frac{\sqrt{C^2-(\beta_{1x}d)^2}}{\beta_{1x}d}, \tag{11.151}$$

respectively, as the modal equations. Note that the right sides of the two equations are the same.

These are transcendental equations and cannot be solved explicitly for $\beta_{1x}d$. Solutions can be found by numerical or graphical methods. We will use a graphical technique because the graph will help us understand the results. We plot both the left and right sides of the modal equations as a function of $\beta_{1x}d$. Wherever the two plots cross we have a solution.

The negative cotangent function and the tangent on the left side are plotted in Fig. 11.34 as solid lines. To actually plot the right side we must choose values for C.

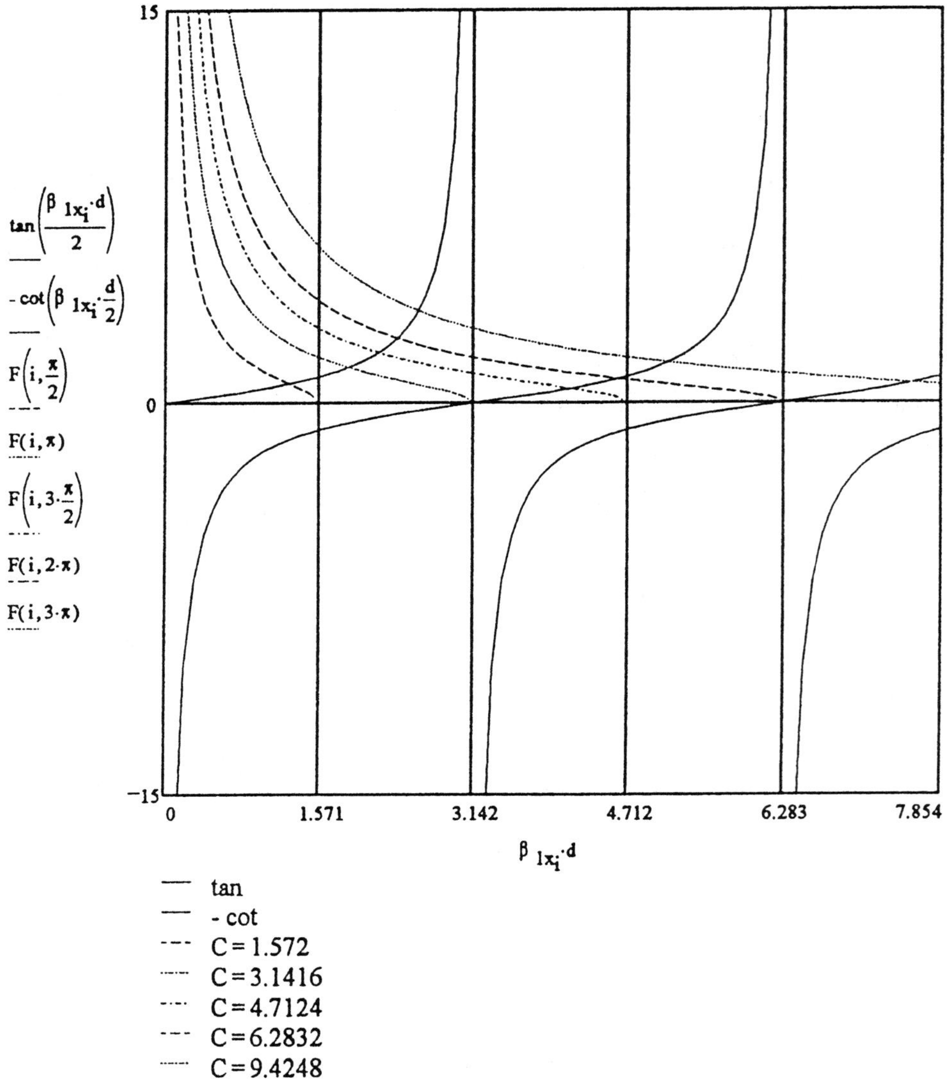

FIGURE 11.34 Plot of the modal equations.

The figure shows the right side plotted for $C = \pi/2$ (dashed), π (dotted), $3\pi/2$ (dash-dot), 2π (dashed), and 3π (dotted). For $C = \pi/2$ there is one intersection giving the value of the mode 0 solution as $\beta_{1x,0}d \approx 0.4\pi$ (0.403π) (more exact values, found by numerical solution, are given in parentheses). For $C = \pi$ there are two intersections giving $\beta_{1x,0}d \approx 0.6\pi$ (0.595π) and giving the value of the mode 1 solution as $\beta_{1x,1}d = \pi$, its cutoff value. For $C = 3\pi/2$ there are again two intersections giving $\beta_{1x,0}d \approx 0.7\pi$ (0.694π) and $\beta_{1x,1}d \approx 1.3\pi$ (1.317π). For $C = 2\pi$ there are three intersections giving $\beta_{1x,0}d \approx 0.8\pi$ (0.754π) and $\beta_{1x,1}d \approx 1.5\pi$ (1.473π) and giving the value of the mode 2 solution as $\beta_{1x,2}d = 2\pi$, its cutoff value. For $C = 3\pi$ there are four intersections giving $\beta_{1x,0}d \approx 0.85\pi$ (0.823π), $\beta_{1x,1}d \approx 1.65\pi(1.633\pi)$, and

$\beta_{1x,2}d \approx 2.4\pi$ (2.407π) and giving the value of the mode 3 solution as $\beta_{1x,3}d = 3\pi$, its cutoff value.

Continuing for different values, we find the cutoff angular frequency of the TE_n mode satisfies $\omega_{cn}\sqrt{\mu_0(\varepsilon_1 - \varepsilon_2)}d = n\pi$ and the cutoff frequency is

$$f_{cn} = \frac{nc_2}{2d\sqrt{(\varepsilon_1/\varepsilon_2) - 1}}, \quad c_2 = \frac{1}{\sqrt{\mu_0\varepsilon_2}}, \quad n = 1, 2, 3, \ldots. \tag{11.152}$$

The crossing of the curves at $\beta_{1x,0}d$ for the TE_0 mode exists regardless of how small we choose $\omega\sqrt{\mu_0(\varepsilon_1 - e_2)}d = C$ to be, and, hence, this is the dominant mode and it is propagating down to zero frequency.

Thus, depending on the value of C, there may be one, two, three, or more values of β_{1x}, which are roots of Eqs. (11.149) and (11.151). The corresponding nth value of β_z is

$$\beta_n = \sqrt{\omega^2\mu_1\varepsilon_1 - \beta_{1x,n}^2}, \quad n = 0, 1, 2, 3, \ldots. \tag{11.153}$$

The x direction attenuation rate α_n into media 2 and 3 for the nth mode is, from Eq. (11.142),

$$\alpha_n = \sqrt{\omega^2\mu_0(\varepsilon_1 - \varepsilon_2) - \beta_{1x,n}^2}. \tag{11.154}$$

Note that at the cutoff frequency for each mode it is the attenuation constant for waves traveling into media 2 and 3 that goes to zero. For frequency less than the cutoff frequency power leaks from the core into the surrounding region; the wave in the core loses power and hence decays as it travels in the z direction. It is no longer useful as a guided wave. Exactly at the cutoff frequency the field extends indefinitely out into the external region on either side of the core dielectric. As frequency increases above cutoff the attenuation constant increases and the field decays more rapidly with distance away from the surface of the core. The wave outside the core is known as a surface wave because its amplitude is significant only along the surface of the dielectric core.

We can now obtain the reflection and transmission coefficients at the $x = 0$ boundary and, using these, the fields in the core and in the surrounding medium. Using Eq. (11.79) we have

$$\Gamma_{TE} = \frac{jX_{02} - R_{01}}{jX_{02} + R_{01}} = \frac{j(\omega\mu_0/\alpha) - (\omega\mu_0/\beta_{1x})}{j(\omega\mu_0/\alpha) + (\omega\mu_0/\beta_{1x})}, \qquad \tau_{TE} = \frac{2jX_{02}}{jX_{02} + R_{01}} = \frac{2j(\omega\mu_2/\alpha)}{j(\omega\mu_2/\alpha) + (\omega\mu_1/\beta_{1x})}.$$

For even modes (TE_n, $n = 0, 2, 4, \ldots$) we multiply numerator and denominator by $-j\alpha/\omega\mu_0$ and use Eq. (11.150) to obtain

$$\Gamma_{TE} = \frac{1 + j(\alpha/\beta_{1x})}{1 - j(\alpha/\beta_{1x})}, \quad \tau_{TE} = \frac{2}{1 - j(\alpha/\beta_{1x})}$$

$$= 1\angle 2\tan^{-1}(\alpha/\beta_{1x}) = \frac{2}{\sqrt{1+(\alpha/\beta_{1x})^2}}\angle \tan^{-1}\left(\frac{\alpha}{\beta_{1x}}\right)$$

$$= 1\angle \beta_{1x}d, \qquad = \frac{2}{\sqrt{1+\tan^2(\beta_{1x}d/2)}}\angle \beta_{1x}d/2$$

$$= 2\cos(\beta_{1x}d/2)\angle \beta_{1x}d/2.$$

For the nth even mode these give

$$\begin{aligned}\underline{E}_{1y,n} &= E_0(e^{-j\beta_{1x,n}x} + \Gamma_{\text{TE}}e^{j\beta_{1x,n}x})e^{-j\beta_n z}\\ &= 2E_0 e^{j\beta_{1x,n}d/2}\cos(\beta_{1x,n}x + \beta_{1x,n}d/2)e^{-j\beta_n z}, \qquad -d < x < 0,\\ &= E_{0n}\cos(\beta_{1x,n}x')e^{-j\beta_n z}, \qquad -d/2 < x' < d/2, \end{aligned} \tag{11.155}$$

where $x' = x + d/2$ measures distance from the center of the core and $E_{0n} = 2E_0\exp(j\beta_{1x,n}d/2)$. The even-order modes are symmetric with respect to the center of the core, since the cosine is an even function. The nth even mode wave transmitted into medium 2 is

$$\begin{aligned}\underline{E}^t_{y,n} &= \tau_{\text{TE}}E_0 e^{-\alpha_n x}e^{-j\beta_n z}\\ &= 2E_0 e^{j\beta_{1x,n}d/2}\cos(\beta_{1x,n}d/2)e^{-\alpha_n x}e^{-j\beta_n z}\\ &= E_{0n}\cos(\beta_{1x,n}d/2)e^{-\alpha_n x}e^{-j\beta_n z}, \qquad x > 0. \end{aligned} \tag{11.156}$$

For odd modes (TE_n, $n = 1, 3, 5, \ldots$) we multiply numerator and denominator by $-\beta_{1x}/\omega\mu_0$ and use Eq. (11.148) to obtain

$$\Gamma_{\text{TE}} = -\frac{1+j(-\beta_{1x}/\alpha)}{1-j(-\beta_{1x}/\mu_1\alpha)} \qquad \tau_{\text{TE}} = \frac{-2j(-\beta_{1x}/\alpha)}{1-j(-\beta_{1x}/\alpha)}$$

$$= -1\angle 2\tan^{-1}(-\beta_{1x}/\alpha) \qquad = \frac{-2j(-\beta_{1x}/\alpha)}{\sqrt{1+(-\beta_{1x}/\alpha)^2}}\angle \tan^{-1}\left(-\frac{\beta_{1x}}{\alpha}\right)$$

$$= -1\angle \beta_{1x}d, \qquad = \frac{-2j\tan(\beta_{1x}d/2)}{\sqrt{1+\tan^2(\beta_{1x}d/2)}}\angle \beta_{1x}d/2$$

$$= -2j\sin(\beta_{1x}d/2)\angle \beta_{1x}d/2.$$

For the nth odd mode these give

$$\begin{aligned}\underline{E}_{1y,n} &= -j2E_0 e^{j\beta_{1x,n}d/2}\sin(\beta_{1x,n}x + \beta_{1x,n}d/2)e^{-j\beta_n z}, \qquad -d < x < 0,\\ &= E_{0n}\sin(\beta_{1x,n}x')e^{-j\beta_n z}, \qquad -d/2 < x' < d/2, \end{aligned} \tag{11.157}$$

where x' is measured from the center of the core and $E_{0n} = -j2E_0\exp(j\beta_{1x,n}d/2)$. The odd-order modes are antisymmetric with respect to the center of the core, since the sine is an odd function. The odd mode wave transmitted into medium 2 is

$$\begin{aligned}\underline{E}^t_{y,n} &= -j2E_0 e^{j\beta_{1x,n}d/2}\sin(\beta_{1x,n}d/2)e^{-\alpha_n x}e^{-j\beta_n z}\\ &= E_{0n}\sin(\beta_{1x,n}d/2)e^{-\alpha_n x}e^{-j\beta_n z}, \quad x > 0.\end{aligned} \tag{11.158}$$

When the modes are close to cutoff the field extends well beyond the core region because α_n is small. In the limit, at cutoff, the field does not decay at all outside the core. When the frequency is much greater than the cutoff frequency, the fields decay very rapidly outside the core, because α_n is large.

11.9.2 Transverse Magnetic (TM) Wave Solutions

From Eqs. (11.98) and (11.99) for TM waves we have

$$R_{01} = R_{1,\mathrm{TM}} = \eta_1\cos\theta_i = \sqrt{\frac{\mu_0}{\varepsilon_1}}\left(\frac{\beta_{1x}}{\omega\sqrt{\mu_0\varepsilon_1}}\right) = \frac{\beta_{1x}}{\omega\varepsilon_1}, \tag{11.159}$$

$$X_{02} = X_{2,\mathrm{TM}} = -\sqrt{\frac{\mu_0}{\varepsilon_2}}\left(\frac{\alpha_{2x}}{\omega\sqrt{\mu_0\varepsilon_2}}\right) = -\frac{\alpha_{2x}}{\omega\varepsilon_2}, \tag{11.160}$$

which, substituted in Eqs. (11.149) and (11.151), gives

$$-\cot\left(\frac{\beta_{1x}d}{2}\right) = \frac{\varepsilon_1\alpha}{\varepsilon_2\beta_{1x}} = \frac{\varepsilon_1\sqrt{\omega^2(\varepsilon_1-\varepsilon_2)d^2-(\beta_{1x}d)^2}}{\varepsilon_2(\beta_{1x}d)} \tag{11.161}$$

$$-\cot\left(\frac{\beta_{1x}d}{2}\right) = \frac{(\varepsilon_1/\varepsilon_2)\sqrt{C^2-(\beta_{1x}d)^2}}{\beta_{1x}d}, \quad C = \omega\sqrt{\mu_0(\varepsilon_1-\varepsilon_2)}\,d \tag{11.162}$$

and

$$\tan\left(\frac{\beta_{1x}d}{2}\right) = \frac{\varepsilon_1\alpha}{\varepsilon_2\beta_{1x}} \tag{11.163}$$

$$\tan\left(\frac{\beta_{1x}d}{2}\right) = \frac{(\varepsilon_1/\varepsilon_2)\sqrt{C^2-(\beta_{1x}d)^2}}{\beta_{1x}d}, \tag{11.164}$$

respectively, as the modal equations. Again note that the right sides of the two equations are identical.

These equations can be solved graphically just as was done for the TE modes. The cutoff frequencies are still given by Eq. (11.152), the phase constant in the z direction is given by Eq. (11.153), and the attenuation constant in the x direction is given by Eq. (11.154). The TM_0 mode also propagates at any frequency down to zero. It is the dominant TM mode.

We can now obtain the reflection and transmission coefficients at the $x = 0$ boundary and, using these, the fields in the core and in the surrounding medium.

Using Eq. (11.79) we have

$$\Gamma_{TM} = \frac{jX_{02} - R_{01}}{jX_{02} + R_{01}} \qquad \tau_{TM} = \frac{2jX_{02}}{jX_{02} + R_{01}}$$

$$= \frac{-j(\alpha/\omega\varepsilon_2) - (\beta_{1x}/\omega\varepsilon_1)}{-j(\alpha/\omega\varepsilon_2) + (\beta_{1x}/\omega\varepsilon_1)}, \qquad = \frac{2j(\omega\varepsilon_2/\alpha)}{j(\omega\varepsilon_2/\alpha) + (\omega\varepsilon_1/\beta_{1x})}.$$

For even modes (TM_n, $n = 0, 2, 4, \ldots$) we multiply numerator and denominator by $-\omega\varepsilon_1/\beta_{1x}$ and use Eq. (11.163) to obtain

$$\Gamma_{TM} = -\frac{1 + j(\varepsilon_1\alpha/\varepsilon_2\beta_{1x})}{1 - j(\varepsilon_1\alpha/\varepsilon_2\beta_{1x})} \qquad \tau_{TM} = -\frac{2j(\varepsilon_1\alpha/\varepsilon_2\beta_{1x})}{1 - j(\varepsilon_1\alpha/\varepsilon_2\beta_{1x})}$$

$$= -1\angle 2\tan^{-1}(\varepsilon_1\alpha/\beta_{1x}) \qquad = -\frac{2j(\varepsilon_1\alpha/\varepsilon_2\beta_{1x})}{\sqrt{1 + (\varepsilon_1\alpha/\varepsilon_2\beta_{1x})^2}} \angle \tan^{-1}\left(\frac{\varepsilon_1\alpha}{\varepsilon_2\beta_{1x}}\right)$$

$$= -1\angle \beta_{1x}d, \qquad = -\frac{2j\tan(\beta_{1x}d/2)}{\sqrt{1 + \tan^2(\beta_{1x}d/2)}} \angle \beta_{1x}d/2$$

$$= -2j\sin(\beta_{1x}d/2) \angle \beta_{1x}d/2.$$

For the nth even mode these give

$$\underline{H}_{1y,n} = -2H_0 e^{j\beta_{1x,n}d/2} \cos(\beta_{1x,n}x + \beta_{1x,n}d/2) e^{-j\beta_n z}, \qquad -d < x < 0$$

$$= H_{0n} \cos(\beta_{1x,n}x') e^{-j\beta_n z}, \qquad -d/2 < x' < d/2, \tag{11.165}$$

where $x' = x + d/2$ measures distance from the center of the core and $H_{0n} = -2H_0 \exp(j\beta_{1x,n}d/2)$. The even-order modes are symmetric with respect to the center of the core. The even mode wave transmitted into medium 2 is

$$\underline{H}^t_{y,n} = -2H_0 e^{j\beta_{1x,n}d/2} \cos(\beta_{1x,n}d/2) e^{-\alpha_n x} e^{-j\beta_n z}$$

$$= H_{0n} \cos(\beta_{1x,n}d/2) e^{-\alpha_n x} e^{-j\beta_n z}, \qquad x > 0. \tag{11.166}$$

For odd modes (TM_n, $n = 1, 3, 5, \ldots$) we multiply numerator and denominator by $j\omega\varepsilon_2/\alpha$ and use Eq. (11.161) to obtain

$$\Gamma_{TM} = \frac{1 + j(-\varepsilon_2\beta_{1x}/\varepsilon_1\alpha)}{1 - j(-\varepsilon_2\beta_{1x}/\varepsilon_1\alpha)} \qquad \tau_{TM} = \frac{2}{1 - j(-\varepsilon_2\beta_{1x}/\varepsilon_1\alpha)}$$

$$= 1\angle 2\tan^{-1}(-\varepsilon_2\beta_{1x}/\varepsilon_1\alpha) \qquad = \frac{2}{\sqrt{1 + \tan^2(\beta_{1x}d/2)}} \angle \tan^{-1}(\beta_{1x}d/2)$$

$$= 1\angle \beta_{1x}d, \qquad = 2\cos(\beta_{1x}d/2) \angle \beta_{1x}d/2.$$

For the nth odd mode these give

$$\underline{H}_{1y,n} = 2jH_0 e^{j\beta_{1x,n}d/2} \sin(\beta_{1x,n}x + \beta_{1x,n}d/2)e^{-j\beta_n z}, \quad -d < x < 0$$

$$= H_{0n} \sin(\beta_{1x,n}x')e^{-j\beta_n z}, \quad -d/2 < x' < d/2, \tag{11.167}$$

where $H_{0n} = 2jH_0 \exp(j\beta_{1x,n}d/2)$. The odd-order modes are antisymmetric with respect to the center of the core. The odd mode wave transmitted into medium 2 is

$$\underline{H}^t_{y,n} = 2jH_0 e^{j\beta_{1x,n}d/2} \sin(\beta_{1x,n}d/2)e^{-\alpha_n x} e^{-j\beta_n z}$$

$$= H_{0n} \sin(\beta_{1x,n}d/2)e^{-\alpha_n x} e^{-j\beta_n z}, \quad x > 0. \tag{11.168}$$

11.9.3 Wave Impedance and Power

For TE modes we define the wave impedance of a forward ($+z$) going wave as

$$Z_{\text{TE}} = -\frac{\underline{E}_y}{\underline{H}_x}.$$

From the Ampère–Maxwell curl equation we have $\underline{H}_x = (1/j\omega\mu_0)(\partial \underline{E}_y/\partial z)$. The z dependence of all the $\underline{E}_y$ components is $\exp(-j\beta_n z)$ and hence the wave impedance of the TE_n mode is

$$Z_{\text{TE}_n} = \frac{\omega\mu_0}{\beta_n}.$$

The time average power density is given by

$$\vec{p}_{f,\text{Avg}} = \vec{a}_z \tfrac{1}{2}\text{Re}(-\underline{E}_y \underline{H}^*_x) = \vec{a}_z \frac{1}{2Z_{\text{TE}_n}} |\underline{E}_y|^2 = \vec{a}_z \frac{\beta_n}{2\omega\mu_0} |\underline{E}_y|^2.$$

The time average power carried by the mode in a width w is found by integrating over a cross section

$$P_{f,\text{Avg}} = \int_{y=-w/2}^{w/2} \int_{x=-\infty}^{\infty} \vec{p}_{f,\text{Avg}} \cdot \vec{a}_z \, dx \, dy.$$

The fields are independent of the y coordinate and the integration over y simply results in multiplication by the width w. The integration over the x coordinate for the symmetric dielectric slab waveguide is twice the integral over the interval from $x = -d/2$ to ∞. Thus

$$P_{f,\text{Avg,TE}_n} = 2\left(\frac{w\beta_n}{2\omega\mu_0}\right)\left(\int_{x=-d/2}^{0} |\underline{E}_{1y,n}|^2 \, dx + \int_{x=0}^{\infty} |\underline{E}^t_{y,n}|^2 \, dx\right).$$

For the even modes we substitute from Eqs. (11.155) and (11.156) to obtain

$$P_{f,\text{Avg,TE}_n} = 2\left(\frac{w\beta_n}{2\omega\mu_0}\right)|E_{0n}|^2\left\{\int_{x'=0}^{d/2}\cos^2(\beta_{1x,n}x')\,dx' + \cos^2\left(\frac{\beta_{1x,n}d}{2}\right)\int_{x=0}^{\infty}e^{-2\alpha_n x}\,dx\right\}$$

$$= \left(\frac{w\beta_n}{\omega\mu_0}\right)|E_{0n}|^2\left\{\frac{1}{2}\left[\frac{d}{2}+\frac{1}{\beta_{1x,n}}\sin\left(\frac{\beta_{1x,n}d}{2}\right)\cos\left(\frac{\beta_{1x,n}d}{2}\right)\right]+\frac{1}{2\alpha_n}\cos^2\left(\frac{\beta_{1x,n}d}{2}\right)\right\}.$$

From Eq. (11.150) we have $\cos(\beta_{1x,n}d/2) = (\beta_{1x,n}/\alpha_n)\sin(\beta_{1x,n}d/2)$, which gives

$$P_{f,\text{Avg,TE}_n} = \left(\frac{w\beta_n}{4\omega\mu_0}\right)|E_{0n}|^2 d_{\text{eff}}, \quad d_{\text{eff}} = d + \frac{2}{\alpha_n}. \tag{11.169}$$

The TE_n modes with n odd give the same result.

The TE_n mode in the dielectric slab waveguide with a core width d carries the same power as the TE_n mode in the parallel plate waveguide with conductor separation d_{eff} when each has the same peak electric field intensity.

For TM modes we define the wave impedance of a forward $(+z)$ going wave as

$$Z_{\text{TM}} = \frac{\underline{E}_x}{\underline{H}_y}.$$

From the Faraday–Maxwell curl equation we have $\underline{E}_x = -(1/j\omega\varepsilon)(\partial \underline{H}_y/\partial z)$. The z dependence of all the $\underline{H}_y$ components is $\exp(-j\beta_n z)$ and hence the wave impedance of the TM_n mode is

$$Z_{\text{TM}_n} = \begin{cases} \dfrac{\beta_n}{\omega\ \varepsilon_1}, & -d < x < 0 \quad \text{or} \quad -\dfrac{d}{2} < x' < \dfrac{d}{2}, \\[2ex] \dfrac{\beta_n}{\omega\varepsilon_2}, & x > 0 \quad \text{and} \quad x < -d \quad \text{or} \quad |x'| > \dfrac{d}{2}. \end{cases}$$

The time average power density is given by

$$\vec{p}_{f,\text{Avg}} = \vec{a}_z\tfrac{1}{2}\text{Re}(\underline{E}_x\underline{H}_y^*) = \vec{a}_z\tfrac{1}{2}Z_{\text{TM}_n}|\underline{H}_y|^2.$$

The time average power carried by the mode in a width w is found by integrating over a cross section. Thus

$$P_{f,\text{Avg,TM}_n} = \frac{1}{2}\times 2\left[\left(\frac{w\beta_n}{\omega\varepsilon_1}\right)\int_{x=-d/2}^{0}|\underline{H}_{1y,n}|^2\,dx + \left(\frac{w\beta_n}{\omega\varepsilon_2}\right)\int_{x=0}^{\infty}|\underline{H}_{y,n}^t|^2\,dx\right].$$

For the even modes we substitute from Eqs. (11.165) and (11.166) to obtain

$$P_{f,\text{Avg,TM}_n} = \left(\frac{w\beta_n}{\omega}\right)|H_{0n}|^2\left\{\frac{1}{\varepsilon_1}\int_{x'=0}^{d/2}\cos^2(\beta_{1x,n}x')\,dx' + \frac{1}{\varepsilon^2}\cos^2\left(\frac{\beta_{1x,n}d}{2}\right)\int_{x=0}^{\infty}e^{-2\alpha_n x}\,dx\right\}$$

$$= \left(\frac{w\beta_n}{4\omega\varepsilon_1}\right)|H_{0n}|^2\left\{2\left[\frac{d}{2}+\frac{1}{\beta_{1x,n}}\sin\left(\frac{\beta_{1x,n}d}{2}\right)\cos\left(\frac{\beta_{1x,n}d}{2}\right)\right]\right.$$

$$\left.+4\left(\frac{\varepsilon_1}{\varepsilon_2}\right)\left(\frac{1}{2\alpha_n}\right)\cos^2\left(\frac{\beta_{1x,n}d}{2}\right)\right\}$$

$$P_{f,\text{Avg, TM}_n} = \left(\frac{w\beta_n}{4\omega\varepsilon_1}\right)|H_{0n}|^2 d_{\text{eff}}. \tag{11.170}$$

The term inside the braces represents d_{eff}. From Eq. (11.163) we have $\cos(\beta_{1x,n}d/2) = (\varepsilon_2\beta_{1x,n}/\varepsilon_1\alpha_n)\sin(\beta_{1x,n}d/2)$, which gives

$$d_{\text{eff}} = d + \frac{2}{\alpha_n}\left[\left(\frac{\varepsilon_2}{\varepsilon_1}\right)\sin^2\left(\frac{\beta_{1x,n}d}{2}\right)+\left(\frac{\varepsilon_1}{\varepsilon_2}\right)\cos^2\left(\frac{\beta_{1x,n}d}{2}\right)\right]$$

$$= d + \frac{2}{\alpha_n}\left[\left(\frac{\varepsilon_2}{\varepsilon_1}\right)\frac{(\varepsilon_1\alpha_n/\varepsilon_2\beta_{1x,n})^2}{1+(\varepsilon_1\alpha_n/\varepsilon_2\beta_{1x,n})^2}+\left(\frac{\varepsilon_1}{\varepsilon_2}\right)\frac{1}{1+(\varepsilon_1\alpha_n/\varepsilon_2\beta_{1x,n})^2}\right]$$

$$d_{\text{eff}} = d + \frac{2}{\alpha_n}\left(\frac{\varepsilon_1}{\varepsilon_2}\right)\left(\frac{\beta_{1x,n}^2+\alpha_n^2}{\beta_{1x,n}^2+(\varepsilon_1/\varepsilon_2)^2\alpha_n^2}\right). \tag{11.171}$$

The TM_n modes with n odd give the same result.

The TM_n mode in the dielectric slab waveguide with a core width d carries the same power as the TM_n mode in the parallel plate waveguide with conductor separation d_{eff} when each has the same peak magnetic field intensity.

PROBLEMS

11.1 **a.** For the electric field intensity found in Example 11.1 find the corresponding magnetic field intensity.

b. Use this magnetic field intensity to obtain the original electric field intensity.

Ans. a. $\underline{\vec{H}} = (1/j0.790)[\vec{a}_z(1.58+j2.50) - j\vec{a}_x 3.41]\times 10^{-3}E_y$.

11.2 In obtaining Eq. (11.9) terms of order $(\sigma/\omega\varepsilon)$ have been retained and terms of order $(\sigma/\omega\varepsilon)^2$ and higher have been neglected.

a. Expand $(1+\sigma/\omega\varepsilon)^{1/2}$, retaining terms of order $(\sigma/\omega\varepsilon)^3$ and hence obtaining more accurate expressions for α and β.

b. Write the expressions for α and β in terms of $(\varepsilon''/\varepsilon')$ instead of $(\sigma/\omega\varepsilon)$.

c. Calculate the percent correction due to the added terms for the material and frequency of Example 11.2.

Ans. a. $\alpha = \dfrac{\sigma}{2}\sqrt{\dfrac{\mu}{\varepsilon}}\left[1-\dfrac{1}{8}\left(\dfrac{\sigma}{\omega\varepsilon}\right)^2\right]$, $\beta = \omega\sqrt{\mu\varepsilon}\left[1+\dfrac{1}{8}\left(\dfrac{\sigma}{\omega\varepsilon}\right)^2\right]$;

c. $1.7\times 10^{-6}\%$.

11.3 For seawater, $\sigma = 4$ S/m and $\varepsilon_r = 72$.

a. At what frequency is $\sigma = \omega\varepsilon$?

b. Calculate α, β, δ_s, λ, v_{ph}, and η at 100 Hz.

c. Repeat at 1 GHz.

d. Repeat at 100 GHz.

e. Comment on the difficulty of underwater communication using electromagnetic waves.

Ans. b. $\alpha = \beta = 0.0397\ \text{m}^{-1}$, $\eta = 9.93 \times 10^{-3}(1 + j)\ \Omega$;
c. $\gamma = 80.9 + j195\ \text{m}^{-1}$; d. $\alpha = 88.9\ \text{m}^{-1}$, $\beta = 17.8 \times 10^3\ \text{m}^{-1}$.

11.4 **a.** Show that

$$\alpha^2 = \tfrac{1}{4}(\gamma + \gamma^*)^2 = \tfrac{1}{2}\omega\mu(g - \omega\varepsilon), \quad \beta^2 = -\tfrac{1}{4}(\gamma - \gamma^*)^2 = \tfrac{1}{2}\omega\mu(g + \omega\varepsilon),$$

and hence obtain Eq. (11.17).

b. Multiplying numerator and denominator of η by $\sqrt{\sigma - j\omega\varepsilon}$, show that $\eta = j\gamma^*/g$ and hence obtain Eq. (11.18).

11.5 **a.** Show that an elliptically polarized wave can be written as the sum of a right and left circularly polarized wave.

b. Show that this is also true for a linearly polarized wave.

11.6 Make a sketch similar to Fig. 11.5, when $\kappa = 2\exp(-j\pi/4)$ and when $\kappa = 2\exp(j\pi/4)$.

11.7 Equations (12.50) and (12.51) give the magnetic field phasors for a TE_{10} mode in a rectangular guide. Find an expression for the value of x at which the magnetic field is circularly polarized.

Ans. $x = \dfrac{a}{\pi}\tan^{-1}\left(\dfrac{\pi}{a\beta}\right)$.

11.8 Verify Eq. (11.40) when $\sigma \neq 0$.

11.9 Equations (12.49) to (12.51) give the electric and magnetic field phasors for a TE_{10} mode in a rectangular guide, with both $\vec{J}_s = 0$ and $\sigma = 0$. For a volume $0 \le x \le a$, $0 \le y \le b$, and $0 \le z \le L$, show that Eq. (11.33) is satisfied.

11.10 **a.** For the polarized wave $\underline{\vec{E}} = |E_0|(\vec{a}_x + \kappa\vec{a}_y)\exp(-j\beta z)$, $\kappa = |\kappa|\exp(j\psi)$, find the magnetic field intensity.

b. Find an expression for the time average power density $\vec{p}_{f,\text{Avg}}$ for this elliptically polarized wave.

c. Assume $\psi = -\pi/2$. Is the electric field intensity right-handed or left-handed polarized? The magnetic field intensity?

Ans. a. $\underline{\vec{H}} = (|E_0|/\eta)(-\vec{a}_x\kappa + \vec{a}_y)\exp(-j\beta z)$;
b. $\vec{p}_{f,\text{Avg}} = \vec{a}_z(|E_0|^2/2\eta)\,(1 + |\kappa|^2)$.

11.11 Assume two lossless media, air or free space and glass ($\varepsilon_r = 4$).

a. For a wave incident from free space onto glass (Case 1), plot θ_t, β_{tz}/β_i, and β_{tx}/β_i versus θ_i.

b. For the wave incident from glass onto free space (Case 2), calculate θ_c and then plot θ_t, β_{tz}/β_i, β_{tx}/β_i, and α_{tx}/β_i versus θ_i.

11.12 **a.** Calculate the maximum deviation of the transmitted phase vector $\vec{\beta}_t$ from the normal to the boundary for a wave incident from air onto copper ($\sigma = 5.8 \times 10^7$) at 10 GHz.

b. Find the frequency at which the deviation equals 1°.

Ans. a. $\psi = 7.9 \times 10^{-3}$ degrees $= 1.38 \times 10^{-4}$ rad.

b. 159×10^3 GHz $= 159 \times 10^{15}$ Hz; this is approximately in the visible light frequency band.

11.13 Show that the time average power density carried by the TE wave, Eqs. (11.63) and (11.64), and by the TM wave, Eqs. (11.65) and (11.66), sums to give the time average power density of the total wave.

Ans. The total time average power density is $\vec{a}_{x'}|E_a|^2/2\eta_1$.

11.14 Show that the time average electric and magnetic energy densities for the TE wave, Eqs. (11.63) and (11.64), and TM wave, Eqs. (11.65) and (11.66), sum to give the time average electric and magnetic energy densities of the total wave.

Ans. For the total wave $w_{e,\text{Avg}} = w_{m,\text{Avg}} = \frac{1}{4}\varepsilon_1|E_a|^2$.

11.15 Plot the polarizing angle $\theta_{ip,\text{TM}}$ and the critical angle θ_c as a function of $\varepsilon_2/\varepsilon_1$ from $\varepsilon_2/\varepsilon_1 = 0.01$ to 100. Use a logarithmic scale for the abscissa.

11.16 We can define a time average power density reflection and transmission coefficient at the boundary between two media as

$$R = \left.\frac{\text{Re}[\underline{E}^r \times \underline{H}^{r*}]\cdot(-\vec{a}_x)}{\text{Re}[\underline{E}^i \times \underline{H}^{i*}]\cdot(\vec{a}_x)}\right|_{x=0}, \qquad T = \left.\frac{\text{Re}[\underline{E}^t \times \underline{H}^{t*}]\cdot(-\vec{a}_x)}{\text{Re}[\underline{E}^i \times \underline{H}^{i*}]\cdot(\vec{a}_x)}\right|_{x=0}.$$

Show that for reflection between two perfect dielectric media, these are given by

$$R_{\text{TE}} = |\Gamma_{\text{TE}}|^2, \quad T_{\text{TE}} = \frac{\eta_1 \cos\theta_t}{\eta_2 \cos\theta_i}|\tau_{\text{TE}}|^2; \quad R_{\text{TM}} = |\Gamma_{\text{TM}}|^2, \quad T_{\text{TM}} = \frac{\eta_1 \cos\theta_i}{\eta_2 \cos\theta_t}|\tau_{\text{TM}}|^2$$

for the TE and TM waves, for Case 1 and for Case 2 with $\theta_i < \theta_c$. Show that for Case 2 with $\theta_i > \theta_c$, $R = 1$ and $T = 0$ for both the TE and TM waves. Show that $R + T = 1$ as required by conservation of time average x-directed power density, for all of these results.

11.17 Plot R and T, as given in Problem 11.16, versus θ_i, when medium 1 is free space and medium 2 is:

a. Glass, $\mu = \mu_0$, $\varepsilon = 2.25\varepsilon_0$, $\sigma = 0$,

b. Distilled water, $\mu = \mu_0$, $\varepsilon = 81\varepsilon_0$, $\sigma = 0$.

11.18 Repeat Problem 11.17, when the two media are reversed.

11.19 For the case of normal incidence, $\theta_i = 0$, show that $\theta_t = 0$ and that

$$\Gamma = \frac{\eta_2 - \eta_1}{\eta_2 + \eta_1}, \quad \tau = \frac{2\eta_2}{\eta_2 + \eta_1}$$

for both the TE and TM cases. In fact, the two cases are identical and the incident, reflected, and transmitted waves are simply TEM to the x direction. Choosing the y direction to be the direction of the incident electric field, show that

$$\vec{\underline{E}}_1 = \vec{\underline{E}}^i + \vec{\underline{E}}^r = \vec{a}_y E_0 (e^{-\gamma_1 x} + \Gamma e^{\gamma_1 x}), \quad \vec{\underline{H}}_1 = \vec{\underline{H}}^i + \vec{\underline{H}}^r = \vec{a}_z \frac{E_0}{\eta_1}(e^{-\gamma_1 x} - \Gamma e^{\gamma_1 x}),$$

$$\vec{\underline{E}}_2 = \vec{\underline{E}}^t = \vec{a}_y \tau E_0 e^{-\gamma_2 x}, \quad \vec{\underline{H}}_2 = \vec{\underline{H}}^t = \vec{a}_z \tau \frac{E_0}{\eta_2} e^{-\gamma_2 x}.$$

11.20 **a.** A plane TEM wave $\vec{\underline{E}}^i = \vec{a}_y E_0 \exp(-j\beta_0 x)$ is normally incident from free space ($x < 0$) onto a good conductor ($x > 0$). The conductor is characterized by ε_0, μ_0, σ, with $\sigma \gg \omega\varepsilon_0$. Show that the transmission coefficient is

$$\tau_1 = 2\sqrt{\frac{\omega\varepsilon_0}{2\sigma}}(1 + j)$$

and that the ratio of transmitted to incident time average power density is

$$T_1 = \frac{p^t_{f,\mathrm{Avg}}}{p^i_{f,\mathrm{Avg}}} = \frac{\mathrm{Re}(\underline{E}^t_y \underline{H}^{t*}_z)}{\mathrm{Re}(\underline{E}^i_y \underline{H}^{i*}_z)} \approx 2^{2/3}\sqrt{\frac{\omega\varepsilon_0}{\sigma}}.$$

b. If the wave is normally incident from the conductor onto free space show that $\tau_2 \approx 2$ and that

$$T_2 \approx 2^{5/2}\sqrt{\frac{\omega\varepsilon_0}{\sigma}}.$$

c. Use these results to show that if the wave $\vec{\underline{E}}^i$ is normally incident from free space on a conductor of thickness d, then the overall power transmission ratio is

$$T_3 \approx T_1 e^{-2d/\delta_s} T_2 = 16\frac{\omega\varepsilon_0}{\sigma} e^{-2d/\delta_s}, \quad \delta_s = \sqrt{\frac{2}{\omega\mu_0\sigma}},$$

when multiple reflections within the conductor are neglected. δ_s is the skin depth in the conductor. This is a good approximation when $d \geq \delta_s$. Evaluate δ_s for copper at 1 MHz and evaluate T_3 for $d = \delta_s/2$, δ_s, and $2\delta_s$.

d. To account for multiple reflections note that the amplitude of the wave transmitted into the conductor is $\tau_1 E_0$, the transmission factor in each passage of the wave through the conductor is $\exp(-\gamma d) = \exp(-d/\delta_s)\exp(-jd/\delta_s)$, and the reflection coefficient at both conductor to free space boundaries is ~ 1. Draw a time–distance diagram for reflections in the conductor and show that the electric field transmitted through the conductor is

$$E^t = \tau_1 E_0 e^{-\gamma d}[1 + e^{-2\gamma d} + (e^{-2\gamma d})^2 + \cdots]\tau_2 = \frac{\tau_1 \tau_2 E_0}{2 \sinh \gamma d}.$$

Show that the overall power transmission ratio is

$$T_4 = \frac{|\tau_1|^2 \ |\tau_2|^2}{4|\sinh \gamma d|^2} = \frac{4(\omega\varepsilon_0/\sigma)}{\sinh^2(d/\delta_s)\cos^2(d/\delta_s) + \cosh^2(d/\delta_s)\sin^2(d/\delta_s)},$$

when multiple reflections are not neglected. Evaluate T_4 for $d = \delta_s/2$, δ_s, and $2\delta_s$ and compare the results with those for T_3.

11.21 A right circularly polarized wave $\vec{\underline{E}}^i = E_0(\vec{a}_y - j\vec{a}_z)\exp(-\gamma_1 x)$ is normally incident on the $x = 0$ boundary between two media. Show that the reflected wave is left circularly polarized for any reflection coefficient Γ at the boundary.

11.22 A plane uniform TEM wave with amplitude $E_0 = 100 \angle 0$ is normally incident from air, $x < 0$, onto a perfect dielectric, $x > 0$ ($\mu = \mu_0$, $\varepsilon = 4\varepsilon_0$). Evaluate the time average incident, reflected, and transmitted power density.

Ans. $p_{i,\text{Avg}} = 13.26$, $p_{r,\text{Avg}} = 1.47$, $p_{t,\text{Avg}} = 11.79\ \text{W/m}^2$.

11.23 It is desired to improve the efficiency of transmission of visible light from air into glass by applying a transparent dielectric coating as a quarter-wave matching transformer. The glass and the dielectric coating are lossless and for the glass $\varepsilon_r = 4$. Assume normal incidence.

a. Calculate the fraction of incident power reflected at the air–glass interface if there is no coating. Is this frequency dependent?

b. The visible spectrum may be taken to be from red light at free-space wavelength of $\lambda_{r0} = 0.76\ \mu\text{m}$ to violet light at $\lambda_{v0} = 0.42\ \mu\text{m}$. Design a dielectric coating to match the glass to the air at the center frequency of the visible band. Your design should specify ε_{rc} for the coating and the thickness d.

c. Calculate the fraction of incident power reflected at the center frequency and for red and violet light. Use the Smith chart or analytic calculations.

Ans. a. $P_{\text{frac}} = p^r_{f,\text{Avg}}/p^i_{f,\text{Avg}} = 1/9 = 0.111$; b. $\varepsilon_{rc} = 2$, $d = 0.0965\mu\text{m}$;
c. $P_{\text{frac}} = 0$ at the center frequency, $P_{\text{frac}} = 0.025$ for red and violet.

11.24 For Example 11.6, show that the x component of time average power density for the incident wave equals the sum of the x component of time average power density for the reflected and transmitted waves, for both the TE and TM waves.

11.25 **a.** For an air-to-glass ($\varepsilon_r = 4$) interface and an incident angle of 45°, calculate the reflection and transmission coefficients for the TE and TM polarizations.

b. Assume the electric field of the incident wave is linearly polarized at 45° from the y axis. Find the angle of polarization from the y axis for the reflected and transmitted waves.

Ans. a. $\Gamma_{TE} = -0.45$, $\tau_{TE} = 0.549$, $\Gamma_{TM} = -0.204$, $\tau_{TE} = 0.796$; b. 24.3° and 55.4° for reflected and transmitted waves, respectively.

11.26 **a.** Let $d = 1$ inch and calculate the cutoff frequencies for the TE_n and TM_n modes, for $n = 1, 2, 3$.

b. Calculate the phase constant, the phase velocity, the group velocity, the wave impedance, and the attenuation due to imperfectly conducting copper walls at $f = 1.5 f_{c1}$ for the TE_1 and TM_1 modes.

11.27 Refer to the transmission line model of Fig. 11.33. Assume a symmetric slab waveguide, $\mu_2 = \mu_3$, $\varepsilon_2 = \varepsilon_3$, and $\alpha_{2x} = \alpha_{3x} = \alpha$. For the TE modes, set the sum of the impedances seen to the right and left of the center line $x' = 0$ to zero and obtain Eq. (11.159). Repeat for the sum of the admittances, to get Eq. (11.160).

11.28 Repeat the previous problem for the TM modes to obtain Eqs. (11.170) and (11.171).

11.29 Equation (11.169) gives the time average power carried by the TE_n mode in the dielectric slab guide for even modes. Show that the same result is obtained for odd modes.

11.30 Equation (11.170) gives the time average power carried by the TM_n mode in the dielectric slab guide for even modes. Show that the same result is obtained for odd modes.

11.31 Sketch the curves for the left and right sides of Eq. (11.151) and find approximate values of $\beta_{1x}d$ when $C = 7.80$.
Ans. Exact values are $\beta_{1x,1}d = 0.792\pi$, $\beta_{1x,2}d = 1.566\pi$, $\beta_{1x,3}d = 2.267\pi$.

11.32 Sketch the curves for the left and right sides of Eq. (11.164) and find approximate values of $\beta_{1x}d$ when $C = 2$, π, 2π, and 7.8. Assume $\varepsilon_1 = 2.25\varepsilon_2$.
Ans. Exact values are $\beta_{1x,1}d = 0.894\pi$, $\beta_{1x,2}d = 1.549\pi$, $\beta_{1x,3}d = 2.145\pi$.

REFERENCES

[1] K. F. Sander, *Microwave Components and Systems*, Addison-Wesley, New York, 1987, p. 146ff.

[2] Robert E. Collin, *Foundations of Microwave Engineering*, McGraw-Hill, New York, 1966, pp. 286–309.

CHAPTER TWELVE

Three-Dimensional Waveguides and Resonators

12.1 INTRODUCTION

In this chapter we will examine the propagation of waves in rectangular and circular waveguides with conducting walls and in dielectric rods. We will also examine resonant cavities with rectangular and circular cross sections.

12.2 TRANSVERSE ELECTRIC AND TRANSVERSE MAGNETIC WAVE SOLUTIONS

We have already been introduced to transverse electric (TE) and transverse magnetic (TM) waves in the last chapter. We will now find a simplified set of Maxwell's equations that are useful in dealing with TE and TM waves on cylindrical structures. This is similar to the derivation for transverse electromagnetic waves (TEM) carried out in Section 9.3.

12.2.1 Transverse Electric Waves

For a TE wave propagating in the z direction, we have, by definition, $E_z = 0$. We can therefore write the TE wave as

$$\underline{\vec{E}}(x, y, z) = \vec{e}_t(x, y)e^{\mp\gamma z}, \quad \vec{e}_t(x, y) = \vec{a}_x e_x(x, y) + \vec{a}_y e_y(x, y), \tag{12.1}$$

$$\underline{\vec{H}}(x, y, z) = [\vec{h}_t(x, y) + \vec{a}_z h_z(x, y)]e^{\mp\gamma z}, \tag{12.2}$$

$$\vec{h}_t(x, y) = \vec{a}_x h_x(x, y) + \vec{a}_y h_y(x, y),$$

where $\vec{e}_t$ and $\vec{h}_t$ are the electric and magnetic fields in the transverse plane and h_z is the longitudinal magnetic field. The z dependence of the wave is given by $\exp(\mp\gamma z)$, where γ is the propagation constant.

We now wish to find a differential equation for h_z, the solution of which, subject to boundary conditions, will give us the longitudinal field in the guide. We further want to find equations that relate h_z to the transverse fields and that relate the transverse fields to each other.

To do this, we return to Maxwell's equations, Eqs. (9.3) and (9.4), and substitute from Eqs. (12.1) and (12.2). All terms involving $\underline{E}_z$ are zero, differentiation with respect to z results in multiplication by $\mp\gamma$, and the exponential $\exp(\mp\gamma z)$ can be canceled from the equations. The results are

$$\pm\gamma e_y = -j\omega\mu h_x \qquad \mp\gamma e_x = -j\omega\mu h_y, \tag{12.3a, b}$$

$$\frac{\partial e_y}{\partial x} - \frac{\partial e_x}{\partial y} = -j\omega\mu h_z \qquad \frac{\partial h_z}{\partial y} \pm \gamma h_y = (\sigma + j\omega\mu)\, e_x, \tag{12.3c, d}$$

$$\mp\gamma h_x - \frac{\partial h_z}{\partial x} = (\sigma + j\omega\mu) e_y \qquad \frac{\partial h_y}{\partial x} - \frac{\partial h_x}{\partial y} = 0. \tag{12.3e, f}$$

Equations (12.3a) and (12.3b) are purely algebraic and can be solved for the electric fields in terms of the magnetic field:

$$e_y = \mp\frac{j\omega\mu}{\gamma} h_x = \mp Z_{\mathrm{TE}} h_x, \tag{12.4a$'$}$$

$$Z_{\mathrm{TE}} = \frac{j\omega\mu}{\gamma}$$

$$e_x = \pm\frac{j\omega\mu}{\gamma} h_y = \pm Z_{\mathrm{TE}} h_y, \tag{12.4b$'$}$$

Using Eq. (12.4b′) to eliminate e_x in Eq. (12.3d) and Eq. (12.4a′) to eliminate e_y in Eq. (12.3e), we obtain

$$\frac{\partial h_z}{\partial y} \pm \gamma h_y = \pm\frac{j\omega\mu(\sigma + j\omega\varepsilon)}{\gamma} h_y = \mp\frac{k^2}{\gamma} h_y$$

$$h_y = \mp\left(\frac{\gamma}{\gamma^2 + k^2}\right)\frac{\partial h_z}{\partial y},$$

$$\mp\gamma h_x - \frac{\partial h_z}{\partial x} = \mp\frac{j\omega\mu(\sigma + j\omega\varepsilon)}{\gamma} h_x = \pm\frac{k^2}{\gamma} h_x \tag{12.5d$'$}$$

$$h_x = \mp\left(\frac{\gamma}{\gamma^2 + k^2}\right)\frac{\partial h_z}{\partial x}, \tag{12.5e$'$}$$

where

$$k = \sqrt{-j\omega\mu(\sigma + j\omega\varepsilon)}.$$

Substituting Eqs. (12.5d′) and (12.5e′) into Eq. (12.3f) gives $0 = 0$, so this equation is satisfied. Finally, substituting Eqs. (12.5d′) and (12.5e′) into Eqs. (12.4b′) and (12.4a′), respectively, and the results into Eq. (12.3c) we obtain

$$\frac{j\omega\mu}{\gamma^2 + k^2}\left(\frac{\partial^2 h_z}{\partial x^2} + \frac{\partial^2 h_z}{\partial y^2}\right) = -j\omega\mu h_z$$

$$\frac{\partial^2 h_z}{\partial x^2} + \frac{\partial^2 h_z}{\partial y^2} + k_c^2 h_z = 0, \tag{12.6}$$

where

$$\gamma^2 + k^2 = k_c^2$$

$$\gamma = \sqrt{k_c^2 - k^2} = \sqrt{k^2 + j\omega\mu(+ j\omega\varepsilon)}. \tag{12.7}$$

Equations (12.6) and (12.7), together, are the modal equation for longitudinal magnetic field intensity h_z.

These results can be written more compactly by using vector notation. Equation (12.6) becomes

$$\nabla_t^2 h_z + k_c^2 h_z = 0, \quad \gamma = \sqrt{k_c^2 - k^2}, \quad k = \sqrt{-j\omega\mu(\sigma + j\omega\varepsilon)}. \tag{12.8}$$

Equations (12.5d′) and (12.5e′) become

$$\vec{h}_t = \mp \left(\frac{\gamma}{k_c^2}\right) \nabla_t h_z. \tag{12.9}$$

Equations (12.4a′) and (12.4b′) become

$$\vec{e}_t = \mp Z_{\mathrm{TE}} \vec{a}_z \times \vec{h}_t, \quad Z_{\mathrm{TE}} = \frac{j\omega\mu}{\gamma}. \tag{12.10}$$

These equations apply to other coordinate systems in addition to the rectangular system from which they were derived. Thus for rectangular coordinates the transverse del operator is

$$\nabla_t = \vec{a}_x \frac{\partial}{\partial x} + \vec{a}_y \frac{\partial}{\partial y},$$

while for cylindrical coordinates it is

$$\nabla_t = \vec{a}_r \frac{\partial}{\partial r} + \vec{a}_\phi \frac{1}{r} \frac{\partial}{\partial \phi}.$$

In summary, the solution of the modal equation, Eq. (12.8), subject to boundary conditions gives the longitudinal field h_z and the value(s) of the propagation constant γ. Equation (12.9) then gives the transverse magnetic field $\vec{h}_t$ and Eq. (12.10) gives the transverse electric field $\vec{e}_t$. This is the complete TE solution.

Before applying this solution to a particular geometry, let us obtain the results for TM waves.

12.2.2 Transverse Magnetic Waves

For a TM wave propagating in the $\pm z$ direction, we have, by definition, $\underline{H}_z = 0$. We can therefore write the TM wave as

$$\underline{\vec{H}}(x, y, z) = \vec{h}_t(x, y)\, e^{\mp \gamma z}, \quad \vec{h}_t(x, y) = \vec{a}_x h_x(x, y) + \vec{a}_y h_y(x, y), \tag{12.11}$$

$$\underline{\vec{E}}(x, y, z) = [\vec{e}_t(x, y) + \vec{a}_z e_z(x, y)]\, e^{\mp \gamma z}, \tag{12.12}$$

$$\vec{e}_t(x, y) = \vec{a}_x e_x(x, y) + \vec{a}_y e_y(x, y),$$

where e_z is the longitudinal electric field.

We could repeat the derivation for TE waves following the same steps. It is easier, however, to use the duality property described in Section 9.2.1. We make these duality substitutions to get the compactly written general forms for TM waves as

$$\nabla_t^2 e_z + k_c^2 e_z = 0, \quad \gamma = \sqrt{k_c^2 - k^2}, \quad k = \sqrt{-j\omega\mu(\sigma + j\omega\varepsilon)}, \tag{12.13}$$

$$\vec{e}_t = \mp \left(\frac{\gamma}{k_c^2}\right) \nabla_t e_z, \tag{12.14}$$

$$\vec{h}_t = \pm\, Y_{\mathrm{TM}} \vec{a}_z \times \vec{e}_t, \quad Y_{\mathrm{TM}} = \frac{\sigma + j\omega\varepsilon}{\gamma}, \quad Z_{\mathrm{TM}} = \frac{\gamma}{\sigma + j\omega\varepsilon}. \tag{12.15}$$

The solution of the modal equation, Eq. (12.13), subject to boundary conditions gives the longitudinal field e_z and the value(s) of the propagation constant γ. Equation (12.14) then gives the transverse electric field $\vec{e}_t$ and Eq. (12.15) gives the transverse magnetic field $\vec{h}_t$. This is the complete TM solution.

These equations give us the tools we need to find solutions in particular geometries.

12.3 SEPARATION OF VARIABLES: SOLUTION OF THE MODAL EQUATION

Equations (12.8) and (12.13) are the modal equations for TE and TM waves, respectively. We will apply these equations to problems where the boundaries are rectangular, as in the rectangular waveguide (Fig. 12.1), and circular, as in the circular waveguide (Fig. 12.2). These equations are of identical form,

$$\nabla_t^2 f + k_c^2 f = 0, \tag{12.16}$$

where f represents the longitudinal field component e_z or h_z. The equation is a second-order homogeneous partial differential equation in two variables, the transverse coordinates. There is a very powerful mathematical technique, called separation of variables, that allows us to convert this partial differential equation into two ordinary differential equations with well known solutions. We will apply this technique to the rectangular and cylindrical coordinates that interest us, but it is applicable in nine other coordinate systems as well, including the spherical system. Separation of variables is also used to solve the three-dimensional Laplace equation, Eq. (6.81), in rectangular, cylindrical, and spherical coordinates.

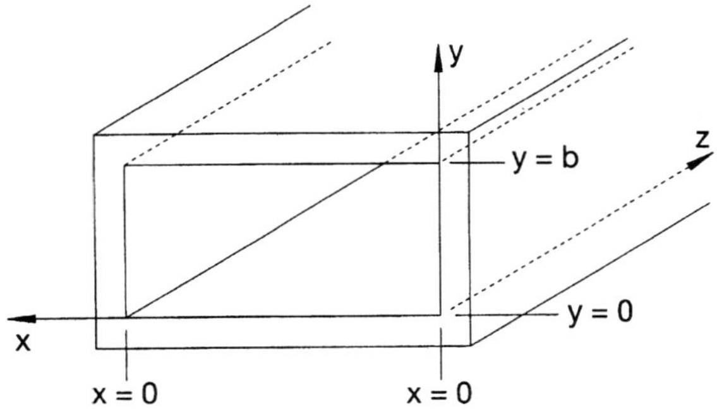

FIGURE 12.1 Rectangular waveguide and coordinate system.

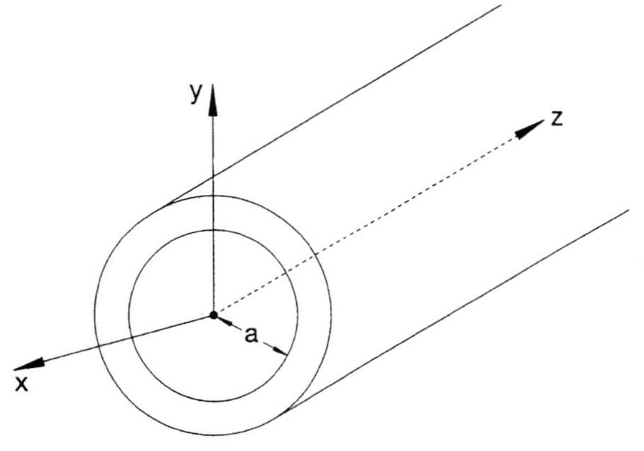

FIGURE 12.2 Circular waveguide and coordinate system.

12.3.1 Separation of Variables in Rectangular Coordinates

We begin with rectangular coordinates. Equation (12.16) is then

$$\frac{\partial^2 f(x,y)}{\partial x^2} + \frac{\partial^2 f(x,y)}{\partial y^2} + k_c^2 f(x,y) = 0.$$

To separate the variables we assume that the unknown function $f(x, y)$ can be written as the product of two unknown functions $X(x)$ and $Y(y)$; thus $f(x, y) = X(x)\,Y(y)$. This is called "separating the variables" because f is a function of both x and y, but X is a function of x alone and Y is a function of y alone.
Substituting in the differential equation we find

$$\frac{d^2X(x)}{dx^2}\,Y(y) + X(x)\,\frac{d^2Y(y)}{dy^2} + k_c^2 X(x)\,Y(y) = 0$$

$$\frac{1}{X(x)}\frac{d^2X(x)}{dx^2} + \frac{1}{Y(y)}\frac{d^2Y(y)}{dy^2} + k_c^2 = 0, \tag{12.17}$$

where we have divided by the product XY.

Two things have happened. First, the partial derivatives have become ordinary derivatives. Second, we have written the equation so that its first term is at most a function of x, its second term is at most a function of y, and the third term is a constant. Now the second and third terms summed together cannot change when x changes, since neither term depends on x. Furthermore, the sum of the first term plus the second and third terms must always be zero. It therefore follows that when x changes the first term cannot change either and it is really a constant, independent of x as well as y. By the same reasoning the second term must also be a constant, independent of y as well as x. We choose the two constants to be $-k_x^2$ and $-k_y^2$, so that Eq. (12.17) becomes three equations:

$$\frac{1}{X(x)}\frac{d^2X(x)}{dx^2} = -k_x^2, \quad \frac{d^2X(x)}{dx^2} + k_x^2 X(x) = 0 \tag{12.18}$$

$$\frac{1}{Y(y)}\frac{d^2Y(y)}{dy^2} = -k_y^2, \quad \frac{d^2Y(y)}{dy^2} + k_y^2 Y(y) = 0 \tag{12.19}$$

$$-k_x^2 - k_y^2 + k_c^2 = 0, \qquad k_c^2 = k_x^2 + k_y^2. \tag{12.20}$$

We now have more, but simpler, equations to solve.

Equations (12.18) and (12.19) are familiar second-order ordinary differential equations and have as their solutions sine and cosine functions. The result for the product $f(x, y) = X(x)Y(y)$ is therefore

$$f(x,y) = (A\cos k_x x + B\sin k_x x)(C\cos k_y y + D\sin k_y y), \tag{12.21}$$

where A, B, C, D, k_x, and k_y are unknown constants. The unknown constants will be determined by boundary conditions as we shall see when we use these results in the rectangular waveguide problem.

An additional comment is warranted. The constants k_x and k_y were chosen so that we would end up with trigonometric functions as the solutions for X and Y. These are the right functions for the waveguide problem that concerns us because we need solutions that go to zero at the values of x and y that correspond to a conducting boundary. For the solution of different problems, the constants could be chosen differently. For example, letting the first term of Eq. (12.18) be equal to $+c_x^2$, instead of $-k_x^2$, we would get

$$\frac{1}{X(x)}\frac{d^2X(x)}{dx^2} = c_x^2, \quad \frac{d^2X(x)}{dx^2} - c_x^2 X(x) = 0, \tag{12.22}$$

which has exponential or hyperbolic functions as its solution. Then the result is

$$f(x,y) = (Ae^{c_x x} + Be^{-c_x x})(C\cos k_y y + D\sin k_y y),$$

or

$$f(x,y) = (A\cosh c_x x + B\sinh c_x x)(C\cos k_y y + D\sin k_y y),$$

$$k_c^2 = -c_x^2 + k_y^2.$$

The physical characteristics of the problem determine which form of the solution is appropriate.

12.3.2 Separation of Variables in Cylindrical Coordinates

In cylindrical coordinates Eq. (12.16) becomes

$$\frac{1}{r}\frac{\partial}{\partial r}\left[r\frac{\partial f(r,\phi)}{\partial r}\right]+\frac{1}{r^2}\frac{\partial^2 f(r,\phi)}{\partial\phi^2}+k_c^2 f(r,\phi)=0.$$

As before, we separate the variables by letting

$$f(r,\phi)=R(r)\,\Phi(\phi).$$

Substituting in the differential equation and dividing by $R\Phi/r^2$ we obtain

$$\left\{\frac{r}{R(r)}\frac{d}{dr}\left[r\frac{dR(r)}{dr}\right]+r^2k_c^2\right\}+\frac{1}{\Phi(\phi)}\frac{d^2\Phi(\phi)}{d\phi^2}=0.$$

We have again eliminated the partial derivatives because each function $R(r)$ and $\Phi(\phi)$ is a function of only one variable. We now use the same argument to separate this result into two equations. The two terms within the braces are at most a function of r alone. The third term is at most a function of ϕ alone. Since the terms within the braces and the last term must sum to zero, each must be equal to a constant. Letting the third term equal $-\nu^2$ we must have the sum of the first two terms equal to ν^2, so that the total sum is zero. This leads to the two equations.

$$\frac{r}{R(r)}\frac{d}{dr}\left[r\frac{dR(r)}{dr}\right]+r^2k_c^2=\nu^2$$

$$\frac{d^2R(r)}{dr^2}+\frac{1}{r}\frac{dR(r)}{dr}+\left(k_c^2-\frac{\nu^2}{r^2}\right)R(r)=0, \tag{12.23}$$

$$\frac{1}{\Phi(\phi)}\frac{d^2\Phi(\phi)}{d\phi^2}=-\nu^2$$

$$\frac{d^2\Phi(\phi)}{d\phi^2}+\nu^2\Phi(\phi)=0. \tag{12.24}$$

We have again converted the second-order partial differential equation into two second-order ordinary differential equations.

The second equation in ϕ has as its solution sine and cosine functions.

The first equation in r is unfortunately more complicated and less familiar. This is known as Bessel's differential equation. Two independent solutions to Bessel's equation are Bessel functions of the first and second kind of order ν, $J_\nu(k_c r)$ and $N_\nu(k_c r)$, respectively. The properties of these two functions are summarized in Appendix C. They each have repeated zeros, which become almost periodic as their arguments get larger. Just like sine and cosine functions, these solutions are suitable to a region in which the field must be zero at the value of r that corresponds to the boundary. The Bessel function of the first kind is finite as its argument goes to zero, but the Bessel function of the second kind is singular (infinite) for zero argument.

The complete solution in cylindrical coordinates is given by

$$f(r, \phi) = [AJ_\nu(k_c r) + BN_\nu(k_c r)]\,[C \cos(\nu\phi) + D \sin(\nu\phi)]. \tag{12.25}$$

The unknown constants A, B, C, D, ν, and k_c will be determined by the boundary conditions for each particular physical problem.

For rectangular coordinates we noted that there was a second form of solution involving exponentials. Similarly, as given in Appendix C, there is a second form of Bessel's equation, which has as its solution functions that approach exponentials as their argument gets large. If we let $k_c^2 = -k_i^2$, then the modified Bessel equation is

$$\frac{d^2R(r)}{dr^2} + \frac{1}{r}\frac{dR(r)}{dr} - \left(k_i^2 + \frac{\nu^2}{r^2}\right)R(r) = 0, \tag{12.26}$$

and its solutions are called modified Bessel functions of the first and second kind of order $\nu, I_\nu(k_i r)$ and $K_\nu(k_i r)$, respectively. The propagation constant, Eq. (12.8) or (12.13), is

$$\gamma = \sqrt{-k_i^2 - k^2} = \sqrt{-k_i^2 + j\omega\mu(\sigma + j\omega\varepsilon)}, \tag{12.27}$$

and the complete solution in cylindrical coordinates is

$$f(r, \phi) = [AI_\nu(k_i r) + BK_\nu(k_i r)]\,[C \cos(\nu\phi) + D \sin(\nu\phi)]. \tag{12.28}$$

12.4 RECTANGULAR WAVEGUIDES

Figure 12.1 shows a rectangular waveguide with inside dimensions $a \times b$, $a \geq b$. The x direction is generally chosen to be along the greater dimension. The conducting walls of the guide are considered to be perfectly conducting in the initial solution to the problem. After the waveguide field solutions or modes are obtained, the attenuation due to finitely conducting walls will be derived. We will first obtain the solution for TE waves and then the solution for TM waves.

12.4.1 Transverse Electric Waves in the Rectangular Guide

For TE waves we must solve Eq. (12.8) subject to the boundary conditions of perfectly conducting walls. The general form of the solution is given by Eq. (12.21), repeated here:

$$h_z(x, y) = (A \cos k_x x + B \sin k_x x)(C \cos k_y y + D \sin k_y y).$$

The perfectly conducting walls require that the tangential electric field must be zero along the walls. Equation (12.10) shows that e_x is proportional to h_y, which in turn is proportional [Eq. (12.9)] to $\partial h_z/\partial y$. Therefore the requirement that $e_x(x,0) = 0$ and $e_x(x,b) = 0$ means that

$$\left.\frac{\partial h_z}{\partial y}\right|_{y=0} = 0 \qquad \left.\frac{\partial h_z}{\partial y}\right|_{y=b} = 0. \tag{12.29a, b}$$

Similarly, Eq. (12.10) shows that e_y is proportional to h_x, which in turn is proportional [Eq. (12.9)] to $\partial h_z/\partial x$. Therefore the requirement that $e_y(0,y) = 0$ and $e_y(a,y) = 0$ means that

$$\left.\frac{\partial h_z}{\partial x}\right|_{x=0} = 0 \qquad \left.\frac{\partial h_z}{\partial x}\right|_{x=a} = 0. \tag{12.30a, b}$$

Equations (12.29a) and (12.29b) and (12.30a) and (12.30b) are the four boundary conditions on h_z.

We now apply these boundary conditions, one at a time. Equation (12.29a) gives

$$\left.\frac{\partial h_z}{\partial y}\right|_{y=0} = k_y(A\cos k_x x + B\sin k_x x)(-C\sin k_y y + D\cos k_y y)\Big|_{y=0} = 0$$

$$k_y(A\cos k_x x + B\sin k_x x)D = 0.$$

For this to be zero for all values of x we must either set A and B both equal to zero, giving the trivial solution $h_z = 0$, or set $D = 0$. We choose the latter so that

$$h_z(x,y) = (A\cos k_x x + B\sin k_x x)(C\cos k_y y).$$

Equation (12.29b) gives

$$\left.\frac{\partial h_z}{\partial y}\right|_{y=b} = k_y(A\cos k_x x + B\sin k_x x)(-C\sin k_y y)\Big|_{y=b} = 0$$

$$k_y(A\cos k_x x + B\sin k_x x)(-C\sin k_y b) = 0.$$

For this to be zero for all values of x we must either set A and B both equal to zero or set $C = 0$, giving the trivial solution $h_z = 0$, or set $\sin k_y b = 0$. We choose the latter. The sine function is zero when its argument is $m\pi$, where m is an integer. Hence $k_y b = m\pi$ and

$$k_y = \frac{m\pi}{b}, \quad m = 0, 1, 2, \ldots. \tag{12.31}$$

So far we have found $D = 0$ and $k_y = m\pi/b$ so that

$$h_z(x,y) = C(A\cos k_x x + B\sin k_x x)\cos\left(\frac{m\pi y}{b}\right).$$

We apply Eq. (12.30a) to get

$$\left.\frac{\partial h_z}{\partial x}\right|_{x=0} = k_x C(-A\sin k_x x + B\cos k_x x)\cos\left.\left(\frac{m\pi y}{b}\right)\right|_{x=0} = 0$$

$$k_x C(B)\cos\left(\frac{m\pi y}{b}\right) = 0,$$

from which we conclude that $B = 0$. This leaves

$$h_z(x, y) = (CA)(\cos k_x x)\cos\left(\frac{m\pi y}{b}\right).$$

We apply the last boundary condition, Eq. (12.30b), to get

$$\left.\frac{\partial h_z}{\partial x}\right|_{x=a} = k_x(AC)(-\sin k_x x)|_{x=a} = 0$$

$$k_x(AC)\sin k_x a = 0.$$

For this to hold without giving the trivial solution we must have $\sin k_x a = 0$ or $k_x a = n\pi$, where n is an integer. Thus

$$k_x = \frac{n\pi}{a}, \quad n = 0, 1, 2, \ldots.$$

We have now applied all the boundary conditions and found all the unknown constants except the product AC. This is the amplitude of the wave, which is determined by the strength of the source producing the wave. For each value of n and m, there is an amplitude that we can call $H_{0,nm}$. Each value of n and m gives a different solution, which is called a mode. The solution for h_z for the TE_{nm} mode is therefore

$$h_{z,nm} = H_{0,nm}\cos\left(\frac{n\pi x}{a}\right)\cos\left(\frac{m\pi y}{b}\right), \quad \begin{array}{l} n = 0, 1, 2, \ldots, \\ m = 0, 1, 2, \ldots, \\ \text{but not} \quad n = m = 0. \end{array} \tag{12.32}$$

There are a doubly infinite number of TE modes, corresponding to the integral values of n and m. We do not allow $n = m = 0$ because this means all the other field components are zero and the wave amplitude is zero.

We now use Eq. (12.20) to find k_c and Eq. (12.8) to find the propagation constant. Thus

$$k_{c,nm}^2 = \left(\frac{n\pi}{a}\right)^2 + \left(\frac{m\pi}{b}\right)^2 \tag{12.33}$$

and

$$\gamma_{nm} = \sqrt{k_{c,nm}^2 - \omega^2 \mu\varepsilon} = \sqrt{\left(\frac{n\pi}{a}\right)^2 + \left(\frac{m\pi}{b}\right)^2 - \omega^2 \mu\varepsilon}\,. \tag{12.34}$$

We see that for frequency low enough so that $\omega^2\mu\varepsilon < k_{c,nm}^2$ the propagation constant is real and the wave attenuates in the $\pm z$ direction. On the other hand, for frequency high enough so that $\omega^2\mu\varepsilon > k_{c,nm}^2$ the propagation constant is imaginary and the wave travels in the $\pm z$ direction. The cutoff frequency at which the transition occurs is

$$f_{c,nm} = \frac{k_{c,nm}}{2\pi\sqrt{\mu\varepsilon}} = c\sqrt{\left(\frac{n}{2a}\right)^2 + \left(\frac{m}{2b}\right)^2}, \quad c = \frac{1}{\sqrt{\mu\varepsilon}}. \tag{12.35}$$

The propagation constant is therefore

$$\gamma_{nm} = \begin{cases} \alpha_{nm} = \sqrt{k_{c,nm}^2 - \omega^2\mu\varepsilon} = \beta\sqrt{\left(\frac{f_{c,nm}}{f}\right)^2 - 1}, & f < f_{c,nm}, \quad (12.36) \\ j\beta_{nm} = j\sqrt{\omega^2\mu\varepsilon - k_{c,nm}^2} = j\beta\sqrt{1 - \left(\frac{f_{c,nm}}{f}\right)^2}, & f > f_{c,nm}, \quad (12.37) \end{cases}$$

where $\beta = \omega\sqrt{\mu\varepsilon} = 2\pi f/c$ is the unguided phase constant in the medium. The guided phase constant β_{nm} is zero at the cutoff frequency and approaches the unguided phase constant as frequency increases.

Figure 12.3 plots the dispersion curve (ω versus β). The slope of a line from the origin to a point on the curve equals the phase velocity. We see that the phase velocity is infinite at the cutoff frequency and approaches c as the frequency increases. The slope of the curve at any point is the group velocity. The group velocity is zero at the cutoff frequency and approaches c as frequency increases.

The phase velocity $v_{\mathrm{ph},nm}$ is given by

$$v_{\mathrm{ph},nm} = \frac{\omega}{\beta_{nm}} = \frac{\omega}{\sqrt{\omega^2\mu\varepsilon - k_{c,nm}^2}} = \frac{c}{\sqrt{1 - (f_{c,nm}/f)^2}}\,. \tag{12.38}$$

The group velocity $v_{\mathrm{g},nm}$ is

$$v_{\mathrm{g},nm} = \left(\frac{\partial\beta_{nm}}{\partial\omega}\right)^{-1} = \left(\frac{\omega\mu\varepsilon}{\beta_{nm}}\right)^{-1} = \frac{c^2}{v_{\mathrm{ph},nm}} = c\sqrt{1 - (f_{c,nm}/f)^2}\,. \tag{12.39}$$

Note that the product of the phase velocity and the group velocity is the unguided velocity of light in the medium squared, $v_{\mathrm{ph},nm}v_{\mathrm{g},nm} = c^2$. The phase and group velocities only have physical meaning for frequency greater than the cutoff frequency. The frequency behavior of these velocities matches the behavior surmised from the dispersion curve.

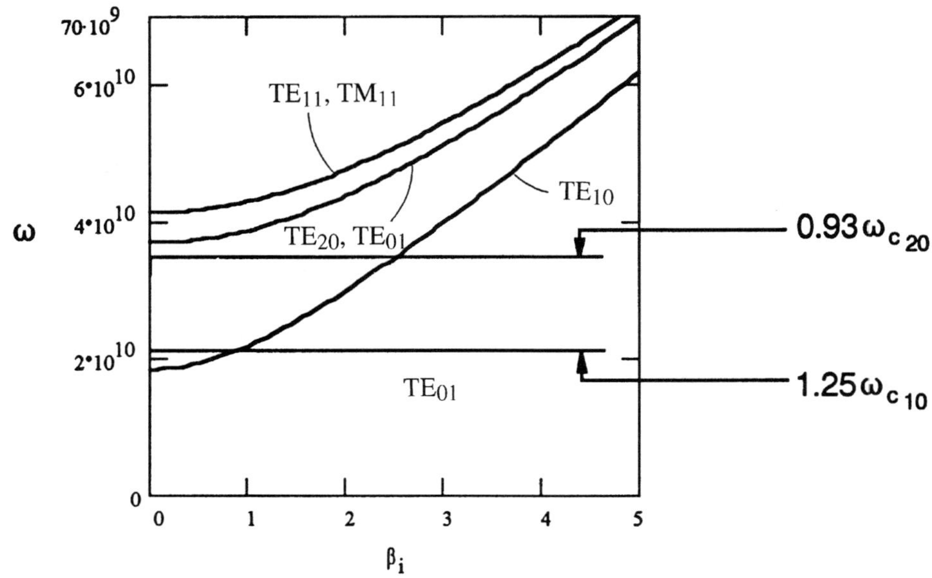

FIGURE 12.3 Dispersion curve or ω versus β diagram for modes in a rectangular waveguide.

Another term that only has physical meaning for frequency greater than the cutoff frequency is the guide wavelength, which is

$$\lambda_{g,nm} = \frac{2\pi}{\beta_{nm}} = \frac{2\pi}{\sqrt{\omega^2\mu\varepsilon - k_{c,nm}^2}} = \frac{\lambda}{\sqrt{1 - (f_{c,nm}/f)^2}}, \tag{12.40}$$

where $\lambda = c/f$ is the unguided wavelength in the medium. The guide wavelength is infinite at the cutoff frequency and approaches the unguided wavelength as frequency increases.

The mode impedance, Eq. (12.10), is

$$Z_{\mathrm{TE}_{nm}} = \frac{j\omega\mu}{\gamma_{nm}} = \begin{cases} \dfrac{j\omega\mu}{\alpha_{nm}} = \dfrac{j\eta}{\sqrt{(f_{c,nm}/f)^2 - 1}}, & f < f_{c,nm}, \quad (12.41) \\[2ex] \dfrac{\omega\mu}{\beta_{nm}} = \dfrac{\eta}{\sqrt{1 - (f_{c,nm}/f)^2}}, & f > f_{c,nm}, \quad (12.42) \end{cases}$$

where $\eta = \sqrt{\mu/\varepsilon}$ is the intrinsic impedance of the medium. When $f < f_{c,nm}$, the mode impedance is inductive reactive and the wave carries no time average power. The mode is evanescent. This is consistent with the result that it attenuates even though there are no losses in the perfect dielectric or the perfectly conducting walls. The inductive nature of the mode impedance indicates that the evanescent TE_{nm} modes predominantly store magnetic energy, $W_{m,\mathrm{Avg}} - W_{e,\mathrm{Avg}} > 0$. When $f > f_{c,nm}$ the mode impedance is resistive and the wave carries time average power without attenuation.

Equation (12.32) gives $h_{z,nm}$. The remaining field components can be found with the help of Eqs. (12.9) and (12.10). The actual phasors are found by multiplying each term by $\exp(\mp\gamma_{nm}z)$. The physical field is then obtained by multiplying by $\exp(j\omega t)$ and taking the real part of the product.

12.4.2 Transverse Magnetic Waves in the Rectangular Guide

For TM waves we must solve Eq. (12.13) subject to the boundary conditions on the perfectly conducting walls. The general solution is

$$e_z(x, y) = (A\cos k_x x + B\sin k_x x)(C\cos k_y y + D\sin k_y y).$$

The boundary condition is that the tangential electric field must be zero along the walls. e_z must satisfy this condition and hence we have

$$e_z(0, y) = 0 \qquad e_z(a, y) = 0 \tag{12.43a, b}$$

$$e_z(x, 0) = 0 \qquad e_z(x, b) = 0. \tag{12.44a, b}$$

Applying Eq. (12.43a) we find

$$e_z(0, y) = A(C\cos k_y y + D\sin k_y y) = 0,$$

from which we conclude that $A = 0$. Using this result and applying Eq. (12.43b) we find

$$e_z(a, y) = B(\sin k_x a)(C\cos k_y y + D\sin k_y y) = 0,$$

from which we conclude that $\sin k_x a = 0$ and $k_x a = n\pi$, where n is an integer. With

$$k_x = \frac{n\pi}{a}, \quad n = 1, 2, 3, \ldots \tag{12.45}$$

we now have

$$e_z(x, y) = B\sin\left(\frac{n\pi x}{a}\right)(C\cos k_y y + D\sin k_y y).$$

We use the boundary condition from Eq. (12.44a) to get

$$e_z(x, y) = B\sin\left(\frac{n\pi x}{a}\right)(C) = 0,$$

from which we conclude that $C = 0$. Using this result and the last boundary condition, Eq. (12.44b), we obtain

$$e_z(x, y) = B\sin\left(\frac{n\pi x}{a}\right)D\sin(k_y b) = 0,$$

from which we conclude that $\sin k_y b = 0$ and $k_y b = m\pi$. The result is

$$k_y = \frac{m\pi}{b}, \quad m = 1, 2, 3, \ldots . \tag{12.46}$$

We have now applied all the boundary conditions and found all the unknown constants except the product BD. This is the amplitude of the wave, which is determined by the strength of the source producing the wave. For each value of n and m, there is an amplitude that we can call $E_{0,nm}$. Each value of n and m gives a different mode solution. The solution for e_z for the TM_{nm} mode is therefore

$$e_{z,nm} = E_{0,nm} \sin\left(\frac{n\pi x}{a}\right) \sin\left(\frac{m\pi y}{b}\right), \quad \begin{array}{l} n = 1, 2, 3, \ldots, \\ m = 1, 2, 3, \ldots . \end{array} \tag{12.47}$$

The separation constants k_x and k_y are the same as they are for the TE modes and therefore $k_{c,nm}$, $f_{c,nm}$, and γ_{nm} are still given by Eqs. (12.33), (12.35) and (12.36) and (12.37), respectively. The mode impedance, Eq. (12.15), is

$$Z_{\mathrm{TM}_{nm}} = \frac{1}{Y_{\mathrm{TM}_{nm}}} = \frac{\gamma}{j\omega\varepsilon} = \begin{cases} \dfrac{\alpha_{nm}}{j\omega\varepsilon} = -j\eta\sqrt{(f_{c,nm}/f)^2 - 1}, & f < f_{c,nm}, \\ \dfrac{\beta_{nm}}{\omega\varepsilon} = \eta\sqrt{1 - (f_{c,nm}/f)^2}, & f > f_{c,nm}. \end{cases} \tag{12.48}$$

The mode impedance is capacitive reactive for $f < f_{c,nm}$ and resistive for $f > f_{c,nm}$. When the frequency is less than the mode's cutoff frequency the wave is evanescent. The capacitive nature of the mode impedance indicates that the evanescent TM_{nm} modes predominantly store electric energy, $W_{m,\mathrm{Avg}} - W_{e,\mathrm{Avg}} < 0$. For frequency greater than the cutoff frequency the wave carries time average power.

12.5 THE DOMINANT TE_{10} MODE IN THE RECTANGULAR WAVEGUIDE

Let us now closely examine Eq. (12.35) for the cutoff frequency of the nmth mode. Assuming that $a > b$ the lowest cutoff frequency will be for the $n = 1$, $m = 0$ mode. Only the transverse electric waves permit one of the indices to be zero and hence this corresponds to the TE_{10} mode. We recall that in a single conductor structure there is no TEM mode. Hence the TE_{10} mode is the dominant mode in the rectangular waveguide.†

In almost all practical situations the signals carried in waveguides are restricted to a frequency range in which only the dominant mode propagates. There are two reasons for this restriction. First, the design and placement of the device used to launch the wave at the input end of the guide or to receive the wave at its output end depends on the character and shape of the field, which is different for each mode. If we

† When $a = b$, then $f_{c,10} = f_{c,01}$. The TE_{10} and TE_{01} modes are then degenerate. Waveguides are not usually constructed with this geometry.

TABLE 12.1 Ratio of Cutoff Frequencies with $a = 2b$

n	1	2	0	1	2	3	3	0
m	0	0	1	1	1	0	1	2
Ratio	1	2	2	2.24	2.83	3	3.61	4
Type	TE	TE	TE	TE, TM	TE, TM	TE	TE, TM	TE

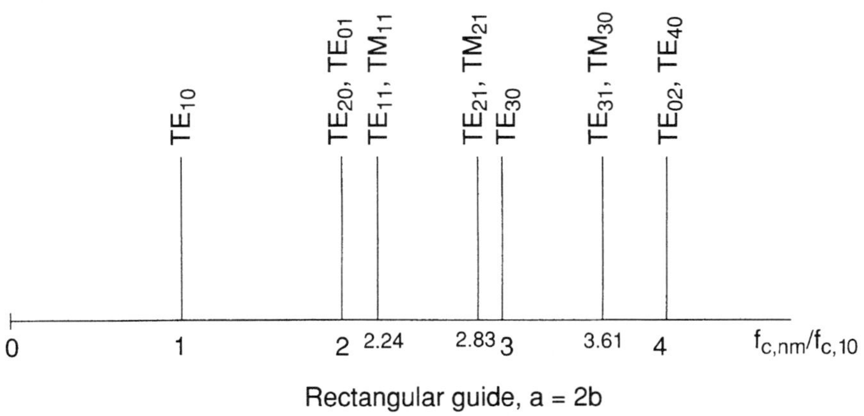

FIGURE 12.4 Cutoff frequencies for the TE and TM modes in a rectangular waveguide with $a = 2b$.

are to design efficient wave launchers and receivers, we must know which mode is carrying the signal and we can only be certain of that if there is only one propagating mode. Second, the signal velocity is different for each mode. This means that if more than one mode carries the signal from the input to the output, each mode will transmit the signal at a different velocity. If we think of packets of information being transmitted, we see that each mode would deliver the packet at a different time, causing distortion and interference with adjacent packets.

Example 12.1 A typical geometry for rectangular waveguides has $a = 2b$. Find the cutoff frequency of the first ten modes as a multiple of the dominant mode cutoff frequency.

Solution The dominant TE_{10} mode's cutoff frequency is $f_{c,10} = c/2a$. Setting $b = a/2$ in Eq. (12.35) and dividing by $c/2a$ we obtain

$$\text{Ratio} = \frac{f_{c,nm}}{f_{c,10}} = \sqrt{n^2 + (2m)^2}, \quad a = 2b.$$

Table 12.1 shows the results of subsitituing values of n and m and Fig. 12.4 shows the results along a frequency line.

We see that for this shape guide there is a frequency band $c/2a < f < c/a$ in which only the dominant mode can propagate. It is generally desirable to allow only dominant mode propagation and hence this sets the maximum theoretical bandwidth for single mode propagation.

As an example, if the inside guide dimensions are 3 by 1.5 inches and the guide interior medium is air, then $f_{c,10} = c/2a \approx (12 \text{ in./ns})/(2 \times 3 \text{ in.}) = 2$ GHz. The TE_{20} and TE_{01} modes begin to propagate at 4 GHz. From 2 to 4 GHz only the dominant TE_{10} mode is a traveling wave.

Let us write out the complete field components for the TE_{10} mode, assuming we are operating above the cutoff frequency, so $\gamma_{10} = j\beta_{10}$. From Eq. (12.32)

$$h_{z,10} = H_{0,10} \cos\left(\frac{\pi x}{a}\right).$$

Using Eq. (12.9) we find

$$\vec{h}_{t,10} = \mp\left(\frac{j\beta_{10}}{k_{c,10}^2}\right)\nabla_t h_{z,10m} = \mp\left(\frac{j\beta_{10}}{(\pi/a)^2}\right)\vec{a}_x\left(-\frac{\pi}{a}\right)H_{0,10}\sin\left(\frac{\pi x}{a}\right)$$

$$= \pm\,\vec{a}_x\left(\frac{j\beta_{10}}{\pi/a}\right)H_{0,10}\sin\left(\frac{\pi x}{a}\right).$$

Using Eq. (12.10) we find

$$\vec{e}_t = \mp\frac{j\omega\mu}{j\beta_{10}}\,\vec{a}_z\times\vec{h}_t = \mp\frac{j\omega\mu}{j\beta_{10}}\,\vec{a}_z\times\left[\mp\,\vec{a}_x\left(\frac{j\beta_{10}}{\pi/a}\right)H_{0,10}\sin\left(\frac{\pi x}{a}\right)\right]$$

$$= -\,\vec{a}_y\left(\frac{j\omega\mu}{\pi/a}\right)H_{0,10}\sin\left(\frac{\pi x}{a}\right).$$

The complete phasors are obtained by multiplying by $\exp(\mp j\beta_{10}z)$:

$$\underline{E}_y = -\left(\frac{j\omega\mu}{\pi/a}\right)H_{0,10}\sin\left(\frac{\pi x}{a}\right)e^{\mp j\beta_{10}z},$$

$$\underline{H}_x = \pm\left(\frac{j\beta_{10}}{\pi/a}\right)H_{0,10}\sin\left(\frac{\pi x}{a}\right)e^{\mp j\beta_{10}z},$$

$$\underline{H}_z = H_{0,10}\cos\left(\frac{\pi x}{a}\right)e^{\mp j\beta_{10}z}.$$

We have now shown that we can use the "big guns" developed earlier in the chapter to obtain the transverse field components from the longitudinal field, at least for the TE_{10} mode. In view of the importance of this mode, let us see if there isn't a simpler way to arrive at the same results.

We know for parallel plates located at $y = 0$ and $y = b$ there is a wave $\underline{E}_y = E_0 \exp(-j\beta z)$. We want a solution like that, except that the field must go to zero at $x = 0$

and $x = a$. It is natural to assume a sine dependence on x in order to get the function to go to zero at those boundaries. Let us guess at a solution with

$$\underline{E}_y = E_0 \sin\left(\frac{\pi x}{a}\right) e^{-j\beta_{10}z}, \tag{12.49}$$

where β_{10} is unknown, and see if it works. "Seeing if it works" means substituting in Maxwell's equations to see if they are satisfied. Using the Faraday–Maxwell curl equation we obtain

$$\begin{aligned}\vec{\underline{H}} &= -\frac{1}{j\omega\mu} \nabla \times \vec{\underline{E}} = -\frac{1}{j\omega\mu}\left(-\vec{a}_x \frac{\partial \underline{E}_y}{\partial z} + \vec{a}_z \frac{\partial \underline{E}_y}{\partial x}\right) \\ &= -\frac{1}{j\omega\mu} E_0 \left[\vec{a}_x\, j\beta_{10} \sin\left(\frac{\pi x}{a}\right) + \vec{a}_z \left(\frac{\pi}{a}\right) \cos\left(\frac{\pi x}{a}\right)\right] e^{-j\beta_{10}z}\end{aligned}$$

or

$$\underline{H}_x = -\frac{\beta_{10}}{\omega\mu} E_0 \sin\left(\frac{\pi x}{a}\right) e^{-j\beta_{10}z}, \tag{12.50}$$

$$\underline{H}_z = -\frac{(\pi/a)}{j\omega\mu} E_0 \cos\left(\frac{\pi x}{a}\right) e^{-j\beta_{10}z}. \tag{12.51}$$

This gives us the full set of field components, but we must ensure that the Ampère–Maxwell equation is satisfied.

$$\begin{aligned}\vec{\underline{E}} &= \frac{1}{j\omega\varepsilon} \nabla \times \vec{\underline{H}} = \frac{1}{j\omega\varepsilon} \vec{a}_y \left(\frac{\partial \underline{H}_x}{\partial z} - \frac{\partial \underline{H}_z}{\partial x}\right) \\ &= \vec{a}_y E_0 \left[\frac{1}{j\omega\varepsilon}\left(\frac{j\beta_{10}^2}{\omega\mu} - \frac{(\pi/a)^2}{j\omega\mu}\right)\right] \sin\left(\frac{\pi x}{a}\right) e^{-j\beta_{10}z}.\end{aligned}$$

For our results to be a solution to Maxwell's equations $\underline{E}_y$ in the last equation must equal our original assumption, Eq. (12.49). This requires that the bracketed term be equal to unity, with the result that

$$\beta_{10} = \sqrt{\omega^2\mu\varepsilon - (\pi/a)^2}. \tag{12.52}$$

If we compare these field components with those previously obtained, we see they are the same with the arbitrary amplitude $E_0 = [-j\omega\mu/(\pi/a)]\, H_{0,10}$. Going from Eq. (12.49) to Eqs. (12.50) and (12.51), remember that Eq. (12.52) is clearly the simplest way to write down the TE_{10} field components. We will use these equations in calculating time average power and losses.

Since "guessing" worked so well, why do we need the formal "big gun" solution? First, "guesses" are easier when you already know the answer. Second, without knowing what all the modes are we could not tell that the TE_{10} mode is the dominant mode. Third, except for TE_{0m} and TE_{n0} modes it is not so simple to guess a field component and then obtain the other field components.

12.5.1 TE_{10} Mode Time Average Power

Let us now calculate the time average power carried by the TE_{10} mode. The time average power density is

$$\vec{p}_{f,\text{Avg}} = \tfrac{1}{2}\text{Re}(\underline{\vec{E}} \times \underline{\vec{H}}^*) = \tfrac{1}{2}\vec{a}_z \left(\frac{\beta_{10}}{\omega\mu}\right) |E_0|^2 \sin^2\left(\frac{\pi x}{a}\right).$$

The power flow is entirely in the z direction, which accords with the fact that power cannot flow into the perfectly conducting walls.

The total time average power carried by the wave is found by integrating over the cross section of the guide.

$$P_{f,\text{Avg}} = \int_S \vec{p}_{f,\text{Avg}} \cdot d\vec{S} = \frac{1}{2}\left(\frac{\beta_{10}}{\omega\mu}\right)|E_0|^2 \int_{y=0}^{b} \int_{x=0}^{a} \sin^2\left(\frac{\pi x}{a}\right) dx\, dy$$

$$= \frac{1}{2}\left(\frac{\beta_{10}}{\omega\mu}\right)\left(\frac{ab}{2}\right)|E_0|^2. \tag{12.53}$$

12.5.2 Attenuation Due to Imperfectly Conducting Walls

We have found the field within the guide, assuming that the walls are perfect conductors. We now use those fields to find the time average power loss per unit length (in the z direction). Recall that the power loss density (W/m^2) at any point on the surface of the conductor is

$$p_{l,\text{Avg}} = \tfrac{1}{2}|H_{\text{tang}}|^2 R_s, \quad R_s = \sqrt{\frac{\omega\mu_c}{2\sigma_c}},$$

where μ_c (usually equal to μ_0) and σ_c are the permeability and conductivity of the conducting walls. The power loss density on each of the four walls is therefore

$$p_{l,\text{Avg}}|_{x=0} = \tfrac{1}{2}|H_z(0,y,z)|^2 R_s = \frac{1}{2}\frac{(\pi/a)^2}{(\omega\mu)^2}|E_0|^2 R_s,$$

$$p_{l,\text{Avg}}|_{x=a} = \tfrac{1}{2}|H_z(a,y,z)|^2 R_s = p_{l,\text{Avg}}|_{x=0},$$

$$p_{l,\text{Avg}}|_{y=0} = \tfrac{1}{2}\left[|H_x(x,0,z)|^2 + |H_z(x,0,z)|^2\right] R_s$$

$$= \frac{1}{2}\left[\left(\frac{\beta}{\omega\mu}\right)^2 \sin^2\left(\frac{\pi x}{a}\right) + \frac{(\pi/a)^2}{(\omega\mu)^2}\cos^2\left(\frac{\pi x}{a}\right)\right]|E_0|^2 R_s,$$

$$p_{l,\text{Avg}}|_{y=b} = \tfrac{1}{2}\left[|H_x(x,b,z)|^2 + |H_z(x,b,z)|^2\right] R_s = p_{l,\text{Avg}}|_{y=0}.$$

To obtain the total power loss per unit length we integrate the power loss density around the perimeter of the guide:

$$P_{l,\text{Avg}} = \int_{y=0}^{b} (p_{l,\text{Avg}}|_{x=0} + p_{l,\text{Avg}}|_{x=a})\, dy + \int_{x=0}^{a} (p_{l,\text{Avg}}|_{y=0} + p_{l,\text{Avg}}|_{y=b})\, dx$$

$$= \frac{1}{2}\left\{2\left[\frac{(\pi/a)^2}{(\omega\mu)^2}\right] b + 2\left[\left(\frac{\beta}{\omega\mu}\right)^2 + \frac{(\pi/a)^2}{(\omega\mu)^2}\right]\frac{a}{2}\right\} |E_0|^2 R_s$$

$$P_{l,\text{Avg}} = \frac{1}{(\omega\mu)^2}\left[\left(\frac{\pi}{a}\right)^2 b + (\omega^2\mu\varepsilon)\frac{a}{2}\right] |E_0|^2 R_s. \tag{12.54}$$

The attenuation constant due to conducting losses in the guide walls is then given by Eq. (10.75):

$$\alpha_c = \frac{P_{l,\text{Avg}}}{2P_{f,\text{Avg}}} = \frac{\dfrac{1}{(\omega\mu)^2}\left[\left(\dfrac{\pi}{a}\right)^2 b + (\omega^2\mu\varepsilon)\dfrac{a}{2}\right] |E_0|^2 R_s}{\left(\dfrac{\beta_{10}}{\omega\mu}\right)|E_0|^2\left(\dfrac{ab}{2}\right)}$$

$$= \frac{1}{\omega\mu\beta_{10}\, ab}\left[2\left(\frac{\pi}{a}\right)^2 b + \omega^2\mu\varepsilon a\right] R_s, \quad R_s = \sqrt{\frac{\omega\mu_c}{2\sigma_c}}. \tag{12.55}$$

This can also be written as

$$\alpha_c = \frac{1}{\eta b\sqrt{1 - (f_{c,10}/f)^2}}\left[1 + \frac{2b}{a}\left(\frac{f_{c,10}}{f}\right)^2\right]\sqrt{\frac{\pi f \mu_c}{\sigma_c}}, \quad \eta = \sqrt{\frac{\mu}{\varepsilon}}, \tag{12.56}$$

to clearly show the frequency dependence.

When the frequency is just greater than the cutoff frequency the attenuation is infinite, due to the term within the radical in the denominator. As frequency increases the attenuation drops rapidly. However, as the frequency increases further and becomes much greater than the cutoff frequency, then the terms $f_{c,10}/f$ can be neglected and the attenuation increases as the square root of frequency. As an example, Fig. 12.5 shows the attenuation plotted as a function of normalized frequency for a copper wall guide with $b = a/2$.

12.5.3 Attenuation Due to Imperfect Dielectric

A lossy dielectric within the guide is characterized by letting $\varepsilon = \varepsilon' - j\varepsilon''$, where $\varepsilon'' \ll \varepsilon'$. The propagation constant (above cutoff) becomes

$$\gamma_{10} = j\sqrt{\omega^2\mu(\varepsilon' - j\varepsilon'') - (\pi/a)^2} = j\sqrt{\omega^2\mu\varepsilon' - (\pi/a)^2}\left[1 - j\frac{\omega^2\mu\varepsilon''}{\omega^2\mu\varepsilon' - (\pi/a)^2}\right]^{1/2}$$

$$\approx j\sqrt{\omega^2\mu\varepsilon' - (\pi/a)^2}\left[1 - j\frac{\omega^2\mu\varepsilon''}{2(\omega^2\mu\varepsilon' - (\pi/a)^2)}\right]$$

$$= \frac{\omega^2\mu\varepsilon''}{2\sqrt{\omega^2\mu\varepsilon' - (\pi/a)^2}} + j\sqrt{\omega^2\mu\varepsilon' - (\pi/a)^2},$$

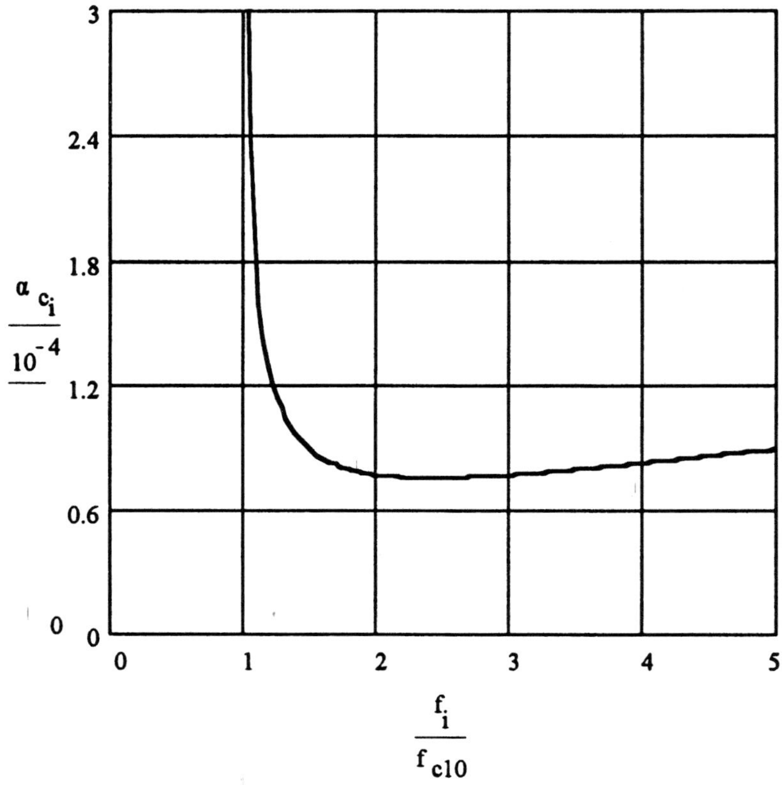

FIGURE 12.5 Attenuation due to wall losses for the TE_{10} mode, $a = 2$ in., $b = 1$ in., and copper walls.

from which we obtain

$$\gamma_{10} = \alpha_d + j\beta_{10}, \quad \alpha_d = \frac{\omega^2 \mu \varepsilon''}{2\beta_{10}}, \quad \beta_{10} = \sqrt{\omega^2 \mu \varepsilon' - \left(\frac{\pi}{a}\right)^2}. \tag{12.57}$$

If both dielectric and conducting wall losses are present, then the total attenuation constant is the sum of the two attenuation constants:

$$\alpha_t = \alpha_c + \alpha_d$$

with ε' replacing ε in the formula for α_c. Generally, conduction losses are much greater than dielectric losses.

12.5.4 Design Considerations

Due to the large conduction losses just above cutoff the usable frequency range for a waveguide has its lower frequency limit about 20% above the cutoff frequency. The upper frequency limit is generally about 5% below the cutoff frequency of the next propagating mode. If $a = 2b$, for example, then the usable frequency band is $1.2f_{c,10} < f < 0.95(2f_{c,10}) = 1.9f_{c,10}$. Looking at Fig. 12.3 we see that the group velocity is fairly constant over this frequency range, so that the dispersion distortion is acceptable unless the guide is very long.

Example 12.2 Design a waveguide for a frequency band with its lower limit at 6.0 GHz. What is the upper frequency limit to allow single mode propagation? Calculate the group velocity at the lower and upper ends of the frequency band.

Solution The cutoff frequency for the guide should be about $f_{c,10} = 6.0/1.2 = 5.0$ GHz. The wider dimension should be $a = c/2f_{c,10} = 3 \times 10^8/2(5 \times 10^9) = 0.0300$ m or 3.00 cm. Taking $b = a/2 = 1.5$ cm, the TE_{20} mode and the TE_{01} mode have a cutoff frequency of 10 GHz. The upper frequency limit will then be $0.95 \times 10 = 9.5$ GHz. The group velocities are

$$\begin{aligned} v_g = c\sqrt{1-(f_{c,10}/f)^2} &= 3\times 10^8\sqrt{1-(5.0/6.0)^2} = 1.66\times 10^8 \text{ m/s} \quad \text{at } 6.0 \text{ GHz} \\ &= 3\times 10^8\sqrt{1-(5.0/9.5)^2} = 2.55\times 10^8 \text{ m/s} \quad \text{at } 9.5 \text{ GHz.} \end{aligned}$$

This appears to be a large change in group velocity, but individual signals do not generally occupy this entire 3.5 GHz band. A TV channel, for example, has a bandwidth of 6 MHz, so that the average fractional change in group velocity over the bandwidth of the TV channel is

$$\frac{\Delta v_g}{v_{g,\text{Average}}} = \frac{(2.55-1.66)}{(2.55+1.66)/2}\left(\frac{6\times 10^6}{3.5\times 10^9}\right) = 0.72\times 10^{-3}.$$

12.5.5 Physical Fields

To obtain the physical field in the guide we multiply the phasor field, Eqs. (12.49), (12.50), and (12.51), by $\exp(j\omega t)$ and take the real part. Assuming for simplicity that E_0 is real, the result is

$$E_y = E_0 \sin\left(\frac{\pi x}{a}\right)\cos(\omega t - \beta_{10} z), \tag{12.58}$$

$$H_x = -\frac{\beta_{10}}{\omega\mu} E_0 \sin\left(\frac{\pi x}{a}\right)\cos(\omega t - \beta_{10} z), \tag{12.59}$$

$$H_z = -\frac{(\pi/a)}{\omega\mu} E_0 \cos\left(\frac{\pi x}{a}\right)\sin(\omega t - \beta_{10} z). \tag{12.60}$$

The z component of the magnetic field intensity is a quarter cycle out of time phase with the other two field components. That is the meaning of the j in Eq. (12.51).

Figure 12.6a shows the electric field intensity within the guide at time $t = 0$. Electric field lines are densest in the center of the guide along $x = a/2$ and vary cosinusoidally in the longitudinal direction from a positive maximum at $z = 0$ to a negative maximum at $z = \lambda_{g,10}/2 = \pi/\beta_{10}$ and so on. As time increases the picture moves in the

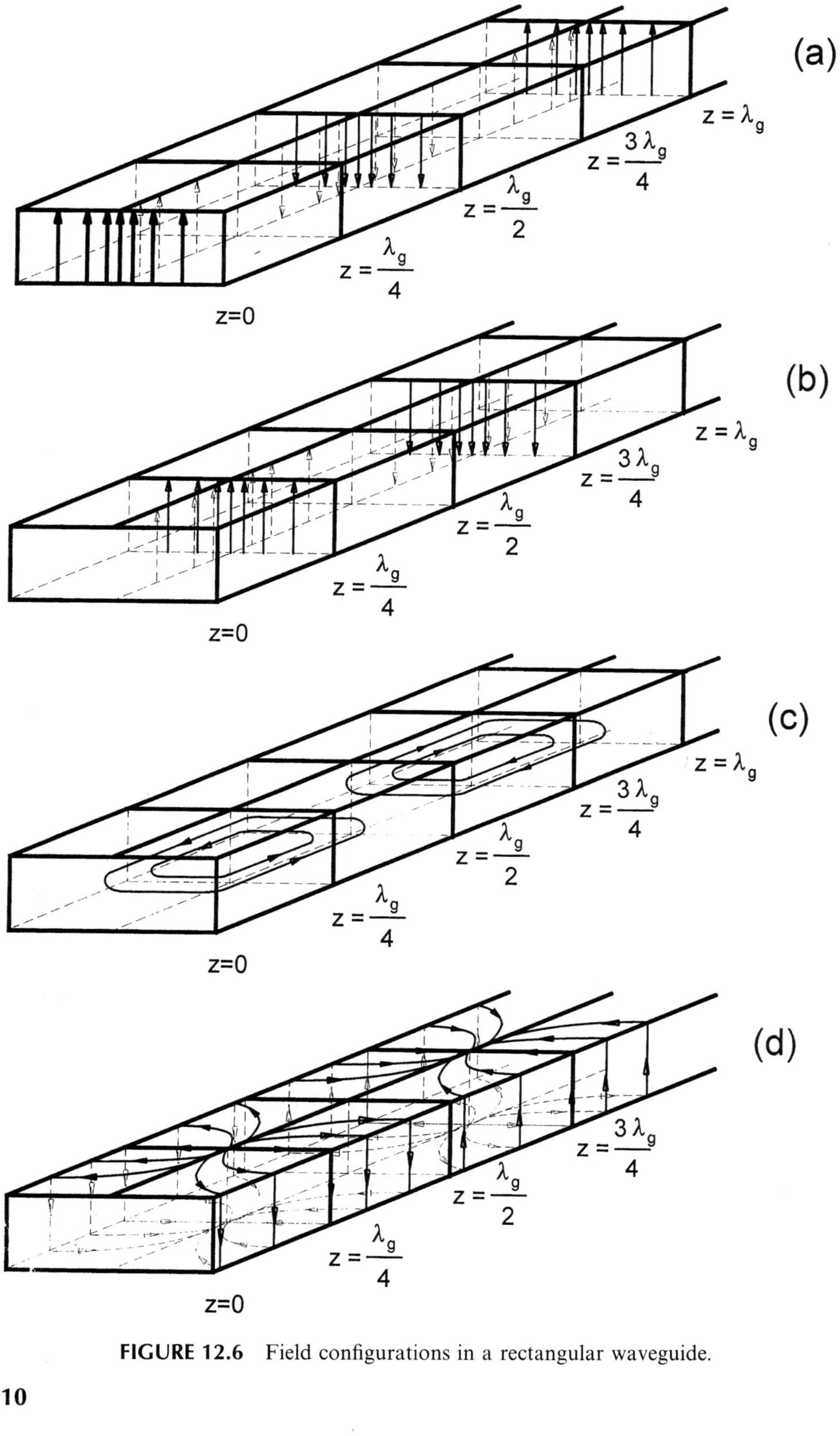

FIGURE 12.6 Field configurations in a rectangular waveguide.

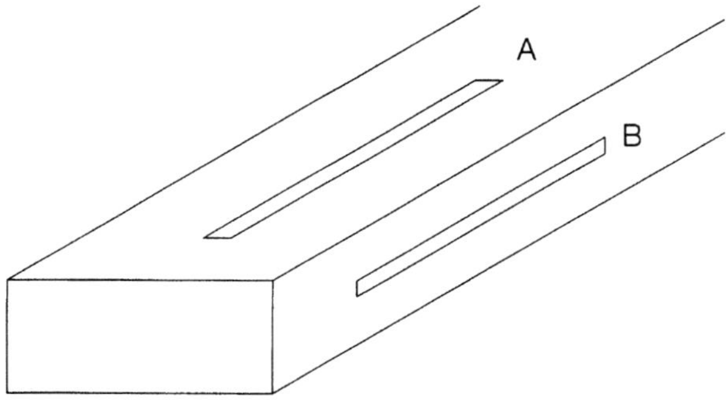

FIGURE 12.7 Slots in a rectangular waveguide.

z direction. Displacement current is proportional to the time rate of change of electric field. Hence the displacement current (Fig. 12.6b) is positive maximum where the electric field lines are increasing most rapidly, at $z = \lambda_{g,10}/4$, and is negative maximum where the electric field lines are decreasing most rapidly, at $z = 3\lambda_{g,10}/4$. Magnetic field lines surround the displacement current in a right-hand rule sense and form closed lines (Fig. 12.6c) centered on $x = a/2$, $z = \lambda_{g,10}/4$ and on $x = a/2$, $z = 3\lambda_{g,10}/4$. Note that where the two sets of closed lines are contiguous they are in the same direction. Conduction current flows on the inner surface of the guide wall (Fig. 12.6d) in accordance with the boundary condition that $\vec{J}_s = \vec{a}_n \times \vec{H}_{\text{tang}}$, where $\vec{a}_n$ points into the guide. The total current is the sum of the displacement and conduction current (Figs. 12.6b and 12.6d), and it flows in a closed continuous circuit. Of course, all the pictures move in the positive $z = 0$ direction with phase velocity $v_{\text{ph},10}$.

Figure 12.6d shows that no current crosses the centerline of the top or bottom face of the guide. Hence a thin slot cut along this line, as shown by slot A in Fig. 12.7, will not disrupt the current or the wave propagation. This is the way a slot is cut to insert a movable electric-field-sensitive probe to measure the field within the guide. On the other hand, cutting a slot in the longitudinal direction on a side wall (slot B in the figure) will break the lines of current, disturb the wave, allow radiation from the slot, and cause reflection of the wave. If slot B were sufficiently long then the side wall would not be a conductor for up- and down-directed currents and the TE_{10} mode could not exist.

12.6 CIRCULAR WAVEGUIDES

Figure 12.2 shows a circular waveguide of radius a. We will find the TE and TM wave solutions assuming that the walls are perfectly conducting. We will then use these solutions to find the attenuation due to imperfectly conducting walls.

12.6.1 Transverse Magnetic Modes in the Circular Waveguide

For a TM wave the longitudinal electric field must satisfy Eq. (12.13), the solution of which is given in cylindrical coordinates in Eq. (12.25). Thus

$$e_z(r, \phi) = [AJ_\nu(k_c r) + BN_\nu(k_c r)][C\cos(\nu\phi) + D\sin(\nu\phi)]. \qquad (12.61)$$

The boundary conditions require that:

1. The field and its derivative with respect to ϕ must be continuous at $\phi = 0$ and $\phi = 2\pi$.
2. The field must be finite everywhere within the guide since there are no sources. In particular, $e_z(0, \phi)$ must be finite.
3. The tangential electric field must be zero along the perfectly conducting wall. Hence $e_z(a, \phi) = 0$.

Condition 1 means that

$$e_z(r,0) = e_z(r,2\pi) \quad \text{or} \quad C = C\cos(2\pi v) + D\sin(2\pi v),$$

$$\left.\frac{\partial e_z(r,\phi)}{\partial \phi}\right|_{\phi=0} = \left.\frac{\partial e_z(r,\phi)}{\partial \phi}\right|_{\phi=2\pi} \quad \text{or} \quad vD = v[-C\sin(2\pi v) + D\cos(2\pi v)].$$

The first equation is satisfied if (a) $C = 0$ and $v = 0, \frac{1}{2}, 1, \frac{3}{2}, \ldots$ or if (b) $v = 0, 1, 2, \ldots$. The second equation is satisfied by possibility (b), but not by (a). Hence $v = n$, $n = 0, 1, 2, 3, \ldots$. Reference to Appendix C, Eq. (C.3), shows that $N_n(0)$ is infinite and hence condition 2 requires that $B = 0$. Equation (12.61) has become

$$e_z = J_n(k_c r)(AC\cos n\phi + AD\sin n\phi). \tag{12.62}$$

The final boundary condition 3 requires that $J_n(k_c a) = 0$ or $k_c a = p_{nm}$, where p_{nm} is the mth zero $m = 1, 2, 3, \ldots$ of J_n. Thus the TM$_{nm}$ mode solution is

$$k_{c,nm} = \frac{p_{nm}}{a}, \quad n = 0, 1, 2, 3, \ldots \quad m = 1, 2, 3, \ldots, \tag{12.63}$$

$$e_{z,nm} = J_n\left(\frac{p_{nm} r}{a}\right)(C_{nm}\cos n\phi + D_{nm}\sin n\phi), \tag{12.64}$$

and

$$\gamma_{nm} = \sqrt{\left(\frac{p_{nm}}{a}\right)^2 - k^2} = \sqrt{\left(\frac{p_{nm}}{a}\right)^2 - \omega^2\mu\varepsilon}. \tag{12.65}$$

There is again a doubly infinite number of modes corresponding to the integers n and m.

The two factors $\cos(n\phi)$ and $\sin(n\phi)$ simply represent the same field configuration but rotated within the guide by an angle of $\pi/2n$ radians ($n = 1, 2, 3, \ldots$). This rotation is possible in a circular guide because the choice of direction for the x axis is arbitrary. For $n = 0$, the sine term is zero and the cosine is constant, with value equal to one, indicating the field is independent of angle. Since these two factors represent the same mode, let us drop the sine term and replace C_{nm} with $E_{0,nm}$, giving

$$e_{z,nm} = E_{0,nm} J_n\left(\frac{p_{nm} r}{a}\right)\cos n\phi. \tag{12.66}$$

TABLE 12.2 Zero p_{nm} of $J_n(p)$

m	1	2	3	4
p_{0m}	2.405	5.520	8.654	11.792
p_{1m}	3.832	7.016	10.173	13.324
p_{2m}	5.136	8.417	11.620	14.796
p_{3m}	6.380	9.761	13.015	16.223

The cutoff frequency for the nmth mode is the frequency for which γ_{nm} goes to zero:

$$f_{c,nm} = \frac{p_{nm}}{2\pi\sqrt{\mu\varepsilon}a} = \frac{cp_{nm}}{2\pi a}. \tag{12.67}$$

For frequencies less than the cutoff frequency the wave attenuates and is evanescent. For frequencies greater than the cutoff frequency the wave propagates without attenuation and carries time average power. Equations (12.36) through (12.40) for the attenuation constant, phase constant, phase velocity, group velocity, and guide wavelength still hold. Table 12.2 gives the first few zero p_{nm} of J_n. Note that $p_{01} = 2.405$ is the smallest value so the TM_{01} mode has the lowest cutoff frequency of all TM modes.

Let us rewrite Eq. (12.66) as a phasor by reinserting the propagation factor for a forward going wave, $\exp(-j\beta_{nm}z)$:

$$\underline{E}_{z,nm} = E_{0,nm} J_n\left(\frac{p_{nm}r}{a}\right)\cos(n\phi)e^{-j\beta_{nm}z}. \tag{12.68}$$

The transverse electric field components are found from Eq. (12.14), which, written for cylindrical propagating modes, is

$$\vec{\underline{E}}_t = -\frac{j\beta}{k_c^2}\left(\vec{a}_r\frac{\partial \underline{E}_z}{\partial r} + \vec{a}_\phi\frac{1}{r}\frac{\partial \underline{E}_z}{\partial \phi}\right).$$

Noting that $dJ_n(qu)/du = q[dJ_n(qu)/d(qu)] = qJ'_n(q)$ we obtain

$$\underline{E}_{r,nm} = -\left(\frac{j\beta_{nm}a}{p_{nm}}\right)E_{0,nm}J'_n\left(\frac{p_{nm}r}{a}\right)\cos(n\phi)e^{-j\beta_{nm}z}, \tag{12.69}$$

$$\underline{E}_{\phi,nm} = \left(\frac{j\beta_{nm}a^2n}{p_{nm}^2r}\right)E_{0,nm}J_{nm}\left(\frac{p_{nm}r}{a}\right)\sin(n\phi)e^{-j\beta_{nm}z}. \tag{12.70}$$

The transverse magnetic field components are found using Eq. (12.15), $\vec{\underline{H}}_t = (\omega\varepsilon/\beta_{nm})\vec{a}_z \times \vec{\underline{E}}_t$. Thus

$$\underline{H}_{r,nm} = -\left(\frac{j\omega\varepsilon a^2n}{p_{nm}^2r}\right)E_{0,nm}J_{nm}\left(\frac{p_{nm}r}{a}\right)\sin(n\phi)e^{-j\beta_{nm}z}, \tag{12.71}$$

$$\underline{H}_{\phi,nm} = -\left(\frac{j\omega\varepsilon a}{p_{nm}}\right) E_{0,nm} J_n'\left(\frac{p_{nm} r}{a}\right) \cos(n\phi) e^{-j\beta_{nm} z}. \tag{12.72}$$

If it is desirable, we can replace the derivative of the Bessel function in the $\underline{E}_r$ and $\underline{H}_\phi$ components with [Eq. (C.7)]

$$J_n'\left(\frac{p_{nm} r}{a}\right) = \frac{na}{p_{nm} r} J_n\left(\frac{p_{nm} r}{a}\right) - J_{n+1}\left(\frac{p_{nm} r}{a}\right).$$

12.6.1.1 Time Average Power and Attenuation for the TM Mode

The time average power density flowing in the guide is in the z direction and is

$$\begin{aligned} p_{f,\text{Avg,TM}_{nm}} &= \tfrac{1}{2} \operatorname{Re}(\underline{E}_r \underline{H}_\phi^* - \underline{E}_\phi \underline{H}_r^*) \\ &= \frac{1}{2} \frac{\omega\varepsilon\beta_{nm} a^2}{p_{nm}^2} |E_{0,nm}|^2 \left\{ \left[J_n'\left(\frac{p_{nm} r}{a}\right)\right]^2 \cos^2 n\phi \right. \\ &\quad \left. + \left(\frac{na}{p_{nm} r}\right)^2 \left[J_n\left(\frac{p_{nm} r}{a}\right)\right]^2 \sin^2 n\phi \right\}. \end{aligned} \tag{12.73}$$

The total power carried by the wave is

$$P_{f,\text{Avg,TM}_{nm}} = \int_{r=0}^{a} \int_{\phi=0}^{2\pi} p_{f,\text{Avg,TM}_{nm}} r \, d\phi \, dr.$$

The integration over ϕ should be done first. When $n = 0$ the sine term is zero and its integral is zero, and the cosine equals unity with an integral of 2π. When $n = 1, 2, \ldots$ the integral of the squared sine or cosine is π. Thus if we let

$$\delta_n = \begin{cases} 1, & n = 0, \\ 2, & n = 1, 2, \ldots, \end{cases} \tag{12.74}$$

we have

$$\begin{aligned} P_{f,\text{Avg,TM}_{nm}} &= \frac{1}{2} \frac{\omega\varepsilon\beta_{nm} a^2}{p_{nm}^2} \left(\frac{2\pi}{\delta_n}\right) |E_{0,nm}|^2 \int_{r=0}^{a} \left\{ \left[J_n'\left(\frac{p_{nm} r}{a}\right)\right]^2 \right. \\ &\quad \left. + \left(\frac{na}{p_{nm} r}\right)^2 \left[J_n\left(\frac{p_{nm} r}{a}\right)\right]^2 \right\} r \, dr. \end{aligned}$$

The remaining integral can be put in the form of the integral of Eq. (C.17) by letting $x = p_{nm} r/a$. Then we have

$$\begin{aligned} \frac{P_{f,\text{Avg,TM}_{nm}}}{C} &= \int_{x_1}^{x^2} x \left\{ [J_n'(x)]^2 + \left(\frac{n}{x}\right)^2 [J_n(x)]^2 \right\} dx \\ &= x J_n(x) J_n'(x) + \frac{x^2}{2} \left\{ [J_n(x)]^2 \left(1 - \frac{n^2}{x^2}\right) + [J_n'(x)]^2 \right\} \Bigg|_{x_1}^{x_2}, \end{aligned}$$

where

$$C = \frac{1}{2}\frac{\omega\varepsilon\beta_{nm}a^4}{p_{nm}^4}\left(\frac{2\pi}{\delta_n}\right)|E_{0,nm}|^2,$$

$x_1 = 0$ and $x_2 = p_{nm}$. At the upper limit $x_2 = p_{nm}$ we have $J_n(p_{nm}) = 0$ and at the lower limit all the terms are zero. Hence

$$\frac{P_{f,\text{Avg},\text{TM}_{nm}}}{C} = \frac{p_{nm}^2}{2}[J_n'(p_{nm})]^2 = \frac{p_{nm}^2}{2}[J_{n+1}(p_{nm})]^2$$

$$P_{f,\text{Avg},\text{TM}_{nm}} = \frac{1}{2}\frac{\omega\varepsilon\beta_{nm}a^4}{p_{nm}^2}\left(\frac{\pi}{\delta_n}\right)|E_{0,nm}|^2[J_{n+1}(p_{nm})]^2. \qquad (12.75)$$

The power loss per unit length is

$$\begin{aligned} P_{l,\text{Avg},\text{TM}_{nm}} &= \frac{1}{2}\int_{\phi=0}^{2\pi} R_s|H_{\text{tang}}(a,\phi,z)|^2 a\,d\phi = \frac{1}{2}R_s\int_{\phi=0}^{2\pi}|H_\phi(a,\phi,z)|^2 a\,d\phi \\ &= \frac{1}{2}\left(\frac{\omega\varepsilon a}{p_{nm}}\right)^2|E_{0,mn}|^2[J_n'(p_{nm})]^2 aR_s\int_{\phi=0}^{2\pi}\cos^2 n\phi\,d\phi \\ &= \frac{1}{2}\left(\frac{\omega\varepsilon a}{p_{nm}}\right)^2|E_{0,mn}|^2[J_{n+1}(p_{nm})]^2\left(\frac{2\pi a}{\delta_n}\right)R_s. \end{aligned} \qquad (12.76)$$

The attenuation constant due to conducting losses in the guide walls is

$$\begin{aligned} \alpha_{c,\text{TM}_{nm}} &= \frac{P_{l,\text{Avg},\text{TM}_{nm}}}{2P_{f,\text{Avg},\text{TM}_{nm}}} = \frac{\omega\varepsilon}{\beta_{nm}a}R_s \\ &= \frac{1}{a\eta\sqrt{1-(f_{c,nm}/f)^2}}\sqrt{\frac{\pi f\mu_c}{\sigma_c}}, \quad \eta = \sqrt{\frac{\mu}{\varepsilon}}. \end{aligned} \qquad (12.77)$$

The last form clearly shows the frequency dependence. Attenuation is very high for frequency just greater than the cutoff frequency, drops rapidly as frequency increases until a minimum is reached, and then rises as the square root of frequency.

Figure 12.8 plots the attenuation constant for several modes in the guide.

12.6.2 Transverse Electric Modes in the Circular Waveguide

For a TE wave the longitudinal magnetic field must satisfy Eq. (12.8), the solution of which is given in cylindrical coordinates in Eq. (12.25). Thus

$$h_z(r,\phi) = [AJ_\nu(k_c r) + BN_\nu(k_c r)][C\cos(\nu\phi) + D\sin(\nu\phi)]. \qquad (12.78)$$

The boundary conditions require that:

1. The field and its derivative with respect to ϕ must be continuous at $\phi = 0$ and $\phi = 2\pi$.

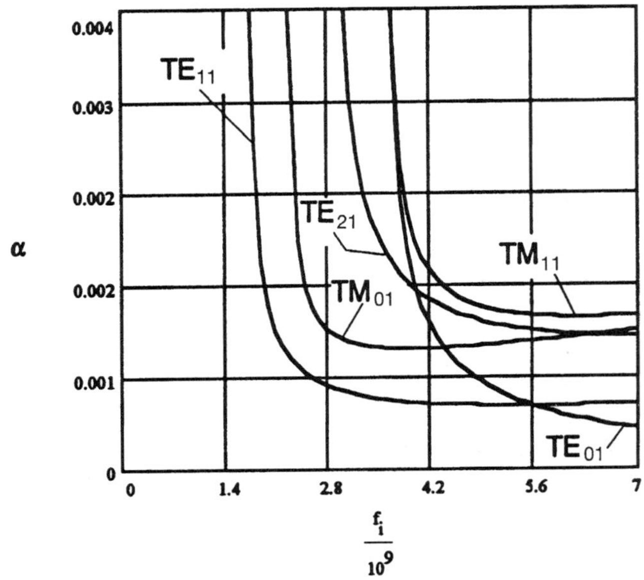

FIGURE 12.8 Wall attenuation in circular waveguide with $a = 2$ in., α in Np/m, and copper walls.

2. The field must be finite everywhere within the guide since there are no sources. In particular, $h_z(0, \phi)$ must be finite.
3. The tangential electric field must be zero along the perfectly conducting wall. The tangential electric field component is e_ϕ, which is proportional to h_r, which is, in turn, proportional to $\partial h_z/\partial r$. Hence we must have $\partial h_z/\partial r = 0$ at $r = a$.

Condition 1 again requires that $\nu = n$, $n = 0, 1, 2, 3, \ldots$. $N_n(0)$ is infinite and hence condition 2 requires that $B = 0$. Equation (12.78) has become

$$h_z = J_n(k_c r)(AC \cos n\phi + AD \sin n\phi). \tag{12.79}$$

The final boundary condition 3 requires that

$$\left.\frac{\partial h_z}{\partial r}\right|_{r=a} = k_c J_n'(k_c a)(AC \cos n\phi + AD \sin n\phi) = 0$$

or $k_c a = p_{nm}'$, where p_{nm}' is the mth zero $m = 1, 2, 3, \ldots$ of J_n'. Thus the TE_{nm} mode solution is

$$k_{c,nm} = \frac{p_{nm}'}{a}, \quad n = 0, 1, 2, 3, \ldots, m = 1, 2, 3, \ldots, \tag{12.80}$$

$$h_{z,nm} = J_n\left(\frac{p_{nm}' r}{a}\right)(C_{nm} \cos n\phi + D_{nm} \sin n\phi), \tag{12.81}$$

TABLE 12.3 Zero p'_{nm} of $J'_n(p')$

m	1	2	3	4
p'_{0m}	3.832	7.016	10.173	13.324
p'_{1m}	1.841	5.331	8.536	11.706
p'_{2m}	3.054	6.076	9.969	13.170
p'_{3m}	4.201	8.015	11.346	14.585

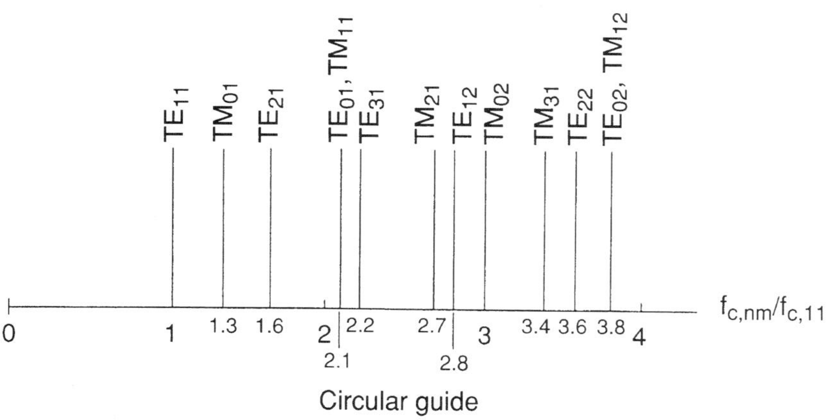

FIGURE 12.9 Cuttoff frequencies for the TE and TM modes in a circular waveguide with $a = 2b$.

and

$$\gamma_{nm} = \sqrt{\left(\frac{p'_{nm}}{a}\right)^2 - k^2} = \sqrt{\left(\frac{p'_{nm}}{a}\right)^2 - \omega^2\mu\varepsilon}. \tag{12.82}$$

The cutoff frequency for the nmth mode is the frequency for which γ_{nm} goes to zero:

$$f_{c,nm} = \frac{p'_{nm}}{2\pi\sqrt{\mu\varepsilon}a} = \frac{cp'_{nm}}{2\pi a}. \tag{12.83}$$

For frequencies less than the cutoff frequency the wave attenuates and is evanescent. For frequencies greater than the cutoff frequency the wave propagates without attenuation and carries time average power. Equations (12.36) through (12.40) for the attenuation constant, phase constant, phase velocity, group velocity, and guide wavelength still hold. Table 12.3 gives the first few zero p'_{nm} of J'_n. Note that $p'_{11} = 1.841$ is the smallest value so the TE_{11} mode has the lowest cutoff frequency of all TE modes. This is also lower than the lowest cutoff frequency for TM modes and hence the TE_{11} mode is the dominant mode in the circular waveguide.

The ratio of the cutoff frequency for the TM_{01} mode to the cutoff frequency for the TE_{11} mode is $2.405/1.841 = 1.3$. This means that the circular guide has a much narrower frequency range for single mode propagation than the rectangular guide. Figure 12.9 shows the ratio of cutoff frequencies in the circular guide and should be compared to Fig. 12.4 for the rectangular guide.

We again note that the difference between the sine and cosine terms is a simple rotation within the circular guide and we can drop the sine term for simplicity. Taking $H_{0,nm}$ as the amplitude of the mode and reinserting the propagation factor $\exp(-j\beta_{nm}z)$ for a forward propagating wave, the longitudinal magnetic field intensity phasor is

$$\underline{H}_{z,nm}(r,\phi,z) = H_{0,nm} J_n\left(\frac{p'_{nm}r}{a}\right)\cos(n\phi)\,e^{-j\beta_{nm}z}. \tag{12.84}$$

The transverse magnetic field components are found from Eq. (12.9), which, written for cylindrical propagating modes, is

$$\vec{\underline{H}}_t = -\frac{j\beta}{k_c^2}\left(\vec{a}_r\frac{\partial \underline{H}_z}{\partial r} + \vec{a}_\phi\frac{1\underline{H}_z}{r\partial\phi}\right).$$

This gives

$$\underline{H}_{r,nm} = -\left(\frac{j\beta_{nm}a}{p'_{nm}}\right) H_{0,nm} J'_n\left(\frac{p'_{nm}r}{a}\right)\cos(n\phi)\,e^{-j\beta_{nm}z}, \tag{12.85}$$

$$\underline{H}_{\phi,nm} = -\left(\frac{j\beta_{nm}a^2n}{p'^2_{nm}r}\right) H_{0,nm} J_n\left(\frac{p'_{nm}r}{a}\right)\sin(n\phi)\,e^{-j\beta_{nm}z}. \tag{12.86}$$

The transverse electric field components are found using Eq. (12.10), $\vec{\underline{E}}_t = -(\omega\mu/\beta)\vec{a}_z \times \vec{\underline{H}}_t$. Thus

$$\underline{E}_{r,nm} = \frac{j\omega\mu a^2 n}{p'^2_{nm}r} H_{0,nm} J_n\left(\frac{p'_{nm}r}{a}\right)\sin(n\phi)\,e^{-j\beta_{nm}z}, \tag{12.87}$$

$$\underline{E}_{\phi,nm} = \frac{j\omega\mu a}{p'_{nm}} H_{0,nm} J'_n\left(\frac{p'_{nm}r}{a}\right)\cos(n\phi)\,e^{-j\beta_{nm}z}. \tag{12.88}$$

Figure 12.10 shows the electric field lines in the transverse plane for the dominant TE_{11} mode. The lines are curved, but they nevertheless resemble the vertical lines of the dominant TE_{10} mode in the rectangular waveguide.

If it is desirable, we can replace the derivative of the Bessel function in the $\underline{H}_r$ and $\underline{E}_\phi$ components by [Eq. (C.7)]

$$J'_n\left(\frac{p_{nm}r}{a}\right) = \frac{na}{p_{nm}r} J_n\left(\frac{p_{nm}r}{a}\right) - J_{n+1}\left(\frac{p_{nm}r}{a}\right).$$

12.6.2.1 *Time Average Power and Attenuation for the TE Mode* The time average power density flowing in the guide is in the z direction and is

$$\begin{aligned} p_{f,\text{Avg},\text{TE}_{nm}} &= \tfrac{1}{2}\,\text{Re}(\underline{E}_r\underline{H}^*_\phi - \underline{E}_\phi\underline{H}^*_r) \\ &= \frac{1}{2}\frac{\omega\mu\beta_{nm}a^2}{p'^2_{nm}}|H_{0,nm}|^2\left\{\left[J'_n\left(\frac{p'_{nm}r}{a}\right)\right]^2\cos^2 n\phi \right. \\ &\quad \left. + \left(\frac{na}{p'_{nm}r}\right)^2\left[J_n\left(\frac{p'_{nm}r}{a}\right)\right]^2\sin^2 n\phi\right\}. \end{aligned} \tag{12.89}$$

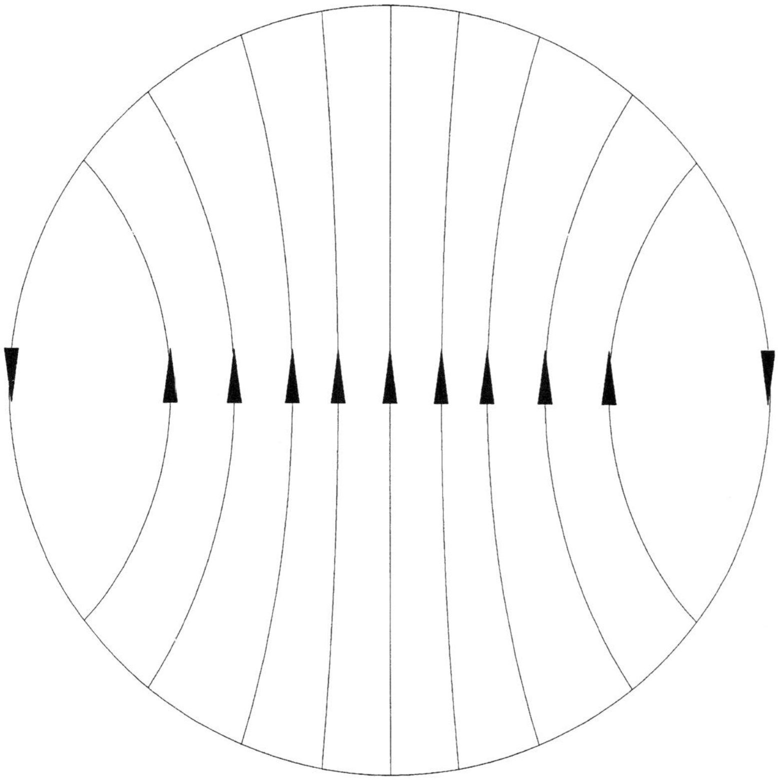

FIGURE 12.10 Electric field lines in the transverse plane for the dominant TE_{11} mode in a circular waveguide.

The total power carried by the wave is

$$P_{f,\,\mathrm{Avg},\,\mathrm{TE}_{nm}} = \int_{r=0}^{a} \int_{\phi=0}^{2\pi} p_{f,\,\mathrm{Avg},\mathrm{TE}_{nm}} r\, d\phi\, dr.$$

The terms within the braces in Eqs. (12.89) and (12.73) are identical, except for the replacement of p_{nm} by p'_{nm}. The integration therefore gives

$$\frac{P_{f,\,\mathrm{Avg},\mathrm{TM}_{nm}}}{C} = xJ_n(x)J'_n(x) + \frac{x^2}{2}\left\{[J_n(x)]^2\left(1 - \frac{n^2}{x^2}\right) + [J'_n(x)]^2\right\}\Bigg|_{x_1}^{x_2},$$

where

$$C = \frac{1}{2}\,\frac{\omega\mu\beta_{nm}a^4}{p'^4_{nm}}\left(\frac{2\pi}{\delta_n}\right)|H_{0,\,nm}|^2,$$

$x_1 = 0$ and $x_2 = p'_{nm}$. At the upper limit $x_2 = p'_{nm}$ we have $J'_n(p'_{nm}) = 0$ and at the lower limit all the terms are zero. Hence

$$\frac{P_{f,\,\mathrm{Avg,TE}_{nm}}}{C} = \frac{p'^2_{nm}}{2}\,[J_n(p'_{nm})]^2\left(1 - \frac{n^2}{p'^2_{nm}}\right),$$

$$P_{f,\,\mathrm{Avg,TE}_{nm}} = \frac{1}{2}\,\frac{\omega\mu\beta_{nm}a^4}{p'^2_{nm}}\left(\frac{\pi}{\delta_n}\right)|H_{0,\,nm}|^2\,[J_n(p'_{nm})]^2\left(1 - \frac{n^2}{p'^2_{nm}}\right). \tag{12.90}$$

The power loss per unit length is

$$\begin{aligned}
P_{l,\,\mathrm{Avg,TE}_{nm}} &= \tfrac{1}{2}\int_{\phi=0}^{2\pi} R_s|H_{\mathrm{tang}}(a,\phi)|^2 a\,d\phi \\
&= \tfrac{1}{2}\,R_s\int_{\phi=0}^{2\pi}[|H_\phi(a,\phi,z)|^2 + |H_z(a,\phi,z)|^2]\,a\,d\phi \\
&= \tfrac{1}{2}\,R_s|H_{0,\,nm}|^2[J_n(p'_{nm})]^2 a\int_{\phi=0}^{2\pi}\left[\left(\frac{\beta_{nm}na}{p'^2_{nm}}\right)^2\sin^2 n\phi + \cos^2 n\phi\right]d\phi
\end{aligned}$$

$$P_{l,\,\mathrm{Avg,TE}_{nm}} = \tfrac{1}{2}\,R_s|H_{0,\,nm}|^2[J_n(p'_{nm})]^2\,a\left[\left(\frac{\beta_{nm}na}{p'^2_{nm}}\right)^2 + 1\right]\left(\frac{2\pi}{\delta_n}\right). \tag{12.91}$$

The attenuation constant due to the imperfectly conducting guide wall is

$$\alpha_{c,\mathrm{TE}_{nm}} = \frac{P_{l,\,\mathrm{Avg,TE}_{nm}}}{2P_{f,\,\mathrm{Avg,\,TE}_{nm}}} = \frac{[(\beta_{nm}na)^2 + (p'_{nm})^4]}{\omega\mu\beta_{nm}a^3\,[(p'_{nm})^2 - n^2]}\,R_s. \tag{12.92}$$

When $n \neq 0$ the attenuation constant shows the usual behavior. It is very large for frequency just greater than the cutoff frequency (due to the β_{nm} in the denominator), decreases rapidly to a minimum, and then rises as the square root of frequency. When $n = 0$, Eq. (12.92) simplifies to

$$\alpha_{c,\mathrm{TE}_{0m}} = \frac{(p'_{0m})^2}{\omega\mu\beta_{0m}a^3}\,R_s = \frac{c(p'_{0m})^2}{(2\pi)^2 f\mu a^3\sqrt{1-(f_{c,0m}/f)^2}}\sqrt{\frac{\pi\mu_c}{f\sigma_c}}, \quad c = \frac{1}{\sqrt{\mu\varepsilon}}. \tag{12.93}$$

The last form shows the frequency behavior explicitly. The attenuation constant is very large for frequency just greater than the cutoff frequency and decreases rapidly as frequency increases. As frequency increases further the attenuation constant continues to decrease, as $f^{-3/2}$. This means that if we excite the TE_{01} mode, for example, at a frequency many times its cutoff frequency, we can attain extremely low attenuation.

This behavior is exciting to the design engineer, who must always seek to reduce losses. The difficulty (and the challenge) is that for frequency many times the cutoff frequency there are many modes that can propagate. We have already discussed the undesirability of multimode propagation, in which power can couple back and forth

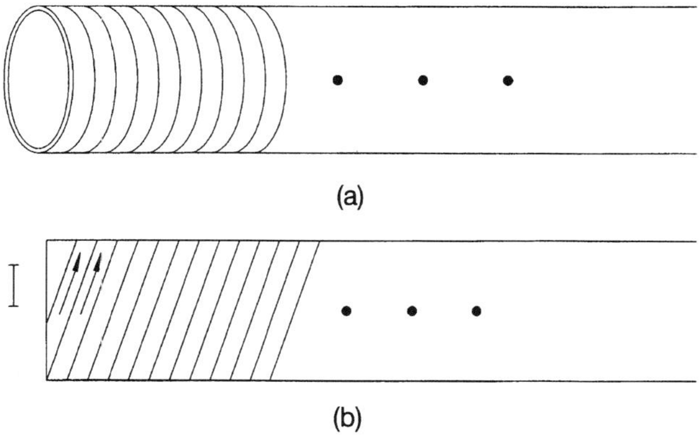

FIGURE 12.11 Circular waveguide fabricated from thin rings of conductors, which are insulated from each other.

to the other modes. Since this low-loss behavior is important, let us rewrite the field components for the TE_{01} mode.

$$\underline{H}_{z,01}(r, \phi, z) = H_{0,01} J_0\left(\frac{3.832r}{a}\right) e^{-j\beta_{01}z}, \tag{12.94}$$

$$\underline{H}_{r,01} = -\left(\frac{j\beta_{01}a}{3.832}\right) H_{0,01} J_0'\left(\frac{3.832r}{a}\right) e^{-j\beta_{01}z}, \tag{12.95}$$

$$\underline{E}_{\phi,01} = \frac{j\omega\mu a}{3.832} H_{0,01} J_0'\left(\frac{3.832r}{a}\right) e^{-j\beta_{01}z}, \tag{12.96}$$

where

$$\beta_{01} = \sqrt{\omega^2\mu\varepsilon - \left(\frac{3.832}{a}\right)^2}. \tag{12.97}$$

For the $n = 0$ modes $\underline{H}_\phi = 0$ and there are no longitudinal (z-directed) currents in the waveguide wall. This is the reason that the attenuation is reduced. The only current in the wall stems from $\underline{H}_z$ evaluated at the wall. This causes a surface current that is purely in the circumferential ($\vec{a}_\phi$) direction. We can suppress other modes by making the circular guide a good circumferential conductor and a poor longitudinal conductor. For example, we can fabricate the guide from thin rings of conductor, which are insulated from each other (Fig. 12.11a), or from a helix with very small pitch (Fig. 12.11b).

12.7 DIELECTRIC ROD WAVEGUIDE

The dielectric rod waveguide is a circular cylinder of radius a and permittivity $\varepsilon_1 = \varepsilon_0\varepsilon_{r1}$ surrounded by a medium with permittivity $\varepsilon_2 = \varepsilon_0\,\varepsilon_{r2}$, where $\varepsilon_1 > \varepsilon_2$

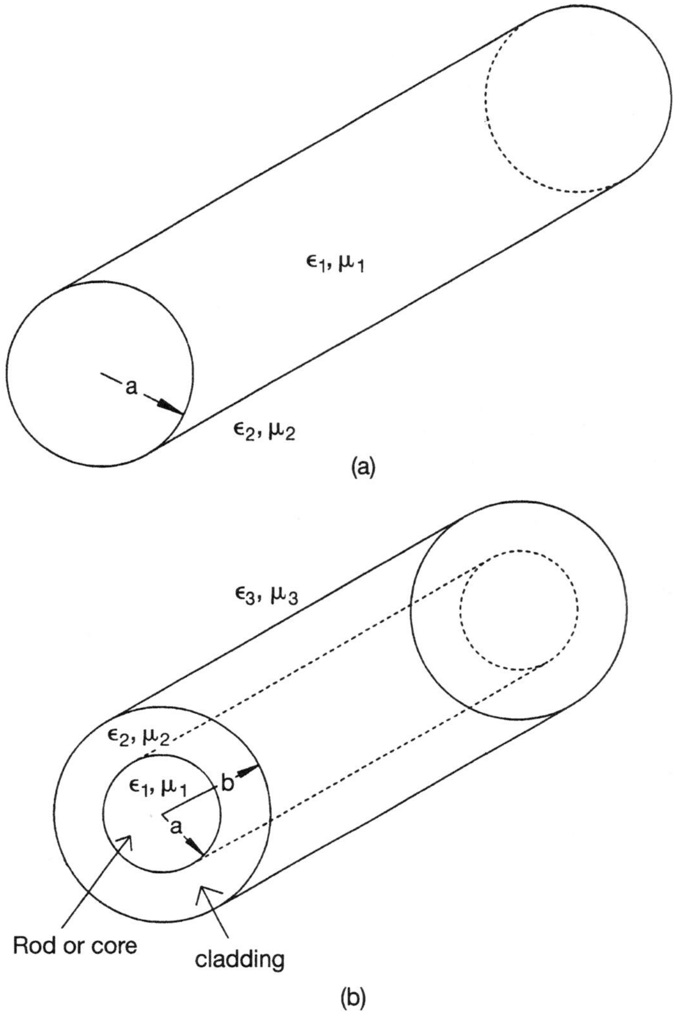

FIGURE 12.12 Dielectric rod waveguide and step-index fiber.

(Fig. 12.12a). This last condition, that the rod be of higher density than the surrounding medium, leads to an internally reflected wave, just as it did for the dielectric slab waveguide (Section 11.9). Indeed, the solutions to the dielectric rod and the dielectric slab are fundamentally the same.

The dielectric rod may also be considered as a simplified model for the step-index fiber, shown in Fig. 12.12b. In this case the rod is called the core and the surrounding material is called the cladding. If the fields are very weak at the cladding–air boundary, then the step-index problem can properly be modeled by the dielectric rod.

We recall that when the dielectric slab successfully guided power the modal solutions were proportional to $\exp(-j\beta_n z)$. The field amplitude decayed exponentially as $\exp(-\alpha_n x)$ outside the slab and no power was transmitted normal to the slab. Exactly at the cutoff frequency for each mode the attenuation constant α_n was zero. Inside the dielectric slab, due to the reflections between boundaries, field amplitudes varied sinusoidally in the transverse dimension. Below the cutoff frequency power leaked out of the slab into the surrounding medium and the wave decayed as it traveled in the z direction. Analogously, for the dielectric rod we seek modes that decay exponentially with radius outside the rod, $\exp(-k_i r)$, and are oscillatory in the radial direction inside the rod. Cutoff will occur at the frequency for which k_i is zero.

Unfortunately, analogies are rarely complete. For the dielectric slab it was possible to separate solutions into TE and TM modes. For the dielectric guide we will find that most (but not all) modes must be hybrid modes, meaning that solutions contain both longitudinal electric and magnetic field intensity components.

12.7.1 Boundary Conditions and Field Solutions

The solutions we are seeking must satisfy the following conditions:

1. The field and its derivatives with respect to ϕ must be continuous at $\phi = 0$ and 2π.
2. The field must be finite on the rod's axis at $r = 0$.
3. The field must decay exponentially far from the rod, $r \gg a$.
4. Tangential electric and magnetic fields must be continuous at the boundary between the rod and the surrounding medium (the core and the cladding). Specifically, this leads to four conditions

$$\underline{E}_{z1}(a, \phi, z) = \underline{E}_{z2}(a, \phi, z), \tag{12.98}$$

$$\underline{H}_{z1}(a, \phi, z) = \underline{H}_{z2}(a, \phi, z), \tag{12.99}$$

$$\underline{E}_{\phi 1}(a, \phi, z) = \underline{E}_{\phi 2}(a, \phi, z), \tag{12.100}$$

$$\underline{H}_{\phi 1}(a, \phi, z) = \underline{H}_{\phi 2}(a, \phi, z), \tag{12.101}$$

   where the subscripts 1 and 2 refer to the field in the dielectric rod or core and to the field in the surrounding medium or cladding, respectively.
5. All fields are proportional to $\exp(-\gamma z)$. The propagation constant in the core must be equal to the propagation constant in the cladding or condition 4 to hold for all values of z.

As a consequence of conditions 1 and 2 the longitudinal fields within the core are

$$e_{z1}(r, \phi) = AJ_0(k_c r) \quad \text{or} \quad e_{z1}(r, \phi) = A \cos(n\phi) J_n(k_c r), \tag{12.102}$$

$$h_{z1}(r, \phi) = BJ_0(k_c r) \quad \text{or} \quad h_{z1}(r, \phi) = B \sin(n\phi) J_n(k_c r), \tag{12.103}$$

for $n = 0$ and $n = 1, 2, 3, \ldots$, respectively. The choice of sine or cosine is arbitrary, but if one is chosen for the electric field then the other must go with the magnetic field. As a consequence of conditions 1 and 3 the fields in the cladding are

$$e_{z2}(r, \phi) = CK_0(k_i r) \quad \text{or} \quad e_{z2}(r, \phi) = C \cos(n\phi) K_n(k_i r), \tag{12.104}$$

$$h_{z2}(r, \phi) = DK_0(k_i r) \quad \text{or} \quad h_{z2}(r, \phi) = D \sin(n\phi) K_n(k_i r), \tag{12.105}$$

where the K's are the modified Bessel functions of the second kind (Section C.4).

Equations (12.9) and (12.10) and Eqs. (12.14) and (12.15) are combined to give the ϕ-directed fields in the core:

$$e_{\phi 1}(r,\phi) = \frac{j\omega\mu_0}{k_c^2}\frac{\partial h_{z1}}{\partial r} - \frac{\gamma}{k_c^2}\frac{1}{r}\frac{\partial e_{z1}}{\partial \phi}, \tag{12.106}$$

$$h_{\phi 1}(r,\phi) = -\frac{\gamma}{k_c^2}\frac{1}{r}\frac{\partial h_{z1}}{\partial \phi} - \frac{j\omega\varepsilon_1}{k_c^2}\frac{\partial e_{z1}}{\partial r}, \tag{12.107}$$

where

$$\gamma^2 = k_c^2 - \omega^2\mu_0\varepsilon_1. \tag{12.108}$$

In the cladding we are using the modified Bessel function so that the wave will decay exponentially to zero for large radius. This means that we must replace k_c by jk_i to get

$$e_{\phi 2}(r,\phi) = -\frac{j\omega\mu_0}{k_i^2}\frac{\partial h_{z2}}{\partial r} + \frac{\gamma}{k_i^2}\frac{1}{r}\frac{\partial e_{z2}}{\partial \phi}, \tag{12.109}$$

$$h_{\phi 2}(r,\phi) = \frac{\gamma}{k_i^2}\frac{1}{r}\frac{\partial h_{z2}}{\partial \phi} + \frac{j\omega\varepsilon_2}{k_i^2}\frac{\partial e_{z2}}{\partial r}, \tag{12.110}$$

where

$$\gamma^2 = -k_i^2 - \omega^2\mu_0\varepsilon_2. \tag{12.111}$$

This system of equations has six unknown constants, A, B, C, D, k_c, and k_i. One amplitude, say, A, is determined by the strength of the exciting source, leaving the other five to be determined by the four conditions, Eqs. (12.98) to (12.101), at $r = a$ and by condition 5, which means the γ's of Eqs. (12.108) and (12.111) are equal.

It will be convenient to work with the dimensionless parameters

$$p = k_c a \quad \text{and} \quad q = k_i a. \tag{12.112}$$

Equating the γ's we obtain

$$p^2 + q^2 = a^2 k_0^2(\varepsilon_{r1} - \varepsilon_{r2}), \quad k_0 = \omega\sqrt{\mu_0\varepsilon_0} = \frac{\omega}{c}. \tag{12.113}$$

The cutoff frequency for any mode occurs when $k_i = 0$, and therefore $q = 0$, so that the field in the surrounding medium does not decay with increasing radius. If we let $\hat{p}$ be the value of p that corresponds to $q = 0$, then Eq. (12.113) gives

$$f_c = \frac{\hat{p}c}{2\pi a\sqrt{\varepsilon_{r1} - \varepsilon_{r2}}}, \quad c = 3\times 10^8 \text{ m/s} \tag{12.114}$$

as the cutoff frequency.

12.7.2 TM and TE Modes when $n = 0$

When the fields are independent of the ϕ coordinate, $n = 0$, we can separate the modes into TM and TE solutions, because the terms $(\partial/\partial\phi)$ in the equations for e_ϕ and h_ϕ are zero.

For the TM modes, we substitute Eqs. (12.102) and (12.104) into Eq. (12.90) and, using Eqs. (12.106) and (12.109), into Eq. (12.100), to obtain

$$AJ_0(p) = CK_0(q),$$

$$-\frac{j\omega\varepsilon_1 a}{p} AJ'_0(p) = \frac{j\omega\varepsilon_2 a}{q} CK'_0(q).$$

Substitute the derivative formulas $J'_0(p) = -J_1(p)$ and $K'_0(q) = -K_1(q)$ and divide the first equation by the second to obtain

$$-\frac{pJ_0(p)}{\varepsilon_{r1}J_1(p)} = \frac{qK_0(q)}{\varepsilon_{r2}K_1(q)} \quad \text{(TM)}. \tag{12.115}$$

Simultaneous solution of Eqs. (12.113) and (12.115) gives the values of p and q and hence k_c and k_i for the TM modes. Before discussing the solution, let us find the similar equation for TE modes.

For the TE modes, we substitute Eqs. (12.103) and (12.105) into Eq. (12.99) and, using Eqs. (12.107) and (12.110), into Eq. (12.101), to obtain

$$BJ_0(p) = DK_0(q),$$

$$\frac{j\omega\mu_0 a}{p} BJ'_0(p) = -\frac{j\omega\mu_0 a}{q} DK'_0(q).$$

We again substitute the derivative formulas and divide the first equation by the second to obtain

$$-\frac{pJ_0(p)}{J_1(p)} = \frac{qK_0(q)}{K_1(q)} \quad \text{(TE)}. \tag{12.116}$$

Simultaneous solution of Eqs. (12.113) and (12.116) gives the values of p and q and hence k_c and k_i for the TE modes.

Solutions in either case are obtained by plotting each pair of equations in the p (ordinate) versus q (abscissa) plane and finding the values at which they cross. To do this, values of p are numerically found from Eqs. (12.115) and (12.116) as q increases from zero. Then Eq. (12.113) is plotted on the same axes and the points of intersection are found. We can find the p axis intercepts as follows. For Eqs. (12.115) and (12.116), the limit of the right side as $q \to 0$ is zero. The left side of either equation is zero at the zeros of J_0, p_{0m}, given in Table 12.2. With $q = 0$ and $\hat{p} = p_{0m}$, Eq. (12.113) gives the cutoff frequency for the TE_{0m} and the TM_{0m} modes as

$$f_{c,0m} = \frac{p_{0m}c}{2\pi a\sqrt{\varepsilon_{r1} - \varepsilon_{r2}}} \qquad c = 3\times 10^8 \text{ m/s}. \tag{12.117}$$

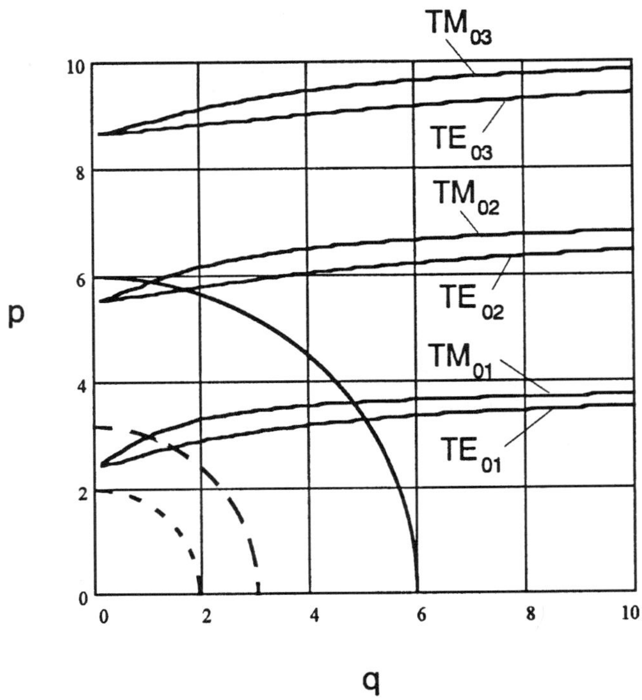

FIGURE 12.13 TE and TM solutions for the modes 1, 2, 3 of a dielectric rod waveguide.

The lowest value is for $p_{01} = 2.405$. This is not, however, the dominant mode in the dielectric rod.

Figure 12.13 plots the TE and TM solutions for the $m = 1, 2, 3$ modes, with $\varepsilon_{r1} = 3\varepsilon_{r2}$. Equation (12.113) is the equation of a circle centered at the origin with radius $C = k_0 a\sqrt{\varepsilon_{r1} - \varepsilon_{r2}}$. The figure shows three arcs of circles. When $C < 2.405$ (dotted) no TE or TM modes can propagate. When $2.405 < C < 5.520$ (dashed) the TE_{01} and the TM_{01} modes propagate, with p and q determined by the intersection. When $5.520 < C < 8.654$ (solid) the $TE_{01}, TE_{02}, TM_{01}$, and TM_{02} modes propagate.

Example 12.3 A dielectric rod has a 2 cm diameter and a relative permeability $\varepsilon_{r1} = 3$. The surrounding medium is air. (a) Calculate the cutoff frequency $f_{c,01}$ for the TM_{01} mode. (b) Plot the longitudinal electric field for this mode at 10 GHz. Assume the electric field intensity on the axis is E_0 V/m. (c) Repeat at 20 GHz.

Solution (a) From Eq. (12.117) we calculate

$$f_{c,10} = \frac{2.405 \times 3 \times 10^8}{2\pi \times 0.01\sqrt{3-1}} = 8.12 \times 10^9 = 8.12 \text{ GHz}.$$

(b) At 10 GHz, the right side of Eq. (12.113) gives $(2\pi \times 10^{10}/3 \times 10^9)(0.01)\sqrt{3-1} = 2.96 \approx 3$. Using the circle with this radius on Fig. 12.13 we get $q \approx 0.8$

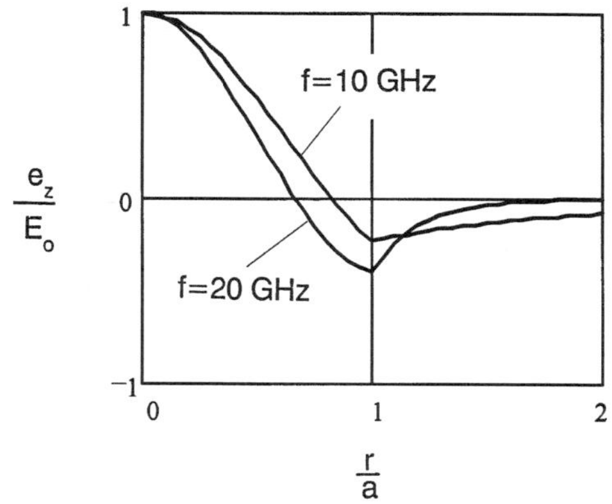

FIGURE 12.14 Normalized longitudinal electric field, TM_{01} mode.

and $p \approx 2.9$. In the rod, $k_c = 2.9/0.01 = 290\ \text{m}^{-1}$ and $e_z(r) = E_0 J_0(290r)$. In the air, $k_i = 0.8/0.01 = 80\ \text{m}^{-1}$ and

$$e_z = E_0 \frac{J_0(2.9)}{K_0(0.8)} K_0(80r).$$

Figure 12.14 shows e_z/E_0 versus r.

(c) At 20 GHz the circle's radius is 6, giving $q \approx 4.8$ and $p \approx 3.6$, corresponding to $k_c = 360\ \text{m}^{-1}$ and $k_i = 480\ \text{m}^{-1}$. We then have

$$\begin{aligned} e_z &= E_0 J_0(360r), \quad r \le a \\ &= E_0 \frac{J_0(3.6)}{K_0(4.8)} K_0(480r), \quad r \ge a. \end{aligned}$$

This is also plotted in Fig. 12.14. The results are similar to those for the dielectric slab waveguide. When the frequency is close to the cutoff frequency, as in (b), the field extends well beyond the core region. When the frequency is much greater than the cutoff frequency, as in (c), the field decays rapidly outside the core.

12.7.3 Hybrid Modes: $n = 1, 2, 3, \ldots$; The Dominant Mode

When n is nonzero, the boundary conditions cannot be satisfied unless the field is a hybrid mode. Substitute Eqs. (12.102) and (12.103) into Eqs. (12.106) and (12.107) and evaluate at $r = a$ to obtain

$$\begin{bmatrix} e_{\phi 1}(a, \phi)/\sin\phi \\ h_{\phi 1}(a, \phi)/\cos\phi \end{bmatrix} = \begin{bmatrix} \dfrac{j\omega\mu_0 a}{p} \dfrac{J_n'(p)}{J_n(p)} & \dfrac{\gamma a n}{p^2} \\ -\dfrac{\gamma a n}{p^2} & -\dfrac{j\omega\varepsilon_1 a}{p} \dfrac{J_n'(p)}{J_n(p)} \end{bmatrix} \begin{bmatrix} BJ_n(p) \\ AJ_n(p) \end{bmatrix}.$$

We substitute Eqs. (12.104) and (12.105) into Eqs. (12.109) and (12.110) and evaluate at $r = a$ to get

$$\begin{bmatrix} e_{\phi 2}(a,\phi)/\sin\phi \\ h_{\phi 2}(a,\phi)/\cos\phi \end{bmatrix} = \begin{bmatrix} -\dfrac{j\omega\mu_0 a}{q}\dfrac{K_n'(q)}{K_n(q)} & -\dfrac{\gamma a n}{q^2} \\ \dfrac{\gamma a n}{q^2} & -\dfrac{j\omega\varepsilon_2 a}{p}\dfrac{J_n'(q)}{J_n(q)} \end{bmatrix} \begin{bmatrix} DK_n(q) \\ CK_n(q) \end{bmatrix}.$$

The two column matrices on the left, involving the ϕ-directed fields at $r = a$, must be equal because of continuity of e_ϕ and h_ϕ, Eqs. (12.100) and (12.101). Similarly, the two column matrices on the right, obtained from the z-directed fields at $r = a$, must be equal because of continuity of e_z and h_z, Eqs. (12.98) and (12.99). Subtract the second from the first to get

$$\begin{bmatrix} 0 \\ 0 \end{bmatrix} = \begin{bmatrix} j\omega\mu_0 a\left(\dfrac{1}{p}\dfrac{J_n'(p)}{J_n(p)} + \dfrac{1}{q}\dfrac{K_n'(q)}{K_n(q)}\right) & \gamma a n\left(\dfrac{1}{p^2}+\dfrac{1}{q^2}\right) \\ -\gamma a n\left(\dfrac{1}{p^2}+\dfrac{1}{q^2}\right) & -j\omega a\left(\dfrac{\varepsilon_1}{p}\dfrac{J_n'(p)}{J_n(p)} + \dfrac{\varepsilon_2}{q}\dfrac{K_n'(q)}{K_n(q)}\right) \end{bmatrix} \begin{bmatrix} BJ_n(p) \\ AJ_n(p) \end{bmatrix}.$$

Since the left side is the zero matrix, we can only have a nontrivial solution for A and B if the determinant of the right-side coefficient matrix is zero. This condition gives

$$\mu_0(\omega a)^2\left[\frac{1}{p}\frac{J_n'(p)}{J_n(p)} + \frac{1}{q}\frac{K_n'(q)}{K_n(q)}\right]\left[\frac{\varepsilon_1}{p}\frac{J_n'(p)}{J_n(p)} + \frac{\varepsilon_2}{q}\frac{K_n'(q)}{K_n(q)}\right] + (\gamma a n)^2\left(\frac{1}{p^2}+\frac{1}{q^2}\right)^2 = 0.$$

This is the characteristic equation, which together with Eq. (12.113) gives the values of p and q. We can put this in a more useful form. Multiply Eqs. (12.108) and (12.111) by a^2, divide the first by p^2 and the second by q^2, and then add the results to get

$$(\gamma a)^2\left(\frac{1}{p^2}+\frac{1}{q^2}\right) = -\mu_0(\omega a)^2\left(\frac{\varepsilon_1}{p^2}+\frac{\varepsilon_2}{q^2}\right).$$

Substitute this in the characteristic equation and divide by $\mu_0(\omega a)^2$ to get

$$\left[\frac{1}{p}\frac{J_n'(p)}{J_n(p)} + \frac{1K_n'(q)}{qK_n(q)}\right]\left[\frac{\varepsilon_1}{p}\frac{J_n'(p)}{J_n(p)} + \frac{\varepsilon_2}{q}\frac{K_n'(q)}{K_n(q)}\right] = n^2\left(\frac{\varepsilon_1}{p^2}+\frac{\varepsilon_2}{q^2}\right)\left(\frac{1}{p^2}+\frac{1}{q^2}\right)$$

$$\frac{\varepsilon_1}{p^2}\left(\frac{J_n'(p)}{J_n(p)}\right)^2 + \frac{(\varepsilon_1+\varepsilon_2)J_n'(p)K_n'(q)}{pqJ_n(p)K_n(q)} + \frac{\varepsilon_2}{q^2}\left(\frac{K_n'(q)}{K_n(q)}\right)^2 = n^2\left(\frac{\varepsilon_1}{p^4} + \frac{\varepsilon_1+\varepsilon_2}{p^2q^2} + \frac{\varepsilon_2}{q^4}\right).$$

Substitute for J'_n from Eq. (C.11) in the first term and for K'_n from Eq. (C.26) in the third therm, cancel common terms, and obtain

$$-\frac{\varepsilon_1 J_{n-1}(p) J_{n+1}(p)}{p^2 J_n^2(p)} + \frac{(\varepsilon_1 + \varepsilon_2) J'_n(p) K'_n(q)}{pq J_n(p) K_n(q)} + \frac{\varepsilon_2 K_{n-1}(q) K_{n+1}(q)}{q^2 K_n^2(q)} = \frac{n^2(\varepsilon_1 + \varepsilon_2)}{p^2 q^2}. \tag{12.118}$$

This result solved simultaneously with Eq. (12.113) gives the solutions for p and q.

Recall that at cutoff for each mode $q = 0$. To find the corresponding values of p we proceed as follows. Multiply both sides of Eq. (12.118) by $q^2 J_n^2(p)$ and take the limit as q approaches zero.

$$\lim_{q \to 0} \left[-\frac{q^2}{p^2} \varepsilon_1 J_{n-1}(p) J_{n+1}(p) + \frac{q}{p} J_n(p)(\varepsilon_1 + \varepsilon_2) J'_n(p) \frac{K'_n(q)}{K_n(q)} + \varepsilon_2 J_n^2(p) \frac{K_{n-1}(q) K_{n+1}(q)}{K_n^2(q)} \right] = \frac{n^2(\varepsilon_1 + \varepsilon_2)}{p^2} J_n^2(p). \tag{12.119}$$

We must now distinguish between $n = 2, 3, 4, \ldots$ and $n = 1$.

12.7.3.1 Hybrid Modes when n = 2, 3, 4, . . .

We now examine the limit, Eq. (12.119), when $n = 2, 3, 4, \ldots$. The first term on the left goes to zero. In the second term

$$\frac{K'_n(q)}{K_n(q)} = \frac{K_{n+1}(q)}{K_n(q)} + \frac{n}{q} \approx -\frac{n!\, 2^n/q^{n-1}}{(n-1)!\, 2^{n-1}/q^n} + \frac{n}{q} = -\frac{2n}{q} + \frac{n}{q} = -\frac{n}{q}. \tag{12.120}$$

In the third term

$$\frac{K_{n-1}(q) K_{n+1}(q)}{K_n^2(q)} \approx \frac{[(n-2)!\, 2^{n-2}/q^{n-1}][n!\, 2^n/q^{n+1}]}{[(n-1)!\, 2^{n-1}/q^n]^2} = \frac{n}{n-1}.$$

The limit becomes

$$J_n(p) \left[-\frac{n}{p} (\varepsilon_1 + \varepsilon_2) J'_n(p) + \frac{n\varepsilon_2}{n-1} J_n(p) \right] = J_n(p) \left[\left(\frac{n}{p} \right)^2 (\varepsilon_1 + \varepsilon_2) J_n(p) \right].$$

This equality is satisfied at the zeros of $J_n(p)$, which we have called p_{nm} (see Table 12.2). It also is satisfied when the term in the brackets on the left equals the term in the brackets on the right. We equate these two terms and substitute for J'_n from Eq. (C.6) to

$$J_{n-1}(p) = \frac{p\varepsilon_2}{(n-1)(\varepsilon_1 + \varepsilon_2)} J_n(p). \tag{12.121}$$

We can call the values of p that satisfy this equation $\bar{p}_{nm}$. These values depend on the permittivities and can be found numerically.

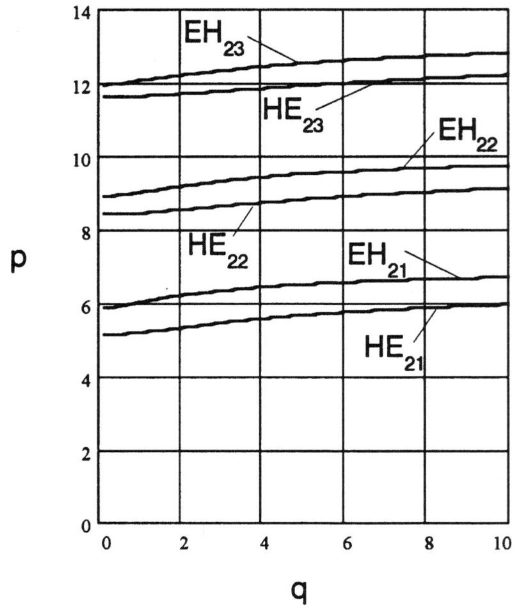

FIGURE 12.15 Solutions for the hybrid $n = 2$ modes for a dielectric rod waveguide with $\varepsilon_r = 3$.

Figure 12.15 plots the solutions for $n = 2$, with $\varepsilon_{r1} = 3\varepsilon_{r2}$. The values of $q = 0$ intercepts are $p_{2,1} = 5.136$, $p_{2,2} = 8.417$, and $p_{2,3} = 11.620$ (from Table 12.2) and $\bar{p}_{2,1} = 5.841$, $\bar{p}_{2,2} = 8.873$, and $\bar{p}_{2,3} = 11.955$ [from numerical evaluation Eq. (12.121)].

Example 12.4 For the rod of Example 12.3, calculate the lowest cutoff frequency for the hybrid $n = 2$ modes.

Solution The smallest p axis intercept in Fig. 12.15 is $p_{2,1} = 5.136$ and the corresponding cutoff frequency is

$$f_{c,\mathrm{HE}_{21}} = \frac{5.136 \times 3 \times 10^8}{2\pi \times 0.01\sqrt{3 - 1}} = 17.34 \times 10^9 = 17.34\ \mathrm{GHz}.$$

12.7.3.2 Hybrid Modes when n = 1; The Dominant Mode We return to Eq. (12.119) and the limit as $q \to 0$, but for $n = 1$. The first term on the left again goes to zero. In the second term Eq. (12.120) is still correct, with $n = 1$. In the third term, however,

$$\lim_{q\to 0} \frac{K_2(q)}{K_1^2(q)} = \frac{2/q^2}{(1/q)^2} = 2.$$

The limit becomes

$$J_1(p)\left[-\frac{1}{p}(\varepsilon_1 + \varepsilon_2)J_1'(p) + 2\varepsilon_2 J_1(p)K_0(q)\right] = J_1(p)\left[\frac{\varepsilon_1 + \varepsilon_2}{p^2} J_1(p)\right].$$

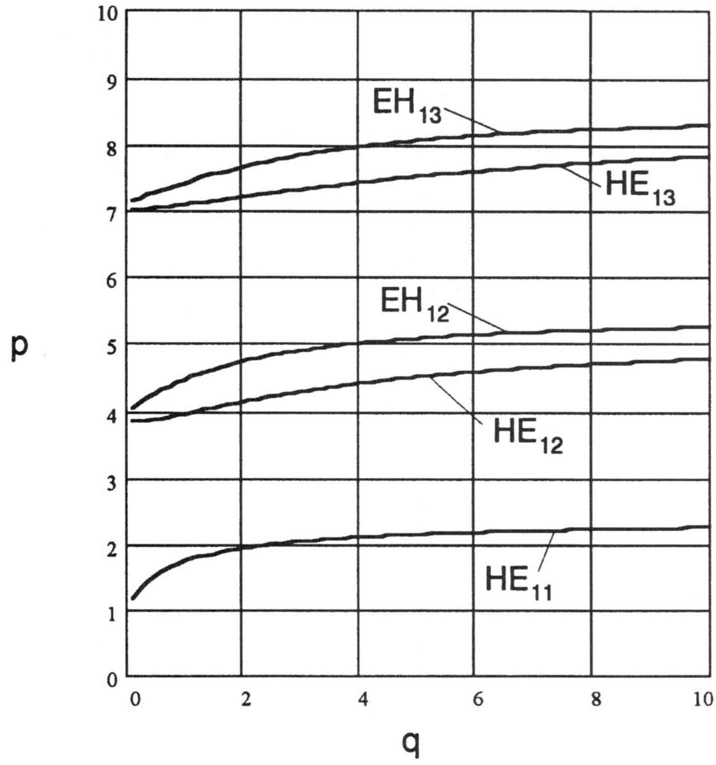

FIGURE 12.16 Solutions for the hybrid $n = 1$ modes for a dielectric rod waveguide with $\varepsilon_r = 3$.

The equality is satisfied at the zeros of $J_1(p)$, p_{1m}, giving the $EH_{1,m+1}$ modes. The equality is also satisfied when the bracketed terms are equal. Equating these terms and using $J_1'(p) = J_0(p) - J_1(p)/p$ we find

$$\frac{(\varepsilon_1 + \varepsilon_2) J_0(p)}{2\varepsilon_2 p J_1(p)} = K_0(q). \tag{12.122}$$

As $q \to 0$ the right side increases without limit. The left side increases without limit at the zeros of $J_1(p)$, p_{1m}, $m = 1, 2, \ldots$, giving the $HE_{1,m+1}$ modes, and also as $p \to 0$, giving the HE_{11} mode. Thus when $n = 1$, the lowest cutoff frequency is zero.

The HE_{11} mode is therefore the dominant mode in the dielectric rod and it propagates down to zero frequency. The next modes to propagate are the TE_{01} and the TM_{01} modes at $p = 2.405$. From $f = 0$ to $f_{c,01} = 2.405c/2\pi a\sqrt{\varepsilon_{r1} - \varepsilon_{r2}}$ only the dominant mode can propagate in the rod.

Figure 12.16 plots the solutions for $n = 1$, with $\varepsilon_{r1} = 3\varepsilon_{r2}$. The $q = 0$ intercepts are $p = 0$, $p_{11} = 3.832$, and $p_{12} = 7.016$. Figure 12.17 combines the solution plots for $n = 0, 1, 2$ to give a complete set of solutions.

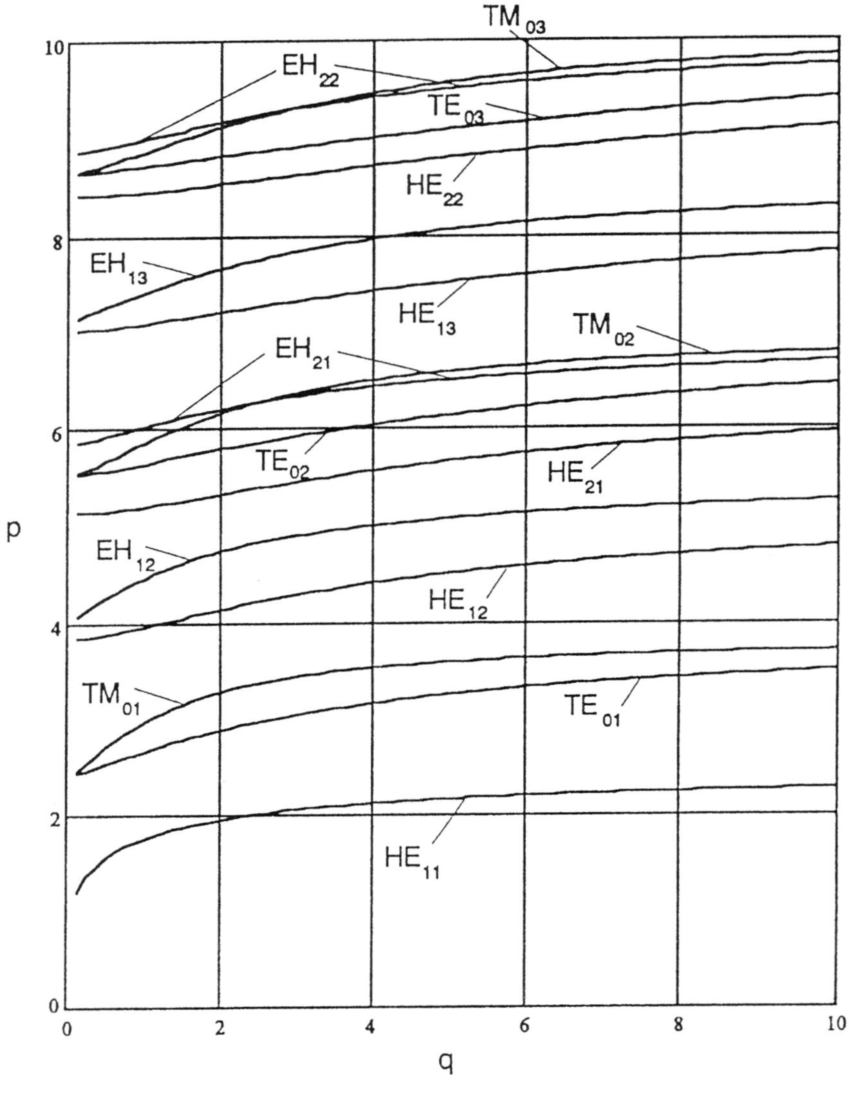

FIGURE 12.17 Dielectric rod mode chart: $\varepsilon_r = 3$, $n = 0, 1, 2$.

Example 12.5 For the dielectric rod of Example 12.3, find the solution corresponding to $p^2 + q^2 = 4$. Evaluate the frequency and plot the longitudinal electric field.

Solution Drawing a circle of radius 2 on Fig. 12.17, we find only the HE_{11} mode is propagating. The solutions from the figure are approximately $q = 1$, $p = 1.7$. More exactly, from Eq. (12.118), $q = 0.983$, $p = 1.742$. Equation (12.113) shows

$$f = \frac{c\sqrt{p^2 + q^2}}{2\pi a\sqrt{\varepsilon_{r1} - \varepsilon_{r2}}} = \frac{3 \times 10^8 \times 2}{2\pi(0.01)\sqrt{3 - 1}} = 6.75 \times 10^9 = 6.75 \text{ GHz}.$$

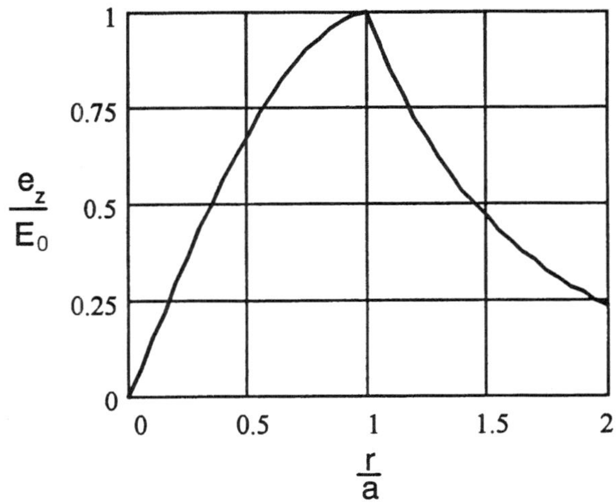

FIGURE 12.18 Longitudinal electric field HE_{11} mode: $q = 0.983$, $p = 1.742$.

From Eqs. (12.102) and (12.104) we have

$$e_{z1} = A \cos\phi \, J_1\left(p \frac{r}{a}\right), \quad r \le a,$$

$$e_{z2} = C \cos\phi \, K_1\left(q \frac{r}{a}\right), \quad r \ge a.$$

We equate these at $r = a$ and normalize to unit amplitude at $\phi = 0$, $r = a$ to obtain

$$e_z = \begin{cases} \cos\phi \, \dfrac{J_1(1.742r/a)}{J_1(1.742)}, & r \le a, \\[2ex] \cos\phi \, \dfrac{K_1(0.983r/a)}{K_1(0.983)}, & r \ge a. \end{cases} \tag{12.123}$$

This result is plotted in Fig. 12.18. Note that the field extends deeply into the free-space region surrounding the core.

12.8 RESONANT CAVITIES

In studying transmission lines we found that a resonance condition was created when both ends of the transmission line were short circuited. We can achieve this result in a waveguide by placing a shorting conducting plate at each end of a length d of waveguide section. The hollow closed resonant conductor that results is known as a resonant cavity. In this section we will find the resonant frequencies for rectangular and circular cavities.

Figure 12.19 shows a cylindrical waveguide shorted with conducting plates at $z = 0$ and $z = d$. At each conducting plate the incident wave will be completely

FIGURE 12.19 Cylindrical waveguide shorted at both ends with conducting plates.

reflected, giving rise to a pure standing wave in which the longitudinal field phasor will be proportional to $\sin \beta z$ or $\cos \beta z$. Equations (12.9), (12.10), (12.14), and (12.15) were derived assuming a z dependence of $\exp(\mp \gamma z)$ and must therefore be revised.

The $\mp \gamma$ in the numerator can be replaced by differentiation of the longitudinal phasor with respect to z. Also, Eqs. (12.9) and (12.10) or Eqs. (12.14) and (12.15) can be combined. The result for TE waves is

$$\vec{\underline{H}}_t = \frac{1}{k_{c,nm}^2} \nabla_t \left(\frac{\partial \underline{H}_z}{\partial z} \right), \quad \vec{\underline{E}}_t = \left(\frac{j\omega\mu}{k_{c,nm}^2} \right) \vec{a}_z \times \nabla_t \underline{H}_z \quad \text{(TE)}, \tag{12.124}$$

and for TM waves it is

$$\vec{\underline{E}}_t = \frac{1}{k_{c,nm}^2} \nabla_t \left(\frac{\partial \underline{E}_z}{\partial z} \right), \quad \vec{\underline{H}}_t = \left(\frac{j\omega\varepsilon}{k_{c,nm}^2} \right) \vec{a}_z \times \nabla_t \underline{E}_z \quad \text{(TM)}. \tag{12.125}$$

The two pairs of equations are duals.

The transverse electric field is tangential to the two conducting end plates and must therefore be zero at $z = 0$ and $z = d$. For the TE wave the transverse electric field has the same z dependence as $\underline{H}_z$ and therefore they must both be proportional to $\sin(q\pi z/d)$, where q is an integer. Equating $q\pi z/d = \beta_{nm} z$ we get

$$\beta_{nm} = \sqrt{k^2 - k_{c,nm}^2} = q\pi/d.$$

Recalling that $k = \omega\sqrt{\mu\varepsilon}$ and solving for the frequency, we have

$$f_{nmq} = \frac{1}{2\pi\sqrt{\mu\varepsilon}} \sqrt{k_{c,nm}^2 + \left(\frac{q\pi}{d} \right)^2} \tag{12.126}$$

as the only values of frequency at which these end conditions are satisfied. The triply infinite set of frequencies f_{nmq} are the resonant frequencies of the cavity. The longitudinal field is

$$\underline{H}_z = H_{0,nmq} h_z \sin\left(\frac{q\pi z}{d} \right), \tag{12.127}$$

where h_z depends on the shape of the cylindrical cross section.

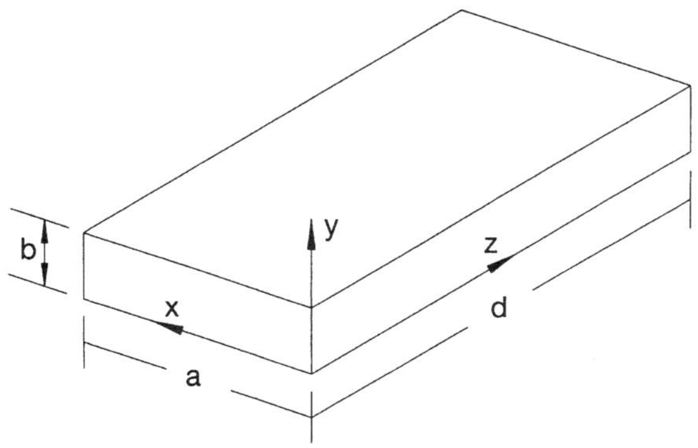

FIGURE 12.20 Rectangular waveguide shorted at both ends with conducting plates.

For TM waves the transverse electric field has z dependence proportional to the derivative of $\underline{E}_z$ with respect to z. If $\underline{E}_z$ is proportional to $\cos(q\pi z/d)$ then $\underline{\vec{E}}_t$ will be proportional to $\sin(q\pi z/d)$, which is zero at the end plates as required. Thus

$$\underline{E}_z = E_{0,nmq}\, e_z \cos\left(\frac{q\pi z}{d}\right), \tag{12.128}$$

where e_z depends on the shape of the cylindrical cross section. We again have $q\pi z/d = \beta_{nm} z$, leading to the same resonant frequencies as Eq. (12.126).

12.8.1 Rectangular Resonant Cavity

Figure 12.20 shows a rectangular waveguide with cross section $a \times b$ and length d. We conventionally assume $a \geq b$ and $d \geq b$. With $k_{c,nm}$ given by Eq. (12.33), we have

$$\begin{aligned} f_{nmq} &= \frac{1}{2\pi\sqrt{\mu\varepsilon}} \sqrt{\left(\frac{n\pi}{a}\right)^2 + \left(\frac{m\pi}{b}\right)^2 + \left(\frac{q\pi}{d}\right)^2} \\ &= c\sqrt{\left(\frac{n}{2a}\right)^2 + \left(\frac{m}{2b}\right)^2 + \left(\frac{q}{2}\right)^2} \end{aligned} \tag{12.129}$$

for the resonant frequencies.

For TE waves using Eqs. (12.32) and (12.127), we find the longitudinal magnetic field intensity

$$\underline{H}_{z,nmq} = H_{0,nmq} \cos\left(\frac{n\pi x}{a}\right) \cos\left(\frac{m\ \pi y}{b}\right) \sin\left(\frac{q\pi z}{d}\right),$$

and using Eq. (12.124) we find the transverse components

$$\underline{H}_{x,nmq} = -\frac{(n\pi/a)(q\pi/d)}{(n\pi/a)^2 + (m\pi/b)^2} H_{0,nmq} \sin\left(\frac{n\pi x}{a}\right) \cos\left(\frac{m\pi y}{b}\right) \cos\left(\frac{q\pi z}{d}\right),$$

$$\underline{H}_{y,nmq} = -\frac{(m\pi/b)(q\pi/d)}{(n\pi/a)^2 + (m\pi/b)^2} H_{0,nmq} \cos\left(\frac{n\pi x}{a}\right) \sin\left(\frac{m\pi y}{b}\right) \cos\left(\frac{q\pi z}{d}\right),$$

$$\underline{E}_{x,nmq} = \frac{j\omega_{nmq}\mu(m\pi/b)}{(n\pi/a)^2 + (m\pi/b)^2} H_{0,nmq} \cos\left(\frac{n\pi x}{a}\right) \sin\left(\frac{m\pi y}{b}\right) \sin\left(\frac{q\pi z}{d}\right),$$

$$\underline{E}_{y,nmq} = \frac{j\omega_{nmq}\mu(n\pi/a)}{(n\pi/a)^2 + (m\pi/b)^2} H_{0,nmq} \sin\left(\frac{n\pi x}{a}\right) \cos\left(\frac{m\pi y}{b}\right) \sin\left(\frac{q\pi z}{d}\right).$$

For these resonant modes we have

$$\left.\begin{aligned} n &= 0, 1, 2, \ldots \\ m &= 0, 1, 2, \ldots \end{aligned}\right\} \quad \text{but not } n = m = 0,$$

$$q = 1, 2, 3, \ldots .$$

The lowest resonant frequency is

$$f_{101} = c\sqrt{\left(\frac{1}{2a}\right)^2 + \left(\frac{1}{2d}\right)^2} \tag{12.130}$$

for the TE_{101} resonant mode.

These equations are derived assuming perfectly conducting walls. When the walls are good, but imperfect, conductors, with conductivity, σ_c, then the fields inside the guide are closely given by these results, but there are losses in the walls. We can find the quality factor of the cavity from $Q_{nmq} = \omega_{nmq} W_{\text{Avg}}/P_{l,\text{Avg}}$. Let us perform these calculations for the fundamental TE_{101} mode. The fields are (dropping the *nmp* subscript)

$$\underline{H}_z = H_0 \cos\left(\frac{\pi x}{a}\right) \sin\left(\frac{\pi z}{d}\right),$$

$$\underline{H}_x = -\left(\frac{a}{d}\right) H_0 \sin\left(\frac{\pi x}{a}\right) \cos\left(\frac{\pi z}{d}\right), \tag{12.131}$$

$$\underline{E}_y = -\frac{j\omega_{101}\mu}{(\pi/a)} H_0 \sin\left(\frac{\pi x}{a}\right) \sin\left(\frac{\pi z}{d}\right).$$

The peak energy stored in the electric field equals the sum of the time average energy stored in the electric and magnetic fields. Hence

$$
\begin{aligned}
W_{\text{Avg}} &= \frac{1}{2}\,\varepsilon \int_V |\underline{E}_y|^2\, dV \\
&= \frac{1}{2}\,\varepsilon \int_{z=0}^{d}\int_{y=0}^{b}\int_{x=0}^{a} \left(\frac{\omega_{101}\mu}{\pi/a}\right)^2 |H_0|^2 \sin^2\left(\frac{\pi x}{a}\right)\sin^2\left(\frac{\pi z}{d}\right) dx\, dy\, dz \\
&= \frac{1}{2}\,\varepsilon \left(\frac{\omega_{101}\mu}{\pi/a}\right)^2 |H_0|^2 \left(\frac{abd}{4}\right) = \frac{1}{2}\,\mu\, |H_0|^2 \left[1+\left(\frac{a}{d}\right)^2\right]\left(\frac{abd}{4}\right). \qquad (12.132)
\end{aligned}
$$

The power loss density is $|H_{\text{tang}}|^2 R_s/2$ and the power loss is the integral over the wall surfaces. On the $x = 0$ wall H_z is tangential. On the $y = 0$ wall H_x and H_z are tangential. On the $z = 0$ wall H_x is tangential. Losses on opposite walls are equal. Thus

$$
\begin{aligned}
P_{L,\text{Avg}} &= \frac{1}{2}\,(2)\left\{\int_{y=0}^{b}\int_{z=0}^{d} |\underline{H}_z(0,y,z)|^2\, dz\, dy + \int_{x=0}^{a}\int_{z=0}^{d} [|\underline{H}_x(x,0,z)|^2 \right. \\
&\qquad \left. + |\underline{H}_z(x,0,z)|^2]\, dz\, dx + \int_{x=0}^{a}\int_{y=0}^{b} |\underline{H}_x(x,y,0)|^2\, dy\, dx\right\} R_s \\
&= \frac{1}{2}\,(2)\,|H_0|^2 \left\{\left(\frac{bd}{2}\right) + \left[\left(\frac{a}{d}\right)^2\left(\frac{ad}{4}\right) + \left(\frac{ad}{4}\right)\right] + \left(\frac{a}{d}\right)^2\left(\frac{ab}{2}\right)\right\} R_s.
\end{aligned}
$$

The cavity Q for the TE_{101} mode is therefore

$$
\begin{aligned}
Q_{\text{TE}_{101}} &= \frac{\omega_{101}\mu[1+(a/d)^2](abd/4)}{2\{(bd/2)+[(a/d)^2(ad/4)+(ad/4)]+(a/d)^2(ab/2)\}R_s} \\
&= \frac{\pi[1+(a/d)^2]^{3/2}}{4\{1+(a/2b)[1+(a/d)^2]+(a/d)^2\}}\,\frac{\eta}{R_s}. \qquad (12.133)
\end{aligned}
$$

The Q of a cubical cavity ($a = b = d$) oscillating in the TE_{101} mode is

$$
Q_{\text{TE}_{101}} = \frac{0.740\eta}{R_s}, \quad a = b = d.
$$

For an air-filled cavity, $\eta = 377\ \Omega$, with copper walls, $R_s = 2.61\times 10^{-7}\sqrt{f_{101}}\ \Omega$, we have $Q = 1.07\times 10^9/\sqrt{f_{101}}$, with $a = c/\sqrt{2}f_{101}$ as the cubical dimension. For example, with a resonant frequency of 10 GHz, $a = 2.12$ cm and $Q = 10{,}700$.

This extremely high value of Q is typical of cavity resonators. A high value of Q allows the design of narrow-band filters and broad-band filters with near-ideal characteristics. Cavity resonators are therefore extensively used in high-quality microwave filters.

Example 12.6 Design an air-filled rectangular cavity to resonate at 10 GHz using a copper waveguide with $a = 0.9$ in. and $b = 0.4$ in. Calculate the cavity Q.

Solution The resonant frequency is

$$10 = 11.8\sqrt{\left(\frac{1}{2(0.9)}\right)^2 + \left(\frac{1}{2d}\right)^2},$$

giving $d = 0.781$ in. For air, $\sqrt{\mu/\varepsilon} = 377\ \Omega$ and for copper $R_s = 2.61 \times 10^{-7}\sqrt{f} = 0.0261\ \Omega$. Substituting in the formula gives $Q = 7830$.

For TM waves, using Eqs. (12.47) and (12.128), we find the longitudinal electric field intensity

$$\underline{E}_{z,nmq} = E_{0,nmq} \sin\left(\frac{n\pi x}{a}\right) \sin\left(\frac{m\pi y}{b}\right) \cos\left(\frac{q\pi z}{d}\right)$$

and (using Eq. 12.125), we get the transverse components

$$\underline{E}_{x,nmq} = -\frac{(n\pi/a)(q\pi/d)}{(n\pi/a)^2 + (m\pi/b)^2} E_{0,nmq} \cos\left(\frac{n\pi x}{a}\right) \sin\left(\frac{m\pi y}{b}\right) \sin\left(\frac{q\pi z}{d}\right),$$

$$\underline{E}_{y,nmq} = -\frac{(m\pi/b)(q\pi/d)}{(n\pi/a)^2 + (m\pi/b)^2} E_{0,nmq} \sin\left(\frac{n\pi x}{a}\right) \cos\left(\frac{m\pi y}{b}\right) \sin\left(\frac{q\pi z}{d}\right),$$

$$\underline{H}_{x,nmq} = -\frac{j\omega_{nmq}\mu(m\pi/b)}{(n\pi/a)^2 + (m\pi/b)^2} E_{0,nmq} \sin\left(\frac{n\pi x}{a}\right) \cos\left(\frac{m\pi y}{b}\right) \cos\left(\frac{q\pi z}{d}\right),$$

$$\underline{H}_{y,nmq} = -\frac{j\omega_{nmq}\mu(n\pi/a)}{(n\pi/a)^2 + (m\pi/b)^2} E_{0,nmq} \cos\left(\frac{n\pi x}{a}\right) \sin\left(\frac{m\pi y}{b}\right) \cos\left(\frac{q\pi z}{d}\right).$$

For these resonant modes we have

$$n = 1, 2, 3, \ldots, \quad m = 1, 2, 3, \ldots, \quad q = 0, 1, 2, \ldots.$$

The lowest resonant frequency is for the TM_{110} mode. Since we have assumed $d \geq b$, $f_{110} \geq f_{101}$.

12.8.1.1 Cavity Coupling A resonant circuit must have energy coupled in and out for it to be useful. Three methods of coupling to a rectangular cavity resonator are illustrated in Fig. 12.21. The coaxial probe (Fig. 12.21a) should be located in a region of strong electric field parallel to the probe for the mode to be excited. For the TE_{101} mode the probe should be in the center of the width of the cavity ($x = a/2$), not too near the ends. When a single probe is used, as shown in solid lines, the cavity can be represented as a resonant impedance loading the coaxial transmission line. When two probes are used, as represented by the solid and dashed lines, the cavity can be represented as a half-wave resonant structure, between input and output coaxial transmission lines. If properly designed this becomes a bandpass filter centered near the resonant frequency of the cavity. The coaxial loop coupling of Fig. 12.21b should be placed so the loop links with strong magnetic fields for the mode to be excited. For the TE_{101} mode the loop should be at $y = b/2$ along the long wall of the cavity (shown

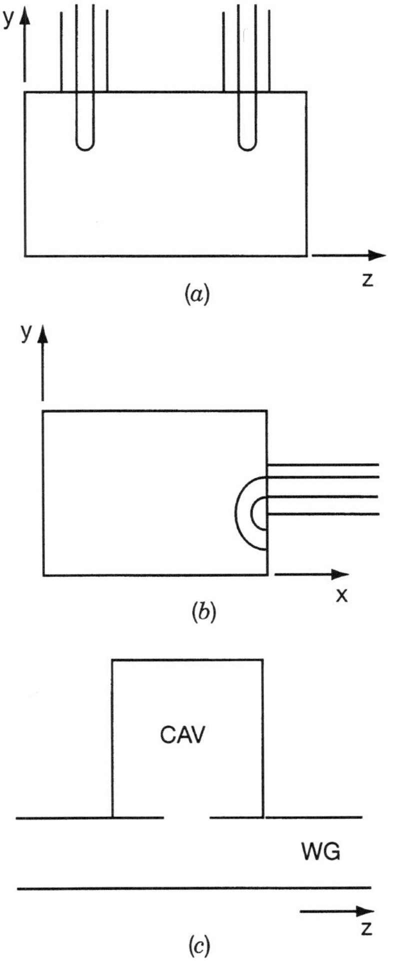

FIGURE 12.21 Three methods of coupling to a rectangular cavity waveguide.

in solid lines) or on the end wall (shown dashed). The cavity can be coupled to a waveguide (WG, Fig. 12.21c) by an aperture or opening in their common wall.

To get some appreciation of the use of a cavity as a narrow bandpass filter, let us look at Fig. 12.22a, which shows a rectangular waveguide with two conducting plates, each with a small circular aperture at its center. We will call f_0 the center frequency of the passband and it should be in the recommended single mode propagation frequency of the guide. The waveguide can be modeled as a transmission line (Fig. 12.22b), and the circular apertures can be modeled as shunt inductors with susceptance

$$B_a = \frac{3ab\lambda_g}{2\pi D^3} Y_g, \quad D \ll 0.8b, \tag{12.134}$$

where λ_g is the guide wavelength and $Y_g = 1/Z_{\text{TE}_{10}} = \beta_{10}/\omega\mu$ is the mode admittance for the TE_{10} mode. At f_0 the value of the guide wavelength, the mode admittance, and

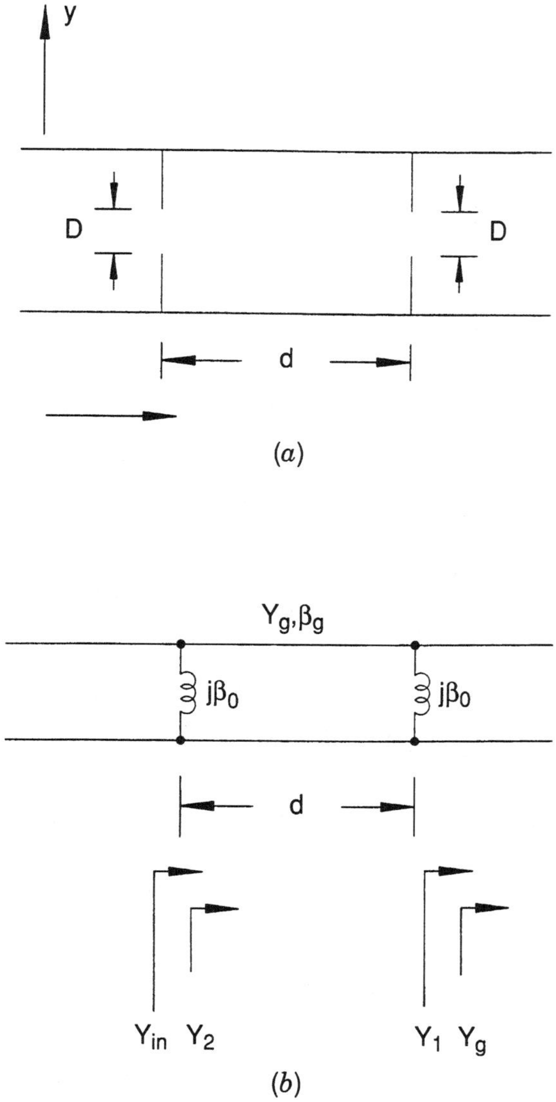

FIGURE 12.22 Narrow bandpass filter realized in a rectangular waveguide.

the aperture susceptance are λ_{g0}, Y_{g0}, and $-B_{a0}$, respectively. The distance d between the two apertures is just slightly less than a half guide wavelength long at f_0, so $d = k\lambda_{g0}/2$, where k is close to but slightly less than unity. The guide to the right is terminated in a matched load $Y_L = Y_g$.

The cavity is loaded with admittance $Y_2 = Y_g - jB_a$ and this admittance is transformed through the distance d to

$$Y_1 = Y_g \left[\frac{Y_2 + jY_g \tan(k\pi\lambda_{g0}/\lambda_g)}{Y_g + jY_2 \tan(k\pi\lambda_{g0}/\lambda_g)} \right]. \tag{12.135}$$

The total input admittance seen by the input waveguide is $Y_{\text{in}} = -jB_a + Y_1$. To achieve full transmission (zero reflection) at f_0, we must have $Y_{\text{in}} = Y_{g0}$, which requires that $Y_1 = Y_{g0} + jB_{a0}$. We equate Y_1 to this value, at f_0, and solve for the tangent term, obtaining $\tan(k\pi) = -2Y_{g0}/B_{a0}$. This gives

$$k = 1 - \frac{1}{\pi}\tan^{-1}\left(\frac{2Y_{g0}}{B_{a0}}\right). \tag{12.136}$$

This value of k causes the inductive susceptance from the aperture on the right to be transformed into an equal capacitive susceptance in parallel with and in resonance with the inductive susceptance of the aperture on the left. As B_{a0}/Y_{g0} increases, $k \to 1$ and the bandwidth of the filter decreases.

Example 12.7 An air-filled rectangular waveguide has dimensions $a = 2.29$ cm by $b = 1.02$ cm. The cutoff frequency for this guide is $f_c = 6.56$ GHz and the recommended frequency range is 8.20 to 12.40 GHz. Calculate the fractional transmitted power versus frequency for $D = 0.7b$, $D = 0.5b$, and $D = 0.3b$. Take the center frequency to be $f_0 = 10.30$ GHz.

Solution At 10.30 GHz $\lambda_{g0} = 0.0378$ m and $Y_{g0} = 0.00205$ S. For the three values of D, the values of B_{a0} are 0.0238, 0.0654, and 0.303 S and the values of k are 0.946, 0.980, and 0.996, respectively. The length of the cavity is $k\lambda_{g0}$, so the three values of D are 0.0358, 0.0370, and 0.0376 m, respectively. We can calculate $Y_{\text{in}} = -jB_a + Y_1$ with the aid of Eq. (12.135). The reflection coefficient and fractional transmitted power can be calculated from

$$\Gamma = \frac{Y_g - Y_{\text{in}}}{Y_g + Y_{\text{in}}}, \quad \frac{P_{\text{out}}}{P_{\text{in}}} = 1 - |\Gamma|^2.$$

The results are plotted in Fig. 12.23. The figure shows the 3 dB or half-power line. The fractional bandwidth for the three aperture diameters is about 5.58×10^{-3}, 0.754×10^{-3} and 0.034×10^{-3}, respectively.

In this example we have completely neglected the losses in the cavity itself. The Q of the resultant filter is due entirely to the external load (the matched load on the output). Including the internal Q of the cavity would lead to slightly greater total Q and hence greater fractional bandwidth.

Generally, microwave filters consist of several resonant structures. For the half-wave cavity resonator this means adding additional apertures along the guide. By properly choosing the value of B_{a0} for each aperture and the value of k for each resonator, different filter characteristics can be obtained [1, 2]. We will not pursue this topic further.

12.8.1.2 Cavity Perturbations: Tuning The reader will have noted that the designs given in the above example require that the cavity be manufactured with small tolerances in its dimensions. In practice, even with the most exacting manufacturing standards, some way to tune the resonant frequency of the cavity is required. Tuning takes the form of small perturbations in the shape of the cavity walls or small perturbations in the material within the cavity.

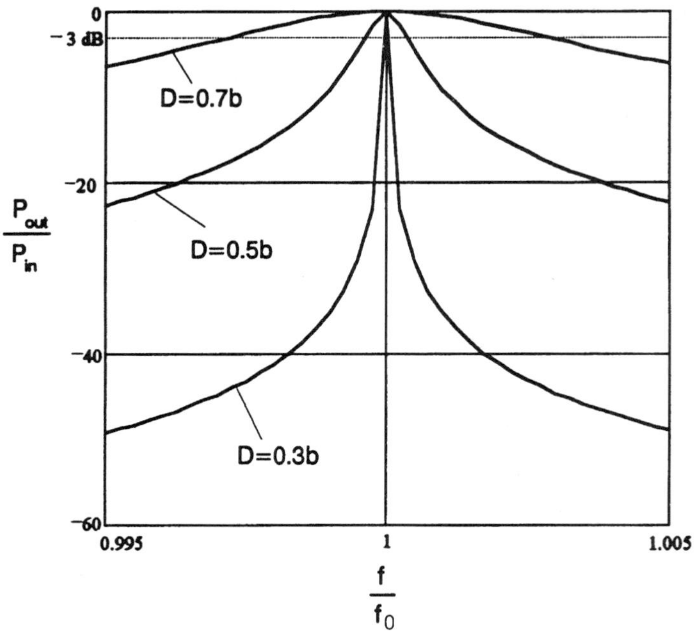

FIGURE 12.23 Insertion loss (P_{out}/P_{in} in dB) versus normalized frequency for a narrow-band filter.

Let us first consider small changes in cavity shape. Figure 12.24a shows an original cavity with volume V, resonant angular frequency f_0 (in some particular mode), and total stored energy W_{Avg}. Figure 12.24b shows the same cavity, but with a small volume ΔV removed due to an indentation in the walls. If $\Delta W_{m,Avg}$ and $\Delta W_{e,Avg}$ are the time average magnetic and electric energy, respectively, originally stored in ΔV, then the new resonant frequency $f_{0,new}$ is, approximately,

$$\frac{f_{0,\text{new}} - f_0}{f_0} \approx \frac{\Delta W_{m,\text{Avg}} - \Delta W_{e,\text{Avg}}}{W_{\text{Avg}}}. \tag{12.137}$$

An indentation at a point of high magnetic field intensity and low electric field intensity, which means that $\Delta W_{m,Avg} > \Delta W_{e,Avg}$, will raise the resonant frequency of the cavity. An indentation at a point of high electric field intensity and low magnetic intensity, meaning that $\Delta W_{m,Avg} < \Delta W_{e,Avg}$, will lower the resonant frequency of the cavity.

Example 12.8 A small screw of radius r_0 is inserted in the center of the top wall (with dimension $a \times d$) so that it projects a length l into the guide. The cavity is air-filled and is resonating in the TE_{101} mode. Calculate the fractional change in resonant frequency.

Solution The fields are given in Eq. (12.131) and the stored energy is given in Eq. (12.132). At the position of the screw the magnetic field intensity is zero and hence

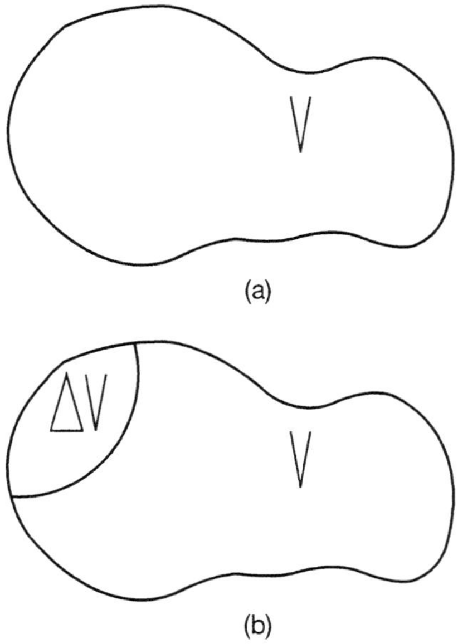

FIGURE 12.24 Illustration of a small perturbation in the volume of a cavity.

$\Delta W_{m,\text{Avg}} = 0$. The time average stored electric energy is approximated by the time average electric energy density $\varepsilon|\underline{E}_y(a/2, b, d/2)^2|/4$ multiplied by the volume of the screw $\pi r_0^2 l$, giving

$$\frac{f_{0,\text{new}} - f_0}{f_0} = -\frac{\frac{1}{4}\varepsilon(2af_0\mu)^2|H_0|^2(\pi r_0^2 l)}{\frac{1}{2}\varepsilon(2af_0\mu)^2|H_0|^2(abd/4)}$$

$$= -2\frac{\pi r_0^2 l}{abd} = -2\frac{\Delta V}{V}.$$

The frequency f_0 is given by Eq. (12.130) with $c = 1/\sqrt{\mu_0\varepsilon_0}$.

Let us next consider a small change in the permittivity or permeability of the material within the cavity. If the original material constants are ε and μ and the perturbed constants are $\varepsilon + \Delta\varepsilon$ and $\mu + \Delta\mu$, then the new resonant frequency is

$$\frac{f_{0,\text{new}} - f_0}{f_0} \approx -\frac{1}{4W_{\text{Avg}}}\int_V (\Delta\mu|\underline{H}|^2 + \Delta\varepsilon|\underline{E}|^2)\, dv. \tag{12.138}$$

This means that any small increase in ε or μ will decrease the resonant frequency of the cavity. Since any large increase in material constants can be considered as a succession of small increases, it follows that any increase in ε or μ will decrease the resonant frequency of the cavity.

When the change in ε and μ is only within a small volume ΔV we obtain

$$\frac{f_{0,\text{new}} - f_0}{f_0} \approx -\frac{1}{W_{\text{Avg}}}\left(\frac{\Delta\mu}{\mu}\Delta W_{m,\text{Avg}} + \frac{\Delta\varepsilon}{\varepsilon}\Delta W_{e,\text{Avg}}\right). \tag{12.139}$$

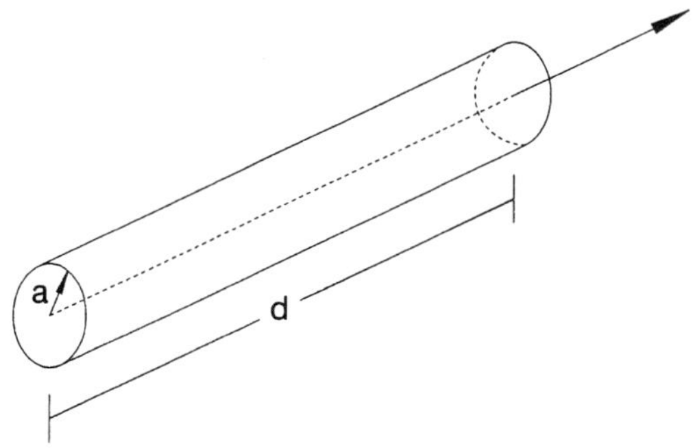

FIGURE 12.25 Circular cylindrical cavity.

Example 12.9 Repeat Example 12.8, but with the screw made of polystyrene, $\varepsilon_r = 2.56$.

Solution With $\varepsilon = \varepsilon_0 + \Delta\varepsilon = \varepsilon_r \varepsilon_0$, we get $\Delta\varepsilon = (\varepsilon_r - 1)\,\varepsilon_0$. The result is

$$\frac{f_{0,\,\text{new}} - f_0}{f_0} = -\,2(\varepsilon_r - 1)\,\frac{\pi r_0^2 l}{abd} = -\,2(\varepsilon_r - 1)\,\frac{\Delta V}{V}.$$

The tuning effect of the polystyrene screw is enhanced by $(\varepsilon_r - 1) = 1.56$ over the tuning effect of the conducting screw.

Any tuning creates additional modes, which extract energy from the desired resonant mode. Tuning therefore increases the losses in the cavity and reduces the Q of the cavity.

12.8.2 Circular Resonant Cavity

Figure 12.25 shows a circular cylindrical cavity with radius a and length d.

For TE modes we have $k_{c,nm} = p'_{nm}/a$ so

$$f_{nmp} = \frac{c}{2\pi}\sqrt{\left(\frac{p'_{nm}}{a}\right)^2 + \left(\frac{q\pi}{d}\right)^2}. \tag{12.140}$$

The longitudinal mangetic field intensity phasor is

$$\underline{H}_{z,nmq} = H_{0,nmq} J_n\left(\frac{p'_{nm} r}{a}\right)\cos(n\phi)\sin\left(\frac{q\pi z}{d}\right)$$

and, using Eq. (12.124), the transverse field phasors are

$$\underline{H}_{r,nmq} = \frac{(q\pi/d)}{(p'_{nm}/a)} H_{0,nmq} J'_n\left(\frac{p'_{nm} r}{a}\right)\cos(n\phi)\cos\left(\frac{q\pi z}{d}\right),$$

$$\underline{H}_{\phi,nmq} = -\,\frac{n(q\pi/d)}{(p'_{nm}/a)^2 r} H_{0,nmq} J_n\left(\frac{p'_{nm} r}{a}\right)\sin(n\phi)\cos\left(\frac{q\pi z}{d}\right),$$

$$\underline{E}_{r,nmq} = \frac{j\omega_{nmq}\mu n}{(p'_{nm}/a)^2} H_{0,nmq} J_n\left(\frac{p'_{nm}r}{a}\right)\sin(n\phi)\sin\left(\frac{q\pi z}{d}\right),$$

$$\underline{E}_{\phi,nmq} = \frac{j\omega_{nmq}\mu}{(p'_{nm}/a)} H_{0,nmq} J'_n\left(\frac{p'_{nm}r}{a}\right)\sin(n\phi)\sin\left(\frac{q\pi z}{d}\right).$$

For TM modes we have $k_{c,nm} = p_{nm}/a$ so

$$f_{nmp} = \frac{c}{2\pi}\sqrt{\left(\frac{p_{nm}}{a}\right)^2 + \left(\frac{q\pi}{d}\right)^2}. \tag{12.141}$$

The longitudinal electric field intensity phasor is

$$\underline{E}_{z,nmq} = E_{0,nmq} J_n\left(\frac{p_{nm}r}{a}\right)\cos(n\phi)\cos\left(\frac{q\pi z}{d}\right)$$

and, using Eq. (12.125), the transverse field phasors are

$$\underline{E}_{r,nmq} = -\frac{(q\pi/d)}{(p_{nm}/a)} E_{0,nmq} J'_n\left(\frac{p_{nm}r}{a}\right)\cos(n\phi)\sin\left(\frac{q\pi z}{d}\right),$$

$$\underline{E}_{\phi,nmq} = \frac{n(q\pi/d)}{(p_{nm}/a)^2 r} E_{0,nmq} J_n\left(\frac{p_{nm}r}{a}\right)\sin(n\phi)\sin\left(\frac{q\pi z}{d}\right),$$

$$\underline{H}_{r,nmq} = -\frac{j\omega_{nmq}\varepsilon n}{(p_{nm}/a)^2} E_{0,nmq} J_n\left(\frac{p_{nm}r}{a}\right)\sin(n\phi)\cos\left(\frac{q\pi z}{d}\right),$$

$$\underline{H}_{\phi,nmq} = -\frac{j\omega_{nmq}\varepsilon}{(p_{nm}/a)} E_{0,nmq} J'_n\left(\frac{p_{nm}r}{a}\right)\cos(n\phi)\cos\left(\frac{q\pi z}{d}\right).$$

For either the TE or TM modes we have $n = 0, 1, 2, \ldots$, $m = 1, 2, 3, \ldots$, and $q = 1, 2, 3, \ldots$. The ratio $2a/d$ determines which mode has the lowest resonant frequency. When $2a/d < 0.985$ the TE_{111} mode has the lowest resonant frequency, but when $2a/d > 0.985$ the TM_{010} mode has the lowest resonant frequency.

Let us again look carefully at the TE_{011} mode, which we already known to be a low-loss mode. The fields are

$$\underline{H}_{z,011} = H_{0,011} J_0\left(\frac{3.832r}{a}\right)\sin\left(\frac{\pi z}{d}\right),$$

$$\underline{H}_{r,011} = \frac{(\pi/d)}{(3.832/a)} H_{0,011} J'_0\left(\frac{3.832r}{a}\right)\cos\left(\frac{\pi z}{d}\right),$$

$$\underline{E}_{\phi,011} = \frac{j\omega_{011}\mu}{(3.832/a)} H_{0,011} J'_0\left(\frac{3.832r}{a}\right)\sin\left(\frac{\pi z}{d}\right).$$

The Q of the cavity is given in Problem 12.15.

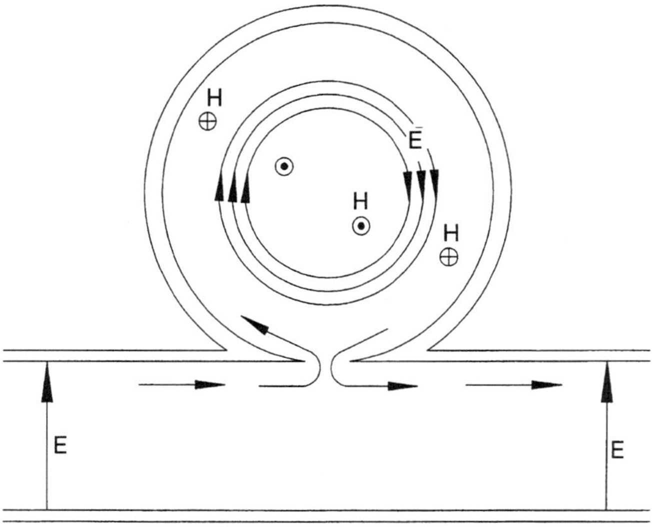

FIGURE 12.26 Coupling between a cylindrical cavity a rectangular waveguide.

On the circular cylindrical wall ($r = a$) $\underline{H}_z$ is tangential, giving rise to circumferential surface current, $\underline{J}_{s\phi}$. On the circular end plates, $\underline{H}_r$ is tangential, again giving rise to circumferential surface current. There are no radial currents and no longitudinal currents, so no current flows across the joint from the circular end plate to the cylindrical walls. If the length d is made adjustable by mounting one end of the cavity on a micrometer screw there is no interruption of the surface currents between the movable end wall and the circular wall of the cavity. As the length d is varied by turning the micrometer screw, the resonant frequency changes. The frequency scale can therefore be calibrated along the length of the cavity. We can mount the cavity on the broad wall of a rectangular waveguide (Fig. 12.26) and use it as a frequency meter or wavemeter to measure the frequency of the dominant TE_{10} mode in the rectangular guide. A thin slot cut from the waveguide to the cavity at the joint provides coupling between the guide and the cavity. When the frequency in the guide matches the cavity resonant frequency, the cavity is excited, there is a small power loss in the cavity, and the transmitted power through the guide dips sharply. Care must be taken not to excite other resonant modes in the cavity, as these will give false readings or, for the TM_{111} mode, which has the same resonant frequency, reduce the Q and hence the sharpness of the power dip.

PROBLEMS

12.1 Laplace's equation in rectangular coordinates is

$$\nabla^2 V = \frac{\partial^2 V}{\partial x^2} + \frac{\partial^2 V}{\partial y^2} + \frac{\partial^2 V}{\partial z^2} = 0.$$

Let $V(x, y, z) = X(x)Y(y)Z(z)$ and obtain the general solution for V. Assume the x and y dependence is harmonic and the z dependence is exponential. What is the relation between the three separation constants? Can the dependence be harmonic in all three coordinates? Can the dependence be exponential in all three coordinates?

Ans. $V = (A \cos k_x x + B \sin k_x x)(C \cos k_y y + D \sin k_y y)$
$[E \exp(k_z z) + F \exp(-k_z z)], \quad k_x^2 + k_y^2 = k_z^2.$

12.2 Laplace's equation in cylindrical coordinates is

$$\nabla^2 V = \frac{1}{r}\frac{\partial}{\partial r}\left(r\frac{\partial v}{\partial r}\right) + \frac{1}{r^2}\frac{\partial^2 V}{\partial \phi^2} + \frac{\partial^2 V}{\partial z^2} = 0.$$

Let $V(r, \phi, z) = R(r)\Phi(\phi)Z(z)$ and obtain the general solution for V. Assume the ϕ dependence is harmonic and the z dependence is exponential.

Ans. $V = [AJ_v(kr) + BN_v(kr)](C \cos v\phi + D \sin v\phi)[E \exp(kz) + F \exp(-kz)].$

12.3 A perfect magnetic conductor is the dual of a perfect electric conductor. At the surface of a perfect magnetic conductor the tangential component of magnetic field intensity must be zero. A rectangular waveguide is constructed with the walls at $x = 0$ and $x = a$ made of perfect electric conductors and the walls at $y = 0$ and $y = b$ made of perfect magnetic conductors. Assume $a > b$.

a. Find $h_z(x, y)$ for the TE modes.

b. Find $e_z(x, y)$ for the TM modes.

c. Find the cutoff frequency for the nmth mode and state which mode is dominant.

Ans. a. $h_{z,nm} = H_{0,nm} \cos(n\pi x/a) \sin(m\pi y/b), \quad n = 0, 1, 2, \ldots, m = 1, 2, \ldots;$

b. $e_{z,nm} = E_{0,nm} \sin(n\pi x/a) \cos(m\pi y/b), \quad n = 1, 2, \ldots, m = 1, 2, \ldots;$

c. $f_{c,nm} = (c/2)\sqrt{(n/a)^2 + (m+b)^2}$, TE_{10} is dominant.

12.4 For an air-filled rectangular waveguide with inside dimensions $a = 0.900$ in. and $b = 0.400$ in., calculate the lowest eight cutoff frequencies and their ratio with the cutoff frequency of the dominant mode. Identify the modes associated with each frequency. Use $c = 11.8 \times 10^8$ in./s. *Note*: The recommended frequency range for this guide is 8.20 to 12.4 GHz.

12.5 For an air-filled rectangular waveguide with inside dimensions $a = 1.580$ cm and $b = 0.790$ cm the recommended frequency range is 12.4 to 18.0 GHz. Calculate the cutoff frequency for the TE_{10} mode and the guide wavelength, phase velocity, group velocity, and attenuation due to conducting wall losses at each end of the recommended frequency band. Assume a brass guide with $\sigma_c = 1.1 \times 10^7$.

Ans. $f_{c10} = 9.49$ GHz, $\beta_g = 167/320$ rad/m, $\lambda_g = 0.0376/0.0196$ m,
$v_{ph} = 4.66/(3.53 \times 10^8)$ m/s, $v_g = 1.93/(2.55 \times 10^8)$ m/s,
$\alpha_c = 0.0550/0.0406$ Np/m or 0.48/0.35 dB/m.

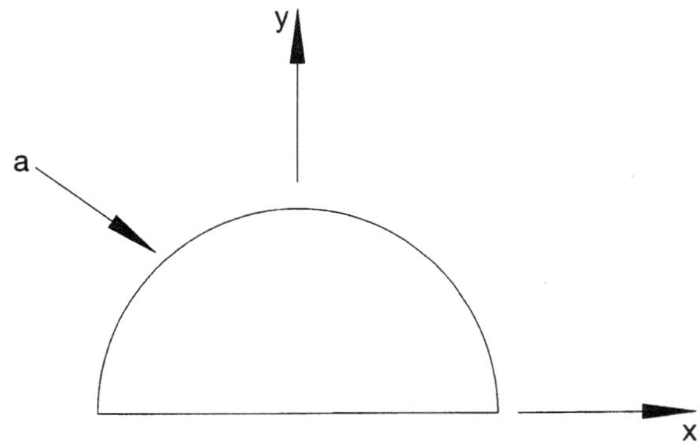

FIGURE 12.27 Problem 12.6.

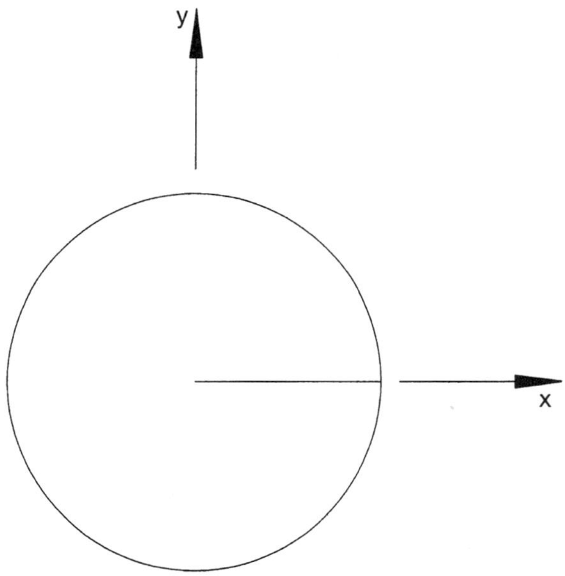

FIGURE 12.28 Problem 12.7.

12.6 Find e_z for the TM modes and h_z for the TE modes in a semicircular guide of radius a (Fig. 12.27). Which is the dominant mode? What is the next mode to propagate? Is the frequency band in which only the dominant mode can propagate larger than for the circular guide?

Ans. $e_{z,nm} = E_{0,nm} J_n(p_{nm} r/a) \sin n\phi, \quad n = 1, 2, \ldots, m = 1, 2, \ldots;$

$h_{z,nm} = H_{0,nm} J_n(p'_{nm} r/a) \cos n\phi, \quad n = 1, 2, \ldots, m = 1, 2, \ldots.$

12.7 Find e_z for the TM modes and h_z for the TE modes in a circular guide of radius a with a conducting baffle in the xz plane extending from $x = 0$ to $x = a$ (Fig. 12.28). Identify the dominant mode and the next three modes to propagate. Is the frequency band in which only the dominant mode can propagate larger than for the circular guide?

Ans. $e_{z,vm} = E_{0,vm} J_v\left(\dfrac{p_{vm} r}{a}\right) \sin v\phi, \quad v = n/2, \quad n = 1, 2, \ldots, m = 1, 2, \ldots;$

$$h_{z,vm} = H_{0,vm} J_v\left(\frac{p'_{vm} r}{a}\right) \cos v\phi, \quad v = n/2, \quad n = 0, 1, 2, \ldots, m = 1, 2, \ldots.$$

These are half-order and integer-order Bessel functions. The half-order Bessel functions and their derivatives have the same zeros as the spherical Bessel functions. $p_{v,m}$ is the mth zero of J_v. $p'_{v,m}$ is the mth zero of the derivative of J_v with respect to its argument.

For $J_{1/2}$: $p_{1/2,1} = 3.14 = \pi$, $p_{1/2,2} = 6.28 = 2\pi$.
For J_1: $p_{1,1} = 3.832$, $p_{1,2} = 7.016$.
For $J_{3/2}$: $p_{3/2,1} = 4.493$, $p_{3/2,2} = 7.725$.
For J_0: $p'_{0,1} = 3.832$, $p'_{0,2} = 7.016$.
For $J_{1/2}$: $p'_{1/2,1} = 1.166$, $p'_{1/2,2} = 4.604$.
For J_1: $p'_{1,1} = 1.841$, $p_{1,2} = 5.331$.
For $J_{3/2}$: $p'_{3/2,1} = 2.461$, $p'_{3/2,2} = 6.029$.

12.8 For the coaxial transmission line the dominant mode is the TEM mode, which propagates down to zero frequency. The upper limit of single mode propagation is when TE and TM modes begin to propagate. Find e_z for the TM modes and h_z for the TE modes and the characteristic equation that determines k_c in each case. The coaxial guide has inner radius a and outer radius b.

Ans. $e_{z,nm} = E_{0,nm}[N_n(k_{c,nm}a) J_n(k_{c,nm}r) - J_n(k_{c,nm}a) N_n(k_{c,nm}r)] \cos n\phi,$
where $k_{c,nm}$ is the mth root of
$J_n(k_c a) N_n(k_c b) - J_n(k_c b) N_n(k_c a) = 0.$
$h_{z,nm} = H_{0,nm} [N'_n[k_{c,nm}a) J_n(k_{c,nm}r) - J'_n(k_{c,nm}a) N_n(k_{c,nm}r)] \cos n\phi,$
where $k_{c,nm}$ is the mth root of
$J'_n(k_c a) N'_n(k_c b) - J'_n(k_c b) N'_n(k_c a) = 0.$

The TE_{11} mode has the lowest cutoff frequency of all TE and TM modes. For this mode the value of $k_{c,1}a$ is approximated with less than 3% error (for $b/a < 15$) by the formula $k_{c,11} = 2/[1 + (b/a)]$.

12.9 Show that

$$\frac{\partial v_g}{\partial f} = -\frac{c^2}{v_g}\left(\frac{f_c}{f}\right)^2 \left(\frac{1}{f}\right)$$

and hence that the fractional change in v_g over a frequency band Δf is

$$\left|\frac{\Delta v_g}{v_g}\right| = \left(\frac{c}{v_g}\right)^2 \left(\frac{f_c}{f}\right)^2 \left|\frac{\Delta f}{f}\right|.$$

Calculate the fractional change if $f_c = 2$ GHz, $\Delta f = 6$ MHz, and $f = 1.5f_c$. Note that 6 MHz is the bandwidth of a TV channel.

Ans. $|\Delta v_g/v_g| = 0.016$ or 1.6%.

12.10 Design a rectangular waveguide for as broad a bandwidth as possible in the dominant mode. The center of the band should be at 3 GHz. The low end f_l should be 20% greater than the cutoff frequency and the high end f_h should be 5% less than the cutoff frequency of the next mode to propagate. Find a, b, f_l, f_h, and the useful bandwidth.

Ans. $a = 7.75$ cm, $b = 3.88$ cm, $f_l = 2.32$ GHz, $f_h = 3.68$ GHz.

12.11 An air-filled rectangular waveguide excited in the dominant mode is loaded with (or feeds) a horn antenna. The input impedance to the antenna is the load impedance on the guide. This impedance is expressed in terms of the fields as $Z_L = Z_{\text{in}} = -\underline{E}_{y,10}/\underline{H}_{x,10}$. In the guide the forward going or incident wave has an amplitude E_0 and the backward going or reflected wave has an amplitude $\Gamma_L E_0$, both at the input plane of the antenna. Find an expression for the reflection coefficient Γ_L in terms of Z_L and $Z_{\text{TE}_{10}} = \omega\mu_0/\beta_{10}$. Evaluate Γ_L if $f = 1.5f_{c,10}$ and $Z_L = 2\eta_0$.

Ans. $\Gamma_L = (Z_L - Z_{\text{TE}_{10}})/(Z_L + Z_{\text{TE}_{10}}), \Gamma_L = 0.197$.

12.12 Design an air-filled rectangular waveguide to operate in the dominant mode over a frequency range from 4 to 5 GHz. The wide dimension should be chosen so that the low frequency is 20% greater than the cutoff frequency. To minimize the attenuation constant the narrow dimension should be chosen as large as possible, consistent with the high frequency being 5% less than the cutoff frequency of the next propagating mode.

Ans. $a = 4.50$ cm, $b = 2.85$ cm.

12.13 The frequency meter of Fig. 12.26 uses the length d of the cavity to measure the frequency of the signal. Plot the resonant frequency versus d for the TE_{011} mode. Let the cavity radius be 2 cm and plot for length $0.25 \le d \le 5$ cm. Is the relation linear? What range of d do you think is most useful: where the frequency varies rapidly with d or where it varies slowly?

12.14 Design a circular cavity resonator in the TE_{011} mode for use as a frequency meter from 2.20 to 3.30 GHz. It is desired to have as much of the frequency band as possible be in the range where the resonant frequency varies slowly with d. Find a value for a and a corresponding range for d. The dimensions of the rectangular waveguide for this frequency range are 8.64 cm by 4.32 cm. *Hint*: Choose a by letting d be very large and letting the resonant frequency be about 0.9(2.20) GHz.

Ans. $a = 9.24$ cm, $d = 5.68$ to 15.6 cm. Other designs are possible.

12.15 Find the expression for the Q of a circular cylindrical cavity resonating in the TE_{011} mode.

REFERENCES

[1] David M. Pozar, *Microwave Engineering*, Addison-Wesley, Reading, MA, 1990, Chap. 9.

[2] Peter A. Rizzi, *Microwave Engineering: Passive Circuits*, Prentice Hall, Englewood Cliffs, NJ, 1988, pp. 4564–500.

CHAPTER THIRTEEN

Microwave Networks

13.1 INTRODUCTION

In your systems, circuit, or network theory courses you have studied the analysis and design of low-frequency circuits consisting of resistors, inductors, capacitors, and electronic devices. Since you are now familiar with wave phenomena you realize that the term "low-frequency circuits" means that the dimensions of the circuit are small in comparison with a wavelength. In such circuits the dimensions of the elements and of the wires that interconnect them are of no consequence.

At microwave or higher frequencies it is desirable to utilize the techniques and knowledge acquired for analysis at low frequency. We have already seen that for TEM waves on two-wire transmission lines we can define voltages and currents. We can also define voltages and currents for TE and TM waves in other transmission channels, such as waveguides. This will allow us to model these channels as transmission lines. It is also possible to model the interconnections of these channels, discontinuities in the channels, microwave devices, and other elements by a combination of transmission lines, resistors, inductors, and capacitors. These models are helpful in analyzing microwave circuits.

13.2 VOLTAGE, CURRENT, AND IMPEDANCE

Our studies have shown us that we can describe electric and magnetic fields on transmission channels in terms of TEM, TE, TM, and hybrid modes. For simplicity we will consider only lossless channels and propagating modes, so $\gamma = j\beta$ and $\alpha = 0$. For each of these modes there are a few common properties.

1. When z is taken as the longitudinal direction in the guide, then the mode phasor field components have $\exp(\mp j\beta z)$ as their z dependence. The upper sign goes with a forward or $+z$ traveling wave and the lower sign goes with a backward or $-z$ traveling wave. Generally, β is a function of frequency, the medium properties, the guide shape and dimensions, and the mode number, but for each mode and frequency it completely describes the wave propagation.

2. The transverse electric field phasor is given in terms of a transverse field $\vec{e}_t$. The transverse magnetic field phasor is given in terms of a transverse field $\vec{h}_t$. $\vec{e}_t$ and $\vec{h}_t$ are functions of the transverse coordinates and they are related by $R_w\vec{h}_t = \vec{a}_z \times \vec{e}_t$, where R_w is the characteristic resistance of the mode. For TEM waves $R_w = \sqrt{\mu/\varepsilon}$, Eq. (9.11), for TE waves $R_w = \omega\mu/\beta$, Eq. (12.10), and for TM waves $R_w = \beta/\omega\varepsilon$, Eq. (12.15).
3. The power carried by each mode is given by an integral of the transverse fields over the transverse plane of the guide.

For any mode, then, the transverse field phasors are

$$\underline{\vec{E}}_t = C^+\vec{e}_t e^{-j\beta z} + C^-\vec{e}_t e^{+j\beta z}, \quad \underline{\vec{H}}_t = C^+\vec{h}_t e^{-j\beta z} - C^-\vec{h}_t e^{+j\beta z}.$$

The time average power flow in the z direction is

$$\begin{aligned} P_{f,\text{Avg}} &= \frac{1}{2}\int_{\text{tr.pl.}} \text{Re}(\underline{\vec{E}}_t \times \underline{\vec{H}}_t^*)\cdot d\vec{S} \\ &= \frac{1}{2R_w}(|C^+|^2 - |C^-|^2)\int_{\text{tr.pl.}} |\vec{e}_t|^2\, dS. \end{aligned} \tag{13.1}$$

For the transmission line model there are voltage and current waves on the line. The transmission line model has a phase constant β identically equal to that of the guide mode and a characteristic resistance R_0. The voltage and current on the line are

$$\underline{V} = V_0^+ e^{-j\beta z} + V_0^- e^{+j\beta z}, \quad \underline{I} = \frac{1}{R_0}(V_0^+ e^{-j\beta z} - V_0^- e^{+j\beta z}).$$

The time average power flow along the line is

$$P_{f,\text{Avg}} = \tfrac{1}{2}\,\text{Re}(\underline{V}\underline{I}^*) = \frac{1}{2R_0}(|V_0^+|^2 - |V_0^-|^2). \tag{13.2}$$

The relation between the model's waves and the guide's waves is given by a constant of proportionality[†] K, such that

$$V_0^+ = KC^+, \quad V_0^- KC^-. \tag{13.3}$$

We have not actually specified either the line characteristic resistance R_0 or the constant K. Whatever choice we make for these, we require that the time average

[†] A comment on the dimensionality of K is required. The units of $\vec{e}_t$ and $\underline{\vec{E}}_t$ are volts per meter and hence C^+ and C^- are dimensionless. The units of $\underline{V}$, V_0^+, and V_0^- are volts. The units of the constant K are therefore volts.

power flow in the guide and the time average power flow in its transmission line model must be the same. Hence equating Eqs. (3.1) and (13.2) gives

$$\frac{1}{2R_w}(|C^+|^2 - |C^-|^2)\int_{\text{tr.pl.}} |\vec{e}_t|^2\, dS = \frac{1}{2R_0}(|V_0^+|^2 - |V_0^-|^2)$$

$$\frac{1}{R_w}\int_{\text{tr.pl.}} |\vec{e}_t|^2\, dS = \frac{K^2}{R_0}$$

or

$$K = \sqrt{\frac{R_0}{R_w}\int_{\text{tr.pl.}} |\vec{e}_t|^2\, dS}. \tag{13.4}$$

We may now choose the value of the transmission line's characteristic resistance R_0 somewhat arbitrarily and then the value of K follows from Eq. (13.4).

There are several common ways to choose this characteristic resistance. We can define it as the ratio of the maximum voltage $\underline{V}$ across the guide to the total longitudinal current $\underline{I}$ along the guide, sometimes called the voltage–current definition of R_0. We can define it in terms of the time average power flow $P_{f,\text{Avg}}$ and the maximum voltage, sometimes called the power–voltage definition. Finally, we can define it in terms of the time average power flow and the total longitudinal current, sometimes called the power–current definition. The formulas for these definitions are

$$R_{0,\text{VI}} = \frac{\underline{V}}{\underline{I}}, \quad R_{0,\text{PV}} = \frac{|\underline{V}|^2}{2P_{f,\text{Avg}}}, \quad \text{and} \quad R_{0,\text{PI}} = \frac{2P_{f,\text{Avg}}}{|\underline{I}|^2}, \tag{13.5}$$

respectively. Let us see what this means for the TE_{10} mode in a rectangular guide.

Example 13.1 For the dominant TE_{10} mode in the rectangular guide, find the transmission line model characteristic resistance for each of the three definitions.

Solution The transverse fields [Eqs. (12.49) and (12.50)] are

$$\underline{E}_y = E_0 \sin\left(\frac{\pi x}{a}\right)e^{-j\beta_{10}z}, \quad \beta_{10} = \sqrt{\omega^2\mu\varepsilon - \left(\frac{\pi}{a}\right)^2},$$

$$\underline{H}_x \frac{\beta_{10}}{\omega\mu} E_0 \sin\left(\frac{\pi x}{a}\right)e^{-j\beta_{10}z}.$$

and the time average power flow [Eq. (12.53)] is

$$P_{f,\text{Avg}} = \frac{1}{2}\left(\frac{\beta_{10}}{\omega\mu}\right)\left(\frac{ab}{2}\right)|E_0|^2.$$

The maximum voltage from the bottom to the top of the guide is

$$\underline{V} = \int_{y=0}^{b} \underline{E}_y(a/2, y, z)\, dy = bE_0 e^{-j\beta_{10} z}.$$

The total longitudinal current is

$$\underline{I} = -\int_{x=0}^{a} \underline{H}_x(x, 0, z)\, dx = \left(\frac{2a}{\pi}\right)\left(\frac{\beta_{10}}{\omega\mu}\right) E_0 e^{-j\beta_{10} z}.$$

The three model characteristic resistances are

$$R_{0,\mathrm{VI}} = \left(\frac{\pi b}{2a}\right)\left(\frac{\omega\mu}{\beta_{10}}\right) = 1.57\left(\frac{b}{a}\right)\left(\frac{\omega\mu}{\beta_{10}}\right), \tag{13.6}$$

$$R_{0,\mathrm{PV}} = \left(\frac{2b}{a}\right)\left(\frac{\omega\mu}{\beta_{10}}\right) = 2\left(\frac{b}{a}\right)\left(\frac{\omega\mu}{\beta_{10}}\right), \tag{13.7}$$

$$R_{0,\mathrm{PI}} = \left(\frac{\pi^2 b}{8a}\right)\left(\frac{\omega\mu}{\beta_{10}}\right) = 1.23\left(\frac{b}{a}\right)\left(\frac{\omega\mu}{\beta_{10}}\right), \tag{13.8}$$

where

$$\left(\frac{\omega\mu}{\beta_{10}}\right) = \frac{\eta}{\sqrt{1 - (f_{c,10}/f)^2}}, \quad f_{c,10} = \frac{1}{2a\sqrt{\mu\varepsilon}}.$$

It should be noted that the three forms of model characteristic resistance are almost identical. Each one is proportional to the mode impedance, $\omega\mu/\beta_{10}$, and each is proportional to the guide height and inversely proportional to the guide width. Even the numerical value of the multiplier is not very different. All three are in fact close to $(b/a)(\omega\mu/\beta_{10})$, which would be the result by all three definitions if the transverse fields were uniform instead of varying sinusoidally across the guide. One of these definitions is no more correct than another and any one can be used as long as the same one is used throughout a single modeling problem.

Example 13.2 Repeat the previous example for the dominant TE_{11} mode in a circular guide.

Solution Figure 12.10 shows the electric field lines for this mode. The maximum voltage across the guide is found by integrating along the vertical field line, which bisects the guide. Thus, letting $p' \equiv p'_{11} = 1.841$ and $\beta \equiv \beta_{11}$ for simplicity, we have

$$\begin{aligned} \underline{V} &= 2\int_{r=0}^{a} \underline{E}_r(r, \pi/2, z)\, dr = \frac{-2j\omega\mu a^2}{(p')^2} H_0 e^{-j\beta z} \int_{r=0}^{a} \frac{1}{r} J_1\left(\frac{p' r}{a}\right) dr \\ &= \frac{2j\omega\mu a^2}{(p')^2} H_0 e^{-j\beta z} F, \quad F = J_1(p') - \int_{u=0}^{p'} J_0(u)\, du. \end{aligned}$$

Using $\underline{H}_\phi$, Eq. (12.86), the longitudinal conduction current flowing in the bottom half of the circular guide is

$$\underline{I} = \int_{\phi=\pi}^{2\pi} \underline{H}_\phi(a, \phi, z) a \, d\phi = \frac{j\beta a^2}{(p')^2} H_0 J_1(p') e^{-j\beta z} \int_{\phi=\pi}^{2\pi} \sin\phi \, d\phi$$

$$= -\frac{2j\beta a^2}{(p')^2} H_0 J_1(p') e^{-j\beta z}.$$

The time average power carried by the wave is given by Eq. (12.90) as

$$P_{f,\text{Avg}} = \frac{1}{2}\frac{\omega\mu\beta a^4}{(p')^4}\left(\frac{\pi}{2}\right) H_0^2 [J_1(p')]^2 [(p')^2 - 1].$$

The resultant model characteristic resistances are

$$R_{0,\text{VI}} = \frac{F}{J_1(p')}\left(\frac{\omega\mu}{\beta_{11}}\right) = 1.38\left(\frac{\omega\mu}{\beta_{11}}\right),$$

$$R_{0,\text{PV}} = \frac{8F^2}{\pi J_1^2(p')[(p')^2 - 1]}\left(\frac{\omega\mu}{\beta_{11}}\right) = 2.03\left(\frac{\omega\mu}{\beta_{11}}\right),$$

$$R_{0,\text{PI}} = \frac{\pi[(p')^2 - 1]}{8}\left(\frac{\omega\mu}{\beta_{11}}\right) = 0.939\left(\frac{\omega\mu}{\beta_{11}}\right),$$

where

$$\left(\frac{\omega\mu}{\beta_{11}}\right) = \frac{\eta}{\sqrt{1 - (f_{c,11}/f)^2}}, \quad f_{c,11} = \frac{p'_{11}}{2\pi\sqrt{\mu\varepsilon}a}.$$

Again, the three forms are the same. All are independent of the guide radius, except as it appears in β, and all are proportional to the wave impedance. Again, any one can be used as long as the same one is used consistently throughout a modeling problem.

When the transmission line model characteristic resistance is determined the voltages and currents on the line can be obtained by using the transmission line analysis of Chapter 10. To return to the waveguide fields use Eq. (13.4) to calculate K and then Eq. (13.3) to obtain C^+ and C^-.

13.3 ONE-PORT, TWO-PORT, AND *N*-PORT JUNCTIONS

As with all highways, waveguides must start somewhere and go somewhere. Also like highways, they may divide or fork into several channels, they may change shape along the way, or they may have various obstacles or barriers along the route. Each of these irregularities can be modeled as a junction involving one or more waveguide channels.

For example, a coaxial cable connected to the signal input of a television set (Fig. 13.1a), a waveguide terminated with a lossy wedge (Fig. 13.1b), or a waveguide

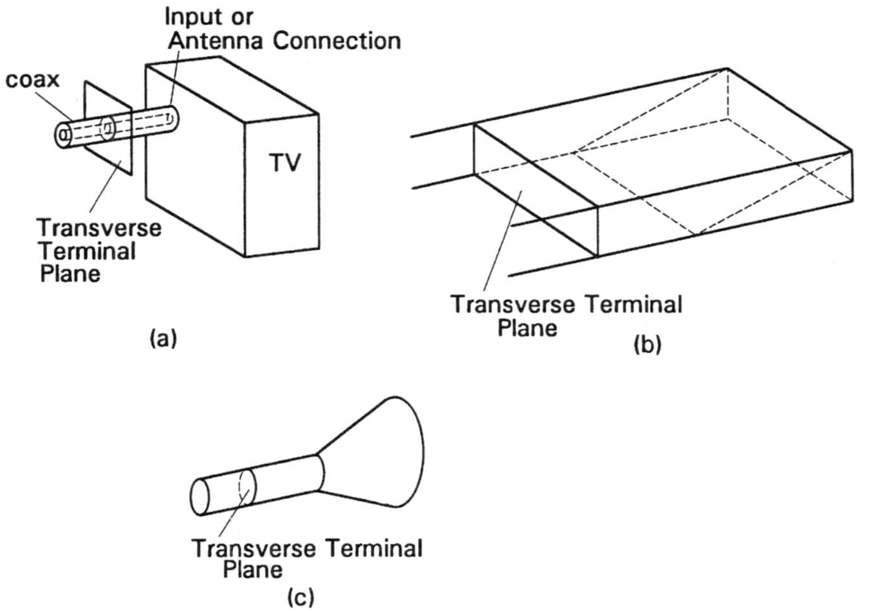

FIGURE 13.1 Example of one-port junctions and their transverse terminal planes.

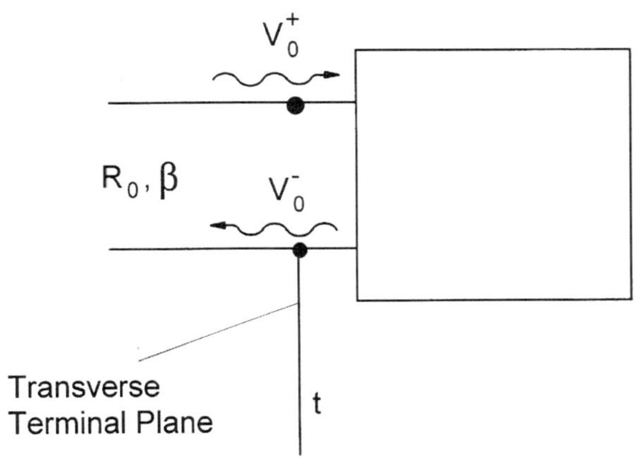

FIGURE 13.2 One-port junction.

terminated in a radiation horn (Fig. 13.1c) can all be modeled as a transmission line terminated in a single-terminal pair box or a one-port junction (Fig. 13.2). The junction represents all of the electromagnetic phenomena to the right of the transverse plane shown in the previous figure. V_0^+ and V_0^- are the amplitudes of the incident and reflected voltage waves at the terminal plane $z = 0$.

Figure 13.3a shows a coaxial cable coupled to a rectangular waveguide. Figure 13.3b shows a rectangular waveguide with a change in height, known as an E-plane step. Figure 13.3c shows a rectangular waveguide with a series of posts inserted into

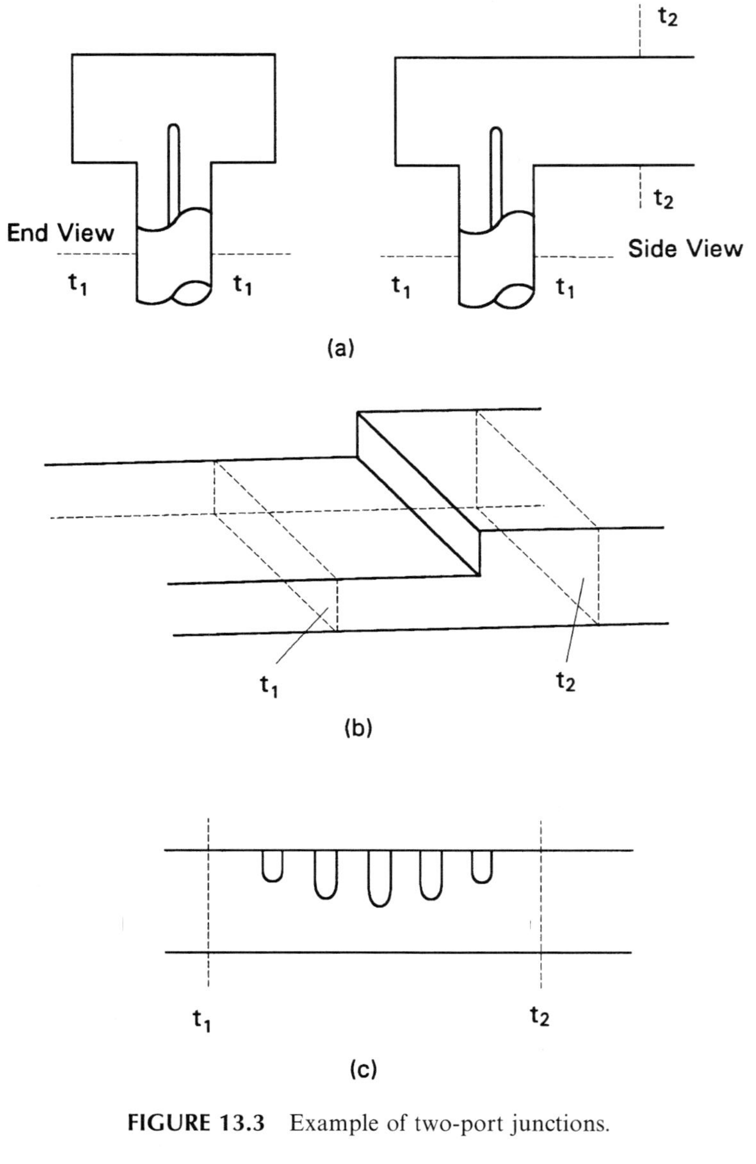

FIGURE 13.3 Example of two-port junctions.

the guide, acting as a frequency filter. These can all be modeled as in Fig. 13.4 with two transmission lines and a two-terminal pair box or a two-port junction, representing the electromagnetic phenomena between the terminal planes shown in the previous figure.

Figure 13.5a shows a microstrip transmission line dividing into two transmission lines, symmetrically, as a simple power divider. This can be modeled as three transmission lines connected to a three-terminal pair or three-port junction as shown in Fig. 13.5b. Figure 13.6a shows a junction between four rectangular waveguides, known as a Magic-Tee, which can be modeled as a four-port junction as in Fig. 13.6b.

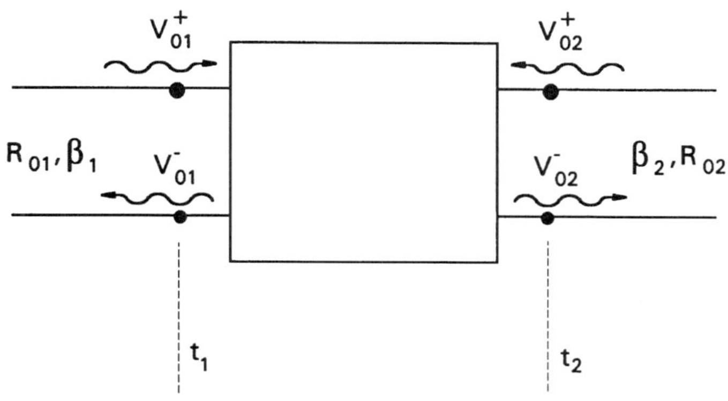

FIGURE 13.4 Equivalent circuit of a two-port junction.

In general, we can represent any set of N waveguides or transmission lines connected together by some physical configuration as an N-port junction connecting together N transmission lines modeling the waveguides (Fig. 13.7). We have already seen how to model the waveguides as transmission lines, each with forward and backward traveling voltage and current waves. We must find a way to mathematically represent the junction so that the relationships between the waves on each line are given.

Before we discuss the representation of the junction, let us recall several things about the waveguides connected to the junction. First, we have found infinite sets of modes can propagate in each guide, but that we generally operate the guide within a frequency range in which only the dominant mode carries time average power and all other modes are evanescent. Due to the complicated boundaries within each junction, many or all of these modes will be generated. For example, in the E-plane step (Fig. 13.3b) when a TE_{10} mode is incident on the junction from the shorter guide, there will be a reflected TE_{10} mode in the shorter guide and a transmitted TE_{10} mode in the taller guide, and, to satisfy the boundary conditions along the plane of the step, all the TE_{1m} evanscent modes will be generated in both guides. Since the evanescent modes decay rapidly away from the junction, we can choose the junction's terminal planes so that only the dominant mode remains at those planes. The representation of the junction includes the effect of all the evanescent modes but gives the relations at the ports between the dominant modes. Second, moving the terminal plane changes the relative phases of the incident and reflected dominant waves at the port. Therefore every description of a junction must be referenced to clearly described terminal planes.

13.4 Z AND Y MATRIX DESCRIPTION OF N-PORT JUNCTIONS

An N-port junction can be represented via its impedance (**Z**) or admittance (**Y**) matrix. Figure 13.8 shows an N-port network with all voltages and currents defined at each

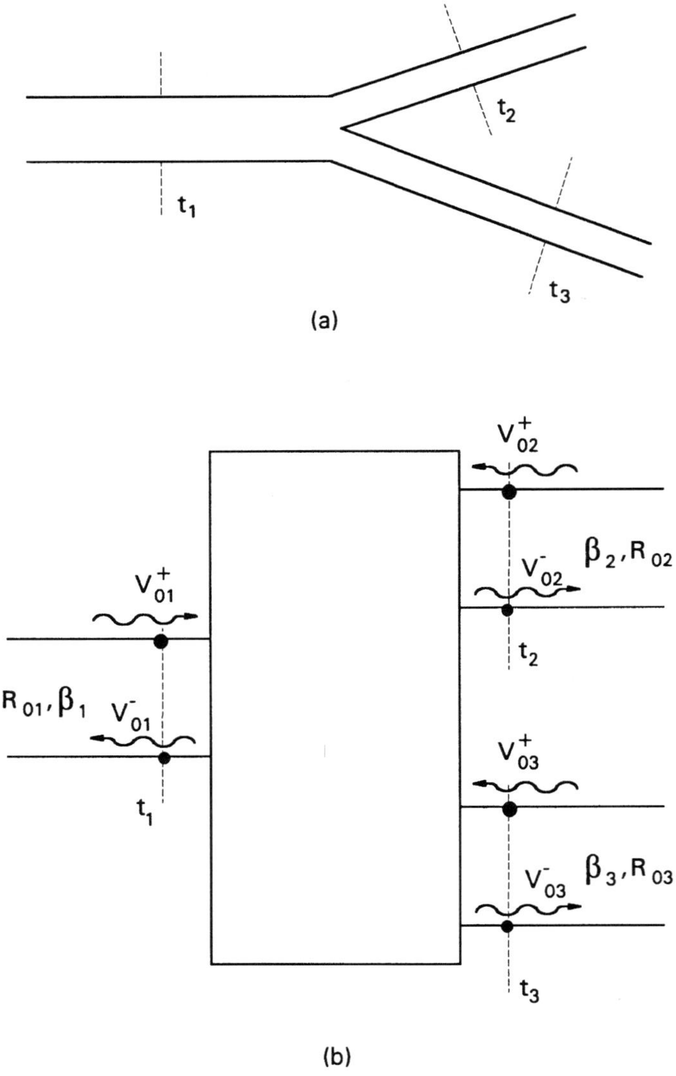

FIGURE 13.5 Equivalent circuit of a three-port junction.

port. The **Z** matrix representation of such a junction is given below:

$$\begin{bmatrix} V_1 \\ V_2 \\ V_3 \\ \vdots \\ V_N \end{bmatrix} = \begin{bmatrix} Z_{11} & Z_{12} & Z_{13} & \cdots & Z_{1N} \\ Z_{21} & Z_{22} & Z_{23} & \cdots & Z_{2N} \\ Z_{31} & Z_{32} & Z_{33} & \cdots & Z_{3N} \\ \vdots & \vdots & \vdots & & \vdots \\ Z_{N1} & Z_{N2} & Z_{N3} & \cdots & Z_{NN} \end{bmatrix} \begin{bmatrix} I_1 \\ I_2 \\ I_3 \\ \vdots \\ I_N \end{bmatrix}. \tag{13.9}$$

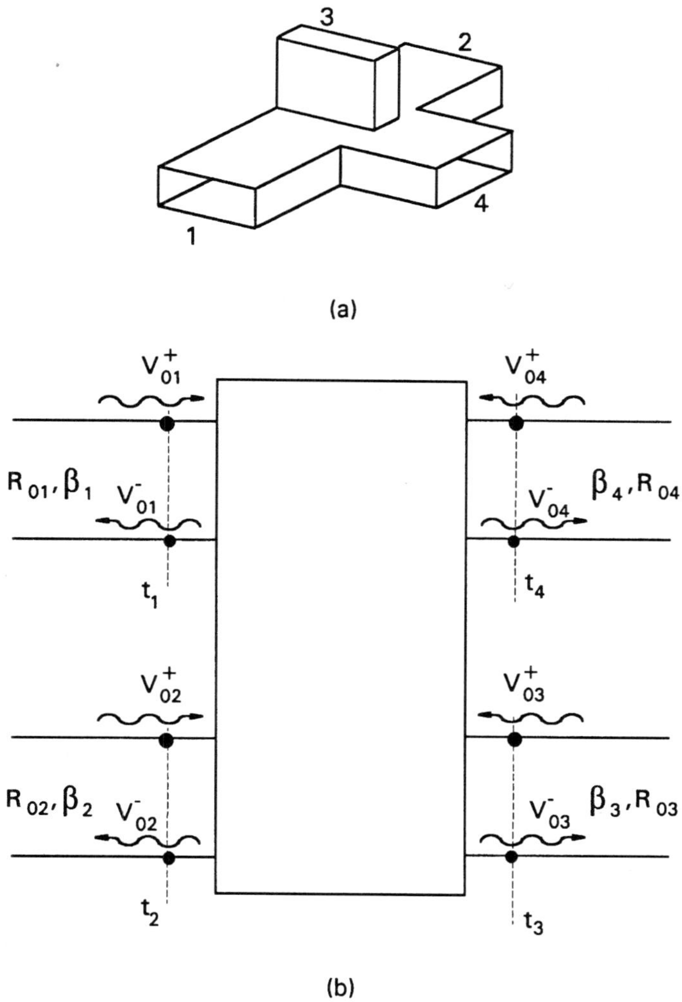

FIGURE 13.6 A Magic-Tee and its four-port circuit represenation.

The Z-parameters are determined as follows:

$$Z_{ij} = \frac{V_i}{I_j}, \quad I_k = 0, \quad k \neq j. \tag{13.10}$$

Determination of the Z-parameters requires the placement of an open circuit to all but one port. Hence the Z-parameters are also known as the open circuit parameters. Complete characterization of an N-port via the Z-parameters requires the determination of N^2 complex parameters or $2 \cdot N^2$ real parameters.

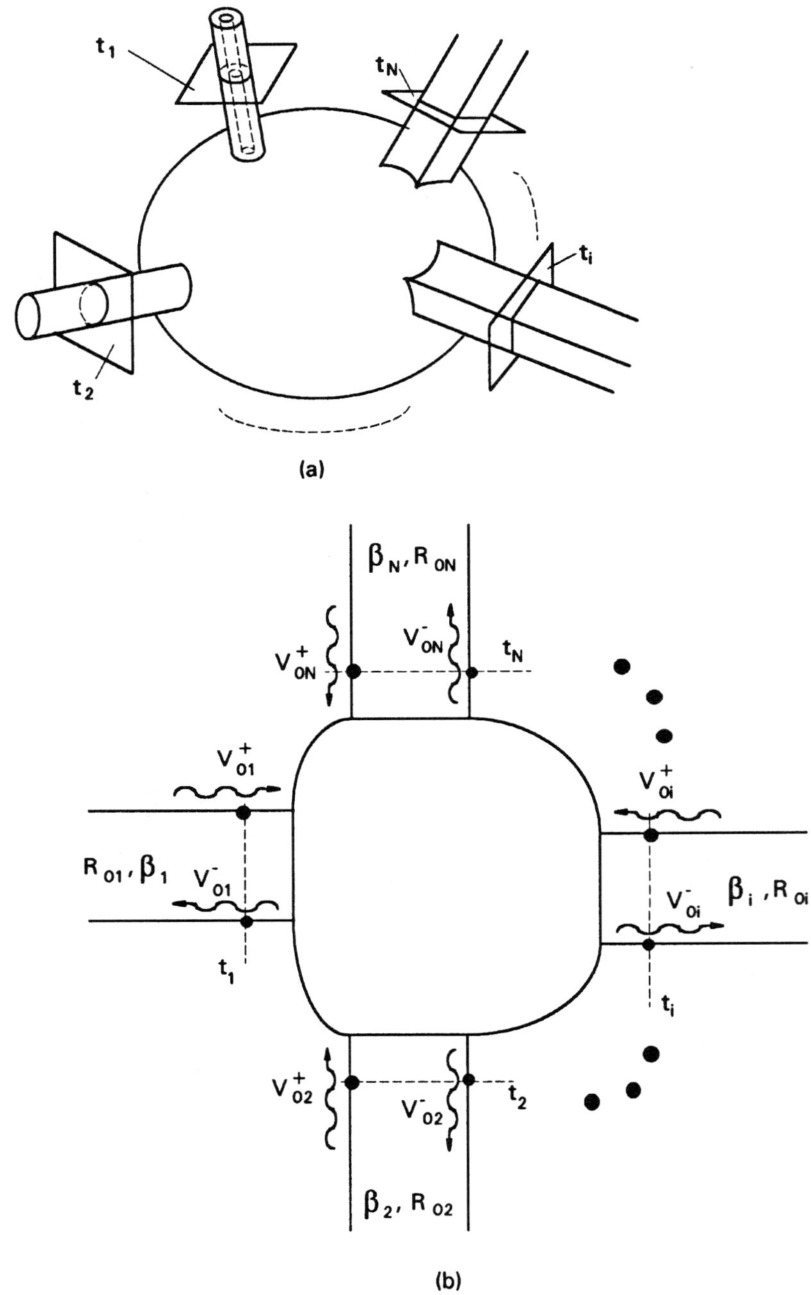

FIGURE 13.7 General N-port and its circuit representation.

Reciprocal networks will have a symmetrical $\mathbf{Z}$ matrix ($Z_{ij} = Z_{ji}$) and require only $N(N + 1)/2$ complex parameters or $N(N + 1)$ real parameters. Lossless networks have purely imaginary Z-parameters ($Z_{ij} = jX_{ij}$). The characterization of a lossless reciprocal network requires the determination of $N(N + 1)/2$ real parameters.

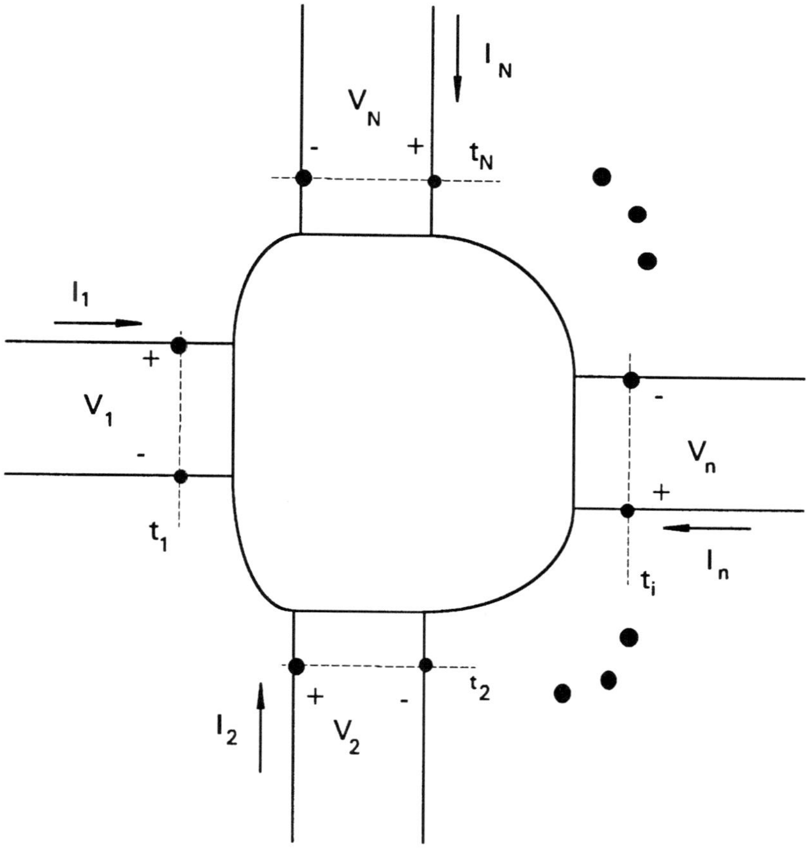

FIGURE 13.8 An N-port junction with voltage and current defined at each port.

The **Y** matrix representation of an N-port network is given below:

$$\begin{bmatrix} I_1 \\ I_2 \\ I_3 \\ \vdots \\ I_N \end{bmatrix} = \begin{bmatrix} Y_{11} & Y_{12} & Y_{13} & \cdots & Y_{1N} \\ Y_{21} & Y_{22} & Y_{23} & \cdots & Y_{2N} \\ Y_{31} & Y_{32} & Y_{33} & \cdots & Y_{3N} \\ \vdots & \vdots & \vdots & & \vdots \\ Y_{N1} & Y_{N2} & Y_{N3} & \cdots & Y_{NN} \end{bmatrix} \begin{bmatrix} V_1 \\ V_2 \\ V_3 \\ \vdots \\ V_N \end{bmatrix}. \tag{13.11}$$

The Y-parameters are determined as follows:

$$Y_{ij} = \frac{I_i}{V_j}, \quad V_k = 0, \quad k \neq j. \tag{13.12}$$

Determination of the Y-parameters requires the placement of a short circuit to all but one port. Hence the Y-parameters are also known as the short circuit parameters.

Reciprocal networks will have a symmetrical **Y** matrix ($Y_{ij} = Y_{ji}$), while lossless networks will have purely imaginary Y-parameters ($Y_{ij} = jB_{ij}$).

The **Z** matrix is the inverse of the **Y** matrix. If one of the matrices is singular, the other matrix representation does not exist. There are networks that do not have a **Z** matrix or a **Y** matrix. Examples of such networks will be given in the next section.

13.4.1 Z and Y Matrices for Two-Port Networks; Series and Parallel Connections of Two-Port Networks

Figure 13.9 shows a two-port network. The Z- and Y-parameters for a shunt admittance and a series impedance are given in Table 13.1 and can easily be obtained by applying the definition of the respective parameter.

The results in Table 13.1 show us that the shunt admittance two-port network does not have a **Y** matrix representation, while the series impedance two-port network does not have a **Z** matrix representation.

Example 13.3 Determine the Z-parameters of the network shown in Fig. 13.12 (T-network).

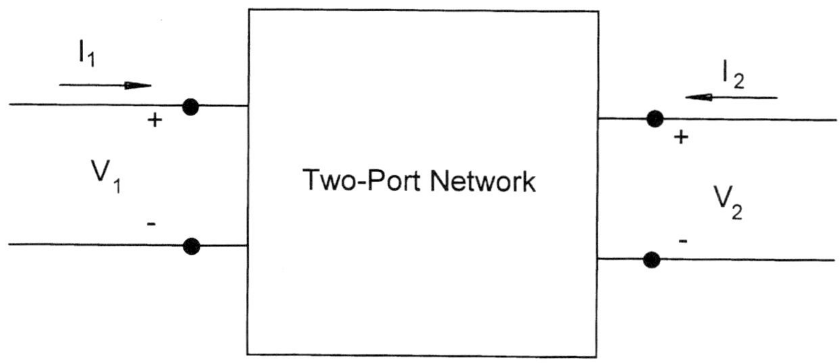

FIGURE 13.9 Two-port network with voltage and current defined at each port.

TABLE 13.1 Z and Y Matrix Elements

		Z				Y			
Network	Schematic	Z_{11}	Z_{12}	Z_{21}	Z_{22}	Y_{11}	Y_{12}	Y_{21}	Y_{22}
Shunt admittance	Fig. 13.10	$1/Y$	$1/Y$	$1/Y$	$1/Y$	—	—	—	—
Series impedance	Fig. 13.11	—	—	—	—	$1/Z$	$1/Z$	$1/Z$	$1/Z$

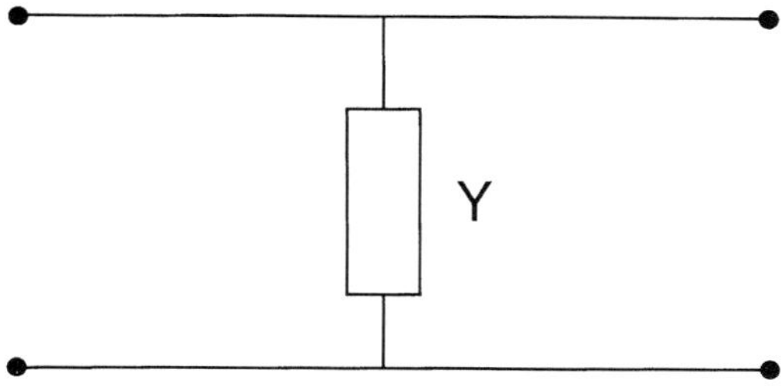

FIGURE 13.10 Circuit representation for a shunt admittance Y.

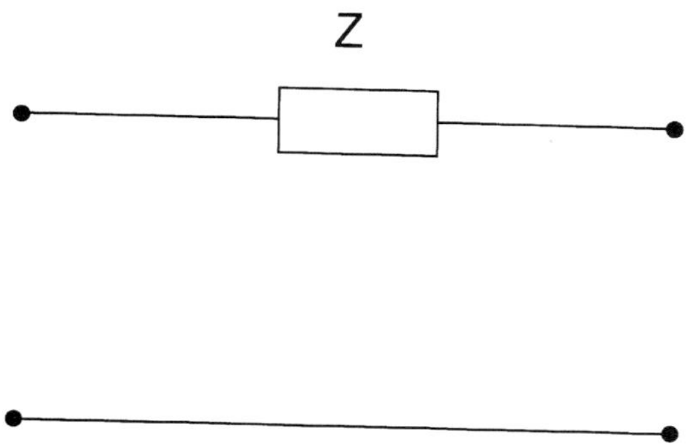

FIGURE 13.11 Circuit representation for a series impedance Z.

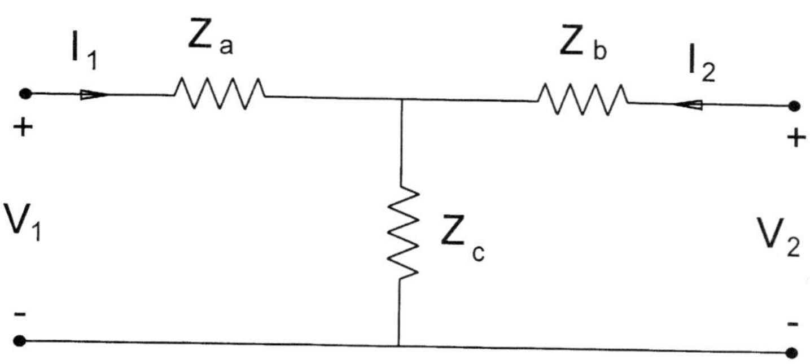

FIGURE 13.12 Circuit representation for a T-network.

Solution Using the definition of the Z-parameters, we find

$$Z_{11} = \left[\frac{V_1}{I_1}\right]_{I_2=0} = Z_a + Z_c,$$

$$Z_{12} = \left[\frac{V_1}{I_2}\right]_{I_1=0} = Z_c,$$

$$Z_{21} = \left[\frac{V_2}{I_1}\right]_{I_2=0} = Z_c,$$

$$Z_{22} = \left[\frac{V_2}{I_2}\right]_{I_1=0} = Z_b + Z_c.$$

From this example, by solving for the impedances $Z_a = Z_{11} - Z_{12}$, $Z_b = Z_{22} - Z_{12}$, and $Z_c = Z_{12}$ as functions of the Z-parameters, one can construct the T-network equivalent for any reciprocal two-port. Figure 13.13 shows the T-network equivalent of any reciprocal two-port for which the Z-parameters exist.

Example 13.4 Determine the Y-parameters of the network shown in Fig. 13.14 (Π-network).

Solution Using the definition of the Y-parameters, we find

$$Y_{11} = \left[\frac{I_1}{V_1}\right]_{V_2=0} = Y_a + Y_c,$$

$$Y_{12} = \left[\frac{I_1}{V_2}\right]_{V_1=0} = -Y_c,$$

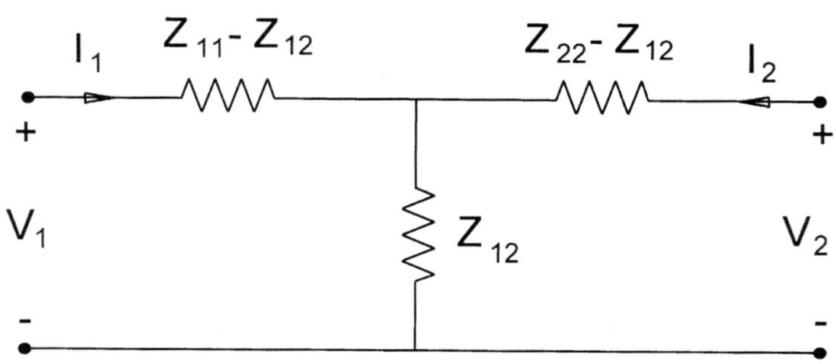

FIGURE 13.13 A T-network equivalent of any reciprocal two-port for which the Z-parameters exist.

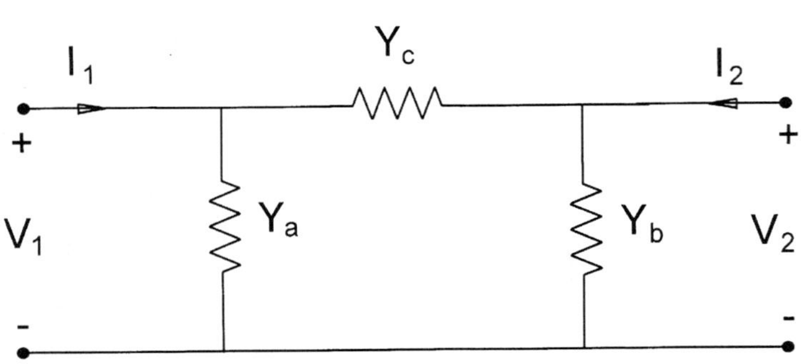

FIGURE 13.14 Circuit representation for a Π-network.

$$Y_{21} = \left[\frac{I_2}{V_1}\right]_{V_2=0} = -Y_c,$$

$$Y_{22} = \left[\frac{I_2}{V_2}\right]_{V_1=0} = Y_b + Y_c.$$

From this example, by solving for the admittances $Y_a = Y_{11} + Y_{12}$, $Y_b = Y_{22} + Y_{12}$, and $Y_c = -Y_{12}$ as functions of the Y-parameters, one can construct the Π-network equivalent for any reciprocal two-port. Figure 13.15 shows the Π-network equivalent of any reciprocal two-port for which the Y-parameters exist.

The series connection of two two-port networks is shown in Fig. 13.16. The relationships among the voltages and currents associated with the two two-port networks are

$$I_1 = I_1^A = I_1^B,$$

$$I_2 = I_2^A = I_2^B,$$

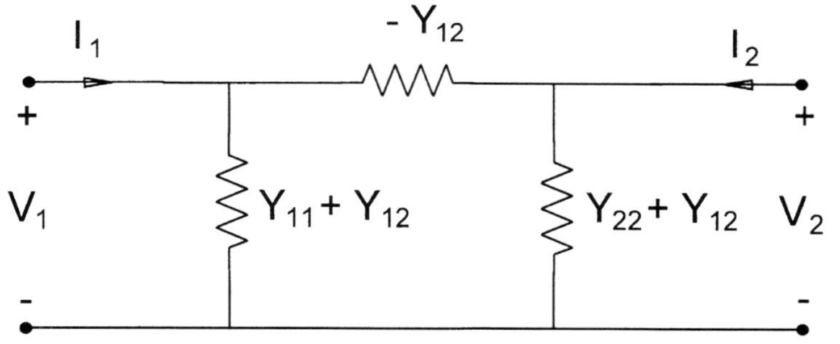

FIGURE 13.15 A Π-network equivalent of any reciprocal two-port for which the Y-parameters exist.

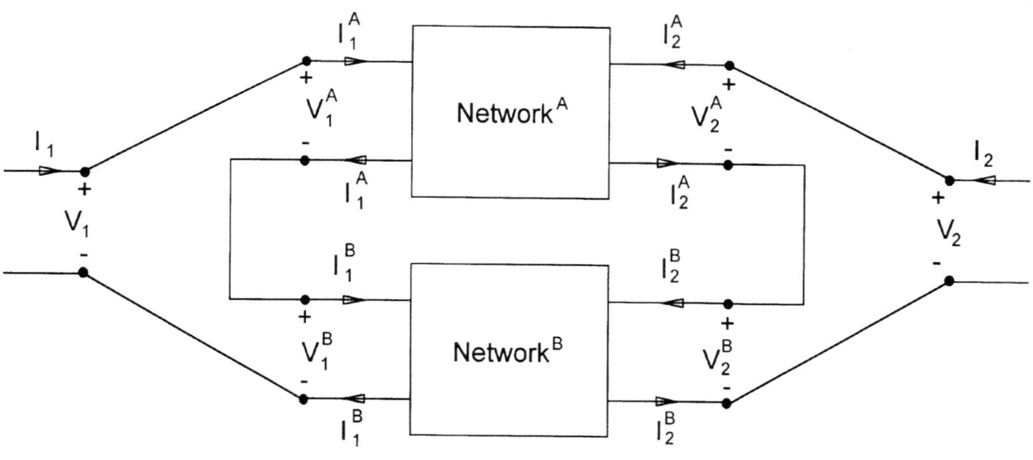

FIGURE 13.16 Series connection of two two-port networks.

$$V_1 = V_1^A + V_1^B,$$
$$V_2 = V_2^A + V_2^B.$$

Expressing these relationships in matrix form we have

$$\begin{bmatrix} V_1 \\ V_2 \end{bmatrix} = \begin{bmatrix} V_1^A + V_1^B \\ V_2^A + V_2^B \end{bmatrix} = \begin{bmatrix} V_1^A \\ V_2^A \end{bmatrix} + \begin{bmatrix} V_1^B \\ V_2^B \end{bmatrix}.$$

Representing each two-port via its **Z** matrix, we find

$$\begin{bmatrix} V_1 \\ V_2 \end{bmatrix} = \begin{bmatrix} Z_{11}^A & Z_{12}^A \\ Z_{21}^A & Z_{22}^A \end{bmatrix}\begin{bmatrix} I_1^A \\ I_2^A \end{bmatrix} + \begin{bmatrix} Z_{11}^B & Z_{12}^B \\ Z_{21}^B & Z_{22}^B \end{bmatrix}\begin{bmatrix} I_1^B \\ I_2^B \end{bmatrix}$$
$$= \left(\begin{bmatrix} Z_{11}^A & Z_{12}^A \\ Z_{21}^A & Z_{22}^A \end{bmatrix} + \begin{bmatrix} Z_{11}^B & Z_{12}^B \\ Z_{21}^B & Z_{22}^B \end{bmatrix}\right)\begin{bmatrix} I_1 \\ I_2 \end{bmatrix}.$$

The **Z** matrix of the series connection of two two-port networks is found by adding the two junctions' **Z** matrices:

$$\mathbf{Z} = \mathbf{Z}^A + \mathbf{Z}^B.$$

The parallel connection of two two-port networks is shown in Fig. 13.17. The relationships among the voltages and currents associated with the two two-port networks are

$$V_1 = V_1^A = V_1^B,$$
$$V_2 = V_2^A = V_2^B,$$

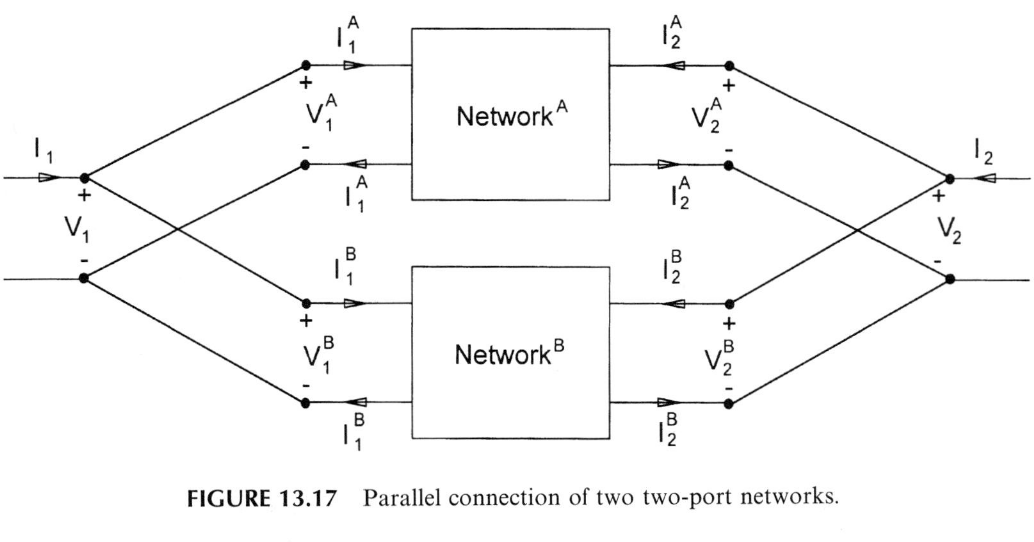

FIGURE 13.17 Parallel connection of two two-port networks.

$$I_1 = I_1^A = I_1^B,$$

$$I_2 = I_2^A + I_2^B.$$

Expressing these relationships in matrix form we have

$$\begin{bmatrix} I_1 \\ I_2 \end{bmatrix} = \begin{bmatrix} I_1^A + I_1^B \\ I_2^A + I_2^B \end{bmatrix} = \begin{bmatrix} I_1^A \\ I_2^A \end{bmatrix} + \begin{bmatrix} I_1^B \\ I_2^B \end{bmatrix}.$$

Representing each two-port via its **Y** matrix, we find

$$\begin{bmatrix} I_1 \\ I_2 \end{bmatrix} = \begin{bmatrix} Y_{11}^A & Y_{12}^A \\ Y_{21}^A & Y_{22}^A \end{bmatrix} \begin{bmatrix} V_1^A \\ V_2^A \end{bmatrix} + \begin{bmatrix} Y_{11}^B & Y_{12}^B \\ Y_{21}^B & Y_{22}^B \end{bmatrix} \begin{bmatrix} V_1^B \\ V_2^B \end{bmatrix}$$

$$= \left(\begin{bmatrix} Y_{11}^A & Y_{12}^A \\ Y_{21}^A & Y_{22}^A \end{bmatrix} + \begin{bmatrix} Y_{11}^B & Y_{12}^B \\ Y_{21}^B & Y_{22}^B \end{bmatrix} \right) \begin{bmatrix} V_1 \\ V_2 \end{bmatrix}.$$

The **Y** matrix of the parallel connection of two two-port networks is found by adding the two junctions' **Y** matrices:

$$\mathbf{Y} = \mathbf{Y}^A + \mathbf{Y}^B.$$

Example 13.5 Determine the **Z** matrix of the network shown in Fig. 13.18. Use the results obtained for series or parallel connected net works.

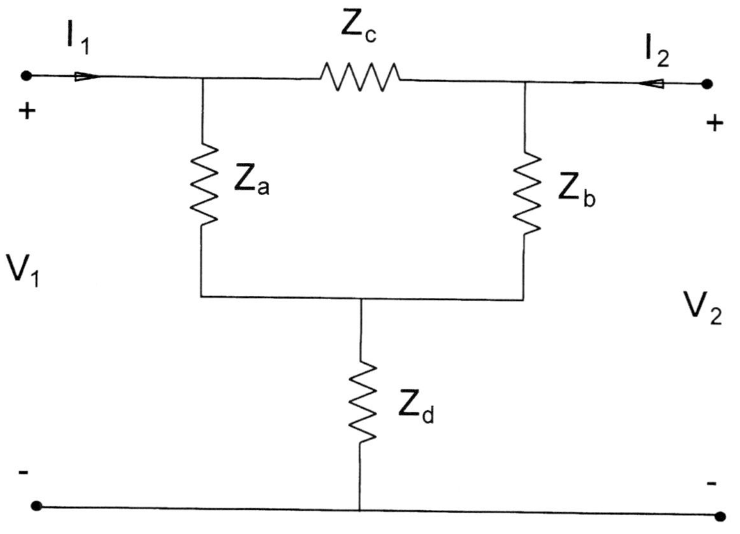

FIGURE 13.18 Network for Example 13.4.

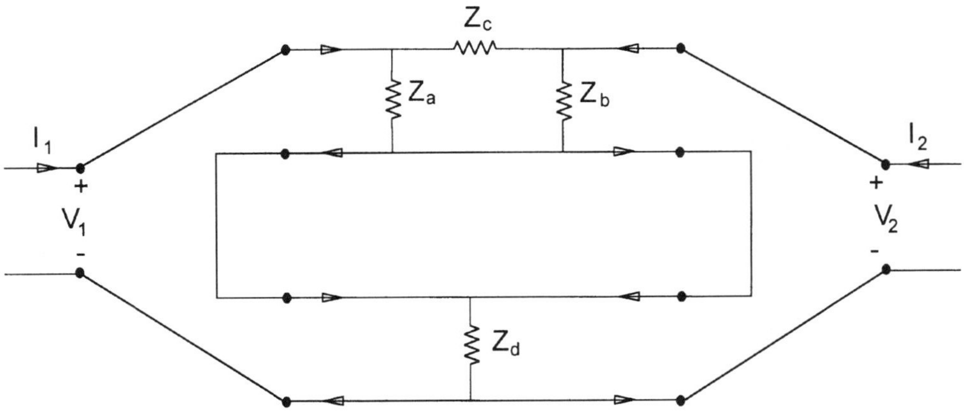

FIGURE 13.19 Network of Fig. 13.18 divided into two series connected subnetworks.

Solution The network under consideration can be divided into two series connected subnetworks as shown in Fig. 13.19. The **Z** matrix of the top network can be shown to be

$$\mathbf{Z}^{\text{top}} = \begin{bmatrix} \dfrac{Z_a(Z_c + Z_b)}{Z_a + Z_b + Z_c} & \dfrac{Z_a Z_b}{Z_a + Z_b + Z_c} \\ \dfrac{Z_a Z_b}{Z_a + Z_b + Z_c} & \dfrac{Z_b(Z_c + Z_a)}{Z_a + Z_b + Z_c} \end{bmatrix}.$$

The **Z** matrix of the bottom network is found in Table 13.1:

$$\mathbf{Z}^{\text{bottom}} = \begin{bmatrix} Z_d & Z_d \\ Z_d & Z_d \end{bmatrix}.$$

Since the two networks are connected in series, the **Z** matrix of the whole network is found by adding the two individual matrices:

$$\mathbf{Z} = \mathbf{Z}^{\text{top}} + \mathbf{Z}^{\text{bottom}} = \begin{bmatrix} \dfrac{Z_a(Z_c + Z_b)}{Z_a + Z_b + Z_c} + Z_d & \dfrac{Z_a Z_b}{Z_a + Z_b + Z_c} + Z_d \\ \dfrac{Z_a Z_b}{Z_a + Z_b + Z_c} + Z_d & \dfrac{Z_b(Z_c + Z_a)}{Z_a + Z_b + Z_c} + Z_d \end{bmatrix}.$$

13.5 VOLTAGE–CURRENT TRANSMISSION (*ABCD*) MATRIX DESCRIPTION OF A TWO-PORT JUNCTION

Figure 13.9 illustrates a two-port junction with voltage $\underline{V}_1$ and current $\underline{I}_1$ at port 1 and voltage $\underline{V}_2$ and current $\underline{I}_2$ at port 2. The junction may be represented by a voltage–current transmission matrix with elements A, B, C, D as

$$\begin{bmatrix} \underline{V}_1 \\ \underline{I}_1 \end{bmatrix} = \begin{bmatrix} A & B \\ C & D \end{bmatrix} \begin{bmatrix} \underline{V}_2 \\ -\underline{I}_2 \end{bmatrix} \tag{13.13}$$

The elements are given in terms of the voltages and currents as

$$A = \left.\frac{\underline{V}_1}{\underline{V}_2}\right|_{\underline{I}_2=0} \qquad C = \left.\frac{\underline{I}_1}{\underline{V}_2}\right|_{\underline{I}_2=0}, \tag{13.14a, b}$$

$$B = \left.\frac{\underline{V}_1}{-\underline{I}_2}\right|_{\underline{V}_2=0} \qquad D = \left.\frac{\underline{I}_1}{-\underline{I}_2}\right|_{\underline{V}_2=0}. \tag{13.15a, b}$$

The elements A and D are dimensionless; B has the units of impedance (Ω); and C has the units of admittance (S). The condition of Eqs. (3.14a, b) that $\underline{I}_2 = 0$ means that port 2 is open circuited. The condition of Eqs. (13.15a, b) that $\underline{V}_2 = 0$ means that port 2 is short-circuited. For a reciprocal junction the elements are related by $AD - BC = 1$.

The $ABCD$ matrix for three simple two ports is given in Table 13.2.

A cascade connection of two two-port junctions is shown in Fig. 13.20. The $ABCD$ matrix of the combined two-port junction is found by matrix multiplication of the two cascaded junction's $ABCD$ matrices. Thus if

$$\begin{bmatrix} \underline{V}_1 \\ \underline{I}_1 \end{bmatrix} = \begin{bmatrix} A_1 & B_1 \\ C_1 & D_1 \end{bmatrix} \begin{bmatrix} \underline{V}_2 \\ -\underline{I}_2 \end{bmatrix} \quad \text{and} \quad \begin{bmatrix} \underline{V}_2 \\ -\underline{I}_2 \end{bmatrix} = \begin{bmatrix} A_2 & B_2 \\ C_2 & D_2 \end{bmatrix} \begin{bmatrix} \underline{V}_3 \\ -\underline{I}_3 \end{bmatrix},$$

TABLE 13.2 ***ABCD* Matrix Elements**

Network	Schematic	A	B	C	D
Shunt admittance	Fig. 13.10	1	0	Y	1
Series impedance	Fig. 13.11	1	Z	0	1
Lossless T-line	Fig. 9.14	$\cos \beta l$	$jR_0 \sin \beta l$	$jG_0 \sin \beta l$	$\cos \beta l$

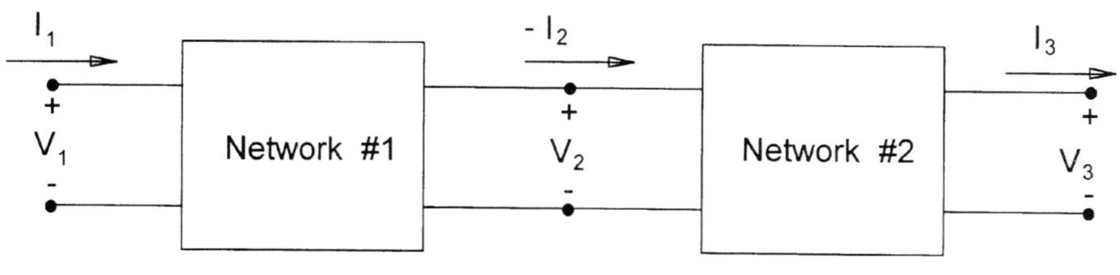

FIGURE 13.20 Cascade connection of two two-port networks.

then

$$\begin{bmatrix} \underline{V}_1 \\ \underline{I}_1 \end{bmatrix} = \begin{bmatrix} A_1 & B_1 \\ C_1 & D_1 \end{bmatrix} \begin{bmatrix} A_2 & B_2 \\ C_2 & D_2 \end{bmatrix} \begin{bmatrix} \underline{V}_3 \\ -\underline{I}_3 \end{bmatrix}$$

$$= \begin{bmatrix} A_1A_2 + B_1C_2 & A_1B_2 + B_1D_2 \\ C_1A_2 + D_1C_2 & C_1B_2 + D_1D_2 \end{bmatrix} \begin{bmatrix} \underline{V}_3 \\ -\underline{I}_3 \end{bmatrix}.$$

We now understand why the matrix is defined in terms of the negative of the current at the second port. With this definition, the output current $-I_2$ from the first junction becomes the input current $+I_2$ to the second junction.

Example 13.6 Find the *ABCD* matrix for a shunt admittance cascaded with a series impedance (Fig. 13.21).

Solution Using the matrices from Table 13.2, we have

$$\begin{bmatrix} A & B \\ C & D \end{bmatrix} = \begin{bmatrix} 1 & Z_1 \\ 0 & 1 \end{bmatrix} \begin{bmatrix} 1 & 0 \\ Y_2 & 1 \end{bmatrix} = \begin{bmatrix} 1 + Z_1Y_2 & Z_1 \\ Y_2 & 1 \end{bmatrix}.$$

Since the connections of different junctions are not cascaded only, series only, or parallel only by combinations of such connections, it becomes important to be able to convert from one type of parameter to another. The best approach in such conversions is to start with the definition of the wanted parameter. Consider the following example that illustrates the approach.

Example 13.7 Determine the *ABCD*-parameter in terms of the *Z*-parameters, for any two-port network.

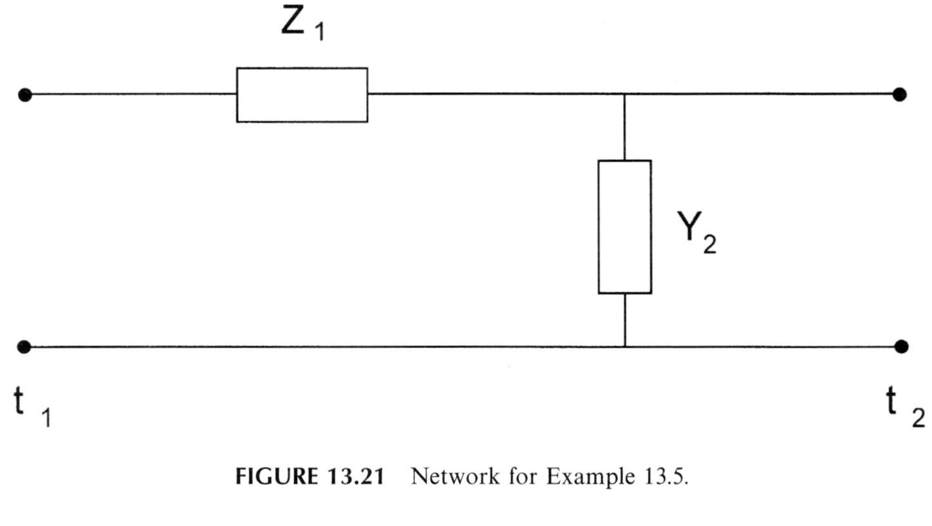

FIGURE 13.21 Network for Example 13.5.

Solution From the definition of the $ABCD$-parameters, the A- and C-parameters are evaluated with the input applied to port 1 and port 2 open-circuited:

$$A = \left[\frac{V_1}{V_2}\right]_{I_2=0} \qquad C = \left[\frac{I_1}{V_2}\right]_{I_2=0}.$$

The same condition $(I_2 = 0$ is now set for the Z-parameter representation of the same network. The voltage–current relationships from the Z-parameter representation become

$$V_1 = Z_{11} I_1 \qquad V_2 = Z_{21} I_1.$$

Therefore the A- and C-parameters become

$$A = \left[\frac{V_1}{V_2}\right]_{I_2=0} = \frac{Z_{11}}{Z_{21}} \qquad C = \left[\frac{I_1}{V_2}\right]_{I_2=0} = \frac{1}{Z_{21}}.$$

The B- and D-parameters are obtained by short circuiting port 2 ($V_2 = 0$):

$$B = \left[\frac{V_1}{-I_2}\right]_{V_2=0} \qquad D = \left[\frac{I_1}{-I_2}\right]_{V_2=0}.$$

The same condition ($V_2 = 0$) is now set for the Z-parameter representation of the same network. The voltage–current relationships from the Z-parameter representation become

$$V_1 = Z_{11} I_1 + Z_{12} I_2 \qquad 0 = Z_{21} I_1 + Z_{22} I_2.$$

Solving for I_1 in terms of I_2, and then substituting in the other equation, we find

$$I_1 = -\frac{Z_{22}}{Z_{21}} I_2 \qquad V_1\left(-Z_{11}\frac{Z_{22}}{Z_{21}} + Z_{12}\right) I_2.$$

The B- and D-parameters become

$$B = \left[\frac{V_1}{-I_2}\right]_{V_2=0} = Z_{11}\frac{Z_{22}}{Z_{21}} - Z_{12} \qquad D = \left[\frac{I_1}{-I_2}\right]_{V_2=0} = \frac{Z_{22}}{Z_{21}}.$$

13.6 SCATTERING MATRIX DESCRIPTION OF AN N-PORT JUNCTION

In Chapters 10 and 11 we studied reflection and transmission when two transmission lines are connected or when a wave is incident on a boundary between two media. We found that we could describe the junction in terms of a reflection and a transmission coefficient. In the terms of a transmission line model and a microwave network these situations are two-port networks. The scattering matrix description of an N-port network is a generalization of the reflection and transmission coefficient concept.

Figure 13.7b shows an N-port junction with a transmission line connected to each port. Along the ith transmission line the voltage and current waves are

$$\underline{V}_i(z_i) = V_{0i}^{+} e^{-j\beta_i z_i} + V_{0i}^{-} e^{j\beta_i z_i},$$

$$\underline{I}_i(z_i) = \frac{1}{R_{0i}} (V_{0i}^{+} e^{-j\beta_i z_i} - V_{0i}^{-} e^{j\beta_i z_i}),$$

where R_{0i} and β_i are the characteristic resistance and phase constant of the ith line, respectively. The longitudinal coordinate z_i has its origin at the terminal plane of the ith port and is increasing or positive toward the port. V_{0i}^{+} gives the amplitude (magnitude and angle) of the incident voltage wave at $z_i = 0$ and V_{0i}^{-} gives the amplitude of the reflected voltage wave at $z_i = 0$ and V_{0i}^{-} gives the amplitude of the reflected voltage wave at $z_i = 0$. We can define a column matrix $\mathbf{V}^{+}$ for the incident waves as the 1 by N matrix whose elements are V_{01}^{+}, $i = 1, 2, \ldots, N$ and a column matrix $\mathbf{V}^{-}$ for the reflected waves as the 1 by N matrix whose elements are V_{0i}^{-}, $i = 1, 2, \ldots, N$. The scattering matrix $\mathbf{S}$ is an N by N matrix that relates the incident wave column matrix to the reflected wave column matrix. Thus

$$\mathbf{V}^{-} = \mathbf{S}\mathbf{V}^{+} \tag{13.16}$$

or, in detail,

$$\begin{bmatrix} V_{01}^{-} \\ V_{02}^{-} \\ V_{03}^{-} \\ \vdots \\ V_{0N}^{-} \end{bmatrix} = \begin{bmatrix} S_{11} & S_{12} & S_{13} & \cdots & S_{1N} \\ S_{21} & S_{22} & S_{23} & \cdots & S_{2N} \\ S_{31} & S_{32} & S_{33} & \cdots & S_{3N} \\ \vdots & \vdots & \vdots & \vdots & \vdots \\ S_{N1} & S_{N2} & S_{N3} & \cdots & S_{NN} \end{bmatrix} \begin{bmatrix} V_{01}^{+} \\ V_{02}^{+} \\ V_{03}^{+} \\ \vdots \\ V_{0N}^{+} \end{bmatrix}. \tag{13.17}$$

Remember that S_{ij} refers to the element in the ith row and the jth column of $\mathbf{S}$.

Let us examine the typical elements of the scattering matrix. A typical element on the diagonal of the matrix is S_{ii}, given by

$$S_{ii} = \frac{V_{0i}^{-}}{V_{0i}^{+}}, \quad V_{0j}^{+} = 0, \quad j \neq i. \tag{13.18}$$

In physical terms this requires that (1) we place a source on the ith line, so that there will be an incident wave on this line; (2) we place a matched load R_{0j} on every other line (the jth lines, $j \neq i$), so that the wave traveling away from the jth port will not be reflected at the jth load; and (3) we measure the ratio of the reflected to the incident wave at the ith port. See Fig. 13.22. Step 2 leaves only the ith port accessible and hence it converts the N-port into a one-port network as shown by the dashed line in this figure. S_{ii} is the reflection coefficient at the ith port, when every other port is terminated in its matched load.

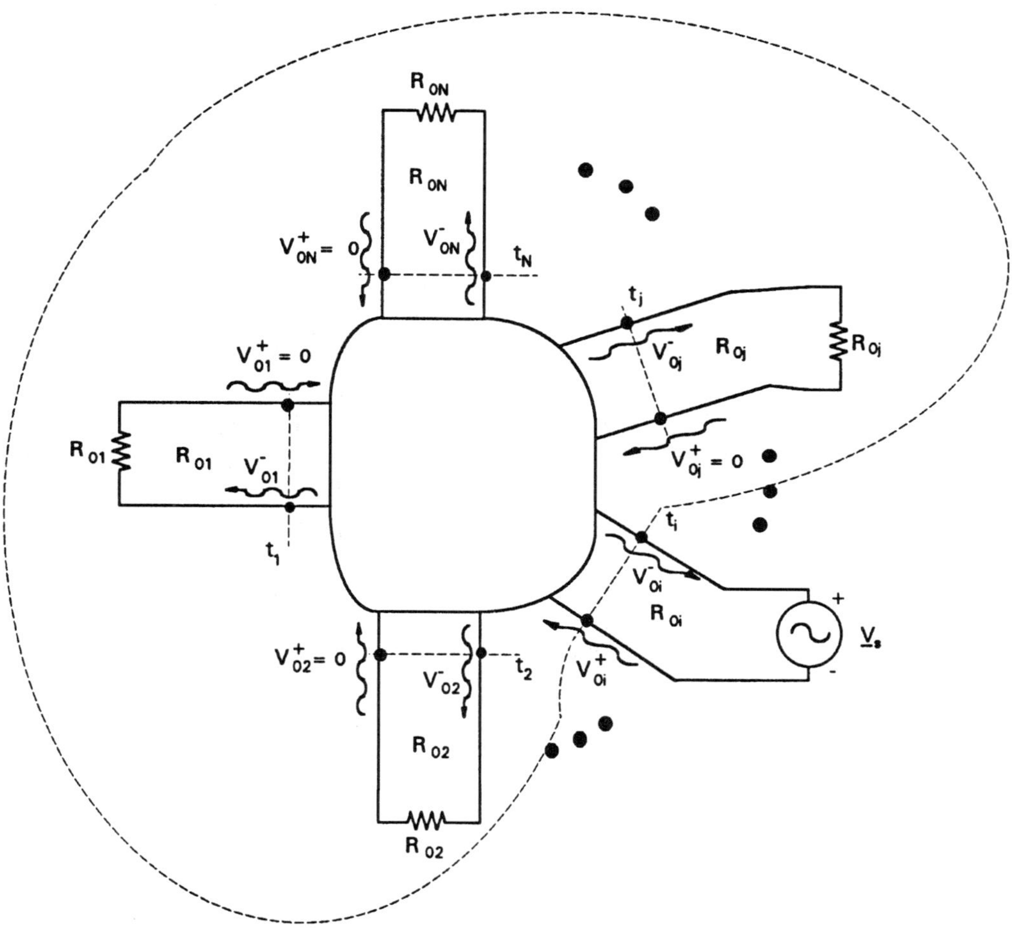

FIGURE 13.22 An N-port network with input signal applied to the $i$the port and all other ports matched. Used to calculate the S_{ii} parameters.

A typical term off the diagonal of the scattering matrix is S_{ij}, $i \neq j$, given by

$$S_{ij} = \frac{V_{0i}^-}{V_{0j}^+}, \quad V_{0i}^+ = 0, \quad i \neq j. \tag{13.19}$$

In physical terms this requires that (1) we place a source on the jth line, so that there will be an incident wave on this line; (2) we place a matched load R_{0i} on every other line (the ith lines, $i \neq j$), so that the wave traveling away from the ith port will not be reflected at the ith load; and (3) we measure the ratio of the transmitted wave V_{0i}^- at the $i$the port to the incident wave V_{0j}^+ at the jth port. See Fig. 13.23. Step 2 leaves only the ith and jth ports accessible and hence it converts the N-port into a two-port network as shown by the dashed line in this figure. S_{ij} is the transmission coefficient from the jth port to the ith port, when every other port is terminated in a matched load.

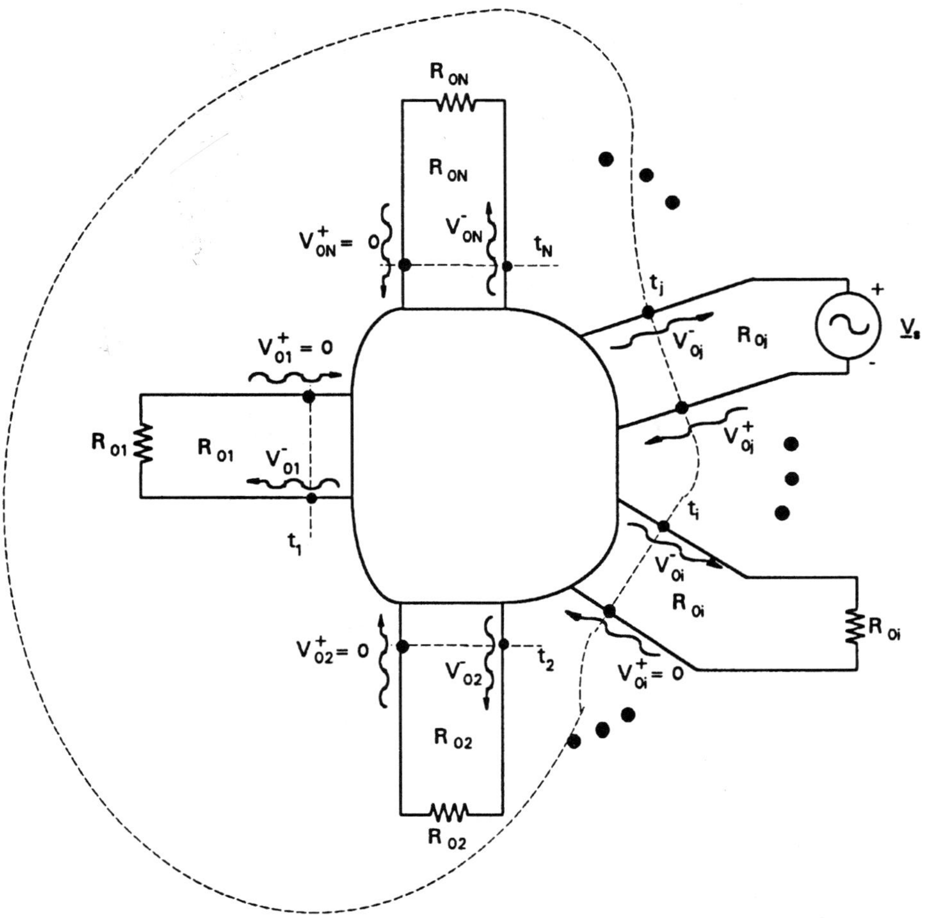

FIGURE 13.23 An N-port network with input signal applied to the jth port and all other ports matched. Used to calculate the S_{ij} parameters.

Example 13.8 Figure 13.24 shows a three-port junction. The lines have characteristics resistances R_{01}, R_{02}, and R_{03} and phase constants β_1, β_2, and β_3, and the *junction contains an internal arbitrary impedance* Z. Find the scattering matrix that represents the junction.

Solution Figure 13.25a shows lines 2 and 3 terminated in their characteristic resistances R_{02} and R_{03}, respectively. The input impedance at port 1 is then $Z_{in} = R_{02} + R_{03} + Z$ and the reflection coefficient at port 1 is

$$S_{11} = \frac{Z_{in} - R_{01}}{Z_{in} + R_{01}} = \frac{R_{02} + R_{03} + Z - R_{01}}{R_{02} + R_{03} + Z + R_{01}}. \tag{13.20}$$

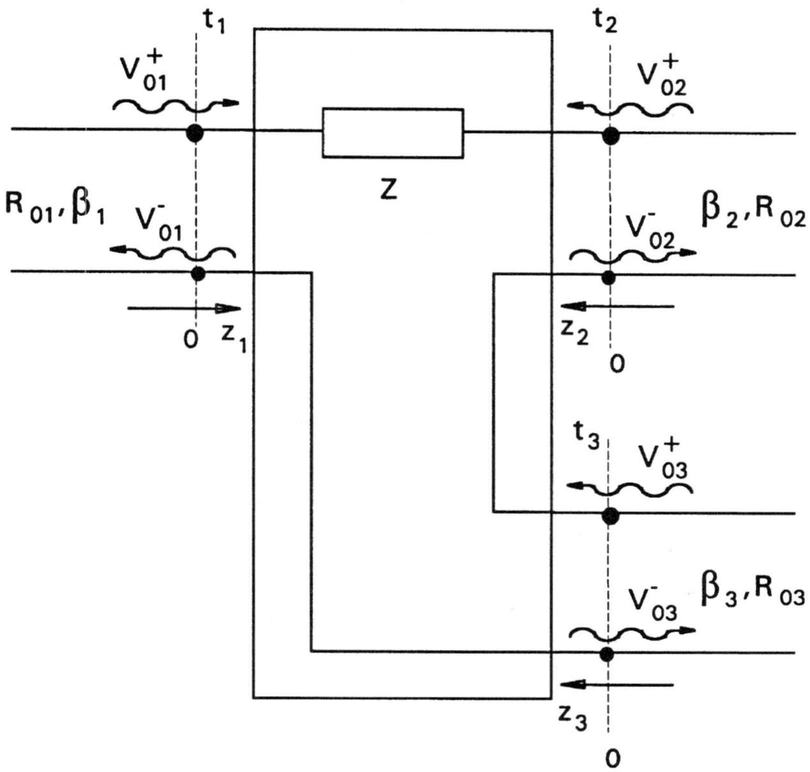

FIGURE 13.24 Three-port junction used in Example 13.8.

Since line 2 is terminated in its characteristic resistance, there is no reflected wave from the load and therefore $V_2^+ = 0$. The total voltage at port 2 equals the voltage of the outward or $-z_2$ going wave, $\underline{V}_2(0) = V_2^-$. At port 1 the voltage is the sum of the incident and the reflected wave, $\underline{V}_1(0) = V_{01}^+ + V_{01}^- = V_{01}^+(1 + S_{11})$. By simple voltage division we therefore have

$$\underline{V}_2(0) = V_{02}^- = \frac{R_{02}}{R_{02} + R_{03} + Z}\underline{V}_1(0) = \frac{E_{02}}{R_{02} + R_{03} + Z}V_{01}^+(1 + S_{11})$$

or, substituting for S_{11},

$$S_{21} = \frac{V_{02}^-}{V_{01}^+} = \frac{2R_{02}}{R_{01} + R_{02} + R_{03} + Z}. \tag{13.21}$$

In the same way

$$S_{31} = \frac{V_{03}^-}{V_{01}^+} = \frac{2R_{03}}{R_{01} + R_{02} + R_{03} + Z}. \tag{13.22}$$

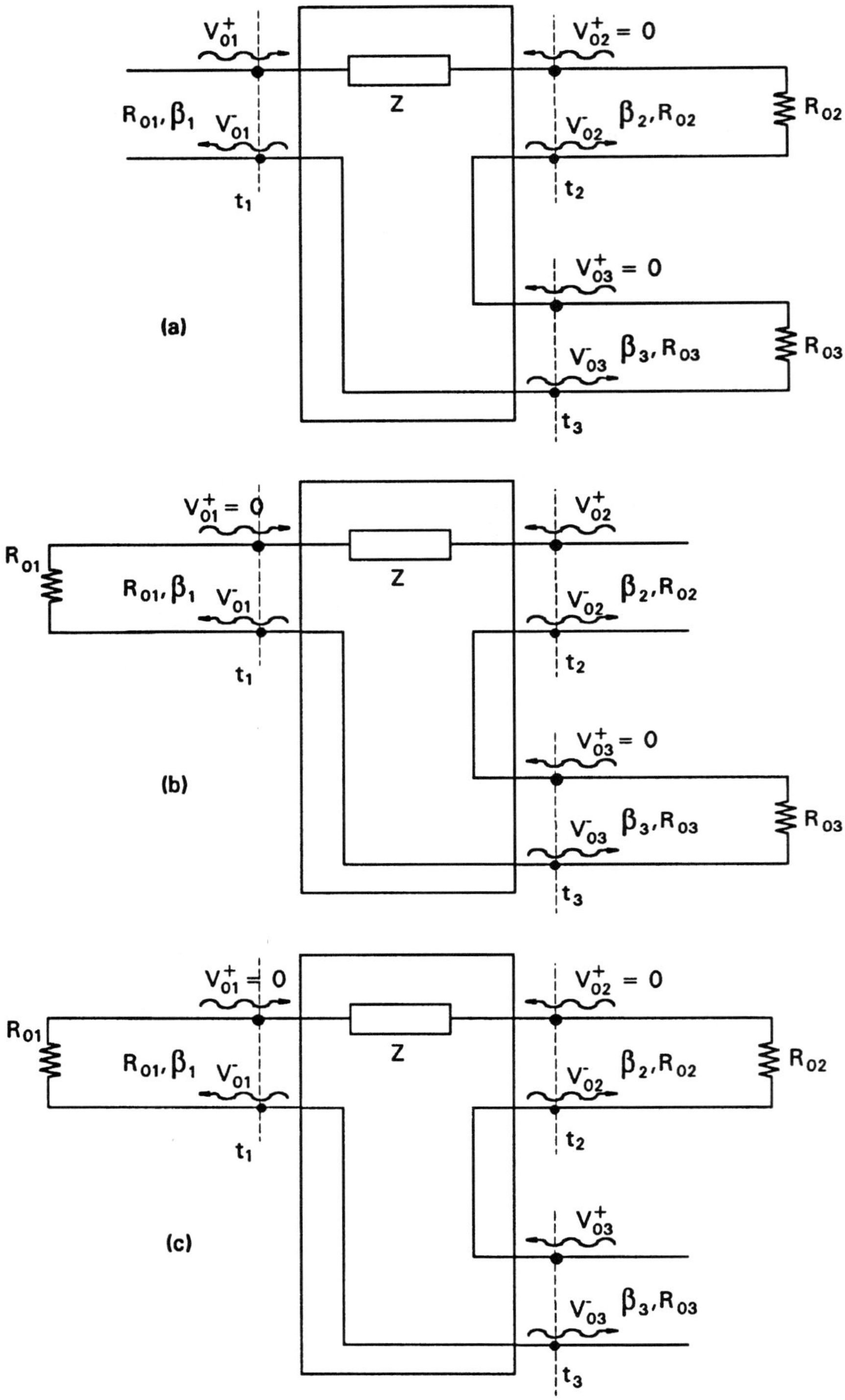

FIGURE 13.25 Circuit configurations used to calculate the *S*-parameters of the three-port in Example 13.8.

The second and third columns are found with the aid of Figs. 13.25b and 13.25c, respectively. The complete scattering matrix is

$$\mathbf{S} = \frac{1}{Z_s}\begin{bmatrix} R_{02} + R_{03} + Z - R_{01} & 2R_{01} & 2R_{01} \\ 2R_{02} & R_{01} + R_{03} + Z - R_{02} & -2R_{02} \\ 2R_{03} & -2R_{03} & R_{01} + R_{02} + Z - R_{03} \end{bmatrix},$$

$$Z_s = R_{01} + R_{02} + R_{03} + Z.$$

You should be sure you understand why two terms contain minus signs.

The scattering matrix is generally the most appropriate description of a junction for three reasons. First, it relates the incident waves to the transmitted waves at the ports and these waves are the natural way to represent the voltage and current along the line. Second, it exists (is nonsingular) for all junctions. Third, and most important, lines usually operate with matched or closely matched loads. It is therefore possible to measure the elements of the scattering matrix exactly as they are defined by Eqs. (3.18) and (13.19). Other descriptions of a junction, such as the *ABCD* matrix description, require the placing of short or open circuits at the ports. These conditions may be impossible to attain or, particularly when the junction contains active elements, may either change the operating point of the junction or destroy it. This third reason makes the scattering matrix description the appropriate one for many junctions even at low frequencies. Most modern microwave measuring equipment is designed to measure the scattering matrix elements.

13.6.1 Symmetry Property of the Scattering Matrix

The scattering matrix for an N-port junction contains N^2 elements. It is natural to ask if there are any relations showing these elements are not independent. Knowing these relations will reduce the number of terms that we must calculate or measure.

Examining the result of our example we see that the off-diagonal terms are related by the equation

$$\frac{S_{ij}}{R_{0i}} = \frac{S_{ji}}{R_{0j}}, \quad i \neq j. \tag{13.23}$$

It is of course dangerous to generalize from a single example. Nevertheless, it can be shown [1] that this result is true for any reciprocal junction. A junction containing only isotropic media is reciprocal.

Equation (13.23) shows the advantage of defining the N characteristic resistances to be equal, $R_{0i} = R_0$, $i = 1, 2, \ldots, N$. When this is true, then,

$$S_{ij} = S_{ji}, \quad i \neq j, \quad \text{when } R_{0i} = R_0,\ i = 1, 2, \ldots, N, \tag{13.24}$$

and the scattering matrix is symmetric.

In either case, knowing S_{ij} determines S_{ji} or vice versa. The number of independent elements of the scattering matrix for a reciprocal junction is therefore N elements on the diagonal plus $(N^2 - N)/2$ independent elements off the diagonal for a total of $N(N+1)/2$ independent elements.

13.6.2 Unitary property of the scattering Matrix

Further properties of the scattering matrix can be derived when the junction is passive and lossless. For a passive and lossless junction the time average power entering the junction from all ports must equal the time average power leaving the junction. Thus

$$2P_{\text{Avg,out}} = \sum_{i=1}^{N} \frac{|V_{0i}^-|^2}{R_{0i}} \sum_{i=1}^{N} \frac{|V_{0i}^+|^2}{R_{0i}} = 2P_{\text{Avg,in}}, \tag{13.25}$$

where

$$V_{0i}^- = \sum_{j=1}^{N} S_{ij} V_{0j}^+. \tag{13.26}$$

The V_{0j}^+ are independent incident voltages and we can choose them as we like. Let us first choose all of them except the kth one to be zero: $V_{0j}^+ = 0$, $j \neq k$. Then Eq. (13.26) gives $V_{0i}^- = S_{ik} V_{0k}^+$, which can be substituted into Eq. (13.25) to give

$$\sum_{i=1}^{N} \frac{|S_{ik}|^2 |V_{0k}^+|^2}{R_{0i}} = \frac{|V_{0k}^+|^2}{R_{0k}} \tag{13.27}$$

$$\sum_{i=1}^{N} \frac{|S_{ik}|^2}{R_{0i}} = \frac{1}{R_{0k}}, \quad k = 1, 2, \ldots, N. \tag{13.28}$$

This result means that if the elements in the kth column are multiplied term by term by their complex conjugates, divided term by term by the characteristic resistance of their row and summed, the result is the reciprocal of the kth characteristic resistance.

A further result may be obtained by choosing all the incident voltages to be zero except the kth and the rth: $V_{0j}^+ = 0$, $j \neq k$, $j \neq r$. Then Eq. (13.26) gives $V_{0i}^- = S_{ik} V_{0k}^+ + S_{ir} V_{0r}^+$, which can be substituted into Eq. (13.25) to give

$$\sum_{i=1}^{N} \frac{1}{R_{0i}} (S_{ik} V_{0k}^+ + S_{ir} V_{0r}^+)(S_{ik} V_{0k}^+ + S_{ir} V_{0r}^+)^* = \frac{|V_{0k}^+|^2}{R_{0k}} + \frac{|V_{0r}^+|^2}{R_{0r}}.$$

Multiplying the terms on the left and using Eq. (13.27), we find

$$\sum_{i=1}^{N} \frac{1}{R_{0i}} [(S_{ik} V_{0k}^+ + S_{ir} V_{0r}^+)^* (S_{ik} V_{0k}^+)^* + (S_{ir} V_{0r}^+)] = 0.$$

We are still free to choose the relative values of V_{0k}^+ and V_{0r}^+. Choosing $V_{0k}^+ = V_{0r}^+ \neq 0$ we find

$$|V_{0r}^+|^2 \sum_{i=1}^{N} \frac{1}{R_{0i}} (S_{ik}S_{ir}^* + (S_{ik}^*S_{ir}) = 0$$

and choosing $V_{0k}^+ = jV_{0r}^+ \neq 0$ we find

$$j|V_{0r}^+|^2 \sum_{i=1}^{N} \frac{1}{R_{0i}} (S_{ik}S_{ir}^* - S_{ik}^*S_{ir}) = 0.$$

We divide each of these results by the nonzero factor multiplying the summation, and then we add to get

$$\sum_{i=1}^{N} \frac{1}{R_{0i}} S_{ik}S_{ir}^* = 0, \quad k, r = 1, 2, \ldots, N, k \neq r. \tag{13.29}$$

This result means that if the elements of any column are multiplied term by term by the complex conjugates of the elements of any other column, each product is divided by the characteristic resistance of its row and summed, the result is zero.

If we define the N characteristic resistances to be equal – $R_{0i} = R_0$, $i = 1, 2, \ldots, N$ – then the results of Eqs. (13.28) and (13.29) are simplified to

$$\sum_{i=1}^{N} |S_{ik}|^2 = 1, \quad k = 1, 2, \ldots, N; \tag{13.30}$$

$$\sum_{i=1}^{N} S_{ik}S_{ir}^* = 0, \quad k, r = 1, 2, \ldots, N, k \neq r. \tag{13.31}$$

For a passive and lossless junction when any column of the scattering matrix is multiplied by its complex conjugate the product is unity and when any column is multiplied by the complex conjugate of any other column the product is zero. A matrix with this property is said to be unitary.

Example 13.9 For the scattering matrix of Example 13.8, let $R_{01} = 10$, $R_{02} = 20$, $R_{03} = 30$, and $Z = j50\ \Omega$. (a) Calculate the elements of the scattering matrix. (b) Show that Eqs. (13.28) and (13.29) are satisfied.

Solution (a) For the values given, $Z_s = 60 + j50$ and

$$\mathbf{S} = \frac{1}{60 + j50} \begin{bmatrix} 40 + j50 & 20 & 20 \\ 40 & 20 + j50 & -40 \\ 60 & -60 & j50 \end{bmatrix}$$

(b) $Z_s Z_s^* = 60^2 + 50^2 = 6100$. Equation (13.28) gives

$$k = 1: \quad \frac{1}{6100}\left(\frac{40^2 + 50^2}{10} + \frac{40^2}{20} + \frac{60^2}{30}\right) = \frac{1}{10},$$

$$k = 2: \quad \frac{1}{6100}\left(\frac{20^2}{10} + \frac{20^2 + 50^2}{20} + \frac{60^2}{30}\right) = \frac{1}{20},$$

$$k = 3: \quad \frac{1}{6100}\left(\frac{20^2}{10} + \frac{40^2}{20} + \frac{50^2}{30}\right) = \frac{1}{30}.$$

Equation (13.29) gives

$$k = 1, r = 2: \quad \frac{1}{6100}\left[\frac{(40 + j50)20}{10} + \frac{40(20 - j50)}{20} + \frac{60(-60)}{30}\right] = 0,$$

$$k = 1, r = 3: \quad \frac{1}{6100}\left[\frac{(40 + j50)20}{10} + \frac{40(-40)}{20} + \frac{60(-j50)}{30}\right] = 0,$$

$$k = 2, r = 3: \quad \frac{1}{6100}\left[\frac{(20)20}{10} + \frac{(20 + j50)(-40)}{20} + \frac{(-60)(-j50)}{30}\right] = 0.$$

13.6.3 Matched *N*-Port

An N-port junction is said to be completely matched when the reflection coefficient at any port is zero if all the other ports are terminated in matched loads. This is equivalent to saying that $S_{ii} = 0$, $i = 1, 2, \ldots, N$. Hence an N-port network is completely matched if all the diagonal elements of its scattering matrix are zero. Conversely, if all N ports are matched the diagonal elements are all equal to zero.

13.6.4 Shift of Port Plane

This elements of the scattering matrix must be derived or measured with terminal planes sufficiently distant from the junction so that the evanescent modes will have decayed to essentially zero amplitude. Once the element values are obtained, however, the terminal plane can be shifted to any convenient location.

For example, in Fig. 13.26, let us assume the scattering matrix $\mathbf{S}$ has been obtained for an N-port junction with the ith port located at the plane t_i. We now wish to find the scattering matrix $\hat{\mathbf{S}}$ when that terminal plane has been shifted a distance d toward the junction to $\hat{t}_i$. This is the equivalent of changing the longitudinal coordinate of the ith line from z_i to $\hat{z}_i = z_i - d$.

In terms of the two longitudinal coordinates the voltage and current on the ith line are

$$\underline{V}_i(z_i) = V_{0i}^+ e^{-j\beta_i z_i} + V_{0i}^- e^{j\beta_i z_i}, \quad \hat{\underline{V}}_i(\hat{z}_i) = \hat{V}_{0i}^+ e^{-j\beta_i \hat{z}_i} + \hat{V}_{0i}^- e^{j\beta_i \hat{z}_i},$$

$$R_{0i}\underline{I}_i(z_i) = V_{0i}^+ e^{-j\beta_i z_i} - V_{0i}^- e^{j\beta_i z_i}, \quad R_{0i}\hat{\underline{I}}_i(\hat{z}_i) = \hat{V}_{0i}^+ e^{-j\beta_i \hat{z}_i} - \hat{V}_{0i}^- e^{j\beta_i \hat{z}_i}.$$

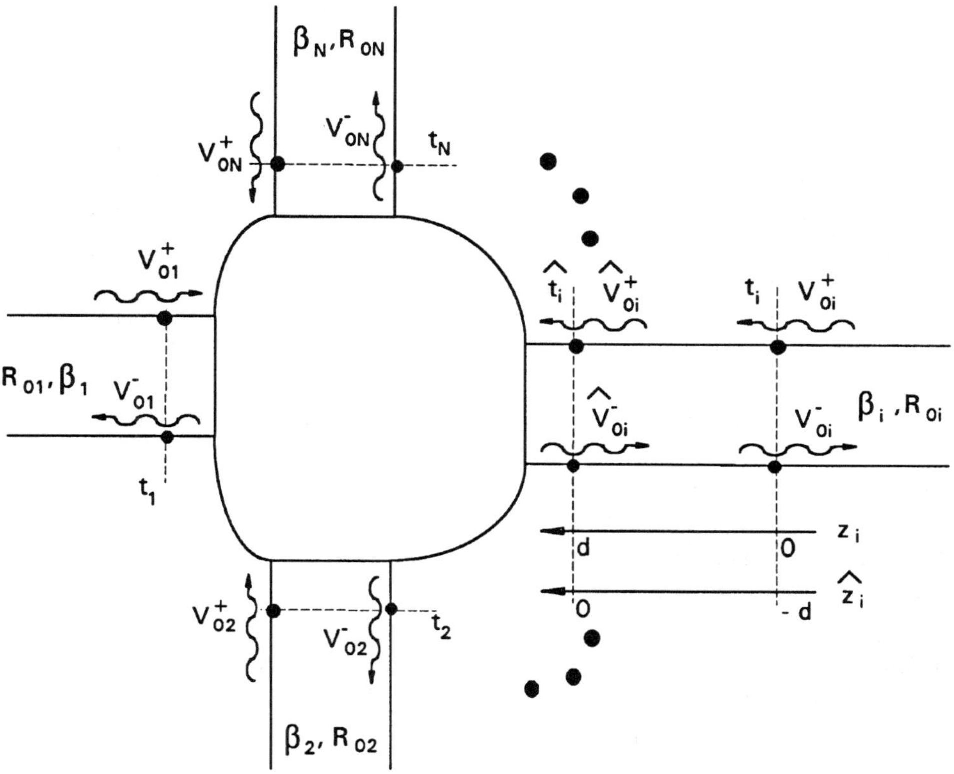

FIGURE 13.26 An N-port junction for which the terminal plane of the ith port has been shifted.

Equating $\hat{\underline{V}}_i(0) = \underline{V}_i(d)$ and $\hat{\underline{I}}_i(0) = \underline{I}_i(d)$, we obtain

$$\hat{V}_{0i}^{+} = V_{0i}^{+} e^{-j\beta_i d}, \quad \hat{V}_{0i}^{-} = V_{0i}^{-} e^{j\beta_i d}.$$

For the ith row of $\hat{\mathbf{S}}$ we have

$$\hat{S}_{ij} = \frac{\hat{V}_{0i}^{-}}{V_{0j}^{+}} = \frac{V_{0i}^{-} e^{j\beta_i d}}{\hat{V}_{0j}^{+}} = S_{ij} e^{j\beta_i d}, \quad j \neq i; \tag{13.32}$$

for the ith column we have

$$\hat{S}_{ji} = \frac{V_{0j}^{-}}{\hat{V}_{0i}^{+}} = \frac{V_{0j}^{-}}{\hat{V}_{0i}^{+} e^{-j\beta_i d}} = S_{ji} e^{j\beta_i d}, \quad j \neq i; \tag{13.33}$$

and for the on-diagonal element we have

$$\hat{S}_{ii} = \frac{\hat{V}_{0i}^{-}}{\hat{V}_{0i}^{+}} = \frac{V_{0i}^{-} e^{j\beta_i d}}{V_{0i}^{+} e^{-j\beta_i d}} = S_{ii} e^{j2\beta_i d}. \tag{13.34}$$

Thus at the shifted terminal plane, the new scattering matrix relation is

$$\begin{bmatrix} V_{01}^- \\ V_{02}^- \\ V_{03}^- \\ \vdots \\ \hat{V}_{0i}^- \\ \vdots \\ V_{0N}^- \end{bmatrix} = \begin{bmatrix} S_{11} & S_{12} & S_{13} & \cdots & S_{1i}e^{j\beta_i d} & \cdots & S_{1N} \\ S_{21} & S_{22} & S_{23} & \cdots & S_{2i}e^{j\beta_i d} & \cdots & S_{2N} \\ S_{31} & S_{32} & S_{33} & \cdots & S_{3i}e^{j\beta_i d} & \cdots & S_{3N} \\ \vdots & \vdots & \vdots & & \vdots & & \vdots \\ S_{i1}e^{j\beta_i d} & S_{i2}e^{j\beta_i d} & S_{i3}e^{j\beta_i d} & \cdots & S_{ii}e^{j2\beta_i d} & \cdots & S_{iN}e^{j\beta_i d} \\ \vdots & \vdots & \vdots & & \vdots & & \vdots \\ S_{N1} & S_{N2} & S_{N3} & \cdots & S_{Ni}e^{j\beta_i d} & \cdots & S_{NN} \end{bmatrix} \begin{bmatrix} V_{01}^+ \\ V_{02}^+ \\ V_{03}^+ \\ \vdots \\ \hat{V}_{0i}^+ \\ \vdots \\ V_{0N}^+ \end{bmatrix}$$

The ith row and the ith column are each multiplied by $\exp(j\beta_i d)$ when the terminal plane is shifted a distance d toward the junction. Since the diagonal element is in both the row and the column it is thereby multiplied by $\exp(j2\beta_i d)$.[†]

Example 13.10 The terminal plane of port 3 in Example 13.9 is shifted by a quarter-wavelength $\lambda_3/4$ toward the junction. Find the new scattering matrix.

Solution The third row and column must be multiplied by $\exp(j\beta_3\lambda_3/4) = \exp(j\pi/2) = j$. The result is

$$\hat{\mathbf{S}} = \frac{1}{60 + j50} \begin{bmatrix} 40 + j50 & 20 & j20 \\ 40 & 20 + j50 & -j40 \\ j60 & -j60 & -j50 \end{bmatrix}.$$

Note that the S_{33} term is multiplied by $(j)(j) = -1$.

13.6.5 Scattering Matrix with a Load Connected to One Port

If a load Z_L is connected to the ith port of an N-port junction (Fig. 13.27), the junction becomes an $(N-1)$-port. It is possible to find the scattering matrix $\hat{\mathbf{S}}$ of the $(N-1)$-port junction in terms of the scattering matrix $\mathbf{S}$ of the original N-port.

The load Z_L can be transformed through the length d to the ith terminal plane with the aid of Eq. (10.7):

$$Z_T = R_{0i} \frac{Z_L + jR_{0i} \tan \beta_i d}{R_{0i} + jZ_L \tan \beta_i d}.$$

[†] This operation is the equivalent of premultiplying and postmultiplying the scattering matrix $\mathbf{S}$ by an $N \times N$ diagonal matrix $\mathbf{P}(P_{ij} = 0,\ i \neq j)$ with $P_{ii} = \exp(j\beta_i d)$, $P_{jj} = 1, j \neq i$. If more than one terminal plane is shifted, so that the ith terminal plane is shifted a distance d_i toward the junction, then the diagonal matrix $\mathbf{P}$ has $P_{ii} = \exp(j\beta_i d_i)$, $i = 1, 2, \ldots, N$. The new scattering matrix is $\hat{\mathbf{S}} = \mathbf{PSP}$.

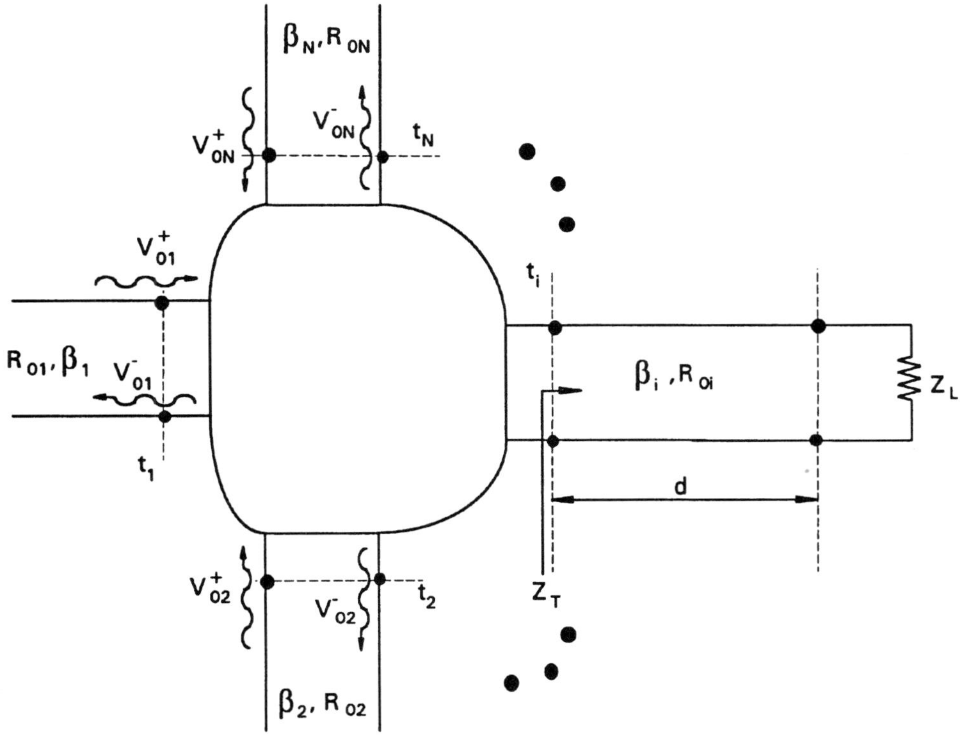

FIGURE 13.27 An N-port with one of its ports terminated with an impednace Z_L.

Z_T is the terminal equivalent impedance connected directly at the ith port. V_0^- is the amplitude of the voltage wave incident on Z_T and V_0^+ is the amplitude of the voltage wave reflected from Z_T. Hence

$$V_{0i}^+ = \Gamma V_{0i}^-, \quad \Gamma = \frac{Z_T - R_{0i}}{Z_T + R_{0i}}.$$

From the original scattering matrix we have

$$V_{0i}^- = \Gamma^{-1} V_{0i}^+ = \sum_{j=1}^{N} S_{ij} V_{0j}^+ = S_{i1} V_{01}^+ S_{i2} V_{02}^+ + \cdots$$
$$+ S_{i,i-1} V_{0,i-1}^+ + S_{ii} V_{0i}^+ + S_{i,i+1} V_{0,i+1}^+ + \cdots + S_{iN} V_{0N}^+.$$

We can solve for V_{0i}^+ and obtain

$$V_{0i}^+ = \frac{\Gamma}{1 - \Gamma S_{ii}} (S_{i1} V_{01}^+ + S_{i2} V_{02}^+ + \cdots + S_{i,i-1} V_{0,i-1}^+ S_{i,i+1} V_{0,i+1}^+ + \cdots + S_{iN}) V_{0N}^+$$
$$= \frac{\Gamma}{1 - \Gamma S_{ii}} \sum_{\substack{j=1 \\ j \neq i}}^{N} S_{ij} V_{0j}^+.$$

Substituting in the scattering matrix to obtain V_{0k}^-, $k \neq i$, we find

$$V_{0k}^- = \sum_{p=1}^{N} S_{kp}V_{0p}^+ = \left(S_{k1} + \frac{S_{ki}\Gamma S_{i1}}{1-\Gamma S_{ii}}\right)V_{01}^+ + \left(S_{k2} + \frac{S_{ki}\Gamma S_{i2}}{1-\Gamma S_{ii}}\right)V_{02}^+ + \cdots$$

$$+ \left(S_{kp} + \frac{S_{ki}\Gamma S_{ip}}{1-\Gamma S_{ii}}\right)V_{0p}^+ + \cdots + \left(S_{kN} + \frac{S_{ki}\Gamma S_{iN}}{1-\Gamma S_{ii}}\right)V_{0N}^+, \quad k,\, p \neq i.$$

The typical term of the new scattering matrix $\hat{\mathbf{S}}$ is therefore

$$\hat{S}_{kp} = S_{kp} + \frac{S_{ki}\Gamma S_{ip}}{1-\Gamma S_{ii}}, \quad k,\, p = 1, 2, \ldots, i-1,$$

$$\hat{S}_{k-1,p-1} = S_{kp} + \frac{S_{ki}\Gamma S_{ip}}{1-\Gamma S_{ii}}, \quad k,\, p = i+1, \ldots, N. \tag{13.35}$$

Example 13.11 A three-port junction has port 3 terminated in a reactive load $Z_T = jX_T$. Find the scattering matrix of the resultant two-port.

Solution The reflection coefficient is $\Gamma = (jX_T - R_3)/(jX_T + R_3) = \exp(j\theta)$. The element of $\hat{\mathbf{S}}$ are

$$\hat{S}_{11} = S_{11} + \frac{S_{13}S_{31}e^{j\theta}}{1 - S_{33}\,e^{j\theta}}, \quad \hat{S}_{12} = S_{12} + \frac{S_{13}S_{32}e^{j\theta}}{1 - S_{33}e^{j\theta}},$$

$$\hat{S}_{21} = S_{21} + \frac{S_{23}S_{31}e^{j\theta}}{1 - S_{33}e^{j\theta}}, \quad \hat{S}_{22} = S_{22} + \frac{S_{23}S_{32}e^{j\theta}}{1 - S_{33}\,e^{j\theta}}.$$

13.6.6 Conversion Between the Z and Y Matrices and the Scattering Matrix

In the process of analyzing complex microwave networks, the need for different characterizations arises. It is therefore necessary to be able to convert from one type of parameter to another. As an example, the S-parameters (which will be dealt with later in the section) of a microwave transistor are known, and what we now need to determine are the S-parameters of the same transistor with a resistor in series with one of its terminals. The series connection of the resistor with the transistor suggests that the Z-parameters would be needed for both devices.

In order to determine the Z-parameters in terms of the S-parameters, the relationships between the incident and reflected voltage waves and the voltages and currents at each port of the microwave junction need to be considered.

$$V_n = V_n^+ + V_n^-, \quad R_{0n}I_n = V_n^+ - V_n^-.$$

Writing these relationships in vector form, we have

$$\mathbf{V} = \mathbf{V}^+ + \mathbf{V}^-, \quad \mathbf{R}_c\mathbf{I} = \mathbf{V}^+ - \mathbf{V}^-,$$

where $\mathbf{R}_c$ is the following diagonal matrix, with the corresponding characteristic resistances as the elements on the main diagonal:

$$\mathbf{R}_c = \begin{bmatrix} R_{01} & 0 & 0 & \dots & 0 \\ 0 & R_{02} & 0 & \dots & 0 \\ 0 & 0 & R_{03} & \dots & 0 \\ \vdots & \vdots & \vdots & \vdots & \vdots \\ 0 & 0 & 0 & \dots & R_{0n} \end{bmatrix}.$$

Using the relationship $\mathbf{V}^- = \mathbf{S}\mathbf{V}^+$ and the identity $\mathbf{V}^+ = \mathbf{U}\mathbf{V}^+$, the above equations become

$$\mathbf{V} = \mathbf{U}\mathbf{V}^+ + \mathbf{S}\mathbf{V}^+ = (\mathbf{U} + \mathbf{S})\mathbf{V}^+, \quad \mathbf{R}_c\mathbf{I} = \mathbf{U}\mathbf{V}^+ - \mathbf{S}\mathbf{V}^+ = (\mathbf{U} - \mathbf{S})\mathbf{V}^+.$$

From the current equation, the vector $\mathbf{V}^+$ is found to be

$$\mathbf{V}^+ = (\mathbf{U} - \mathbf{S})^{-1}\mathbf{R}_c\mathbf{I}.$$

The relationship between the voltage vector and the current vector is obtained by substituting the expression of the vector $\mathbf{V}^+$ into the voltage equation:

$$\mathbf{V} = (\mathbf{U} + \mathbf{S})(\mathbf{U} - \mathbf{S})^{-1}\mathbf{R}_c\mathbf{I}.$$

From the Z-parameter characterization $\mathbf{V} = \mathbf{Z}\mathbf{I}$. Therefore

$$\mathbf{Z} = (\mathbf{U} + \mathbf{S})(\mathbf{U} - \mathbf{S})^{-1}\mathbf{R}_c. \tag{13.36}$$

In the case when all characteristic resistances are equal ($R_{01} = R_{02} = \cdots = R_{0N} = R_0$), the $\mathbf{Z}$ matrix becomes

$$\mathbf{Z} = R_0(\mathbf{U} + \mathbf{S})(\mathbf{U} - \mathbf{S})^{-1}. \tag{13.37}$$

Example 13.12 Determine the $\mathbf{Z}$ matrix for a microwave transistor, shown in Fig. 13.28, with the following $\mathbf{S}$ matrix:

$$\mathbf{S} = \begin{bmatrix} 0.7\angle -142^\circ & 0.1\angle 55^\circ \\ 3.5\angle 127^\circ & 0.8\angle -50^\circ \end{bmatrix}.$$

Solution MathCad will be used to solve this example. The $\mathbf{S}$ matrix and the unit matrix have to be defined first:

$$\mathbf{S} := \begin{bmatrix} 0.7 \cdot e^{-142j\cdot(\pi/180)} & 0.1 \cdot e^{55j\cdot(\pi/180)} \\ 3.5 \cdot e^{127j\cdot(\pi/180)} & 0.8 \cdot e^{-50j\cdot(\pi/180)} \end{bmatrix} \qquad \mathbf{U} := \begin{pmatrix} 1 & 0 \\ 0 & 1 \end{pmatrix}.$$

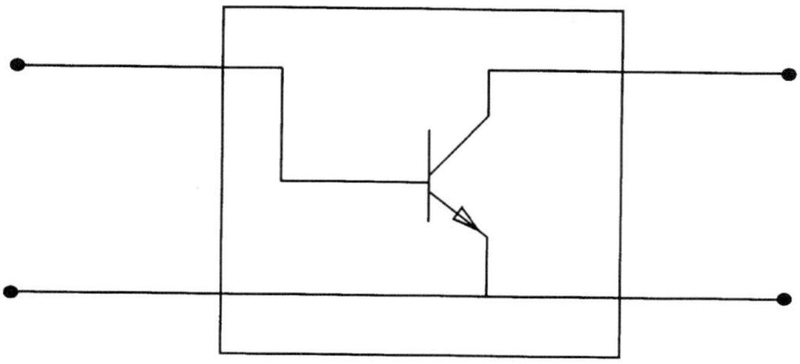

FIGURE 13.28 Two-port representation of a transistor with the emitter terminal common to the input and output.

The **Z** matrix is then defined as

$$\mathbf{Z} := 50 \cdot (\mathbf{U} + \mathbf{S}) \cdot (\mathbf{U} - \mathbf{S})^{-1}.$$

The **Z** matrix elements are viewed by typing **Z** followed by the equal sign. The Mathcad result is

$$\mathbf{Z} = \begin{pmatrix} 4.168 - 2.651j & 6.935 + 0.073j \\ 72.583 + 231.62j & 36.941 - 70.095j \end{pmatrix}.$$

Using a similar procedure, the **S** matrix can be determined from the **Z** matrix as follows:

$$\mathbf{S} = (\mathbf{Z} - \mathbf{R}_c)(\mathbf{Z} + \mathbf{R}_c)^{-1}. \tag{13.38}$$

The conversion from Y-parameters to S-parameters and vice versa is performed as follows:

$$\mathbf{Y} = \mathbf{R}_c^{-1}(\mathbf{U} - \mathbf{S})(\mathbf{U} + \mathbf{S})^{-1}, \tag{13.39}$$

$$\mathbf{S} = (\mathbf{U} - \mathbf{R}_c\mathbf{Y})(\mathbf{U} + \mathbf{R}_c\mathbf{Y})^{-1}. \tag{13.40}$$

In the case of equal characteristic resistances ($R_{01} = R_{02} = \cdots = R_{0N} = R_0$), the conversion expressions become

$$\mathbf{S} = (\mathbf{Z} - R_0\mathbf{U})(\mathbf{Z} + R_0\mathbf{U})^{-1}, \tag{13.41}$$

$$\mathbf{Y} = \frac{1}{R_0}(\mathbf{U} - \mathbf{S})(\mathbf{U} + \mathbf{S})^{-1}, \tag{13.42}$$

$$\mathbf{S} = (\mathbf{U} - R_0\mathbf{Y})(\mathbf{U} + R_0\mathbf{Y})^{-1}. \tag{13.43}$$

The conversion equations can be memorized quite easily by using an analogy with the relaionships between the impedance and the reflection coefficient:

$$\Gamma = \frac{Z - R_0}{Z + R_0} \qquad Z = R_0 \frac{1 + \Gamma}{1 - \Gamma}.$$

Example 13.13 The microwave transistor from the previous example will be connected as shown in Fig. 13.29. Determine the **S** matrix of the new circuit.

Solution The microwave transistor's emitter and the resistance R_e are connected in series. The network can be divided into two two-port networks, which are connected in series, as shown in Fig. 13.30. The **Z** matrix of the whole network is then determined as the sum of the **Z** matrices of the microwave transistor and the two-port network containing the resistance R_e.

$$\mathbf{Z} = \begin{bmatrix} 4.168 - 2.651j & 6.935 + 0.073j \\ 72.583 + 231.62j & 36.941 - 70.095j \end{bmatrix} + \begin{bmatrix} 10 & 10 \\ 10 & 10 \end{bmatrix}$$
$$= \begin{bmatrix} 14.168 - 2.651j & 16.935 + 0.073j \\ 82.583 + 231.62j & 46.941 - 70.095j \end{bmatrix}.$$

Using MathCad, the **S** matrix of the entire network is then calculated as

$$\mathbf{S} = \begin{bmatrix} -0.092 - 0.531j & 0.081 + 0.152j \\ -1.676 + 1.849j & 0.669 - 0.561j \end{bmatrix} = \begin{bmatrix} 0.5389 \angle -99.829^\circ & 0.1722 \angle 61.947^\circ \\ 2.496 \angle 132.19^\circ & 0.873 \angle -39.982^\circ \end{bmatrix}.$$

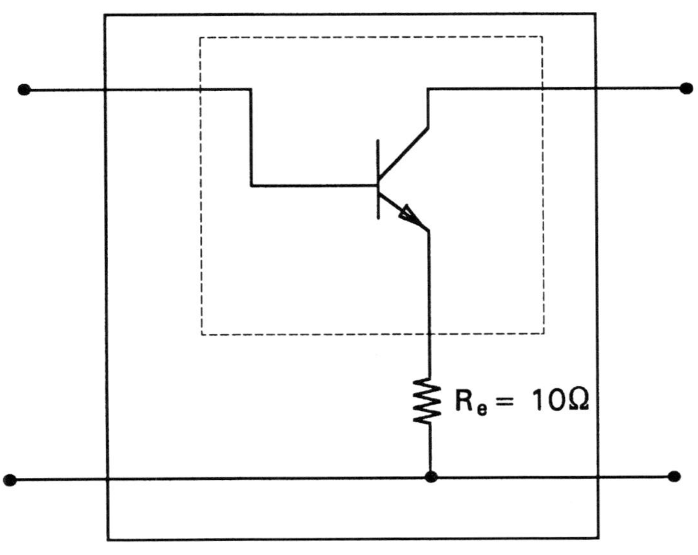

FIGURE 13.29 Two-port representation of a transistor with a series resistor $R_e = 10\ \Omega$ in the emitter terminal.

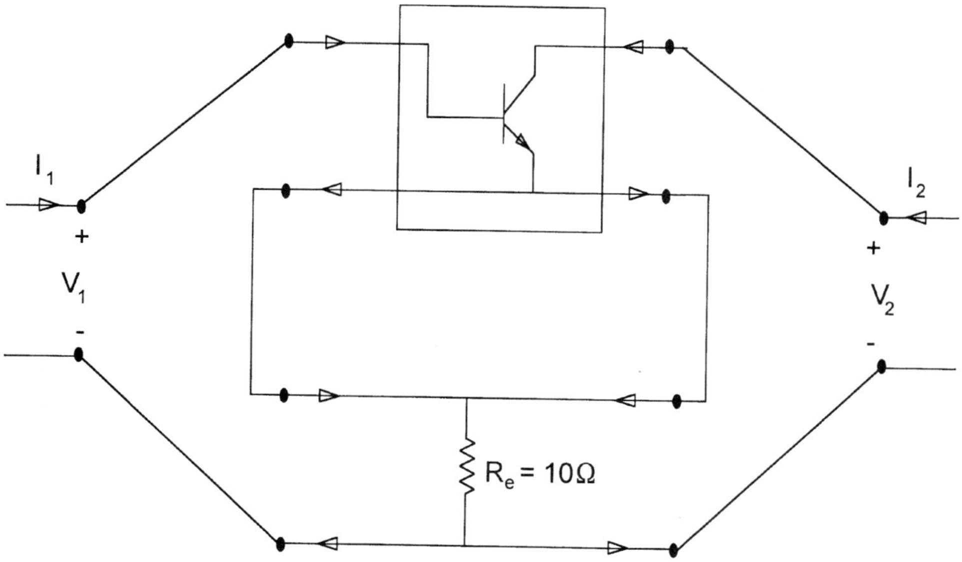

FIGURE 13.30 Network of Fig. 13.29 divided into two series connected subnetworks.

13.6.7 Conversion from the *ABCD* Matrix to the Scattering Matrix

A two-port junction can be described in terms of its scattering matrix or its *ABCD* matrix. At the ports we have

$$\underline{V}_1 = V_{01}^+ + V_{01}^-, \quad \underline{V}_2 = V_{02}^+ + V_{02}^-,$$

$$\underline{I}_1 = \frac{1}{R_{01}}(V_{01}^+ - V_{01}^-), \quad \underline{I}_2 = \frac{1}{R_{02}}(V_{02}^+ - V_{02}^-).$$

Substituting these in Eqs. (13.18) and (13.19) and rearranging terms, we can find [2]

$$\mathbf{S} = \frac{\begin{bmatrix} (B - CR_{01}R_{02}) + (AR_{02} - DR_{01}) & 2R_{01}(AD - BC) \\ 2R_{02} & (B - CR_{01}R_{02}) - (AR_{02} - DR_{01}) \end{bmatrix}}{(B + CR_{01}R_{02} \ + AR_{02} + DR_{01})}. \tag{13.44}$$

13.7 EXAMPLES OF THE SCATTERING MATRIX FOR PASSIVE DEVICES

In this section we will describe a few passive microwave devices. Several other devices are given in the problems. This presentation is to illustrate the principles we have presented, not to be all inclusive. Additional information is available elsewhere [3].

13.7.1 One-Port Devices

The scattering matrix for a one-port device has only a single element S_{11}, which is the reflection coefficient Γ at the terminal plane.

For a matched load $S_{11} = \Gamma = 0$. Matched loads can be achieved in waveguides with either a lossy wedge (Fig. 13.1b) or a resistive card (Fig. 13.31) and in microstrip line with a lumped resistor connected from the line to the ground plane. Microstrip resistors can also be fabricated by depositing a highly resistive line on the dielectric substrate. For high-resistivity conductors with good dielectric the per unit length transmission line parameters satisfy $R \gg \omega L$ and $\omega C \gg G$. We then have $Z_0 \approx \sqrt{R/j\omega C}$ and $\gamma \approx \sqrt{j\omega^{\phi\gamma}}$. For a shorted resistive transmission line of length d we have $Z_{in} = Z_0 \tan h(\gamma d) \approx Z_0 \gamma d = Rd$, if $|\gamma d| \ll 1$. By making $Rd = R_0$ a match can be achieved [4].

For a short circuit $S_{11} = \Gamma = -\exp(j2\beta d)$, where d is the distance from the plane of the short to the terminal plane of the port. A fixed position short can be achieved simply by terminating the waveguide in a conducting plate, with care taken to ensure good electrical contact.

Movable shorts in waveguides can be achieved by a sliding spring contact along the guide (Fig. 13.32a). Unfortunately, this does not generally result in a reliably low-resistance connection. For example, a total normalized resistance $\hat{R} = 0.02$ between the guide walls and the shorting block results in a reflection coefficient $|\Gamma| = 0.96$, instead of unity, and a power loss of almost 8%. This is generally not acceptable.

A better design, called a choke-type adjustable short, is shown in Fig. 13.32b for a rectangular guide. In this design there is no contact between the front of the plunger and the guide wall. The plunger length l_1 and the folded length l_2 are both a quarter-wavelength long, $l_1 = l_2 = \lambda_{g0}/4$, at the center of the guide frequency band. An equivalent transmission line circuit is shown in Fig. 13.33. The folded length is terminated in a short circuit, which is transformed to an open circuit at the contact point. This means that no current flows through the contact resistance R. The plunger length transforms the open circuit into a short circuit at the face of the plunger.

Off the center frequency the guide wavelength changes and the two lines are no longer a quarter-wavelength long. We can find the reflection coefficient with the aid of the equivalent circuit. The guide impedances R_0 are proportional to $(b/a)(\omega\mu/\beta_g)$. Since β_g depends on the guide width a, but not the guide height, it is the same for the

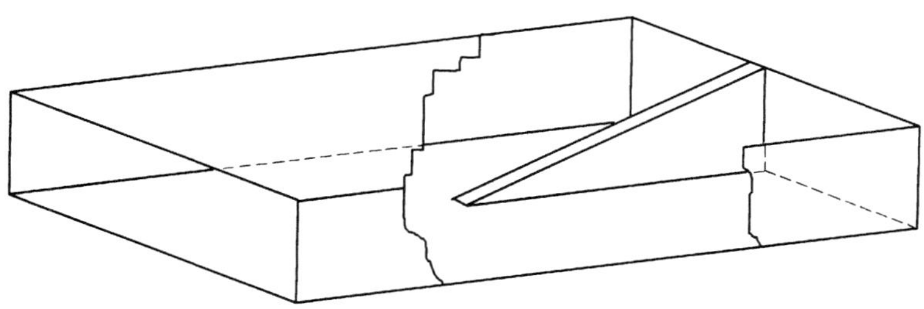

FIGURE 13.31 Matched load achieved by terminating the waveguide with a resistive card.

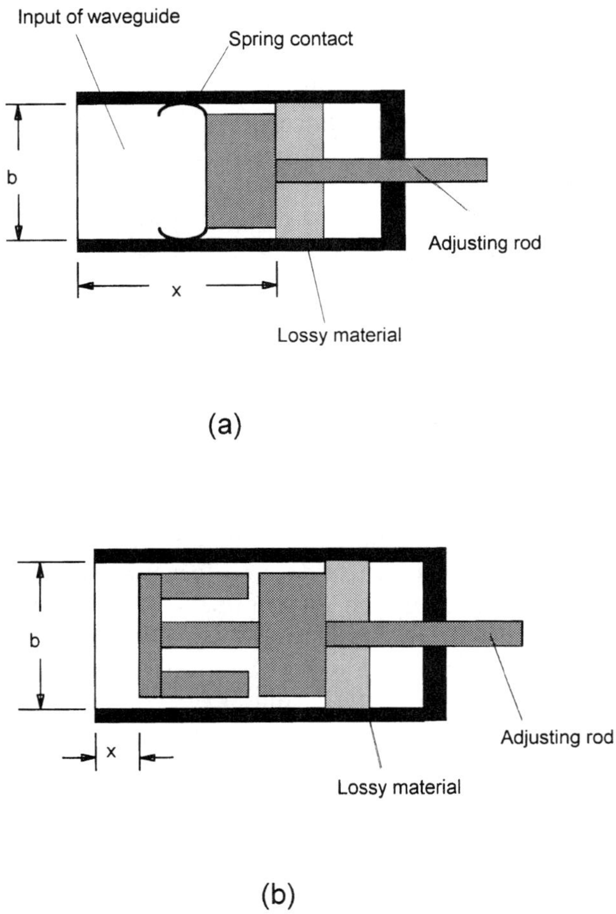

FIGURE 13.32 Side view of adjustable waveguide short circuits of the (a) contacting type and (b) noncontacting type.

guide, the plunger length, and the folded length. Normalizing all impedances with respect to the characteristic resistance R_{0g} of the guide of height b, we have

$$\hat{R}_{01} = \frac{R_{01}}{R_{0g}} = \frac{b_1}{b}, \quad \hat{R}_{02} = \frac{R_{02}}{R_{0g}} = \frac{b_2}{b}, \quad \hat{R} = \frac{R}{R_{0g}}, \quad \beta_g l_1 = \beta_g l_2 = \frac{\pi \lambda_{g0}}{2\lambda_g} = r,$$

$$\hat{Z}_{\text{in},2} = j\left(\frac{R_{02}}{R_{0g}}\right)\tan\left(\frac{\pi\lambda_{g0}}{2\lambda_g}\right) = j\left(\frac{b_2}{b}\right)\tan r,$$

$$Z_{\text{in},1} = R_{01}\frac{(Z_{\text{in},2} + R) + jR_{01}\tan(\pi\lambda_{g0}/2\lambda_g)}{R_{01} + j(Z_{\text{in},2} + R)\tan(\pi\lambda_{g0}/2\lambda_g)}$$

$$\hat{Z}_{\text{in},1} = \left(\frac{b_1}{b}\right)\left[\frac{j(b_2/b)\tan r + \hat{R} + j(b_1/b)\tan r}{(b_1/b) - (b_2/b)\tan^2 r + j\hat{R}\tan r}\right].$$

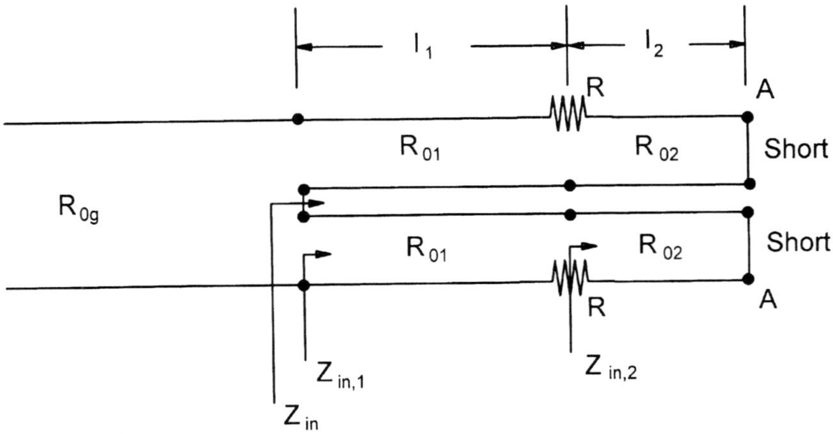

FIGURE 13.33 Equivalent transmission line circuit for the sliding short in Fig. 13.32b.

The total normalized input impedance is twice this value because the two halves of the choke are in series. Dividing the numerator and denominator by $(j\tan r)^2$ to remove the singularity at $r = r_0 = \pi/2$ gives

$$\hat{Z}_{\text{in}} = -j2\left(\frac{b_1}{b}\right)\cot r\frac{(b_1/b) + (b_2/b) - j\hat{R}\cot r}{(b_2/b) - j\hat{R}\cot r - (b_1/b)\cot^2 r}. \tag{13.45}$$

Note that $r_0 = \pi/2$, $\cot r_0 = 0$ and $Z_{\text{in}} = 0$ independent of the contact resistance $\hat{R}$. For a typical rectangular waveguide operating within its recommended frequency range, the value of r will range from $r_{\text{min}} = 0.6(\pi/2)$ to $r_{\text{max}} = 1.4(\pi/2)$. As shown in the following example, the power loss is much less than for the sliding contact adjustable short.

Example 13.14 For X-band waveguide $a = 0.900$ in. and $b = 0.400$ in. and the recommended frequency range is 8.20 to 12.4 GHz. Let $b_1 = 0.020$ in. and $b_2 = 0.100$ in. Find $l_1 = l_2$ and the input impedance, reflection coefficient, and percent lost power at each end of the frequency range. Assume $\hat{R} = 0.01$.

Solution The cutoff frequency is $f_c = c/2a = 11.8/(2 \times 0.9) = 6.56$ GHz and the center frequency is $f_0 = (8.20 + 12.4)/2 = 10.3$ GHz. The guide wavelength is calculated as

$$\lambda_g|_{12.4} = \frac{11.8/12.4}{\sqrt{1 - (6.56/12.4)^2}} = 1.12 \text{ in.}, \quad \lambda_g|_{10.3} = 1.49 \text{ in.}, \quad \lambda_g|_{8.2} = 2.40 \text{ in.}$$

The guide wavelength at the center frequency is therefore $\lambda_{g0} = 1.49$ in. and $l_1 = l_2 = 0.372$ in. The range of r is from $(\pi/2)(1.49/1.12) = 2.09$ rad to

$(\pi/2)(1.49/2.40) = 0.975$ rad. For the larger value of r we have $\cot r = -0.572$, which we substitute in Eq. (13.45) to get

$$\hat{Z}_{\text{in}} = -j2(0.05)(-0.572)\frac{0.3 + j0.00572}{0.25 + j0.00572 - 0.0155}$$

$$\approx j2(0.05)(0.572)(0.3)(1 + j0.0191)(1 - j0.0244)/0.235$$

$$\approx j(0.0730)(1 - j0.0053) = 0.387 \times 10^{-3} - j0.0730.$$

The reflection coefficient is

$$\Gamma = \frac{0.387 \times 10^{-3} - j0.0730 - 1}{0.387 \times 10^{-3} - j0.0730 + 1} \approx -\left(\frac{1 - 0.387 \times 10^{-3}}{1 + 0.387 \times 10^{-3}}\right)\left(\frac{1 + j0.0730}{1 - j0.0730}\right)$$

$$\approx -[1 - 2(0.387 \times 10^{-3})]\angle 2(0.0730) = 0.999226\angle 172^\circ.$$

The percent of power reflected is $100(1 - |\Gamma|^2) = 0.15\%$ at 12.4 GHz.

The smaller value of r gives $\Gamma = 0.99869\angle -170^\circ$ and 0.26% power reflected at 8.2 GHz. This is a satisfactory design for all practical purposes.

13.7.2 Two-Port Devices

Many two-port devices can be modeled as two cascaded transmission lines with a shunt reactance at their junction (Fig. 13.34). Before discussing specific devices let us obtain the scattering matrix for this general situation. It is convenient to use admittances because of the parallel connections.

With line 2 terminated in its characteristic admittance G_{02} the total admittance terminating line 1 is $G_{02} + jB$. The reflection coefficient, Eq. (10.33), is

$$S_{11} = \Gamma = -\left[\frac{G_{02} + jB - G_{01}}{G_{02} + jB + G_{01}}\right]$$

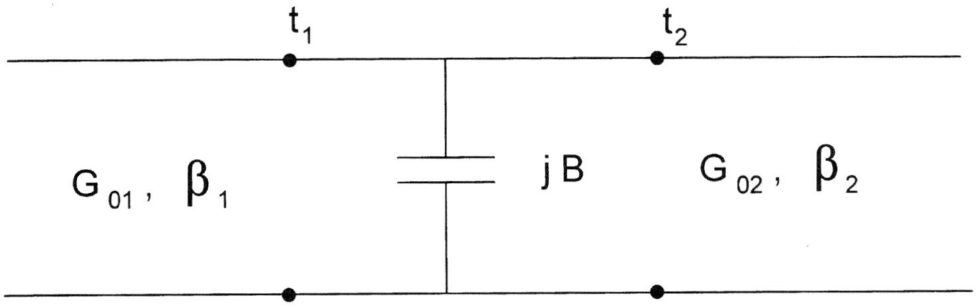

FIGURE 13.34 Two transmission lines with a shunt reactance at the junction.

and the transmission coefficient is

$$S_{21} = \tau = 1 + \Gamma = \frac{2G_{01}}{G_{02} + jB + G_{01}}.$$

The terms S_{22} and S_{12} can be found in the same way or by interchanging the subscripts "1" and "2." The resultant scattering matrix is

$$\mathbf{S} = \frac{1}{G_{01} + G_{02} + jB} \begin{bmatrix} G_{01} - G_{02} - jB & 2G_{02} \\ 2G_{01} & G_{02} - G_{01} - jB \end{bmatrix}. \tag{13.46}$$

We can now look at some specific devices.

13.7.2.1 Inductive Iris in a Rectangular Waveguide Figure 13.35 shows a symmetric inductive iris or window in a guide with characteristic admittance G_0. This iris maintains uniformity of the structure with respect to the y coordinate. On the conducting surface of the obstacle the transverse electric field must be zero and at the aperture in the obstacle both the transverse electric and magnetic fields must be continuous. With a TE_{10} wave incident on the obstacle there will be reflected and transmitted TE_{10} waves. In order to satisfy the boundary conditions and keep the fields independent of the y coordinate, evanescent TE_{n0} modes are generated on each side of the boundary. These evanescent modes store predominantly magnetic energy [Eq. (12.41)] and the boundary can therefore be represented as a shunt inductor. This means that B in Eq. (13.46) is negative.

Figure 13.36 plots $-(B/G_0)(a/\lambda_g)$ as a function of d/a, with a/λ as a parameter [5]. Similar results are available for an asymmetric iris and for symmetric and asymmetric

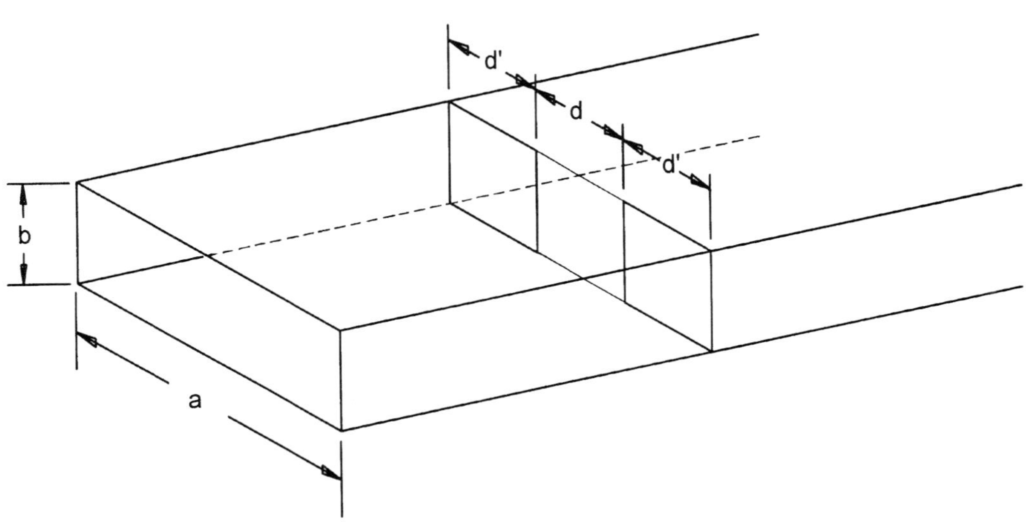

FIGURE 13.35 Symmetric inductive iris or window in a waveguide.

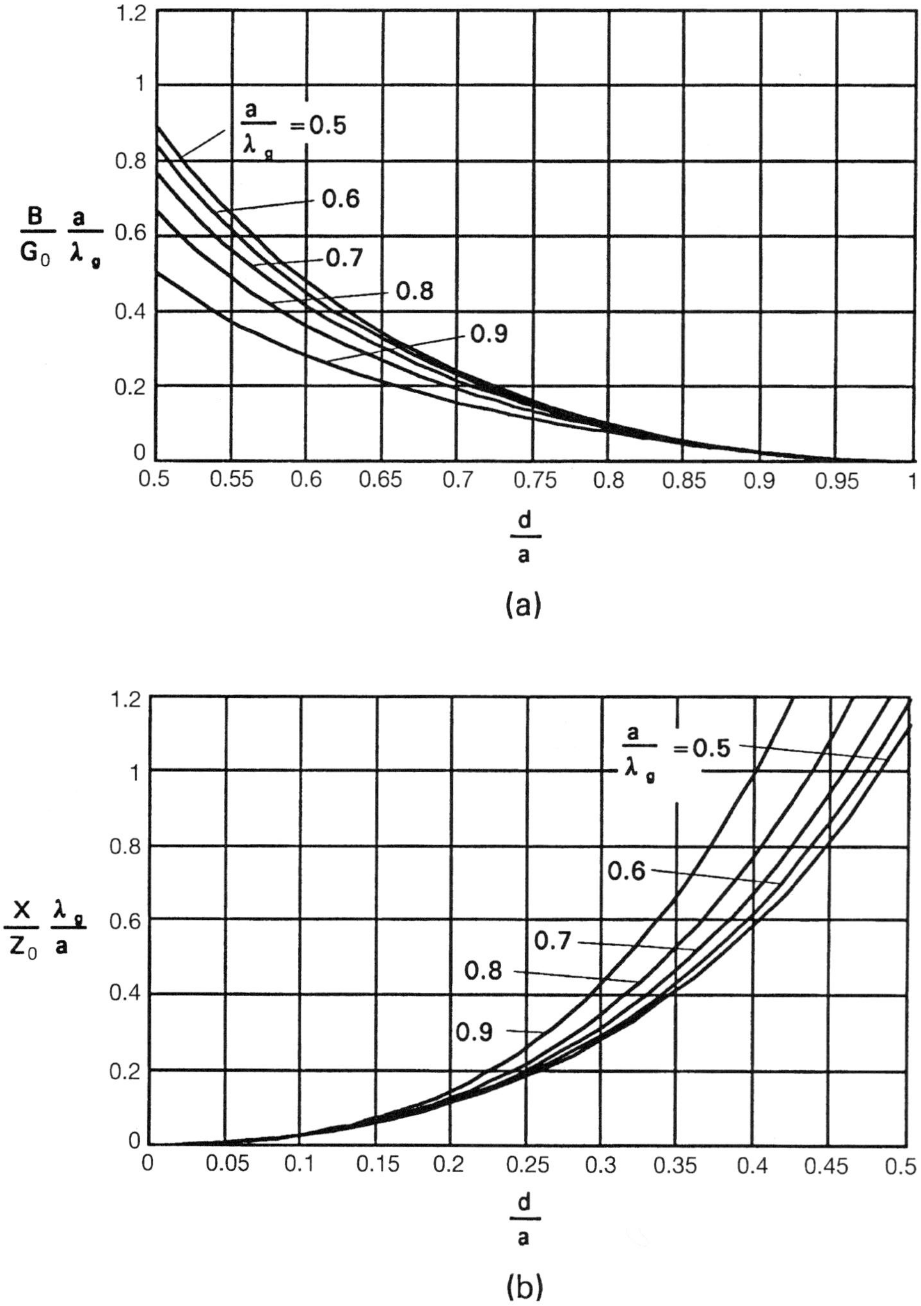

FIGURE 13.36 Plot of $-(B/G_0)(a/\lambda_g)$ versus d/a.

obstacles. Note that with $G_{01} = G_{02} = G_0$ the scattering matrix Eq. (13.46) becomes

$$\mathbf{S} = \frac{1}{2 + j\hat{B}} \begin{bmatrix} -j\hat{B} & 2 \\ 2 & -j\hat{B} \end{bmatrix}, \quad \hat{B} = \frac{B}{G_0}.$$

The iris can be used to match a load to the guide as in single stub tuning (Section 10.7.2).

Example 13.15 A horn antenna is connected as a load on an X-band waveguide (0.900 in. by 0.400 in.) and is found to have a normalized input admittance $\hat{Y}_a = 4 - j3$ at 10.0 GHz. Design an inductive iris to match the antenna to the guide at this frequency.

Solution The cutoff frequency is $f_c = 11.8/(2 \times 0.9) = 6.56$ GHz. At 10.0 GHz the wavelength is $\lambda = 11.8/10.0 = 1.18$ in. The guide wavelength is $\lambda_g = 1.18/\sqrt{1 - (6.56/10.0)^2} = 1.56$ in. We enter the Smith chart in Fig. 13.37 at $\hat{Y}_\alpha$ (point A)

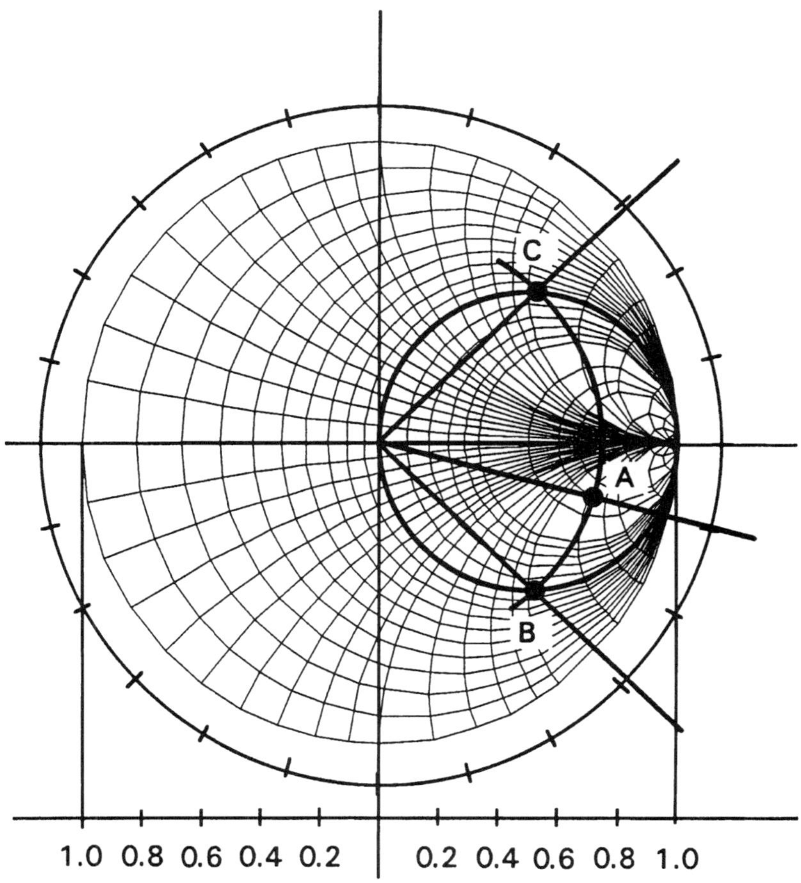

FIGURE 13.37 Smith chart solution to Example 13.15.

and rotate until we reach the $G/G_0 = 1$ circle at points B or C. At point B the susceptance is negative (inductive), requiring a positive (capacitive) susceptance to make a match, so we reject this point. At point C the susceptance is positive (capacitive), requiring a negative (inductive) susceptance for the match. This is what we need. At point C, $\hat{Y} = 1 + j2.1$, so we want the iris to give us a normalized susceptance of $\hat{B} = -2.1$. Then $-(B/G_0)(a/\lambda_g) = 2.1(0.9/1.56) = 1.21$. We have $a/\lambda = 0.9/1.18 = 0.76 \approx 0.8$, so, using Fig. 13.36, we get $d/a = 0.445$ or $d = 0.400$ in. From the Smith chart, we rotated $(0.5 - 0.270 + 0.189) = 0.419$ guide wavelengths toward the generator to get from the antenna to the position at which the iris should be located. This corresponds to a distance of $l = 0.654$ in. Our design is complete.

13.7.2.2 Capacitive Iris in a Rectangular Waveguide Figure 13.38 shows a symmetric capacitive iris with zero thickness in a guide with characteristic admittance G_0. This iris maintains uniformity of the structure with respect to the x coordinate. On the conducting surface of the obstacle the transverse electric field must be zero and at the aperture in the obstacle both the transverse electric and magnetic fields must be continuous. With a TE_{10} wave incident on the obstacle there will be reflected and transmitted TE_{10} waves. In order to satisfy the boundary conditions evanescent TE_{1m} and TM_{1m} modes are generated on each side of the boundary. The evanescent TM_{1m} modes store electric energy [Eq. (12.48)] and the evanescent TE_{1m} modes store magnetic energy, but the electric energy storage of the TM_{1m} modes predominates. The boundary can therefore be represented as a shunt capacitor. This means that B in Eq. (13.46) is positive.

Figure 13.39 plots $(B/G_0)(\lambda_g/b)$ as a function of d/b, with b/λ_g as a parameter [6]. Similar results are available for the asymmetric iris and for symmetric and asymmetric obstacles.

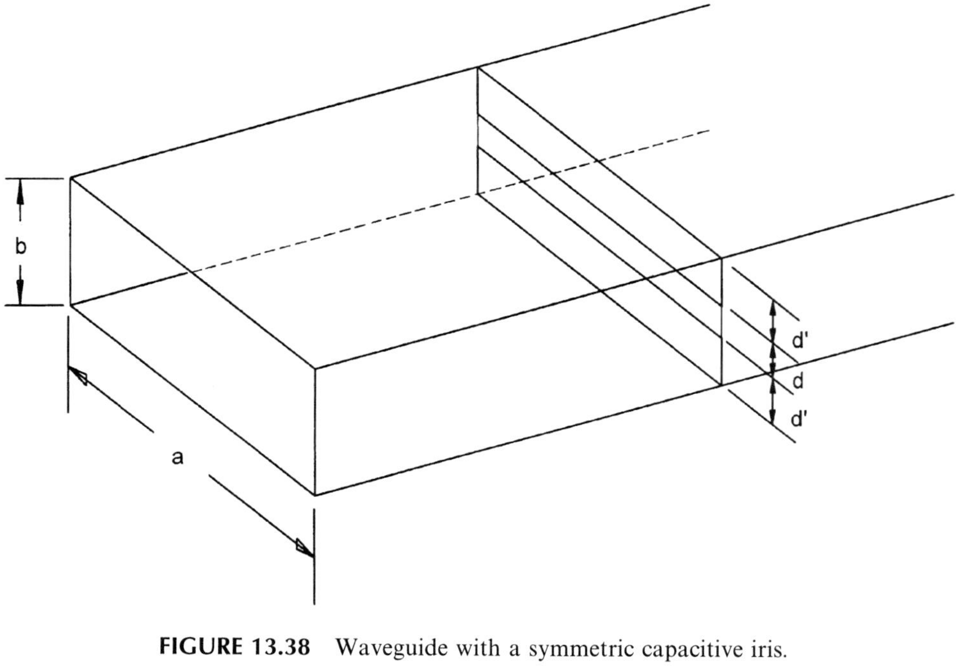

FIGURE 13.38 Waveguide with a symmetric capacitive iris.

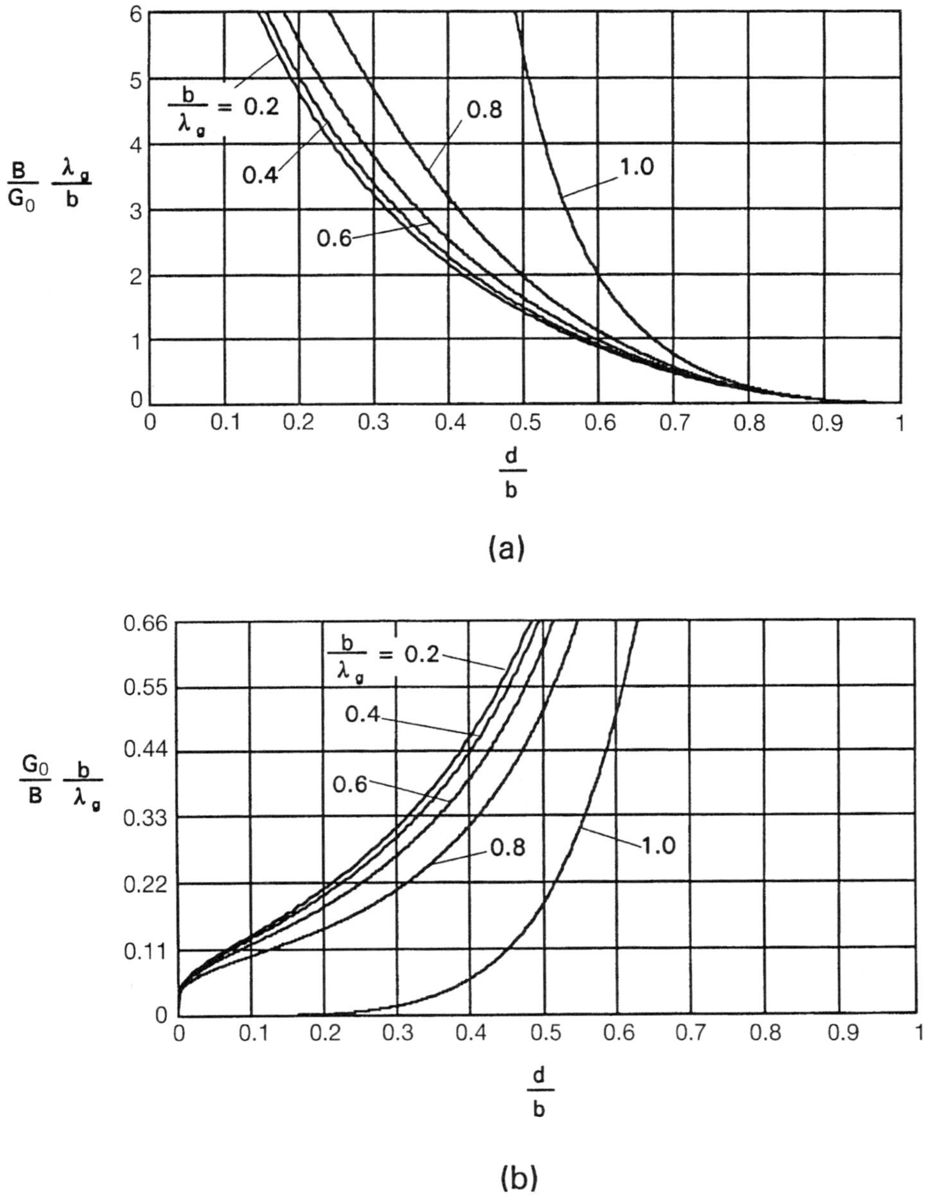

FIGURE 13.39 Plot of $-(B/G_0)(\lambda_g/b)$ versus d/b.

Example 13.16 Repeat the previous example, but use a capacitive iris to do the matching.

Solution This time we can use the solution at point B in Fig. 13.37. At point B, $\hat{Y} = 1 - j2.1$, so we want the diaphragm to give us a normalized susceptance of $\hat{B} = 2.1$. Then $(B/G_0)(\lambda_g/b) = 2.1(1.56/0.4) = 8.19$. We have $b/\lambda_g = 0.4/1.56 = 0.26$, so, using Fig. 13.36, we get $d/b = 0.09$ or $d = 0.036$ in. From the Smith chart, we rotated

$0.311 - 0.270 = 0.041$ guide wavelengths toward the generator to get from the antenna to the position of the diaphragm. This corresponds to a distance of $l = 0.064$ in. We again have a design.

Let us briefly compare the two designs. The capacitive design places the diaphragm closer to the load than the inductive design, which increases the bandwidth of the match. On the other hand, the very small opening of the capacitive design makes the susceptance very sensitive to this dimension. Further study is required before a design choice can be made.

13.7.2.3 Change in Height in the Rectangular Waveguide Figure 13.40 shows a symmetrical change in the height of a rectangular waveguide. Figure 13.41 gives $(B/G_{01})(\lambda_g/b_1)$ as a function of b_2/b_1 with b_1/λ_g as a parameter [7]. In this situation $G_{02}/G_{01} = b_1/b_2$.

Example 13.17 A 0.900 in. wide guide is changed in height from $b_1 = 0.600$ in. to $b_2 = 0.300$ in. as shown in Fig. 13.40. Evaluate the scattering matrix of the junction at 10.00 GHz.

Solution The scattering matrix elements of Eq. (13.46) can be normalized with respect to G_{01}, giving

$$\mathbf{S} = \frac{1}{1 + (b_1/b_2) + j(B/G_{01})} \begin{bmatrix} 1 - (b_1/b_2) - j(B/G_{01}) & 2(b_1/b_2) \\ 2 & (b_1/b_2) - 1 - jB/G_{01} \end{bmatrix}.$$

At 10.00 GHz, $\lambda_g = 1.56$ in. (see Example 13.15), so $b_2/b_1 = 0.50$ and $b_1/\lambda_g = 0.385$. Using Fig. 13.41 we have $(B/G_{01})(\lambda_g/b_1) = 0.85$. Then $B/G_{01} = 0.33$ and the scattering

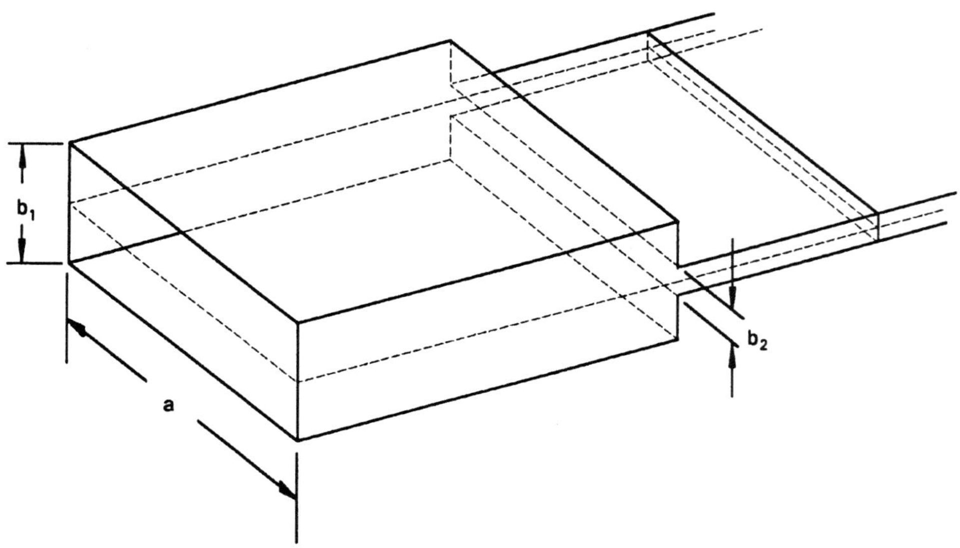

FIGURE 13.40 Symmetrical change in height of a rectangular waveguide.

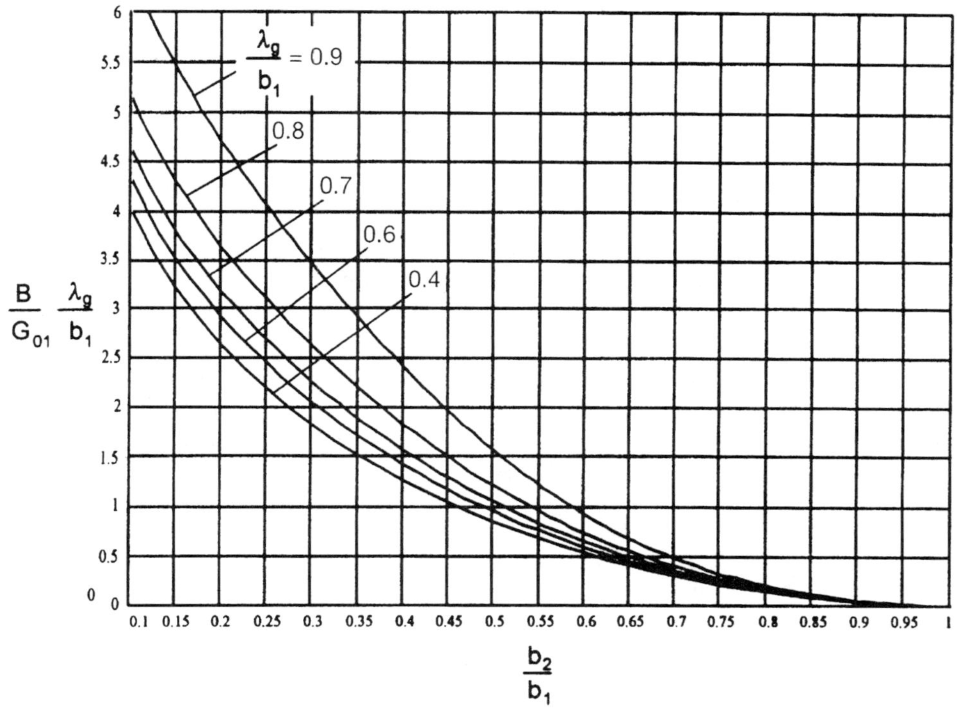

FIGURE 13.41 Normalized shunt susceptance for change in height of rectangular waveguide.

matrix is evaluated as

$$\mathbf{S} = \frac{1}{1 + 2.00 + j0.33} \begin{bmatrix} 1 - 2 - j0.33 & 4 \\ 2 & 2.00 - 1 - j0.33 \end{bmatrix}$$

$$= \begin{bmatrix} 0.349 \angle -168^\circ & 1.325 \angle -6.3^\circ \\ 0.663 \angle -6.3^\circ & 0.349 \angle -24.54^\circ \end{bmatrix}.$$

The result satisfies both the symmetry property, Eq. (13.23), and the unitary properties, Eqs. (13.28) and (13.29).

The guide can be split in half by a horizontal conducting plane (shown dashed in Fig. 13.40). The upper half of the guide (or the lower half of the guide) is an asymmetrical junction of two guides. We will use a circumflex (hat) to designate the dimensions and parameters of the upper half of the guide. The capacitors $\hat{C}$ representing the junction in each half of the split guide are in series to give the capacitance C of the original junction. Hence $\hat{C} = 2C$ and the new susceptance $\hat{B}$ equals twice the old susceptance B. If λ_g in Fig. 13.41 is replaced by $\lambda_g/2$, then the plots show $(\hat{B}/\hat{G}_{01})(\lambda_g/2\hat{b}_1)$ as a function of $\hat{b}_2/\hat{b}_1$ with $2\hat{b}_1/\lambda_g$ as a parameter.

13.7.3 Three-Port Device: Power Divider

Figure 13.42 shows the Wilkinson 3 dB power divider, realized in microstrip. The tops of the lines are shown, with the dielectric layer and ground plane underneath. The structure is symmetric about the dashed centerline. The resistor R can be fabricated as a film or a lumped resistor. As we shall see, with the two lines having a characteristic resistance $\sqrt{2}R_0$ and length $l = \lambda_0/4$ at the center frequency f_0 and with $R = 2R_0$, the divider is matched: $S_{11} = S_{22} = S_{33} = 0$ at f_0. Symmetry then guarantees that a signal incident on port 1 will split equally and in phase to ports 2 and 3.

Let us find the scattering matrix elements at the center frequency.

First consider an input voltage at port 1 with ports 2 and 3 matched. Ports 2 and 3 will be at the same voltage so no current will flow in the resistor. The resistor can therefore be removed from the circuit (Fig. 13.43). The loads R_0 are transformed by the quarter-wavelength line into two resistors $(\sqrt{2}R_0)^2/R_0 = 2R_0$ in parallel at port 1. The total load seen looking into port 1 is R_0 and that port is matched ($S_{11} = 0$). The line voltage along each quarter-wave line is

$$V(z) = V_0(e^{-j\beta z} + \Gamma_0 e^{j\beta z}), \quad \Gamma_0 = \frac{R_0 - \sqrt{2}R_0}{R_0 + \sqrt{2}R_0} = \frac{1 - \sqrt{2}}{1 + \sqrt{2}}.$$

The voltage at port 2 or 3 relative to the voltage at port 1 is

$$S_{21} = \frac{V(0)}{V(-\lambda_0/4)} = \frac{V_0(1 + \Gamma_0)}{V_0(j - j\Gamma_0)} = \frac{-j}{\sqrt{2}}.$$

This is also the value of S_{12}, S_{13}, and S_{31} due to the symmetry of the circuit and the scattering matrix.

Next, consider matched voltage sources with "even" voltages $V_e = V_g/2$ and "odd" voltages $V_o = V_g/2$ at ports 2 and 3 with port 1 matched (Fig. 13.44a). The total source

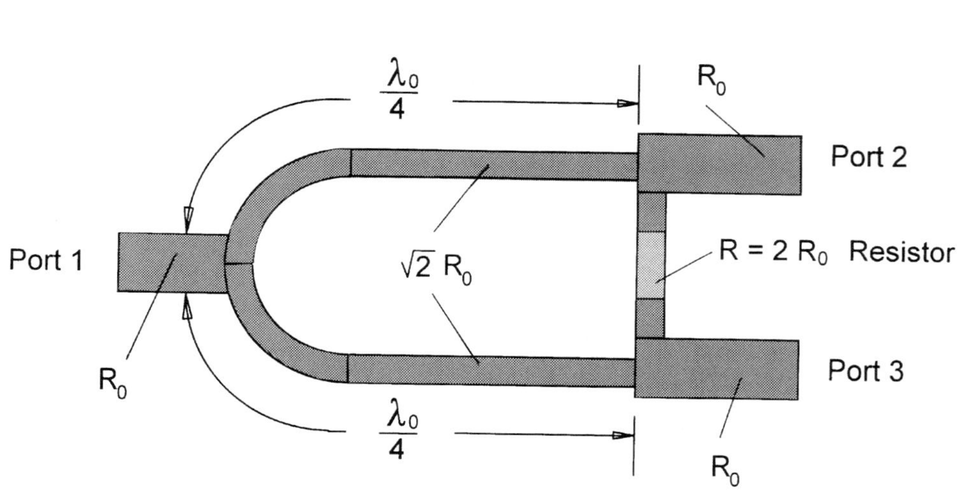

FIGURE 13.42 Wilkinson power divider.

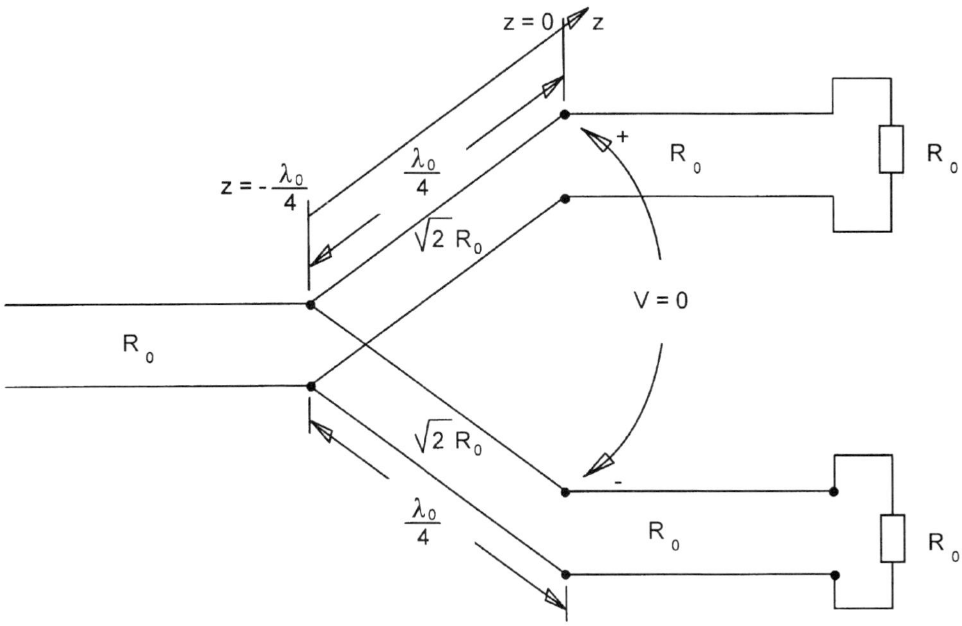

FIGURE 13.43 Transmission line power divider with the resistor removed.

voltage at port 2 is $V_{s2} = V_e + V_o = V_g$, while the total source voltage at port 3 is $V_{s3} = V_e - V_o = 0$. The load resistor at port 1 is shown as two resistors $2R_0$ in parallel to maintain symmetry.

With only the even voltages applied, the current crossing the centerline is zero and the circuit can be split as shown in Fig. 13.44b. The load $2R_0$ is transformed by the quarter-wave line to $(\sqrt{2}R_0)^2/2R_0 = R_0$, giving a reflection coefficient $S_{22,e} = 0$. The voltage at port 2 and at port 3 is $V_g/4$.

With only the odd voltages applied, the centerline will be at zero voltage, resulting in the equivalent circuit of Fig. 13.44c. The short at port 1 is transformed by the quarter-wave line into an open at port 2, so that only $R/2 = R_0$ appears as a load looking into port 2. The reflection coefficient is therefore $S_{22,o} = 0$. The voltage at port 2 is $V_{2o} = V_g/4$ and the voltage at port 3 is $-V_g/4$.

With both even and odd voltages applied, the reflection coefficient at port 2 is $S_{22} = S_{22,o} + S_{22,e} = 0$ and, because of the symmetry of the circuit, $S_{33} = 0$. The total voltage at port 2 is $\underline{V}_2 = V_{o2}^+ = V_g/2$ and the total voltage at port 3 is $\underline{V}_3 = V_{o3}^- = 0$. Hence $S_{32} = 0$ and from the symmetry of the circuit we have $S_{23} = S_{32} = 0$. Ports 2 and 3 are matched and ports 2 and 3 are isolated from each other at f_0. At f_0 the scattering matrix is

$$\mathbf{S} = \begin{bmatrix} 0 & -j/\sqrt{2} & -j/\sqrt{2} \\ -j/\sqrt{2} & 0 & 0 \\ -j/\sqrt{2} & 0 & 0 \end{bmatrix}. \qquad (13.47)$$

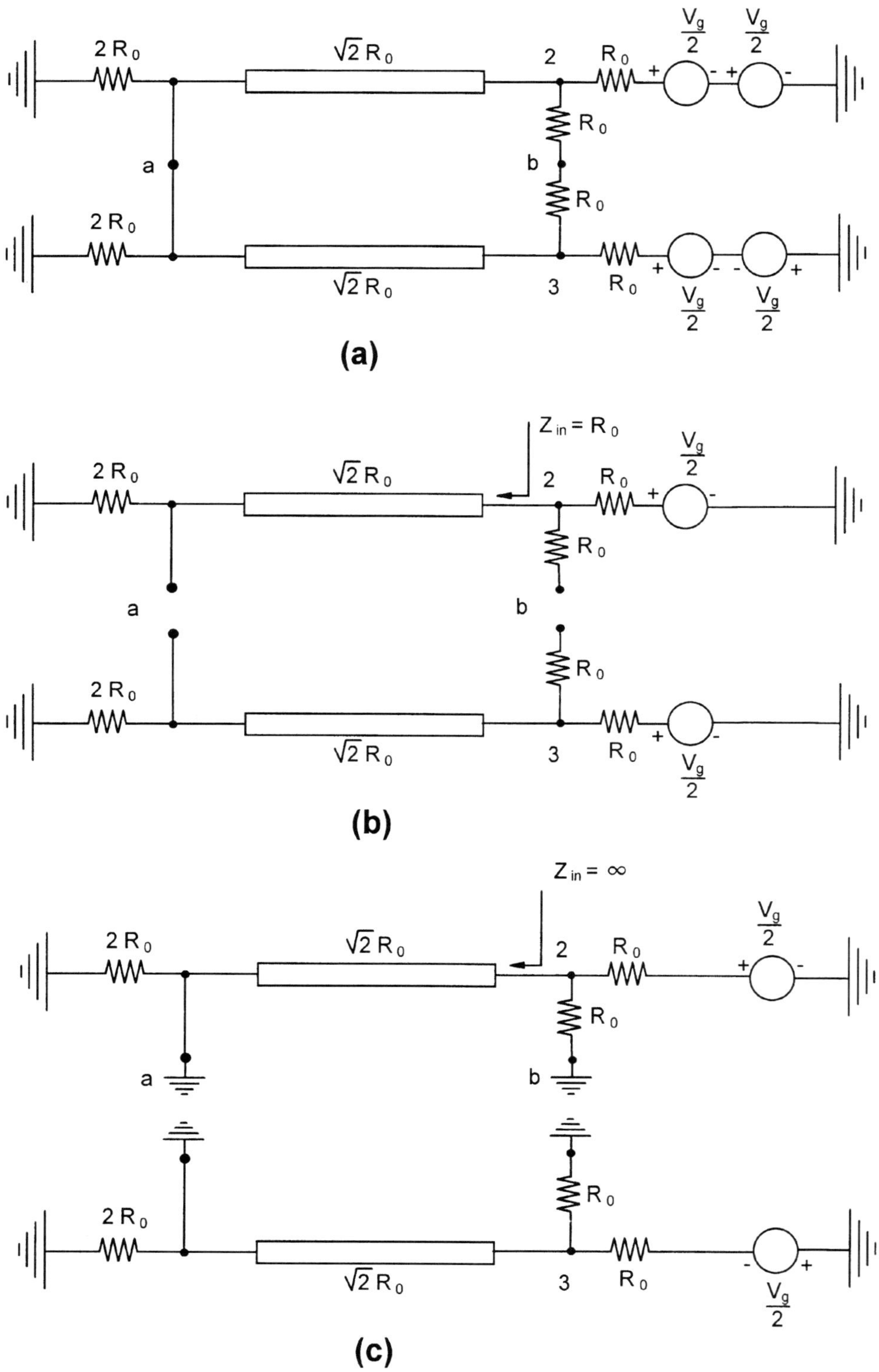

FIGURE 13.44 Even- and odd-mode equivalent circuits of the power divider.

At any other frequency the two lines of length $l = \lambda_0/4$ are not a quarter-wavelength long. The three ports are not perfectly matched, the power coupled from port 1 to ports 2 and 3 is not exactly one-half and ports 2 and 3 are not perfectly isolated. Using the same method of analysis as above, we could show that

$$S_{11} = \frac{j\sqrt{2}\cot\theta}{4 - j3\sqrt{2}\cot\theta}, \quad \theta = \beta l = \left(\frac{2\pi}{\lambda}\right)\left(\frac{\lambda_0}{4}\right) = \left(\frac{\pi}{2}\right)\left(\frac{f}{f_0}\right),$$

$$S_{21} = S_{12} = S_{31} = S_{13} = \frac{-j2\sqrt{2}\operatorname{cosec}\theta}{4 - j3\sqrt{2}\cot\theta},$$

$$S_{22} = S_{33} = \frac{\cot^2\theta}{8 - 3\cot^2\theta - j8\sqrt{2}\cot\theta},$$

$$S_{23} = S_{32} = \frac{-2(\cot^2\theta + j\sqrt{2}\cot\theta)}{8 - 3\cot^2\theta - j8\sqrt{2}\cot\theta}. \tag{13.48}$$

The parameters for a coupler are usually stated in decibel measure as follows:

$$\begin{aligned} &\text{Return loss}_1 = -20\log|S_{11}|, \quad &&\text{Return loss}_2 = -20\log|S_{22}|, \\ &\text{Coupling} = -20\log|S_{21}|, &&\text{Isolation} = -20\log|S_{23}|. \end{aligned} \tag{13.49}$$

Figure 13.45 plots these quantities over a 100% $[f = f_0(1 \pm 0.5)]$ bandwidth. For most applications this design is not suitable for a system requiring greater than 40% bandwidth.

13.7.4 Four-Port Device: Directional Coupler

We will examine the directional coupler as an example of a four-port device. Directional couplers are frequently used in measuring systems to monitor power. There are many designs for directional couplers in waveguide, coax, and microstrip [8]. We will first discuss the general properties of couplers and then find the scattering matrix for a branch line coupler.

13.7.4.1 Definition and Properties Figure 13.46 shows a schematic illustration of a directional coupler.† An ideal directional coupler has two properties: (1) power into any port is coupled into the opposite two ports but is not coupled into the adjacent port; and (2) all four ports are matched.‡

† The numbering of the ports is not standardized and is different in the various texts and references.

‡ It is easy to use the unitary property of the scattering matrix to show that if property 1 is satisfied and if any two ports are matched, then the other two ports must also be matched.

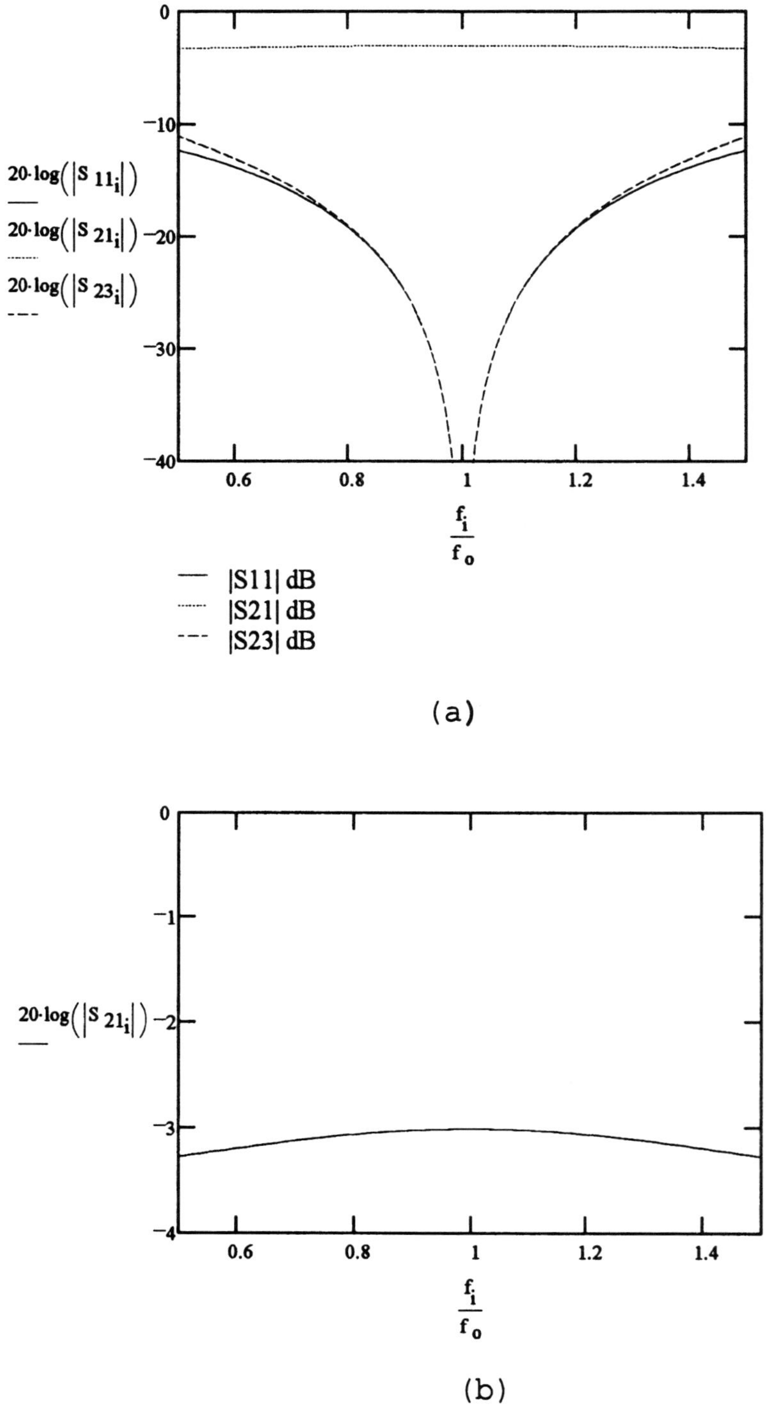

FIGURE 13.45 Wilkinson power divider *S*-parameters.

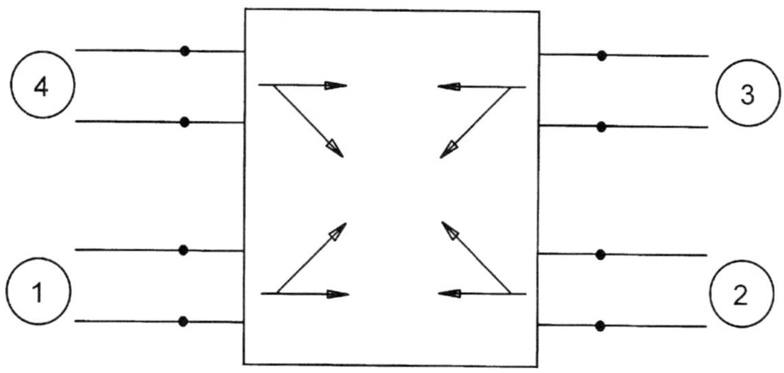

FIGURE 13.46 Schematic representation of a directional coupler.

The first property means that $S_{14} = S_{41} = 0$ and $S_{23} = S_{32} = 0$ and the second means that $S_{11} = S_{22} = S_{33} = S_{44} = 0$. The scattering matrix is therefore given by

$$\mathbf{S} = \begin{bmatrix} 0 & S_{12} & S_{13} & 0 \\ S_{12} & 0 & 0 & S_{24} \\ S_{13} & 0 & 0 & S_{34} \\ 0 & S_{24} & S_{34} & 0 \end{bmatrix}. \tag{13.50}$$

The remaining nonzero elements are not independent because of the unitary property of the scattering matrix. As is usually the case in practice, assume that each of the four lines connected to the ports has the same characteristic resistance. Taking the product of column 1 and the conjugate of column 4 and the same for columns 2 and 3, we find

$$S_{12}S_{24}^* = -S_{13}S_{34}^*, \tag{13.51}$$

$$S_{12}S_{13}^* = -S_{24}S_{34}^*. \tag{13.52}$$

We divide the first equation by the second to get $S_{13}S_{13}^* = S_{24}S_{24}^*$ or

$$|S_{13}| = |S_{24}| \equiv K_a. \tag{13.53}$$

We take the conjugate of Eq. (13.51) and multiply by Eq. (13.52) to get $S_{12}S_{12}^* = S_{34}S_{34}^*$ or

$$|S_{12}| = |S_{34}| \equiv K_b. \tag{13.54}$$

Equation (13.53) means that the magnitude of the transmission from port 1 to port 3 (and from port 3 to 1) is the same as the magnitude of the transmission from port 2 to port 4 (and from port 4 to 2). Equation (13.54) means that the magnitude of the transmission from port 1 to port 2 (and from port 2 to 1) is the same as the magnitude of the transmission from port 3 to port 4 (and from port 4 to 3).

Taking the product of the first column of **S** with its conjugate we find

$$|S_{12}|^2 + |S_{13}|^2 = 1,$$
$$K_b^2 + K_a^2 = 1. \tag{13.55}$$

The scattering matrix has been determined to be

$$\mathbf{S} = \begin{bmatrix} 0 & K_b \angle \phi_b & K_a \angle \phi_a & 0 \\ K_b \angle \phi_b & 0 & 0 & K_a \angle \phi_c \\ K_a \angle \phi_a & 0 & 0 & K_b \angle \phi_d \\ 0 & K_a \angle \phi_c & K_b \angle \phi_d & 0 \end{bmatrix}. \tag{13.56}$$

Just as K_a and K_b are not independent, neither are the angles. Equations (13.51) and (13.52) each require that

$$\phi_d = \phi_a + \phi_c - \phi_b \pm \pi. \tag{13.57}$$

The values of the angles can be adjusted by shifting the terminal planes at the various ports. For example, by shifting ports 1, 2, and 4 we can arbitrarily choose ϕ_a, ϕ_b, and ϕ_c, respectively. Equation (13.57) then gives the value of ϕ_d.

Equations (13.55), (13.56), and (13.57) together give the scattering matrix of the ideal coupler.

Directional couplers are usually characterized by their coupling and directivity. Letting P^i be the incident power into port 1, P^c be the coupled power out of port 3, and P^d the power out of the isolated port 4, we define the coupling and the directivity in decibels to be

$$\text{Coupling} = C = 10\log\left(\frac{P^i}{P^c}\right) = -20\log|S_{13}|, \tag{13.58}$$

$$\text{Directivity} = D = 10\log\left(\frac{P^c}{P^d}\right) = -20\log|S_{14}|. \tag{13.59}$$

For an ideal coupler the directivity is infinite. The return loss at each port is also an important parameter. Designating the incident and the reflected power at the jth port as P_j^i and P_j^r, respectively, then the return loss at that port is

$$\text{Return loss}_j = 10\log\left(\frac{P_j^i}{P_j^r}\right) = -20\log|S_{jj}|.$$

For an ideal coupler the return loss is infinite at each port.

Figure 13.47 shows a directional coupler used to monitor the power from a transmitter to an antenna. The power meter is matched to the line (i.e., has an input resistance R_0). Power reflected by the antenna is absorbed by the matched load on port 4 and by the transmitter.

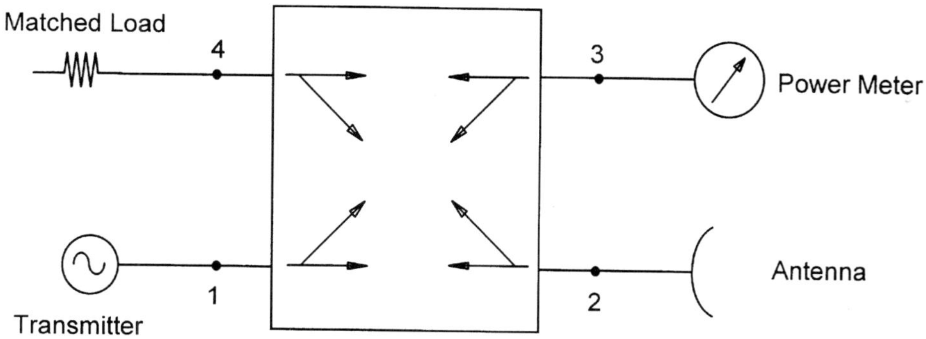

FIGURE 13.47 Directional coupler used to monitor the power transmitted to an antenna.

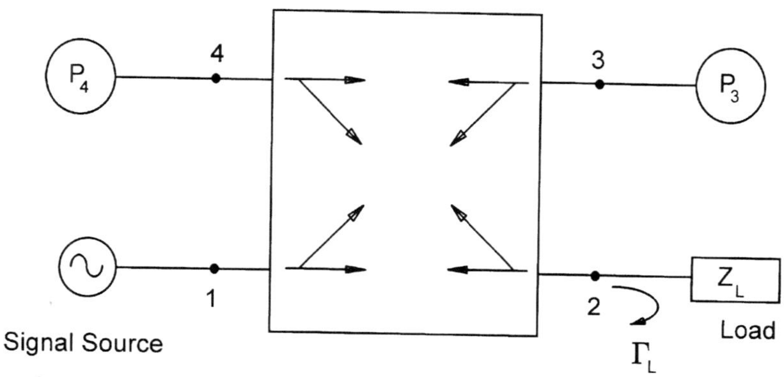

FIGURE 13.48 Directional coupler used to measure the magnitude of the reflection coefficient of the load Z_L.

Figure 13.48 shows a directional coupler used to measure the magnitude of the reflection coefficient at the load Z_L. The powers measured by the matched power meters ($V_3^+ = V_4^+ = 0$) are calculated as

$$P_3 = \frac{1}{2}\frac{|V_3^-|^2}{R_0}, \quad P_4 = \frac{1}{2}\frac{|V_4^-|^2}{R_0},$$

where

$$V_3^- = S_{13}V_1^+ + S_{34}V_4^+ = S_{13}V_1^+,$$
$$V_4^- = S_{24}V_2^+ + S_{34}V_3^+ = S_{24}V_2^+.$$

But $V_2^+ = \Gamma_L V_2^-$. Therefore V_4^- becomes

$$V_4^- = S_{24}\Gamma_L V_2^- = S_{24}\Gamma_L(S_{12}V_1^+ + S_{24}V_4^+) = S_{24}\Gamma_L S_{12}V_1^+.$$

Substituting the expressions for V_3^- and V_4^- into the power equations, and taking the ratio of powers at ports 4 and 3, the solution for the magnitude of the reflection coefficient is

$$|\Gamma|^2 = K_b^2 \frac{P_4}{P_2}.$$

Example 13.18 Determine a calibration procedure that would allow you to use a directional coupler, with unknown characteristics (but with a very good directivity), to perform measurements of the magnitude of the reflection coefficient. (*Hint*: Calibration procedures involve the use of known standards, with known reflection coefficients.)

Solution If the unknown load is replaced by a load with a known reflection coefficient, like a short circuit [$\Gamma(\text{short}) = -1$], the power ratio will allow the determination of the unknown coefficient K_b:

$$1 = K_b^2 \frac{P_4(\text{short})}{P_3(\text{short})}.$$

Therefore

$$K_b = \sqrt{\frac{P_3(\text{short})}{P_4(\text{short})}}.$$

The magnitude of the unknown reflection coefficient Γ_L is then determined as

$$|\Gamma_L| = \sqrt{\frac{P_3(\text{short})}{P_4(\text{short})} \cdot \frac{P_4(\Gamma_L)}{P_3(\Gamma_L)}}.$$

13.7.4.2 Branch Line Coupler Figure 13.49 illustrates a branch line coupler, realized in microstrip. The tops of the lines are shown, with the dielectric layer and ground plane underneath. The device is symmetric about the centerline shown in the figure. Both the through line, with characteristic resistance R_{01}, and the branch line, with characteristic resistance R_{02}, are a quarter-wavelength long at the center design frequency f_0. We have assumed that each port is connected to a transmission line with characteristic resistance R_0.

We can analyze the structure by connecting matched loads to ports 2 and 3 and matched voltage sources at ports 1 and 4 (Fig. 13.50a). At port 1, $\underline{V}_{s1} = \underline{V}_g/2 + \underline{V}_g/2 = \underline{V}_g$. At port 4, $\underline{V}_{s4} = \underline{V}_g/2 - \underline{V}_g/2 = 0$. With only the even voltages applied, no current will cross the centerline and each half of the coupler becomes a two-port as shown in Fig. 13.50b. Since no current crosses the centerline, the half branch lines are open-circuited. For this two-port the reflection coefficient at port 1 is Γ_e and the transmission coefficient from port 1 to port 2 is τ_e. With only the odd voltages applied, the centerline is at zero voltage and each half of the coupler becomes a two-port as shown in Fig. 13.50c. Since the voltage at the centerline is zero, the half branch lines

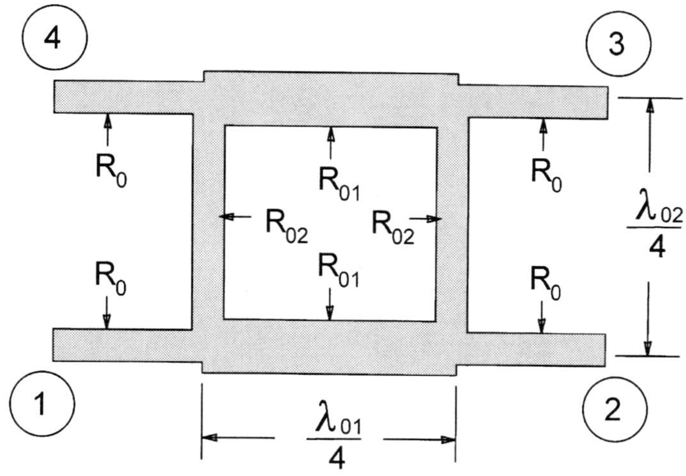

FIGURE 13.49 Top view of a branch line coupler realized in microstrip.

are short-circuited. For this two-port the reflection coefficient at port 1 is Γ_o and the transmission coefficient from port 1 to port 2 is τ_o. With both the even and odd voltages, the voltage at port 4 is zero and it is connected to a matched load. Since ports 2, 3, and 4 are matched the voltage at each of these ports is simply equal to the outgoing voltage wave at the port, which is the sum of the outgoing wave with the even and the odd source connected. Hence

$$V_{o1}^- = \tfrac{1}{2}(\Gamma_e + \Gamma_o)V_{o1}^+ = S_{11}V_{o1}^+, \quad V_{o2}^- = \tfrac{1}{2}(\tau_e + \tau_o)V_{o1}^+ = S_{21}V_{o1}^+,$$

$$V_{o4}^- = \tfrac{1}{2}(\Gamma_e - \Gamma_o)V_{o1}^+ = S_{41}V_{o1}^+, \quad V_{o3}^- = \tfrac{1}{2}(\tau_e - \tau_o)V_{o1}^+ = S_{31}V_{o1}^+.$$

Recalling that $S_{ij} = S_{ji}$ and utilizing the symmetry of the branch line coupler ($S_{22} = S_{33} = S_{44} = S_{11}$, $S_{34} = S_{21}$, $S_{32} = S_{41}$, and $S_{31} = S_{24}$), we obtain the scattering matrix

$$\mathbf{S} = \begin{bmatrix} (\Gamma_e + \Gamma_o)/2 & (\tau_e + \tau_o)/2 & (\tau_e - \tau_o)/2 & (\Gamma_e - \Gamma_o)/2 \\ (\tau_e + \tau_o)/2 & (\Gamma_e + \Gamma_o)/2 & (\Gamma_e - \Gamma_o)/2 & (\tau_e - \tau_o)/2 \\ (\tau_e - \tau_o)/2 & (\Gamma_e - \Gamma_o)/2 & (\Gamma_e + \Gamma_o)/2 & (\tau_e + \tau_o)/2 \\ (\Gamma_e - \Gamma_o)/2 & (\tau_e - \tau_o)/2 & (\tau_e + \tau_o)/2 & (\Gamma_e + \Gamma_o)/2 \end{bmatrix}. \tag{13.60}$$

For this matrix to match with that of the ideal coupler, Eq. (13.56), at the design center frequency we must have $\Gamma_e = \Gamma_o = 0$. It remains to solve for the even and odd reflection and transmission coefficients.

For both the even and the odd excitation the equivalent circuit from port 1 to port 2 is a shunt susceptance jB_s, cascaded with a transmission line R_{01}, $\theta = \beta l = (\pi/2)(f/f_0)$, cascaded with a shunt susceptance jB_s (Fig. 13.51). Using the *ABCD*

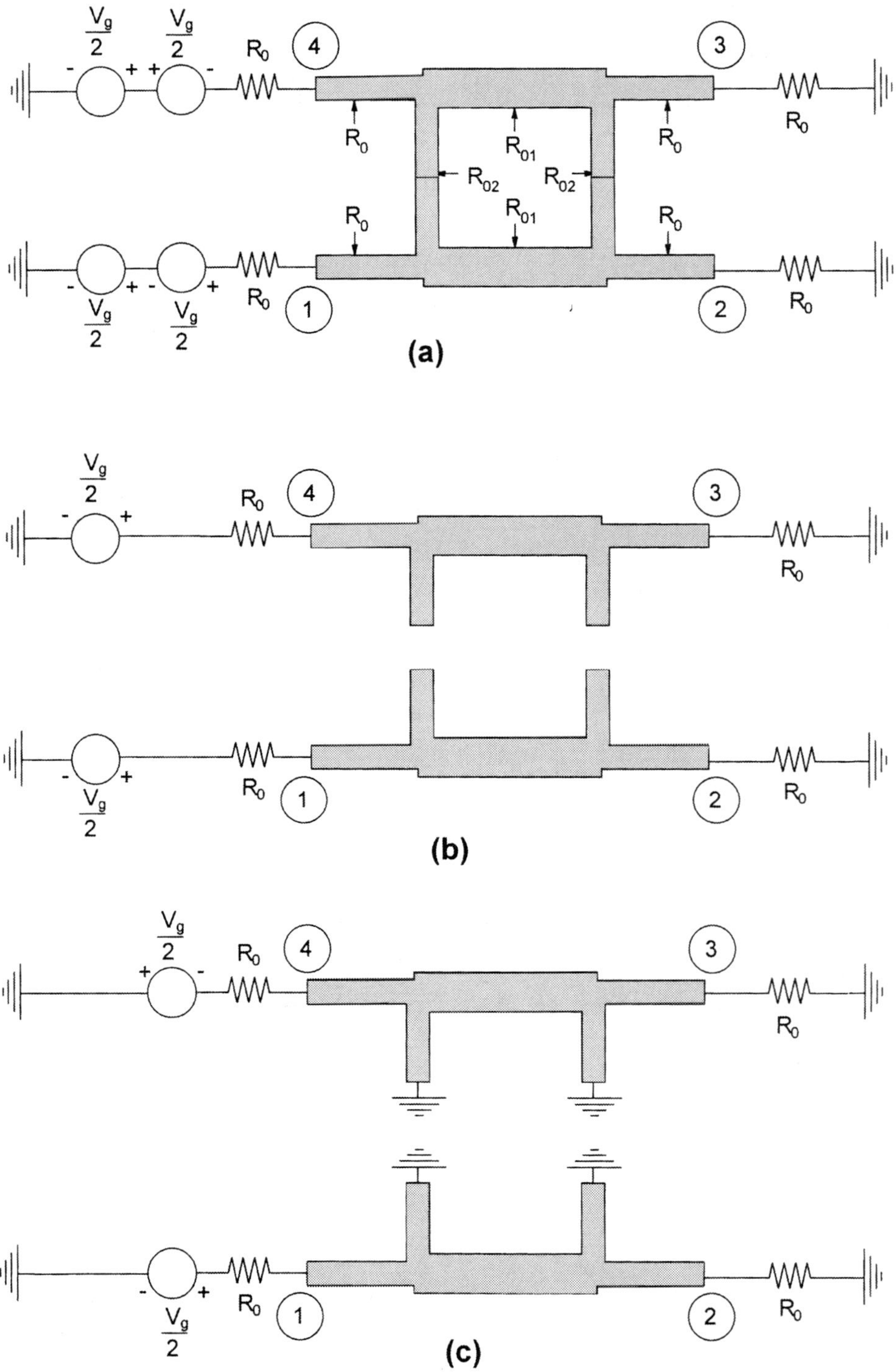

FIGURE 13.50 Even- and odd-mode equivalent circuits for a branch line coupler.

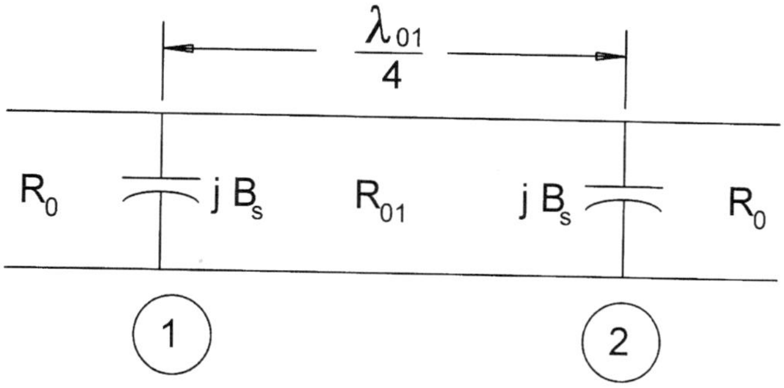

FIGURE 13.51 Equivalent circuit from port 1 to port 2 for both even and odd excitations of a branch line coupler.

matrices from Table 13.2, we have

$$\begin{bmatrix} V_1 \\ I_1 \end{bmatrix} = \begin{bmatrix} 1 & 0 \\ jB_s & 1 \end{bmatrix} \begin{bmatrix} \cos\theta & jR_{01}\sin\theta \\ jG_{01}\sin\theta & \cos\theta \end{bmatrix} \begin{bmatrix} 1 & 0 \\ jB_s & 1 \end{bmatrix} \begin{bmatrix} V_2 \\ -I_2 \end{bmatrix}. \tag{13.61}$$

The reflection coefficient and the transmission coefficient are given by S_{11} and S_{21}, respectively, in Eq. (13.44), used with $R_{01} = R_{02} = R_0$. Thus

$$\Gamma = \frac{B - CR_0^2 + (A - D)R_0}{B + CR_0^2 + (A + D)R_0}, \tag{13.62}$$

$$\tau = \frac{2R_0}{B + CR_0^2 + (A + D)R_0}. \tag{13.63}$$

For even and odd excitations we have

$$B_{se} = G_{02}\tan(\theta/2), \quad B_{so} = -G_{02}\cot(\theta/2), \tag{13.64a, b}$$

respectively.

At the design frequency f_0, $\theta = \pi/2$, $\cos\theta = 0$, $\sin\theta = 1$, $B_{se} = G_{02}$ and $B_{so} = -G_{02}$. Carrying out the matrix multiplication of Eq. (13.61) we obtain

$$A_e = D_e = -\frac{R_{01}}{R_{02}}, \quad B_e = jR_{01}, \quad C_e = jR_{01}\left(\frac{1}{R_{01}^2} - \frac{1}{R_{02}^2}\right),$$

for even excitation and

$$A_o = D_o = \frac{R_{01}}{R_{02}}, \quad B_e = jR_{01}, \quad C_e = jR_{01}\left(\frac{1}{R_{01}^2} - \frac{1}{R_{02}^2}\right),$$

for odd excitation. With $A = D$ the numerator of the reflection coefficient, Eq. (13.62), is $B - CR_0^2$. Setting this to zero we obtain

$$R_{02} = \frac{R_0 \; R_{01}}{\sqrt{R_0^2 \; - R_{01}^2}} \tag{13.65}$$

for both even and odd excitation. Substituting this for R_{02}, the transmission coefficient becomes

$$\tau_e = \frac{1}{R_0}\left(-\sqrt{R_0^2 - R_{01}^2} - jR_{01}\right), \quad \tau_o = \frac{1}{R_0}\left(\sqrt{R_0^2 - R_{01}^2} - jR_{01}\right),$$

giving

$$\frac{1}{2}(\tau_e + \tau_o) = -j\frac{R_{01}}{R_0}, \quad \frac{1}{2}(\tau_e - \tau_o) = -\sqrt{1 - \left(\frac{R_{01}}{R_0}\right)^2}.$$

The scattering matrix is therefore

$$\mathbf{S} = \begin{bmatrix} 0 & -j(R_{01}/R_0) & -\sqrt{1-(R_{01}/R_0)^2} & 0 \\ -j(R_{01}/R_0) & 0 & 0 & -\sqrt{1-(R_{01}/R_0)^2} \\ -\sqrt{1-(R_{01}/R_0)^2} & 0 & 0 & -j(R_{01}/R_0) \\ 0 & -\sqrt{1-(R_{01}/R_0)^2} & -j(R_{01}/R_0) & 0 \end{bmatrix} \tag{13.66}$$

at the design frequency f_0.

The coupling, Eq. (13.58), is

$$\text{Coupling} = C = -\,20\log|S_{13}| = -10\log\left[1 - \left(\frac{R_{01}}{R_0}\right)^2\right]. \tag{13.67}$$

When we specify the coupling C we specify R_{01} and, from Eq. (13.65), R_{02}. Table 13.3 gives the required value of R_{01}/R_0 and R_{02}/R_0 for several couplings. Branch line couplers cannot generally be designed for weaker coupling than 10 dB because the high characteristic resistance of the branch line cannot be achieved. Figure 13.52

TABLE 13.3 Branch Line Coupler

C (dB)	R_{01}/R_0	R_{02}/R_0
3	$1/\sqrt{2} = 0.707$	1
6	0.865	1.727
10	0.949	3.00

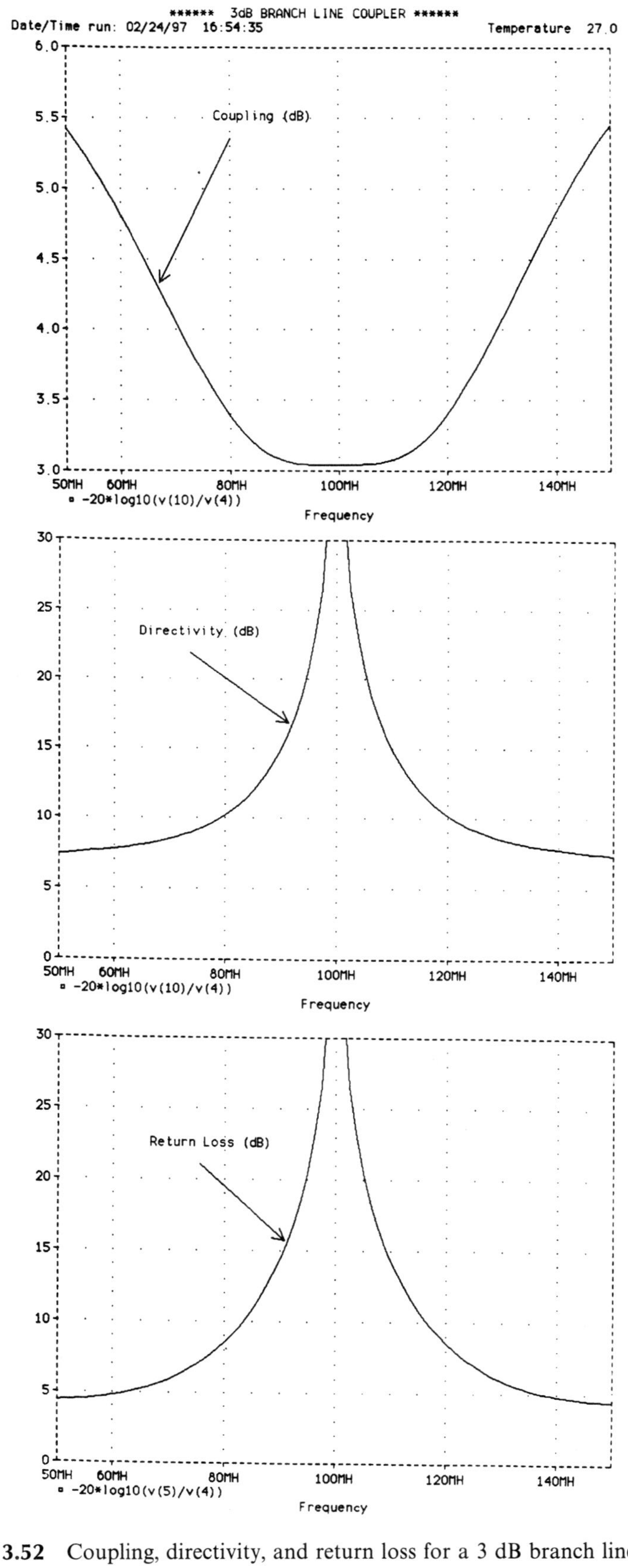

FIGURE 13.52 Coupling, directivity, and return loss for a 3 dB branch line coupler.

shows the coupling, directivity, and return loss over a 50% bandwidth for a 3 dB coupler.

13.8 *S*-PARAMETER MEASUREMENTS

Determination of the *S*-parameters, as their definitions imply, requires the connection of matched terminations at different ports and then the measurement of the reflection and transmission coefficients throughout the network. The actual measurements of the magnitude and phase of the reflection and the transmission coefficients are performed with the use of vector network analyzers. An example of a scalar network analyzer, for the measurement of the magnitude of the reflection coefficient only, is the directional coupler from the previous section. The measurement of both magnitude and phase angle requires more sophisticated equipment and data manipulation. This section gives an introduction to the network analysis systems.

13.8.1 Traditional Vector Network Analyzers

The most common vector network analyzer used throughout the industry is the type manufactured by Hewlett Packard or Wiltron. This type of vector network analyzer is capable of measuring both the magnitude and phase angle of the *S*-parameters of any two-port or multiport devices for a wide frequency range, as well as performing time domain reflectometry measurements (used for cable and transmission line diagnosis, etc.). The measurement of the magnitude and, more importantly, the measurement of the phase angle of the *S*-parameters is performed by down-converting the very-high-frequency signals to an IF frequency range, where the measurement of the phase angle can be performed accurately. In order to account for the imperfections in the system, a calibration procedure would have to precede any measurement. This type of vector network analyzer is very accurate but requires "near-ideal" and, hence, expensive parts.

13.8.2 Six-Port Network Analyzers

The concept of six-port measurements stemmed from the idea of Engen and Hoer to use power measurements to determine the phase of the reflection coefficient or any other *S*-parameter. The idea was born from the need for a measurement technique that required very few "near-ideal" components.

The real advantage of the six-port reflectometer, and six-port vector analyzer in general, over the traditional vector analyzers is that it is easier to measure amplitude, which is the square root of the power, than to measure both amplitude and phase.

The phase information is extracted by manipulating the power equations.

Suppose an unknown vector, $\vec{a}_2$, has to be determined with respect to a known reference vector $\vec{a}_1$ – see Fig. 13.53 – using only absolute value or power measurements. The phase angle between the two vectors $\vec{a}_2$ and $\vec{a}_1$ can be determined by measuring the magnitude of the sum and the difference of the two vectors as well as

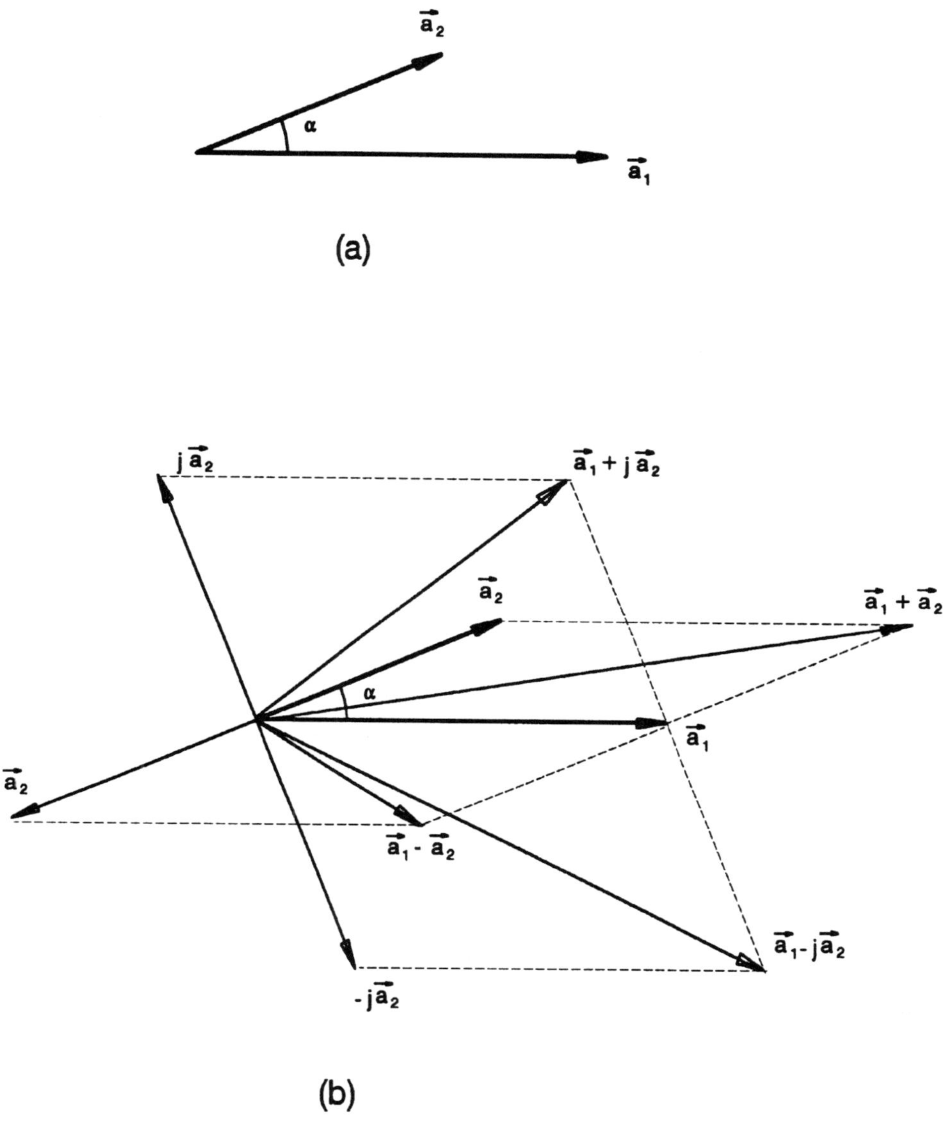

FIGURE 13.53 (a) Unknown vector $\vec{a}_2$ and reference vector $\vec{a}_1$ and (b) the vector operations that lead to the determination of the angle α.

the sum and the difference of the reference vector $\vec{a}_1$ and the unknown vector shifted by 90°.

$$|\vec{a}_1 + \vec{a}_2|^2 = |\vec{a}_1|^2 + |\vec{a}_2|^2 + 2|\vec{a}_1\vec{a}_2|\cos\alpha,$$

$$|\vec{a}_1 - \vec{a}_2|^2 = |\vec{a}_1|^2 + |\vec{a}_2|^2 - 2|\vec{a}_1\vec{a}_2|\cos\alpha,$$

$$|\vec{a}_1 + j\vec{a}_2|^2 = |\vec{a}_1|^2 + |\vec{a}_2|^2 - 2|\vec{a}_1\vec{a}_2|\sin\alpha,$$

$$|\vec{a}_1 - j\vec{a}_2|^2 = |\vec{a}_1|^2 + |\vec{a}_2|^2 + 2|\vec{a}_1\vec{a}_2|\sin\alpha.$$

The angle α is completely determined by finding both $\cos\alpha$ and $\sin\alpha$ from the above equations:

$$\cos\alpha = \frac{|\vec{a}_1 + \vec{a}_2|^2 - |\vec{a}_1 - \vec{a}_2|^2}{4\,|\vec{a}_1\vec{a}_2|},$$

$$\sin\alpha = \frac{|\vec{a}_1 - j\vec{a}_2|^2 - |\vec{a}_1 + j\vec{a}_2|^2}{4\,|\vec{a}_1\vec{a}_2|}.$$

The amplitude of the unknown vector is determined as follows:

$$|\vec{a}_2| = \sqrt{\tfrac{1}{2}\left(|\vec{a}_1 - \vec{a}_2|^2 + |\vec{a}_1 + \vec{a}_2|^2 - 2|\vec{a}_1|^2\right)}.$$

Even though there are networks that allow the above operations to be performed, a general six-port network can be used to determine the value of any unknown reflection coefficient.

The determination of any reflection coefficient requires the use of at least a five-port network. It can be shown that for a general four-port network, as depicted in Fig. 13.54, the ratio of powers measured at ports 4 and 3 is given by

$$\frac{P_4}{P_3} = A_4 \frac{|1 - q_4\Gamma_L|^2}{|1 - q_3\Gamma_L|^2},$$

where the constants A_4, q_3, and q_4 are functions of the S-parameters of the four-port network. The reason for taking the power ratio is to make the measurements independent of the input power level. From the above equation, we could, at best, determine the magnitude of the reflection coefficient, which makes the four-port a candidate for a scalar reflectometer (see Example 13.18).

It seems that we need another power equation in order to be able to determine the complex reflection coefficient. A five-port reflectometer is shown in Fig. 13.55. We

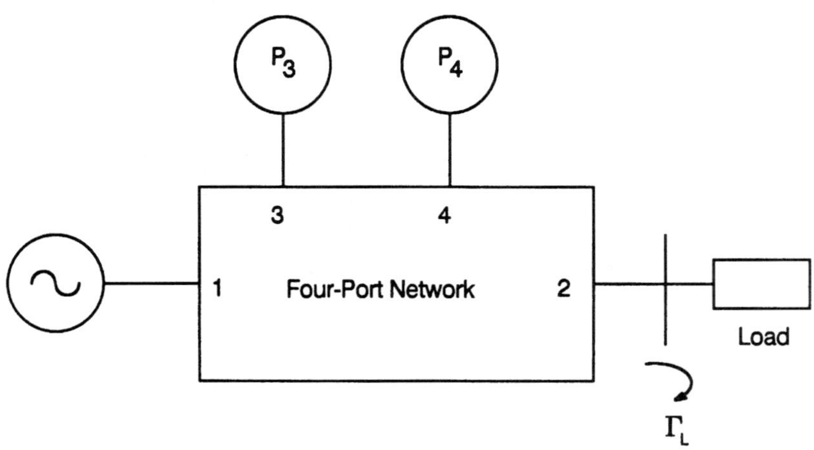

FIGURE 13.54 General four-port network with signal generator, power meters, and an unknown load.

now have two power ratios that we can manipulate:

$$\frac{P_4}{P_3} = A_4 \frac{|1 - q_4\Gamma_L|^2}{|1 - q_3\Gamma_L|^2},$$

$$\frac{P_5}{P_3} = A_5 \frac{|1 - q_5\Gamma_L|^2}{|1 - q_3\Gamma_L|^2}.$$

It can be shown that the above equations represent circles in the Γ plane. The solution for the reflection coefficient is the intersection of the two circles. But two different circles intersect at most at two points (at one point if they are tangent). Thus there are two possible solutions for Γ_L. If the loads that are tested are passive loads, the five-port reflectometer can be used to measure the reflection coefficient, by eliminating the solution outside the unit circle. Since there is a degree of uncertainty in the solution for the reflection coefficient, the need to increase the number of ports becomes evident.

The six-port reflectometer is shown in Fig. 13.56. We now have four power-measuring ports. The power ratio equations for the six-port reflectometer are given below:

$$\frac{P_4}{P_3} = A_4 \frac{|1 - q_4\Gamma_L|^2}{|1 - q_3\Gamma_L|^2},$$

$$\frac{P_5}{P_3} = A_5 \frac{|1 - q_5\Gamma_L|^2}{|1 - q_3\Gamma_L|^2},$$

$$\frac{P_6}{P_3} = A_6 \frac{|1 - q_6\Gamma_L|^2}{|1 - q_3\Gamma_L|^2}.$$

The solution for the reflection coefficient is found as the intersection of the three circles that the above equations represent in the Γ plane. Figure 13.57 shows the three

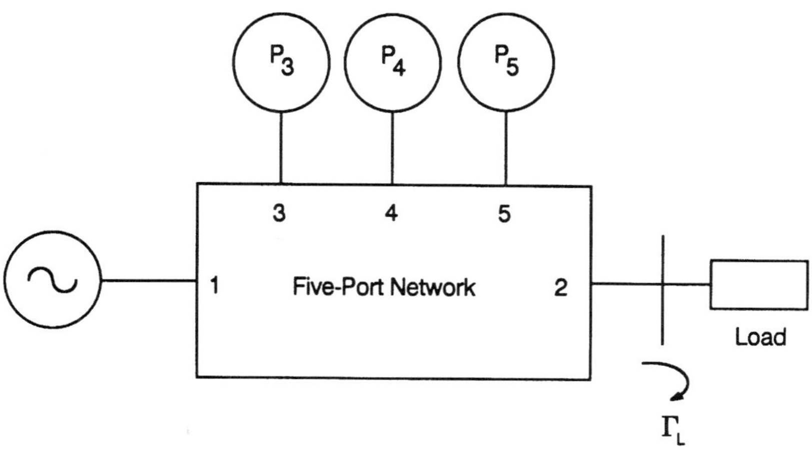

FIGURE 13.55 General five-port network with signal generator, power meters, and an unknown load.

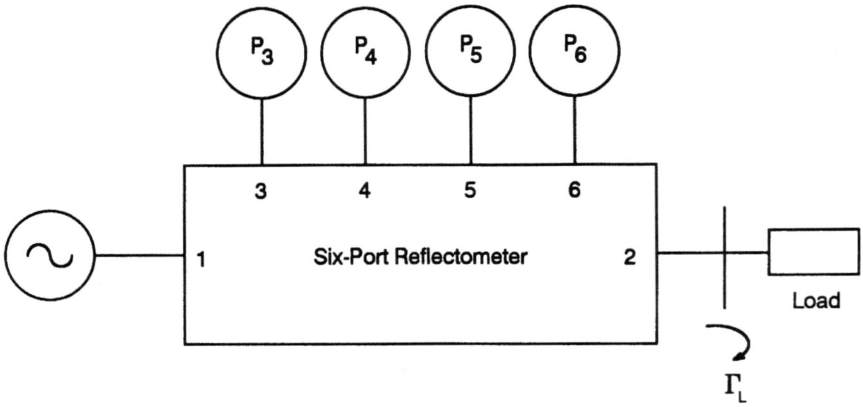

FIGURE 13.56 General six-port reflectometer with signal generator, power meters, and an unknown load.

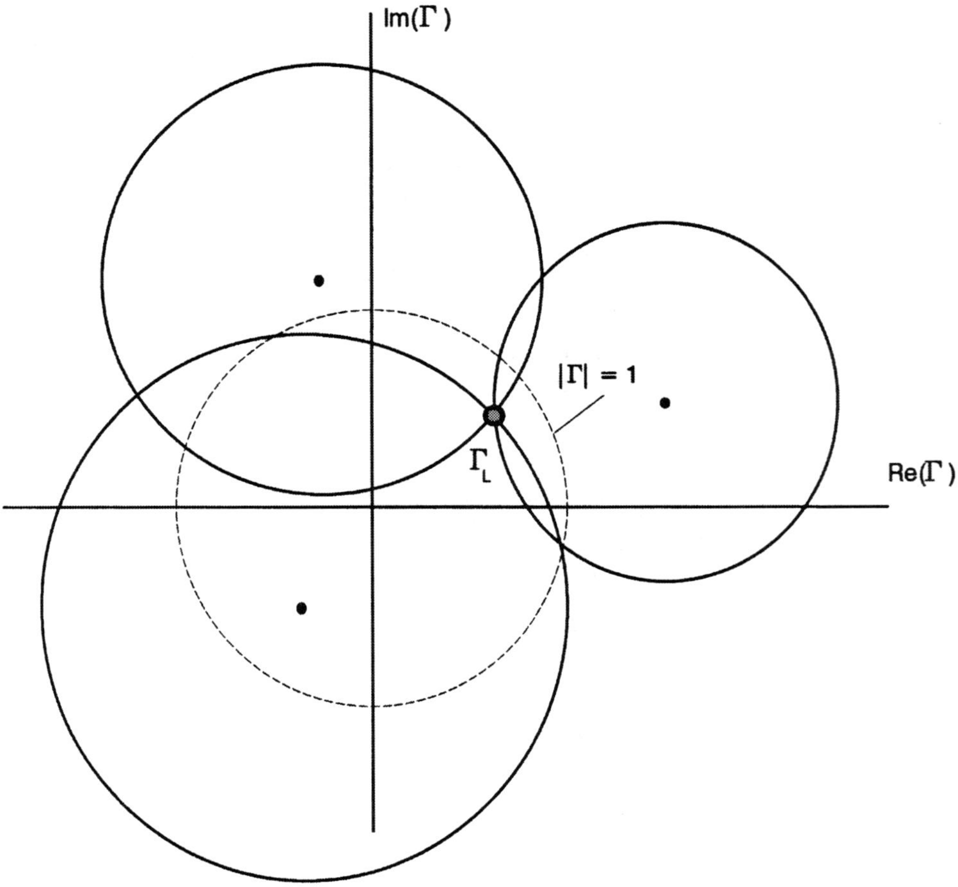

FIGURE 13.57 Three power ratio circles and their intersection, which gives the solution for the unknown reflection coefficient.

circles and their intersection, which gives the solution for the unknown reflection coefficient.

The solution for Γ_L requires knowledge of the three real constants A_4, A_5, and A_6 and of the four complex constants q_3, q_4, q_5, and q_6. Determination of all these constants is performed through a calibration procedure. The calibration is performed by connecting at least four standard terminations at the measurement port and then finding the solution to the nonlinear system thus obtained. The calibration procedure is relatively involved and is the only price paid for the gain in hardware simplicity.

In order to measure transmission coefficients, two six-port networks are used—hence the name dual six-port network analyzer.

Six-port network analyzers are very accurate and less expensive than the traditional vector network analyzers, but they require complex software for their calibration.

PROBLEMS

13.1 For the TEM mode in a coaxial line, obtain the three characteristic resistances defined by Eq. (13.5).

Ans. $R_{0,\mathrm{VI}} = R_{0,\mathrm{PV}} = R_{0,\mathrm{PI}} = (\eta/2\pi)\ln(b/a)$.

13.2 For the TE_{11} mode in a circular guide, obtain the three characteristic resistances defined by Eq. (13.5). *Hints*: Define the voltage along the vertical field line shown in Fig. 12.10,

$$\underline{V} = 2\int_{r=0}^{a} \underline{E}_r(r, \pi/2, z)\, dr.$$

Define the current as the total longitudinal conduction current in the bottom half of the guide,

$$\underline{I} = -\int_{\phi=\pi}^{2\pi} \underline{H}_\phi(a, \phi, z)a\, d\phi.$$

The time average power carried by the wave is given by Eq. (12.90). Show that the resultant model characteristic resistances are

$$R_{0,\mathrm{VI}} = \frac{F}{J_1(p')}\left(\frac{\omega\mu}{\beta_{11}}\right) = 1.38\left(\frac{\omega\mu}{\beta_{11}}\right), \quad F = J_1(p') - \int_{u=0}^{p'} J_0(u)\, du,$$

$$R_{0,\mathrm{PV}} = \frac{8F^2}{\pi J_1^2(p')[(p')^2 - 1]}\left(\frac{\omega\mu}{\beta_{11}}\right) = 2.03\left(\frac{\omega\mu}{\beta_{11}}\right),$$

$$R_{0,\mathrm{PI}} = \frac{\pi([p')^2 - 1]}{8}\left(\frac{\omega\mu}{\beta_{11}}\right) = 0.939\left(\frac{\omega\mu}{\beta_{11}}\right),$$

where $p' = p'_{11} = 1.841$ and

$$\left(\frac{\omega\mu}{\beta_{11}}\right) = \frac{\eta}{\sqrt{1-(f_{c,11}/f)^2}}, \quad f_{c,11} = \frac{p'}{2\pi\sqrt{\mu\varepsilon a}}.$$

The three forms are the same. All are independent of the guide radius, except as it appears in β, and all are proportional to the wave impedance.

13.3 The definitions of characteristic resistance given in Eq. (13.5) are not the only ones possible. For example, we may arbitrarily define $R_0 = 50\ \Omega$ and require that the power match with the actual power carried by the wave.

13.4 Solve Example 13.5 by decomposing the network into two parallel networks.

13.5 Show that the *ABCD* matrix for the three two-port networks of Table 13.2 are as given.

13.6 Obtain the *ABCD* matrix for a transmission line (R_0, β) with shunt admittances Y_1 and Y_2 (Fig. 13.58). Specialize the result to the case $Y_1 = Y_2 = jB$.

Ans. $A = \cos\beta l + jR_0 Y_2 \sin\beta l, \quad B = jR_0 \cos\beta l,$

$C = (Y_1 + Y_2)\cos\beta l + j(R_0 Y_1 Y_2 + G_0)\sin\beta l,$

$D = \cos\beta l + jR_0 Y_1 \sin\beta l.$

13.7 Express the *ABCD*-parameters in terms of the *Y*-parameters.

13.8 Express the *Z*-parameters in terms of the *ABCD*-parameters.

13.9 Express the *Y*-parameters in terms of the *ABCD*-parameters.

13.10 Construct a MathCad file that will accept as input the *Z*-, *Y*-, or *ABCD*-parameters of a network and outputs all other parameters (*Z*, *Y*, *ABCD*). Provisions must be made for networks that do not have **Z** or **Y** matrix.

13.11 Obtain the scattering matrix for the shunt admittance shown in the first line of Table 13.2. Assume the transmission lines to the left and right have characteristic resistances R_{01} and R_{02}, respectively. Use two methods: (a) directly from the defining equations for the scattering matrix elements, Eqs. (13.18) and (13.19); and (b) by converting the *ABCD* matrix as in Eq. (13.44). Verify that

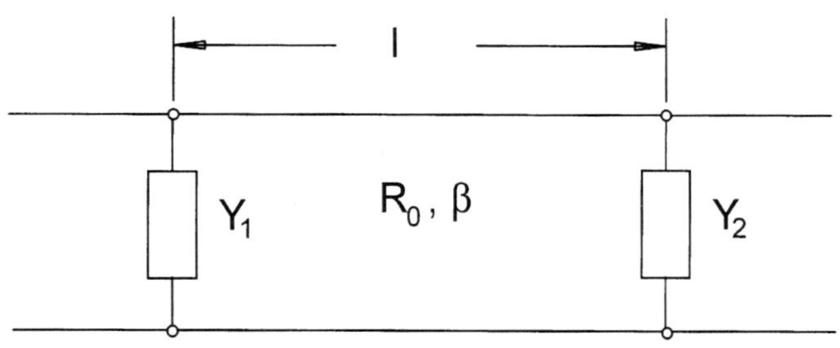

FIGURE 13.58 Problem 13.6.

the symmetry property is satisfied and that if $Y = jB$ the unitary property is also satisfied.

Ans. Den $= R_{01} + R_{02} + YR_{01}R_{02}$, S_{11}Den $= R_{02} - R_{01} - YR_{01}R_{02}$, S_{21}Den $= 2R_{02}$.

13.12 Obtain the scattering matrix for the series admittance shown in the second line of Table 13.2. Assume the transmission lines to the left and right have characteristic resistances R_{01} and R_{02}, respectively. Use two methods: (a) directly from the defining equations for the scattering matrix elements, Eqs. (13.18) and (13.19); and (b) by converting the *ABCD* matrix as in Eq. (13.44). Verify that the symmetry property is satisfied and that if $Z = jX$ the unitary property is also satisfied.

Ans. Den $= Z + R_{01} + R_{02}$, S_{11}Den $= Z + R_{01} - R_{02}$, S_{12}Den $= 2R_{01}$.

13.13 Obtain the scattering matrix for the transmission line shown in the third line of Table 13.2. Assume the transmission lines to the left and right have characteristic resistances R_{01} and R_{02}, respectively. Use two methods: (a) directly from the defining equations for the scattering matrix elements, Eqs. (13.18) and (13.19); and (b) by converting the *ABCD* matrix as in Eq. (13.44). Verify that the symmetry and unitary properties are satisfied.

Ans. Den $= R_0(R_{01} + R_{02}) + j(R_0^2 + R_{01}R_{02})\tan \beta l$,
S_{11}Den $= R_0(R_{01} - R_{02}) + j(R_0^2 - R_{01}R_{02})\tan \beta l$,
S_{21}Den $= 2R_0R_{02}/\cos \beta l$.

13.14 Find the scattering matrix for a transmission line (R_0, β) of length l when $R_{01} = R_{02} = R_0$.

Ans. $S_{11} = S_{22} = 0$, $S_{21} = S_{12} = \exp(-j\beta l)$.

13.15 Consider a lossless, passive reciprocal three-port connected to lines with the same characteristic resistance, $R_{01} = R_{02} = R_{03} \equiv R_0$. Using the symmetry and unitary properties show that the three-port cannot be completely matched. *Hint*: Assume that $S_{11} = S_{22} = S_{33} = 0$ and show that a contradiction results.

13.16 Prove that the matrix products in Eqs. (13.37), (13.41), (13.42), and (13.43) commute.

13.17 Use MathCad to verify the results in Example 13.8 by first finding the Y-parameters and then using the conversion equations between the **Y** and **S** matrices to obtain the S-parameters. *Hint*: you may have to evaluate separately the expressions involved in the conversion.

13.18 Determine the S-parameters of the power divider obtained by removing the resistor from the Wilkinson power divider. Verify that this power divider is not matched at all ports. Determine its **Z** and/or **Y** matrices.

13.19 Use the unitary property to show that for a directional coupler if power into any port is coupled into the two opposite ports, but is not coupled into the adjacent port, and if any two ports are matched, then the remaining two ports are also matched.

13.20 Determine the configuration of a scalar network analyzer that uses as centerpiece a directional coupler. The scalar network analyzer should be able to measure the magnitude of the S-parameters of any two-port network. Design the calibration procedure for the scalar analyzer, making the measurement procedure independent of the characteristics of the hardware involved.

13.21 Show that the power ratio equations in Section 13.8.2 represent circles and determine their centers and radii.

REFERENCES

[1] Roger F. Harrington, *Time-Harmonic Electromagnetic Fields*, McGraw-Hill, New York, 1961, p. 400 and preceding.

[2] R. W. Beaty and D. M. Keat, "Relationships between different kinds of network parameters, not assuming reciprocity or equality of the waveguide of transmission line characteristic impedances," *Proc. IEEE*, Jan. 1964, p. 84.

[3] Peter A. Rizzi, *Microwave Engineering: Passive Circuits*, Prentice Hall, Englewood Cliffs, NJ, 1988.

[4] L. Young and H. Sobol (Eds.), *Advances in MICROWAVES*, Vol. 8, Academic Press, New York, 1974, p. 158.

[5] N. Marcovitz, *Waveguide Handbook*, McGraw-Hill, New York, 1951, pp. 221–224.

[6] N. Marcovitz, *Waveguide Handbook*, McGraw-Hill, New York, 1951, pp. 218–221.

[7] N. Marcovitz, *Waveguide Handbook*, McGraw-Hill, New York, 1951, p. 309.

[8] Peter A. Rizzi, *Microwave Engineering: Passive Circuits*, Prentice Hall, Englewood Cliffs, NJ, 1988, pp. 358–364, 367–390.

CHAPTER FOURTEEN

Radiation and Antennas

14.1 INTRODUCTION

We have seen that fixed charges are the source of electrostatic fields and that steady currents are the source of magnetostatic fields. We have also seen that time-varying charges and currents act as sources in the wave equations, Eqs. (5.43) and (5.44). A time harmonically varying charge and the inseparable time harmonically varying current are the sources of time harmonically varying electromagnetic fields. These fields, as we have seen, propagate in space and transfer energy from one location to another.

In this chapter we will find the field of an oscillating charge dipole or current element. This is the basic radiating element. When the current distribution is known on an antenna structure, the fields can be found by resolving the current into current elements of the proper amplitude and relative phase and then integrating over the complete distribution.

We will also study some examples of how radiating elements are arranged or combined so as to focus the radiation in desired directions. If we are interested in point-to-point communication, increasing the focusing power or gain of the antenna decreases the power that must be radiated by the transmitter in order to maintain the same power at the receiver. Thus if the transmitting and receiving antennas each have their power gains increased by a factor of 10, the transmitted power can be reduced by a factor of 100, an important advantage.

14.2 FIELD OF AN OSCILLATING CURRENT ELEMENT

For static electric fields the point charge was the basic element from which we found the field due to any charge distribution. For static magnetic fields the current element played that role. For time harmonic currents we need to find the field of an oscillating current element, from which we can calculate, by addition or integration, the fields due to any known current distribution.

We will find the field of an oscillating current element in two steps. First, we will find the field due to a time harmonic current lying along the positive z axis. Second, we will superimpose a second semi-infinite line of current, so as to cancel all but a short current element at the origin. These are the same steps we took in Section 8.2.1 to find the field of the steady current element.

14.2.1 Field of a Time Harmonic Current on a Semi-infinite Wire

We begin by finding the field due to a current on a semi-infinite wire located along the positive z axis (Fig. 14.1). Looking back at Sections 4.5.3 and 4.5.4, we know that if the current is constant with time the magnetic field intensity is

$$H_\phi = \frac{I_0(1 + \cos\theta)}{4\pi R \sin\theta}.$$

For a time harmonic current along the wire the magnetic field intensity must conform to this result as $\omega \to 0$. Clearly, at any frequency, the circular symmetry of the source requires that the fields are independent of ϕ. Let us guess that at any frequency the dependence on θ is the same as at zero frequency and, therefore, that for the time harmonic current the magnetic field intensity phasor will be

$$\vec{\underline{H}} = \vec{a}_\phi \underline{H}_\phi = \vec{a}_\phi \frac{I_0(1 + \cos\theta)}{4\pi R \sin\theta} F(R), \tag{14.1}$$

where $F(R) \to 1$ as $\omega \to 0$. We will test our guess by substituting this formula in Maxwell's equations: if Maxwell's equations are satisfied, our guess was correct; otherwise, we must start over.

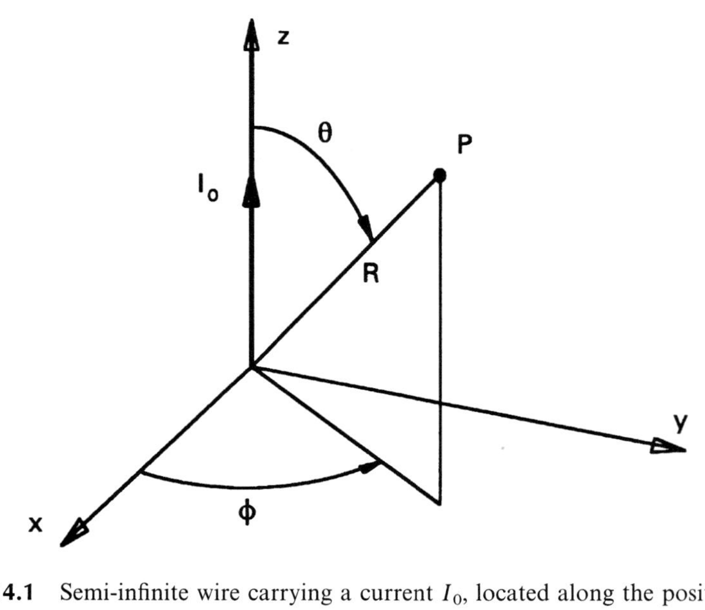

FIGURE 14.1 Semi-infinite wire carrying a current I_0, located along the positive z axis.

To find the electric field intensity, we substitute in $\vec{\underline{E}} = (\nabla \times \vec{\underline{H}})/j\omega\varepsilon_0$, giving

$$\vec{\underline{E}} = \frac{1}{j\omega\varepsilon_0}\left[\vec{a}_R \frac{1}{R\sin\theta}\frac{\partial}{\partial\theta}(\sin\theta\,\underline{H}_\phi) - \vec{a}_\theta \frac{1}{R}\frac{\partial}{\partial R}(R\underline{H}_\phi)\right] \tag{14.2}$$

$$= \frac{-I_0}{4\pi j\omega\varepsilon_0}\left[\vec{a}_R \frac{F}{R^2} + \vec{a}_\theta \frac{1+\cos\theta}{R\sin\theta}\frac{dF}{dR}\right]. \tag{14.3}$$

We now substitute this field in $\vec{\underline{H}} = -(\nabla \times \vec{\underline{E}})/j\omega\mu_0$, giving

$$\vec{\underline{H}} = -\vec{a}_\phi \frac{1}{j\omega\mu_0}\frac{1}{R}\left[\frac{\partial}{\partial R}(R\underline{E}_\theta) - \frac{\partial \underline{E}_R}{\partial\theta}\right]$$

$$= -\vec{a}_\phi\left(\frac{I_0}{\omega^2\mu_0\varepsilon_0}\right)\left(\frac{1+\cos\theta}{4\pi R\sin\theta}\right)\frac{d^2F}{dR^2}. \tag{14.4}$$

Equating this result to the originally assumed $\vec{\underline{H}}$, Eq. (14.1), shows that the original guess satisfies Maxwell's equations if $-(1/\omega^2\mu_0\varepsilon_0)(d^2F/dR^2) = F$ or

$$\frac{d^2F}{dR^2} + \omega^2\mu_0\varepsilon_0 F = 0. \tag{14.5}$$

Solutions to this equation are

$$F = K^+ e^{-j\beta R} + K^- e^{j\beta R}, \quad \beta = \omega\sqrt{\mu_0\varepsilon_0}.$$

The first term represents a spherical wave traveling radially outward from the origin. The second term represents a spherical wave traveling radially inward toward the origin. The second term might be generated if, at some distance R_0, there were a discontinuity in the medium causing a portion of the outward wave to be reflected. In our case, infinite free space, the medium is everywhere uniform and hence we choose $K^- = 0$. Taking $K^+ = 1$ to maintain the correct amplitude of the field at zero frequency and substituting F in Eqs. (14.4) and (14.3), we have

$$\underline{H}_\phi = \frac{I_0(1+\cos\theta)}{4\pi R\sin\theta} e^{-j\beta R}, \quad \beta = \omega\sqrt{\mu_0\varepsilon_0}, \tag{14.6}$$

and

$$\underline{E}_R = \frac{-I_0}{4\pi j\omega\varepsilon_0 R^2} e^{-j\beta R}, \tag{14.7}$$

$$\underline{E}_\theta = \frac{\eta I_0(1+\cos\theta)}{4\pi R\sin\theta} e^{-j\beta R} = \eta\underline{H}_\phi, \quad \eta = \sqrt{\frac{\mu_0}{\varepsilon_0}}. \tag{14.8}$$

Our original guess for $\vec{\underline{H}}$ was correct and these are the fields surrounding the current shown in Fig. 14.1.

Taking the ratio of the magnitudes of these electric field components we find

$$\left|\frac{\underline{E}_R}{\underline{E}_\theta}\right| \alpha \frac{1}{\omega\varepsilon_0\eta R} = \frac{1}{\beta R} = \frac{\lambda}{2\pi R}.$$

Thus the magnitude of the radial component is much less than the magnitude of the θ component as the distance from the origin increases to one or more wavelengths. This is characteristic of radiated spherical waves.

The phasor current in the wire is found from the magnetic field intensity

$$\underline{I}(z) = \lim_{\theta\to 0}\int_{\phi=0}^{2\pi} [\underline{H}_\phi(R, \theta)R\sin\theta]\,|_{R=z}\, d\phi = I_0 e^{-j\beta z} \tag{14.9}$$

and the physical current is

$$I(z, t) = I_0 \cos(\omega t - \beta z).$$

Note that the current does not go to zero at $z = 0$, the end of the wire. This is not physically realistic. At $\theta = 0$ the radial electric field is along the wire and impedes the current flow. To drive the current against this force there must be per unit length voltage sources $-\underline{E}_R$ distributed along the wire. These sources become increasingly large as $R = z \to 0$. This is another indication that the physical situation is unrealistic.

14.2.2 Field of a Time Harmonic Current Element Located at the Origin

As our second step, we superimpose a second equal but opposite current over the first current, so that the total current is zero except over $0 < z < d$ (Fig. 14.2). The second current is $-I_0\exp(-j\beta z) = -I_0\exp(-j\beta d)\exp(-j\beta z')$, where $z' = z - d$. From Eq. (14.6), the total magnetic field intensity is

$$\underline{H}_\phi = \frac{I_0(1+\cos\theta)}{4\pi R\sin\theta}e^{-j\beta R} - \frac{I_0 e^{-j\beta d}(1+\cos\theta')}{4\pi R'\sin\theta'}e^{-j\beta R'}. \tag{14.10}$$

For $d \ll R$ and $d \ll \lambda = 2\pi/\beta$,

$$R' = \sqrt{R^2 + d^2 - 2Rd\cos\theta} \approx R[1 - (d/R)\cos\theta], \tag{14.11}$$

$$\begin{aligned}\cos\theta' &= \frac{z-d}{R'} \approx \frac{z-d}{R}\left[1 - \left(\frac{d}{R}\right)\cos\theta\right]^{-1} \\ &\approx \left(\cos\theta - \frac{d}{R}\right)\left[1 + \left(\frac{d}{R}\right)\cos\theta\right] \approx \cos\theta - \frac{d}{R}(1 - \cos^2\theta) \\ &= \cos\theta - \frac{d}{R}\sin^2\theta,\end{aligned} \tag{14.12}$$

$$e^{-j\beta R'} \approx \exp(j\beta d\cos\theta)\exp(-j\beta R) \approx (1 + j\beta d\cos\theta)e^{-j\beta R}, \tag{14.13}$$

$$e^{-j\beta d} \approx 1 - j\beta d, \tag{14.14}$$

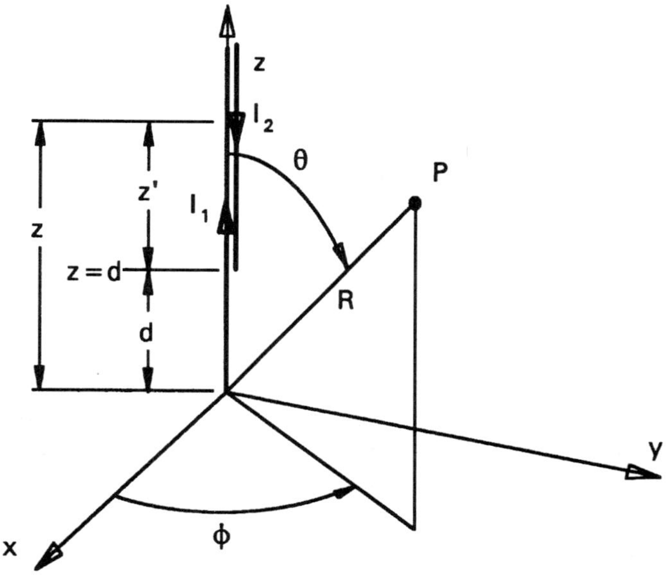

FIGURE 14.2 Current element located at the origin made of two semi-infinite wires carrying equal but opposite currents.

correct to terms of order d/R and d/λ. Noting that $R' \sin\theta' = R \sin\theta$,

$$\underline{H}_\phi \approx \frac{I_0}{4\pi R \sin\theta} [(1+\cos\theta) - (1 - j\beta d)(1+\cos\theta')(1 + j\beta d \cos\theta)]\, e^{-j\beta R}$$

$$\approx \frac{I_0}{4\pi R \sin\theta} [\cos\theta - \cos\theta' + j\beta d(1+\cos\theta')(1-\cos\theta)]\, e^{-j\beta R}.$$

In this expression we obtain $\cos\theta - \cos\theta'$ from Eq. (14.12) and we let $(1+\cos\theta') \approx (1+\cos\theta)$. This latter approximation introduces an error of order d/R, but because of the $j\beta d$ multiplier, the error is actually of second order, $d^2/\lambda R$. Replacing $1 - \cos^2\theta$ with $\sin^2\theta$, we find

$$\underline{H}_\phi = \frac{j\beta I_0 d}{4\pi R}\left(1 + \frac{1}{j\beta R}\right) \sin\theta\, e^{-j\beta R}, \quad \beta = \omega\sqrt{\mu_0 \varepsilon_0}. \tag{14.15}$$

This is the result we have been seeking.

The electric field intensity is found from Eq. (14.7):

$$\underline{E}_R = \frac{\eta I_0 d}{2\pi R^2}\left(1 + \frac{1}{j\beta R}\right) \cos\theta\, e^{-j\beta R}, \quad \eta = \sqrt{\frac{\mu_0}{\varepsilon_0}}, \tag{14.16}$$

$$\underline{E}_\theta = \frac{j\eta\beta I_0 d}{4\pi R}\left[1 + \frac{1}{j\beta R} + \frac{1}{(j\beta R)^2}\right] \sin\theta\, e^{-j\beta R}. \tag{14.17}$$

These give us the complete field of a time harmonic current element $I_0 d$ located at the origin in free space and pointing positively along the z axis.

When more than one current element is present we can find the complete field by summing the fields due to each element. When a line current is present we replace $I_0 d$ with $\underline{I}\, d\vec{l}$ and integrate over the line. When a surface current is present we replace $I_0 d$ with $\underline{\vec{J}}_s\, dS$ and integrate over the surface. When the current is distributed throughout a volume we replace $I_0 d$ with $\underline{\vec{J}}\, \mathrm{dv}$ and integrate over the volume. In principle this is easy, but in practice it may be difficult.

14.2.3 Near or Quasi-stationary Field

Let us examine the field in the region of space where $\beta R = 2\pi R/\lambda \ll 1$. We can be in this region by holding the frequency and hence the wavelength fixed and reducing the distance of the field point from the origin until the inequality is satisfied. Alternately, we can hold the distance fixed and let the frequency be reduced—and hence the wavelength is increased—until the inequality is satisfied. The first implies a near field and the second implies a quasi-stationary field, but in either case the result is the same.

In either case we have $\beta R = 2\pi R/\lambda \to 0$ and $\exp(-j\beta R) \to 1$, so that the results corresponding to Eqs. (14.15), (14.16), and (14.17), respectively, are

$$\underline{H}_\phi \to \frac{I_0 d}{4\pi R^2} \sin\theta, \tag{14.18}$$

$$\underline{E}_R \to \frac{q_0 d}{2\pi\varepsilon R^3} \cos\theta, \quad q_0 = \frac{I_0}{j\omega}, \tag{14.19}$$

$$\underline{E}_\theta \to \frac{q_0 d}{4\pi\varepsilon R^3} \sin\theta. \tag{14.20}$$

The magnetic field intensity approaches that of a steady current element of moment $I_0 d$ located at the origin along the z axis, Eq. (8.8). The equation of continuity requires that current I_0 at the upper end of the element of the length d give rise to an accumulation of charge $q_0 = I_0/j\omega$. Similarly, at the lower end of the element the charge $-q_0$ accumulates. These point charges form a dipole of moment $q_0 d$ located at the origin along the z axis and therefore the electric field intensity found here is the same as that of Eq. (6.33).

14.2.4 Far or Radiation Field

We now examine the field in the region of space where $\beta R = 2\pi R/\lambda \gg 1$. Again, this can imply sufficiently increased distance at fixed frequency or sufficiently increased frequency (and reduced wavelength) at fixed distance. The field in this region of space is known as the far or radiation field. Neglecting terms in $1/\beta R$ or $1/(\beta R)^2$, Eqs. (14.15),

(14.16), and (14.17) become

$$\underline{H}_\phi = \frac{j\beta I_0 d}{4\pi R} \sin\theta \, e^{-j\beta R},$$

$$\underline{E}_R = \frac{\eta I_0 d}{2\pi R^2} \cos\theta \, e^{-j\beta R}, \quad \eta = \sqrt{\frac{\mu_0}{\varepsilon_0}},$$

$$\underline{E}_\theta = \frac{j\eta\beta I_0 d}{4\pi R} \sin\theta \, e^{-j\beta R}.$$

The ratio of the magnitudes of the radial and θ components of the electric field intensity is

$$\left|\frac{\underline{E}_R}{\underline{E}_\theta}\right| = \frac{2}{\beta R} \cot\theta.$$

Thus, in the far field, the field is substantially in the θ direction, except near the axis of the current element where $\cot\theta$ is very large.

In summary, the far field of the current element is given by

$$\underline{H}_\phi = \frac{j\beta I_0 d}{4\pi R} \sin\theta \, e^{-j\beta R}, \quad \underline{E}_\theta = \eta \underline{H}_\phi. \tag{14.21}$$

This is a transverse electromagnetic (TEM) spherical wave propagating radially outward from the origin. The electric and magnetic field intensities are in time phase. Both field intensities are proportional to $\sin\theta$ and hence are maximum in the equatorial plane ($\theta = \pi/2$) and go to zero along the axis of the element ($\theta = 0$).

14.2.5 Time Average Power Density and Radiated Power

The time average power density in the far field is

$$\vec{p}_{f,\text{Avg}} = \tfrac{1}{2}\operatorname{Re}(\underline{\vec{E}} \times \underline{\vec{H}}^*) = \vec{a}_R \tfrac{1}{2}\operatorname{Re}(\underline{E}_\theta \underline{H}_\phi^*)$$

$$= \vec{a}_R \tfrac{1}{2}\eta\left(\frac{\beta d}{4\pi R}\right)^2 I_0 I_0^* \sin^2\theta. \tag{14.22}$$

The power density is focused or directed so that it is greatest in the equatorial plane and goes to zero along the element's axis. Figure 14.3 shows the power density in any plane that includes the z axis, normalized to unity at $\theta = \pi/2$.

We have assumed that the medium in which the wave is propagating is lossless. As a consequence, the total time average power flow outward through any closed surface surrounding the current element $P_{f,\text{Avg}}$ must be the same, independent of the surface's size or shape. We can therefore evaluate $P_{f,\text{Avg}}$ most easily by integrating over a sphere

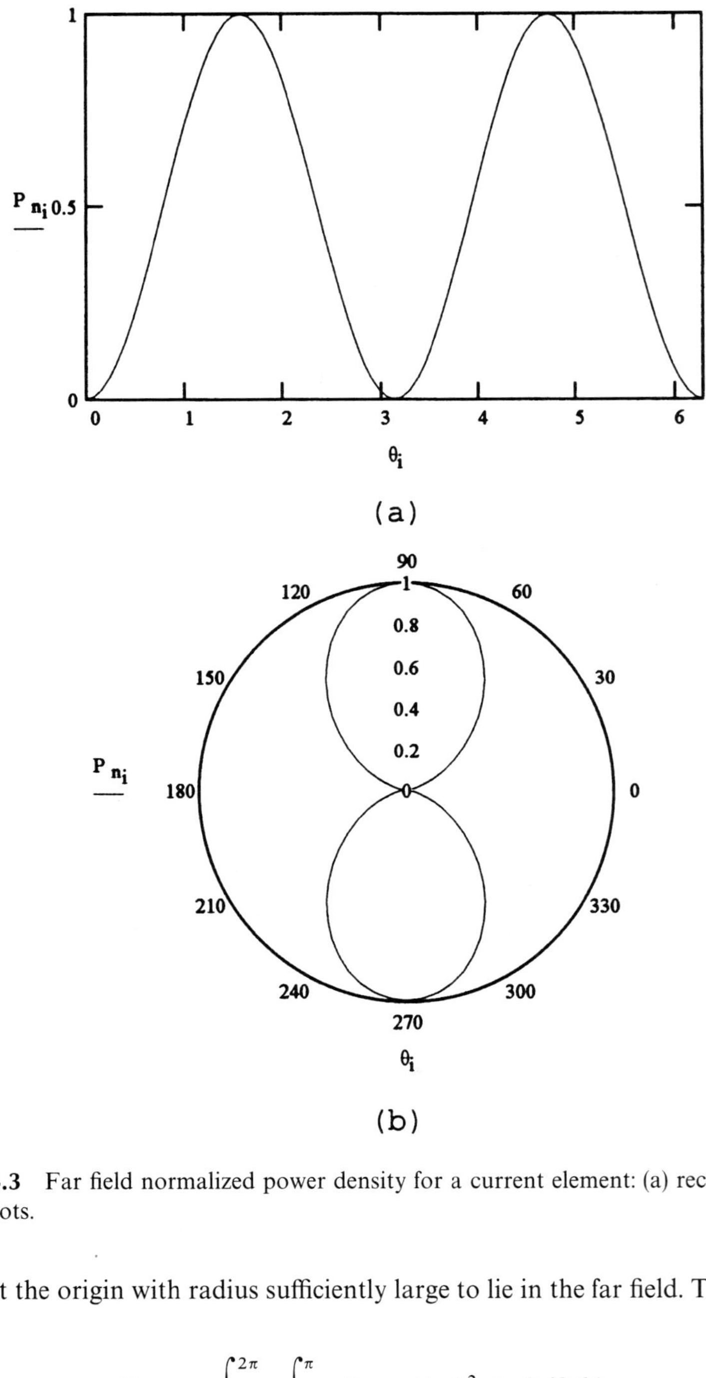

FIGURE 14.3 Far field normalized power density for a current element: (a) rectangular and (b) polar plots.

centered at the origin with radius sufficiently large to lie in the far field. The result is

$$P_{f,\text{Avg}} = \int_{\phi=0}^{2\pi} \int_{\theta=0}^{\pi} \vec{p}_{f,\text{Avg}} \cdot \vec{a}_R R^2 \sin\theta \, d\theta \, d\phi$$

$$= \tfrac{1}{2}\eta \left(\frac{\beta d}{4\pi}\right)^2 I_0 I_0^*(2\pi) \int_0^{\pi} \sin^3\theta \, d\theta$$

$$= \frac{\eta}{12\pi}(\beta d)^2 I_0 I_0^* . \tag{14.23}$$

This is the time average radiated power.

Let us find the radiated power by a different method. In the vicinity of the current element we can expand the fields in a power series in βR. Thus

$$e^{-j\beta R} = 1 - j\beta R - \tfrac{1}{2}(\beta R)^2 + j\tfrac{1}{6}(\beta R)^3 + \tfrac{1}{24}(\beta R)^4 + \cdots$$

and Eqs. (14.15), (14.16), and (14.17) become

$$\underline{H}_\phi = \frac{I_0 d}{4\pi R^2}[1 + \tfrac{1}{2}(\beta R)^2 - j\tfrac{1}{3}(\beta R)^3 + \cdots]\sin\theta,$$

$$\underline{E}_R = \frac{\eta I_0 d}{j2\pi\beta R^3}[1 + \tfrac{1}{2}(\beta R)^2 - j\tfrac{1}{3}(\beta R)^3 - \tfrac{1}{8}(\beta R)^4 + \cdots]\cos\theta,$$

$$\underline{E}_\theta = \frac{\eta I_0 d}{j4\pi\beta R^3}[1 - \tfrac{1}{2}(\beta R)^2 + j\tfrac{2}{3}(\beta R)^3 + \tfrac{3}{8}(\beta R)^4 + \cdots]\sin\theta.$$

The first terms are the near or quasi-static fields, Eqs. (14.18), (14.19), and (14.20). The electric field in the z direction is $\underline{E}_z = \underline{E}_R\cos\theta - \underline{E}_\theta\sin\theta$, giving

$$\begin{aligned}\underline{E}_z = \frac{\eta I_0 d}{j4\pi\beta R^3}&[(2\cos^2\theta - \sin^2\theta) + (\beta R)^2(1 - \tfrac{1}{2}\sin^2\theta)\\ &- j\tfrac{2}{3}(\beta R)^3 - \tfrac{1}{8}(\beta R)^4(1 + \sin^2\theta) + \cdots].\end{aligned} \tag{14.24}$$

To drive the current in the element requires a voltage $\underline{V} = -\underline{E}_z(R,\theta)d|_{R\to 0}$. The time average power supplied by this voltage source is

$$P_{\text{Avg}} = \tfrac{1}{2}\operatorname{Re}(\underline{V}I_0^*) = \tfrac{1}{2}\operatorname{Re}(-\underline{E}_z I_0^* d)|_{R\to 0}.$$

Examining $\underline{E}_z$, Eq. (14.24), we note that there is a j in the denominator factor outside the bracket so that the first and second terms of the series are 90° out of phase with the current and do not contribute any time average power. This is also true for the fourth term, but, more importantly, this term and all higher terms in the series approach zero as $R \to 0$ and hence they do not contribute any time average power. The remaining third term gives

$$\begin{aligned}P_{\text{Avg}} &= \tfrac{1}{2}\left(\frac{\eta I_0 d}{j4\pi\beta R^3}\right)[j\tfrac{2}{3}(\beta R)^3](I_0^* d)\\ &= \frac{\eta}{12\pi}(\beta d)^2 I_0 I_0^*,\end{aligned}$$

which is the same result as that of Eq. (14.23).

It was noted that the first and second terms of $\underline{E}_z$ do not contribute time average power. They do indicate reactive power and, since they are inversely proportional to R and R^2, they imply infinite reactive power. The first term corresponds to the infinite electric energy stored in the field of the electric dipole and the second to the infinite magnetic energy stored in the field of the current element. Clearly, one cannot physically build and drive an ideal current element.

14.3 ANTENNA PARAMETERS

Various parameters are used to describe the radiation from an antenna. This section gives the definitions of the most important parameters and applies them to the current element. In subsequent sections they will be used to describe different antennas.

14.3.1 Antenna Patterns

Antenna field patterns are plots of the relative electric field strength in the far field region. Ideally, the antenna field pattern should be a three-dimensional plot of the strength of the electric field intensity, normalized to the peak value, as a function of the θ and ϕ coordinates. To avoid drawing three-dimensional diagrams the patterns are shown as planar sections of the three-dimensional figure. Most commonly, E-plane and H-plane patterns, called the principal plane patterns, are shown. The E-plane pattern gives the normalized strength of the electric field in the plane that contains the radius vector from the antenna to the point of maximum electric field and the electric field intensity vector. The H-plane pattern gives the normalized strength of the electric field in the plane that contains the radius vector from the antenna to the point of maximum electric field and the magnetic intensity vector. The two planes are normal to each other.

For the current element, with far fields given by Eq. (14.21), the peak value of the electric field intensity occurs at $\theta = \pi/2$ for any value of ϕ. The E plane is thus the $\phi = C$ plane and the electric field strength varies as $\sin\theta$ in this plane (Fig. 14.4a). The H plane is the $\theta = \pi/2$ plane in which the electric field strength is constant (independent of ϕ) (Fig. 14.4b). The three-dimensional field pattern (Fig. 14.4c) is the surface of revolution generated by rotating the pattern of Fig. 14.4a about the z axis.

The antenna radiation or power (density) pattern gives the radiated time average power, normalized with respect to its peak value. Again, these plots are usually given in the principal planes defined above.

The power density for the current element is given by Eq. (14.22). The E-plane power pattern drawn in any $\phi = C$ plane is a plot of $\sin^2\theta$ (Fig. 14.4d). The H-plane power pattern drawn in the $\theta = \pi/2$ plane is again a unit circle.

The half-power beam width is usually given in each principal plane. It is the angle between the points on the field pattern in which the electric field intensity is reduced by $1/\sqrt{2}$ from its peak value or the angle between points on the power pattern in which the power density is reduced by one-half of the peak value. These are also called the 3 dB widths. The smaller the half-power beam width the more focused or directive is the antenna pattern.

For the current element, in the E plane the normalized field intensity falls by $1/\sqrt{2}$ (and the power density falls by 1/2) at $\theta = 45^\circ$ and 135°. The E plane half-power beam

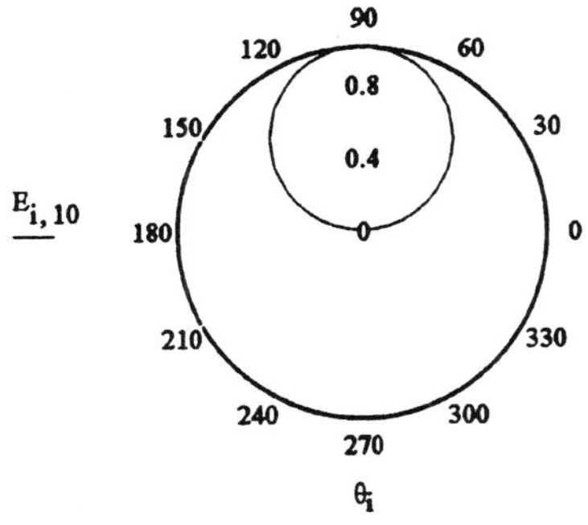

(a) E-plane field pattern

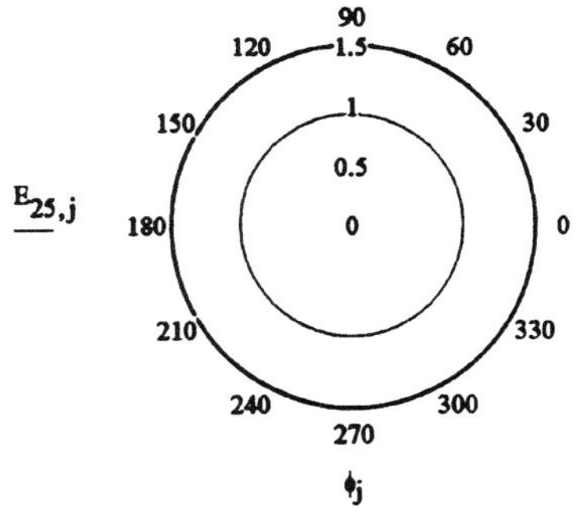

(b) H-plane field pattern

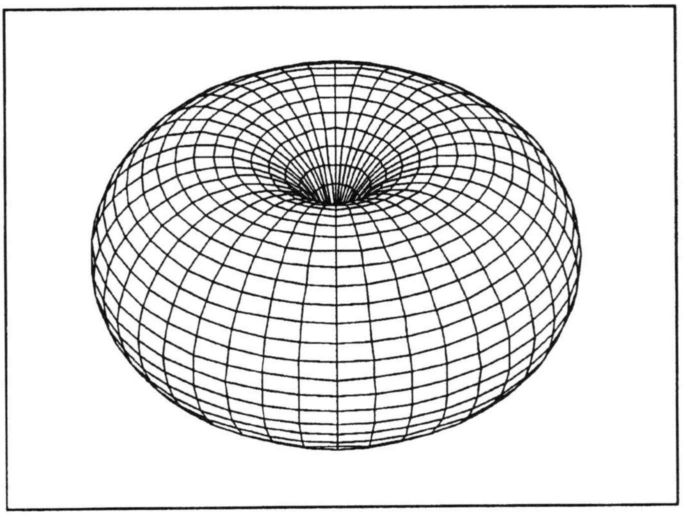

X,Y,Z

(c) Three-dimensional field pattern

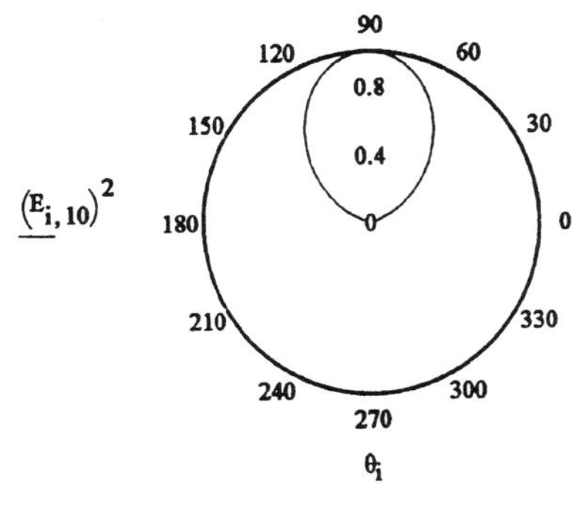

(d) E-plane power pattern

FIGURE 14.4 Field and power patterns for a current element.

width is therefore $135^\circ - 45^\circ = 90^\circ$. The H plane does not show a beam width. The pattern is moderately focused in the E plane and unfocused or omnidirectional in the H plane.

14.3.2 Directivity

For any antenna the time average power density in the far field $\vec{p}_{f,\mathrm{Avg}}$ will be in the radial direction, will be a function of the θ and ϕ coordinates, and will be inversely proportional to R^2. The total time average power radiated through any closed surface surrounding the antenna is $P_{f,\mathrm{Avg}}$. Since the medium is lossless, this can be found most easily by integrating the power density over a spherical surface of radius R, with surface area $S = 4\pi R^2$. The average power density radiated through the sphere is therefore $P_{f,\mathrm{Avg}}/S$. We define the directional directivity $D(\theta, \phi)$ as the power density radiated in any direction divided by the average power density:

$$D(\theta, \phi) = \frac{\vec{p}_{f,\mathrm{Avg}} \cdot \vec{a}_R}{P_{f,\mathrm{Avg}}/S} = \frac{4\pi R^2 \vec{p}_{f,\mathrm{Avg}} \cdot \vec{a}_R}{P_{f,\mathrm{Avg}}}. \tag{14.25}$$

The maximum directional directivity or, simply, the directivity D is the maximum value of the directional directivity:

$$D = D(\theta, \phi)_{\max} = \frac{(\vec{p}_{f,\mathrm{Avg}} \cdot \vec{a}_R)_{\max}}{P_{f,\mathrm{Avg}}/S} = \frac{4\pi R^2 (\vec{p}_{f,\mathrm{Avg}} \cdot \vec{a}_R)_{\max}}{P_{f,\mathrm{Avg}}}. \tag{14.26}$$

For the current element we have already found $\vec{p}_{f,\mathrm{Avg}}$ and $P_{f,\mathrm{Avg}}$, Eqs. (14.22) and (14.23), so

$$D(\theta, \phi) = \frac{4\pi R^2 [(\eta/2)(\beta d/4\pi R)^2 I_0 I_0^* \sin^2\theta]}{(\eta/12\pi)(\beta d)^2 I_0 I_0^*} = \tfrac{3}{2}\sin^2\theta. \tag{14.27}$$

The maximum value is at $\theta = \pi/2$ and thus $D = \frac{3}{2}$.

Calculation of the directivity frequently involves integrations that are either difficult or impossible to perform analytically. We can make an estimate of the directivity from the half-power angles around the peak power density. Draw two orthogonal planes through the beam with their line of intersection lying along the direction of the peak power density (Fig. 14.5). Letting Θ_{HP} be the half-power angle in one plane and Φ_{HP} be the half-power angle in the other plane, we can estimate the total power through a sphere of radius R as $P_{f,\mathrm{Avg}} \approx (p_{f,\mathrm{Avg}})_{\max} R^2 \Theta_{\mathrm{HP}} \Phi_{\mathrm{HP}}$. Substituting this estimate in the denominator of Eq. (14.26) gives

$$D \approx \frac{4\pi}{\Theta_{\mathrm{HP}}\Phi_{\mathrm{HP}}} = \frac{41{,}253}{\Theta^\circ_{\mathrm{HP}}\Phi^\circ_{\mathrm{HP}}}. \tag{14.28}$$

If there are two equal beams in the pattern, a factor of 2 occurs in the denominator. For the current element, one half-power angle is 90° and the other is 360°, so we would estimate $D \approx 1.27$ compared with the exact value of 1.5 found above.

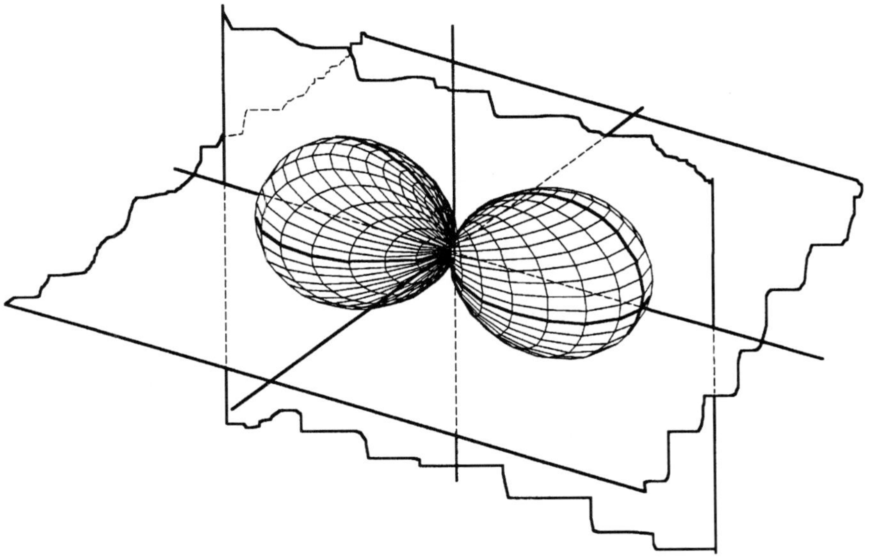

FIGURE 14.5 Power pattern with two orthogonal planes through the beam.

14.3.3 Radiation Resistance

Assuming no losses, the time average power delivered to the antenna must equal the time average power radiated by the antenna, $P_{f,\text{Avg}}$. The time average power delivered to the antenna can be written as $P = I_0 I_0^* R_r/2$, where I_0 is the amplitude of the current at the antenna terminals and R_r is called the radiation resistance. Thus we have

$$R_r = \frac{2P_{f,\text{Avg}}}{I_0 I_0^*}. \tag{14.29}$$

The radiation resistance is the resistive part of the impedance seen looking into the lossless antenna.

For the current element, using Eq. (14.23),

$$R_r = \frac{\eta}{6\pi}(\beta d)^2.$$

14.3.4 Input Impedance and Efficiency

The input impedance at the terminals of the antenna has three parts. The first is the radiation resistance R_r, which represents the power radiated from the antenna. The second is the loss resistance R_{loss}, which represents the heating losses within the antenna. The third is the reactive term X_a, which represents the magnetic and electric energy stored in the vicinity of the antenna. Hence

$$Z_{\text{in}} = R_r + R_{\text{loss}} + jX_a.$$

The total time average power input to the antenna is the heat loss power $I_0 I_0^* R_{\text{loss}}/2$ plus the power radiated $I_0 I_0^* R_r/2$. The antenna efficiency is the radiated power divided by the total power

$$\text{eff} = \frac{R_r}{R_{\text{loss}} + R_r}. \tag{14.30}$$

For the current element we have already noted that the magnetic and electric energy stored in the vicinity of the element is infinite and hence we cannot evaluate the reactive part of the input impedance. Unless we attribute a finite length and wire radius to the current element it is also impossible to evaluate the heat loss. We will therefore defer additional discussion of efficiency until we consider the short thin linear antenna (Section 14.4.3).

14.4 THIN LINEAR ANTENNAS

Let us examine the radiation fields from a thin wire fed at its center by a time harmonic voltage source, which is very small in size (Fig. 14.6). We will consider that the wire is so thin that it can be modeled as a line current along the part of the z axis covered by the wire. This idealization of the thin wire will allow us to calculate the radiated fields far from the wire but will not yield good results for the fields close to a wire of finite diameter.

14.4.1 Filamentary Current

Our first task is to find the line current $\underline{I}(z)$ along the wire. There are two basic ground rules. First, the current must go to zero at the ends of the wire, $\underline{I}(L-) = \underline{I}(-L+) = 0$, so that no point charge $\underline{I}(L-)/j\omega$ or $\underline{I}(-L+)/j\omega$ will accumulate at either end of the wire.† Second, the current through the voltage source at the center of the wire must be continuous, $\underline{I}(0+) = \underline{I}(0-)$.

First consider the wire along the positive z axis, from $z = 0$ to $z = L$. We have already obtained the exact field of a semi-infinite current filament along the positive z axis (Fig. 14.1) and found that the current phasor is a progressive wave, $I_0 \exp(-j\beta z)$, Eq. (14.9). If we add a second filamentary current $-I_0 \exp(-j\beta z)$ along the positive z axis starting at $z = L$ (Fig. 14.7a), then the net current is zero for $z > L$. The magnetic field intensity can be found by summing the field, Eq. (14.6), with the same formula, but with I_0, θ and R replaced by $-I_0 \exp(-j\beta L)$, θ_1, and R_1, respectively. Unfortunately, these two currents leave a current $\underline{I}(L-) = I_0 \exp(-j\beta L)$ at the upper end of the wire, violating the condition that $\underline{I}(L-) = 0$. To eliminate this we add a third filamentary current $-I_0 \exp(-j\beta L)\exp[j\beta(z-L)]$ along the z axis from $z = L$ to $z = -\infty$ (Fig. 14.7b). The total magnetic field intensity is now the previous sum plus an additional field, Eq. (14.6), with I_0, θ, and R replaced by $-I_0 \exp(-j\beta L)$, $\pi - \theta_1$, and R_1, respectively. Finally, add a fourth filamentary

† The "+" or "−" after the L implies the limit as $\delta \to 0$ of $L \pm \delta$. This notation continues in what follows.

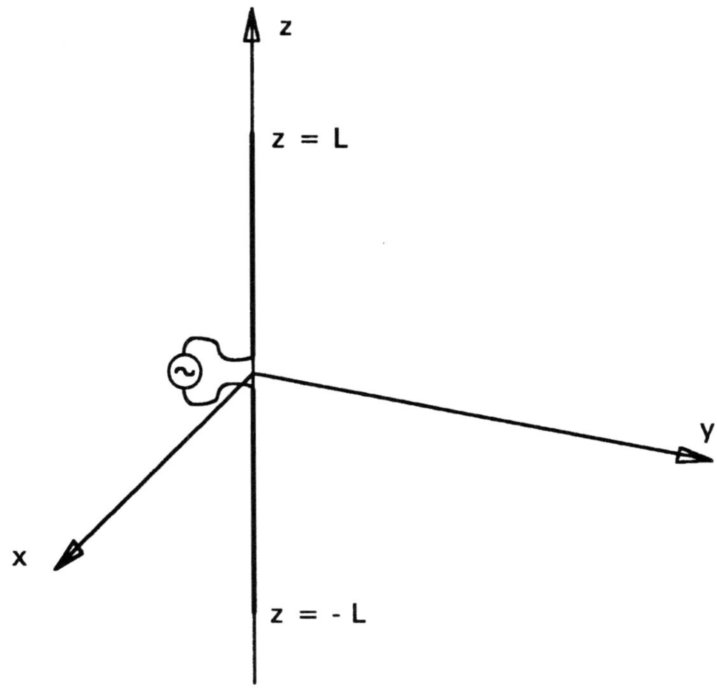

FIGURE 14.6 Thin wire fed at its center by a time harmonic source.

current $I_0 \exp(-j\beta L)\exp[j\beta(z - L)]$ along the negative z axis (Fig. 14.7c), so that the net current there is zero. The final total magnetic field intensity is the previous sum plus an additional field, Eq. (14.6), with I_0 and θ replaced by $I_0 \exp(-j2\beta L)$ and $\pi - \theta$, respectively. The sum of the four filamentary currents shown in Fig. 14.7c is the correct current along the positive z axis.

Adding the two filamentary currents that flow in the region $0 < z < L$ we obtain the current in the top wire,

$$\underline{I}(z) = I_0 e^{-j\beta z} - I_0 e^{-j\beta L} e^{j\beta(z-L)} = I_0 e^{-j\beta L}[e^{j\beta(L-z)} - e^{-j\beta(L-z)}]$$
$$= I_m \sin\beta(L - z), \quad 0 < z < L, \tag{14.31}$$

where $2jI_0 \exp(-j\beta L)$ has been replaced by I_m. Not surprisingly, the current on the "open circuited" upper wire is a standing wave with a node at $z = L$. The current at the voltage source is $\underline{I}(0) = I_m \sin\beta L$.

The current in the lower wire must also be a standing wave, but with a node at $z = -L$, and the current at the voltage source must be continuous. Hence

$$\underline{I}(z) = I_m \sin\beta(L + z), \quad -L < z < 0. \tag{14.32}$$

Equations (14.31) and (14.32) give the current along the filamentary wire driven by a voltage source at the center.

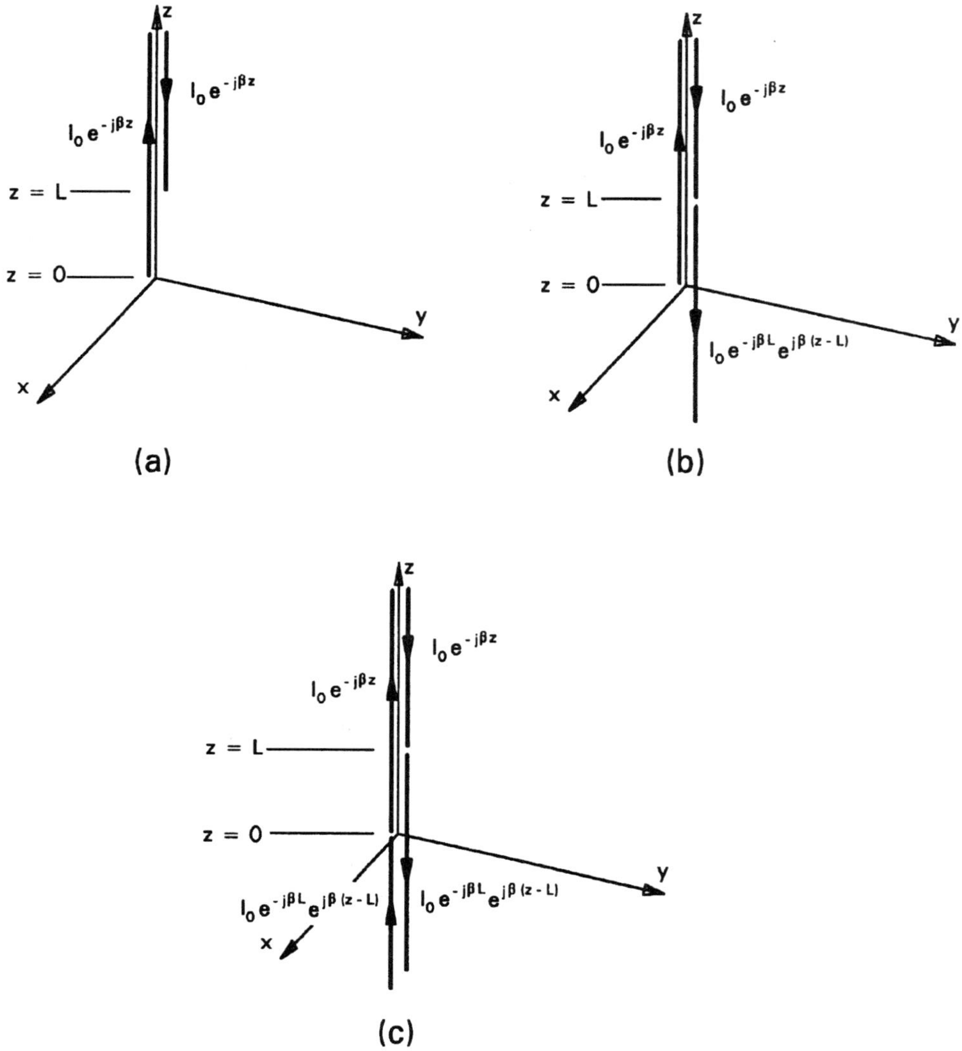

FIGURE 14.7 Construction of correct current distribution for the top wire in Fig. 14.6.

14.4.2 Far Field—Power Density and Power

The far field due to this current can be obtained from Eq. (14.21) with the help of Fig. 14.8. The far field due to the current element $\underline{I}(z')\,dz'$ is

$$d\underline{H}_\phi = \frac{j\beta \underline{I}(z')\,dz'}{4\pi R} \sin\theta \, e^{-j\beta(R - z'\cos\theta)}$$

$$= \frac{j\beta}{4\pi R} \sin\theta \, [\underline{I}(z') e^{j\beta z'\cos\theta} dz'] \, e^{-j\beta R}. \qquad (14.33)$$

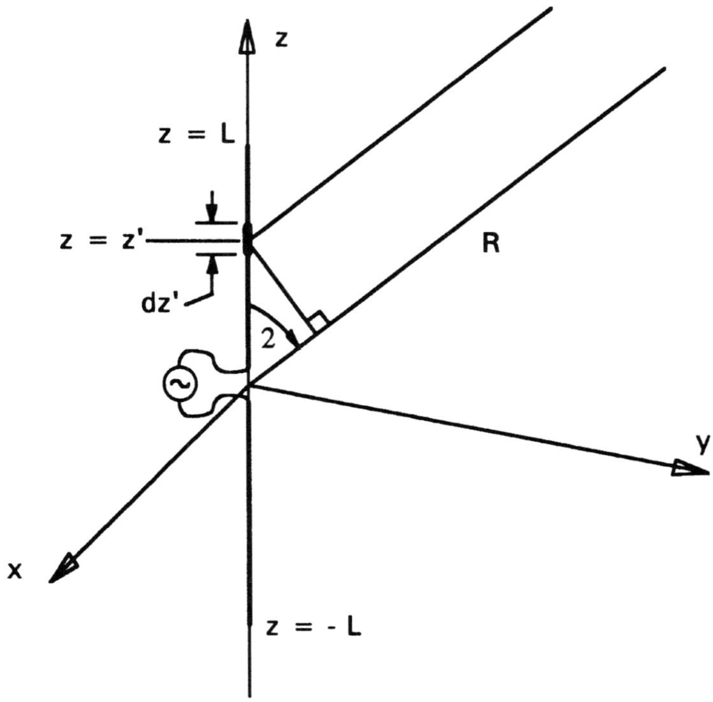

FIGURE 14.8 Thin wire center-fed antenna with current element along the wire used to determine its far fields.

The total field is found by substituting for $\underline{I}(z')$ from Eqs. (14.31) and (14.32) and integrating over the wire,

$$\underline{H}_\phi = \frac{j\beta I_m}{4\pi R}\sin\theta\left[\int_{z'=0}^{L}\sin\beta(L-z')\,e^{j\beta z'\cos\theta}\,dz' + \int_{z'=-L}^{0}\sin\beta(L+z')\,e^{j\beta z'\cos\theta}\,dz'\right]e^{-j\beta R}.$$

The integration is tedious, but not difficult. In the first integral let $L - z' = u$ and in the second let $L + z' = u$. The bracketed integral becomes

$$[\cdots] = e^{j\beta L\cos\theta}\int_0^L e^{-j\beta u\cos\theta}\sin\beta u\,du + e^{-j\beta L\cos\theta}\int_0^L e^{j\beta u\cos\theta}\sin\beta u\,du.$$

Both integrals are in the standard form

$$\int e^{au}\sin bu\,du = \frac{e^{au}}{a^2+b^2}(a\sin bu - b\cos bu),$$

with $a = \mp j\beta\cos\theta$ in the first and second integrals, respectively, and $b = \beta$ in both. In either case $a^2 + b^2 = \beta^2\sin^2\theta$. Performing the integration, substituting the limits, and canceling and combining terms gives

$$[\cdots] = \frac{2}{\beta\sin^2\theta}[\cos(\beta L\cos\theta) - \cos\beta L]$$

and therefore

$$\underline{H}_\phi = \frac{jI_m}{2\pi R}\left[\frac{\cos(\beta L\cos\theta) - \cos\beta L}{\sin\theta}\right]e^{-j\beta R}, \quad \underline{E}_{c\theta} = \eta\underline{H}_\phi. \tag{14.34}$$

These are the magnetic and electric field intensities in the far field $R \gg L$, $R \gg \lambda$.

The time average power density is

$$\begin{aligned}\vec{p}_{f,\text{Avg}}(\theta, \phi) &= \vec{a}_R \tfrac{1}{2}\,\text{Re}\,(\underline{E}_\theta\underline{H}_\phi^*)\\ &= \vec{a}_R\frac{\eta I_m I_m^*}{8\pi R^2}\left[\frac{\cos(\beta L\cos\theta) - \cos\beta L}{\sin\theta}\right]^2\end{aligned} \tag{14.35}$$

and the total time average power is

$$\begin{aligned}P_{f,\text{Avg}} &= \int_{\phi=0}^{2\pi}\int_{\theta=0}^{\pi}\vec{p}_{f,\text{Avg}}\cdot\vec{a}_R\,R^2\sin\theta\,d\theta\,d\phi\\ &= \frac{\eta I_m I_m^*}{4}\int_0^\pi\frac{[\cos(\beta L\cos\theta) - \cos\beta L]^2}{\sin\theta}\,d\theta.\end{aligned} \tag{14.36}$$

We will not perform this last integral in this general form [1] but will look at some special cases.

14.4.3 Short Antenna

When the antenna is short, $\beta L \ll 1$ or $L < 0.1\lambda$, then the sinusoidal current distribution along the wire, Eqs. (14.31) and (14.32), is approximately linear,

$$\underline{I}(z) = I_m\sin\beta(L - |z|) \approx I_m\beta(L - |z|) = I_0(I - |z|/L),$$

where $I_0 = I_m\beta L$ is the input current at $z = 0$. When substituting this in Eq. (14.33), we note that $\beta z'\cos\theta \le \beta L \ll 1$ and hence $\exp(j\beta L\cos\theta) \approx 1$. This means that waves arriving at a distant point from any point along the short antenna are in phase. The integration of $d\underline{H}_\phi$ over the length becomes a simple integration of $\underline{I}(z')$,

$$\begin{aligned}\underline{H}_\phi &= \frac{j\beta}{4\pi R}\sin\theta\left[\int_0^L I_0\left(1 - \frac{z'}{L}\right)dz' + \int_{-L}^0 I_0\left(1 + \frac{z'}{L}\right)dz'\right]e^{-j\beta R}\\ &= \frac{J\beta I_0 L}{4\pi R}\sin\theta\,e^{-j\beta R}, \quad \underline{E}_\phi = \eta\underline{H}_\phi.\end{aligned} \tag{14.37}$$

Comparing this result with Eq. (14.23), we see that the short antenna of total length $2L$ has the same far field as the current element of length d. The time average radiated power is

$$P_{f,\text{Avg}} = \frac{\eta}{12\pi}(\beta L)^2 I_0 I_0^*.$$

The antenna patterns and directivity are the same as those for the current element. The radiation resistance is $R_r = (\eta/6\pi)(\beta L)^2$. The antenna capacitance can be given approximately by [2]

$$C_{\text{ant}} = \frac{\pi\varepsilon L}{\ln(2L/a) - 1 - \ln 2}$$

and the antenna inductance is approximately [3]

$$L_{\text{ant}} = \frac{\mu L}{3\pi}\left(\ln\frac{2L}{a} - \frac{11}{6}\right).$$

The loss resistance can be estimated by finding the ohmic power loss. If R_w is the wire resistance per unit length, then the power loss is

$$\begin{aligned} \mathrm{P}_L &= \tfrac{1}{2} R_w \int_{-L}^{L} |I(z)|^2\, dz = R_w I_0 I_0^* \int_0^L \left(1 - \frac{z}{L}\right)^2 dz \\ &= \tfrac{1}{3} L R_w I_0 I_0^* \end{aligned}$$

This power loss must equal $\frac{1}{2} R_{\text{loss}} I_0 I_0^*$ and therefore $R_{\text{loss}} = (\frac{2}{3}L)R_w$. For high frequency the wire resistance is the surface resistance divided by the circumference, $R_w = R_z/2\pi a$, giving

$$R_{\text{loss}} = \frac{L}{3\pi a} R_s, \quad R_s = \sqrt{\frac{\omega\mu}{2\sigma}}.$$

Figure 14.9 shows the equivalent circuit of the antenna.

Example 14.1 Evaluate the radiation resistance, the loss resistance, the efficiency, and the input impedance for an antenna with $L = 1$ m and $a = 2$ mm, at (a) $f = 30$ MHz and (b) $f = 3$ MHz. The antenna is fabricated of copper ($R_s = 2.61 \times 10^{-7}\sqrt{f}$) and is operating in free space.

Solution (a) At 30 MHz the wavelength is $3 \times 10^8/30 \times 10^6 = 10$ m. The ratio $2L/a = 1000$. The surface resistance is $R_s = 1.43 \times 10^{-3}\,\Omega$ and

$$R_{\text{loss}} = \frac{1 \times 1.43 \times 10^{-3}}{3\pi \times 2 \times 10^{-3}} = 0.076\,\Omega.$$

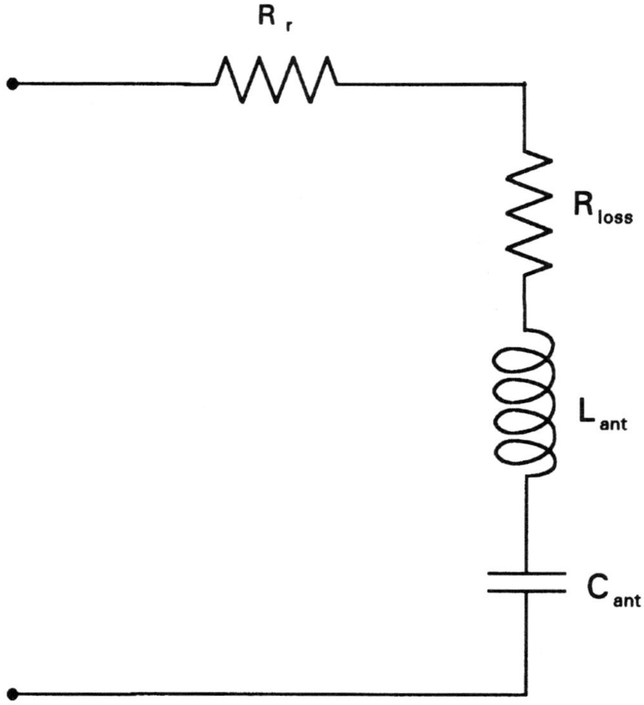

FIGURE 14.9 Equivalent circuit of the short antenna.

The radiation resistance is

$$R_r = \frac{120\pi}{6\pi}(2\pi)^2\left(\frac{L}{\lambda}\right)^2 = 80\pi^2\left(\frac{L}{\lambda}\right)^2 = 0.8\pi^2 = 7.90\ \Omega.$$

The capacitive reactance is

$$\frac{1}{\omega C_{\text{ant}}} = \frac{\ln(1000) - 1 - \ln 2}{2\pi \times 30 \times 10^6 \times \pi \times (1/36\pi) \times 10^{-9} \times 1} = \frac{5.21}{5.24 \times 10^{-3}} = 996\ \Omega,$$

and the inductive reactance is

$$\omega L_{\text{ant}} = 2\pi \times 30 \times 10^6 \left(\frac{4\pi \times 10^{-7} \times 1}{3\pi}\right)\left(\ln 1000 - \frac{11}{6}\right) = 127\ \Omega.$$

The input impedance is

$$Z_{\text{in}} = (R_r + R_{\text{loss}}) + j\left(\omega L_{\text{ant}} - \frac{1}{\omega C_{\text{ant}}}\right) = 7.98 - j869\ \Omega,$$

and the efficiency is

$$\text{eff} = \frac{7.90}{7.90 + 0.08} = 0.99.$$

While the efficiency is good, the high input reactance makes it difficult to match the generator to the antenna.

(b) At 3 MHz the wavelength is 100 m. The loss resistance decreases by a factor of $\sqrt{10}$ to $R_{\text{loss}} = 0.024\ \Omega$. The radiation resistance decreases by $(1/10)^2$ to $R_r = 0.079\ \Omega$. The capacitive reactance increases by a factor of 10 to $1/\omega C_{\text{ant}} = 9960\ \Omega$ and the inductive reactance decreases by a factor of 10 to $\omega L_{\text{ant}} = 12.7\ \Omega$. The input impedance is now

$$Z_{\text{in}} = 0.103 - j9950\ \Omega$$

and the efficiency is

$$\text{eff} = \frac{0.079}{0.079 + 0.024} = 0.77,$$

a lower value. More importantly, the extremely high input reactance will make it very difficult to match the generator to the antenna.

At either frequency the antenna could be matched to the transmission line that excites it by adding an inductive reactance in series with the antenna. This has two difficulties. First, the inductor will itself have ohmic losses, which will reduce the efficiency of the antenna system. Second, the resonance condition between the inductor and the short dipole means that the input reactance to the series combination will change rapidly as signal frequency varies, yielding a narrow bandwidth. An additional difficulty with the short dipole is that the transmission line that excites it must have an unrealistically low characteristic resistance for a proper match.

14.4.4 Half-Wave Dipole, $2L = \lambda/2$

When $L = \lambda/4$ the total length of the antenna is a half-wavelength, $2L = \lambda/2$, and it is called a half-wave dipole. We now have $\beta L = \pi/2$ and Eqs. (14.31) and (14.32) for the current along the wire become

$$\underline{I}(z) = I_m \cos \beta z, \tag{14.38}$$

where I_m is the input current. Equation (14.34) for the far field becomes

$$\underline{H}_\phi = \frac{jI_m}{2\pi R}\left[\frac{\cos\left(\frac{1}{2}\pi\cos\theta\right)}{\sin\theta}\right]e^{-j\beta R}, \quad \underline{E}_\theta = \eta \underline{H}_\phi. \tag{14.39}$$

The time average power density is

$$\vec{p}_{f,\text{Avg}} = \vec{a}_R \tfrac{1}{2} \frac{\eta I_m I_m^*}{(2\pi R)^2} \left[\frac{\cos(\frac{1}{2}\pi \cos\theta)}{\sin\theta} \right]^2. \tag{14.40}$$

The time average radiated power is

$$P_{f,\text{Avg}} = \tfrac{1}{2} \frac{\eta I_m I_m^*}{(2\pi R)^2} \int_{\phi=0}^{2\pi} \int_{\theta=0}^{\pi} \left[\frac{\cos(\frac{1}{2}\pi \cos\theta)}{\sin\theta} \right]^2 R^2 \sin\theta \, d\theta \, d\phi \tag{14.41}$$

$$= \frac{\eta I_m I_m^*}{4\pi} \int_0^{\pi} \frac{\cos^2(\frac{1}{2}\pi \cos\theta)}{\sin\theta} \, d\theta = \frac{\eta I_m I_m^*}{2\pi} \int_0^{\pi/2} \frac{\cos^2(\frac{1}{2}\pi \cos\theta)}{\sin\theta} \, d\theta.$$

Some manipulation is required to transform the integrals into a standard form. First, use $\cos^2 x = \frac{1}{2}(1 + \cos 2x)$ and then let $u = \cos\theta$ to get

$$P_{f,\text{Avg}} = \frac{\eta I_m I_m^*}{4\pi} \int_0^{\pi/2} \frac{1 + \cos(\pi \cos\theta)}{\sin^2\theta} \sin\theta \, d\theta = \frac{\eta I_m I_m^*}{4\pi} \int_0^1 \frac{1 + \cos \pi u}{1 - u^2} \, du.$$

Next, use $1/(1 - u^2) = \frac{1}{2}[1/(1 + u) + 1/(1 - u)]$. In the resultant second integral let $u = -u'$ and then combine the two integrals to get

$$P_{f,\text{Avg}} = \frac{\eta I_m I_m^*}{8\pi} \int_{-1}^{1} \frac{1 + \cos \pi u}{1 + u} \, du = \frac{\eta I_m I_m^*}{8\pi} \int_0^2 \frac{1 - \cos \pi v}{v} \, dv,$$

where the last form is obtained by letting $v = 1 + u$. Finally, letting $\pi v = t$ we have

$$P_{f,\text{Avg}} = \frac{\eta I_m I_m^*}{8\pi} \int_0^{2\pi} \frac{1 - \cos t}{t} \, dt. \tag{14.42}$$

This integral occurs frequently in antenna theory, has been studied extensively, and is given in terms of a related tabulated integral called the cosine integral†, [1]Ci(x) [4],

$$\text{Ci}(x) = -\int_x^{\infty} \frac{\cos t}{t} \, dt.$$

† There is a related integral called the sine integral,

$$\text{Si}(x) = \int_0^x \frac{\sin t}{t} \, dt.$$

The integral itself is sometimes designated as Cin(x) and sometimes as $S_1(x)$ and is given by

$$\text{Cin}(x) = S_1(x) = \int_0^x \frac{1 - \cos t}{t}\, dt = C + \ln x + \int_x^\infty \frac{\cos t}{t}\, dt$$

$$= C + \ln x - \text{Ci}(x), \quad C = 0.577\cdots$$

$$\approx C + \ln x - \frac{\sin x}{x}, \quad x \gg 1.$$

Evaluating at $x = 2\pi$ gives $\text{Cin}(2\pi) = 2.436$ and hence

$$P_{f,\text{Avg}} = \frac{\eta I_m I_m^*}{8\pi}(2.436) = 36.54 I_m I_m^*.$$

The E-plane antenna and power patterns (Fig. 14.10) are plots of the bracketed expressions in Eqs. (14.39) and (14.40), respectively. The maximum power density occurs at $\theta = \pi/2$ and is $\frac{1}{2}(\eta I_m I_m^*)/(2\pi R)^2$. The directivity is this maximum multiplied by $4\pi R^2$ and divided by $P_{f,\text{Avg}}$:

$$D = 4/2.436 = 1.64.$$

This is compared with $D = 1.5$ for the current element and the short dipole. The increased directivity is only a factor of $1.64/1.5 = 1.09$ or 0.4 dB. The radiation resistance is

$$R_r = 2P_{f,\text{Avg}}/I_m I_m^* = 2(36.54) = 73.1\ \Omega.$$

For a cylindrical wire of radius a the ohmic power loss is

$$P_L = \tfrac{1}{2} R_w \int_{-L}^{L} I_m I_m^* \cos^2 \beta z\, dz = \left(\frac{1}{2}\right)\left(\frac{R_s}{2\pi a}\right) I_m I_m^* L$$

and hence $R_{\text{loss}} = R_z \lambda / 8\pi a$. The input reactance is very difficult to calculate and depends on the radius a of the wire. If the antenna is shortened just slightly, so the overall length is 0.48λ instead of 0.5λ, then the input reactance is resonant and close to zero [5]. This shortening has very little effect on the current distribution in the wire and hence little effect on the antenna patterns, the directivity, or the radiation resistance.

The real advantage of the half-wave dipole over the short dipole is not the negligibly increased directivity, but the greater ease of matching the antenna to the transmission line that feeds it. First, no series matching inductor is needed and, second, the characteristic resistance of the transmission line should be the more feasible value of about 73 Ω. Also, because of the large radiation resistance, the loss resistance will be comparatively small and the efficiency high.

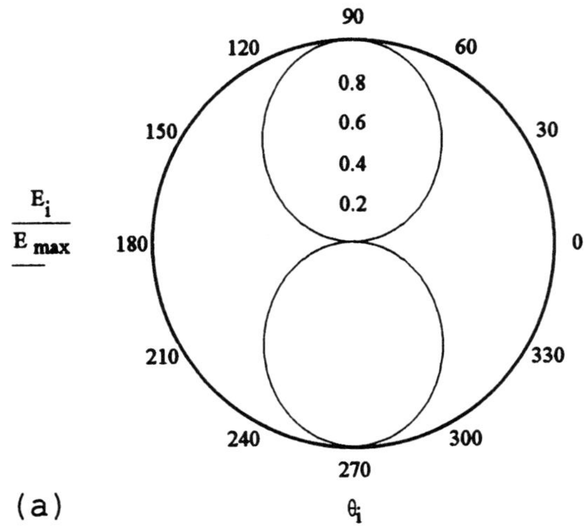

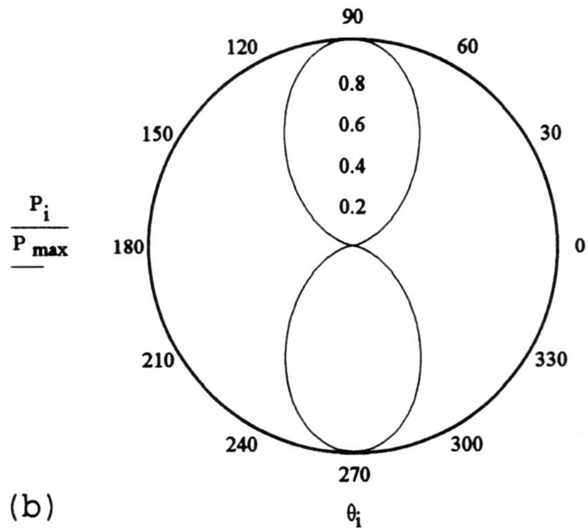

FIGURE 14.10 Normalized (a) E-plane field and (b) power patterns for a half-wave dipole.

14.4.5 Full-Wave Dipole, $2L = \lambda$

When $L = \lambda/2$ the total length of the antenna is a full-wavelength, $2L = \lambda$, and it is called a full-wave dipole. We now have $\beta L = \pi$ and Eqs. (14.31) and (14.32) for the current along the wire become

$$\underline{I}(z) = I_m \sin \beta |z|, \tag{14.43}$$

where I_m is the peak current. Equation (14.34) for the far field becomes

$$\underline{H}_\phi = \frac{jI_m}{\pi R}\left[\frac{\cos^2(\frac{1}{2}\pi \cos\theta)}{\sin\theta}\right], \quad \underline{E}_\phi = \eta \underline{H}_\phi. \tag{14.44}$$

The time average power density is

$$\vec{p}_{f,\text{Avg}} = \vec{a}_R \, \tfrac{1}{2} \frac{\eta I_m I_m^*}{(\pi R)^2} \left[\frac{\cos^2(\frac{1}{2}\pi \cos\theta)}{\sin\theta} \right]^2. \tag{14.45}$$

To find the time average power radiated we repeat the integration of Eq. (14.44) through Eq. (14.36), giving

$$P_{f,\text{Avg}} = \frac{\eta I_m I_m^*}{4\pi} \int_0^{2\pi} \frac{(1-\cos t)^2}{t}\, dt.$$

We can write $(1-\cos t)^2 = 2(1-\cos t) - (1-\cos 2t)/2$ to get

$$\begin{aligned} P_{f,\text{Avg}} &= \frac{\eta I_m I_m^*}{4\pi} \left[2 \int_0^{2\pi} \frac{1-\cos t}{t}\, dt - \tfrac{1}{2} \int_0^{4\pi} \frac{1-\cos 2t}{(2t)}\, d(2t) \right] \\ &= \frac{\eta I_m I_m^*}{4\pi} [2\,\text{Cin}(2\pi) - \tfrac{1}{2}\,\text{Cin}(4\pi)]. \end{aligned}$$

Using $\eta = 120\pi$, $\text{Cin}(2\pi) = 2.436$, and $\text{Cin}(4\pi) = 3.107$, we find

$$P_{f,\text{Avg}} = 99.55 I_m I_m^*. \tag{14.46}$$

The E-plane antenna and power patterns (Fig. 14.11) are plots of the bracketed expressions in Eqs. (14.44) and (14.45), respectively. The maximum power density occurs at $\theta = \pi/2$ and is $\frac{1}{2}(\eta I_m I_m^*)/(\pi R)^2$. The directivity is this maximum multiplied by $4\pi R^2$ and divided by $P_{f,\text{Avg}}$:

$$D = 2.41. \tag{14.47}$$

This is compared with $D = 1.5$ for the current element and the short dipole and $D = 1.64$ for the half-wave dipole. The increased directivity over the current element is a factor of $2.41/1.5 = 1.61$ or 2.1 dB.

The radiation resistance referred to the current maximum I_m is $2P_{f,\text{Avg}}/I_m I_m^* = 199.1\ \Omega$. The input current is $\underline{I}(0) = 0$, so the radiation resistance referred to the input current is infinite. When the finite radius of the wire is accounted for, the radiation resistance is found to be large, but not infinite. Measurements [6] (indicate that the reactive part of the input impedance is resonant in the range $0.8 < 2L/\lambda < 0.9$. For example, when $L/a = 100$, the reactance is zero at $2L/\lambda \approx 0.8$ and $R_r \approx 1200\ \Omega$; when $L/a = 400$, the reactance is zero at $2L/\lambda \approx 0.9$ and $R_r \approx 1700\ \Omega$. The increasing Q as L/a increases means that the bandwidth is reduced.

The small increases in gain over the current element for the half-wave and full-wave dipoles indicate that more complicated antenna structures are required for substantial increases in gain.

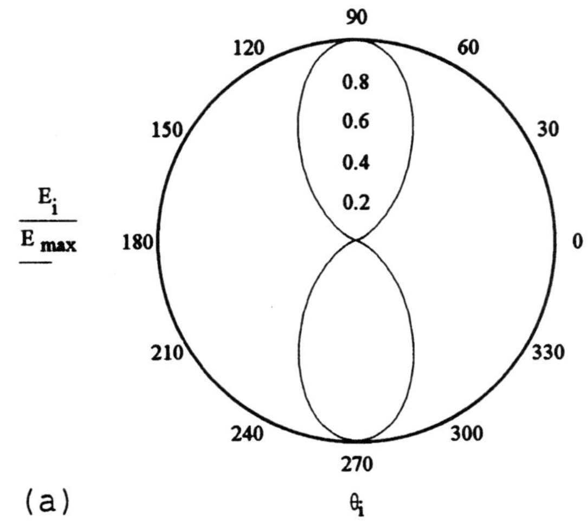

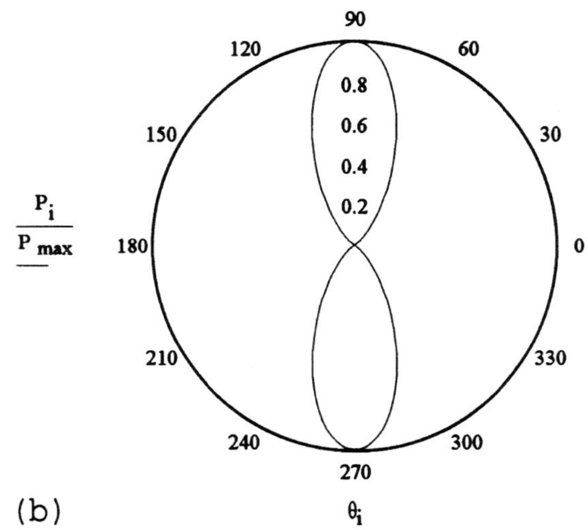

FIGURE 14.11 Normalized (a) E-plane field and (b) power patterns for a full-wave dipole.

14.5 ANTENNA ARRAYS

An antenna array consists of a number of identical antennas or antenna elements, each with the same orientation, distributed in space. By properly choosing the relative position of the antennas and the relative magnitude and phase of their excitations, it is possible to shape the overall radiation pattern and increase the directivity in a chosen direction.

Antenna arrays are very important in wireless communication applications, in which sophisticated arrays are installed even on the hand-held sets to improve the communication efficiency and allow for scanning in order to locate the best cell for improved transmission.

14.5.1 Pattern Multiplication

Let us assume we have N identical antenna elements, each with the same orientation, located about the origin at source positions $\vec{R}_n$, $n = 0, \ldots, N-1$ (Fig. 14.12). A single element of this type and orientation located at the origin with input current I_0 would give rise to a far zone or radiation field of

$$\underline{\vec{E}}(\vec{R}) = V_0 \vec{f}(\theta, \phi) \frac{e^{-j\beta R}}{4\pi R}, \quad \eta \underline{\vec{H}} = \vec{a}_r \times \underline{\vec{E}}, \tag{14.48}$$

where $\vec{R}$ is the position vector of the field point located at (R, θ, ϕ). For example, for the current element located along the z axis, Eq. (14.48), we have $V_0 = j\beta\eta I_0 d$ and $\vec{f} = \vec{a}_\theta \sin\theta$. The far field due to the array of elements is

$$\underline{\vec{E}} = \sum_{n=0}^{N-1} C_n V_0 \vec{f}(\theta_n, \phi_n) \frac{e^{j\beta|\vec{R} - \vec{R}_n}}{4\pi|\vec{R} - \vec{R}_n|}, \tag{14.49}$$

where C_n is a dimensionless complex number giving the strength and phase of the nth element relative to V_0.

Since the field point is in the far field, $|\vec{R}|$ is much greater than the largest dimension of each antenna element and the wavelength and, also, $|\vec{R}| \gg |\vec{R}_n|$. Because of the last inequality, the lines from each element to the field point are considered to be parallel, making the θ_n and ϕ_n angles with respect to each element equal to θ and ϕ. The term $|\vec{R} - \vec{R}_n|$ appearing in the denominator is taken to be R. Only the differences in amplitude and phase of individual antenna element excitations designated by C_n and the difference in propagation phase delay from the element to the field point included in $\exp(j\beta|\vec{R} - \vec{R}|)$ need to be considered.

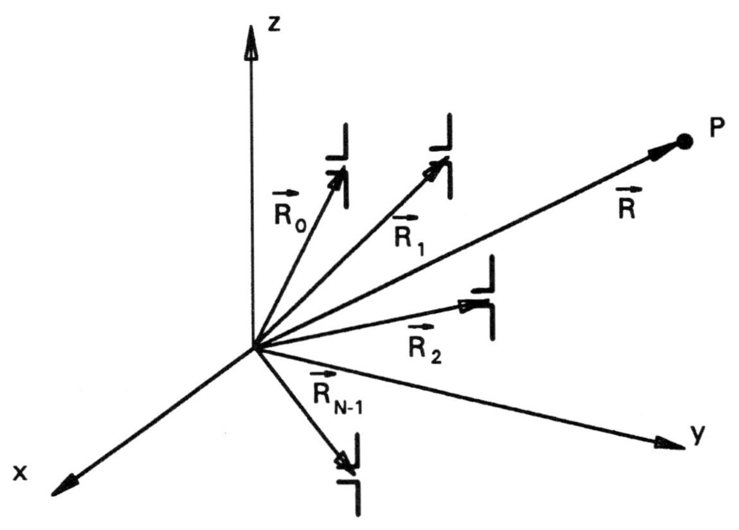

FIGURE 14.12 Array of N identical antennas.

Using $|\vec{R} - \vec{R}_n| \approx R - \vec{a}_R \cdot \vec{R}_n$ we have

$$\underline{\vec{E}} = V_0 \vec{f}(\theta, \phi) \frac{e^{-j\beta R}}{4\pi R} \sum_{n=0}^{N-1} C_n e^{-j\beta \vec{a}_R \cdot \vec{R}_n}$$

$$= V_0 \vec{f}(\theta, \phi) F(\theta, \phi) \frac{e^{-j\beta R}}{4\pi R}, \quad F(\theta, \phi) = \sum_{n=0}^{N-1} C_n e^{-j\beta \vec{a}_R \cdot \vec{R}_n}. \tag{14.50}$$

The term $\vec{f}(\theta, \phi)$ is called the element factor, as it is a characteristic of the individual antenna element in the array, and the summation $F(\theta, \phi)$ is called the array factor, as it depends only on the relative amplitude and phase excitation and position of the elements in the array.

The total field $\underline{\vec{H}}$ from the array of antennas is the product of the field from an individual element, $E_0 \vec{f} \exp(-j\beta R)/4\pi R$, and the array factor F. This is the principle of pattern multiplication: the field pattern of the array is the product of the element pattern and the array pattern.

Similarly, the power density pattern is given by $|V_0|^2 |\vec{f}|^2 |F|^2/\eta$ normalized so as to have a maximum value of unity. This also obeys the principle of pattern multiplication, consisting of the product of the element and the array factor magnitudes squared.

The principle of pattern multiplication depends on each element in the array being identical to every other element. In reality, when identical physical elements are arranged in an array the fields of each element will affect the fields of every other element. This effect, known as mutual coupling, means that the individual elements will not radiate identical field patterns. Generally, this effect is small and the principle of pattern multiplication gives good results for the field and power patterns.

14.5.2 Two-Element Arrays

Consider the array formed by two sources placed at a distance d along the x axis, symmetrically about the origin, as shown in Fig. 14.13. Taking element 1 to be the phase reference and assuming that the two sources are similar and of equal amplitude, the far zone radiation fields are given by

$$\underline{\vec{E}}_1(\vec{R}_1) = V_0 \vec{f}(\theta, \phi) \frac{e^{-j\beta R_1}}{4\pi R_1},$$

$$\underline{\vec{E}}_2(\vec{R}_2) = V_0 \vec{f}(\theta, \phi) \frac{e^{-j(\beta R_2 + \theta)}}{4\pi R_2}.$$

Making the same assumption that $R_1 \approx R_2 \approx R$ as far as the amplitude is concerned, and substituting $R_1 = R - (d/2)\cos(\psi)$ and $R_2 = R + (d/2)\cos(\psi)$ in the phase component, the total field due to the two elements is given by

$$\underline{\vec{E}}(\vec{R}) = V_0 \vec{f}(\theta, \phi) \frac{1}{4\pi R_1} [e^{-j\beta[R - (d/2)\cos(\psi)]} + e^{-j\beta[R + (d/2)\cos(\psi)] + \theta}].$$

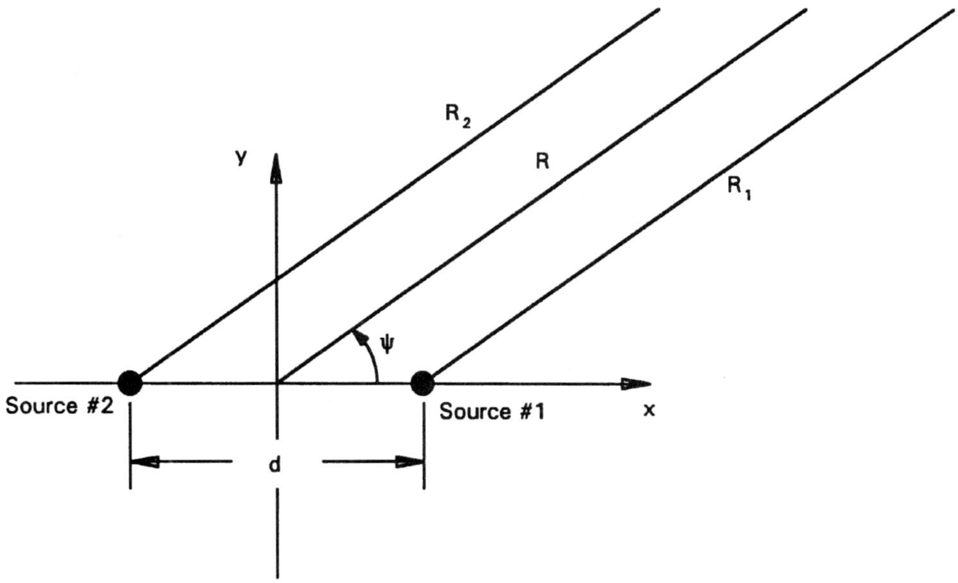

FIGURE 14.13 Array of two identical sources centered at the origin.

The array factor for the two-element array becomes

$$F = e^{-j(\beta R+\theta/2)}[e^{j[\beta(d/2)\cos(\psi)+\theta/2]} + e^{-j[\beta(d/2)\cos(\psi)+\theta/2]}]$$

$$= 2e^{-j(\beta R+\theta/2)}\cos\left(\beta\frac{d}{2}\cos(\psi)+\frac{\theta}{2}\right).$$

The absolute value of the array factor is given by

$$|F| = 2\left|\cos\left(\beta\frac{d}{2}\cos(\psi)+\frac{\theta}{2}\right)\right|.$$

If the two sources are in phase the amplitude of the array factor is then given by

$$|F| = 2\left|\cos\left(\beta\frac{d}{2}\cos(\psi)\right)\right|.$$

For sources that are 180° out of phase, the amplitude of the array factor is

$$|F| = 2\left|\cos\left(\beta\frac{d}{2}\cos(\psi)+90^\circ\right)\right| = 2\left|\sin\left(\beta\frac{d}{2}\cos(\psi)\right)\right|.$$

If the sources are in quadrature ($\theta = 90^\circ$), then the result is

$$|F| = 2\left|\cos\left(\beta\frac{d}{2}\cos(\psi)+45^\circ\right)\right|.$$

The radiation pattern of such arrays depends on the spacing of the elements as well as the phase difference between them. It is possible to synthesize a radiation pattern by using a single-element pair array or combinations of such pairs with the help of the pattern multiplication principle.

14.5.2.1 Pattern Synthesis Pattern synthesis is a very important process in the design of antenna arrays. An antenna engineer may be required to design an array that would have maximum radiation in certain directions and nulls in other directions, thus avoiding interference with other transmitters or receivers. The knowledge of specifications on the maxima and nulls of an array pattern to be synthesized represents enough information for an array factor to be defined and an array configuration to be generated.

Since the array factor of a pair of elements has two variables, d and θ, the patterns that could be synthesized with such an array could only satisfy a couple of constraints.

Example 14.2 Design a uniform array of similar sources along the x axis that exhibits a maximum at $\psi = 45°$, and a null at $\psi = 135°$, as shown in Fig. 14.14. Determine the distance between the elements and the relative phase difference.

Solution The array factor for a two-element array is given by

$$|F| = 2\left|\cos\left(\beta \frac{d}{2} \cos(\psi) + \frac{\theta}{2}\right)\right|.$$

The maximum condition requires $|F| = \text{maximum}$ at $\psi = 45°$. This condition is satisfied if

$$\left|\cos\left(\beta \frac{d}{2} \cos(45°) + \frac{\theta}{2}\right)\right| = 1.$$

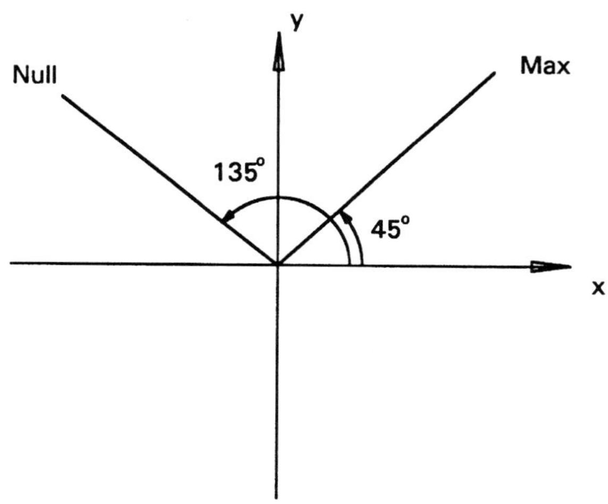

FIGURE 14.14 Specifications in Example 14.2.

The argument of the cosine has to take the following values:

$$\beta \frac{d}{2} \cos(45^\circ) + \frac{\theta}{2} = n\pi$$

$$\frac{\pi d}{\lambda} \cdot \frac{1}{\sqrt{2}} + \frac{\theta}{2} = n\pi.$$

The null condition requires that

$$\beta \frac{d}{2} \cos(135^\circ) + \frac{\theta}{2} = \frac{\pi}{2} + m\pi$$

$$\frac{\pi d}{\lambda}\left(-\frac{1}{\sqrt{2}}\right) + \frac{\theta}{2} = \frac{\pi}{2} + m\pi.$$

Selecting $n = 0$, $m = -1$, the values of d and θ are as follows:

$$d = \frac{\sqrt{2}}{4}\lambda, \quad \theta = -\frac{\pi}{2}.$$

The array factor becomes

$$|F| = 2\left|\cos\left(\frac{\pi\sqrt{2}}{4}\cos(\psi) - \frac{\pi}{4}\right)\right|.$$

The array factor is plotted in Fig. 14.15.

The following example will require more than one pair of elements in the array. The array will be constructed using the pattern multiplication principle. A primary array is designed to satisfy some of the constraints, while the secondary pattern will satisfy the rest of the constraints. Redundancy in the constraint satisfied by the two patterns may be needed to satisfy the overall pattern description. The final spatial arrangement of the array elements is accomplished by taking two primary arrays and placing them with their centers on each of the elements of the secondary array, as shown in Fig. 14.16.

Example 14.3 Design a uniform array of similar sources along the x axis that exhibits a broad maximum from $\psi = -30^\circ$ to $\psi = 30^\circ$ and has nulls at $\psi = 135^\circ$ and $\psi = 60^\circ$, as shown in Fig. 14.17. Determine the number of elements, the distances among them, and the relative phase differences. Plot the array factor.

Solution Based on the number of constraints, an array of four elements is selected, which will be the result of the pattern multiplication of two two-element arrays. One of the arrays will be the primary while the other will be considered the secondary array.

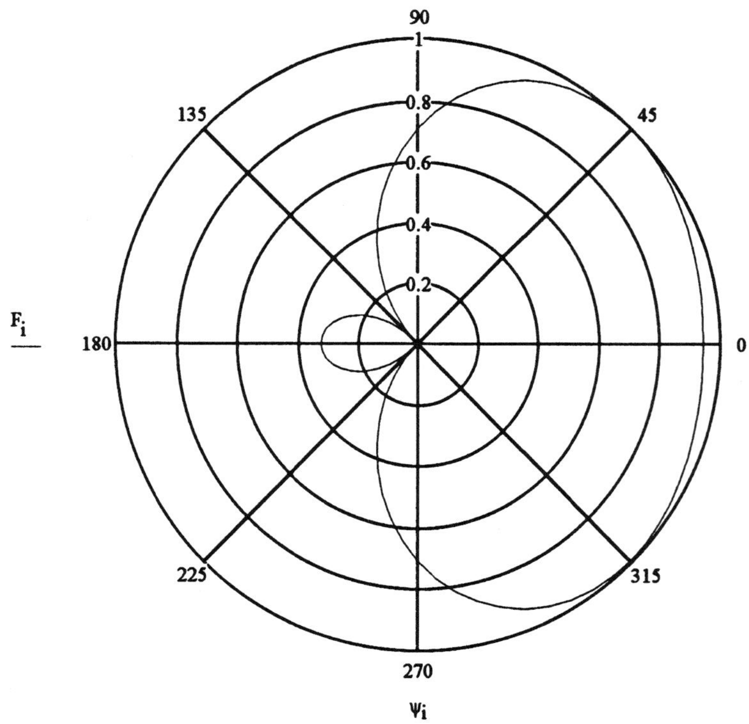

FIGURE 14.15 Array factor plot in Example 14.2.

The primary array is selected to satisfy the following conditions:

- Broad maximum from $\psi = -30^\circ$ to $\psi = 30^\circ$
- Null at $\psi = 135^\circ$

The secondary array is selected to satisfy the following conditions:

- Maximum at $\psi = 0^\circ$
- Null at $\psi = 60^\circ$

As it is with any engineering design, past experience comes in handy in tackling any new task. We notice that the design from the previous example exhibited a null at $\psi = 135^\circ$ and relatively flat maximum from $\psi = -45^\circ$ to $\psi = 45^\circ$. Therefore the primary array is selected to have $d_1 = \sqrt{2}\lambda/4$, $\theta_1 = -\pi/2$, with an array factor given by

$$|F_1| = 2\left|\cos\left(\frac{\pi\sqrt{2}}{4}\cos(\psi) - \frac{\pi}{4}\right)\right|.$$

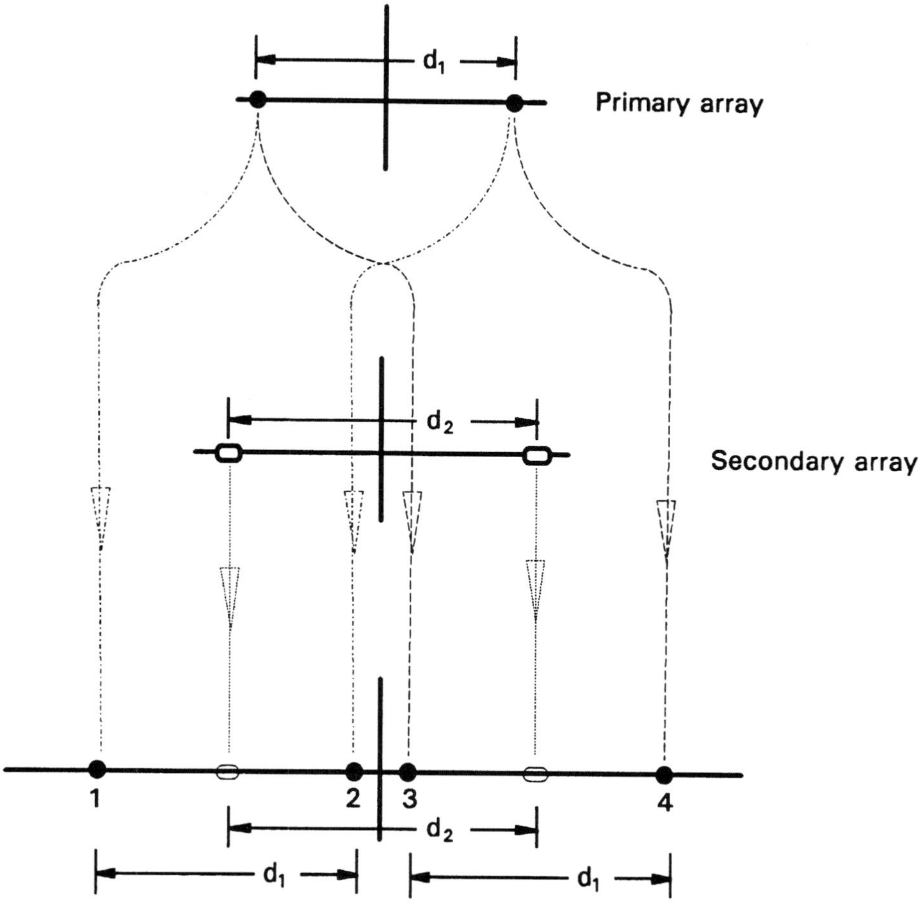

FIGURE 14.16 Construction of a four-element array from two two-element arrays using the pattern multiplication principle.

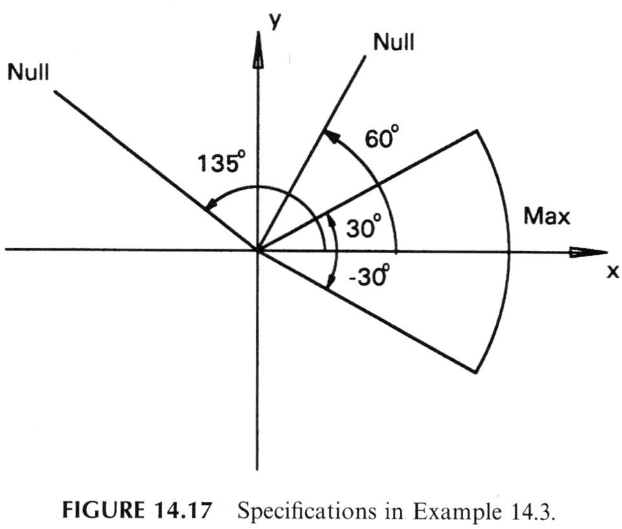

FIGURE 14.17 Specifications in Example 14.3.

The secondary array design follows these steps.

- Maximum at $\psi = 0°$:

$$\left|\cos\left(\frac{\pi d_2}{\lambda}\cos(0°) + \frac{\theta_2}{2}\right)\right| = 1$$

$$\frac{\pi d_2}{\lambda}\cdot 1 + \frac{\theta_2}{2} = n\pi\,.$$

- Null at $\psi = 60°$:

$$\left|\cos\left(\frac{\pi d_2}{\lambda}\cos(60°) + \frac{\theta_2}{2}\right)\right| = 0$$

$$\frac{\pi d_2}{\lambda}\cdot\frac{1}{2} + \frac{\theta_2}{2} = \frac{\pi}{2} m\pi\,.$$

Selecting $n = 0$, $m = -1$, the distance and the phase angle difference between the elements of the secondary array are

$$d_2 = \lambda, \quad \theta_2 = -2\pi\,.$$

The array factor for the secondary array is

$$|F_2| = 2|\cos(\pi\cos(\psi) - \pi)|\,.$$

The overall array factor is obtained by taking the product of the two array factors:

$$|F| = |F_1|\cdot|F_2| = 4\left|\cos\left(\frac{\pi\sqrt{2}}{4}\cos(\psi) - \frac{\pi}{4}\right)\right|\cdot|\cos(\pi\cos(\psi) - \pi)|\,.$$

The plot of the array factor is shown in Fig. 14.18. Figure 14.19 depicts the physical arrangement of the elements of the primary array (a), secondary array (b), and overall array (c).

For more complex pattern synthesis applications, larger arrays have to be used, or other techniques might have to be employed.

14.5.3 Uniform Linear Arrays

In a linear array the elements lie along a straight line, which we can take to be the x axis. If we further assume that the elements are spaced a distance d apart, that they are excited by equal-amplitude excitations, and that the phase change from one element to the next is the same all along the array, then we have a uniform linear array (Fig. 14.20). These conditions are $\vec{R}_n = \vec{a}_x nd' |C_n| = C_0$, and $\angle C_n - nd\alpha$, where α (rad/m) is the phase shift per unit length along the array line and $d\alpha$ is the phase difference between adjacent elements. These last two conditions combine to give $C_n = C_0\exp(jnd\alpha)$.

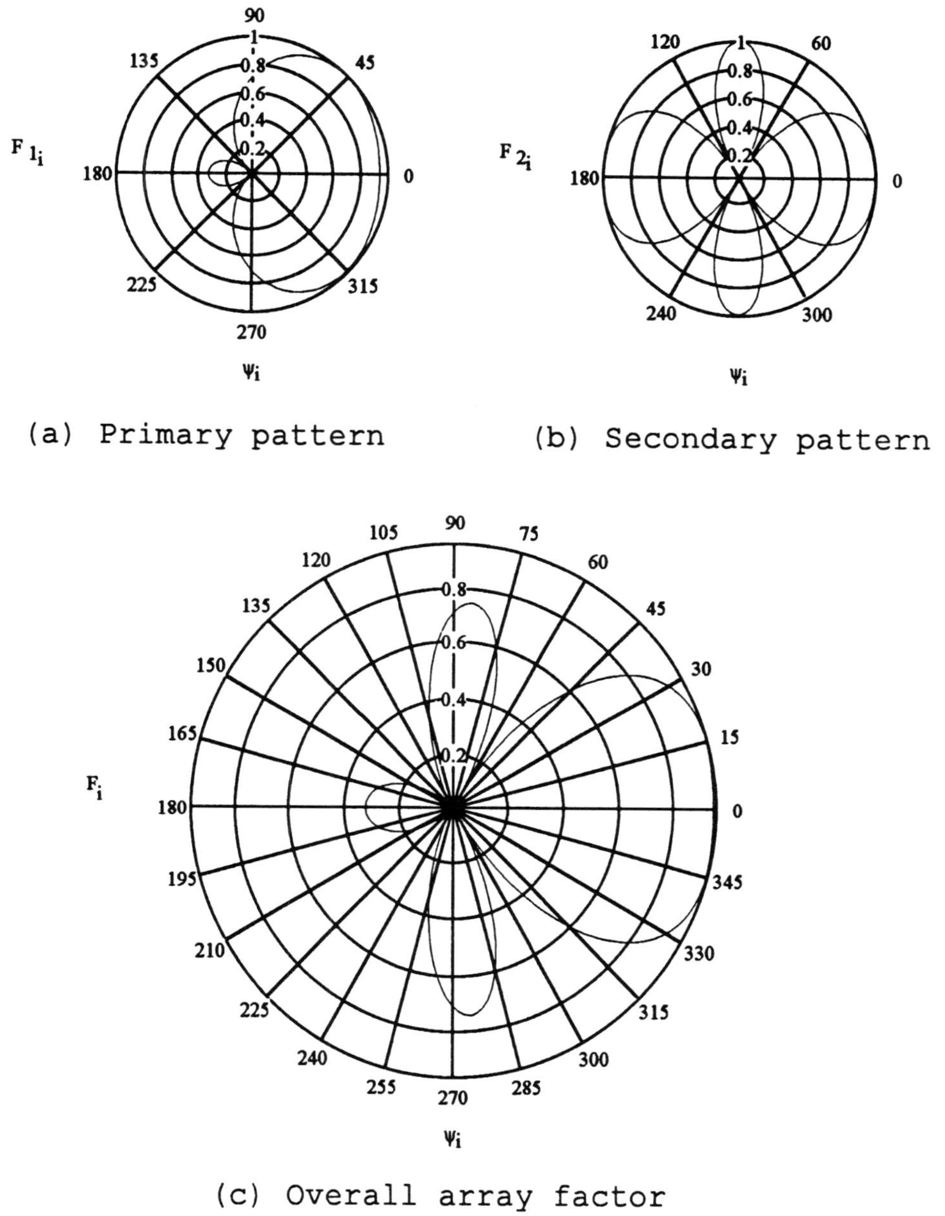

FIGURE 14.18 Array factor plot in Example 14.3.

The position vector $\vec{R}$ makes an angle ψ with the x axis. Then $\vec{a}_R \cdot \vec{a}_x = \cos\psi = \sin\theta\cos\phi$ [Eq. (2.18)] and $\vec{a}_R \cdot \vec{R}_n = nd\cos\psi = nd\sin\theta\cos\phi$. The array factor is

$$F = C_0 \sum_{n=0}^{N-1} e^{njd(\alpha + \beta\cos\psi)} = C_0 \sum_{n=0}^{N-1} (e^{ju})^n, \quad u = d(\alpha + \beta\cos\psi). \tag{14.51}$$

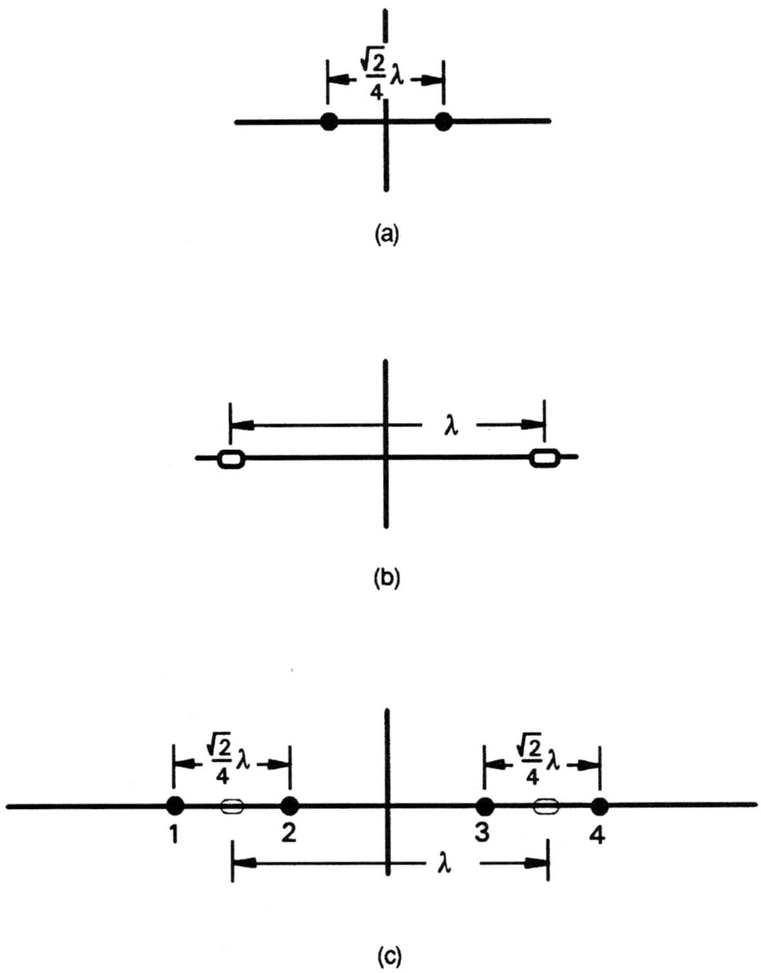

FIGURE 14.19 Physical arrangement of the elements of the (a) primary, (b) secondary, and (c) overall arrays in Example 14.3.

We sum the geometric series, $1 + Z + Z^2 + \cdots + Z^{N-1} = (1 - Z^N)/(1 - Z)$, to obtain

$$F = C_0 \frac{1 - e^{jNu}}{1 - e^{ju}} = C_0 \left(\frac{e^{jNu/2}}{e^{ju/2}} \right) \left(\frac{e^{jNu/2} - e^{-jNu/2}}{e^{ju/2} - e^{-ju/2}} \right)$$

$$= C_0 \frac{\sin \frac{1}{2} Nu}{\sin \frac{1}{2} u} \angle (N-1)(u/2). \tag{14.52}$$

The magnitude of the array factor

$$|F| = C_0 \left| \frac{\sin \frac{1}{2} Nu}{\sin \frac{1}{2} u} \right| = C_0 \left| \frac{\sin \frac{1}{2} Nd(\alpha + \beta \cos \psi}{\sin \frac{1}{2} d(\alpha + \beta \cos \psi} \right| \tag{14.53}$$

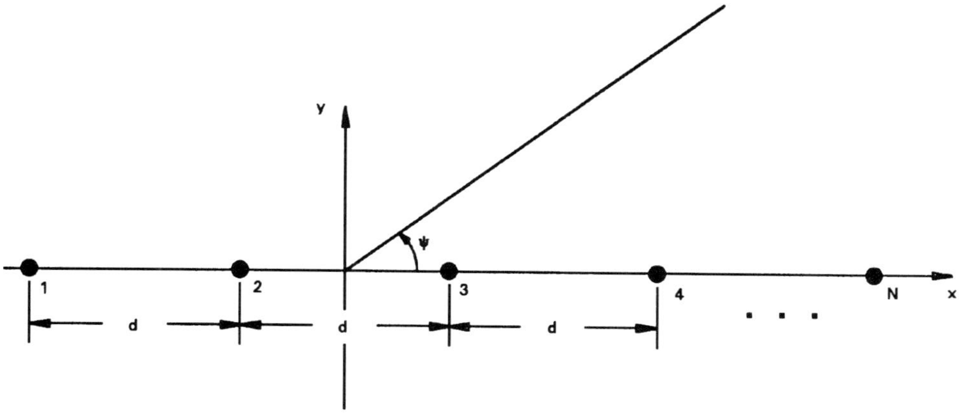

FIGURE 14.20 Uniform linear array.

is plotted in Fig. 14.21. The function has a maximum peak at $u = 0, \pm 2\pi, \pm 4\pi, \ldots$, with the maximum value being N, corresponding to the in-phase addition of the field from each of the N elements of the array. This is the main lobe or main beam of the array pattern. The main beam at $u = 0$ points in the direction

$$\alpha + \beta \cos\psi = 0, \quad \psi = \cos^{-1}(-\alpha/\beta), \tag{14.54}$$

with respect to the x axis, which is the array axis.

Zeros or nulls of the array factor occur at $u = \pm 2\pi/N, \pm 4\pi/N, \ldots, \pm 2(N-1)\pi/N$, so the beam width of the main lobe is $4\pi/N$. The main lobe is surrounded by side lobes. The first side lobes, on either side of the main lobe, have their peaks approximately at $u = 3\pi/N$ and a value of $|1/\sin(3\pi/2N)|$. The field amplitude gain of the main beam over the first side lobes is therefore $N|\sin(3\pi/2N)|$. For large values of N (say, eight or more), the sine function will closely equal its argument and the field amplitude gain of the main beam over the first side lobes will approach $3\pi/2 = 4.71$ or 13.5 dB.

The value of $\cos\psi$ is limited to ± 1 and hence the physically possible range of u is $d(\alpha \pm \beta)$, which is called the physical space (also known as the visible region or visible space). We normally want only one main beam to be present in the array pattern, so the physical space should not extend to the next main lobe on either side of the main beam. This requires that, when $\alpha > 0$, we should have $d(\alpha + \beta) < 2\pi$ or $d/\lambda < 1 - d\alpha/2\pi$; when $\alpha < 0$, we should have $-2\pi < d(-|\alpha| - \beta)$ or $d/\lambda < 1 - d|\alpha|/2\pi$; and when $\alpha = 0$, we should have $d/\lambda < 1$.

Two important cases of the uniform linear array are considered next.

14.5.3.1 Uniform Linear Broadside Array A linear array that produces the main beam in the plane perpendicular to the line of the array is called a broadside array. To accomplish this we must have $u = \alpha + \beta \cos\psi = 0$ when $\psi = \pi/2$. Since $\cos\psi$ is then zero, we get $\alpha = 0$. This means that the elements are all in phase and therefore add constructively in the $\psi = \pi/2$ plane broadside to the array. Figure 14.22 shows the

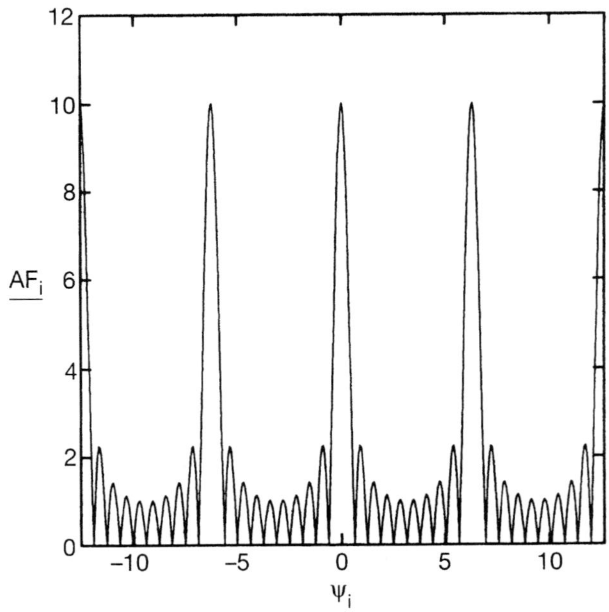

FIGURE 14.21 Array factor for a uniform linear array ($N = 10$).

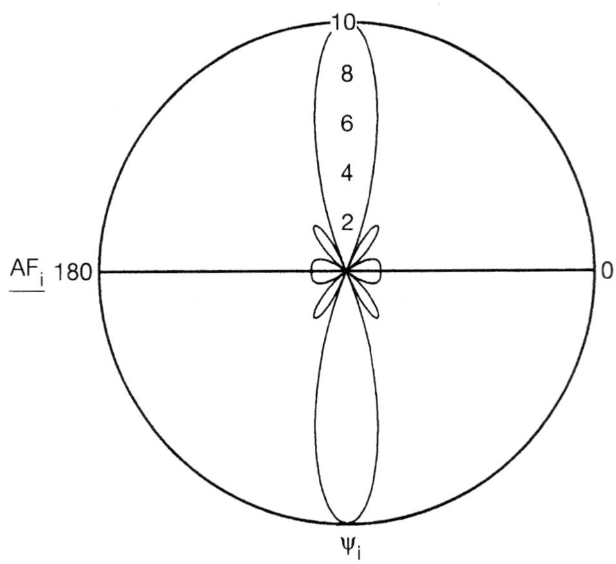

FIGURE 14.22 Array factor for a uniform linear broadside array ($N = 10$).

field pattern in any plane that includes the line of the array. The three-dimensional pattern is a figure of revolution about this line.

We can get a measure of the directivity by finding the beam width of the main beam. The nulls adjacent to the main beam occur at $u = d\beta \cos(\pi/2 \mp \Delta\psi/2) = \mp 2\pi/N$, where $\Delta\psi$ is the beam width about the center of the main beam at $\psi = \pi/2$. This gives

$$\cos\left(\frac{\pi}{2} \mp \frac{\Delta\psi}{2}\right) = \pm \sin\left(\frac{\Delta\psi}{2}\right) = \pm \frac{\lambda}{Nd}$$

or

$$\Delta\psi = 2 \sin^{-1}\left(\frac{\lambda}{Nd}\right) \approx \frac{2\lambda}{Nd}. \tag{14.55}$$

The last form is valid when N is large. The length of the array is $L = (N-1)d \approx Nd$, when N is large, so the angle between nulls is $\Delta\psi \approx 2\lambda/L$. The half-power angle is about one-half of this, so $\Delta\psi_{\text{HP}} \approx \lambda/Nd \approx \lambda/L$. Using Eq. (14.28), the directivity of the array is about

$$D = \frac{4\pi}{2\pi(\lambda/Nd)} = \frac{2Nd}{\lambda} = \frac{2L}{\lambda}.$$

The directivity of the array of elements depends on the radiation element factor $\vec{f}$ as well as the array factor F. If the element is half-wave dipole parallel to the z axis then the field pattern for the array of half-wave dipoles (recall that $\cos\psi = \sin\theta\cos\phi$) is

$$\left|\frac{\underline{E}_\theta}{\underline{E}_{\theta,\,\max}}\right| = \frac{1}{N}\left|\frac{\cos(\frac{1}{2}\pi\cos\theta)}{\sin\theta}\right|\left|\frac{\sin[(N\pi d/\lambda)\sin\theta\cos\phi]}{\sin[(\pi d/\lambda)\sin\theta\cos\phi]}\right|. \tag{14.56}$$

The beams for the dipole are in the $\theta = \pi/2$ direction, coinciding with the xy plane. The main beam for the array is in the $\phi = \pi/2$ plane, coinciding with the yz plane. The product results in two main beams pointing along the positive and negative y axes.

The directivity is

$$D = D\left(\frac{\pi}{2}, \frac{\pi}{2}\right) = \frac{4\pi N^2}{I},$$
$$I = \int_{\phi=0}^{2\pi}\int_{\theta=0}^{\pi}\left[\frac{\cos(\frac{1}{2}\pi\cos\theta)}{\sin\theta}\right]^2\left[\frac{\sin[(N\pi d/\lambda)\sin\theta\cos\phi]}{\sin[(\pi d/\lambda)\sin\theta\cos\phi]}\right]^2 \sin\theta\, d\theta\, d\phi. \tag{14.57}$$

The integral is complicated, but it can be performed numerically.

Let us use Eq. (14.28) to estimate the directivity. The dipole has a half-power angle of 78° or 1.36 rad and we have estimated the array to have a half-power angle of λ/Nd in an orthogonal plane. There are two main beams and hence

$$D \approx \frac{4\pi}{2 \times 1.36 \times (\lambda/Nd)} = \frac{4.6Nd}{\lambda} = 4.6\frac{L}{\lambda}. \tag{14.58}$$

For example, a 20-element array of half-wave dipoles with a spacing $d = \lambda/2$ would have a directivity of about 46 or 17 dB.

14.5.3.2 Uniform Linear End-Fire Array An array that produces a beam along its axis is called an end-fire array. From Eq. (14.51), to make $u = 0$ when $\psi = 0$ we must have $\alpha = -\beta$. This means that the wave radiated along the x axis from each element will arrive in phase and hence add constructively. Figure 14.23 shows the field pattern in any plane that includes the line of the array. The pattern is a figure of revolution about this line.

If we again take the elements to be half-wave dipoles parallel to the z axis, then the field pattern is

$$\left|\frac{\underline{E}_\theta}{\underline{E}_{\theta,\,\max}}\right| = \frac{1}{N}\left|\frac{\cos(\frac{1}{2}\pi\cos\theta)}{\sin\theta}\right|\left|\frac{\sin[(N\pi d/\lambda)(\sin\theta\cos\phi - 1])}{\sin[(\pi d/\lambda)(\sin\theta\cos\phi - 1)]}\right|. \tag{14.59}$$

To have only one main beam in the physical space we must have $d/\lambda < 1 - d|\alpha|/2\pi = 1 - d/\lambda$ or $d/\lambda < 1/2$.

The first nulls around the main beam occur at $u = \pm 2\pi/N = d\beta(\cos\psi - 1)$ or $\cos\psi(1 \pm \lambda/Nd)$. Since $\cos\psi \le 1$, only the minus sign is a physical solution. With N large ψ will be close to zero so we can approximate $\cos\psi$ by the first two terms of its MacLaurin series expansion

$$\cos\psi \approx 1 - \tfrac{1}{2}\psi^2 = 1 - \frac{\lambda}{Nd},$$

$$\psi = \sqrt{\frac{2\lambda}{Nd}}.$$

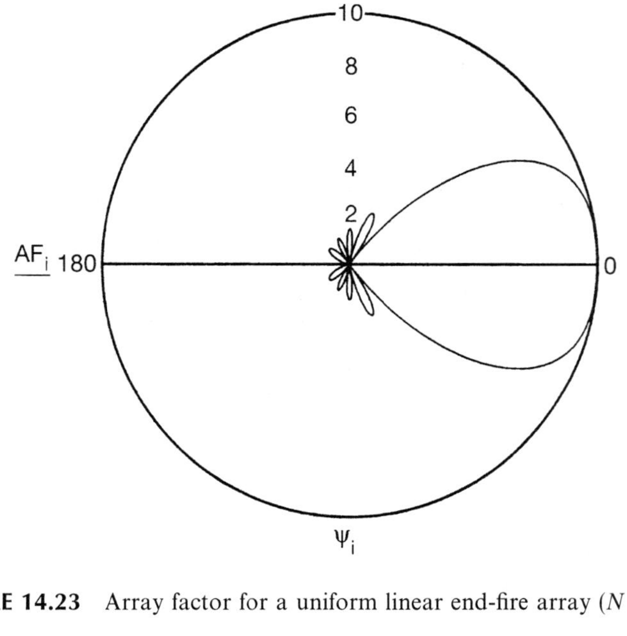

FIGURE 14.23 Array factor for a uniform linear end-fire array ($N = 10$).

The total angle between nulls is twice this. Again estimating the half-power angle as half the angle between nulls, we have

$$\Theta_{HP} = \Phi_{HP} = \sqrt{\frac{2\lambda}{Nd}} = \sqrt{\frac{2\lambda}{L}}.$$

Within this angle the dipole field pattern will be essentially constant at its peak, so this is the half-power angle of the array and dipole together. Using Eq. (14.28) we estimate the directivity as

$$D = \frac{4\pi}{(\sqrt{2\lambda/Nd})^2} = \frac{2\pi Nd}{\lambda} = \frac{2\pi L}{\lambda}.$$

For example, a 40-element array of half-wave dipoles with a spacing $d = \lambda/4$ would have a directivity of about 63 or 18 dB. This array and the broadside array that closed the previous section have the same length.

The exact directivity is given by

$$D = D\left(\frac{\pi}{2}, 0\right) = \frac{4\pi N^2}{I},$$
$$I = \int_{\phi=0}^{2\pi} \int_{\theta=0}^{\pi} \left[\frac{\cos(\frac{1}{2}\pi\cos\theta)}{\sin\theta}\right]^2 \left\{\frac{\sin[(N\pi d/\lambda)(\sin\theta\cos\phi - 1)]}{\sin[(\pi d/\lambda)(\sin\theta\cos\phi - 1)]}\right\}^2 \sin\theta \, d\theta \, d\phi. \tag{14.60}$$

Again, the integral can only be evaluated numerically.

Improved performance can be obtained from arrays with nonuniform amplitude distributions. Examples of such arrays have amplitudes that are proportional to either binomial or Tchebysheff polynomial coefficients, to name only two of the most commonly used distributions. These arrays have improved parameters in terms of side lobe levels and directivity. The reader is encouraged to consult Ref. [7] and [8] for a more detailed analysis of such nonuniform arrays.

14.6 SPECIAL ANTENNA CONFIGURATIONS

This section presents an overview of horn antennas and of the different types of reflector antennas. The purpose of this section is to introduce the different configurations of horns and reflectors and describe their performance. For a detailed analysis of these antennas, the reader is encouraged to consult the references presented at the end of the chapter.

14.6.1 Horn Antennas

The horn antennas are some of the most widely used feed elements in microwave antennas, with applications in satellite communications, radio astronomy, point-to-point terrestrial communications, and so on. The horn antennas are easy to construct and have large gain, with superior overall performance to other types of antennas. Typically, horns are used as calibration and performance measurement standards for other antennas.

Typical configurations of the antennas are shown in Fig. 14.24. The horns represent flared waveguides that taper from waveguide dimensions to a larger aperture. The dimensions of the aperture and the type of flaring are determining factors in the overall performance of the horn.

The electric field at the aperture of a rectangular horn antenna can be approximated to have the same shape as the corresponding component of the dominant TE_{10} mode. This approximation neglects the effects of currents on the exterior surfaces. The field is then given by

$$E_y(x, y) = E_{0,10} \sin\left(\frac{\pi x}{A}\right),$$

where A is the dimension of the horn aperture in the x direction.

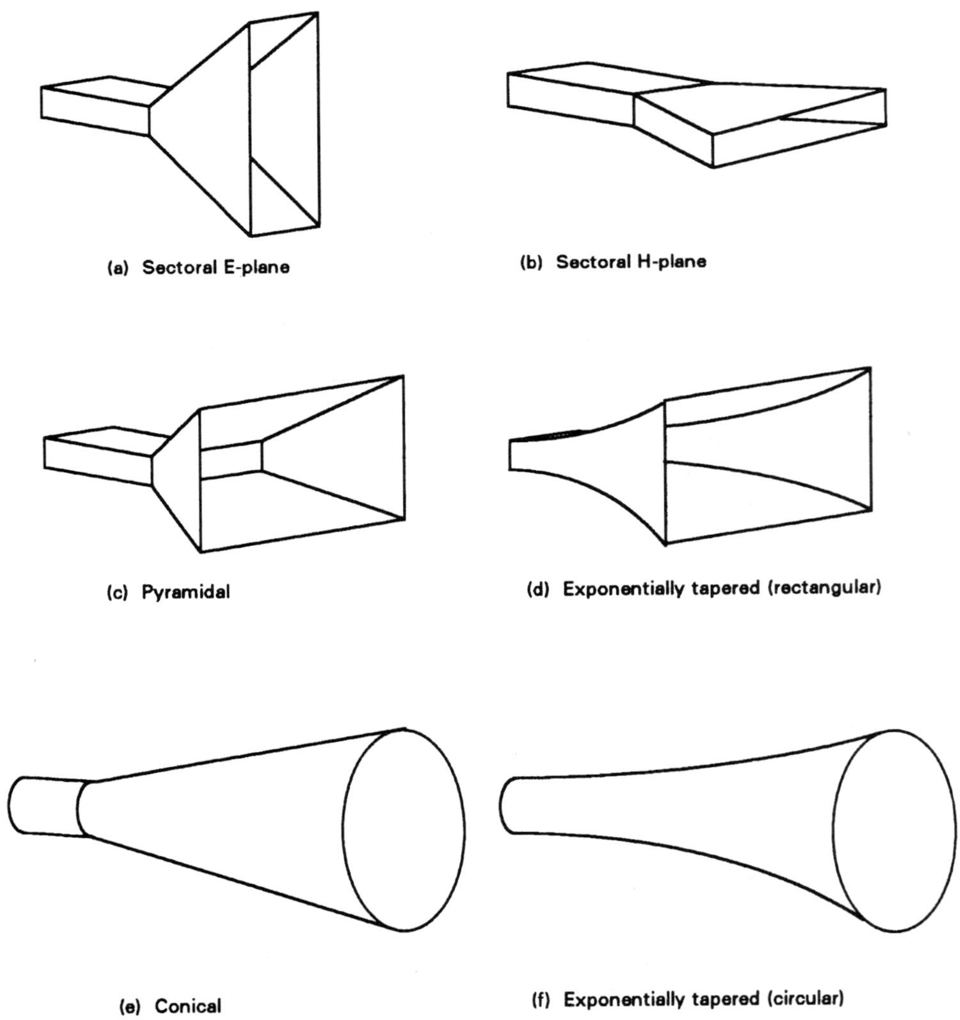

FIGURE 14.24 Rectangular and circular horn antennas.

A similar expression is obtained for the circular waveguide horn antenna, in which the dominant mode is TE_{11} [9].

Rigorous derivations of the electromagnetic radiation fields are presented by Balanis [10].

Improvements in the performance of the horns are accomplished by changes in the construction. Ridge horns, which could have a single or double ridge, have increased bandwidth by lowering the dominant mode cutoff frequency [10].

Reduction in the side lobes may be accomplished by inserting septum plates in the horn, and reduction in cross-polarization and edge diffraction as well as increased symmetry of the pattern and aperture efficiency can be realized by the use of corrugations (grooves) in the walls of the horn [9, 10].

Figure 14.25 shows the different structures mentioned above.

14.6.2 Reflector Antennas

Reflectors are used to modify the radiation characteristics of elemental antennas. Reflector antennas can take a variety of shapes ranging from a flat sheet to parabolic geometry as shown in Fig. 14.26. The type of reflector as well as the position of the feed element (elemental antenna) are determining factors in the shaping of the radiation pattern.

14.6.2.1 Sheet Reflectors The sheet reflectors are large planar reflecting surfaces placed in the proximity of an elemental antenna. The position of the element with respect to the reflector, the polarization of the feed element, and size of the reflecting surface are controlling factors for the radiation pattern, directivity, bandwidth, and impedance. Larger directivities can be achieved by positioning the feed element closer to the reflector, at the cost of reduced bandwidth.

The analysis of such structures reduces itself to the application of the method of images.

The case of an isotropic source and a sheet reflector is depicted in Fig. 14.27. The array formed by the isotropic feed element and its image has the following array factor:

$$F(\psi) = 2j\sin\left(d\frac{2\pi}{\lambda}\cos(\psi)\right).$$

The physical radiation pattern is the one obtained for $-\pi/2 < \psi < \pi/2$.

In cases in which the reflector surface cannot be considered infinite, the analysis follows the geometrical theory of diffraction [11, 12]. The space around the reflector is then divided into several regions, which include the direct radition fields, the surface reflected fields, and the fields due to diffraction.

14.6.2.2 Corner Reflector The corner reflector provides an improvement in directivity and gain over the sheet reflector. In the case of the sheet reflector, the image method resulted in only one image element. The corner reflector yields multiple image

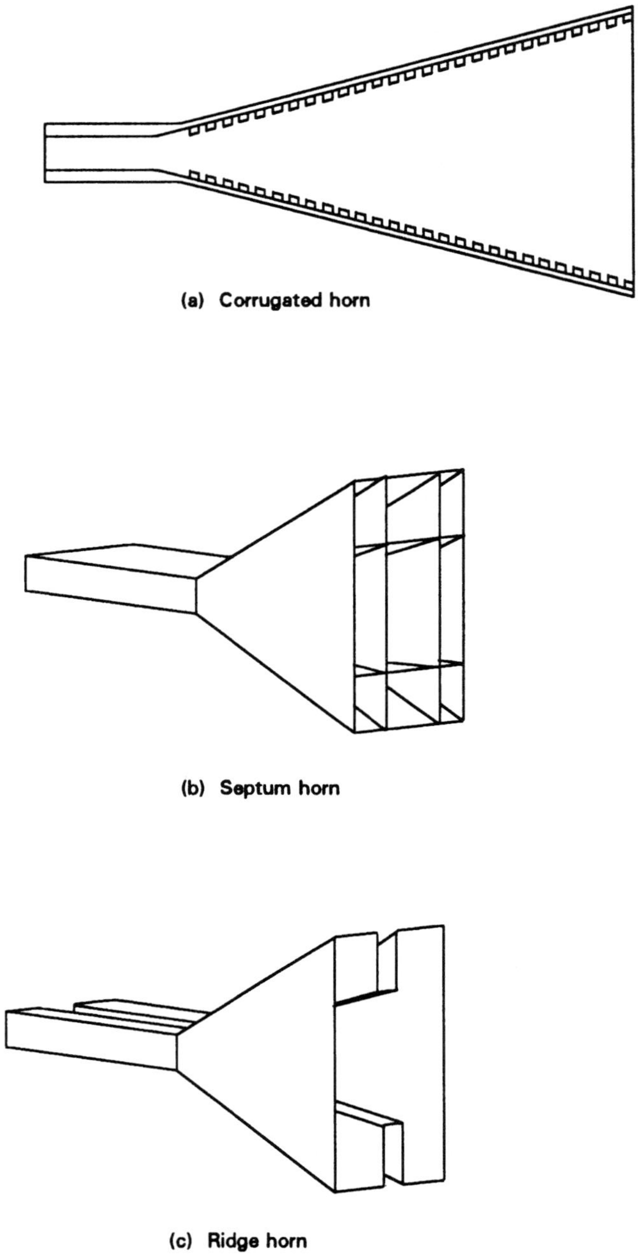

FIGURE 14.25 (a) Corrugated, (b) septum, and (c) ridge horn antennas.

elements. The number of image elements is calculated as

$$n = \frac{360^\circ}{\alpha} - 1,$$

where α is the angle of the corner, which could take values of 90°, 60°, 45°, and so on. The sheet reflector can be considered as a corner reflector with a corner angle of 180°.

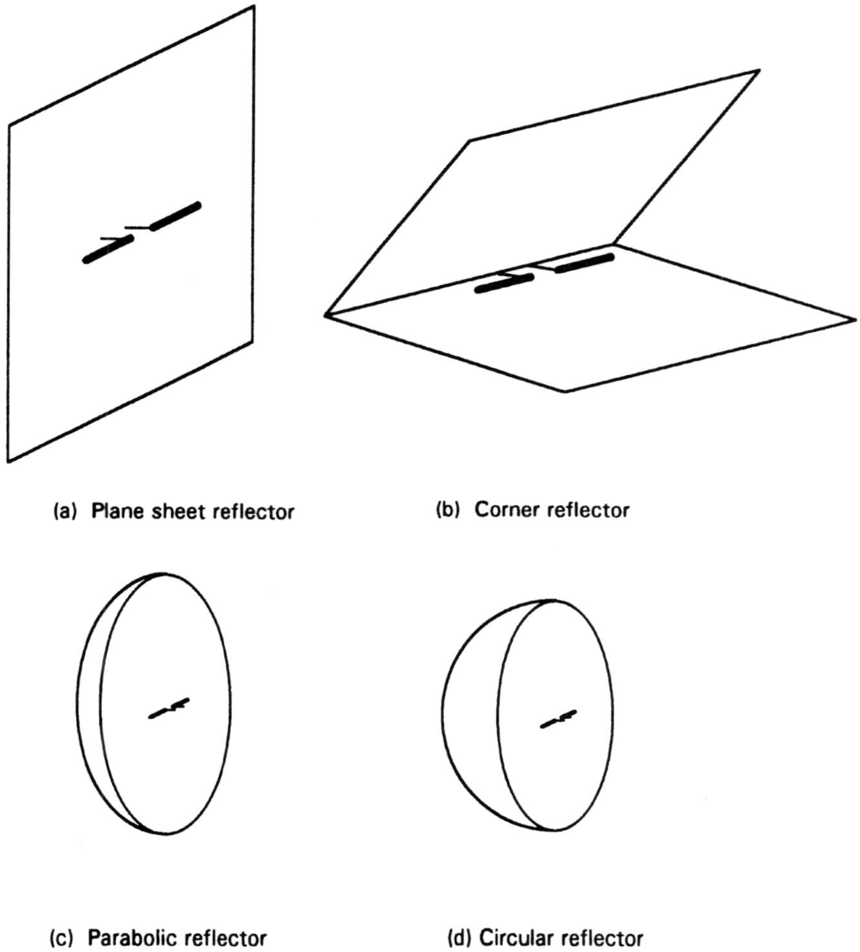

FIGURE 14.26 Different geometries of reflectors.

The most commonly used configuration is a 90° corner reflector. This type of reflector, as a passive element, returns any incident wave back in the same direction. The method of images leads to the analysis of an array of four elements (the feed element and the three image elements) as shown in Fig. 14.28. The array factor for the elements 1 and 3, which are in phase, becomes

$$F_{1,3}(\psi) = 2\cos(d\beta\cos(\psi)) = 2\cos\left(d\frac{2\pi}{\lambda}\cos(\psi)\right),$$

where d represents the distance of the feed element to the corner of the reflector.

The array factor for elements 2 and 4, which are also in phase with each other but out of phase by 180° with element 1, is given by

$$F_{2,4}(\psi) = -2\cos(d\beta\cos(\psi + 90°)) = -2\cos\left(d\frac{2\pi}{\lambda}\sin(\psi)\right).$$

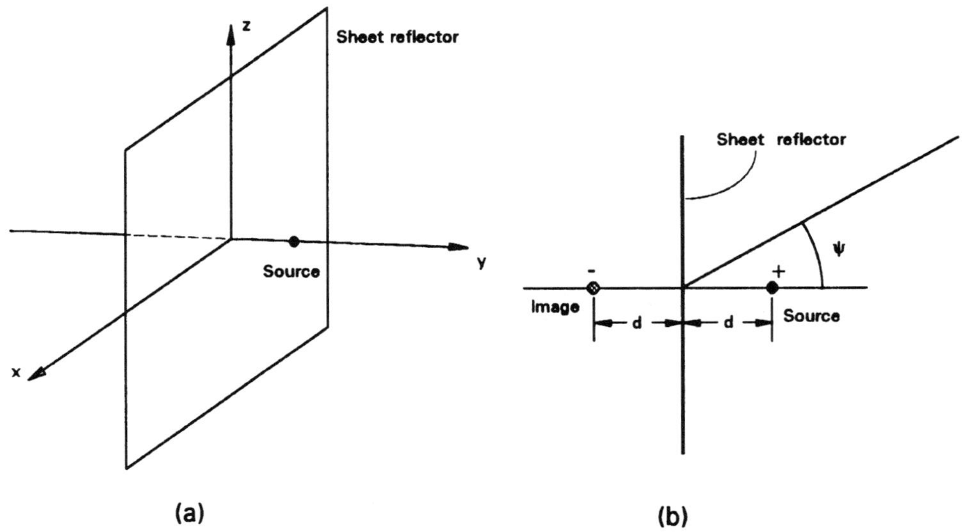

FIGURE 14.27 (a) Sheet reflector with an isotropic source and (b) the image obtained from the method of images.

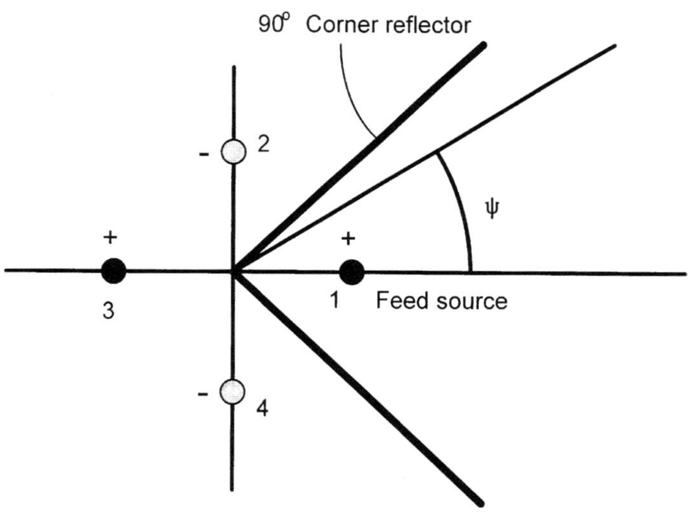

FIGURE 14.28 A 90° corner reflector with feed source and its three images.

The total array factor then becomes

$$F(\psi) = 2\left[\cos\left(d\,\frac{2\pi}{\lambda}\cos(\psi)\right) - \cos\left(d\,\frac{2\pi}{\lambda}\sin(\psi)\right)\right].$$

Example 14.4 Determine the array factor due to the source and its images resulting from a 60° corner reflector, using an isotropic feed element.

Solution The method of images suggests the existence of five images aside from the feed element, positioned as shown in Fig. 14.29. The elements marked with a plus are in phase, while the elements marked with a minus are 180° out of phase with respect to the feed element. The elements are paired as follows: 1 and 4, 5 and 2, and 3 and 6. According to Section 14.5.2, the array pattern of elements 1 and 4—two isotropic sources positioned a distance $2d$ apart, which are out of phase by 180°—is given by

$$F_{1,4}(\psi) = 2j\sin(d\beta\cos(\psi)) = 2j\sin\left(d\frac{2\pi}{\lambda}\cos(\psi)\right).$$

The array factors for the other two pairs of elements are obtained from the above equation by changing the angle that the position vector makes with the axis of the array. For elements 5 and 2, the angle ψ is replaced by $120^\circ + \psi$. For elements 3 and 6, the angle ψ is replaced by $240^\circ + \psi$. The corresponding array factors are

$$F_{5,2}(\psi) = 2j\sin(d\beta\cos(120^\circ + \psi)) = 2j\sin\left(d\frac{2\pi}{\lambda}\cos(120^\circ + \psi)\right),$$

$$F_{3,6}(\psi) = 2j\sin(d\beta\cos(240^\circ + \psi)) = 2j\sin\left(d\frac{2\pi}{\lambda}\cos(240^\circ + \psi)\right).$$

The total array factor is obtained by adding the three components

$$\begin{aligned} F(\psi) &= 2j\left[\sin\left(d\frac{2\pi}{\lambda}\cos(\psi)\right) + \sin\left(d\frac{2\pi}{\lambda}\cos(120^\circ + \psi)\right) + \sin\left(d\frac{2\pi}{\lambda}\cos(240^\circ + \psi)\right)\right] \\ &= 2j\left[\sin\left(d\frac{2\pi}{\lambda}\cos(\psi)\right) - \sin\left(d\frac{2\pi}{\lambda}\cos(\psi - 60^\circ)\right) - \sin\left(d\frac{2\pi}{\lambda}\cos(\psi + 60^\circ)\right)\right]. \end{aligned}$$

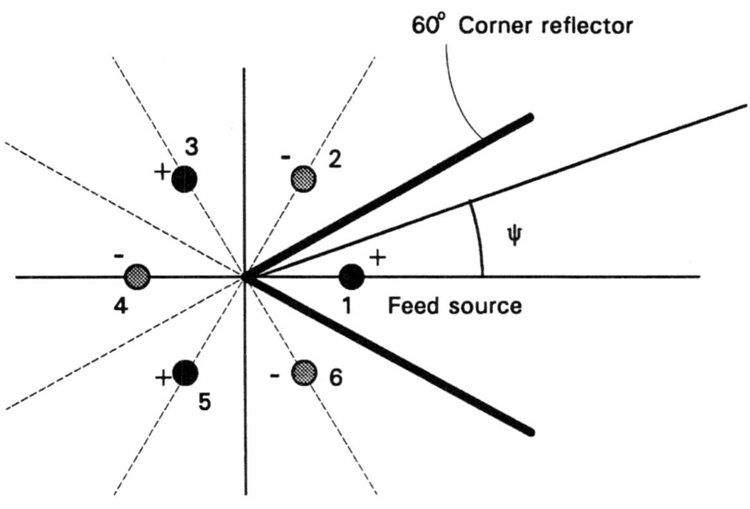

FIGURE 14.29 A 60° corner reflector with feed source and its five images.

14.6.2.3 Parabolic Reflector The parabolic reflector can take the shape of a cylindrical parabolic reflector or of a paraboloid reflector. The cylindrical parabolic reflector transforms a cylindrical wave into a plane wave at its aperture, while the paraboloid reflector transforms a spherical wave into a uniform plane wave at its aperture. The two types of parabolic reflectors are shown in Fig. 14.30.

A parabola, shown in Fig. 14.31, is defined as the locus of points that are equidistant from a point, called the focus, and a line, called the directrix. It can be shown that the wave arriving at the plane PP′ is a plane wave.

If the feed source is isotropic, then the direct radiation from the feed element may interfere with the reflected radiation from the parabolic reflector. The problem can

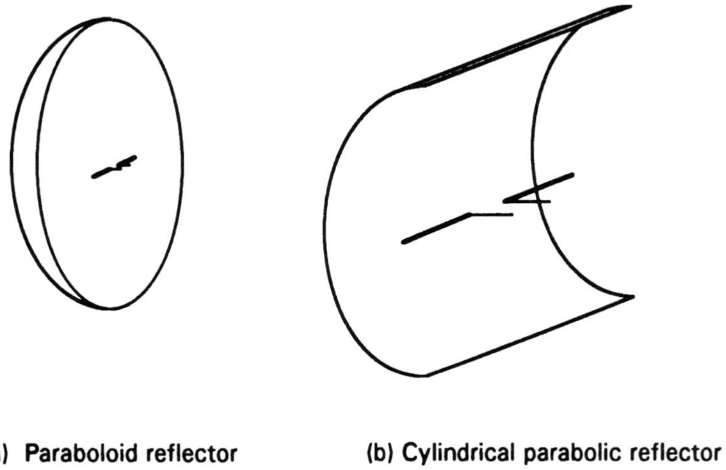

FIGURE 14.30 (a) Paraboloid and (b) cylindrical parabolic reflectors.

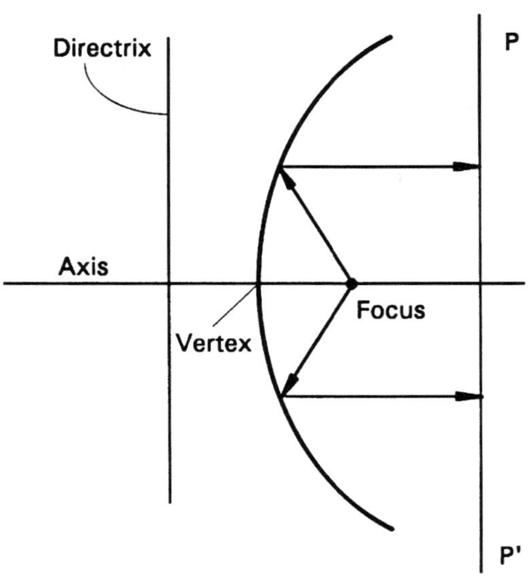

FIGURE 14.31 Parabola geometry.

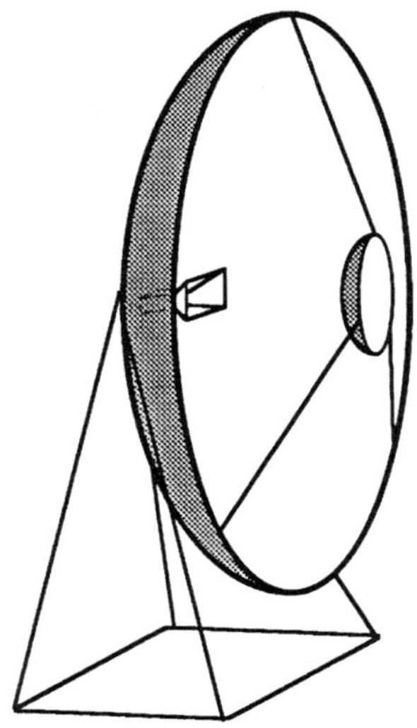

FIGURE 14.32 Cassegrain reflector with a horn as the feed element.

be remedied either by positioning the feed element (the focus of the parabola) at a distance that is an odd number of $\lambda/4$ from the vertex of the parabola, or by using directional feed elements that have a hemispherical radiation pattern. A detailed analysis of these reflectors is presented in Refs. [13, 14].

An improvement over the regular parabolic reflector is the Cassegrain reflector, which positions the feed element at the vertex of the main parabola and uses an additional parabolic subreflector in place of the usual feed element, as shown in Fig. 14.32. The Cassegrain reflector allows for the shaping of the radiation pattern of the feed element, providing a more uniform field distribution over the aperture and an increased efficiency. The Cassegrain subreflector must be large compared to the wavelength, yet small enough compared to the main reflector in order to reduce the blockage of the wavefront. These represent some of the limitations of the Cassegrain reflector.

PROBLEMS

14.1 Calculate the E and H fields of a time harmonic current element located at the origin at a distance $R = 0.0001\lambda$, $R = 3\lambda$, and $R = 100\lambda$, for $\theta = \pi/2$. Use the exact and approximate formulas and compare the results. State in each case how appropriate it is to use the approximate equations. Use MathCad to solve this problem.

14.2 Determine the half-power beam widths for an antenna that has a far field electric field intensity proportional to:

a. $\sin(\theta)\sin(\phi)$.

b. $\sin^2(\theta)\sin^2(\phi)$.

c. $\sin^3(\theta)\sin^3(\phi)$.

14.3 Determine the approximate directivity of the antennas in Problem 14.2.

14.4 Calculate the approximate antenna capacitance and inductance if the length of the short antenna is (a) $L = 10a$, (b) $L = 100a$, and (c) $L = 1000a$, where $a = 2$ mm.

14.5 Determine the loss resistance for the antennas in Problem 14.4 at a frequency of 30 MHz if the wire is made of copper.

14.6 Plot the array factor for a two-element uniform array separated by a distance (a) $d = \lambda/4$, (b) $d = \lambda/2$, and (c) $d = \lambda$, when the elements are in phase.

14.7 Plot the array factor for a two-element uniform array separated by a distance (a) $d = \lambda/4$, (b) $d = \lambda/2$, and (c) $d = \lambda$, when the elements are 180° out of phase.

14.8 Plot the array factor for a two-element uniform array separated by a distance (a) $d = \lambda/4$, (b) $d = \lambda/2$, and (c) $d = \lambda$, when the elements are in quadrature.

14.9 Design an array of elements along the x axis with an array factor that exhibits a maximum at $\psi = 0°$ and a null at $\psi = 120°$. Determine the number of elements, spacing between the elements, and relative phase angles.

14.10 Design an array with an array factor that exhibits a maximum at $\psi = 60°$ and nulls at $\psi = 0°$ and $\psi = 120°$. Determine the number of elements, spacing between the elements, and relative phase angles.

14.11 Design an array with an array factor that exhibits a broad maximum between $\psi = -45°$ and $\psi = 45°$ and nulls at $\psi = 90°$ and $\psi = 150°$. Determine the number of elements, spacing between the elements, and relative phase angles.

14.12 Determine the number of half-wave dipoles, needed in a linear broadside array to obtain a directivity of 100, when the spacing between the elements is (a) $\lambda/4$, (b) $\lambda/2$, and (c) λ.

14.13 Use MathCad to evaluate the exact directivity for the arrays in Problem 14.12.

14.14 Determine the number of half-wave dipoles needed in a linear end-fire array to obtain a directivity of 100, when the spacing between the elements is (a) $\lambda/4$, (b) $\lambda/2$, and (c) λ.

14.15 Use MathCad to evaluate the exact directivity for the arrays in Problem 14.14.

14.16 Determine the half-power beam widths for the arrays in Problem 14.14.

14.17 Plot the radiation patterns for the arrays in Problem 14.12 using MathCad.

14.18 Plot the radiation patterns for the arrays in Problem 14.14 using MathCad.

14.19 Show that, if the phase is varied linearly, the array factor pattern can be directed at different angles. This is the principle of electronic scanning (as opposed to mechanical scanning) for phased arrays. Plot the array factor pattern for different phase angles and note the shifting of the main lobe.

14.20 Use MathCad to plot the radiation pattern of a 90° corner reflector excited by an isotropic source positioned away from the corner at a distance of (a) $\lambda/8$, (b) $\lambda/4$, (c) $\lambda/2$, and (d) λ.

14.21 Repeat Problem 14.10 for a 60° corner reflector.

14.22 Determine the expression of the radiation pattern for a 45° corner reflector. Plot the radiation pattern for the distances between the corner and the isotropic feed element given in Problem 14.20.

REFERENCES

[1] Edward A. Wolff, *Antenna Analysis*, Wiley, New York, 1966, p. 40.

[2] Sergai A. Schelkunoff and Harald T. Friis, *Antennas, Theory and Practice*, Wiley, New York, 1952, p. 306.

[3] Sergai A. Schelkunoff and Harald T. Friis, *Antennas, Theory and Practice*, Wiley, New York, 1952, p. 312.

[4] Eugene Jahnke and Fritz Emde, *Tables of Functions with Formulae and Curves*, 4th ed., Dover Publications, New York, 1945.

[5] Robert E. Collin, *Antennas and Radiowave Propagation*, McGraw-Hill, New York, 1985, pp. 37–38.

[6] Ibid.

[7] Robert E. Collin, *Antennas and Radiowave Propagation*, McGraw-Hill, New York, 1985, pp. 235–260.

[8] J. D. Kraus, *Antennas*, 2nd ed., McGraw-Hill, New York, 1988, pp. 159–175.

[9] J. D. Kraus, *Antennas*, 2nd ed., McGraw-Hill, New York, 1988, pp. 644–659.

[10] C. A. Balanis, *Antenna Theory, Analysis and Design*, Harper & Row, New York, 1992, pp. 532–586.

[11] C. A. Balanis, *Antenna Theory, Analysis and Design*, Harper & Row, New York, 1992, pp. 502–522.

[12] J. D. Kraus and K. R. Carver, *Electromagnetics*, 2nd ed., McGraw-Hill, New York, 1973, pp. 464–468.

[13] J. D. Kraus, *Antennas*, 2nd ed., McGraw-Hill, New York, 1988, pp. 561–619.

[14] C. A. Balanis, *Antenna Theory, Analysis and Design*, Harper & Row, New York, 1992, pp. 604–646.

APPENDIX A

Material Constants

TABLE A.1 Relative Dielectric Permittivity

Material	Relative Permittivity, ε_r
Air	1.0006
Polystyrene foam	1.03
Wood	1.7
Paper	2–4
Teflon	2.1
Polyethylene	2.26
Rubber	2.5–3
Polystyrene	2.54
Lucite	2.56
Plexiglass	2.6–3.5
Earth (dry–wet)	3–30
Snow	3.3
Nylon	3.5–4
Quartz	3.78
Silica	3.8
Glass	4–10
Ice	4.2
Bakelite	4.8–4.9
Asbestos fiber	4.8
Pyrex glass	5
Mica	5.4
Porcelain	6
Neoprene	6.7
Ethyl alcohol	24–25
Glycerol	42.5
Seawater	70
Distilled water	81
Titanium dioxide	90–100
Barium titanate	1200

TABLE A.2 Relative Magnetic Permeability

Material	Relative Permeability, μ_r
	Diamagnetic
Gold	0.99996
Bismuth	0.999983
Copper	0.999991
Water	0.9999912
Silver	0.999998
Paraffin	0.9999994
Vacuum	1.0
	Paramagnetic
Air	1.00000036
Aluminium	1.000021
Nickel chloride	1.00004
Tungsten	1.00008
Platinum	1.00029
Palladium	1.00082
Liquid oxygen	1.0035
	Ferromagnetic
Cobalt	250
Nickel	600
Mild steel	2,000
Iron	5,000
Mu-metal	100,000
Supermalloy	800,000–1,000,000

TABLE A.3 Conductivity (S/m)

Material	σ
Silver	6.17×10^{7}
Copper	5.8×10^{7}
Gold	4.1×10^{7}
Aluminium	3.53×10^{7}
Tungsten	1.8×10^{7}
Brass	1.5×10^{7}
Nickel	1.38×10^{7}
Bronze	1×10^{7}
Iron	1×10^{7}
Solder	7×10^{6}
Lead	4.8×10^{6}
Mercury	1×10^{6}
Graphite	7×10^{4}
Seawater	4
Germanium (intrinsic)	2.2
Silicon (intrinsic)	1.56×10^{-3}
Fresh water	10^{-3}
Earth (dry–wet)	10^{-5}–3×10^{-2}
Distilled water	2×10^{-4}
Bakelite	10^{-9}
Porcelain	10^{-10}
Glass	10^{-10}–10^{-14}
Mica	10^{-11}–10^{-15}
Diamond	2×10^{-3}
Petroleum oil	10^{-14}
Rubber	10^{-15}
Polystyrene	10^{-16}
Quartz	1.3×10^{-18}
Fused quartz	4×10^{-18}

APPENDIX B

Introduction to MathCad

MathCad is a very versatile mathematical software package from the MathSoft Company. It allows for evaluation of sophisticated mathematical expressions numerically as well as symbolically. Manipulation of matrices and matrix operations is done with ease and elegance. The graphical capabilities of MathCad are also impressive, as it allows for both 2D and 3D plots. The version used in this book is MathCad 4.0 or 5.0 + for Windows. This appendix presents a reference guide for the quick application of MathCad to electromagnetic applications, and it, by any means, tries to replace the user's guide provided by the manufacturer. This appendix will allow beginning users to become familiar with and apply MathCad to the examples and homework problems in the text.

B.1 VARIABLE DEFINITION

As the program is invoked, an empty sheet appears, where expressions to be evaluated and almost any mathematical task can be written just as it would be done on a piece of paper. The screen has a menu bar, which can be activated either by clicking on the selected item with the mouse or by typing the underlined letter in the menu title while pressing the Alt key, tool bar, and a strip of buttons (operator palette) on the left edge of the window. There are five sets of function buttons, which allow the quick typing of different mathematical functions, as well as a comprehensive set of Greek symbols, and which can be scrolled through by clicking the top number button above the strip (the numbers range from 1 through 5).

Values can be assigned to variables by typing the variable name followed by the define (or assign) operator (:=) and the value of the variable. The define operator is invoked by pressing the colon (:) key. Right after the colon key is pressed MathCad displays a black rectangle, which is called a placeholder. Numbers or algebraic expressions can be typed. The placeholder is also displayed after each mathematical operator, requiring an entry to be made for the missing operand.

$$A := 4$$

If the variable is defined as an expression of other variables, all the other variables have to be previously defined, which means anywhere above or to the left of the expression, since MathCad reads and evaluates expressions from top to bottom and left to right. For example,

$$x := 4$$

$$y := -2$$

$$A := 3 \cdot \sqrt{x} + y$$

The value of the variable A is viewed by typing the variable followed by the equal sign. MathCad will type the answer after the equal sign.

$$A = 4$$

Expressions are entered as you would on a sheet of paper. For example, the expression

$$x^3 + \frac{\sqrt{y}}{z^{-1} - 3 \cdot z^2 + 3}$$

is entered with the following key sequence:

x ^ 3 [SPACE BAR] + \y [↑] [↑]/z ^ − 1 [SPACE BAR] − 3. z ^ 2 [↑] [↑] + 3

After "x ^ 3" the [SPACE BAR] is pressed in order to select the expression and the "+" sign to appear in the right place. If the [SPACE BAR] is not pressed, the "+" sign will appear right next to 3 in the exponent. The "\" key is an accelerator key for the square root symbol. The same can be accomplished by selecting the square root symbol from the first left palette. An expression or a part of an expression can be selected (or enclosed in a selection box) by using the [↑] key as many times as needed, when the cursor (the vertical bar that appears in an expression or equation edit mode) is active inside the expression. In the above key sequence the [↑] was used twice in order to select the $\sqrt{y}$. The [SPACE BAR] key could have been used as well. The last [SPACE BAR] can be replaced by three [↑]. One will select the 1, the second will select the − 1, and the third will select z^{-1}. A little practice with editing will familiarize the user with the key sequences needed for any kind of mathematical expression.

In some array cases, the index variable has to be defined as a range variable. The range variable is defined by typing the variable name followed by the define (or assign) operator, then entering the first value in the range followed by a comma and the second value in the range, then by typing semicolon (;) followed by the last value in the range:

$$k := 0, 1; 19$$

This kind of range variable can be used to define either a function,

$$v(k) := 2000 \cdot \left(\frac{k}{1 - (1+k)^{-5}} \right)$$

or a vector,

$$V_k := \frac{k^2}{2^k}$$

In the case of the function, it can be evaluated at other than the already defined values for k. In the case of the vector, k sets the index of each element of the vector.

B.2 VECTORS AND MATRICES: DEFINITIONS AND OPERATIONS

Vectors and matrices are defined either by defining each of their entries, as described above, or by using the *Math* menu and selecting *Matrices*. The first definition requires the insertion of the index or indices as subscripts of the vector or matrix. The *indices for vectors and matrices start at* 0 and they appear as subscripts to the variable name. Indices are inserted by typing the square bracket ([ ) key after the variable name, and more than one index are separated by commas. This should not be confused with the literal subscript that any variable can have. The literal subscript is defined by typing the period (.) key after the variable. In the next example, E_x is the name of the vector:

$$E_{x_k} := 25 \sin\left(k \cdot \frac{\pi}{20} \right)$$

This type of array definition lends itself to applications where large vectors and matrices are needed. The following 20 by 20 matrix is easily defined:

$$i := 0, 1, .. 19$$

$$j := 0, 1 .. 19$$

$$M_{i,j} := \sin\left(i \cdot \frac{\pi}{20} \right) \cdot \cos\left(j \cdot \frac{\pi}{20} \right)$$

The typing sequence that created the name of the variable and its indices is

$$M[i, j.$$

Vectors and matrices can also be defined by selecting *Matrices* from the *Math* menu. As a result of the selection, a dialogue box appears, where the number of columns and rows can be entered. For example, the selection, of a 3×1 matrix will produce the following display:

$$V_x := \begin{pmatrix} \blacksquare \\ \blacksquare \\ \blacksquare \end{pmatrix}$$

The black rectangles are the placeholders where the elements of the matrix have to be defined. Clicking each placeholder with the mouse allows the entering of the numerical value or algebraic expression that corresponds to each matrix element. This procedure may be very time consuming for large matrices.

MathCad allows a number of vector and matrix operations to be performed easily. Addition and subtraction are performed by using the addition (+) and subtraction (−) operators:

$$V_1 := \begin{pmatrix} 1 \\ 2 \\ 4 \end{pmatrix} \qquad V_2 := \begin{pmatrix} 2 \\ 3 \\ -1 \end{pmatrix}$$

$$V_{\text{sum}} := V_1 + V_2$$

$$V_{\text{sum}} = \begin{pmatrix} 3 \\ 5 \\ 3 \end{pmatrix}$$

Multiplication is performed using the regular multiplication (·) operator sign. Care must be taken when multiplying matrices so that the operation makes sense.

If two vectors are separated by the regular multiplication sign, MathCad will perform the dot product of the two vectors.

$$\begin{pmatrix} 2 \\ 3 \\ -2 \end{pmatrix} \cdot \begin{pmatrix} -1 \\ 5 \\ 4 \end{pmatrix} = 5$$

The cross product of two vectors is performed by selecting the cross product key from the second left operator palette. The two vectors are entered in the two placeholders.

$$\begin{pmatrix} 2 \\ 3 \\ -2 \end{pmatrix} \times \begin{pmatrix} -1 \\ 5 \\ 4 \end{pmatrix} = \begin{pmatrix} 22 \\ -6 \\ 13 \end{pmatrix}$$

The determinant of a matrix is evaluated by using the "absolute value" symbol, which can be selected from the left operator palette.

$$M := \begin{pmatrix} 2 & 1 & -3 \\ 3 & -2 & 0 \\ -4 & 0 & 0 \end{pmatrix}$$

$$|M| = 24$$

The inverse of a matrix can be calculated by raising the matrix to the power -1 (using the "^" key):

$$M := \begin{pmatrix} 2 & 1 & -3 \\ 3 & -2 & 0 \\ -4 & 0 & 0 \end{pmatrix}$$

$$M^{-1} = \begin{pmatrix} 0 & 0 & -0.25 \\ 0 & -0.5 & -0.375 \\ -0.333 & -0.167 & -0.292 \end{pmatrix}$$

MathCad allows the symbolic evaluation of matrix operations. When the product of two matrices is to be evaluated symbolically, the matrices have to be explicitly entered in the expression. The *Smart Math* feature has to be selected either from the *Math* menu or by clicking the mouse on the top "graduation hat" button on the top of the window. Select the complete expression by clicking the mouse on the expression and using the up-arrow key repeatedly to enclose the expression in a blue rectangle. Then the *Simplify* feature has to be selected from the *Symbolic* menu.

$$\begin{pmatrix} 2 \cdot x \cdot y \\ 4 \cdot x \cdot z^2 \\ -7 \cdot y^3 \cdot z \end{pmatrix} \times \begin{pmatrix} x \\ 2 \cdot y \\ 3 \cdot z \end{pmatrix}$$

After simplification, the displayed result is

$$\begin{pmatrix} 12 \cdot x \cdot z^3 + 14 \cdot y^4 \cdot z \\ -7 \cdot y^3 \cdot z \cdot x - 6 \cdot x \cdot y \cdot z \\ 4 \cdot x \cdot y^2 - 4 \cdot x^2 \cdot z^2 \end{pmatrix}$$

Some matrix operations may be performed by using the *Matrix Operations* function from the *Symbolic* menu.

B.3 OTHER USEFUL FUNCTIONS

Some of the more useful functions in numerical methods applications are the "*if*" and "*until*" functions. The finite difference method examples in this text were implemented with the help of these two functions.

The "*if*" function returns a value or expression that depends on the status of a logic condition:

$$A := if(condition, true\_value, false\_value)$$

The variable A will be assigned the value *true_value* if the *condition* is true, and it will be assigned the value *false_value* if the *condition* is false. The following example shows how an array can be defined using the "*if*" function:

$$i := 0..24$$

$$H_i := if(i > 4, if(i < 20, 1, 0),0)$$

For more information and detailed reference to MathCad commands and operations, the reader is encouraged to consult the *MathCad User's Manual*, provided with the software.

APPENDIX C

Bessel Functions

This appendix presents some basic information about a group of functions called Bessel functions. These functions arise from the solution of a second-order ordinary differential equation called Bessel's equation, just as the trigonometric functions "sine" and "cosine" and the exponential functions arise from the solution of a second-order ordinary differential equation. Bessel functions have been studied extensively, their values are both tabulated and easily computed, and their properties are well known.

C.1 BESSEL'S EQUATION: SERIES SOLUTION AND ASYMPTOTIC FORMS

Bessel's equation is

$$x\frac{d}{dx}\left(x\frac{dy}{dx}\right) + (x^2 - v^2)y = 0. \tag{C.1}$$

This is a second-order ordinary differential equation; x is the independent variable and v is a parameter, either of which can have any value—real, imaginary, or complex. The solution is $y(x, v)$, which is a function of the independent variable and the parameter.

Since this is a second-order ordinary differential equation there must be two independent solutions. To find these solutions we could substitute a power series in x with unknown coefficients and find the required relations between the coefficients to satisfy the differential equation [1]. The results are

$$J_v(x) = \sum_{m=0}^{\infty} \frac{(-)^m x^{2m+v}}{m!\,(m+v)!\,2^{2m+v}},$$

$$J_{-v}(x) = \sum_{m=0}^{\infty} \frac{(-)^m x^{2m-v}}{m!\,(m-v)!\,2^{2m-v}},$$

as can be verified by direct substitution. These two independent solutions are called Bessel functions of the first kind of order ν and $-\nu$, respectively. Note that ν is written as a subscript instead of as a variable of the function; this is purely a matter of custom.

The cases where $\nu = n$, an integer, are of particular importance to us and these cases result in a problem. When $\nu = n$, then

$$\frac{1}{(m-n)!} = 0, \quad \text{for } m < n,$$

and hence

$$J_{-n}(x) = \sum_{m=n}^{\infty} \frac{(-)^m x^{2m-n}}{m!\,(m-n)!\,2^{2m-n}}.$$

Letting $p = m - n$, we obtain

$$J_{-n}(x) = \sum_{p=0}^{\infty} \frac{(-1)^p x^{2p+n}}{(p+n)!\,p!\,2^{2p+n}} = (-)^n J_n(x).$$

Thus when $\nu = n$ the two solutions are no longer independent and a different independent solution is needed. Since J_ν and $J_{-\nu}$ are independent solutions, we can form a new solution as a linear combination of the two. In particular, let

$$N_\nu(x) = \frac{J_\nu(x)\cos\nu\pi - J_{-\nu}(x)}{\sin\nu\pi}.$$

This solution to Bessel's equation is called a Bessel function of the second kind or a Neumann function of order ν. As $\nu \to n$, the numerator and denominator both approach zero, but their ratio, which could be evaluated by L'Hôpital's rule, defines $N_n(x)$. Thus, with N_ν defined as above, we have

$$N_n(x) = \lim_{\nu \to n} N_\nu(x).$$

For small values of x we have the asymptotic expressions

$$J_0(x) \approx 1, \tag{C.2}$$

$$N_0(x) \approx \frac{2}{\pi}\ln\left(\frac{\gamma x}{2}\right), \quad \gamma = 1.781\ldots, \quad \ln\gamma = 0.5772\ldots, \tag{C.3}$$

$$J_\nu(x) \approx \frac{1}{\nu!}\left(\frac{x}{2}\right)^\nu, \tag{C.4}$$

$$N_\nu(x) \approx -\frac{(\nu-1)!}{\pi}\left(\frac{2}{x}\right)^\nu, \quad \mathscr{R}(\nu) > 0. \tag{C.5}$$

For large values of x

$$\left.\begin{aligned} J_\nu(x) &\approx \sqrt{\frac{2}{\pi x}}\cos\left(x - \frac{\pi}{4} - \frac{\nu\pi}{2}\right), \\ N_\nu(x) &\approx \sqrt{\frac{2}{\pi x}}\sin\left(x - \frac{\pi}{4} - \frac{\nu\pi}{2}\right), \end{aligned}\right\} \quad -\pi < \angle x < \pi.$$

Thus, as $x \to 0$, $J_0 \to 1$, $J_\nu \to 0$, $N_0 \to \infty$, and $N_\nu \to \infty$. For $x \to \infty$, the functions become oscillatory with period 2π, within an envelope that decreases as $1/\sqrt{x}$. Bessel's equations appear in cylindrical problems with x representing kr. The J functions are suitable for regions that include the z axis, while the N functions are suitable for regions that either exclude the z axis or that have line sources located on the z axis. For large r the oscillatory nature of the solutions suggests the representation of reflections of cylindrical waves at some boundary.

Figures C.1 and C.2 plot J_n and N_n for $n = 0, 1, 2, 3$.

Other useful solutions are obtained by linear combinations of these functions. In particular,

$$H_\nu^{(1)}(x) = J_\nu(x) + jN_\nu(x),$$

$$H_\nu^{(2)}(x) = J_\nu(x) - jN_\nu(x),$$

are called Hankel functions of the first and second kind of order ν, respectively. For large values of x

$$H_\nu^{(1)} \approx \sqrt{\frac{2}{j\pi x}}\, j^{-\nu} e^{jx}, \quad H_\nu^{(2)} \approx \sqrt{\frac{j2}{\pi x}}\, j^{-\nu} e^{-jx}.$$

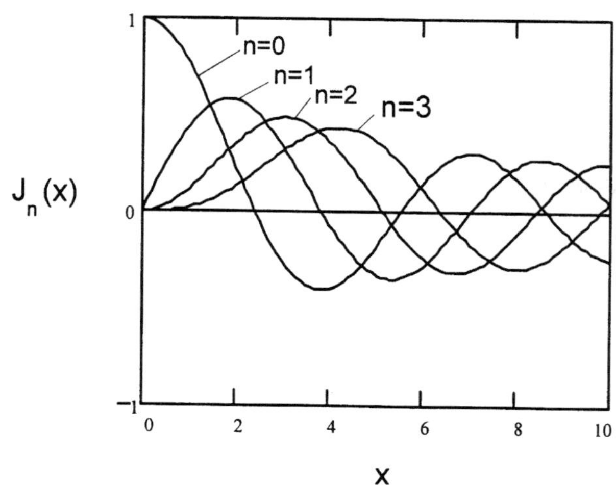

Figure C.1 Bessel functions of the first kind.

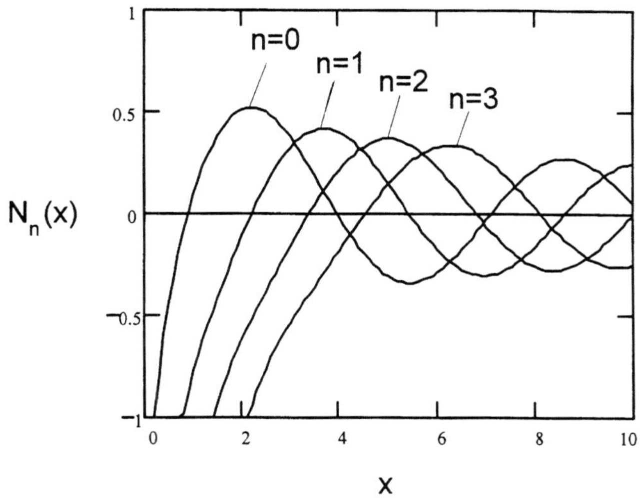

Figure C.2 Bessel functions of the second kind.

Recalling that x represents kr in cylindrical wave problems, the Hankel function of the first kind represents an outward traveling cylindrical wave and the Hankel function of the second kind represents an inward traveling cylindrical wave.

C.2 DERIVATIVE FORMULAS AND RECURSION RELATIONS

By direct differentiation of the series for J_ν we can show that

$$\frac{dJ_\nu(x)}{dx} \equiv J'_\nu(x) = J_{\nu-1}(x) - \frac{\nu}{x} J_\nu(x)$$

$$= - J_{\nu+1}(x) \frac{\nu}{x} J_\nu(x).$$

N_ν, $H_\nu^{(1)}$, and $H_\nu^{(2)}$ give identical results. Hence, letting $Z_\nu(x)$ represent any solution to Bessel's equation, we have

$$Z'_\nu(x) = + Z_{\nu-1}(x) - \frac{\nu}{x} Z_\nu(x), \tag{C.6}$$

$$Z'_\nu(x) = - Z_{\nu+1}(x) - \frac{\nu}{x} Z_\nu(x) \tag{C.7}$$

as formulas for the derivative with respect to the argument. Adding gives

$$2Z'_\nu(x) = Z_{\nu-1}(x) - Z_{\nu+1}(x). \tag{C.8}$$

Subtracting Eq. (C.7) from Eq. (C.6) gives

$$\left(\frac{2v}{x}\right)Z_v(x) = Z_{v-1}(x) + Z_{v+1}(x) \tag{C.9}$$

or, replacing v by $v - 1$,

$$Z_v(x) = \frac{2(v-1)}{x} Z_{v-1}(x) - Z_{v-2}(x) \tag{C.10}$$

as recurrence relations.

We will also use

$$[Z_v'(x)]^2 = -Z_{v-1}(x)Z_{v+1}(x) + \left(\frac{v}{x}\right)^2 Z_v^2(x), \tag{C.11}$$

which is obtained from squaring Eq. (C.8), subtracting Eq. (C.9) squared, and rearranging terms.

C.3 INTEGRAL RELATIONS

Several integrals involving the Bessel functions are useful to us.

Let $Z_p(\alpha x)$ be any Bessel function solution to

$$x\frac{d}{dx}\left[x\frac{dZ_p(\alpha x)}{dx}\right] + (\alpha^2 x^2 - p^2)Z_p(\alpha x) = 0$$

and $\bar{Z}_q(\beta x)$ be any solution to

$$x\frac{d}{dx}\left[x\frac{d\bar{Z}_q(\beta x)}{dx}\right] + (\beta^2 x^2 - q^2)\bar{Z}_q(\beta x) = 0.$$

Multiplying the first equation by $\bar{Z}_q(\beta x)$, which we will write as $\bar{Z}$, and the second equation by $Z_p(\alpha x)$, which we will write as Z, and then subtracting gives

$$\frac{1}{x}[(\alpha^2x^2 - p^2) - (\beta^2x^2 - q^2)]Z\bar{Z} = Z\frac{d}{dx}\left(x\frac{d\bar{Z}}{dx}\right) - \bar{Z}\left(x\frac{dZ}{dx}\right)$$

$$\left[(\alpha^2 - \beta^2)x - \frac{p^2 - q^2}{x}\right]Z\bar{Z} = \frac{d}{dx}\left[\left(x\frac{d\bar{Z}}{dx}\right) - \bar{Z}\left(x\frac{dZ}{dx}\right)\right].$$

Multiply by dx gives a perfect differential on the right side so that integrating gives

$$\int\left[(\alpha^2 - \beta^2)x - \frac{p^2 - q^2}{x}\right]Z_p(\alpha x)\bar{Z}_q(\beta x)\,dx = xZ_p(\alpha x)\frac{d\bar{Z}_q(\beta x)}{dx} - x\bar{Z}_q(\beta x)\frac{dZ_p(\alpha x)}{dx}$$

$$= \beta x Z_p(\alpha x)\bar{Z}_{q-1}(\beta x) - \alpha x Z_{p-1}(\alpha x)\bar{Z}_q(\beta x) + (p - q)Z_p(\alpha x)\bar{Z}_q(\beta x)$$

after substitution of Eq. (C.6) for the derivatives.

If $\alpha = \beta$, but $q \neq p$, then

$$\int \frac{1}{x} Z_p(\alpha x)\bar{Z}_q(\alpha x)\,dx = \alpha x \frac{Z_{p-1}(\alpha x)\bar{Z}_q(\alpha x) - Z_p(\alpha x)\bar{Z}_{q-1}(\alpha x)}{p^2 - q^2} - \frac{Z_p(\alpha x)\bar{Z}_q(\alpha x)}{p+q}. \tag{C.12}$$

If $q = p$, but $\alpha \neq \beta$, then

$$\int xZ_p(\alpha x)\bar{Z}_p(\beta x)\,dx = \frac{\beta x Z_p(\alpha x)\bar{Z}_{p-1}(\beta x) - \alpha x Z_{p-1}(\alpha x)\bar{Z}_p(\beta x)}{\alpha^2 - \beta^2}. \tag{C.13}$$

If the solutions are of the same type this becomes

$$\int xZ_p(\alpha x)Z_p(\beta x)\,dx = \frac{\beta x\, Z_p(\alpha x)Z_{p-1}(\beta x) - \alpha x\, Z_{p-1}(\alpha x)Z_p(\beta x)}{\alpha^2 - \beta^2}.$$

$$= \frac{N(\alpha, \beta, x)}{D(\alpha, \beta)}. \tag{C.14}$$

As $\beta \to \alpha$ the numerator $N(\alpha, \beta, x)$ and the denominator $D(\alpha, \beta)$ each approach zero. Using L'Hôpital's rule

$$\int x\,[Z_p(\alpha x)]^2\,dx = \left.\frac{\partial N/\partial\beta}{\partial D/\partial\beta}\right|_{\beta=\alpha}$$

gives

$$\int x\,[Z_p(\alpha x)]^2\,dx = \frac{x^2}{2}\left\{[Z_p(\alpha x)]^2\left(1 - \frac{p^2}{\alpha^2 x^2}\right) + [Z_p'(\alpha x)]^2\right\} \tag{C.15}$$

$$= \frac{x^2}{2}\{[Z_p(\alpha x)]^2 - Z_{p-1}(\alpha x)Z_{p+1}(\alpha x)\}. \tag{C.16}$$

There is one last integral that will be useful when we calculate the power carried by a mode in a circular waveguide.

$$I = \int_{x_1}^{x_2} x\left\{[J_n'(x)]^2 + \frac{n^2}{x^2}[J_n(x)]^2\right\}dx$$

$$= \int_{x_1}^{x_2} x\,[J_n'(x)]^2\,dx + n^2\int_{x_1}^{x_2}\frac{1}{x}[J_n(x)]^2\,dx.$$

We will integrate the first integral by parts

$$u = xJ_n'(x) \qquad dv = J_n'(x)\,dx$$

$$du = \frac{d}{dx}\left[x\frac{dJ_n'(x)}{dx}\right]dx \qquad = \frac{dJ_n(x)}{dx}dx = dJ_n(x)$$

$$= \frac{1}{x}(n^2 - x^2)J_n(x)\,dt \qquad v = J_n(x)$$

giving

$$\int_{x_1}^{x_2} x[J_n'(x)]^2\,dx = xJ_n(x)J_n'(x)|_{x_1}^{x_2} - \int_{x_1}^{x_2} \frac{1}{x}(n^2 - x^2)[J_n(x)]^2\,dx.$$

Returning to our original integral we now have

$$I = xJ_n(x)J_n'(x)|_{x_1}^{x_2} + \int_{x_1}^{x_2} x[J_n(x)]^2\,dx,$$

which can be integrated with the aid of Eq. (C.15):

$$I = xJ_n(x)J_n'(x) + \frac{x^2}{2}\left\{[J_n(x)]^2\left(1 - \frac{n^2}{x^2}\right) + [J_n'(x)]^2\right\}\Bigg|_{x_1}^{x_2}. \tag{C.17}$$

C.4 MODIFIED BESSEL FUNCTIONS

If $x = ju$, Bessel's equation becomes

$$u\frac{d}{du}\left(u\frac{dy}{du}\right) - (u^2 + \nu^2)y = 0,$$

which is known as the modified Bessel equation. Independent solutions known as the modified Bessel functions of the first and second kind of order ν are

$$I_\nu(u) = \sum_{m=0}^{\infty} \frac{u^{\nu+2m}}{m!(\nu+m)!2^{\nu+2m}},$$

$$K_\nu(u) = \frac{\pi}{2\sin\nu\pi\,[I_{-\nu}(u) - I_\nu(u)]}, \quad K_n(u) = \lim_{\nu\to n} K_\nu(u).$$

These functions are real for real values of u. Their relation to the Bessel functions is analogous to the relation of the exponential functions to the trigonometric functions. The modified Bessel functions are related to the Bessel functions by

$$I_\nu(u) = j^\nu J_\nu(-ju),$$

$$K_\nu(u) = \frac{\pi}{2}(-j)^{\nu+1}[J_\nu(-ju) - jN_\nu(-ju)] = \frac{\pi}{2}(-j)^{\nu+1}H_\nu^{(2)}(-ju).$$

For small u

$$I_0(u) \approx 1, \quad K_0(u) \approx -\ln\left(\frac{\gamma u}{2}\right),$$

$$I_\nu(u) \approx \frac{u^\nu}{\nu!2^\nu}, \quad K_\nu(u) \approx \frac{(\nu-1)!2^{\nu-1}}{u^\nu}; \tag{C.18}$$

and for large x

$$I_\nu(u) \approx \frac{1}{\sqrt{2\pi u}}\, e^{u}, \quad K_\nu(u) \approx \sqrt{\frac{\pi}{2u}}\, e^{-u}, \tag{C.19}$$

indicating usefulness in representing evanescent cylindrical waves.

Derivative formulas for the modified Bessel functions are

$$I'_\nu(u) = I_{\nu+1}(u) + \frac{\nu}{u}\, I_\nu(u) \tag{C.20}$$

$$I'_\nu(u) = I_{\nu-1}(u) - \frac{\nu}{u}\, I_\nu(u) \tag{C.21}$$

and

$$K'_\nu(u) = -\, K_{\nu+1}(u) + \frac{\nu}{u}\, K_\nu(u) \tag{C.22}$$

$$K'_\nu(u) = -\, K_{\nu-1}(u) - \frac{\nu}{u}\, K_\nu(u). \tag{C.23}$$

Recurrence relations are

$$I_\nu(u) = -\frac{2(\nu-1)}{u}\, I_{\nu-1}(u) + I_{\nu-2}(u) \tag{C.24}$$

and

$$K_\nu(u) = \frac{2(\nu-1)}{u}\, K_{\nu-1}(u) + K_{\nu-2}(u). \tag{C.25}$$

We will also use

$$[K'_\nu(x)]^2 = K_{\nu-1}(x) K_{\nu+1}(x) + \left(\frac{\nu}{x}\right)^2 K_\nu^2(x), \tag{C.26}$$

which is obtained in the same way as the similar formula for the Bessel functions.

REFERENCES

[1] Francis Begnaud Hildebrand, *Advanced Calculus for Engineers*, Publisher, Cambridge, MA, 1960, Chap. 4.

Index

WILEY SERIES IN MICROWAVE AND OPTICAL ENGINEERING

KAI CHANG, Editor
Texas A&M University

FIBER-OPTIC COMMUNICATION SYSTEMS, Second Edition • *Govind P. Agrawal*

COHERENT OPTICAL COMMUNICATIONS SYSTEMS • *Silvello Betti, Giancarlo De Marchis and Eugenio Iannone*

HIGH-FREQUENCY ELECTROMAGNETIC TECHNIQUES: RECENT ADVANCES AND APPLICATIONS • *Asoke K. Bhattacharyya*

COMPUTATIONAL METHODS FOR ELECTROMAGNETICS AND MICROWAVES • *Richard C. Booton, Jr.*

MICROWAVE RING CIRCUITS AND ANTENNAS • *Kai Chang*

MICROWAVE SOLID-STATE CIRCUITS AND APPLICATIONS • *Kai Chang*

DIODE LASERS AND PHOTONIC INTEGRATED CIRCUITS • *Larry Coldren and Scott Corzine*

MULTICONDUCTOR TRANSMISSION-LINE STRUCTURES: MODAL ANALYSIS TECHNIQUES • *J. A. Brandão Faria*

PHASED ARRAY-BASED SYSTEMS AND APPLICATIONS • *Nick Fourikis*

FUNDAMENTALS OF MICROWAVE TRANSMISSION LINES • *Jon C. Freeman*

MICROSTRIP CIRCUITS • *Fred Gardiol*

HIGH-SPEED VLSI INTERCONNECTIONS: MODELING, ANALYSIS, AND SIMULATION • *A. K. Goel*

PHASED ARRAY ANTENNAS • *R. C. Hansen*

HIGH-FREQUENCY ANALOG INTEGRATED CIRCUIT DESIGN • *Ravender Goyal (ed.)*

MICROWAVE APPROACH TO HIGHLY IRREGULAR FIBER OPTICS • *Huang Hung-Chia*

NONLINEAR OPTICAL COMMUNICATION NETWORKS • *Eugenio Iannone, Antonio Mecozzi, Francesco Matera, and Marina Settembre*

FINITE ELEMENT SOFTWARE FOR MICROWAVE ENGINEERING • *Tatsuo Itoh, Giuseppe Pelosi and Peter P. Silvester (eds.)*

SUPERCONDUCTOR TECHNOLOGY: APPLICATIONS TO MICROWAVE, ELECTRO-OPTICS, ELECTRICAL MACHINES, AND PROPULSION SYSTEMS • *A. R. Jha*

OPTICAL COMPUTING: AN INTRODUCTION • *M. A. Karim and A. S. S. Awwal*

INTRODUCTION TO ELECTROMAGNETIC AND MICROWAVE ENGINEERING • *Paul R. Karmel, Gabriel D. Colef, and Raymond L. Camisa*

MILLIMETER WAVE OPTICAL DIELECTRIC INTEGRATED GUIDES AND CIRCUITS • *Shiban K. Koul*

MICROWAVE DEVICES, CIRCUITS AND THEIR INTERACTION • *Charles A. Lee and G. Conrad Dalman*

ADVANCES IN MICROSTRIP AND PRINTED ANTENNAS • *Kai-Fong Lee and Wei Chen (eds.)*

OPTOELECTRONIC PACKAGING • *A. R. Mickelson, N. R. Basavanhally, and Y. C. Lee (eds.)*

ANTENNAS FOR RADAR AND COMMUNICATIONS: A POLARIMETRIC APPROACH • *Harold Mott*

INTEGRATED ACTIVE ANTENNAS AND SPATIAL POWER COMBINING • *Julio A. Navarro and Kai Chang*

FREQUENCY CONTROL OF SEMICONDUCTOR LASERS • *Motoichi Ohtsu (ed.)*

SOLAR CELLS AND THEIR APPLICATIONS • *Larry D. Partain (ed.)*

ANALYSIS OF MULTICONDUCTOR TRANSMISSION LINES • *Clayton R. Paul*

INTRODUCTION TO ELECTROMAGNETIC COMPATIBILITY • *Clayton R. Paul*

INTRODUCTION TO HIGH-SPEED ELECTRONICS AND OPTOELECTRONICS • *Leonard M. Riaziat*

NEW FRONTIERS IN MEDICAL DEVICE TECHNOLOGY • *Arye Rosen and Harel Rosen (eds.)*

NONLINEAR OPTICS • *E. G. Sauter*

FREQUENCY SELECTIVE SURFACE AND GRID ARRAY • *T. K. Wu (ed.)*

ACTIVE AND QUASI-OPTICAL ARRAYS FOR SOLID-STATE POWER COMBINING • *Robert A. York and Zoya B. Popović (eds.)*

OPTICAL SIGNAL PROCESSING, COMPUTING AND NEURAL NETWORKS • *Francis T. S. Yu and Suganda Jutamulia*